Wetlands

Wetlands

SECOND EDITION

William J. Mitsch
School of Natural Resources
The Ohio State University
Columbus, Ohio

James G. Gosselink
Center for Coastal, Energy,
and Environmental Resources
Louisiana State University
Baton Rouge, Louisiana

JOHN WILEY & SONS, INC.

New York Chichester Weinheim Brisbane Singapore Toronto

This book is printed on acid-free paper. ♾

Published simultaneously in Canada.

Library of Congress Cataloging-in-Publication Data:

Mitsch, William J.
Wetlands / William J. Mitsch and James G. Gosselink. — 2nd ed.
p. cm.
Includes bibliographical references (p.) and index.
ISBN 0-471-28437-8
1. Wetland ecology—United States. 2. Wetlands—United States.
3. Wetland conservation—United States. I. Gosselink, James G.
II. Title.
QH104.M57 1993
574.5'26325'0973—dc20 92-36129
CIP

Printed in the United States of America

10 9 8

Contents

Part 2
The Wetland Environment

Part 3
Coastal Wetland Ecosystems

Part 5
Management of Wetlands

Preface

It hardly seems like seven years since we wrote the first edition of *Wetlands*. Much has happened in wetland ecology and management since the mid-1980s. Wetlands were not yet "discovered" then, except by a modest number of biological and physical scientists and resource managers. Very few politicians and citizens had heard of wetlands. As an illustration of this, a graduate student from another department came up to us not long ago and asked why, in her search of television news stories, she had not uncovered any reference to wetlands before 1989. Today it is hard to find anyone who has not heard of these ecosystems, their values to society, and the controversies in which they are sometimes involved. Wetlands moved from the eleventh page of the sports section (the outdoor column) of the daily newspaper to the business section and sometimes to the front page. The subject of wetland protection has been debated on the editorial pages of *USA Today* and *The Wall Street Journal*, and in the halls of the U.S. Congress. Grade school children now read about wetlands and discuss animals that live in the swamps and marshes. Developers, landscape designers, and farmers now know and often grudgingly discuss the laws and regulations that protect wetlands. Engineers, soil scientists, and geologists have tried to learn quickly about wetlands when they found that they had to define them and even sometimes to build them. The famous wetland policy of "no net loss" expounded by former U.S. President Bush and modified by his over-zealous White House Council on Competitiveness brought wetland protection and definition into the national environmental spotlight. Environmental organizations such as the Audubon Society, the Environmental Defense Fund, Friends of the Earth, and the World Wildlife Fund have joined long-time wetland friends such as Ducks

Unlimited in championing wetland conservation. The Ramsar Convention and the North American Waterfowl Management Plan have brought wetland preservation to the international political arena.

We have watched and participated in this phenomenal growth in the interest in wetlands, a growth fueled by the newly recognized functions and values of wetlands, by their rapid disappearance in the lower 48 states, and by the environmental awareness that has swept the world. We have also proudly participated in regional, national, and international debates and discussions on wetland science and wetland policy. We have seen a generation of wetland scientists and managers become involved professionally in this new field of wetland ecology with a little help at times from a textbook called *Wetlands*.

Shortly after the first edition appeared, it became clear to us that a second edition was needed because wetland science and its literature were developing and maturing so rapidly. There are now several wetland conferences per year; the INTECOL conference has grown exponentially to the point where it set an attendance record for wetland meetings last September. In North America, the *Society of Wetland Scientists* is approaching, in membership numbers, the size of the longer-standing *Ecological Society of America*. The SWS journal, *Wetlands*, has been joined by *Wetlands Ecology and Management*. Wetland papers have appeared with increasing frequency not only in ecological literature but also in engineering and geological journals. It is our guess that the wetlands literature has more than doubled since our first edition.

To capture the essence of this rapid growth in the understanding of wetlands, while trying to keep both the simplicity and the attention to detail that made the first edition appreciated by many, we have carefully modified this book. We have improved or replaced almost all the figures and tables and have enhanced and upgraded the text throughout the book. We kept the simple yet successful outline of the first edition by first presenting general wetland history (Part 1) and principles (Part II on hydrology, biogeochemistry, adaptations, and succession), followed by seven ecosystem chapters on coastal (Part III) and inland (Part IV) wetland types, and ending with a section on management of wetlands (Part V). Upon examination of the details, readers will see these major changes:

- a new chapter on **wetland creation and restoration** (Chapter 17) with the latest principles, approaches, techniques, and design aspects for constructing or restoring wetlands for habitat replacement, salt marsh restoration, forested wetland restoration, and water quality improvement
- greater emphasis on **international ecology and management of wetlands** throughout the book, with more information on the global extent of wetlands, their role in the biosphere (e.g., the Greenhouse Effect), the use and ecology of wetlands in other countries, and international efforts to protect wetlands
- extensive and comprehensive rewriting of the basic chapters on **wetland hydrology and biogeochemistry**, in order to include new methodologies and

greater detail on wetland processes. The chapter on wetland hydrology (Chapter 4) now incorporates pulsing hydroperiods, several additional examples of wetland water budgets, more equations for calculating hydrologic flows of wetlands, greater discussion on wetland evapotranspiration, and new data on specific effects of wetland hydrology on wetland vegetation and productivity. The chapter on biogeochemistry (Chapter 5) now includes methods for identifying mineral wetland soils and measuring their redox potential, new diagrams—all drawn in a similar format—for nitrogen, carbon, sulfur, and phosphorus cycling in wetlands, new information on generation of both methane and hydrogen sulfide from wetlands, and updates of wetlands as sources, sinks, and transformers of nutrients.

- a revision of Chapter 7 on **wetland succession**, which includes more discussion of paleoecology of wetlands, seed banks, and landscape-scale development of wetlands
- inclusion of major new material on **western United States riparian wetlands** (in Chapter 14) as examples of wetlands in arid environments. This material includes riparian system geomorphology at regional, intra-riparian, and trans-riparian scales
- discussion of **new research areas in wetlands,** including seed banks and dispersal, landscape ecology, mapping and modeling, floating marshes, functional replacement, wetland soils, wetlands and agriculture, the global role of northern peatlands, and the success of mitigation wetlands
- more emphasis on the **fauna of wetlands**, including waterfowl trends in North America, wildlife and fish use of wetlands, and management of wetlands for wildlife
- more discussion on **quantifying values to wetlands**, including updates on approaches to valuation of wetlands such as WET, willingness-to-pay, replacement value, and emergy (embodied energy)
- more emphasis on the **human interface with wetlands** (e.g., social, economic, political, and legal aspects) and how we now value and protect wetlands. This includes an update on all the new laws, regulations, and approaches that govern wetlands in United States, such as "no net loss," Swampbuster, wetland delineation, and jurisdictional wetlands. This material is presented in its historical context, and the importance of a scientific basis for political decisions on wetlands is stressed throughout.
- an **update of our already extensive bibliography** with over 800 *new* entries, compared to fewer than 700 citations in the first edition. The total bibliography now shows almost 1,500 references.
- comprehensive **summaries of chapters,** now given as abstracts at the beginning of each chapter
- dozens of new **photographs and line drawings** of wetlands and research results with wetlands

While making these changes, we have attempted to keep the book useful for its two intended purposes, namely, (1) to provide a comprehensive reference for scientists, engineers, and planners involved in the ecology and management of wetlands; and (2) to serve as a textbook for university courses in wetland ecology. Since the book appeared in 1986, we have noticed its adoption for use in newly formed wetland ecology courses at a number of colleges and universities. To those professors and instructors who have used the book, we are grateful. We do not take this support for granted and we hope that you will find the new edition even more useful for your teaching. We hear from students, too, who have told us that the extensive literature in the book has helped in many term papers, theses, and dissertations. We hope that the new edition will provide an even better entry into the vast wetlands literature. We continue to believe that the book is best for upper-level undergraduate and graduate courses and that some background in ecology is almost essential.

As with the first edition, we had the assistance of many individuals. We greatly appreciate the support from Van Nostrand Reinhold, especially our copy editor, Kevin Callahan, and our marketing manager, Alex Padro. Judy Brief and Mark Licker were extremely encouraging in this project while they were with VNR. As before, we keenly appreciate our colleagues who took time from busy schedules to provide useful reviews of both the first edition and new chapter drafts. This list includes Kathy Ewel, Carol Johnston, Robert Knight, Lyndon Lee, Irv Mendelssohn, Ramesh Reddy, and Michael Scott. We also thank those who took the time to provide insightful book reviews of the first edition as well as material and suggestions for the second, including Thomas Baugh, Andy Clewell, Bob Costanza, Julie Cronk, David Dennis, Elaine Evers, Stephen Faulkner, Siobhan Fennessy, Carl Folke, Nora Gappa, Jack Henry, Laura Foster Huenneke, Paul Keddy, Ronald Keil, Joe Larson, Greg Limscombe, Ariel Lugo, Ed Maltby, John Marshall, Bill Martin, Bill Niering, Jonathan Phillips, Brian Reeder, Carol Reschke, Cheryl Runyon, Christine Samuel, Charles Sasser, Mike Smart, Judy Stout, Gordon Thayer, Ralph Tiner, Robert Twilley, Mary T. and Michael Vogel, Ben Wu, and Joy Zedler. Valerie Atlas of D.C. Comics, Harold Marcus and Ron Anderson of Macbeth Division of Kollmorgen Instruments Corporation, and Linda Boardman of Ocean Spray Cranberries, Inc. generously supplied key prints from their companies. We received clerical, drafting, and other assistance from Terry Buckley, Mashriqui Hassan, Janice Johnson, Rebecca Mitsch, and Renée Wilson. We also would like to thank our graduate students and associates at the School of Natural Resources at Ohio State University and the Coastal Ecology Institute at Louisiana State University both for the help that they provided and for allowing us the time to complete this major undertaking.

When writing a book of this type, we often feel a special kinship with the authors whose work we are citing. It was with particular sadness that we learned in the past two years about the deaths of two prominent wetland ecologists who made major contributions to the field and thus to this book. We thus dedicate this book to the memory of William E. Odum, University of Virginia, and Ralph E. Good, Rutgers University. Two finer wetland scientists and colleagues would be hard to find.

As always, we appreciate the support that we received from our families, particularly from our wives, Ruthmarie H. Mitsch and Jean V. Gosselink, who supported us and persevered with us during the long days and months we devoted to writing two editions of this book.

William J. Mitsch
Columbus, Ohio

James G. Gosselink
Baton Rouge, Louisiana

PART *1*

INTRODUCTION

Wetlands—Their History, Science, and Management

1

Wetlands are a major feature of the landscape in almost all parts of the world. Although many cultures have lived among and even depended on wetlands for centuries, the modern history of wetlands is fraught with misunderstanding and fear. Wetlands disappeared at an alarming rate in the United States and elsewhere until recently. Now their multiple values are recognized, and wetland protection is the norm. Wetlands have properties that are not adequately covered by present terrestrial and aquatic ecology, suggesting that there is a case to be made for wetland science as a unique multidiscipline encompassing many fields, including ecology, chemistry, hydrology, and engineering. Wetlands are unique because of their hydrologic conditions and their role as ecotones between terrestrial and aquatic systems. Wetland management, as the applied side of wetland science, requires an understanding of the scientific aspects of wetlands balanced with legal, institutional, and economic realities. As the interest in wetlands grows, so too do professional societies, journals, and literature concerned with wetlands.

Wetlands are among the most important ecosystems on Earth. In the great scheme of things, the swampy environment of the Carboniferous Period produced and preserved many of the fossil fuels on which we now depend. In more recent biological and human time periods, wetlands are valuable as sources, sinks, and transformers of a multitude of chemical, biological, and genetic materials. Although the value of wetlands for fish and wildlife protection has been known for several decades, some of the other benefits have been identified more recently.

Wetlands are sometimes described as "the kidneys of the landscape" because of the functions that they perform in hydrologic and chemical cycles and because they function as the downstream receivers of wastes from both natural and human sources. They have been found to cleanse polluted waters, prevent floods, protect shorelines, and recharge groundwater aquifers.

Wetlands have also been called "biological supermarkets" for the extensive food chain and rich biodiversity they support. They play major roles in the landscape by providing unique habitats for a wide variety of flora and fauna. Now that we have become concerned about the health of our entire planet, wetlands are being described by some as carbon dioxide sinks and climate stabilizers on a global scale.

These values of wetlands are now being recognized and translated into wetland protection laws, regulations, and management plans. Wetlands have been drained, ditched, and filled throughout history but never as quickly or as effectively as was undertaken in the United States beginning in the mid-1800s. Since then more than half of the nation's original wetlands have been drained. More recently wetlands have become the *cause célèbre* for conservation-minded people and organizations in the United States and throughout the world in part because they had been disappearing at alarming rates and in part because their loss represents an easily recognizable loss of nature to economic "progress." Many scientists, engineers, lawyers, and regulators are now finding it both useful and necessary to become specialists in wetland ecology and wetland management in order to understand, preserve, and even reconstruct these fragile ecosystems. This book is for these aspiring wetland specialists as well as for those who would like to know more about the structure and function of these unique ecosystems. It is a book about wetlands—how they work and how we manage them.

THE GLOBAL EXTENT OF WETLANDS

Wetlands include the swamps, bogs, marshes, mires, fens, and other wet ecosystems found throughout the world under many names (see Chapters 2 and 3 for terms and definitions). They are ubiquitous, that is, they are found on every continent except Antarctica and in every clime from the tropics to the tundra. Although any estimate of the extent of wetlands in the world is difficult and depends on an accurate definition, the most commonly used approximation is that more than 6 percent of the land surface of the world, or 8.6 million km^2, is wetland (Fig. 1–1 and Table 1–1). Matthews and Fung (1987) estimated fewer wetlands in the world (5.3 million km^2), close to the 5.0 million km^2 of global "peatlands" suggested by Finlayson and Moser (1991). Wetlands are found in arid regions as inland salt flats; in humid, cool regions as bogs, fens, and tundra; along rivers and streams as riparian wetland forests and back swamps; and along temperate, subtropical, and tropical coastlines as salt marshes and mangrove

Table 1–1. Estimated Area of Wetlands in the World by Climatic Zone

Zone	Climate	Wetland Area km^2 x 1,000	Percent of Total Land Area
Polar	Humid; semihumid	200	2.5
Boreal	Humid; semihumid	2,558	11.0
Subboreal	Humid	539	7.3
	Semiarid	342	4.2
	Arid	136	1.9
Subtropical	Humid	1,077	17.2
	Semiarid	629	7.6
	Arid	439	4.5
Tropical	Humid	2,317	8.7
	Semiarid	221	1.4
	Arid	100	0.8
World Total		8,558	6.4

Source: Based on data from Bazilevich et al., 1971; Maltby and Turner, 1983

swamps. According to the data in Table 1–1, almost 56 percent of the estimated total wetland area in the world is found in tropical and subtropical regions (4.8 million km^2). The remaining 3.8 million km^2 is mostly boreal peatland. Gorham (1991) estimated 3.5 million km^2 of boreal and subarctic peatlands in Russia, Canada, the USA, and Fennoscania combined.

HISTORY AND WETLANDS

There is no way to estimate how much impact humans have had on the global extent of wetlands except to observe that in developed and heavily populated regions of the world, the impact has ranged from significant to total (Maltby, 1986). But the importance of wetland environments to the development and sustenance of cultures throughout human history is unmistakable. Since early civilization many cultures have learned to live in harmony with wetlands and have benefited economically from surrounding wetlands, whereas other cultures quickly drained the landscape. The ancient Babylonians, Egyptians, and the Aztec in what is now Mexico, developed special systems of water delivery involving wetlands. Mexico City, in fact, is the site of a wetland/lake that disappeared during the past 400 years as a result of human influence (Fig. 1–2). Major cities in the United States such as Chicago and the capital, Washington, D.C., stand on sites that were in part wetlands. Many of the large airports—Boston, New Orleans, and J. F. Kennedy in New York, to name a few—are situated on former wetlands.

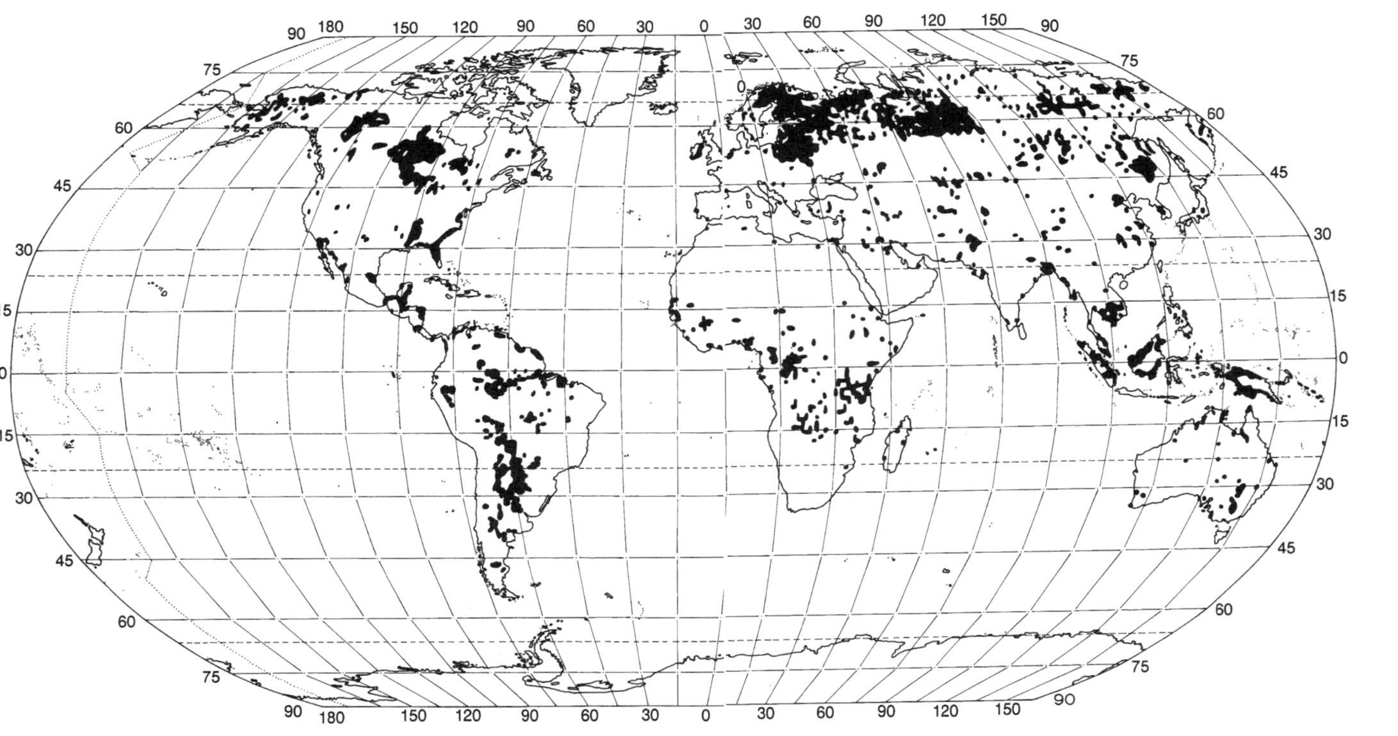

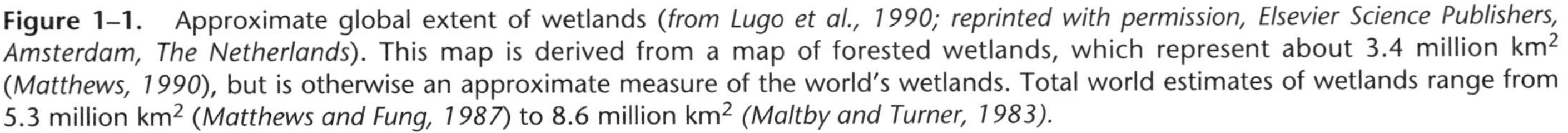

Figure 1–1. Approximate global extent of wetlands (*from Lugo et al., 1990; reprinted with permission, Elsevier Science Publishers, Amsterdam, The Netherlands*). This map is derived from a map of forested wetlands, which represent about 3.4 million km^2 (*Matthews, 1990*), but is otherwise an approximate measure of the world's wetlands. Total world estimates of wetlands range from 5.3 million km^2 (*Matthews and Fung, 1987*) to 8.6 million km^2 *(Maltby and Turner, 1983)*.

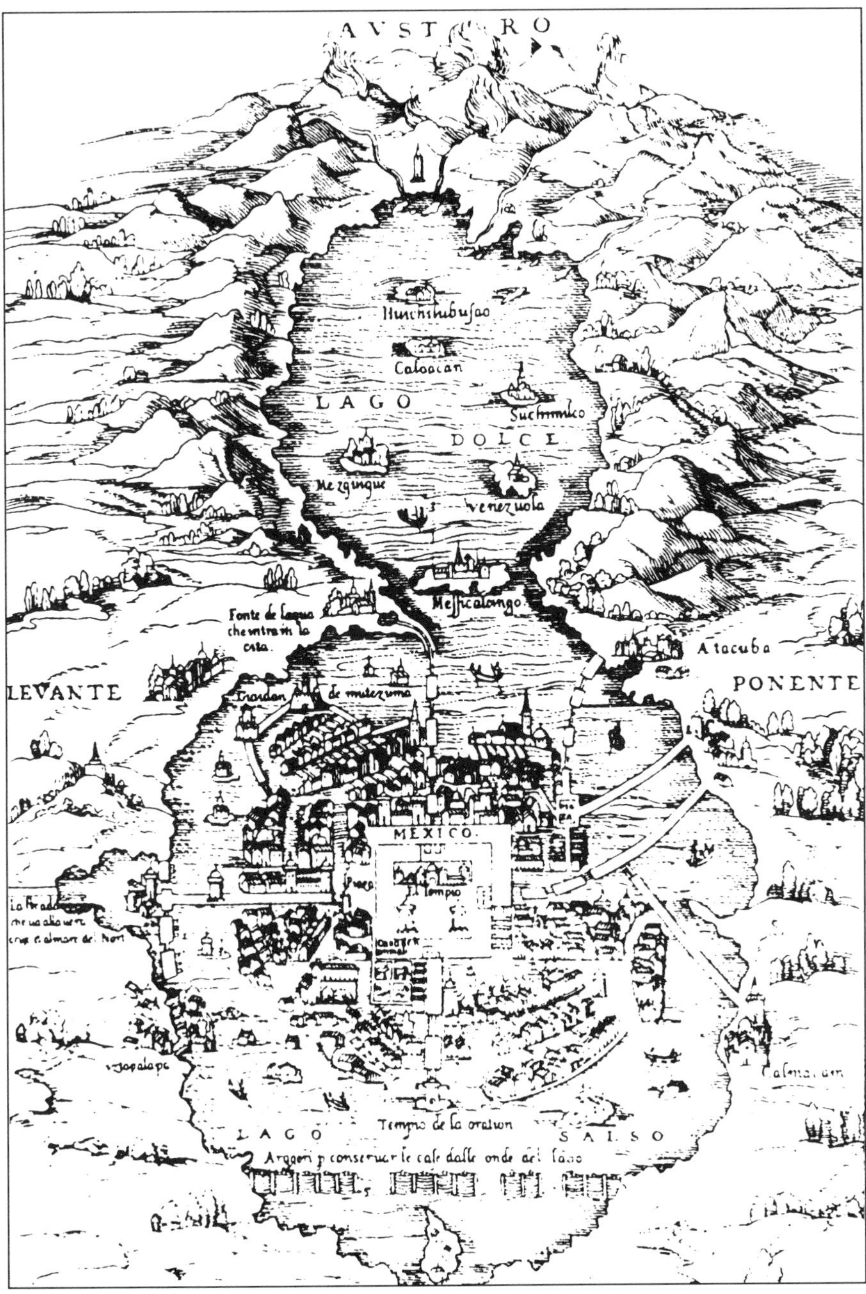

Figure 1–2. A plan of the current city of Mexico City, c. 1556, by Giovanni Battista Ramusio, *El conquistador anónimo. (From "Mexico—Esplendores de treinta siglos," reprinted by The Metropolitan Museum of Art, New York, and Amigos de las Artes de Mexico, Los Angeles)*

WETLANDS HAVE BEEN PART OF THE HISTORY OF MANY CULTURES

Figure 1–3a. The Camargue, in the Rhone River delta region of southern France, is a historically important wetland region in Europe where Camarguais have lived since the Middle Ages. *(Photograph by Tom Nebbia, Horseshoe, N.C., reprinted by permission)*

Figure 1–3b. A Cajun lumberjack camp in the Atchafalaya swamp of coastal Louisiana. American Cajuns are descendants of the French colonists of Acadia (present-day Nova Scotia, Canada) who moved to the Louisiana delta in the last half of the 18th century and flourished within the bayou wetlands. *(Photograph courtesy of Louisiana Collection, Tulane University Library, New Orleans, Louisiana)*

Coles and Coles (1989) refer to the people who live in proximity to wetlands and whose culture is linked to them as *wetlanders*. For example, the Camarguais of southern France (Fig. 1–3a), the Cajuns of Louisiana (Fig. 1–3b), and the Marsh Arabs of southern Iraq (Fig. 1–3c) still live in harmony with wetlands after hundreds of years. Domestic wetlands such as rice paddies feed an estimated half of the world's population (Fig. 1–3d), and countless plant and animal products are harvested from wetlands in countries such as China (Fig. 1–3e, f). Cranberries are harvested from bogs, and the industry continues to thrive today in the United States (Fig. 1–3g). The Russians and the Irish have mined their peatlands for several centuries, using peat as a source of energy. Mangrove wetlands are important for timber, food, and tannin in many countries throughout Indo-Malaysia, East Africa, and Central and South America. For centuries, salt marshes in northern Europe and the British Isles (and later in New England) were used for grazing, hay production, fences, and thatching for roofs. Reeds from wetlands are still used for fencing and thatching today in Romania, Iraq, Japan, and China. The production of fish in systems integrated in shallow ponds or rice paddies developed several thousands of years ago in China and Southeast Asia, and crayfish harvesting is still practiced in the wetlands of Louisiana and the Philippines.

With all of these values and uses, not to mention the aesthetics of a landscape in which water and land often provide a striking panorama, one would expect wetlands to have a history of being revered by humanity; this has certainly not always been the case. Wetlands have been depicted as sinister and forbidding, as

Figure 1–3c. The Marsh Arabs of southern Iraq have lived for centuries on artificial islands in marshes at the confluence of the Tigris and Euphrates rivers. *(Photograph by Nik Wheeler, Los Angeles, California, reprinted by permission)*

Figure 1–3d. Rice production and water buffalo are supported by the wetland environment in many Asian countries, such as this location in Thailand. *(Photograph by Phillip Moore, Louisiana State University, reprinted by permission)*

Figure 1–3e. Interior wetlands in Weishan county, Shandong province, China, where approximately 60,000 people live amid wetland-canal systems, harvesting aquatic plants for food and fiber. *(Photograph by W. J. Mitsch)*

Figure 1–3f. Wetland plants such as *Zizania latifolia* are harvested and sold in markets such as this one in Suzhou, Jiangsu province, China, where these and several other aquatic plants are cooked and served as vegetables. *(Photograph by W. J. Mitsch)*

Figure 1–3g. Cranberry wet harvesting is done by flooding bogs in several regions of the United States. The cranberry plant (*Vaccinium macrocarpon*) is native to bogs and marshes of the northern United States and was first cultivated in Massachusetts. It is now also an important fruit crop in Wisconsin, New Jersey, Washington, and Oregon. *(Reproduced with the permission of Ocean Spray Cranberries, Inc., Lakeville-Middleboro, Massachusetts)*

having little economic value throughout most of history. For example, Dante's *Divine Comedy* describes a marsh of the Styx in the upper Hell as the final resting place for the wrathful:

> Thus we pursued our path round a wide arc of that ghast pool,
> Between the soggy marsh and arid shore,
> Still eyeing those who gulp the marish [marsh] foul.

Centuries later Carl Linneaus, crossing the Lappland peatlands, compared the experience to that same Styx of Hell:

> Shortly afterwards began the muskegs, which mostly stood under water; these we had to cross for miles; think with what misery, every step up to our knees. The whole of this land of the Lapps was mostly muskeg, *hinc vocavi Styx*. Never can the priest so describe hell, because it is no worse. Never have poets been able to picture Styx so foul, since that is no fouler.
>
> — *Carl Linnaeus, 1732*

In the 18th century an Englishman who surveyed the Great Dismal Swamp on the Virginia–North Carolina border and is credited with naming it described the wetland as

> [a] horrible desert, the foul damps ascend without ceasing, corrupt the air and render it unfit for respiration. . . . Never was Rum, that cordial of Life, found more necessary than in this Dirty Place.
>
> — *Colonel William Byrd III (1674–1744) "Historie of the Dividing Line Betwixt Virginia and North Carolina" in* The Westover Manuscripts, *written 1728–1736, Petersburg, Va.; E. and J. C. Ruffin, printers, 1841, 143p.*

Even those who study and have been associated with wetlands have been belittled in literature:

> Hardy went down to botanise in the swamp, while Meredith climbed towards the sun. Meredith became, at his best, a sort of daintily dressed Walt Whitman: Hardy became a sort of village atheist brooding and blaspheming over the village idiot.
>
> — *G. K. Chesterton (1874–1936), Chapter 2 in* The Victorian Age in Literature, *Henry Holt and Company, New York, 1913*

Our English language is filled with words that suggest negative images of wetlands. We get *bogged down* in detail; we are *swamped* with work. Even the mythical *bogeyman*, the character featured in stories that frighten children in many countries, may be associated with European bogs. The sinister and foreboding depiction of wetlands by Hollywood reached its nadir with movies such as the classic *Creature from the Black Lagoon* (1954) and a comic-book-turned-

Figure 1–4. A sinister image of wetlands, especially swamps, is still portrayed in popular media such as comic books and Hollywood movies, although the man-turned-plant *Swamp Thing* shown here is a hero as he fights injustice and even toxic pollution in Louisiana swamps. (SWAMP THING DARK GENESIS™ and *copyright © 1991 by DC Comics, Inc., reprinted with permission*)

Figure 1–5. Agricultural and urban development has transformed half of the wetlands in the United States to uplands since presettlement times. For example, in the Everglades in South Florida, some of the landscape has transformed from natural marshes, swamps, and sloughs typical of the Everglades (*top*) to high-density housing surrounding artificial ponds (*bottom*). *(Photographs by W. J. Mitsch)*

cult-movie, *Swamp Thing* (1982) and its sequel *Return of the Swamp Thing* (1989). But even Swamp Thing, the man/monster depicted in Fig. 1–4, evolved in the 1980s from a feared creature to a protector of wetlands and biodiversity.

Prior to the mid-1970s, the drainage and destruction of wetlands were accepted practices in the United States and were even encouraged by specific government policies. Wetlands were replaced by agricultural fields and by commercial and residential development (Fig. 1–5). Had those trends continued, the resource would be in danger of extinction. Only through the combined activities of hunters and anglers, scientists and engineers, and lawyers and environmentalists have wetlands been elevated to the level of respect in public policy that they deserve. U.S. government interest in wetlands was first reflected in activities such as the sale of federal "duck stamps" to waterfowl hunters that began in 1934 (Fig. 1–6). More than 1.4 million hectares (3.5 million acres) of wetlands were preserved through this program alone during the period 1934–1984. The federal government now supports a variety of other wetland-protection programs. Individual states have also enacted wetland protection laws or have used existing statutes to preserve these valuable resources. But as long as wetlands remain more difficult to stroll through than a forest and more difficult to cross by boat than a lake, they will remain a misunderstood ecosystem to many people.

WETLAND SCIENCE AND WETLAND SCIENTISTS

Wetlands have also been an enigma to scientists. They are difficult to define precisely, not only because of their great geographical extent, but also because of the wide variety of hydrologic conditions in which they are found. Wetlands are usually found at the interface of terrestrial ecosystems, such as upland forests and grasslands, and aquatic systems such as deep lakes and oceans (Fig. 1–7), making them different from each yet highly dependent on both. Because they combine attributes of both aquatic and terrestrial ecosystems but are neither, wetlands have fallen between the cracks of the scientific disciplines of terrestrial and aquatic ecology.

A specialization in the study of wetlands is often termed *wetland science* or *wetland ecology,* and those who carry out such investigations are called *wetland scientists* or *wetland ecologists.* The term *mire ecologist* has also been used. Some have suggested that the study of all wetlands be termed *telmatology* (Victor Masing, oral communication), a term now used by some scientists to refer to the study of northern peatlands. No matter what the field is called, it is apparent that there are several good reasons for treating wetland ecology as a distinct field of study.

1. Wetlands have unique properties that are not adequately covered by present ecological paradigms.

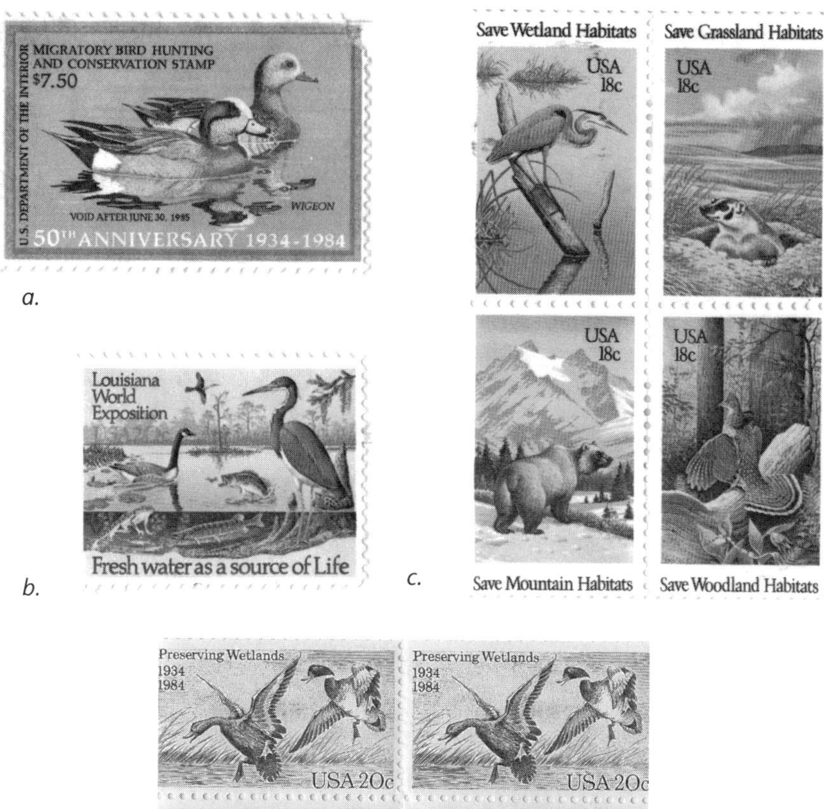

Figure 1–6. Federal involvement in wetlands has been reflected in stamps such as: *a.* "Duck Stamps," which must be purchased by any waterfowl hunter over the age of 16 in the United States. Revenue has been used to acquire and protect wetlands used for hunting; *b.* A wetland panorama for the Louisiana World Exposition of 1984; *c.* A wetland stamp as part of a habitat preservation commemorative stamp set (1979); and *d.* A wetland stamp honoring the 50th anniversary of the duck stamp (1984).

2. Wetland studies have begun to identify some common properties of seemingly disparate wetland types.
3. Wetland investigations require a multidisciplinary approach or training in a number of fields not routinely studied as a unit.
4. There is a great deal of interest in formulating sound policy for the regulation and management of wetlands. These regulations and management approaches need a strong scientific underpinning integrated as wetland ecology.

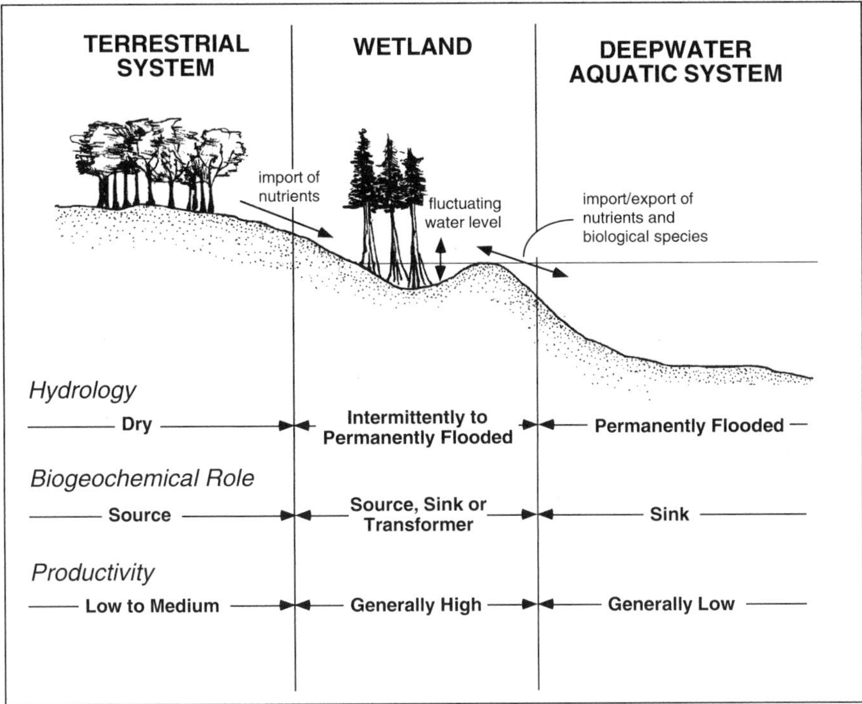

Figure 1–7. Wetlands are often located at the ecotones between dry terrestrial systems and permanently flooded deepwater aquatic systems such as rivers, lakes, estuaries, or oceans. As such, they have an intermediate hydrology, a biogeochemical role as source, sink, or transformer of chemicals, and generally high productivity if they are open to hydrologic and chemical fluxes.

A growing body of evidence suggests that the unique characteristics of wetlands—standing water or waterlogged soils, anoxic conditions, and plant and animal adaptations—may provide some common ground for study that is neither terrestrial ecology nor aquatic ecology. Wetlands provide opportunities for testing "universal" ecological theories and principles involving succession and energy flow, which were developed for aquatic or terrestrial ecosystems. They also provide an excellent laboratory for the study of principles related to transition zones, ecological interfaces, and ecotones.

Wetlands are often ecotones, that is, transition zones, between uplands and deepwater aquatic systems (Figure 1–7). This niche in the landscape allows wetlands to function as organic exporters or inorganic nutrient sinks. Also, this transition position often leads to high biodiversity in wetlands, which "borrow" species from both aquatic and terrestrial systems. Some wetlands, because of their connections to both upland and aquatic systems, have the distinction of being cited as among the most productive ecosystems on Earth.

Our knowledge of different wetland types such as those discussed in this book is for the most part isolated in distinctive literatures and scientific circles. One set of literature deals with coastal wetlands, another with forested wetlands, and still another with freshwater marshes and peatlands. Nevertheless, a number of investigators (e.g., Gosselink and Turner, 1978; H. T. Odum, 1982, 1984; Brown and Lugo, 1982; Nixon, 1988; Madden et al., 1988; Lugo et al., 1988, 1990) have analyzed the properties and functions common to all wetlands. This is probably one of the most exciting areas for wetland research because there is so much to be learned. Comparisons of these wetlands have shown the importance of hydrologic flow-through for the maintenance and productivity of these ecosystems. The anoxic biochemical processes that are common to all wetlands provide another area for comparative research and pose many questions: What are the roles of different wetland types in local and global chemical cycles? How do the activities of humans influence these cycles in various wetlands? What are the synergistic effects of hydrology, chemical inputs, and climatic conditions on wetland productivity? How can plant and animal adaptations to stress be compared in various wetland types?

The true wetland ecologist must be an ecological generalist because of the number of sciences that bear on those ecosystems. Because hydrologic conditions are so important in determining the structures and functions of the wetland ecosystem, a wetland scientist should be well versed in surface and groundwater hydrology. The shallow water environment means that chemistry—particularly for water, sediments, soils, and water-sediment interactions—is an important science. Similarly, questions about wetlands as sources, sinks, or transformers of chemicals require investigators to be versed in a number of biological and chemical techniques. The identification of wetland vegetation and animals requires botanical and zoological skills, and backgrounds in microbial biochemistry and soil science contribute significantly to the understanding of the anoxic environment. Understanding adaptations of wetland biota to the flooded environment requires backgrounds in both biochemistry and physiology. If wetland scientists are to become more involved in the management of wetlands, some engineering techniques, particularly for wetland hydrologic control or wetland creation, need to be learned. Finally, a holistic view of these complex ecosystems can be achieved only through an understanding of the principles of ecology, which can lead to synthesis through the application of techniques developed in ecosystem ecology and systems analysis.

Many scientists now study wetlands. Only a relatively few pioneers, however, investigated these systems in any detail prior to the 1960s. Most of the early scientific studies dealt with classical botanical surveys or investigations of peat structure. Early work on coastal salt marshes and mangroves was published by scientists such as Valentine J. Chapman (1938; 1940), John Henry Davis (1940; 1943), John M. Teal (1958; 1962), Lawrence R. Pomeroy (1959), and Eugene Odum (1961). Henry Chandler Cowles (1899), Edgar N. Transeau (1903),

Herman Kurz (1928), A. P. Dachnowski-Stokes (1935), R. L. Lindeman (1941; 1942), and Eville Gorham (1956; 1961; 1967) are among the investigators in North America who made early contributions to the study of freshwater wetlands, particularly northern peatlands. A number of early scientific studies of peatland hydrology were also produced, particularly in Europe and the Soviet Union. Only later did investigators such as Teal, Pomeroy, Eugene and H. T. Odum, and their colleagues and students begin to use modern ecosystem approaches in wetlands studies. Several research centers devoted to the study of wetlands have since been established in the United States, including the Sapelo Island Marine Institute in Georgia, the Center for Coastal, Energy, and Environmental Resources at Louisiana State University, the Center for Wetlands at the University of Florida, and the Pacific Estuarine Research Laboratory at San Diego State University. In addition, a professional society now exists, the *Society of Wetland Scientists,* which has among its goals providing a forum for the exchange of ideas within wetland science and developing wetland science as a distinct discipline. INTECOL, the *International Association of Ecology,* sponsors a major international wetland conference every four years somewhere in the world.

WETLAND MANAGERS AND WETLAND MANAGEMENT

Just as there are wetland scientists who are researching the details and workings of wetlands, so too are there those who are involved, by choice or by vocation, in some of the many aspects of wetland management. Those individuals, whom we call *wetland managers,* are engaged in activities that range from waterfowl production to wastewater treatment. They must be able to balance the scientific aspects of wetlands with a myriad of legal, institutional, and economic constraints to provide optimum wetland management. The management of wetlands has become so important because these ecosystems are still being drained or encroached on by agricultural enterprises and urban areas. The simple act of being able to identify the boundaries of wetlands has become an important skill for a new type of wetland technician called a *wetland delineator.*

Private organizations such as Ducks Unlimited, Inc., and the Nature Conservancy have protected wetlands by purchasing thousands of hectares of wetlands throughout North America. Through a treaty known as the *Ramsar Convention* and an agreement jointly signed by the United States and Canada in 1986 called the *North American Waterfowl Management Plan,* wetlands are now being protected primarily for their waterfowl value on an international scale. In 1988, a federally sponsored National Wetlands Policy Forum (1988) in the United States raised public and political awareness of wetland loss and recommended a policy of "no net loss" of wetlands. This recommendation has stimulated widespread interest in wetland restoration and creation to replace lost wetlands. Subsequently a National Research Council report (1992) called for the fulfillment of an ambitious goal of gaining 4 million hectares (10 million acres) of wetlands by the year 2010 largely through the reconversion of crop and pasture land. Wetland creation for specific functions is an exciting new area of wetland management that needs trained specialists and may eventually stem the tide of loss and lead to an increase in this important resource.

THE WETLAND LITERATURE

The emphasis on wetland science and management has been demonstrated by a veritable flood of reports, scientific studies, and conference proceedings since the mid- to the late 1970s. The citations in this book are only the tip of the iceberg of the literature on wetlands, much of which has been published since the mid-1980s. Two journals, *Wetlands* and *Wetlands Ecology and Management*, are now published to disseminate scientific and management papers on wetlands, and several other scholarly journals publish frequent papers on wetlands. Several proceedings have been published from conferences on wetlands held in the United States (e.g., Good et al., 1978; Kusler and Montanari, 1978; Greeson et al., 1979; Johnson and McCormick, 1979; Clark and Benforado, 1981; Hook et al., 1988; Sharitz and Gibbons, 1989; Kusler and Kentula, 1990; Gosselink et al., 1990b) and throughout the world (Logofet and Luckyanov, 1982; Gopal et al., 1982; Pokorny et al., 1987; Mitsch et al., 1988; Lefeuvre, 1989; Patten, 1990; Maltby et al., 1992; Mitsch, 1994). Beautiful coffee-table books and articles containing color photographs have been developed by Thomas (1976) and Dennis (1988) on U.S. swamps; by Niering (1985), Littlehales and Niering (1991), and Mitchell et al. (1992) on North American wetlands; by McComb and Lake (1990) on Australian wetlands; and by Finlayson and Moser (1991) on wetlands of the world.

Government agencies have contributed to the wetland literature. The U.S. Fish and Wildlife Service has been involved in the classification and inventory of wetlands (Cowardin et al., 1979; Frayer et al., 1983; Dahl, 1990; Dahl and Johnson, 1991) and has published a series of community profiles on various regional wetlands. The U.S. Environmental Protection Agency has been interested in the impact of human activity on wetlands (Darnell, 1976; Adamus and Brandt, 1990) and in wetlands as possible systems for the control of water pollution (U.S. Environmental Protection Agency, 1983; Kentula et al., 1992; Olson, 1992). Along with the U.S. Army Corps of Engineers, the agency is now in the center of the wetland definition debate in the United States (see Environmental Defense Fund and World Wildlife Fund, 1992). Interest in wetlands has been expressed by the U.S. Congress, which supported summary studies of wetlands, including *Wetland Management* (Zinn and Copeland, 1982) and *Wetlands: Their Use and Regulation* (Office of Technology Assessment, 1984), and sponsored several wetland management bills in the early 1990s, most in response to changing wetland regulations.

Definitions of Wetlands

2

Wetlands have many distinguishing features, the most notable of which are the presence of standing water, unique wetland soils, and vegetation adapted to or tolerant of saturated soils. Wetlands are not easily defined, however, especially for legal purposes, because they have a considerable range of hydrologic conditions, because they are found along a gradient at the margins of well-defined uplands and deepwater systems, and because of their great variation in size, location, and human influence. Common definitions have been used for centuries and are frequently used and misused today. Formal definitions have been developed by several federal agencies in the United States, by scientists in Canada, and by an international treaty known as the Ramsar Convention. These definitions include considerable detail and are used for both scientific and management purposes. No absolute answer to "What is a wetland?" should be expected, but legal definitions involving wetland protection are becoming increasingly comprehensive.

The most common questions that the uninitiated ask about wetlands are "Now, what exactly *is* a wetland?" or "Is that the same as a swamp?" These are surprisingly good questions, and it is not altogether clear that they have been answered completely by wetland scientists and managers. Wetland definitions and terms are many and are often confusing or even contradictory. Nevertheless, definitions are important for both the scientific understanding of these systems and for their proper management.

Defining *wetlands* has become difficult in part because less science and more politics have often been used in the definition when large economic stakes are involved in determining what the term *wetlands* means. Defining the boundaries of wetlands, referred to as "delineation" in the United States, became important when society began to recognize the value of these systems (see Chapter 15) and began to translate that recognition into laws to protect society from further wetland loss (see Chapter 16). In the nineteenth century, when the drainage of wetlands was the norm, a wetland definition was unimportant because it was considered desirable to produce upland from wetlands by draining them. Even as the value of wetlands was being recognized in the early 1970s, there was little interest in precise definitions until it was realized that a better accounting of the remaining wetland resources in this country was needed and definitions were necessary to achieve that inventory. When national and international laws and regulations pertaining to wetland preservation began to be written in the late 1970s and afterward, the need for precision became even greater as individuals recognized that definitions were having an impact on what they could or could not do with their land. The definition of a wetland, then, took on greater importance than the definition of almost any other ecosystem. But just as an estimate of the boundary of a forest or desert or grassland is based on scientifically defensible criteria, so too should the definition of wetlands be based on scientific measures to as great a degree as possible. What society chooses to do with wetlands, once the definition has been chosen, remains a political decision.

DISTINGUISHING FEATURES OF WETLANDS

We can easily identify a coastal salt marsh, with its great uniformity of cordgrass and its maze of tidal creeks, as a wetland. A cypress swamp, with majestic trees festooned with Spanish moss and standing in knee-deep water, provides an unmistakable image of a wetland. A northern sphagnum bog, surrounded by tamarack trees that quake as people trudge by, is another easily recognized wetland. All of those sites have several features in common. All have shallow water or saturated soil, all accumulate organic plant materials that decompose slowly, and all support a variety of plants and animals adapted to the saturated conditions. Wetland definitions, then, often include three main components:

1. Wetlands are distinguished by the presence of water, either at the surface or within the root zone.
2. Wetlands often have unique soil conditions that differ from adjacent uplands.
3. Wetlands support vegetation adapted to the wet conditions *(hydrophytes)* and conversely are characterized by an absence of flooding-intolerant vegetation.

THE PROBLEM OF WETLAND DEFINITION

Although the ideas of shallow water or saturated conditions, unique wetland soils, and vegetation adapted to wet conditions are fairly straightforward, combining these three factors to obtain a precise definition is difficult because of a number of characteristics that distinguish wetlands from other ecosystems yet make them less easy to define (Zinn and Copeland, 1982; Environmental Defense Fund and World Wildlife Fund, 1992):

1. Although water is present for at least part of the time, the depth and duration of flooding vary considerably from wetland to wetland and from year to year. Some wetlands are continually flooded, whereas others are flooded only briefly at the surface or even just below the surface. Similarly, because fluctuating water levels can vary from season to season and year to year in the same wetland type, the boundaries of wetlands cannot be always determined by the presence of water at any one time.
2. Wetlands are often at the margins between deep water and terrestrial uplands and are influenced by both systems. This ecotone position has been suggested by some as evidence that wetlands are mere extensions of either the terrestrial or aquatic ecosystem or both and have no separate identity. Most wetland scientists, however, see emergent properties in wetlands not contained in either upland or deepwater systems.
3. Wetland species (plants, animals, and microbes) range from those that have adaptions to live in either wet or dry conditions (*facultative*) to those adapted to only a wet environment (*obligate*), making difficult their use as wetland indicators.
4. Wetlands vary widely in size, ranging from small prairie potholes of a few hectares in size to large expanses of wetlands several hundreds of square kilometers in area. Although this range in scale is not unique to wetlands, the question of scale is important for their conservation. Wetlands can be lost in large parcels or, more commonly, one small piece at a time in a process called *cumulative loss*. Are wetlands better defined functionally on a large scale or in small parcels?
5. Wetland location can vary greatly, from inland to coastal wetlands and from rural to urban regions. Whereas most ecosystem types, for example, forests or lakes, have similar ecosystem structure and function, there are great differences among different wetland types such as coastal salt marshes, inland pothole marshes, and forested bottomland hardwoods. In rural areas, wetlands are likely to be associated with farmlands, whereas wetlands in urban areas are often subjected to the impact of extreme pollution and altered hydrology.
6. Wetland condition, or the degree to which the wetland is influenced by humans, varies greatly from region to region and from wetland to wetland. Many wetlands can easily be drained and turned into dry lands by human

intervention; similarly, altered hydrology or increased runoff can cause wetlands to develop where they were not found before. Some animals such as beavers, muskrats, and alligators can play a role in developing wetlands. Because wetlands are so easily disturbed, it is often difficult to identify them after such disturbances; this is the case, for example, with wetlands that have been farmed for a number of years.

As described by R. L. Smith (1980), "Wetlands are a half-way world between terrestrial and aquatic ecosystems and exhibit some of the characteristics of each." They form part of a continuous gradient between uplands and open water. As a result, in any definition the upper and the lower limits of wetland excursion are arbitrary boundaries. Consequently, few definitions adequately describe all wetlands. The problem of definition usually arises at the edges of wetlands, toward either wetter or drier conditions. How far upland and how infrequently should the land flood before we can declare that it is not a wetland? At the other edge, how far can we venture into a lake, pond, estuary, or ocean before we leave a wetland? Does a floating mat of vegetation define a wetland? What about a submerged bed of rooted vascular vegetation?

The frequency of flooding is the variable that has made the definition of wetlands particularly controversial. Some classifications include seasonally flooded bottomland hardwood forests, whereas others exclude them because they are dry for most of the year. Because wetland characteristics grade continuously from aquatic to terrestrial, there is no single, universally recognized definition of what a wetland is. This lack has caused confusion and inconsistencies in the management, classification, and inventorying of wetland systems, but considering the diversity of types, sizes, location, and conditions of wetlands in this country, inconsistencies should be no surprise.

FORMAL DEFINITIONS

Precise wetland definitions are needed for two distinct interest groups: (1) wetland scientists and (2) wetland managers and regulators. The wetland scientist is interested in a flexible yet rigorous definition that facilitates classification, inventory, and research. The wetland manager is concerned with laws or regulations designed to prevent or control wetland modification and thus needs clear, legally binding definitions. Because of these differing needs, different definitions have evolved for the two groups. The discrepancy between the regulatory definition of "jurisdictional" wetlands and other definitions in the United States has meant, for example, that maps developed for wetland-inventory purposes cannot be used for regulating wetland development. This is a source of considerable confusion to regulators and landowners.

Early U.S. Definition—Circular 39

One of the earliest definitions of the term *wetlands,* one that is still frequently used today by both wetland scientists and managers, was presented by the U.S. Fish and Wildlife Service in 1956 in a publication that is frequently referred to as Circular 39 (Shaw and Fredine, 1956):

> The term "wetlands" . . . refers to lowlands covered with shallow and sometimes temporary or intermittent waters. They are referred to by such names as marshes, swamps, bogs, wet meadows, potholes, sloughs, and river-overflow lands. Shallow lakes and ponds, usually with emergent vegetation as a conspicuous feature, are included in the definition, but the permanent waters of streams, reservoirs, and deep lakes are not included. Neither are water areas that are so temporary as to have little or no effect on the development of moist-soil vegetation.

The Circular 39 definition (1) emphasized wetlands that were important as waterfowl habitats, and (2) included 20 types of wetlands that served as the basis for the main wetland classification used in the United States until the 1970s (see Chapter 18). It thus served the limited needs of both the wetland manager and the wetland scientist.

U.S. Scientific Definition—Fish and Wildlife Service

Perhaps the most comprehensive definition of wetlands was adopted by wetland scientists in the U.S. Fish and Wildlife Service in 1979, after several years of review. The definition was presented in a report entitled *Classification of Wetlands and Deepwater Habitats of the United States* (Cowardin et al., 1979):

> Wetlands are lands transitional between terrestrial and aquatic systems where the water table is usually at or near the surface or the land is covered by shallow water. . . . Wetlands must have one or more of the following three attributes: (1) at least periodically, the land supports predominantly hydrophytes, (2) the substrate is predominantly undrained hydric soil, and (3) the substrate is nonsoil and is saturated with water or covered by shallow water at some time during the growing season of each year.

This definition is still one of the most widely accepted by wetland scientists in the United States today. Designed for the scientist as well as the manager, it is broad, flexible, and comprehensive and includes descriptions of vegetation, hydrology, and soil. It has its main utility in scientific studies and inventories and generally has been more difficult to apply to the management and regulation of wetlands. It has also been accepted as the official definition of wetlands by India and has been used in proposed wetland legislation by some states in the

United States. Like the Circular 39 definition, this definition serves as the basis for a detailed wetland classification and an updated and comprehensive inventory of wetlands in the United States. That classification and inventory are described in more detail in Chapter 18.

Canadian Wetland Definition

Canadians, who deal with vast areas of inland northern peatlands, have developed a specific national definition of wetlands. At a workshop of the Canadian National Wetlands Working Group, Zoltai (1979) defined wetlands as ". . . areas where wet soils are prevalent, having a water table near or above the mineral soil for the most part of the thawed season, supporting a hydrophylic vegetation." Tarnocai (1979), at the same workshop, presented the definition used in the Canadian Wetland Registry, an inventory and data bank on Canadian wetlands. The definition is similar to Zoltai's:

> Wetland is defined as land having the water table at, near, or above the land surface or which is saturated for a long enough period to promote wetland or aquatic processes as indicated by hydric soils, hydrophytic vegetation, and various kinds of biological activity which are adapted to the wet environment.

These definitions emphasize wet soil conditions, particularly during the growing season. The last definition was presented as a formal definition of Canadian wetlands in the book *Wetlands of Canada* written by the National Wetlands Working Group (1988) after more than a decade of planning and refinement. In that publication Zoltai (1988) notes that "wetlands include waterlogged soils where in some cases the production of plant materials exceeds the rate of decomposition." He describes the wet and dry extremes of wetlands as:

- shallow open waters, generally less than 2 m; and
- periodically inundated areas only if waterlogged conditions dominate throughout the development of the ecosystem.

The International Definition

The International Union for the Conservation of Nature and Natural Resources (IUCN) in the Convention on Wetlands of International Importance Especially as Waterfowl Habitat, better known as the *Ramsar Convention*, adopted the following definition of wetlands (Navid, 1989; Finlayson and Moser, 1991):

> areas of marsh, fen, peatland or water, whether natural or artificial, permanent or temporary, with water that is static or flowing, fresh, brackish, or salt including areas of marine water, the depth of which at low tide does not exceed 6 meters.

The definition, which was adopted at the first meeting of the convention in Ramsar, Iran, in 1971, states that wetlands

> may incorporate riparian and coastal zones adjacent to the wetlands and island or bodies of marine water deeper than six meters at low tide lying within the wetlands.

This definition does not include vegetation or soil and extends wetlands to water depths of 6 m or more, well beyond the depth usually considered wetlands in the United States and Canada. Navid (1989) suggests that this definition could be interpreted to include "a wide variety of habitat types including rivers, coastal areas, and even coral reefs."

Legal Definitions

A U.S. government regulatory definition of wetlands is found in the regulations used by the U.S. Army Corps of Engineers for the implementation of a dredge-and-fill permit system required by Section 404 of the 1977 Clean Water Act Amendments. The latest version of that definition is given as follows:

> The term "wetlands" means those areas that are inundated or saturated by surface or ground water at a frequency and duration sufficient to support, and that under normal circumstances do support, a prevalence of vegetation typically adapted for life in saturated soil conditions. Wetlands generally include swamps, marshes, bogs, and similar areas. (33 CFR328.3(b); 1984)

This definition replaced a 1975 definition that stated that "those areas that normally are characterized by the prevalence of vegetation that *requires* saturated soil conditions for growth and reproduction" (42 *Fed. Reg.* 37128, July 19, 1977; italics added) because the Corps of Engineers found that the old definition excluded "many forms of truly aquatic vegetation that are prevalent in an inundated or saturated area, but that do not require saturated soil from a biological standpoint for their growth and reproduction." The terms "normally" in the old definition and "that under normal circumstances do support" in the new definition were intended "to respond to situations in which an individual would attempt to eliminate the permit review requirements of Section 404 by destroying the aquatic vegetation. . . ." (quotes from 42 *Fed. Reg.* 37128, July 19, 1977). The need to revise the 1975 definition illustrates how difficult it has been to develop a legally useful definition that also accurately reflects the ecological reality of a wetland site.

This legal definition of wetlands has been debated in the courts in several cases, some of which have become landmark cases. In one of the first court tests of wetland protection, the Fifth Circuit of the U.S. Court of Appeals ruled in 1972, in *Zabel* v. *Tabb,* that the U.S. Army Corps of Engineers has the right to refuse a permit for filling of a mangrove wetland in Florida. In 1975, in *Natural Resources Defense Council* v. *Callaway,* wetlands were included in the category "waters of the United States," as described by the Clean Water Act. Prior to that time, the Corps of Engineers regulated dredge-and-fill activities (Section 404 of the Clean Water Act) for navigable waterways only; since that decision, wetlands have been legally included in the definition of waters of the United States. In 1985, the question of regulation of wetlands reached the U.S. Supreme Court for the first time. The court upheld the broad definition of wetlands to include groundwater in *United States* v. *Riverside Bayview Homes, Inc.* (Want, 1990).

The Army Corps of Engineers' definition cited above emphasizes only one indicator, vegetative cover, to determine the presence or absence of a wetland. It is difficult to include soil information and water conditions in a wetland definition when its main purpose is to determine jurisdiction for regulatory purposes and there is little time to examine the site in detail. The wetland manager is interested in a definition that allows the rapid identification of a wetland and the degree to which it has been or could be altered. He or she is also interested in the delineation of wetland boundaries; establishing boundaries is facilitated by defining the wetland simply, according to the presence or absence of certain species of vegetation or aquatic life.

Since 1989 the term *jurisdictional wetland* has been used for legally defined wetlands in the United States under the above definition to delineate those areas that are under the jurisdiction of Section 404 of the Clean Water Act or the Swampbuster Provision of the Food Security Act. Several federal manuals spelling out specific methodologies for identifying wetlands have been written or proposed, all indicating that the three criteria for wetlands, namely wetland hydrology, wetland soils, and hydrophytic vegetation, must be present. But the manuals have differed in prescribing ways in which these three criteria can be proved in the field. These aspects of wetland management are discussed in more detail in Chapter 16.

Choice of a Definition

A wetland definition that will prove satisfactory to all users has not yet been developed because the definition of wetlands depends on the objectives and the field of interest of the user. Lefor and Kennard (1977), in a review of the many definitions used for inland wetlands in northeastern United States, showed that different definitions can be formulated by the geologist, soil scientist, hydrologist, biologist, systems ecologist, sociologist, economist, political scientist, public health scientist, and lawyer. This variance is a natural result of the differences in emphasis in the definer's training and a result of the different ways in which individual disciplines deal with wetlands. For ecological studies and inventories, the 1979 U.S. Fish and Wildlife Service definition has been and should continue to be applied to wetlands in the United States. When wetland management, particularly regulation, is necessary, the U.S. Army Corps of Engineers' definition, as modified, is probably most appropriate. But just as important as the precision of the definition of a wetland is the consistency with which it is used. That is the difficulty that we face when science and legal issues meet, as they often do, in resource management questions such as wetland conservation versus wetland drainage. Applying a comprehensive definition in a uniform and fair way requires a generation of well-trained wetland scientists and managers armed with a fundamental understanding of the processes that are important and unique to wetlands.

Wetland Types and Wetland Resources of North America

3

Wetlands are numerous and diverse in North America, but drastic changes have occurred since presettlement times. An estimated 53 percent of the original wetlands in the lower 48 states has been lost because of drainage and other human activities. On a regional basis, the greatest wetland losses are occurring in the Lower Mississippi Alluvial Plain and the prairie pothole region of the northcentral states. Estimates of original wetlands in the lower 48 states vary around 89 million hectares (220 million acres). The most accurate estimate of present wetland area is 42 million hectares (104 million acres), of which 80 percent are inland wetlands and 20 percent are coastal. Current measures of the wetland area in Alaska are about 69 million hectares (170 million acres). By contrast, Canada has an estimated 127 million hectares (314 million acres) of wetlands.

The historical terminology of wetlands has been confusing and often contradictory. In this book we classify wetlands as seven major types: tidal salt marshes, tidal freshwater marshes, mangrove wetlands, northern peatlands, inland marshes, southern deepwater swamps, and riparian wetlands. In general, regional wetland areas are heterogeneous mosaics of several of these classes. Historical and ecological characteristics of some important regional wetlands of the United States and Canada, several of which have been threatened or eliminated by human development, are discussed here.

North America has always had an abundance and a diversity of wetlands. The wetlands that exist now, however, may represent only a fraction of those seen by pioneers as they advanced across the continent about 200 years ago. Soggy marshes and wet meadows must have evoked the interest of the first settlers as they passed through the hills and mountains of the East and entered the level prairie of the Midwest. Peatlands and prairie potholes had to be a common sight to those who settled in the north country in what is now Michigan, Wisconsin, Minnesota, the Dakotas, and the central Canadian Provinces. Those who traveled on the Mississippi River or rivers in the southeastern Coastal Plain must have marveled at the mammoth structures of many of the virgin cypress swamps and bottomland hardwood forests that lined those rivers. Indeed, salt marshes and mangroves probably represented inhospitable barriers to many of the explorers and settlers as they arrived in the New World. All of those wetlands undoubtedly played an important role in the development of North America. In the early period of settlement, wetlands were accepted as part of the landscape, and there was little desire or capability to change their hydrologic conditions to any great degree. As described by Shaw and Fredine (1956):

> The great natural wealth that originally made possible the growth and development of the United States included a generous endowment of shallow-water and waterlogged lands. The original inhabitants of the New World had utilized the animals living among these wet places for food and clothing, but they permitted the land to essentially remain unchanged.

As the westward movement slowed and towns, villages, cities, and farms began to be expanded on a regional scale, wetlands were increasingly viewed as wastelands that should be drained for reasons as varied as disease control and agricultural expansion. There followed an unfortunate yet understandable period in the country's growth, when wetland drainage and destruction became the accepted norm. The Swamp Lands Acts of 1849, 1850, and 1860 set the tone for that period by encouraging much drainage in 15 of the interior and West Coast states. Coastal wetlands in the populated Northeast were drained and dissected to accommodate urban development. Many old growth forested bottomlands and swamps in the Southeast were lumbered for their valuable products, reflecting little concern for their regeneration for the enjoyment and use of future generations. From the middle of the nineteenth century to the middle of the twentieth century, the United States went through a period in which wetland removal was not questioned. Indeed, it was considered the proper thing to do.

WETLAND TERMS AND TYPES

A number of common terms have been used over the years to describe different types of wetlands (Gore, 1983b). The history of the use and misuse of these words has often revealed a decidedly regional or at least continental origin. Although the lack of standardization of terms is confusing, many of the old terms are rich in meaning to those familiar with them. They often bring to mind vivid images of specific kinds of ecosystems that have distinct vegetation, animals, and other characteristics. The following are some of the more popular terms used to describe particular kinds of wetlands:

Bog. A peat-accumulating wetland that has no significant inflows or outflows and supports acidophilic mosses, particularly sphagnum.

Bottomland. Lowlands along streams and rivers, usually on alluvial floodplains that are periodically flooded. These are usually forested and in the Southeast are often sometimes called bottomland hardwood forests.

Fen. A peat-accumulating wetland that receives some drainage from surrounding mineral soil and usually supports marshlike vegetation.

Marsh. A frequently or continually inundated wetland characterized by emergent herbaceous vegetation adapted to saturated soil conditions. In European terminology a marsh has a mineral soil substrate and does not accumulate peat.

Mire. Synonymous with any peat-accumulating wetland (European definition).

Moor. Synonymous with peatland (European definition). A highmoor is a raised bog, whereas a lowmoor is a peatland in a basin or depression that is not elevated above its perimeter.

Muskeg. Large expanses of peatlands or bogs; particularly used in Canada and Alaska.

Peatland. A generic term of any wetland that accumulates partially decayed plant matter.

Playa. Term used in southwestern United States for marshlike ponds similar to potholes (see below) but with a different geologic origin.

Pothole. Shallow, marshlike pond, particularly as found in the Dakotas and central Canadian provinces.

Reedswamp. Marsh dominated by *Phragmites* (common reed); term used particularly in eastern Europe.

Slough. A swamp or shallow lake system in the northern and midwestern United States. A slowly flowing shallow swamp or marsh in southeastern U.S.

Swamp. Wetland dominated by trees or shrubs (U.S. definition). In Europe a forested fen or reedgrass-dominated wetland is often called a swamp, for example, reedswamp.

Vernal pool. Shallow, intermittently flooded wet meadow, generally dry for most of the summer and fall.

Wet Meadow. Grassland with waterlogged soil near the surface but without standing water for most of the year.

Wet Prairie. Similar to a marsh but with water levels usually intermediate between a marsh and a wet meadow.

Each of these terms has a specific meaning to the initiated, and many are still widely used both by scientists and laypersons. It is often difficult, however, to

Table 3–1. Comparison of Terms Used to Describe Similar Inland Nonforested Freshwater Wetlands

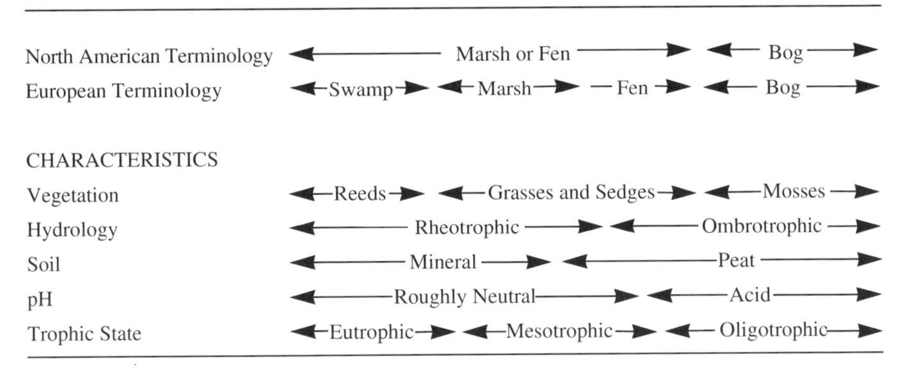

North American Terminology	← Marsh or Fen →			← Bog →
European Terminology	← Swamp →	← Marsh →	— Fen →	← Bog →
CHARACTERISTICS				
Vegetation	← Reeds →	← Grasses and Sedges →		← Mosses →
Hydrology	← Rheotrophic →			← Ombrotrophic →
Soil	← Mineral →		← Peat →	
pH	← Roughly Neutral →			← Acid →
Trophic State	← Eutrophic →	← Mesotrophic →		← Oligotrophic →

convey the same meaning to an international scientific community because some languages have no direct equivalents for certain words that define wetlands. For example, the word *swamp* has no direct equivalent in Russian because there are few forested wetlands there. On the other hand, *bog* can easily be translated because bogs are a common feature of the landscape there.

Table 3–1 illustrates the confusion in terminology that occurs because of different regional or continental uses of terms for similar types of wetlands. In North American terminology, nonforested inland wetlands are often casually classified either as peat-forming, low-nutrient acid bogs or as marshes. European terminology, which is much older, is also much richer and distinguishes at least four different kinds of wetlands from mineral-rich reed beds, called swamps, to wet grassland marshes, to fens, and finally to bogs or moors. In some sources (e.g., Moore, 1984), all of these are considered mires. According to others, mires are limited to peat-building wetlands. The European classification is based on the amount of surface water and nutrient inflow (rheotrophy), type of vegetation, pH, and peat-building characteristics. Two points can be made: First, the physical and biotic characteristics grade continuously from one of these wetland types to the next; hence any classification is to an extent arbitrary; second, the same term may refer to different systems in different regions. For example, a European swamp is dominated by reeds. In the United States the term *swamp* is almost always used to connote a forested wetland.

In this book we use a simple wetland classification scheme based on commonly used terms. The classification scheme used in the United States, as part of the National Wetlands Inventory (Cowardin et al., 1979), is much more formal and all-encompassing, but it is also much too complex to use conveniently here. (The National Wetlands Inventory is described in Chapter 18.) Seven major types of wetlands in the United States can be divided into two major groups: coastal and inland (Table 3–2). This way of classifying wetlands uses terminology that can be easily recognized and understood. These classes of wetlands are

Table 3–2. Types of Wetlands

Wetland Types Used In This Book	*National Wetlands Inventory Equivalent*[a]	*Book Chapter*
Coastal Wetland Ecosystems		
Tidal Salt Marshes	Estuarine intertidal emergent, haline	8
Tidal Freshwater Marshes	Estuarine intertidal emergent, fresh	9
Mangrove Wetlands	Estuarine intertidal forested and shrub, haline	10
Inland Wetland Ecosystems		
Freshwater Marshes	Palustrine emergent	11
Northern Peatlands	Palustrine moss-lichen	12
Southern Deepwater Swamps	Palustrine forested and scrub-shrub	13
Riparian Wetlands	Palustrine forested and scrub-shrub	14

[a]Cowardin et al., 1979

generally recognizable ecosystems about which extensive research literature is available. Regulatory agencies also deal with these systems, and management strategies and regulations have been developed for these wetland types. It is recognized that there are other types of wetlands such as inland saline marshes, scrub-shrub swamps, and even red maple swamps that may "fall between the cracks" in this simple wetland classification, but the seven classes cover most wetlands currently found in North America.

Coastal Wetlands

Several types of wetlands in the coastal areas are influenced by alternate floods and ebbs of tides. Near coastlines the salinity of the water approaches that of the ocean, whereas further inland the tidal effect can remain significant even when the salinity is that of freshwater. If the accounting includes all coastal counties in the conterminous United States, coastal wetlands cover 11.1 million hectares (27.4 million acres); forested and scrub-shrub wetlands make up 63 percent of this total, salt marshes about 16 percent, and fresh marshes 17 percent (Field et al., 1991).

Tidal Salt Marshes

Salt marshes (Fig. 3–1) are found throughout the world along protected coastlines in the middle and high latitudes. In the United States, salt marshes are often dominated by the grass *Spartina* in the low intertidal zone and the rush *Juncus* in the upper intertidal zone. Plants and animals in these systems have adapted to

Figure 3–1. Tidal salt marsh. *(Photograph by Charles E. Sasser)*

the stresses of salinity, periodic inundation, and extremes in temperature. Salt marshes are most prevalent in the United States along the Eastern Coast from Maine to Florida and on into Louisiana and Texas along the Gulf of Mexico. They are also found in narrow belts on the West Coast of the United States and along much of the coastline of Alaska. The area and types of coastal wetland depend on where the inland boundary of the coast is drawn. If intertidal wetlands alone are considered the area is about 2 million hectares (5 million acres), of which 1.7 million hectares are salt marshes (Field et al., 1991).

Tidal Freshwater Marshes

Inland from the tidal salt marshes but still close enough to the coast to experience tidal effects (Fig. 3–2), these wetlands, dominated by a variety of grasses and by annual and perennial broad-leaved aquatic plants, are found primarily along the Middle and South Atlantic coasts and along the coasts of Louisiana and Texas. An estimated 164,000 hectares (405,000 acres) of this type of wetland exist along the Atlantic coast of the United States (W. E. Odum et al., 1984). Field et al. (1991) estimated that tidal freshwater marshes range from 85,000 to 261,000 hectares (211,000 to 645,000 acres) for the entire conterminous United States. The uncertainty in the estimate relates to where the line is drawn between tidal and nontidal areas in determining how much of the extensive Gulf Coast freshwater wetlands are tidal. Tidal freshwater marshes can be described as intermediate in the continuum from coastal salt marshes to freshwater marshes. Because they are tidally influenced but lack the salinity stress

Figure 3–2. Tidal freshwater marsh.

of salt marshes, tidal freshwater marshes have often been reported to be very productive ecosystems, although a considerable range in productivity has been measured in them.

Mangrove Wetlands

The tidal salt marsh gives way to the mangrove swamp (Fig. 3–3) in subtropical and tropical regions of the world. The word *mangrove* refers to both the wetland itself and to the salt-tolerant trees that dominate those wetlands. In the United States, mangrove wetlands are limited primarily to the southern tip of Florida, although small mangrove stands are scattered as far north as Louisiana and Texas. Approximately 287,000 hectares (709,000 acres) of this wetland exist in the United States, a small fraction of the 14 million hectares (35 million acres) found worldwide (Finlayson and Moser, 1991). The mangrove wetland is generally dominated by the red mangrove tree (*Rhizophora*) and the black mangrove tree (*Avicennia*) in Florida. Like the salt marsh, the mangrove swamp requires protection from the open ocean and occurs in a wide range of salinity and tidal influence.

Inland Wetlands

On an areal basis most of the wetlands of the United States and Canada are not located along the coastlines but are found in the interior regions. Frayer et al.

Figure 3–3. Mangrove wetland.

(1983) estimated that 32 million hectares (79.4 million acres) or about 80 percent of the total wetlands in the lower 48 states are inland. It is difficult to put these wetlands into simple categories. Our simplified scheme divides them into two groups that are decidedly regional—northern peatlands and southern deepwater swamps, and into two other categories that are found throughout the climatic zones—freshwater marshes and riparian ecosystems. These divisions roughly parallel the divisions that persist in both the scientific literature and the specializations of wetland scientists.

Figure 3–4. Inland freshwater marsh.

Freshwater Marshes

This category includes a diverse group of wetlands characterized by (1) emergent soft-stemmed aquatic plants such as cattails, arrowheads, pickerel-weed, reeds, and several species of grasses and sedges, (2) a shallow water regime, and (3) generally shallow peat deposits (Fig. 3–4). These wetlands are ubiquitous in North America and are estimated to cover about 7 million hectares (17 million acres) in the United States. Major regions where marshes dominate include the prairie pothole region of the Dakotas, the Great Lake coastal marshes, and the Everglades of Florida. They occur in isolated basins, as fringes around lakes, and in sluggish streams and rivers.

Northern Peatlands

As defined here, northern peatlands include the deep peat deposits of the north temperate regions of North America (Fig. 3–5). In the United States these systems are limited primarily to Wisconsin, Michigan, Minnesota, and the glaciated Northeast, although similar peat deposits, called pocosins, are found on the Coastal Plain of the Southeast. There are also mountaintop bogs in the Appalachian Mountains of West Virginia. The total U.S. area of marshes and peatlands (excluding Alaska) is estimated at 9.8 million hectares (24.3 million acres). Minnesota, containing an estimated 2.7 million hectares, has the largest peatland area in the United States (Glaser, 1987). The much more extensive peatlands of Canada cover an estimated 111 million hectares (275 million acres; Zoltai, 1988). Bogs and fens, the two major types of peatlands, occur as thick peat deposits in old lake basins or as blankets across the landscape. Many of

Figure 3–5. Northern peatland. (*Photograph by Curtis Richardson*)

these lake basins were formed by the last glaciation, and the peatlands are considered to be a late stage of a "filling-in" process. There is a wealth of European scientific literature on this wetland type, much of which has influenced the more recent North American literature on the subject. Bogs are noted for their nutrient deficiency and waterlogged conditions and for the biological adaptations to these conditions such as carnivorous plants and nutrient conservation.

Southern Deepwater Swamps

These are freshwater woody wetlands of the southeastern United States that have standing water for most if not all of the growing season (Fig. 3–6). These swamps occur in a variety of nutrient and hydrologic conditions and are normally dominated by various species of cypress (*Taxodium*) and gum/ tupelo (*Nyssa*). These wetlands can occur as isolated cypress domes fed primarily by rainwater or as alluvial swamps that are flooded annually by adjacent streams and rivers.

Figure 3–6. Southern deepwater swamp.

Riparian Forested Wetlands

Extensive tracts of riparian wetlands, which occur along rivers and streams, are occasionally flooded by those bodies of water but are otherwise dry for varying portions of the growing season (Fig. 3–7). Riparian forests and deepwater swamps combined constitute the most extensive class of wetlands in the United States, covering about 22.3 million hectares (55 million acres; Dahl and Johnson, 1991). In the southeastern United States, riparian ecosystems are referred to as bottomland hardwood forests. They contain a diverse vegetation that varies along gradients of flooding frequency. Riparian wetlands also occur in arid and semiarid regions of the United States, where they are often a conspicuous feature of the landscape in contrast with the surrounding nonforested vegetation. Riparian ecosystems are generally considered to be more productive than the adjacent uplands because of the periodic inflow of nutrients, especially when flooding is seasonal rather than continuous.

Figure 3–7. Riparian forested wetland during flood stage.

THE STATUS OF WETLANDS IN THE UNITED STATES

Major regions of wetlands in the United States are shown in Figure 3–8. Regional diversity and the lack of unanimity about the definition of a wetland, as described in Chapter 2, make it difficult to inventory the wetland resources of the United States. Nevertheless, attempts have been made to find out how many wetlands there are in the United States and to determine the rate at which they are changing. Based on a review of several studies that have been made of wetland trends in the United States, two general statements can be made: (1) Estimates of the area of wetlands in the United States, while they vary widely, are becoming more accurate and (2) most studies indicate a rapid rate of wetland loss in the United States, at least prior to the mid-1970s.

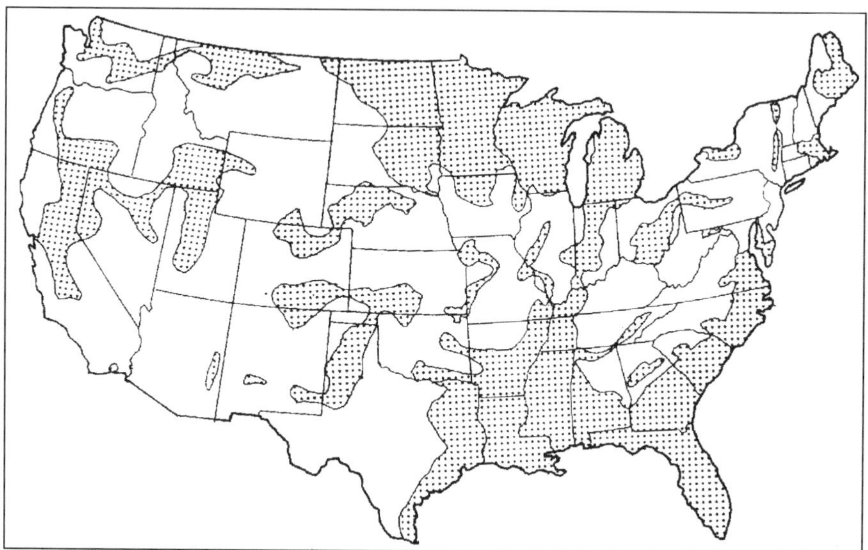

Figure 3–8. Distribution of wetlands in conterminous United States. *(National Wetlands Inventory, U.S. Fish and Wildlife Service, St. Petersburg, FL)*

Wetland Area Estimates

Historical estimates of wetland area in the 48 conterminous states are given in Table 3–3. The numbers vary widely for several reasons. First, the purposes of the inventories varied from study to study. Early wetland censuses, e.g., Wright (1907) and Gray et al. (1924), were based on lands suitable for drainage for agriculture. Later inventories of wetlands (Shaw and Fredine, 1956) were concerned with waterfowl protection. Only within the last decade have wetland inventories considered all of the values of these ecosystems. Second, the definition and classification of wetlands varied with each study, ranging from simple terms to complex hierarchical classifications. Third, the methods available for estimating wetlands changed over the years or varied in accuracy. Remote sensing from aircraft and satellites is one example of a technique for wetland studies that was not generally available or used before the 1970s. Fourth, early estimates were often based on fragmentary records. Finally, in a number of instances the borders of geographical or political units changed between censuses, leading to gaps or overlaps in data.

The first studies of wetland abundance, by the U.S. Department of Agriculture, took place in the early twentieth century. A 1906 inventory was the result of a request from Congress to determine "the amount and location of swamp and overflow lands in the United States that can be reclaimed for agriculture" (Wright, 1907, as quoted in Shaw and Fredine, 1956). That study, which did not include intertidal coastal wetlands, estimated 32 million hectares

Table 3–3. Estimates of Wetland Area in the United States at Different Times

Date	*Wetland Area, million hectares*[a]	*Reference*
Presettlement	87	Roe and Ayres (1954)
Presettlement	51	Soil Conservation Service (cited in Shaw and Fredine, 1956)
Presettlement	60–75	Office of Technology Assessment (1984)
Presettlement	86.2	USDA estimate, in Dahl (1990)
Presettlement	89.5	Dahl (1990)
1906	32[b]	Wright (1907)
1922	37 (total) 3 (tidal) 34 (inland)	Gray et al. (1924)
1940	39.4[c]	Whooten and Purcell (1949)
1954	30.1[d] (total) 3.8 (coastal) 26.3 (inland)	Shaw and Fredine (1956)
1954	43.8 (total) 2.3 (estuarine) 41.5 (inland)	Frayer et al. (1983)
1974	40.1 (total) 2.1 (estuarine) 38.0 (inland)	Ibid.; Tiner (1984)
mid-1970s	42.8[e] (total) 2.2 (estuarine) 40.6 (inland)	Dahl and Johnson (1991)
mid-1980s	41.8[e] (total) 2.2 (estuarine) 39.6 (inland)	Dahl and Johnson (1991)

[a]For 48 conterminous states unless otherwise noted.
[b]Does not include tidal wetlands or eight public land states in West.
[c]Outside of organized drainage enterprises.
[d]Only included wetlands important for waterfowl.
[e]Based on estimates of NWI classes for vegetated estuarine and palustrine wetlands.

(79 million acres) of wetlands in the United States. Of that area 21.3 million hectares (52.7 million acres), or two thirds of the total, were found to be "not fit for cultivation, even in favorable years, unless cleared or protected" (Shaw and Fredine, 1956). A second inventory, conducted in 1922 (Gray et al., 1924), found 37 million hectares (91.5 million acres) of wetlands, including 3 million hectares (7.4 million acres) of tidal marshes and the remainder composed of inland wetlands.

One of the first wetland surveys in the United States that was undertaken based on their habitat values rather than on value for agriculture was carried out in 1954 and published by the U.S. Fish and Wildlife Service two years later as Circular 39 (Shaw and Fredine, 1956). That survey, which relied on a classification scheme of 20 wetland types (described in Chap. 18), estimated that the nation had 30.1 million hectares (74 million acres) of wetlands that were important to waterfowl. The study further identified 25.7 million hectares (63.5 million acres) as inland freshwater wetlands, 0.6 million hectares (1.6 million acres) as inland saline wetlands, 2.1 million hectares (5.3 million acres) as coastal saline wetlands, and 1.6 million hectares (4.0 million acres) as coastal freshwater wetlands. The major shortcoming of the Circular 39 survey was that it considered only wetlands that were important for waterfowl and thus failed to consider a large portion of wetlands in the United States. Nevertheless, it is still referred to today and represents a benchmark summary of waterfowl wetlands for the 48 conterminous states.

A later assessment of wetland abundance in the United States, called the National Wetland Trends Study (NWTS), was conducted by the U.S. Fish and Wildlife Service (Frayer et al. 1983). That study used statistical analysis of wetland data derived from detailed mapping for the National Wetlands Inventory to estimate the total coverage of wetlands in the lower 48 states. The results indicated that, in the mid-1950s, there were 43.8 million hectares (108 million acres) of wetlands in the United States, about 45 percent more than the estimate by the Circular 39 study (Shaw and Fredine, 1956). This difference reflects the fact that the NWTS study was interested in all wetlands, whereas the Circular 39 study was interested only in wetlands that were important to waterfowl. The NWTS study also estimated that by the mid-1970s wetlands in the United States had decreased to 40 million hectares (99 million acres). Of those totals, all were inland wetlands except for 2.3 million hectares (5.6 million acres) of estuarine wetlands in the 1950s and 2.1 million hectares (5.2 million acres) of estuarine wetlands in the 1970s. More recent Fish and Wildlife Service analyses of mid-1970s and mid-1980s data (Dahl and Johnson, 1991) show higher estimates for the 1970s and a continuing slow decline to about 42 million hectares (103 million acres) in the mid-1980s.

Presettlement Wetland Estimates

Several estimates have been made of the area of original natural wetlands that existed in the United States in presettlement times. The U.S. Soil Conservation Service estimated that there were originally 51 million hectares (127 million acres) in the lower 48 states (Shaw and Fredine, 1956). Roe and Ayres (1954) included drained wet soils along with swamps and marshes to estimate 87 million hectares (215 million acres) of wetlands in presettlement times. These estimates often have been compared with existing wetland acreage to determine losses. The U.S. Congress Office of Technology Assessment interpreted prior wetland and wet soil data to suggest a range of 60 to 75 million hectares (149 to 185 million acres), although the authors admit that the estimates "are limited by the lack of good data on the amount of land that has been drained or otherwise reclaimed and the relationship between wetlands and wet soils" (Office of Technology Assessment, 1984). Two recent analyses (see Dahl, 1990) give estimates close to Ayres's original figure of 87 million hectares. Given the uncertainty of original records, these are likely to be as definitive as is possible.

State-by-State Distribution

The extent of wetlands by state from the mid-1950s survey of Shaw and Fredine (1956) and from recent estimates by the National Wetlands Inventory is shown in Table 3–4. Although the methods and accuracy of the surveys are not directly comparable, the numbers do give an indication of which areas of the country are dominated by wetlands important for wildlife. The data presented by the National Wetlands Inventory (see Table 3–4) show that, in general, previous estimates of wetlands in individual states were low. For example, in the National Wetland Inventory, the wetland estimates in Alabama, Arizona, Colorado, Connecticut, Illinois, Indiana, Iowa, Kansas, Maine, Massachusetts, Montana, Nebraska, New Hampshire, New Jersey, New Mexico, New York, Ohio, Oklahoma, Oregon, Pennsylvania, Rhode Island, South Dakota, Texas, Vermont, Virginia, West Virginia, and Wyoming are more than double those of previous studies. A few states, notably Arizona, New Hampshire, West Virginia, and Wyoming show a ten-fold or more increase in wetland area between the two studies.

Including Alaska in wetlands estimates of the United States changes the national numbers altogether. The recent estimate by Dahl (1990) of 69 million hectares of wetlands in Alaska alone is 50 percent greater than the total estimate for the lower 48 states. Thus the inclusion of Alaska in wetland surveys of the United States more than doubles the wetland area.

Wetland Changes

Estimates of the rate of wetland loss in the United States have been published in a number of studies. Table 3–4 shows the loss by state; 53 percent of the wetlands in the conterminous United States have been lost since the late 1700s. Shaw and Fredine (1956) gave evidence of the magnitude of wetland loss in the United States from 1850 to 1950 by demonstrating that there was a 46 percent loss of wetlands in seven states that were covered by the Swamp Land Act of 1850. That study also cited a Department of Agriculture estimate that "in the country as a whole, 45 million acres [18 million hectares] were reclaimed by a combination of clearing, drainage, and flood control on land in publicly organized drainage and flood-control enterprises" (Shaw and Fredine, 1956). Based on an estimated 51.4 million hectares (127 million acres) of original wetlands in the United States, Shaw and Fredine (1956) suggested that 35 percent of the wetlands in the United States were lost from primitive times to the 1950s (Table 3–5).

Table 3–4. Surveys of Area of Wetlands by State in the United States[a]

State	Original Wetlands Circa 1780[b] hectares x 1,000	Original Wetlands Circa 1780[b] acres x 1,000	1954 Survey[c] hectares x 1,000	1954 Survey[c] acres x 1,000	National Wetlands Inventory, mid-1980s[b] hectares x 1,000	National Wetlands Inventory, mid-1980s[b] acres x 1,000	Change (percent)[d]
Alabama	3,063	7,568	650	1,598	1,531	3,784	-50
Alaska	68,799	190,000	—	—	68,799	170,000	-0.1
Arizona	377	931	11.5	28	243	600	-36
Arkansas	3,986	9,849	1,532	3,785	1,119	2,764	-72
California	2,024	5,000	226	559	184	454	-91
Colorado	809	2,000	164	404	405	1,000	-50
Connecticut	271	670	9.5	23	70	173	-74
Delaware	194	480	53	131	90	223	-54
Florida	8,225	20,325	6,955	17,185	4,467	11,038	-46
Georgia	2,769	6,843	2,396	5,920	2,144	5,298	-23
Hawaii	24	59	—	—	21	52	-12
Idaho	355	877	44	109	156	386	-56
Illinois	3,323	8,212	173	427	508	1, 255	-85
Indiana	2,266	5,600	115	283	304	751	-87
Iowa	1,620	4,000	56	138	171	422	-89
Kansas	340	841	83	204	176	435	-48
Kentucky	634	1,566	110	273	121	300	-81
Louisiana	6,554	16,195	3,904	9,647	3,555	8,784	-46
Maine	2,614	6,460	154	381	2,104	5,199	-19
Maryland	668	1,650	117	290	178	440	-73
Massachusetts	331	818	94	232	238	588	-28
Michigan	4,533	11,200	1,302	3,217	2,259	5,583	-50
Minnesota	6,100	15,070	2,042	5,045	3,521	8,700	-42
Mississippi	3,995	9,872	1,048	2,589	1,646	4,067	-59
Missouri	1,960	4,844	153	377	260	643	-87

Montana	464	1,147	76	187	340	840	-27
Nebraska	1,178	2,911	263	650	771	1,906	-35
Nevada	197	487	78	192	96	236	-52
New Hampshire	89	220	5.5	14	81	200	-9
New Jersey	607	1,500	109	270	370	916	-39
New Mexico	291	720	20	48	195	482	-33
New York	1,037	2,562	86	213	415	1,025	-60
North Carolina	4,488	11,090	1,641	4,055	2,300	5,690	-44
North Dakota	1,994	4,928	616	1,523	1,008	2,490	-49
Ohio	2,024	5,000	40	98	195	483	-90
Oklahoma	1,150	2,843	113	280	384	950	-67
Oregon	915	2,262	191	473	564	1,394	-38
Pennsylvania	456	1,127	21	53	202	499	-56
Rhode Island	42	103	10	25	26	65	-37
South Carolina	2,596	6,414	1,367	3,377	1,885	4,659	-27
South Dakota	1,107	2,735	304	752	720	1,780	-35
Tennessee	784	1,937	335	828	318	787	-59
Texas	6,475	16,000	1,514	3,741	3,080	7,612	-52
Utah	325	802	475	1,174	226	558	-30
Vermont	138	341	15	38	89	220	-35
Virginia	748	1,849	219	541	435	1,075	-42
Washington	546	1,350	94	233	380	938	-31
West Virginia	54	134	1.5	4	41	102	-24
Wisconsin	3,966	9,800	1,129	2,791	2,157	5,331	-46
Wyoming	809	2,000	12	30	506	1,250	-38
Total Wetlands	158,395	391,388	—	—	111,060	274,426	-30
Total "Lower 48"	89,491	221,130	30,126	74,439	42,240	104,374	-53

[a]Surveys used different techniques and should not be directly compared.
[b]Dahl, 1990
[c]Shaw and Fredine, 1956
[d]from original wetlands (1780s) to mid-1980s

In a study of wetland losses in the United States from the 1950s to the 1970s, Frayer et al. (1983) estimated a net loss of more than 3.7 million hectares (9.1 million acres) (8.5% loss), or an average annual loss of 185,000 hectares (460,000 acres) (Table 3–5). The losses continued into the 1980s at a reduced rate (Tables 3–5 and 3–6). Over the 20-year interval from the mid-1950s to the mid-1970s, wetland area equivalent to the combined size of Massachusetts, Connecticut, and Rhode Island was lost. Palustrine emergent wetlands and palustrine forested wetlands were hardest hit. They disappeared at the rate of 7 percent and 4.8 percent per decade, respectively. In the decade after the mid-1970s, inland freshwater marshes (palustrine emergent wetlands) sustained no additional losses,

Table 3–5. Estimates of Wetland Losses in the Conterminous United States

	Wetland Loss			
Period	*hectares x 1,000*	*acres x 1,000*	*Percent*	*Reference*
Total Wetlands				
Presettlement–1950s	18,000	45,000	35	Shaw and Fredine (1956)
Presettlement–1980s	47,300	116,800	53	Dahl (1990)
1922–1954			0.2/yr	Zinn and Copeland (1982)
1954–1970s			0.50-0.65/yr	Ibid.
1950s–1970s	3,700 (185/yr)	9,150 (460/yr)	8.5 (0.4/yr)	Frayer et al. (1983)
1970s–1980s[a]	1,057	2,611	2.5	Dahl and Johnson (1991)
Southern Bottomland Hardwood Forests				
1883–1991	6,500	16,100	77	The Nature Conservancy (1992)
1940–1975	2,300 (65/yr)	5,700 (160/yr)	16 (0.45/yr)	Turner et al. (1981)
1960–1975	2,600 (175/yr)	6,500 (431/yr)	18 (1.2/yr)	Ibid.
Coastal Wetlands				
1922–1954	260 (8.1/yr)	650 (20/yr)	6.5 (0.2/yr)	Gosselink and Baumann (1980)
1954–1974	370 (19/yr)	920 (47/yr)	9.9 (0.5/yr)	Ibid.
1950s–1970s	146 (7.3/yr)	360 (18/yr)	—	Tiner (1984)
1970s–1980s[b]	29 (2.9/yr)	71 (7.1/yr)	1.7 (0.15/yr)	Dahl and Johnson (1991)

[a]Details given in Table 3–6
[b]Vegetated estuarine emergent wetlands

but forested wetland loss increased slightly to 5.0 percent per decade (Table 3–6).

Several studies have described losses of particular types of wetlands. Gosselink and Baumann (1980) estimated a loss rate of coastal wetlands of 8,100 hectares/year (20,000 acres/yr) between 1922 and 1954 and a higher loss rate of 19,000 hectares/year (47,000 acres/yr) between 1954 and the 1970s. By comparison, Tiner (1984) estimated that 7,300 hectares/year (18,000 acres/yr) of estuarine wetlands were lost from the 1950s to 1970s, and Dahl and Johnson (1991) measured an annual loss of 2,900 hectares (7,100 acres) from the mid-1970s to the mid-1980s. Thus for coastal wetlands the loss rates declined as public understanding of the value of wetlands increased and wetland regulation became more stringent.

The bottomland hardwood forests of the lower Mississippi River alluvial floodplain provide another sobering example. These forests covered approximately 8.5 million hectares (21 million acres) prior to European settlement. Now fewer than 2 million hectares (4.9 million acres) remain, more than 95 percent of them in Louisiana, Mississippi, and Arkansas (Nature Conservancy, 1992). The

Table 3–6. Conterminous United States Wetland Area, 1980s, and Estimates of Wetland Loss in the Previous Decade

Wetland Class		*Wetland Total Area mid-1980s*	*Wetland Loss Since mid-1970s*	
National Wetlands Inventory[a]	*This Book*	*hectares x 1,000*	*hectares x 1,000*	*Percent*
Estuarine intertidal emergent	Tidal salt marshes and tidal freshwater marshes	1,649	28.7	1.7
Estuarine intertidal forested and shrub	Mangrove wetlands	287	0	0
Palustrine emergent	Inland freshwater marshes and northern peatlands	9,928	89	(0.9)[b]
Palustrine forested and shrub[c]	Southern deepwater swamps and riparian wetlands	27,152	1,442	5.0
Total palustrine and estuarine wetlands[d]		41,780	1,057	2.5

[a]The National Wetlands Inventory classification system on which these data are based has no exact equivalence with the wetland types considered in this book.

[b]Indicates gain.

[c]This classification includes 6 million hectares of inland shrub wetlands.

[d]Includes nonvegetated areas and is not the sum of preceding rows. Palustrine shrub wetlands sustained almost no net loss.

Source: Based on data fom Dahl and Johnson, 1991

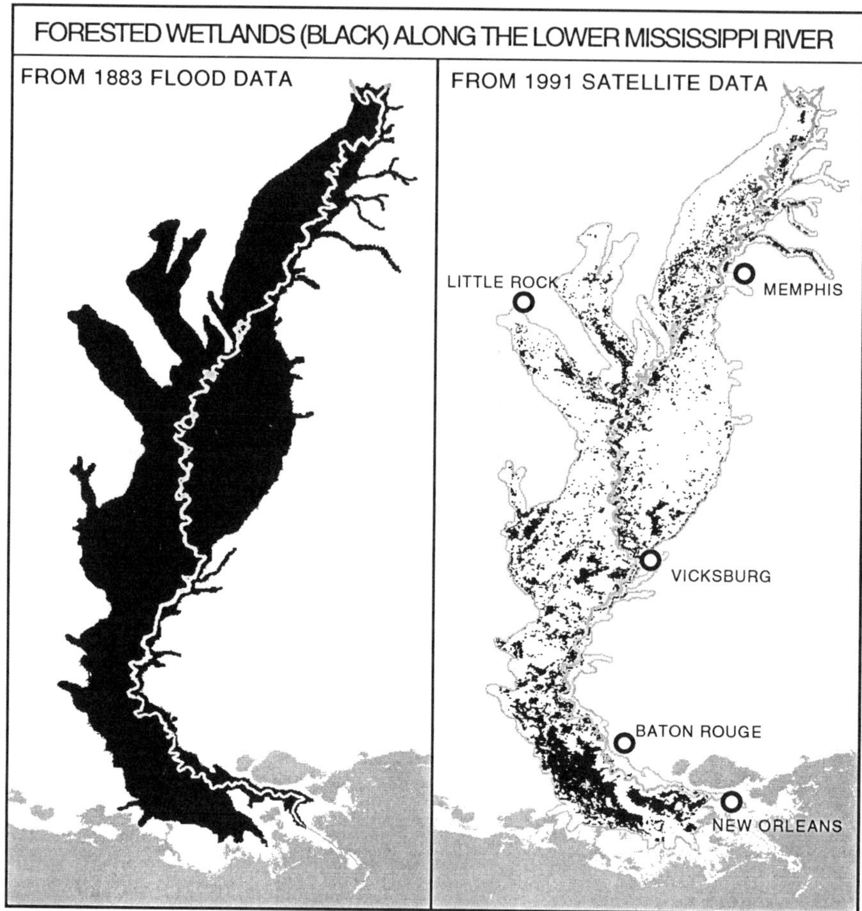

Figure 3–9. Historic and present distribution of bottomland wetland forests (black) in the Mississippi River alluvial floodplain. The 1991 map was developed from remotely sensed advanced Very High Resolution data, which has a spatial resolution of 1.1 km. (*From The Nature Conservancy, 1992; copyright © 1992 by The Nature Conservancy, Baton Rouge, LA, reprinted with permission*)

23 percent that have survived are seriously fragmented and have lost many of their original functions (Fig. 3–9).

By themselves estimates of net wetland losses provide an incomplete picture of the dynamics of change. A more complete picture would show that human activities converted millions of hectares of wetland from one class to another. Through those conversions some wetlands classes increased in area at the expense of other types (Fig. 3–10). Considering the period from the mid-1970s to the mid-1980s, swamps and forested riparian wetlands suffered the greatest loss, 1.4 million hectares (3.4 million acres). Although 800,000 hectares (2 mil-

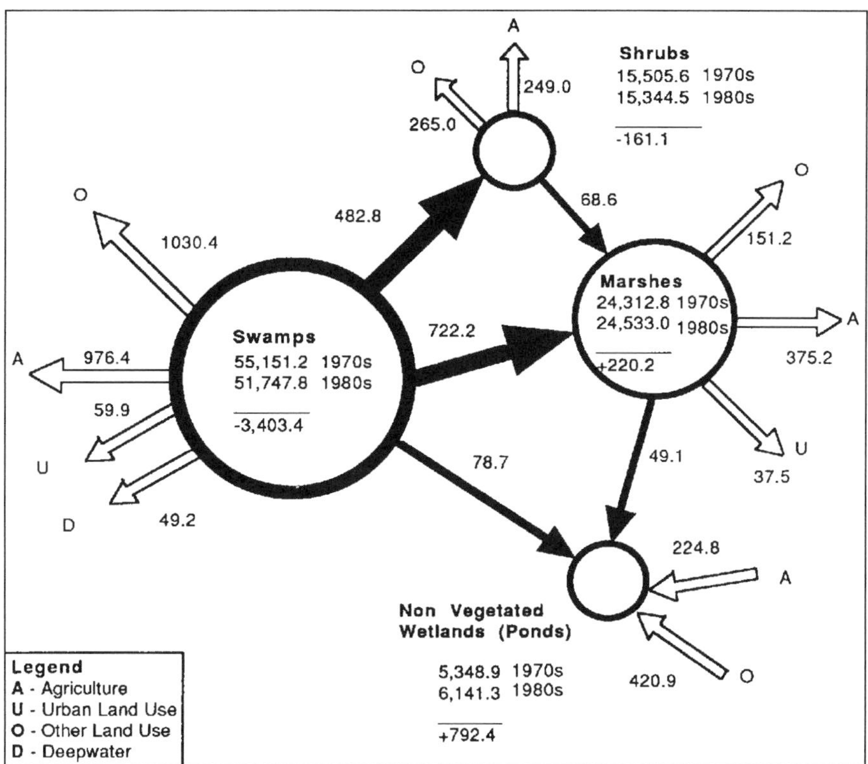

Figure 3–10. Wetland conversions in the conterminous United States, mid-1970s to mid-1980s (all numbers in thousands of acres; 100 acres = 40.5 ha). The figure shows how misleading the net change figures are. For example, although there was a net gain in freshwater marshes, it occurred despite destruction of about 500,000 acres, because 722,000 acres of swamps were converted to freshwater marsh. See text for further discussion. (*From Dahl and Johnson, 1991*)

lion acres) were converted to agricultural and other land uses, large areas were converted to other wetland types: 292,000 hectares to marshes, 195,000 hectares to shrubs, and 32,000 hectares to nonvegetated wetlands. Although shrub wetlands lost 208,000 hectares to agriculture and other nonwetland uses, this was almost offset by the conversion of forested wetlands to shrubs, leaving a net loss of 65,000 hectares. The net gain of 89,000 hectares of marshes occurred despite a loss of 213,000 hectares to agriculture and other land uses because 320,000 hectares of swamps and shrub wetlands changed to marshes (Dahl and Johnson, 1991; in this report, based on the National Wetlands Inventory classification scheme, shrub-scrub wetlands are considered a separate class of palustrine wetlands; in our book, we consider them with swamps and riparian wetlands).

WETLANDS OF CANADA

Canada has about three times the area of wetlands found in the lower 48 states. By one estimate, Canada has about 127.2 million hectares (314 million acres) of wetland, or about 14 percent of the country (Zoltai, 1988). Most of that area (111.3 million hectares, or about 275 million acres) is defined as peatlands. The greatest concentration of Canadian wetlands can be found in the provinces of Manitoba and Ontario. The National Wetlands Working Group (1988) estimated that there were 22.5 million hectares and 29.2 million hectares, respectively, of wetlands in these two provinces, or about 41 percent of the total wetlands of Canada. Much of this total is boreal forested peatlands as bogs and fens, but there are also many deltaic and shoreline marshes and floodplain swamps in the region (Zoltai et al., 1988).

SOME MAJOR REGIONAL WETLANDS OF NORTH AMERICA

Many regions in the United States and Canada support or once supported large, contiguous wetlands or many smaller and more numerous wetlands. These are called regional wetlands in this book. They are often large, heterogeneous wetland areas such as the Okefenokee Swamp in Georgia and Florida that defy categorization as one type of wetland ecosystem. Regional wetlands can also be large expanses that support wetlands of a similar type such as the Prairie Pothole region of the Dakotas and Minnesota in the United States and Manitoba, Saskatchewan, and Alberta in Canada or the Pocosins region of North Carolina. Some regional wetlands such as the Great Dismal Swamp at the Virginia–North Carolina border have been drastically altered since presettlement times, whereas others such as the Great Kankakee Marsh of northern Indiana and Illinois and the Great Black Swamp of northwestern Ohio have almost disappeared as a result of extensive drainage programs.

Each of these regional wetland areas has had a significant influence on the folklore and development of its region, and many have benefited from major investigations by wetland scientists. These studies have taught us much about wetlands and have identified much of their intrinsic value. Descriptions of some of the regional wetlands, shown in Figure 3–11, are given below. The reader is referred to Chapters 8 through 14 and to the scientific literature cited below for a more complete discussion of the structures and functions of these wetlands. For those who wish a more complete description of some of these wetlands and several others in the United States, Thomas (1976) provides an excellent photojournalistic interpretation, and the National Audubon Society has published a colorful field guide to inland wetlands of the United States (Niering, 1985). Littlehales and Niering (1991) have combined to produce beautifully illustrated descriptions of North American wetlands. The National Wetlands Working Group (1988) provides a particularly comprehensive description of major regional wetlands in Canada.

Figure 3–11. Location of some major regional wetlands in North America, as discussed in text.

The Everglades and Big Cypress Swamp

The southern tip of Florida, from Lake Okeechobee southward to the Florida Bay, harbors one of the unique regional wetlands in the world. The region encompasses three major types of wetlands in its 34,000 km^2 (13,000 mi^2): the Everglades, the Big Cypress Swamp, and the coastal mangroves and glades. The water that passes through the Everglades on its journey from Lake Okeechobee to the sea is often called a river of grass, although it is often only centimeters in depth and 80 km wide. The Everglades is dominated by saw grass (*Cladium jamaicense*), which is actually a sedge, not a grass. The expanses of saw grass, which can be flooded by up to a meter of water in the wet season (summer) and burned in a fire in the dry season (winter/spring), are interspersed with deeper water sloughs and tree islands, or *hammocks*, that support a vast diversity of tropical and subtropical plants, including hardwood trees, palms, orchids, and other air plants. To the west of the saw grass Everglades is the Big Cypress Swamp, called *big* because of its great expanse, not because of the size of the trees. The swamp is dominated by cypress interspersed with pine flatwoods and wet prairie. It receives about 125 cm of rainfall per year but does not receive major amounts of overland flow as the Everglades does. The third major wetland type, mangroves, forms impenetrable thickets where the saw grass and cypress swamps meet the saline waters of the coastline.

Numerous popular books and articles, including the classic *The Everglades: River of Grass* by Marjory Stoneman Douglas (1947), have been written about the Everglades and its natural and human history. Davis (1940, 1943) gives some of the earliest and best descriptions of the plant communities in southern Florida. An extensive study of the functional aspects of wetland ecosystems in the Fahkahatchee Strand of the Big Cypress Swamp was published by the U.S. Environmental Protection Agency (Carter et al., 1973), and the findings of several years of ecological studies that have taken place in Corkscrew Swamp Sanctuary also have been published (Duever et al., 1984; Duever, 1984).

Since about half of the original Everglades has been lost to agriculture (the Everglades Agricultural Area) in the north and to urban development in the east, concern for the remaining Everglades has been extended to the quality and quantity of water delivered to the Everglades through a series of canals and water conservation areas (Koch and Reddy, 1992; Gunderson and Loftus, 1993). Several plans for improving the water quality as it leaves the agricultural areas have been suggested, some in extensive court proceedings. There is also concern about the loss of habitat and the suitable hydroperiod for declining populations of wading birds such as the wood stork (*Mycteria americana*) and the white ibis (*Eudocimus albus*) (Walters et al., 1992). North of the Everglades there is a renewed effort to restore the ecological functions of the Kissimmee River, including many of its backswamp areas (Loftin et al., 1990). This river feeds Lake Okeechobee, which, in turn, spills over to the Everglades.

The Okefenokee Swamp

The Okefenokee Swamp in southeastern Georgia and northeastern Florida is a 1,750 km^2 (680 mi^2) mosaic of several different types of wetland communities. It is believed to have been formed in the Pleistocene or later when ocean water was impounded and isolated from the receding sea by a sand ridge (now referred to as the Trail Ridge) that kept water from flowing directly toward the Atlantic. The swamp forms the headwaters of two river systems: the Suwannee River, which flows southwest through Florida to the Gulf of Mexico, and the St. Mary's River, which flows southward and then eastward to the Atlantic Ocean.

Much of the swamp is now part of the Okefenokee National Wildlife Refuge, established in 1937 by Congress. The Okefenokee is named for an Indian word meaning "Land of Trembling Earth" because of the numerous vegetated floating islands that dot the wet prairies. Auble et al. (1982) have identified six major wetland communities in the Okefenokee Swamp: (1) pond cypress forest, (2) emergent and aquatic bed prairie, (3) broad-leaved evergreen forest, (4) broad-leaved shrub wetland, (5) mixed cypress forest, and (6) black gum forest. Pond cypress (*Taxodium distichum* var. nutans), black gum (*Nyssa sylvatica* var. biflora), and various evergreen bays (e.g., *Magnolia virginiana*) are found in slightly elevated areas where water and peat deposits are shallow. Open areas, which are called prairies, include lakes, emergent marshes of *Panicum* and *Carex,* floating-leaved marshes of water lilies (e.g., *Nuphar* and *Nymphaea*), and bladderwort (*Utricularia*). Fires that actually burn peat layers are an important part of this ecosystem and have recurred in a 20- to 30-year cycle when water levels became very low (Cypert, 1961). Many believe that the open prairies represent early successional stages, maintained by burning and logging, of what would otherwise be a swamp forest.

A more complete description of the Okefenokee Swamp is contained in a book compendium edited by Cohen et al. (1984) and in numerous papers (Wright and Wright, 1932; Cypert, 1961, 1972; Schlesinger and Chabot, 1977; Schlesinger, 1978; Auble et al., 1982; Bosserman, 1983a, 1983b).

The Pocosins

Pocosins are evergreen shrub bogs found on the Atlantic Coastal Plain from Virginia to northern Florida. These wetlands are particularly dominant in North Carolina, where an estimated 3,700 km^2 (1,450 mi^2) remained undisturbed or only slightly altered in 1980, whereas 8,300 km^2 (3,200 mi^2) were drained for other land uses between 1962 and 1979 (Richardson et al., 1981). The word *pocosin* comes from the Algonquin phrase for "swamp on a hill." In successional progression and in nutrient-poor acid conditions, pocosins resemble bogs typical of much colder climes and, in fact, were classified as bogs in the 1954 National Wetland Survey (Shaw and Fredine, 1956). Richardson et al. (1981) described the typical pocosin ecosystem in North Carolina as being dominated by evergreen shrubs (*Cyrilla racemiflora, Magnolia virginiana, Persea borbonia, Ilex glabra, Myrica heterophylla,* and *Smilax laurifolia*) and pine (*Pinus serotina*). Pocosins are found "growing on water-logged, acid, nutrient poor, sandy or peaty soils located on broad, flat topographic plateaus, usually removed from large streams and subject to periodic burning" (Richardson et al., 1981). Draining and ditching for agriculture and forestry have affected pocosins in North Carolina. Proposed peat mining and phosphate mining could cause serious losses of these wetlands.

Summaries of the ecological, economic, and legal aspects of pocosin management are included in *Pocosin Wetlands,* edited by Richardson (1981), and in a review paper by Richardson (1983). Descriptions of the extent and phytosociology of these wetlands have been published by Wells (1928), Woodwell (1956), Wilson (1962), and Kologiski (1977).

The Great Dismal Swamp

The Great Dismal Swamp is one of the northernmost "southern" swamps on the Atlantic Coastal Plain and one of the most studied and romanticized wetlands in the United States. The swamp covers approximately 850 km^2 (330 mi^2) in southeastern Virginia and northeastern North Carolina near the urban sprawl of Norfolk–Newport News–Virginia Beach. Its size once extended over 2,000 km^2 (770 mi^2) (F. P. Day, 1982). The swamp has been severely affected by human activity during the past 200 years. Draining, ditching, logging, and fire played a role in diminishing its size and altering its ecological communities (Berkeley and Berkeley, 1976). The Great Dismal Swamp was once primarily a magnificent bald cypress-gum swamp that contained extensive stands of Atlantic white cedar (*Chamaecyparis thyoides*). Although remnants of those communities still exist today, much of the swamp is dominated by red maple (*Acer rubrum*), and mixed hardwoods are found in drier ridges (F. P. Day, 1982). In the center of the swamp lies Lake Drummond, a shallow, tea-colored, acidic body of water. The source of water for the swamp is thought to be underground along its western edge as well as surface runoff and precipitation. Drainage occurred in the Great Dismal Swamp early as 1763 when a corporation called the Dismal Swamp Land Company, which was owned in part by George Washington, built a canal from the western edge of the swamp to Lake Drummond to establish farms in the basin. In general, that effort, like several others in the ensuing years, failed. Timber companies, however, found economic reward in the swamp by harvesting the cypress and cedar for shipbuilding and other uses. One of the last timber companies that owned the swamp, the Union Camp Corporation, gave almost 250 km^2 of the swamp to the federal government to be maintained as a national wildlife refuge (Dabel and Day, 1977).

An extensive literature describes the history and ecology of the Great Dismal Swamp (e.g., Berkeley and Berkeley, 1976; Whitehead, 1972). The structural and functional characteristics of the vegetation of the swamp have been described in several studies from Old Dominion University (Dabel and Day, 1977; Day and Dabel, 1978; McKinley and Day, 1979; Atchue et al., 1982, 1983; Gomez and Day, 1982; Train and Day, 1982; Day, 1982, 1987; Day et al., 1988; Atchue et al., 1983; Laderman, 1989; Powell and Day, 1991). Also, at least one book, *The Great Dismal Swamp* (Kirk, 1979), describes ecological and historical aspects of this important wetland.

The Big Rivers of the South Atlantic Coast

The Atlantic Coastal Plain, extending from North Carolina to the Savannah River in Georgia, is a land dominated by forested wetlands and marshes and cut by large rivers that drain the Piedmont and cross the Coastal Plain in a northwest–southeast direction to the ocean. These rivers include the Roanoke, Chowan, Little Pee Dee, Great Pee Dee, Lynches, Black, Santee, Congaree, Altamaha, Cooper, Edisto, Combahee, Coosawhatchie, and Savannah as well as a host of smaller tributaries. Extensive bottomland hardwood forests and cypress swamps line these rivers and spread into the lowlands between them. Interspersed in these forests are hundreds of Carolina Bays, small elliptical-shaped lakes of uncertain origin surrounded by or overgrown with marshes and forested wetlands (Knight et al., 1984). The origin of these lake-wetland complexes, of which there are more than 500,000 along the eastern Coastal Plain, has been suggested to be meteor showers, wind, or groundwater flow (D. C. Johnson, 1942; H. T. Odum, 1951; Prouty, 1952; Savage, 1983). Along the coast freshwater tides on the lower rivers formerly overflowed extensive forests, but many of these were cleared in the early 1800s to establish rice plantations. Most of the rice plantations have since been abandoned, and the former fields are now extensive freshwater marshes that have become a paradise for ducks and geese. The estuaries at the mouths of the rivers support the most extensive salt marshes on the Southeast Coast.

In 1825 Robert Mills wrote of Richland County, South Carolina: "What clouds of miasma, invisible to sight, almost continually rise from these sinks of corruption, and who can calculate the extent of its pestilential influence?" (quoted by Dennis, 1988). At that time only 10,000 hectares of the 163,000 hectares county were being cultivated. Almost all the rest was a vast, untouched swamp. Our appreciation of those swamps has changed dramatically since that time, and parts of that swamp are now the Congaree Swamp National Monument and the Francis Beidler Forest. Both preserves contain extensive stands of cypress more than 500 years old that escaped the logger's ax in the late 1800s.

Excellent descriptions of the wetlands of this region were written by Savage (1956, 1983), Cely (1974), Gaddy, (1978), Wharton (1978), Porcher (1981), Sharitz and Gibbons (1982), and Dennis (1988).

The Prairie Potholes

A significant number of small wetlands, primarily freshwater marshes, are found in a 780,000 km^2 (300,000 mi^2) region in the states of North and South Dakota and Minnesota and in the Canadian provinces of Manitoba, Saskatchewan, and Alberta (Fig. 3–12). These wetlands, called *prairie potholes,* were formed by glacial action during the Pleistocene. This region is considered one of the most important wetland regions in the world because of its numerous shallow lakes

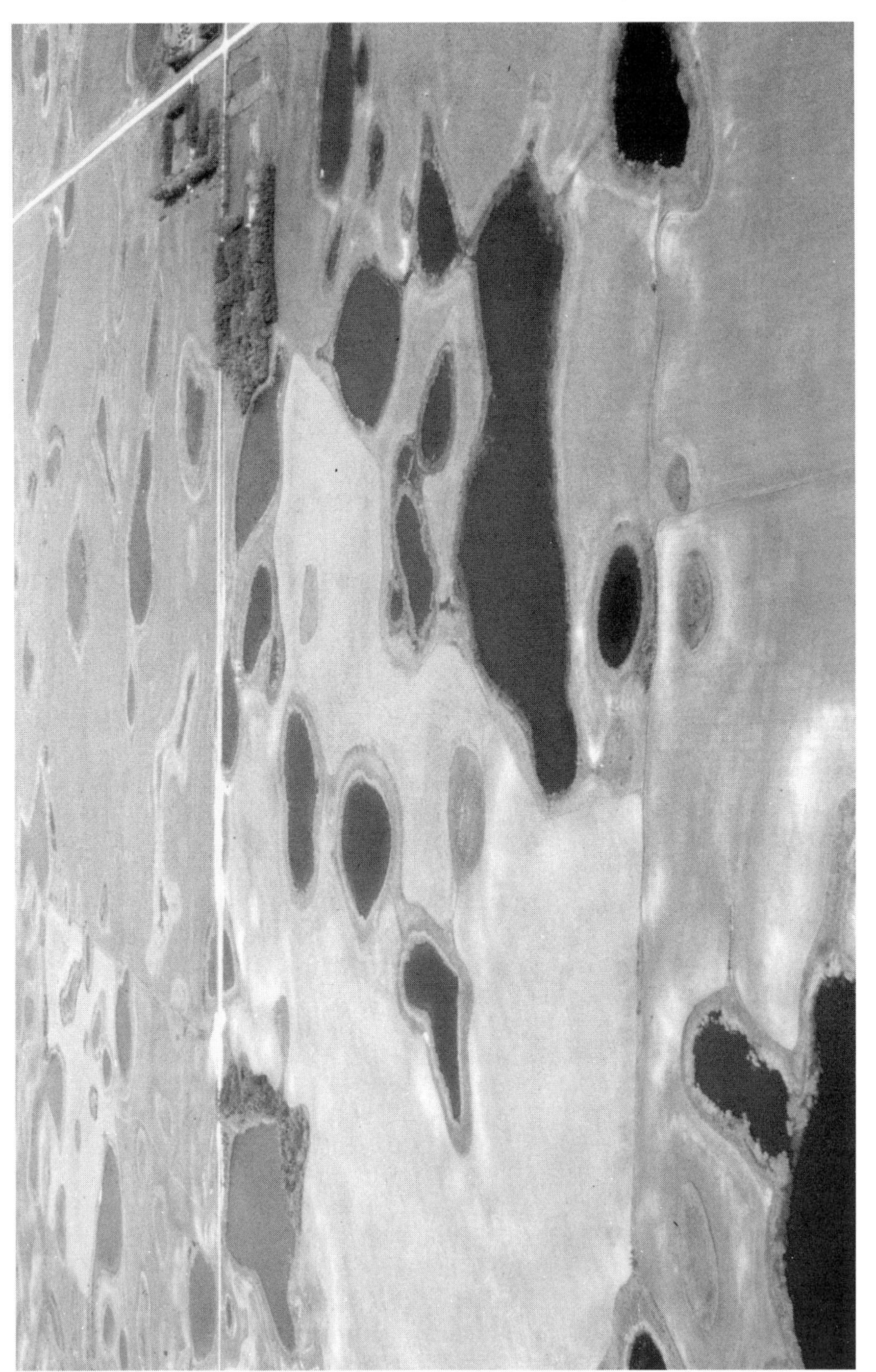

and marshes, its rich soils, and its warm summers (Weller, 1981). It is believed that 50 percent to 75 percent of all the waterfowl produced in North America in any given year comes from this region (Leitch and Danielson, 1979; Ogaard et al., 1981). Many of the prairie potholes have been drained or altered for agriculture. More than half of the original 80,000 km^2 of wetlands may have been drained (Leitch, 1989). An estimated 500 km^2 (190 mi^2) of prairie pothole wetlands in North Dakota, South Dakota, and Minnesota were lost between 1964 and 1968 alone (Weller, 1981). But major efforts to protect the remaining prairie potholes are progressing. Almost 4,000 km^2 (1,500 mi^2) of wetlands have been purchased under the U.S. Fish and Wildlife Service Waterfowl Production Area program in North Dakota alone since the early 1960s (Weller, 1981). The Nature Conservancy and other private foundations have also purchased many wetlands in the region.

Much of the literature on prairie potholes has centered on their role in the breeding and feeding of migratory waterfowl and other nongame species (e.g., Stewart and Kantrud, 1973; Kantrud and Stewart, 1977). A selected annotated bibliography on this subject has been prepared by Ogaard et al. (1981). A wetland classification, described in more detail in Chapter 18, was developed by Stewart and Kantrud (1971, 1972) for the prairie pothole region. The social and economic implications of draining prairie pothole wetlands were described by Leitch and Danielson (1979), Leitch (1989), and Leitch and Ekstrom (1989). The ecology of the marshes of this region is discussed in books written by Weller (1987) and edited by van der Valk (1989).

The Nebraska Sandhills

South of the prairie pothole region is an irregular-shaped region of 51,800 km^2 in northern Nebraska described as "the largest stabilized dune field in the Western Hemisphere" (Novacek, 1989). These Nebraska sandhills represent an interesting and sensitive coexistence of wetlands, agriculture, and a very important aquifer. The area was originally mixed-grass prairie composed of thousands of small wetlands in the interdunal valleys. Much of the region is now used for farming and rangeland agriculture, and many of the wetlands in the region have been preserved even though the vegetation is often harvested for hay or grazed by cattle. The Ogallala Aquifer is an important source of water for the region and is recharged to a significant degree through overlying dune sands and to some extent through the wetlands. It has been estimated that there are 558,000

Figure 3–12 (*opposite page*). Oblique aerial view of prairie pothole wetlands, showing many small ponds surrounded by wetlands, in the middle of large agricultural fields. (*File photograph, U.S. Fish and Wildlife Service, Northern Prairie Wildlife Research Center, Jamestown, North Dakota*)

hectares (1.4 million acres) of wetlands in the Nebraska sandhills; many of these wetlands are wet meadows that contain water levels determined by local water table levels. The wetlands in the region have been threatened by agricultural development, especially pivot irrigation systems that cause a lowering of the local water tables despite increased wetland flooding in the vicinity of the irrigation systems. Like the prairie potholes to the north, the Nebraska sandhill wetlands are important breeding grounds for numerous waterfowl, including about 2 percent of the mallard duck breeding population in the north–central flyway (Novacek, 1989).

The Great Kankakee Marsh

For all practical purposes, this wetland no longer exists, although until about 100 years ago, it was one of the largest marsh-swamp basins in the interior United States. Located primarily in northwestern Indiana and northeastern Illinois, the Kankakee River Basin is 13,700 km^2 (5,300 mi^2) in size, including 8,100 km^2 (3,140 mi^2) in Indiana, where most of the original Kankakee Marsh was located. From the river's source to the Illinois line, a direct distance of only 120 km (75 mi), the river originally meandered through 2,000 bends along 390 km (240 mi), with a nearly level fall of only 8 cm/km (5 in/mi). Numerous wetlands, primarily wet prairies and marshes, remained virtually undisturbed until the 1830s, when settlers began to enter the region. The naturalist Charles Bartlett (1904) described the wetland as follows:

> More than a million acres of swaying reeds, fluttering flags, clumps of wild rice, thick-crowding lily pads, soft beds of cool green mosses, shimmering ponds and black mire and trembling bogs—such is Kankakee Land. These wonderful fens, or marshes, together with their wide-reaching lateral extensions, spread themselves over an area far greater than that of the Dismal Swamp of Virginia and North Carolina.

The Kankakee region was considered a prime hunting area until the wholesale clearing of the land for cropland and pasture began in the 1850s. The Kankakee River and almost all of its tributaries in Indiana were channelized into a straight ditch in the late nineteenth century and early twentieth century. In 1938 the Kankakee River in Indiana was reported to be one of the largest drainage ditches in the United States; the Great Kankakee Swamp was essentially gone by then.

Some historical and scientific literature exists for the Great Kankakee Marsh. Early accounts of the region were given by Bartlett (1904) and Meyer (1935). Scientists have studied riparian forested wetlands in the Illinois portion of the wetland, where some preservation was achieved (Mitsch et al., 1979b, c; Mitsch and Rust, 1984).

The Great Black Swamp

Another vast wetland of the Midwest that has practically ceased to exist is the Great Black Swamp in what is now northwestern Ohio, although several wetlands remain as marshes managed for waterfowl at the western end of Lake Erie. The Great Black Swamp was once a combination of marshland and forested swamps that extended about 160 km long and 40 km wide in a southwesterly direction from the lake and covered an estimated 4,000 km^2. The bottom of an ancient extension of Lake Erie, the Black Swamp was named for the rich, black muck that developed in areas where drainage was poor as a result of several ridges that existed perpendicular to the direction of the flow to the lake (Herdendorf, 1987). There are numerous accounts of the difficulty that early settlers and armies (especially during the War of 1812) had in negotiating this region, and few towns of significant size have developed in the location of the original swamp. One account of travel through the region in the early 1800s suggested that "man and horse had to travel mid-leg deep in mud" for three days just to cover a distance of only 50 km (Kaatz, 1955). As with many other wetlands in the Midwest, state and federal drainage acts led to the rapid drainage of this wetland until little of it was left by the beginning of the twentieth century. Only one small example of an interior forested wetland and several coastal marshes (about 150 km^2) remain of the original western Lake Erie wetlands. Descriptions of some of these remaining Great Lakes coastal wetlands are presented in Prince and D'Itri (1985), Herdendorf (1987), Krieger et al. (1992), and Mitsch (1989, 1992b).

Canada's Central and Eastern Province Wetlands

The southern Ontario wetlands include those along the shorelines of the Great Lakes and along the St. Lawrence lowlands and several river valleys (Glooschenko and Grondin, 1988). The marshes of this region, especially those along the Great Lakes in Ontario and in the St. Lawrence lowlands of Ontario and Quebec, are important habitats for migratory waterfowl. Several of the wetlands along the Great Lakes in southern Ontario, including Long Point and Point Pelee on Lake Erie and the St. Clair National Wildlife Area on Lake St. Clair, and along the Upper St. Lawrence River in eastern Ontario and southwestern Quebec, including Cap Tourmente National Wildlife Area and Lac Saint-François, are on Canada's list of Ramsar wetland sites of international waterfowl significance (see Chapter 16). The peatlands of northern Ontario and Manitoba are extensive regions that are used less by waterfowl and more by a wide variety of mammals, including moose, wolf, beaver, and muskrats. Wild rice (*Zizania palustris*), a common plant in littoral zones of boreal lakes, is often harvested for human consumption (Glooschenko and Grondin, 1988). Some of the boreal wetlands are mined for peat that is used for horticultural purposes or fuel.

The Mississippi River Delta

As the Mississippi River reaches the last phase of its journey to the Gulf of Mexico in southeastern Louisiana, it enters one of the most wetland-rich regions of the world. The total area of marshes, swamps, and shallow coastal lakes covers more than 36,000 km^2 (14,000 mi^2). Much of the richness is found in the Atchafalaya River Basin, a distributary of the Mississippi River that serves as both a flood-relief valve for the Mississippi River and a potential captor of its main flow. The Atchafalaya Basin by itself is the third-largest continuous wetland area in the United States (Thomas, 1976) and contains 30 percent of all the bottomland hardwoods in the entire lower Mississippi alluvial valley (Nature Conservancy, 1992). The river passes through this narrow 4,700 km^2 (1,800 mi^2) basin for 190 km (120 mi), supplying water for 1,700 km^2 (650 mi^2) of bottomlands and cypress-tupelo swamps and another 260 km^2 (100 mi^2) of permanent bodies of water (Hern et al., 1980). The Atchafalaya Basin, contained within a system of artificial and natural levees, has had a controversial history of human intervention characterized by dredging, channelization, and oil and gas production.

Another frequently studied freshwater wetland area in the Delta is the Barataria Bay estuary, Louisiana, an interdistributary basin of the Mississippi River that is now isolated from the Mississippi River by a series of flood-control levees. This basin, 6,500 km^2 (2,500 mi^2) in size, contains 700 km^2 (270 mi^2) of wetlands, including cypress-tupelo swamps, bottomland hardwood forests, marshes, and shallow lakes (Conner and Day, 1976).

As the Mississippi River distributaries reach the sea, salt marshes replace the freshwater wetlands. These salt marshes, some of the most extensive and productive in the United States, depend on the influx of fresh water, nutrients, and organic matter from upstream swamps. The total amount of freshwater and saltwater wetlands is decreasing at a rapid rate in coastal Louisiana, amounting to a total wetland loss of 66 km^2 (25 mi^2) per year attributed to both natural and artificial causes (Dunbar, 1990; Dunbar et al., 1992).

Numerous studies have been published on the Delta wetlands and freshwater swamps (Penfound and Hathaway, 1938; Montz and Cherubini, 1973; Conner and Day, 1976, 1982, 1987; J. W. Day et al., 1977; Conner et al., 1981; Kemp and Day, 1984) and coastal marshes (Gosselink, 1984; Conner and Day, 1987). The land loss in coastal Louisiana was most recently described by Britsch and Kemp (1990), Dunbar (1990), and Dunbar et al. (1992).

San Francisco Bay

One of the most altered and most urbanized wetland areas in the United States is San Francisco Bay in northern California. The marshes surrounding the bay covered more than 2,200 km^2 (850 mi^2) when the first European settlers arrived.

Almost 95 percent of that area has since been destroyed, leaving 125 km^2 (50 mi^2) of tidal marsh, some of which was created recently (Josselyn, 1983).

The ecological systems that make up San Francisco Bay range from deep, open water to salt and brackish marshes. The salt marshes are dominated by Pacific cordgrass (*Spartina foliosa*) and pickleweed (*Salicornia virginica*), and the brackish marshes support bulrushes (*Scirpus* spp.) and cattails (*Typha* spp.). Soon after the beginning of the Gold Rush in 1849, the demise of the bay's wetlands began. Industries such as agriculture and salt production first used the wetlands, clearing the native vegetation and diking and draining the marsh. At the same time other marshes were developing in the bay as a result of rapid sedimentation. The sedimentation was caused primarily by upstream hydraulic mining. Sedimentation and erosion continue to be the greatest problems encountered in the remaining tidal wetlands.

An excellent summary of the ecology and management of San Francisco Bay was written by Josselyn (1983), and the ecology of the bay region was described by a number of researchers, including MacDonald (1977), Mahall and Park (1976a, b, c), and Balling and Resh (1983).

PART 2

THE WETLAND ENVIRONMENT

Hydrology of Wetlands

4

Hydrologic conditions are extremely important for the maintenance of a wetland's structure and function, although simple cause and effect relationships are difficult to establish. Hydrologic conditions affect many abiotic factors, including soil anaerobiosis, nutrient availability, and in coastal wetlands, salinity. These, in turn, determine the flora and fauna that develop in a wetland. Finally, completing the cycle, biotic components are active in altering the wetland hydrology. The hydroperiod, or hydrologic signature of a wetland, is the result of the balance between inflows and outflows of water (called the water budget), the soil contours in the wetland, and the subsurface conditions. The hydroperiod can have dramatic seasonal and year-to-year variations, yet it remains the major determinant of wetland function. Major hydrologic inflows include precipitation, flooding rivers, surface flows, groundwater, and tides in coastal wetlands. Simple hydrologic measurements, a water budget approach, and concepts such as turnover time in wetland studies can contribute to a better understanding of specific wetlands. Hydrology affects the species composition and richness, primary productivity, organic accumulation, and nutrient cycling in wetlands. Generally, productivity is highest in wetlands that have highest flow-through of water and nutrients or in wetlands with pulsing hydroperiods. Decomposition in wetlands is slower in anaerobic standing water than it is under dry conditions. Although many wetlands are organic exporters, this cannot be generalized even within one wetland type. Nutrient cycling is enhanced by hydrology-mediated inputs, and nutrient availability is often increased by reduced conditions in wetland substrates.

The hydrology of a wetland creates the unique physiochemical conditions that make such an ecosystem different from both well-drained terrestrial systems and deepwater aquatic systems. Hydrologic pathways such as precipitation, surface runoff, groundwater, tides, and flooding rivers transport energy and nutrients to and from wetlands. Water depth, flow patterns, and duration and frequency of flooding, which are the result of all of the hydrologic inputs and outputs, influence the biochemistry of the soils and are major factors in the ultimate selection of the biota of wetlands. Biota ranging from microbial communities to vegetation to waterfowl are all constrained or enhanced by hydrologic conditions. An important point about wetlands—one that is often missed by ecologists who begin to study these systems, is this: *Hydrology is probably the single most important determinant of the establishment and maintenance of specific types of wetlands and wetland processes.* An understanding of rudimentary hydrology should be in the repertoire of any wetland scientist.

THE IMPORTANCE OF HYDROLOGY IN WETLANDS

Ecological Processes and Hydrology

Wetlands are transitional between terrestrial and open-water aquatic ecosystems (see Chapter 1). They are transitional in terms of spatial arrangement, for they are usually found between uplands and aquatic systems. They are also transitional in the amount of water they store and process. Wetlands represent the aquatic edge of many terrestrial (emergent) plants and animals; they also represent the terrestrial edge of many aquatic (submersed) plants and animals. Hence small changes in hydrology can result in significant biotic changes. A conceptual model of the role of hydrology in wetlands is shown in Figure 4–1. Hydrologic conditions can directly modify or change chemical and physical properties such as nutrient availability, degree of substrate anoxia, soil salinity, sediment properties, and pH. Except in nutrient-poor bogs, water inputs are the major source of nutrients to wetlands; water outflows often remove biotic and abiotic material from wetlands as well. These modifications of the physiochemical environment, in turn, have a direct impact on the biotic response in the wetland (Gosselink and Turner, 1978). When hydrologic conditions in wetlands change even slightly, the biota may respond with massive changes in species composition and richness and in ecosystem productivity. When the hydrologic pattern remains similar from year to year, a wetland's structural and functional integrity may persist for many years.

Biotic Control of Wetland Hydrology

Just as many other ecosystems exert feedback (cybernetic) control of their physical environments, wetland ecosystems are not simply passive to their hydrologic conditions. The feedback loop in Figure 4–1 shows that biotic components of

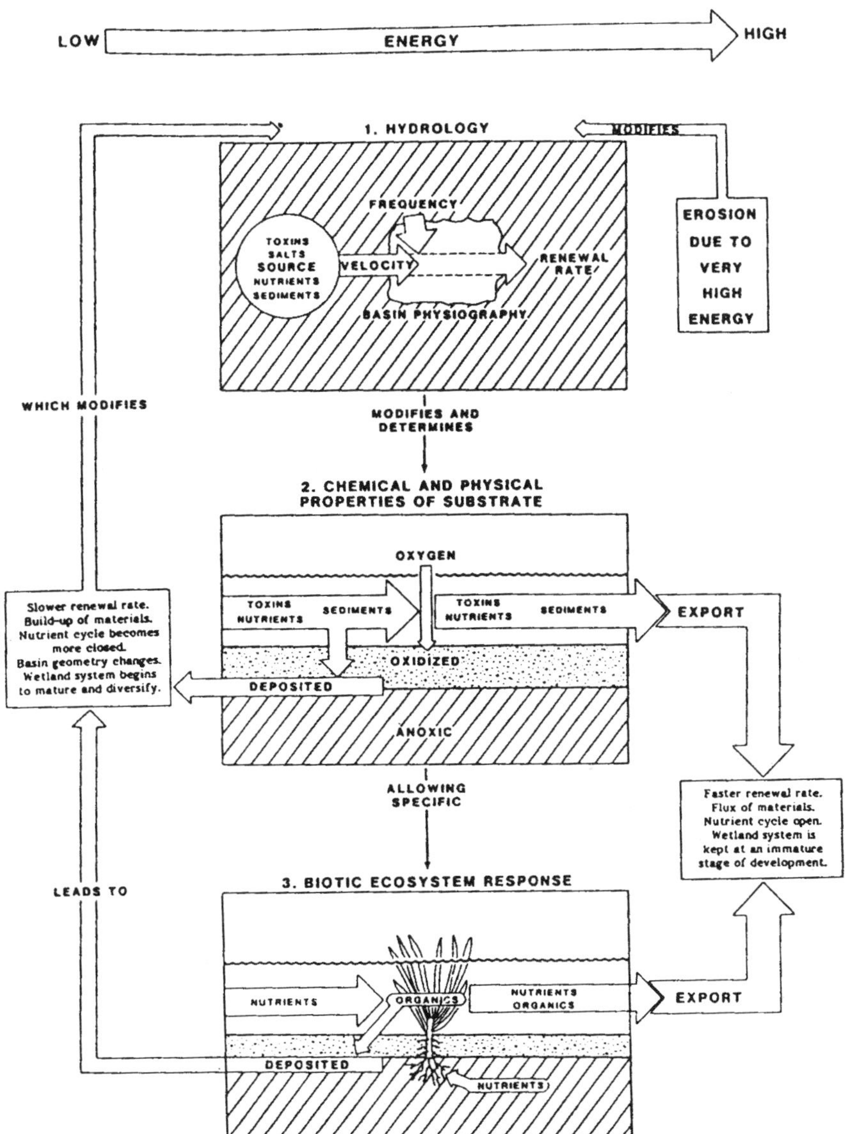

Figure 4–1. Conceptual model showing the direct and indirect effects of hydrology on wetlands. *(From Wicker et al., 1982, after Gosselink and Turner, 1978)*

wetlands, mainly vegetation, can control their water conditions through a variety of mechanisms, including peat building, sediment trapping, nutrient retention, water shading, and transpiration. Many marshes and some riparian wetlands accumulate sediments or organic peat, thereby eventually decreasing the duration and frequency with which they are flooded. Wetland vegetation influences hydrologic conditions by binding sediments to reduce erosion, by trapping sediment, by interrupting water flows, and by building peat deposits (Gosselink, 1984). Bogs build peat to the point at which they are no longer influenced at the surface by the inflow of mineral waters. Some trees in some southern swamps save water by their deciduous nature, their seasonal shading, and their relatively slow rates of transpiration.

Animals contribute to hydrologic modifications and subsequent changes in wetlands (Fig. 4–2; see for example Naiman, 1988). The exploits of beavers (*Castor canadensis*) in much of North America in both creating and destroying wetland habitats are well known. They build dams on streams, backing up water across great expanses, creating wetlands where none existed before, and possibly even altering global carbon biogeochemistry (Naiman et al., 1991). American alligators (*Alligator mississippiensis*) are known for their role in the Florida Everglades in constructing "gator holes" that serve as oases for fish, turtles, snails, and other aquatic animals during the dry season. In all of these cases, the biota of the ecosystem have contributed to their own survival by influencing the ecosystem's hydrology.

Studies of Wetland Hydrology

Until recently, the importance of hydrology in wetland function contrasted markedly with the paucity of published research on the subject. Most early wetland investigations that dealt with hydrology explored the relationships between hydrologic variables (usually water depth) and wetland productivity (e.g., Conner and Day, 1976; Mitsch and Ewel, 1979) or species composition (e.g., Heinselman, 1963, 1970; McKnight et al. 1981; Huffman and Forsythe, 1981). There have been several review papers on various aspects of the hydrology of wetlands, but many of them were published only recently (e.g., Linacre, 1976; Gosselink and Turner, 1978; Carter et al., 1979; Bedinger, 1981; Ingram, 1983; Carter, 1986; Carter and Novitzki, 1988; Winter, 1988; Siegel, 1988a; O'Brien, 1988; Duever, 1988, 1990; Kadlec, 1989; Winter and Llamas, 1993); few comprehensive studies have described in detail the hydrologic characteristics within specific wetland types. An exception to this has been the study of northern peatlands, for which a wealth of literature exists, including work in the former Soviet Union (e.g., Romanov, 1968; Ivanov, 1981), in the British Isles (Ingram et al., 1974; Ingram, 1982; Gilman, 1982), and in North America (e.g., Bay, 1967, 1969; Boelter and Verry, 1977; Verry and Boelter, 1979; Wilcox et al., 1986; Siegel, 1988b). Some of the more notable hydrology studies for other types of wetlands in the United States have included salt marshes (Hemond and Burke, 1981; Hemond and Fifield, 1982), cypress swamps (R. C. Smith, 1975; Heimburg, 1984), and large-scale wetland complexes (Rykiel, 1977, 1984; Hyatt and Brook, 1984).

Figure 4–2. Two animals, beaver (*top*) and alligator (*bottom*), that can significantly modify hydrology and subsequent chemical and physical properties of wetlands. (*Top photo copyright © 1980 by Alvin E. Staffen, reprinted with permission. Bottom photo copyright © 1991 by David M. Dennis, reprinted with permission.*)

WETLAND HYDROPERIOD

The *hydroperiod* is the seasonal pattern of the water level of a wetland and is like a hydrologic signature of each wetland type. It defines the rise and fall of a wetland's surface and subsurface water. It characterizes each type of wetland, and the constancy of its pattern from year to year ensures a reasonable stability for that wetland. The hydroperiod is an integration of all inflows and outflows of water, but it is also influenced by physical features of the terrain and by proximity to other bodies of water. Many terms are used to describe qualitatively a wetland's hydroperiod. Table 4–1 gives several definitions that have been suggested by the U.S. Fish and Wildlife Service. For wetlands that are not subtidal or permanently flooded, the amount of time that wetland is in standing water is called the *flood duration*, and the average number of times that a wetland is flooded in a given period is known as the *flood frequency*. Both terms are used to describe periodically flooded wetlands such as coastal salt marshes and riparian wetlands.

Some typical hydroperiods for very different wetlands are shown in Fig. 4–3. For a cypress dome in north-central Florida (Fig. 4–3a), the ecosystem has standing water during the wet summer season and dry periods in the late autumn and early spring. A coastal salt marsh has a hydroperiod of semidiurnal flooding and dewatering superimposed on a twice-monthly pattern of spring and ebb tides (Fig. 4–3b). There are also hydroperiods that have less pronounced seasonal

Table 4–1. Definitions of Wetland Hydroperiods

Tidal Wetlands

Subtidal—permanently flooded with tidal water

Irregularly Exposed—surface exposed by tides less often than daily

Regularly Flooded—alternately flooded and exposed at least once daily

Irregularly Flooded—flooded less often than daily

Nontidal Wetlands

Permanently Flooded—flooded throughout the year in all years

Intermittently Exposed—flooded throughout the year except in years of extreme drought

Semipermanently Flooded—flooded in the growing season in most years

Seasonally Flooded—flooded for extended periods during the growing season, but usually no surface water by end of growing season

Saturated—substrate is saturated for extended periods in the growing season, but standing water is rarely present

Temporarily Flooded—flooded for brief periods in the growing season, but water table is otherwise well below surface

Intermittently Flooded—surface is usually exposed with surface water present for variable periods without detectable seasonal pattern.

Source: After Cowardin et al., 1979

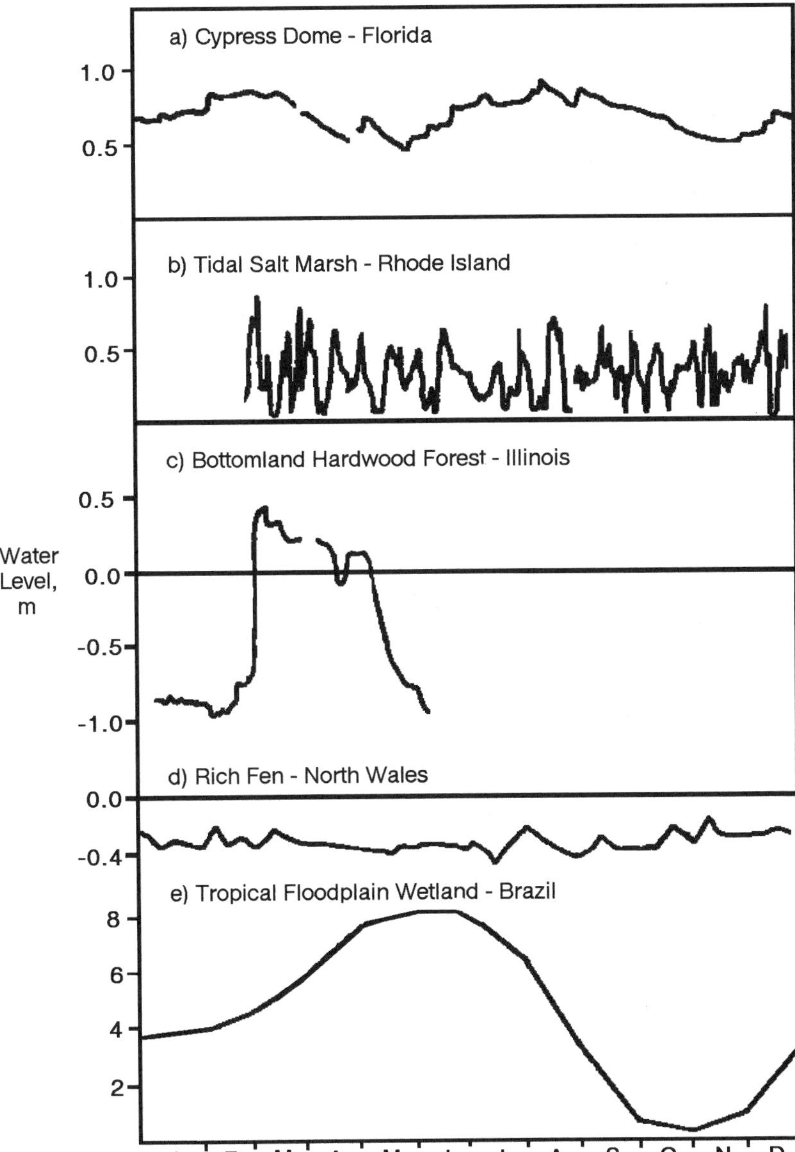

Figure 4–3. Hydroperiods for several different wetlands: *a.* cypress dome in northcentral Florida; *b.* New England salt marsh (*after Nixon and Oviatt, 1973*); *c.* bottomland hardwood forest along Kankakee River in northeastern Illinois (*from Mitsch et al., 1979b*); *d.* peatland (fen) in northern Wales (*from Gilman, 1982*); *e.* Amazon floodplain forested wetland at confluence of Amazon and Negro Rivers, Manaus, Brazil (*from Junk, 1982*). Vertical scale indicates water depth relative to soil surface.

fluctuations, as in the below-ground water level of many bogs and fens (Fig. 4–3d). Low-order riverine wetlands respond sharply to local rainfall events rather than to general seasonal patterns. For example, the hydroperiod of many bottomland hardwood forests (Fig. 4–3c) on low-order streams is sudden and relatively short seasonal flooding due to local precipitation and thawing conditions followed by a rapid drop of the water level. On the other hand, a high-order river is more influenced by seasonal patterns of precipitation throughout a large watershed rather than local precipitation (Junk et al., 1989). The annual fluctuation of water in the tropical floodplain forest near Manaus, Brazil, at the confluence of the Rio Negro and Rio Amazon is a more predictable seasonal pattern that includes a tremendous seasonal fluctuation of almost 8 m because of the flooding rivers (Fig. 4–3e).

Year-to-Year Fluctuations

The hydroperiod, of course, is not the same each year but varies statistically according to climate and antecedent conditions. Great variability can be seen from year to year for some wetlands, as illustrated in Figure 4–4 for a prairie pothole regional wetland in Canada and the Big Cypress Swamp/Everglades region of south Florida. In the pothole region, a wet-dry cycle of 10 to 20 years is seen; spring is almost always wetter than fall but depths vary significantly from year to year (Fig. 4–4a). Figure 4–4b illustrates cases of an even seasonal rainfall pattern for the Everglades in 1957–1958, which caused a fairly stable hydroperiod through the year, and a significant dry season in 1970–1971, which caused the hydroperiod to vary about 1.5 m between high and low water.

Pulsing Water Levels

Water levels in most wetlands (all of the hydroperiods shown in Figure 4–3 except for the rich fen) are generally not stable but fluctuate seasonally (high-order riparian wetlands) daily or semi-daily (types of tidal wetlands) or unpredictably (wetlands in low-order streams and coastal wetlands with wind-driven tides). In fact, wetland hydroperiods that show the greatest differences between

Figure 4–4 (*opposite page*). Year-to-year fluctuations in wetland hydroperiod: *a.* spring (May 1) and fall (October 31) water depths for 1962–86 for shallow open-water wetland in prairie pothole region of southwestern Saskatchewan, Canada, and *b.* wet and dry season hydrographs for Big Cypress Swamp region near the Everglades, Florida. (a. *from Kantrud et al., 1989, as adapted from J. B. Millar, 1971, redrawn with permission of J. B. Millar; b. from Freiberger, 1972 as cited in Duever, 1988; copyright © 1988 by Elsevier Science Publishers, Amsterdam, reprinted by permission*)

a.

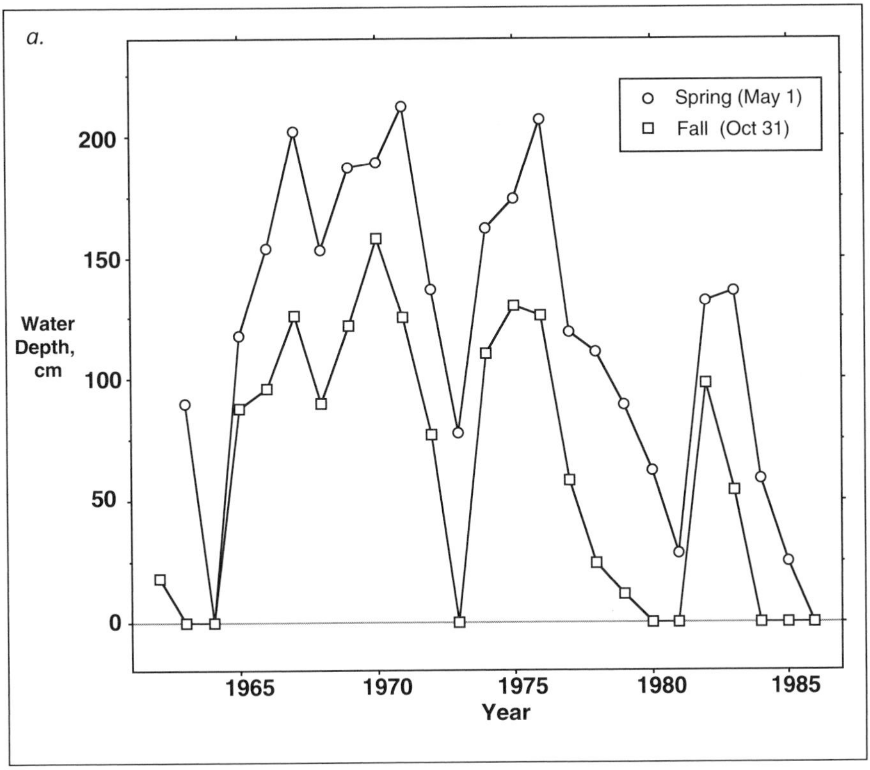

○ Spring (May 1)
□ Fall (Oct 31)

Water Depth, cm

200
150
100
50
0

1965 1970 1975 1980 1985

Year

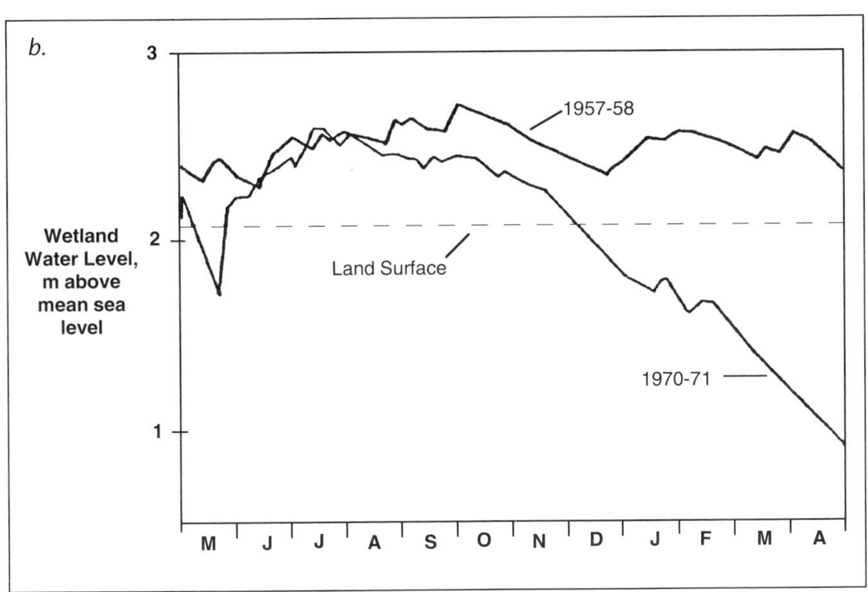

high and low water levels such as those seen in riverine wetlands are often caused by flooding "pulses" that occur seasonally or periodically (Junk et al., 1989; see Fig. 4–3c, e). These pulses nourish the riverine wetland with additional nutrients and carry away detritus and waste products. Pulse-fed wetlands are often the most productive wetlands and are the most favorable for exporting materials, energy, and biota to adjacent ecosystems (see Specific Effects of Hydrology on Wetlands in this chapter). Despite that obvious fact, many wetland managers, especially those who manage wetlands for waterfowl, often manage for stable water levels. Fredrickson and Reid (1990) stated that "Because the goal of many [wetland] management scenarios is to counteract the effects of seasonal and long-term droughts, a general tendency is to restrict water level fluctuations in managed wetlands. This misconception is based on the fact that most wetland wildlife requires water for most stages in their life cycles." Kushlan (1989) suggested that because the avian fauna that use wetlands often possess adaptations to fluctuating water levels, the active manipulation of water levels may be appropriate in artificially managed wetlands. A seasonally fluctuating water level, then, is the rule, not the exception, in most wetlands.

THE OVERALL WETLAND WATER BUDGET

The hydroperiod, or hydrologic state of a given wetland, can be summarized as being a result of the following factors:

1. the balance between the inflows and outflows of water
2. the surface contours of the landscape
3. subsurface soil, geology, and groundwater conditions

The first condition defines the water budget of the wetland, whereas the second and the third define the capacity of the wetland to store water. The general balance between water storage and inflows and outflows, illustrated in Figure 4–5, is expressed as

$$\Delta V/\Delta t = P_n + S_i + G_i - ET - S_o - G_o \pm T \tag{4.1}$$

where

V = volume of water storage in wetlands

$\Delta V/\Delta t$ = change in volume of water storage in wetland per unit time, t

P_n = net precipitation

S_i = surface inflows, including flooding streams

G_i = groundwater inflows

ET = evapotranspiration

S_o = surface outflows
G_o = groundwater outflows
T = tidal inflow (+) or outflow (-)

The average water depth $\bar{d}$, at any one time, can further be described as

$$\bar{d} = V/A \tag{4.2}$$

where A = wetland surface area.

Thus each of the terms in Equation 4.1 can be expressed in terms of depth, per unit time, e.g., cm/yr, or in terms of volume per unit time, e.g., m^3/day.

Examples of Water Budgets

Equation 4.1 serves as a useful summary of the major hydrologic components of any wetland water budget. Examples of hydrologic budgets for several wetlands are shown in Figure 4–6. The terms in the equation, however, vary in importance according to the type of wetland observed; furthermore, not all terms in

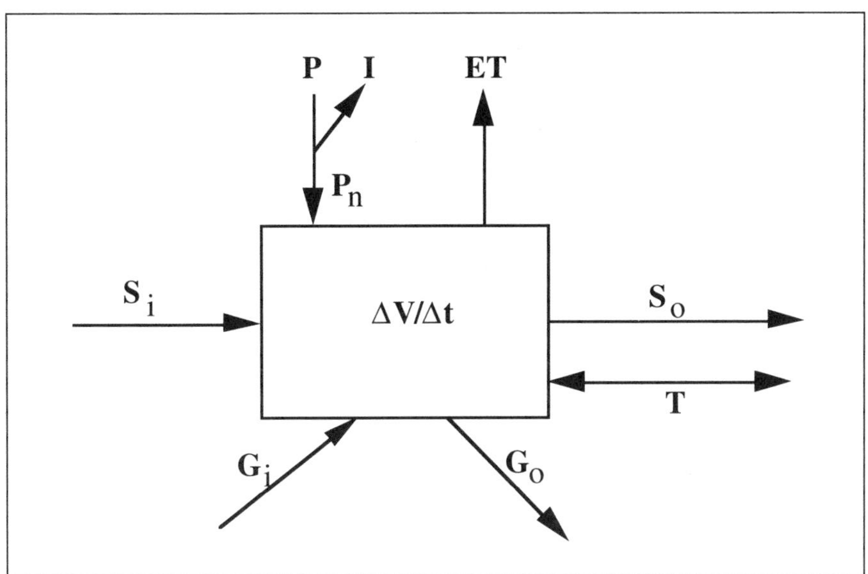

Figure 4–5. Generalized water budget for a wetland corresponding to terms in Equation 4.1. P = precipitation, ET = evapotranspiration, I = interception, P_n = net precipitation, S_i = surface inflow, S_o = surface outflow, G_i = groundwater inflow, G_o = groundwater outflow, $\Delta V/\Delta t$ = change in storage per unit time, T = tide or seiche.

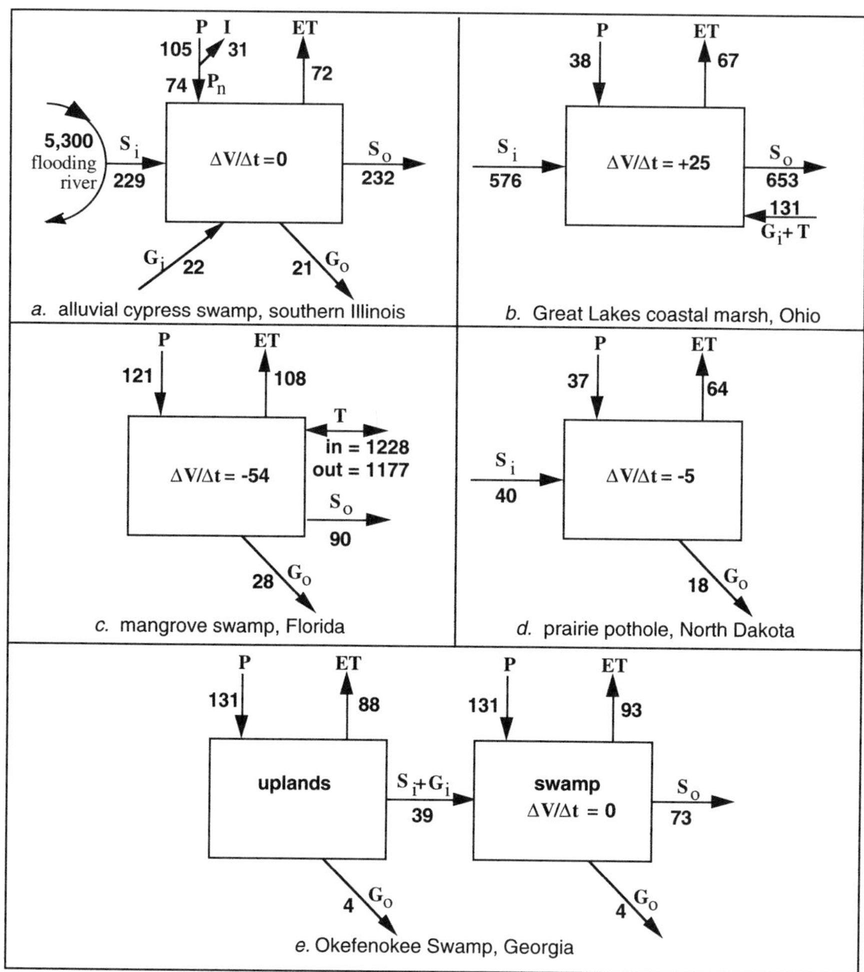

Figure 4–6. Annual water budget for several wetlands including (*above*) *a.* an alluvial cypress swamp in southern Illinois (*after Mitsch, 1979*); *b.* a Lake Erie coastal marsh in northern Ohio (March through September, 1988 only, during a drought year; *after Mitsch and Reeder, 1992*); *c.* a black mangrove swamp in southwestern Florida (*after Twilley, 1982, as cited in S. Brown, 1990*); *d.* prairie pothole marshes in North Dakota (average of 10 wetlands; after Shjeflo, 1968, *as cited in Winter, 1989*); *e.* the Okefenokee Swamp watershed in Georgia (*after Rykiel, 1984*); and (*opposite page*) *f.* a rich fen in northern Wales (*after Gilman, 1982*); *g.* the Green Swamp region of central Florida (*after Pride et al., 1966 as cited in Carter et al., 1979*); *h.* Thoreau's Bog, Concord, Massachusetts (*after Hemond, 1980, as cited in Brown, 1990*); *i.* a pocosin swamp (*average of 3 years; after Richardson, 1983 as cited in Brown, 1990*). All values are expressed in cm/yr unless otherwise noted. See Figure 4–5 for symbol definitions.

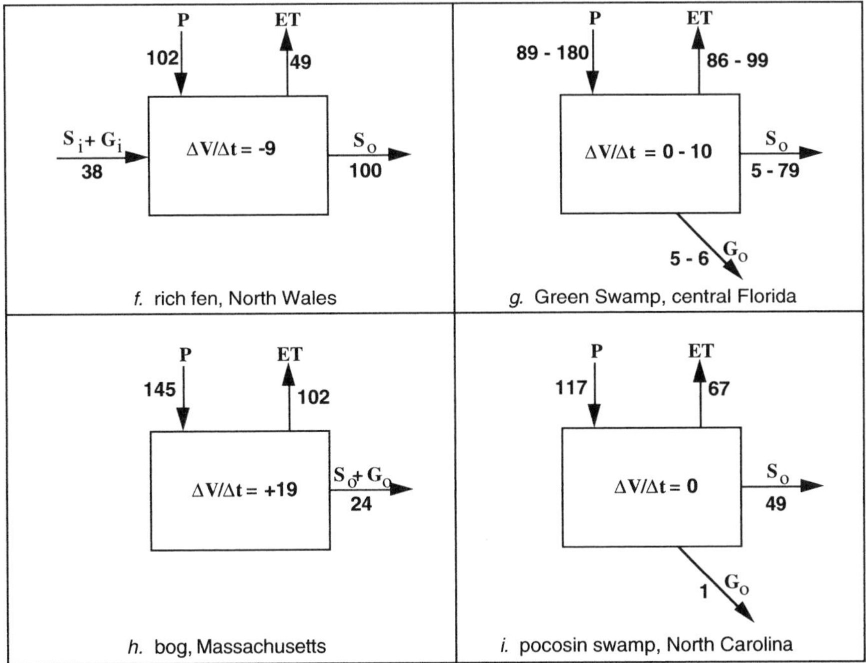

the hydrologic budget apply to all wetlands (Table 4–2). There is a great variability in certain flows, particularly in surface inflows and outflows, depending on the openness of the wetlands. An alluvial cypress swamp in southern Illinois received a gross inflow of floodwater from one flood that was more than 50 times the gross precipitation for the entire year (Fig. 4–6a). Even the net surface inflow from that flood (the water left behind after the flooding river receded) was three times the precipitation input for the entire year. Surface and groundwater inflows to a coastal Lake Erie marsh in Ohio were estimated to be almost 20 times the precipitation for a major part of a drought year (Fig. 4–6b), and tides contributed 10 times the precipitation to a black mangrove swamp in Florida (Fig. 4–6c). In contrast to these inflow-dominated wetlands, surface inflow is approximately equal to the precipitation inflow in the prairie pothole marshes of North Dakota (Fig. 4–6d), considerably less than precipitation for the Okefenokee Swamp in Georgia (Fig. 4–6e) and a rich fen in North Wales (Fig. 4–6f) and essentially nonexistent in the upland Green Swamp of central Florida (Fig. 4–6g), a bog in Massachusetts (Fig. 4–6h), and a pocosin wetland of North Carolina (Fig. 4–6i). In most of these examples, the change in storage is small or zero, indicating that the water level at the end of the study period (usually an annual cycle) is close to where it was at the beginning of the study period.

Table 4–2. Major Components of Hydrologic Budgets for Wetlands

Component	*Pattern*	*Wetlands Affected*
Precipitation	Varies with climate although many regions have distinct wet and dry seasons	All
Surface Inflows and Outflows	Seasonally, often matched with precipitation pattern or spring thaw; can be channelized as streamflow or nonchannelized as runoff; includes river flooding of alluvial wetlands	Potentially all wetlands except ombrotrophic bogs; riparian wetlands, including bottomland hardwood forests and other alluvial wetlands, are particularly affected by river flooding
Groundwater	Less seasonal than surface inflows and not always present	Potentially all wetlands except ombrotrophic bogs and other perched wetlands
Evapotranspiration	Seasonal with peaks in summer and low rates in winter. Dependent on meterorological, physical, and biological conditions in wetlands	All
Tides	One to two tidal periods per day; flooding frequency varies with elevation	Tidal freshwater and salt marshes; mangrove swamps

Residence Time

A generally useful concept of wetland hydrology is that of the *renewal rate* or *turnover rate* of water, defined as the ratio of throughput to average volume within the system:

$$t^{-1} = Q_t/V \tag{4.3}$$

where

t^{-1} = renewal rate (1/time)

Q_t = total inflow rate

V = average volume of water storage in wetland

Few measurements of renewal rates have been made in wetlands, although it is a frequently used parameter in limnological studies. Chemical and biotic properties are often determined by the openness of the system, and the renewal rate is an index of this since it indicates how rapidly the water in the system is replaced. The reciprocal of the renewal rate is the *residence time* (t), (sometimes called *retention time* by engineers, for constructed wetlands; see Chapter 17) which is a measure of the average time that water remains in the wetland. Recent evidence, however, suggests that the theoretical residence time, as calculated by Equation 4.3, is often much longer than the actual residence time as water flows through a wetland because of non-uniform mixing. Because there are often parts of a wetland where waters are not well mixed, the theoretical residence time (t) estimate should be used with caution when estimating the hydrodynamics of wetlands.

PRECIPITATION

Wetlands occur most extensively in regions where *precipitation,* a term that includes rainfall and snowfall, is in excess of losses such as evapotranspiration and surface runoff. Exceptions to this generality occur where surface inflows are seasonally abundant or tides are prevalent such as coastal salt marshes or in arid regions such as the western United States, where riparian wetlands depend more on river flow and less on local precipitation. Precipitation generally has well-defined yearly patterns, although variations among years may be great. An almost uniform pattern of precipitation exists for eastern North America because of the heavy influence of both cold and warm fronts and summer convective storms. The northern Great Plains experience a summer peak in precipitation. Relatively less precipitation occurs in the winter because of the cold continental high pressure that recedes northward in the summer. By contrast, parts of the West Coast have a Mediterranean-type climate characterized by wet winters and pronounced dry summers. The northern extremes of Canada show more uniform patterns of precipitation, but overall amounts are small. The precipitation pattern in Florida shows a decidedly wet season in summer caused by the almost daily convective storms.

The fate of precipitation that falls on wetlands with forested, shrub or emergent vegetation is shown in Figure 4–7. When some of the precipitation is retained by the vegetation cover, particularly in forested wetlands, the amount that actually passes through the vegetation to the water or substrate below is called *throughfall.* The amount of precipitation that is retained in the overlying vegetation canopy is called *interception.* Interception depends on several factors, including the total amount of precipitation, the intensity of the precipitation, and the character of the vegetation, including the stage of vegetation development, the type of vegetation, e.g., deciduous or evergreen, and the strata of the vegetation, e.g., tree, shrub, or emergent macrophyte. The percent of precipitation that is intercepted in forests varies between 8 and 35 percent. One review cites a median value of 13 percent for several studies of deciduous forests and 28 percent for coniferous forests (Dunne and Leopold, 1978). The water budget in

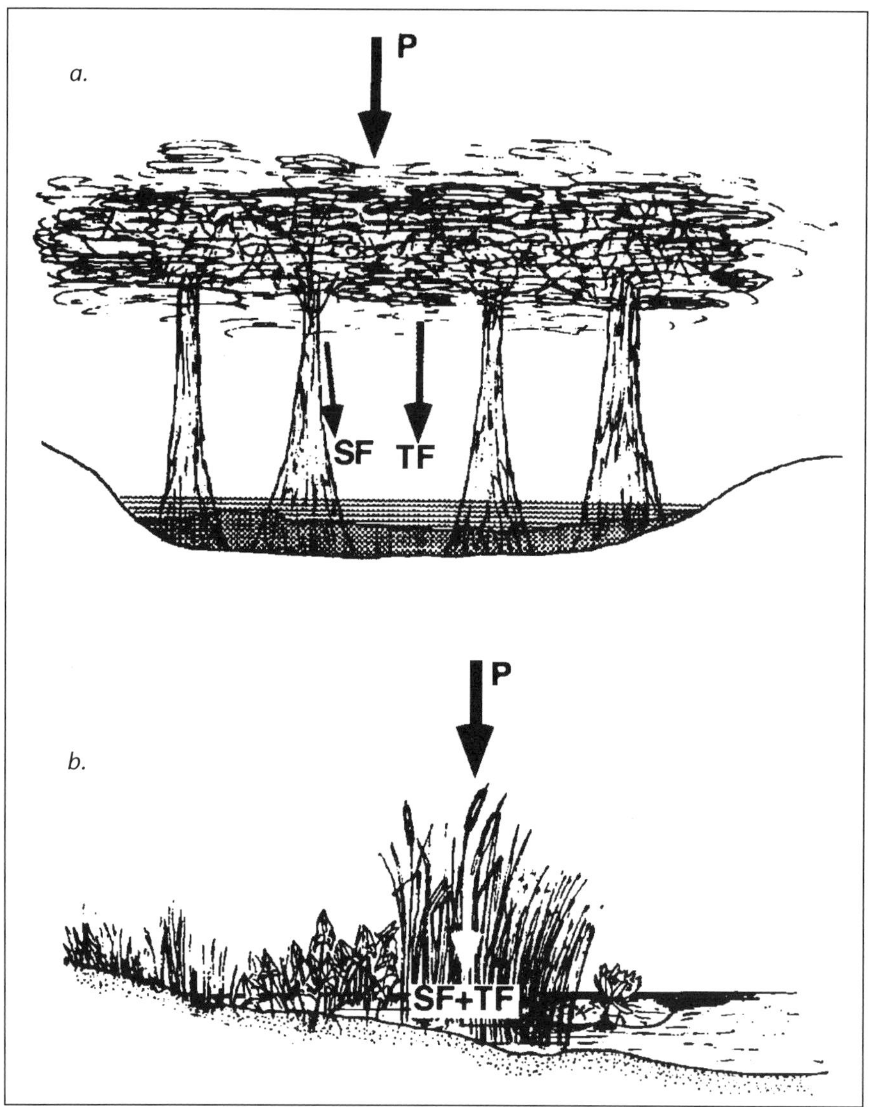

Figure 4–7. Fate of precipitation in *a.* a forested wetland and *b.* a marsh. P = precipitation; TF = throughfall; SF = stemflow.

Figure 4–6a illustrates that 29 percent of precipitation in a forested wetland was intercepted by a canopy dominated by *Taxodium distichum,* a deciduous conifer.

Little is known about the interception of precipitation by emergent macrophytes, but it probably is similar to that measured in grasslands or croplands. Essentially, in those systems, interception at maximum growth can be as high as that in a forest (10–35 percent of gross precipitation). On an annual basis the percent intercepted would be expected to be much less in nonforested wetlands than in forested wetlands because of the dormancy of herbaceous plants (macrophytes in emergent wetlands) in winter. It follows that replacing one wetland type with another (e.g., a marsh for a forested wetland) may not completely replace the former wetland's hydrologic function. On the other hand, an interesting hypothesis about interception and the subsequent evaporation of water from leaf surfaces is that, because the same amount of energy is required whether water evaporates from the surface of a leaf or is transpired by the plant, the evaporation of intercepted water is not "lost" because it may reduce the amount of transpiration loss that occurs (Dunne and Leopold, 1978). This argues that wetlands with high and low interception may be similar in overall water loss to the atmosphere.

Another term related to precipitation, *stemflow,* refers to water that passes down the stems of the vegetation (Fig. 4–7). This flow is generally a minor component of the water budget of a wetland. For example, Heimburg (1984) found that stemflow was, at maximum, 3 percent of throughfall in cypress dome wetlands in north-central Florida.

These terms are related in a simple water balance as follows:

$$P = I + TF + SF \tag{4.4}$$

where

P = total precipitation

I = interception

TF = throughfall

SF = stemflow

The total amount of precipitation that actually reaches the water surface or substrate of a wetland is called the net precipitation (P_n) and is defined as

$$P_n = P - I \tag{4.5}$$

Combining Equations 4.4 and 4.5 yields the most commonly used form for estimating net precipitation in wetlands

$$P_n = TF + SF \tag{4.6}$$

SURFACE FLOWS

Wetlands can be receiving systems for surface water flows (inflows), or surface water streams can originate in wetlands to feed downstream systems (outflows). Surface outflows are found in many wetlands that are in the upstream reaches of

Table 4–3. Description and Hydrologic Response Coefficients for Estimating Direct Runoff from Forested Watersheds in Eastern United States

	Watershed Area, A_w (ha)	*Mean Elevation (m)*	*Mean Slope (%)*	*Soil texture*[a]	*Forest Type*[b]	*Hydrologic Response Coefficient R_p*
Coweeta 2, N.C.	13	850	30	SL	OH	0.04
Coweeta 18, N.C.	13	820	32	SL	OH	0.05
Coweeta 14, N.C.	62	880	21	SL	OH	0.05
Coweeta 21, N.C.	24	990	34	SL	OH	0.06
Bent Ck 7, N.C.	297	940	22	SL	OH	0.06
Coweeta 8, N.C.	760	950	22	SL	MH	0.07
Union 3, S.C.	9	170	7	SC	P	0.08
Coweeta 28, N.C.	146	1,200	33	SL	MH	0.10
Copper Basin 2, Tenn.	36	580	27	SL	OH	0.10
Leading Ridge 1, Pa.	123	370	19	TL	MH	0.11
Dilldown Ck, Pa.	619	580	4	SL	SO	0.12
Fernow 4, W.Va.	39	820	18	TL	MH	0.14
Coweeta 36, N.C.	46	1,300	47	SL	MH	0.15
Burlington Bk, Conn.	1,067	270	3	SS	NH	0.17
Hubbard Brook 4, N.H.	36	600	26	NL	NH	0.18

[a]SL, sandy loam; SC, sandy clay; TL, silt loam; SS, stony sand; NL, stony loam.
[b]OH, oak hickory; MH, mixed hardwoods; NH, northern hardwoods; SO, scrub oak; P, pine
Source: From R. Lee, 1980, after Hewlett and Hibbert, 1967

a watershed. Often these wetlands are important water flow regulators for downstream rivers. Some wetlands have surface outflows that develop only when their water stages exceed certain levels.

Surface Inflows

Wetlands are subjected to surface inflows of several types. *Overland flow* is nonchannelized sheet flow that usually occurs during and immediately following rainfall or a spring thaw or as tides rise in coastal wetlands. If a wetland is influenced by a drainage basin, channelized *streamflow* may enter the wetland during most or all of the year. Often wetlands are an integrated part of a stream or river; for example, as instream freshwater marshes. Wetlands that form in wide shallow expanses of river channels are greatly influenced by the seasonal streamflow patterns of the river. Coastal saline and brackish wetlands are also significantly influenced by freshwater runoff and streamflow that contribute nutrients and energy to the wetland and often ameliorate the effects of soil salinity and anoxia. Wetlands can also receive surface inflow from seasonal or episodic pulses of flood flow from adjacent streams and rivers that may otherwise not be connected hydrologically with the wetland.

Surface runoff from a drainage basin into a wetland is usually difficult to estimate without a great deal of data. Nevertheless, it is often one of the most important sources of water in a wetland's hydrologic budget. The direct runoff component of streamflow refers to rainfall during a storm that causes an immediate increase in streamflow. An estimate of the amount of precipitation that results in direct runoff, or *quickflow*, from an individual storm can be determined from the following equation:

$$S_i = R_p \cdot P \cdot A_w \tag{4.7}$$

where

S_i = direct surface runoff to wetland, m^3 per storm event

R_p = hydrologic response coefficient

P = average precipitation in watershed, m

A_w = area of watershed draining into wetland, m^2

This equation states that the flow is proportional to the volume of precipitation (P x A_w) on the watershed feeding the wetland in question. The values of R_p, which represent the fraction of precipitation in the watershed that becomes direct surface runoff, range from 4 percent to 18 percent for small watersheds in the eastern United States (R. Lee, 1980); a summary of values for certain conditions of slope, soil, and forest type are shown in Table 4–3.

While Equation 4.7 predicts the entire direct runoff caused by a storm event, in some cases wetland scientists and managers might be interested in calculating the peak runoff (*flood peak*) into a wetland caused by a specific rainfall event. Although this is generally a difficult calculation for large watersheds, a formula with the unlikely name of the *rational runoff method* is a widely accepted and useful way to predict peak runoff for watersheds of less than 80 hectares (200 acres). The equation is given by

$$S_{i(pk)} = 0.278\ CIA_w \tag{4.8}$$

where

$S_{i(pk)}$ = peak runoff into wetland (m^3/sec)

Table 4–4. Values of the Rational Runoff Coefficicient, C, Used to Calculate Peak Runoff

	C
Urban Areas	
Business areas: high-value districts	0.75–0.95
neighborhood districts	0.50–0.70
Residential areas: single-family dwellings	0.30–0.50
multiple-family dwellings	0.40–0.75
suburban	0.25–0.40
Industrial areas: light	0.50–0.80
heavy	0.60–0.90
Parks and cemeteries	0.10–0.25
Playgrounds	0.20–0.35
Unimproved land	0.10–0.30
Rural Areas	
Sandy and gravelly soils: cultivated	0.20
pasture	0.15
woodland	0.10
Loams and similar soils: cultivated	0.40
pasture	0.35
woodland	0.30
Heavy clay soils; shallow soils over bedrock:	
cultivated	0.50
pasture	0.45
woodland	0.40

Source: From Dunne and Leopold, 1978

C = rational runoff coefficient (see Table 4–4)

I = rainfall intensity (mm/hr)

A_w = area of watershed draining into wetland, km^2

The coefficient C, which ranges between 0 and 1 (Table 4–4), depends on the upstream land use. Concentrated urban areas have a coefficient ranging from 0.5 to 0.95, and rural areas have lower coefficients that greatly depend on soil types, with sandy soils lowest (C = 0.1–0.2) and clay soils highest (C = 0.4–0.5).

Channelized streamflow into and out of wetlands is described simply as the product of the cross-sectional area of the stream (A) and the average velocity (V) and can be determined through stream velocity measurements in the field:

$$S_i \text{ or } S_o = A_x \cdot V \tag{4.9}$$

where

S_i, S_o = surface channelized flow into or out of wetland m^3/sec

A_x = cross sectional area of the stream, m^2

V = average velocity, m/sec

The velocity can be determined in a number of ways, ranging from velocity meters that are hand-held at various locations in the stream cross section to the floating orange technique, where the velocity of a floating orange or similar fruit (which is 90 percent or more water and therefore floats but just beneath the water surface) is timed as it goes downstream. If a continuous or daily record of stream-

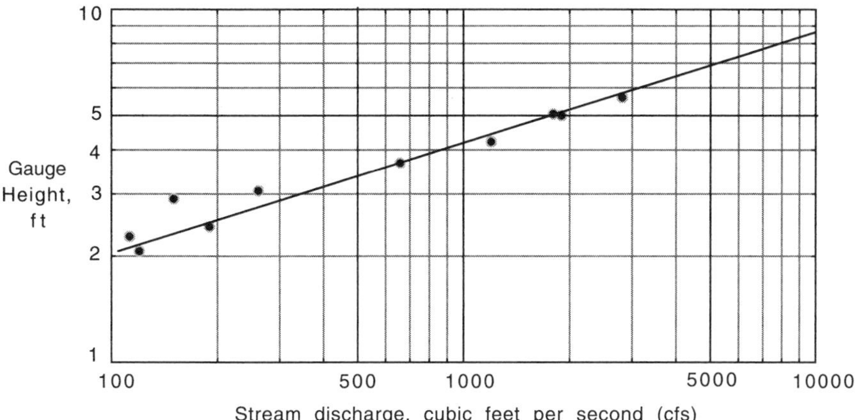

Figure 4–8. Rating curve for streamflow determination as a function of stream stage. This example is from New Fork River at Boulder, Wyoming *(redrawn from Dunne and Leopold, 1978; copyright © W.H. Freeman and Company, redrawn with permission)*. 100 cfs = 2.832 m^3/sec.

flow is needed, then a *rating curve* (Fig. 4–8), a plot of instantaneous streamflow as measured with Equation 4.9 versus stream elevation or stage, is useful. If this type of rating curve is developed for a stream (the basis of most hydrologic streamflow gauging stations operated by the United States Geological Survey), then a simple measurement of the stage in the stream can be used to determine the streamflow. Caution should be taken in using this approach for streams flowing into wetlands to ensure that no "backwater effect" of the wetland's water level will affect the stream stage at the point of measurement.

When an estimate of surface flow into or out of a riverine wetland is needed and no stream velocity measurements are available, the *Manning Equation* can often be used if the slope of the stream and a description of the surface roughness are known:

$$S_i \; or \; S_o = \frac{A_x \; R^{2/3} \; s^{1/2}}{n} \tag{4.10}$$

where

n = roughness coefficient (Manning coefficient) (see Table 4–5)

R = hydraulic radius, m (cross-sectional area divided by wetted perimeter)

s = channel slope, dimensionless

Examples of roughness coefficients are given in Table 4–5. Although these coefficients have not generally been determined as part of wetland studies, they can often be applied to streamflow in and out of wetlands. The relationship is particularly useful for estimating streamflow where velocities are too low to measure directly and to estimate flood peaks from high-water marks on ungauged streams (Lee, 1980). These circumstances are common in wetland studies.

Floods and Riparian Wetlands

A special case of surface inflow occurs in wetlands that are in floodplains adjacent to rivers or streams and are occasionally flooded by those rivers or streams. These ecosystems are often called *riparian wetlands* (Chap. 14). The flooding of

Table 4–5. Roughness Coefficients (n) for Manning Equation Used to Determine Streamflow in Natural Streams and Channels

Stream Conditions	*Manning Coefficient, n*
Straightened earth canals	0.02
Winding natural streams with some plant growth	0.035
Mountain streams with rocky streambed	0.040-0.050
Winding natural streams with high plant growth	0.042-0.052
Sluggish streams with high plant growth	0.065
Very sluggish streams with high plant growth	0.112

Source: After Chow, 1964, and R. Lee, 1980

these wetlands varies in intensity, duration, and number of floods from year to year, although the probability of flooding is fairly predictable. In the eastern and midwestern United States and in much of Canada, a pattern of winter or spring flooding caused by rains and sudden snowmelt is often observed (Fig. 4–9). When river flow begins to overflow onto the floodplain, the streamflow is referred to as *bankfull discharge*. A hydrograph of a stream that flooded its

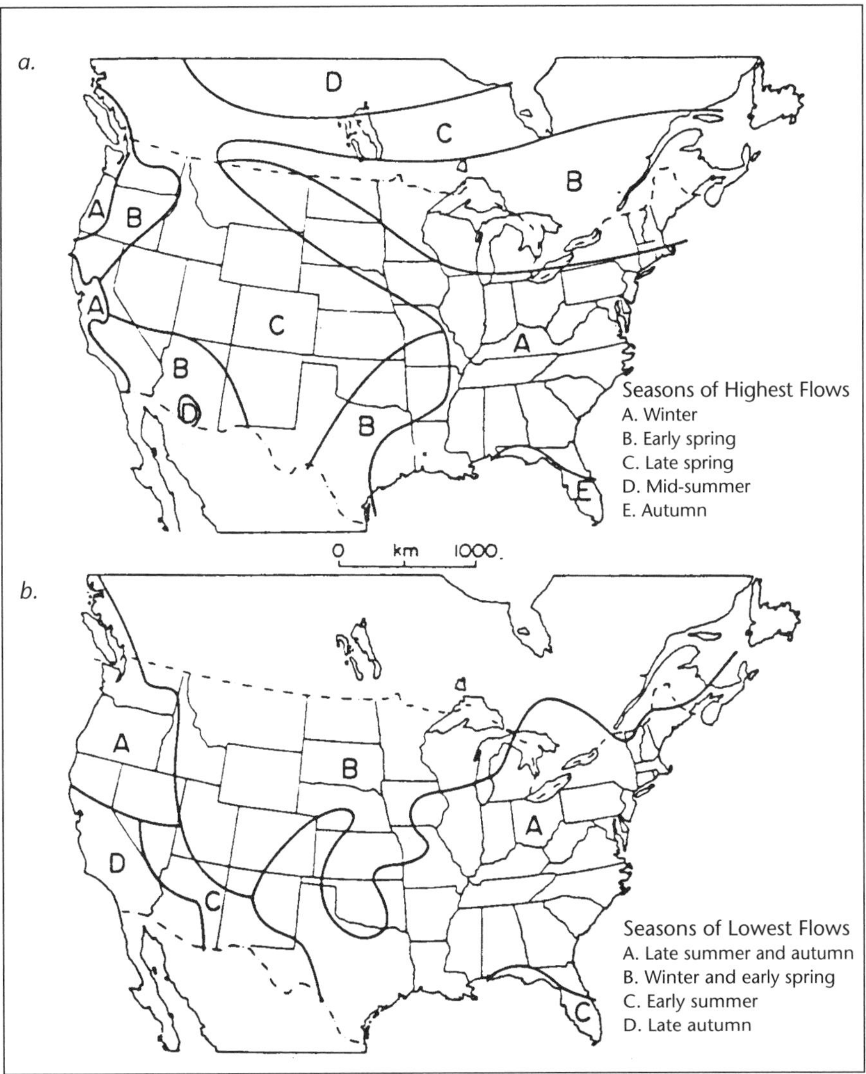

Figure 4–9. Periods of *a.* maximum and *b.* minimum streamflow in North America. *(From Beaumont, 1975; copyright © Blackwell Scientific Publications, reprinted with permission)*

riparian wetlands above bankfull discharge is shown in Figure 4–10. Riparian floodplains in many parts of the United States have recurrence intervals between 1 and 2 years for bankfull discharge, with an average of approximately 1.5 years (Fig. 4–11) (Leopold et al., 1964). The *recurrence interval* is the average interval between the recurrence of floods at a given level or greater flood (Linsley and Franzini, 1979). The inverse of the recurrence interval is the average probability of flooding in any one year. Figure 4–11 indicates that a stream will overflow its banks onto the adjacent riparian forest with a probability of 1/1.5, or 67 percent; this means that these rivers, on the average, overflow their banks in two of three years. Figure 4–11 also demonstrates that twice bankfull discharge occurs at recurrence intervals of between five and ten years; this flow, however, results in only a 30 percent greater depth over bankfull depth on the floodplain.

Surface Outflow

When it is confined to a channel, surface outflow from wetlands can be determined with the general equations for surface flow (see Equations 4.9 and 4.10 above). When a continuous record is desirable, a rating curve related to stream stage, as described above, can be developed. The outflow can also be estimated to be a function of the water level in the wetland itself according to the equation:

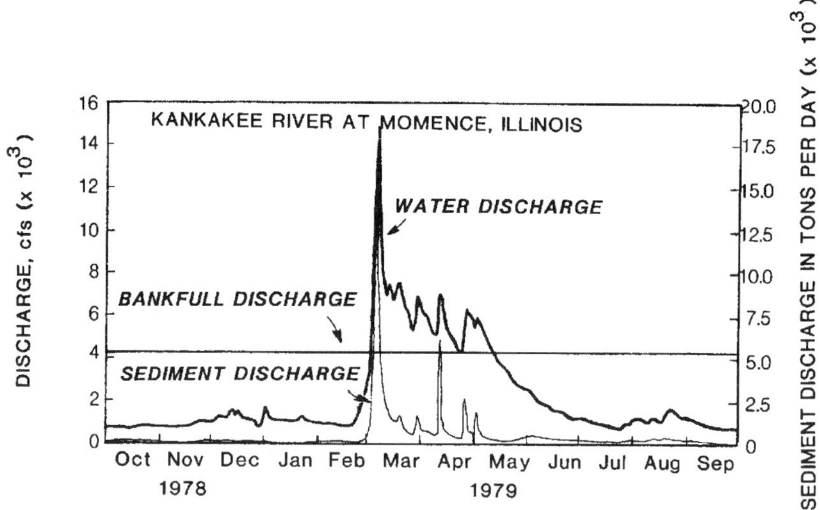

Figure 4–10. River hydrograph from northeastern Illinois, indicating bankfull discharge when riparian wetland is flooded, and sediment load of river. 1000 cfs = 28.32 m^3/sec. *(After Bhowmik et al., 1980)*

$$S_o = x\,L^y \tag{4.11}$$

where

S_o = surface outflow

L = wetland water level (cm above a control structure such as weir)

x,y = calibration coefficients

If a control structure such as a rectangular or V-notched weir is used to measure the outflow from a wetland, standard equations of the form of Equation 4.11 can be obtained from water measurement manuals (e.g., U.S. Department of Interior, 1984).

GROUNDWATER

Recharge-Discharge Wetlands

Groundwater can heavily influence some wetlands, whereas in others it may have hardly any effect at all (Carter, 1986; Carter and Novitzki, 1988). The recharge-discharge function of wetlands on groundwater resources has often

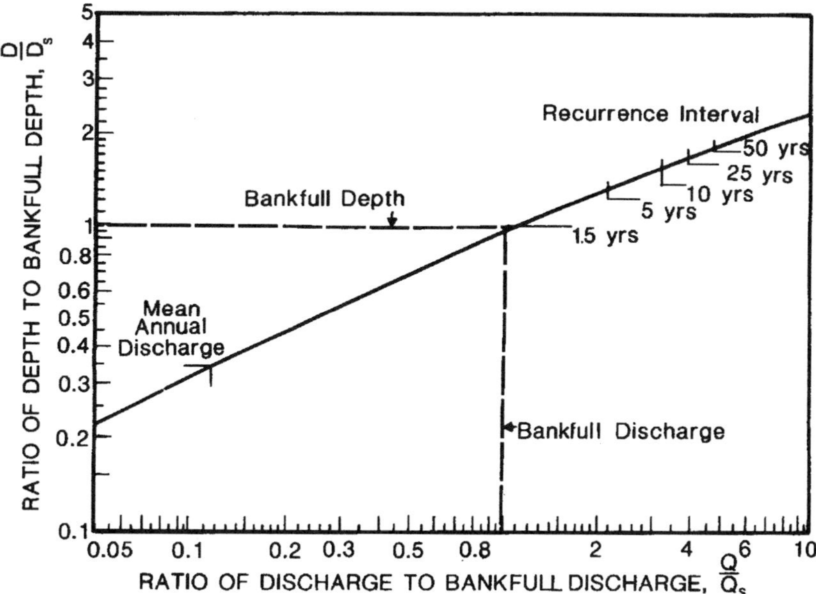

Figure 4–11. Relationships among streamflow (discharge), stream depth, and recurrence interval for streams in midwestern and southern United States. *(After Leopold et al., 1964)*

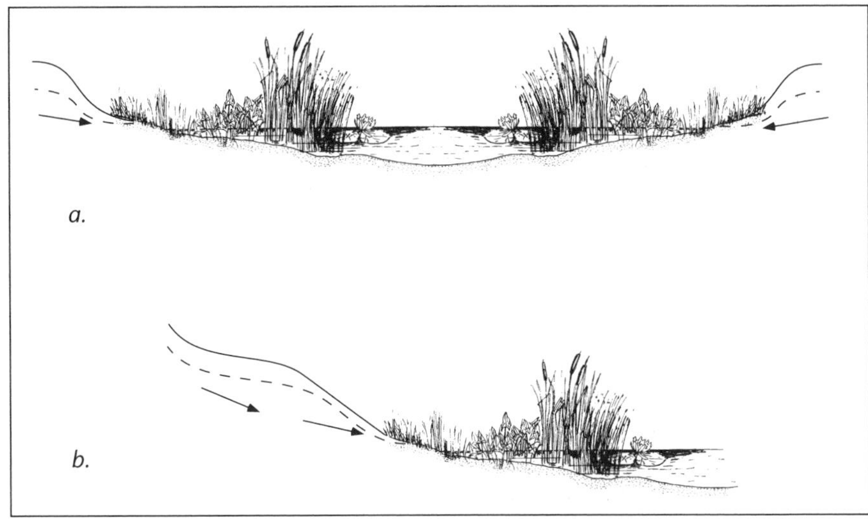

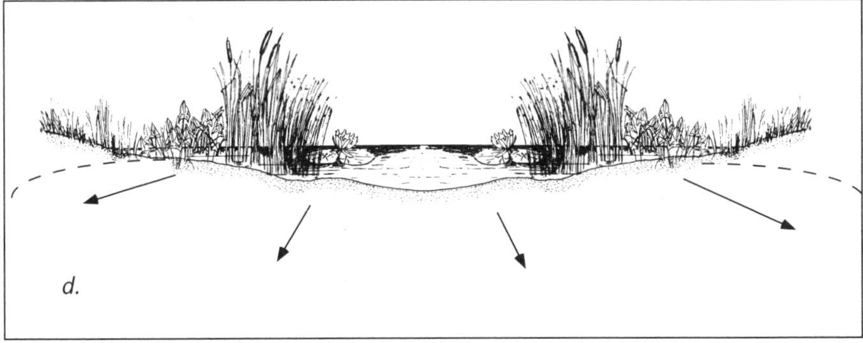

Figure 4–12. Possible wetland discharge-recharge interchanges with wetlands, including (*above*) *a.* marsh as groundwater depression wetland or "discharge" wetland; *b.* groundwater "spring" or "seep" wetland or groundwater slope wetland at base of steep slope; *c.* floodplain wetland fed by groundwater; *d.* groundwater "recharge" wetland; (*opposite page*) *e.* perched wetland or surface water depression wetland; *f.* groundwater flow through tidal wetland. *(Some terminology after Novitzki, 1979)*

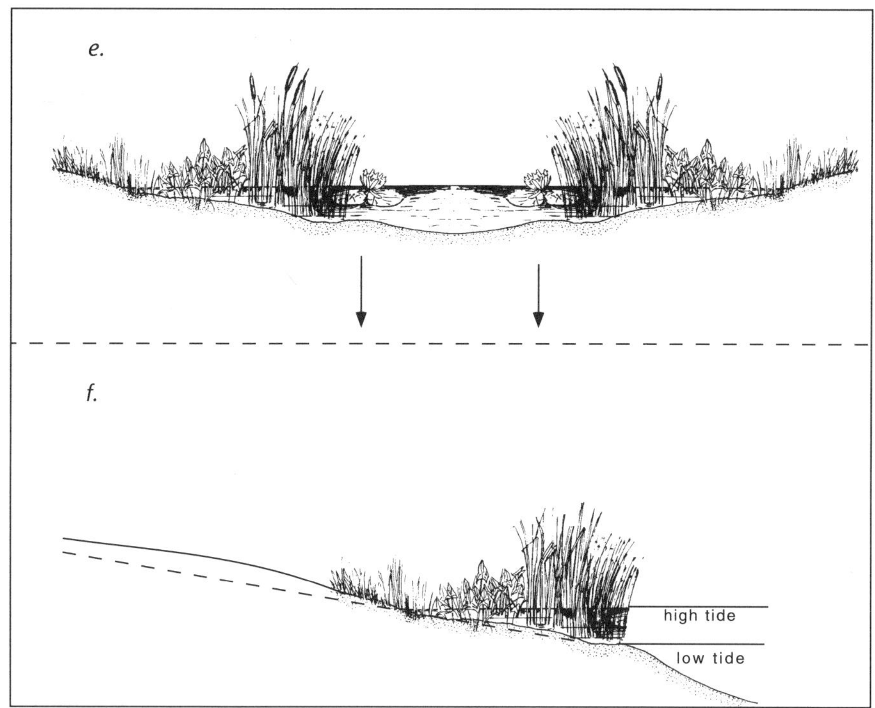

been cited as one of the most important attributes of wetlands, but it does not hold for all wetland types; nor is there sufficient experience with site specific studies to make many generalizations (Siegel, 1988a). Groundwater inflows result when the surface water (or groundwater) level of a wetland is lower hydrologically than the water table of the surrounding land (called a *discharge wetland* by geologists who generally view their water budget from a groundwater, not a wetland, perspective). Wetlands can intercept the water table in such a way that they have only inflows and no outflows, as shown for a prairie marsh in Figure 4–12a. Another type of discharge wetland is called a *spring* or *seep* wetland and is often found at the base of steep slopes where the groundwater surface intersects with the land surface (Fig. 4–12b). This type of wetland often discharges excess water downstream, usually as surface water (Novitzki, 1979; Winter, 1988). A wetland can have both inflows and outflows of groundwater, as shown in the riparian wetland in Figure 4–12c.

When the water level in a wetland is higher than the water table of its surroundings, groundwater will flow out of the wetland (called a *recharge wetland* Fig. 4–12d). When a wetland is well above the groundwater of the area, the wetland is referred to as being perched (Fig. 4–12e). This type of wetland, also referred to as a "surface water depression wetland" by Novitzki (1979), loses water only through infiltration into the ground and through evapotranspiration. Tidally influenced wetlands often have significant groundwater inflows that can

Table 4–6. Typical Hydraulic Conductivity for Wetland Soils Compared with Other Soil Materials

Wetland or Soil Type	*Hydraulic Conductivity cm/sec x 10^{-5}*	*Reference*
Northern Peatlands		
Highly Humified Blanket Bog, U.K.	0.02-0.006	Ingram, 1967
Fen, U.S.S.R.		
slightly decomposed	500	Romanov, 1968
moderately decomposed	80	
highly decomposed	1	
Carex fen, U.S.S.R.		
0-50 cm deep	310	Romanov, 1968
100-150 cm deep	6	
North American Peatlands (general)		
fibric	>150	Verry and Boelter, 1979
hemic	1.2-150	
sapric	<1.2	
Coastal Salt Marsh		
Great Sippewissett Marsh, Mass. (vertical conductivity)		Hemond and Fifield, 1982
0-30 cm deep	1.8	
high permeability zone	2,600	
sand-peat transition zone	9.4	
Non-Peat Wetland Soils		
Cypress Dome, Florida		
clay with minor sand	0.02-0.1	Smith, 1975
sand	30	
Okefenokee Swamp Watershed, Georgia	2.8-834	Hyatt and Brook, 1984
Mineral Soils (general)		
Clay	0.05	Linsley and Franzini, 1979
Limestone	5.0	
Sand	5000	

Source: Partially after Rycroft et al., 1975

reduce soil salinity and keep the wetland soil wet even during low tide (Fig. 4–12f).

A final type of wetland, one that is fairly common, is very little influenced by or influences groundwater. Because wetlands often occur where soils have poor permeability, the major source of water can be restricted to surface water runoff, with losses occurring only through evapotranspiration and other surface outflows. This type of wetland often has fluctuating hydroperiods and intermittent flooding (e.g., some prairie potholes, Fig. 4–4a), and standing water is dependent on seasonal surface inflows. If, on the other hand, such a wetland were to be influenced by groundwater, its water level would be better buffered against dramatic seasonal changes or at least it will be semipermanently flooded (Winter, 1988).

Darcy's Law

The flow of groundwater into, through, and out of a wetland is often described by *Darcy's Law*, an equation familiar to groundwater hydrologists. This law states that the flow of groundwater is proportional to (1) the slope of the piezometric surface, or the hydraulic gradient, and (2) the hydraulic conductivity, or *permeability*, the capacity of the soil to conduct water flow. In equation form, Darcy's Law is given as

$$G = k \cdot a \cdot s \tag{4.12}$$

where

G = flow rate of groundwater (volume per unit time)

k = hydraulic conductivity or permeability (length per unit time)

a = groundwater cross-sectional area perpendicular to the direction of flow

s = hydraulic gradient (slope of water table or piezometric surface)

Despite the importance of groundwater flows in the budgets of many wetlands, there is a poor understanding of groundwater hydraulics in wetlands, particularly in those that have organic soils. Table 4–6 gives some typical values of hydraulic conductivity from wetland studies, while Figure 4–13 shows the normal range of hydraulic conductivity for wetland peat as a function of fiber content. The hydraulic conductivity can be predicted for some peatland soils from their bulk density or fiber content, both of which can easily be measured. In general, the conductivity of organic peat decreases as the fiber content decreases through the process of decomposition. Water can pass through fibric, or poorly decomposed, peats a thousand times faster than it can through more decomposed sapric peats (Verry and Boelter, 1979). The type of plant material that makes up the peat is also important. Peat composed of the remains of grasses and sedges such as *Phragmites* and *Carex,* for example, is more permeable than the remains of most mosses, including sphagnum (Ingram, 1983). Rycroft et al. (1975) properly note that hydraulic conductivity of peat can vary over 9 to 10 orders of magnitude, between 10^{-8} and 10^{2} cm/sec. They also note that there has been disagreement over methods for measuring hydraulic conductivity and about whether Darcy's Law applies to flow through organic peat (Hemond and Goldman, 1985; Kadlec, 1989).

When groundwater flows into wetlands, it can often be an important source of

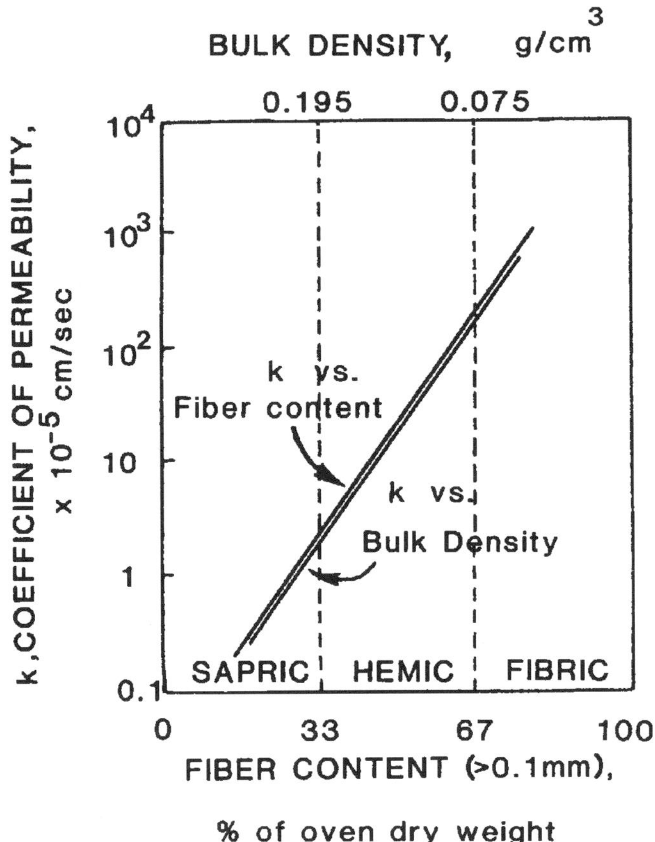

Figure 4–13. Permeability of peatland soil as a function of fiber content and bulk density. (*After Verry and Boelter, 1979; copyright © 1979 American Water Resources Association, reprinted with permission*)

nutrients and dissolved minerals. This is particularly true in the early stages of peatland development and in many coastal marshes. Fresh groundwater can influence coastal wetlands by lowering salinity, particularly at the inland edges of the wetland.

EVAPOTRANSPIRATION

The water that vaporizes from water or soil in a wetland (evaporation), together with moisture that passes through vascular plants to the atmosphere (transpiration), is called *evapotranspiration*. The meteorological factors that affect evaporation and transpiration are similar as long as there is adequate moisture, a condition that almost always exists in most wetlands. The rate of evapotranspiration is proportional to the difference between the vapor pressure at the water surface (or at the leaf surface) and the vapor pressure in the overlying air. This is described in a version of *Dalton's Law*:

$$E = c\,f(u)\,(e_w - e_a) \tag{4.13}$$

where

E = rate of evaporation

c = mass transfer coefficient

$f(u)$ = function of windspeed, u

e_w = vapor pressure at surface, or saturation vapor pressure at wet surface

e_a = vapor pressure in surrounding air

Evaporation and transpiration are enhanced by meteorological conditions such as solar radiation or surface temperature that increase the value of the vapor pressure at the evaporation surface or by factors such as decreased humidity or increased wind speed that decrease the vapor pressure of the surrounding air. This equation assumes an adequate supply of water for capillary movement in the soil or for access by rooted plants. When the water supply is limited (not a frequent occurrence in wetlands), evapotranspiration is limited as well. Transpiration can also be physiologically limited by certain plants through the closing of leaf stomata despite adequate moisture during periods of stress such as anoxia.

Empirical Estimates of Wetland Evapotranspiration

Evapotranspiration can be determined with any number of empirical equations that use easily measured meteorological variables or by various direct measures. One of the most frequently used empirical equations for evapotranspiration from terrestrial ecosystems, which has been applied with some success to wetlands, is the *Thornthwaite Equation* for potential evapotranspiration (Chow, 1964):

$$ET_i = 16\,(10T_i/I)^a \tag{4.14}$$

where

ET_i = potential evapotranspiration for month i, mm/mo

T_i = mean monthly temperature, °C

I = local heat index $= \sum_{i=1}^{12} (T_i/5)^{1.514}$

$a = (0.675 \times I^3 - 77.1 \times I^2 + 17{,}920 \times I + 492{,}390) \times 10^{-6}$

This equation was used to determine evapotranspiration from the Okefenokee Swamp in Georgia by Rykiel (1977, 1984). For a 26-year period examined in that study, average evapotranspiration ranged from 21 mm/mo in December to 179 mm/mo in July. Kadlec et al. (1988) tested the Thornthwaite Equation on wetland evapotranspiration in Michigan and Nevada and found it to underpredict actual evapotranspiration, especially in the arid Nevada site.

A second empirical relationship that has had many applications in hydrologic and agricultural studies but relatively few in wetlands is the *Penman Equation* (Penman, 1948; Chow, 1964). This equation, based on both Dalton's Law and the energy budget approach, is given as

$$ET = \frac{\Delta\ H\ +\ 0.27\ E_a}{\Delta\ +\ 0.27} \tag{4.15}$$

where

ET = evapotranspiration, mm/day

Δ = slope of curve of saturation vapor pressure *vs.* mean air temperature, mm Hg/°C

H = net radiation, cal/cm^2-day

= $R_t\,(1 - a) - R_b$

R_t = total shortwave radiation

a = albedo of wetland surface

R_b = effective outgoing longwave radiation = $f(T^4)$

E_a = term describing the contribution of mass-transfer to evaporation

= $0.35\ (0.5 + 0.00625\ u)\ (e_w - e_a)$

u = wind speed 2 m above ground, km/day

e_w = saturation vapor pressure of water surface at mean air temperature, mm Hg

e_a = vapor pressure in surrounding air, mm Hg

The Penman Equation was compared with the pan evaporation (multiplied by 0.8 factor) and other methods at natural enriched fens in Michigan and constructed wetlands in Nevada by Kadlec et al. (1988). They found that the Penman Equation, like the Thornthwaite Equation, generally underpredicted evapotranspiration from the Michigan wetland (Fig. 4–14) but agreed within a few percent with other measurement techniques for the Nevada wetlands.

Another empirical relationship for describing summer evapotranspiration was developed by Scheffe (1978) and was described by Hammer and Kadlec (1983). The equation, which was used individually for sedges, willow, leatherleaf, and cattail vegetation covers, is

$$ET = \alpha + \beta B + \delta C + \gamma D + \lambda E \tag{4.16}$$

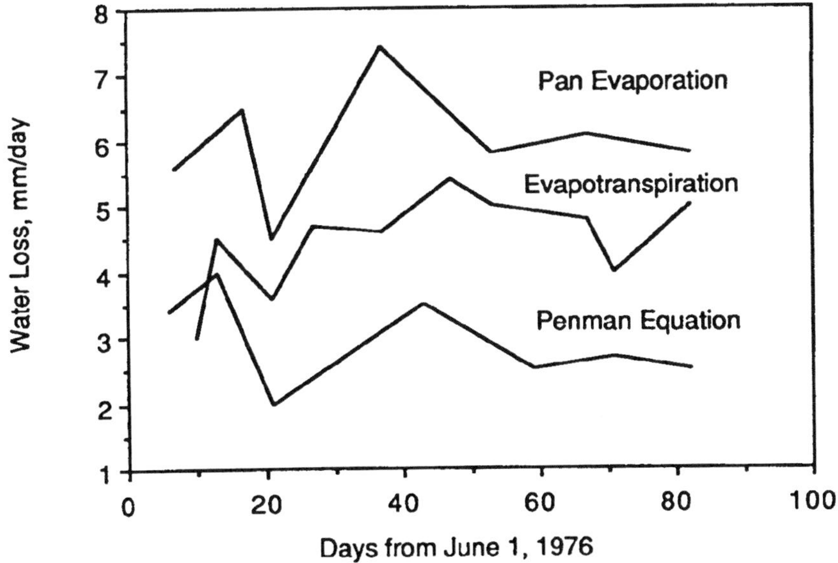

Figure 4–14. Comparison of evapotranspiration measured by diurnal change in water level with pan evaporation measurements and calculations from Penman equation for Houghton Lake, Michigan, enriched fen (sedge site) in summer 1976. (*From Kadlec et al., 1988; copyright © by Donald D. Hook, reprinted with permission*)

where

$\alpha, \beta, \delta, \gamma, \lambda$ = correlation coefficients

B = incident radiation (measured by pyranograph)

C = air temperature

D = relative humidity

E = wind speed

The equation gives estimates that are better than some more frequently used evapotranspiration relationships, although when the results of using this model were compared to actual measurements, the radiation term was shown to dominate (Hammer and Kadlec, 1983).

Because of the many meteorological and biological factors that affect evapotranspiration, none of the many empirical relationships, including the Thornthwaite, Penman, and Hammer and Kadlec Equations, is entirely satisfactory for estimating wetland evapotranspiration. Lee (1980) cautions that there is "no reliable method of estimating evapotranspiration rates based on simple weather-element data or potential evapotranspiration." Nevertheless, these equations of potential evapotranspiration offer the most cost-effective first approximations for estimating water loss. Furthermore, when applied to wetlands, which are only rarely devoid of an adequate water supply, they may be more reliable than their applications to upland terrain, where evapotranspiration can be limited by a lack of soil water.

Direct Measurement of Wetland Evapotranspiration

Several direct measurement techniques can be used in wetlands to determine evapotranspiration. Over fairly uniform areas, it is possible to determine evapotranspiration from heat and water balances through the plant canopy (Hsu et al., 1972). Evapotranspiration from wetlands has also been calculated from measurements of the increase in water vapor in air flowing through vegetation chambers (S. L. Brown, 1981) and from observing the diurnal cycles of groundwater or surface water in wetlands (Mitsch et al., 1977; Heimburg, 1984; Ewel and Smith, 1992). This latter method, described in Figure 4–15, can be calculated as follows:

$$ET = S_y\,(24\,h \pm s) \qquad (4.17)$$

where

ET = evapotranspiration, mm/day

S_y = specific yield of aquifer (unitless)

= 1.0 for standing water wetlands

< 1.0 for groundwater wetlands

h = hourly rise in water level from midnight to 4:00 A.M., mm/hr

s = net fall (+) or rise (-) of water table or water surface in one day

The pattern assumes active "pumping" of water by vegetation during the day and a constant rate of recharge equal to the midnight-to-4:00-A.M. rate. This method also assumes that evapotranspiration is negligible around midnight and that the water table around this time approximates the daily mean. The water level is usually at or near the root zone in many wetlands, a necessary condition for this method to measure evapotranspiration accurately (Todd, 1964).

Effects of Vegetation on Wetland Evapotranspiration

A question about evapotranspiration from wetlands, which does not elicit a uniform answer in the literature, is, "Does the presence of wetland vegetation increase or decrease the loss of water over that which would occur from an open body of

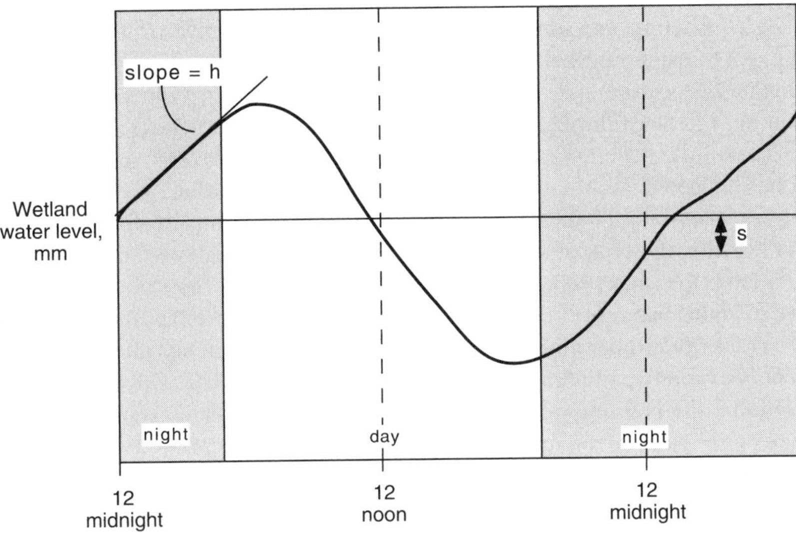

Figure 4–15. Diurnal water fluctuation in some wetlands as it is used to calculate evapotranspiration with Equation 4.17. *(After Todd, 1964)*

water?" Data from individual studies are conflicting. Obviously, the presence of vegetation retards evaporation from the water surface, but the question is whether the transpiration of water through the plants equals or exceeds the difference (Kadlec et al., 1988; Kadlec, 1989). Eggelsmann (1963) found evaporation from bogs in Germany to be generally less than that from open water except during wet summer months. In studies of evapotranspiration from small bogs in northern Minnesota, Bay (1967) found it to be 88 percent to 121 percent of open-water evaporation. Eisenlohr (1976) found 10 percent lower evapotranspiration from vegetated prairie potholes than from nonvegetated potholes in North Dakota. Hall et al. (1972), through a series of measurements and calculations, estimated that a stand of vegetation in a small New Hampshire wetland lost 80 percent more water than did the open water in the wetland. In a forested pond cypress dome in north-central Florida, Heimburg (1984) found that swamp evapotranspiration was about 80 percent of pan evaporation during the dry season (spring and fall) and as low as 60 percent of pan evaporation during the wet season (summer). S. L. Brown (1981) found that transpiration losses from pond cypress wetlands were lower than evaporation from an open water surface even with adequate standing water.

In the arid West, it has been a long-standing practice to conserve water for irrigation and other uses by clearing riparian vegetation from streams. In this environment where groundwater is often well below the surface but within the rooting zone of deep-rooted plants, trees "pump" water to the leaf surface and actively transpire even when little evaporation occurs at the soil surface.

The conflicting measurements and the difficulty of measuring evaporation and evapotranspiration led Linacre (1976) to conclude that neither the presence of wetland vegetation nor the type of vegetation had major influences on evaporation rates, at least during the active growing season. Bernatowicz et al. (1976) also found little difference in evapotranspiration among several species of vegetation. This general unimportance of vegetation-species variation on overall wetland water loss is probably a reasonable conclusion for most wetlands, although it is clear that the type of wetland ecosystem and the season are important considerations. Ingram (1983), for example, found that fens have about 40 percent more evapotranspiration than do treeless bogs and that evaporation from the bogs is less than potential evapotranspiration in the summer and greater than potential evapotranspiration in the winter. Furthermore, H. T. Odum (1984) concluded that the draining of Florida cypress swamps and their "replacement with either open water or other kinds of vegetation may decrease available water, increasing frequency of drought, raising microclimate temperatures in summer, and reducing productivity of natural and agricultural ecosystems."

TIDES

The periodic and predictable tidal inundation of coastal salt marshes, mangroves, and freshwater tidal marshes is a major hydrologic feature of these wetlands. The tide acts as a stress by causing submergence, saline soils, and soil anaerobiosis; it acts as a subsidy by removing excess salts, reestablishing aerobic conditions, and providing nutrients. Tides also shift and alter the sediment patterns in coastal wetlands, causing a uniform surface to develop.

Typical tidal patterns for several coastal areas in the Atlantic and Gulf coasts of the United States are shown in Figure 4–16a. Seasonal as well as diurnal patterns exist in the tidal rhythms. Annual variations of mean monthly sea level are as great as 25 cm (Fig. 4–16b). Tides also have significant bimonthly patterns because they are generated by the gravitational pull of the moon and, to a lesser extent, the sun. When the sun and the moon are in line and pull together, which occurs almost every two weeks, *spring tides,* or tides of the greatest amplitude, develop. When the sun and the moon are at right angles, *neap tides,* or tides of least amplitude, occur. Spring tides occur roughly at full and new moons, whereas neap tides occur during the first and third quarters.

Tides vary more locally than regionally. The primary determinant is the coastline configuration. In North America, tidal amplitudes vary from less than 1 meter along the Texas Gulf Coast to several meters in the Bay of Fundy in Nova Scotia. Tidal amplitude can actually increase as one progresses inland in some funnel-shaped estuaries (W. E. Odum et al., 1984).

Typically on a rising tide, water flows up tidal creek channels until the channels are bankfull. It overflows first at the upstream end, where tidal creeks break

a.

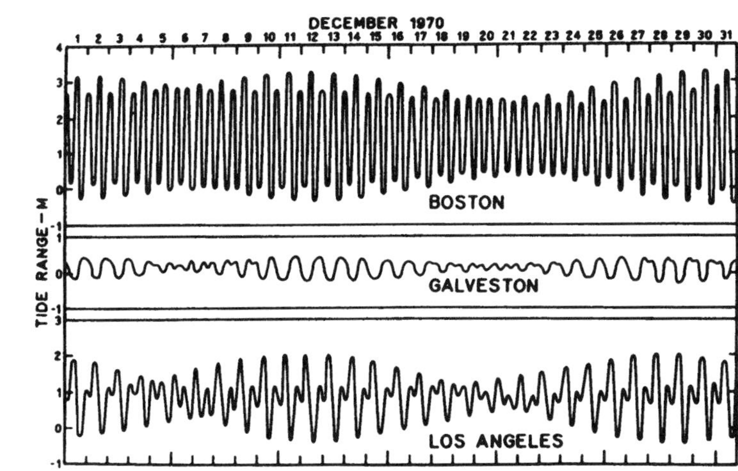

b.

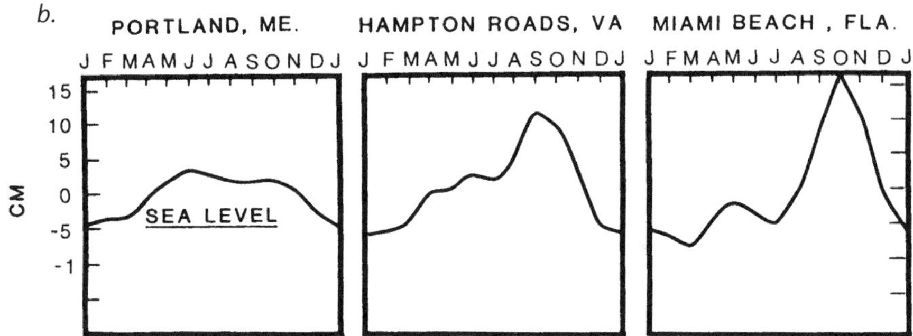

Figure 4–16. *a.* Daily pattern of tides, and *b.* seasonal changes in mean monthly sea level for several locations in North America. *(After Emery and Uchupi, 1972)*

up into small creeklets that lack natural levees. The overflowing water spreads back downstream over the marsh surface. On falling tides the flows are reversed. At low tides water continues to drain through the natural levee sediments into adjacent creeks because these sediments tend to be relatively coarse; in the marsh interior, where sediments are finer, drainage is poor and water is often impounded in small depressions in the marsh.

SPECIFIC EFFECTS OF HYDROLOGY ON WETLANDS

The effects of hydrology on wetland structure and function can be described with a complicated series of cause and effect relationships. A conceptual model that shows the general effects of hydrology in wetland ecosystems was shown in

Figure 4–1. The effects are shown to be primarily on the chemical and physical aspects of the wetlands, which, in turn, affect the biotic components of the ecosystem. The biotic components, in turn, have a feedback effect on hydrology. Several principles underscoring the importance of hydrology in wetlands can be elucidated from the studies that have been conducted to date. These principles, discussed below, are as follows:

1. Hydrology leads to a unique vegetation composition but can limit or enhance species richness.
2. Primary productivity and other ecosystem functions in wetlands are often enhanced by flowing conditions and a pulsing hydroperiod and are often depressed by stagnant conditions.
3. Accumulation of organic material in wetlands is controlled by hydrology through its influence on primary productivity, decomposition, and export of particulate organic matter.
4. Nutrient cycling and nutrient availability are both significantly influenced by hydrologic conditions.

Species Composition and Diversity

Hydrology is a two-edged sword for species composition and diversity in wetlands. It acts as a limit or a stimulus to species richness, depending on the hydroperiod and physical energies. At a minimum, the hydrology acts to select water-tolerant vegetation in both freshwater and saltwater conditions. Of the thousands of vascular plants that are on Earth, relatively few have adapted to waterlogged soils. (These adaptations are discussed in more detail in Chapter 6.) Although it is difficult to generalize, many wetlands that sustain long flooding durations have lower species richness in vegetation than do less frequently flooded areas. Waterlogged soils and the subsequent changes in oxygen content and other chemical conditions significantly limit the number and the types of rooted plants that can survive in this environment. McKnight et al. (1981), in describing the effects of water on species composition in a riparian wetland, stated that "In general, as one goes from the hydric [wet] to the more mesic [dry] bottomland sites, the possible combinations or mixtures of species increases." Bedinger (1979), in reviewing the literature of flooding effects on tree species, attributes the effects to the following factors:

1. Different species have different physiological responses to flooding.
2. Large trees show greater tolerance to flooding than do seedlings.
3. Plant establishment depends on the tolerance of the seeds to flooding.
4. Plant succession depends on the geomorphic evolution of the floodplain such as by sediment deposition or stream downcutting.

Heinselman (1970) found a change in vegetation richness for seven different hydrologically defined conditions of northern peatlands. He noted an increase in diversity, as measured by the number of species, as the flowthrough conditions increased (Table 4–7). In this case, the flowing water can be thought of as a stimulus to diversity, probably caused by its ability to renew minerals and reduce anaerobic conditions.

Hydrology also stimulates diversity when the action of water and transported sediments creates spatial heterogeneity, opening up additional ecological niches (Gosselink and Turner, 1978). When rivers flood riparian wetlands or when tides rise and fall on coastal marshes, erosion, scouring, and sediment deposition sometimes create niches that allow diverse habitats to develop. On the other hand, flowing water can also create a very uniform surface that might cause monospecific stands of *Typha* or *Phragmites* to dominate a freshwater marsh or *Spartina* to dominate a coastal marsh. Keddy (1992) likens water level fluctuations in wetlands to fires in forests. They eliminate one growth form of vegetation (e.g., woody plants) in favor of another (e.g., herbaceous species) and allow regeneration of species from buried seeds (see Chapter 7).

Primary Productivity

In general, the "openness" of a wetland to hydrological fluxes is probably one of the most important determinants of potential primary productivity. For example, peatlands that have flow-through conditions (fens) have long been known to be more productive than stagnant raised bogs (Moore and Bellamy, 1974; see Chapter 12). A number of studies have found that wetlands in stagnant (nonflowing) or continuously deep water have low productivities, whereas wetlands that are in slowly flowing strands or are open to flooding rivers have high productivities. Brinson et al. (1981a) summarized the results of many of these studies by describing the net biomass production of forested freshwater wetlands in order of greatest to least productivity:

flowing water swamps > sluggish flow swamps > stillwater swamps

The relationship between hydrology and ecosystem primary productivity has been investigated most extensively for forested wetlands (e.g., Conner and Day, 1976; Mitsch and Ewel, 1979; S. L. Brown, 1981). A general relationship was developed by Mitsch and Ewel (1979) for cypress productivity as a function of hydrology in Florida. That study concluded that

> Cypress-hardwood associations, found primarily in riverine and flowing strand systems, have the most productive cypress trees. The short hydroperiod favors both root aeration during the long dry periods and elimination of water-intolerant species during the short wet periods. The

Table 4–7. The Relationship Between the Hydrologic Regime and Species Richness in Northern Minnesota Peatlands

	Species Present						
	Tree	*Shrub*	*Field herbs*	*Grasses and ferns*	*Ground layer*	*Total*	*Flow Conditions*
1. Rich swamp forest	6	16	28	11	10	71	Good surface flow; minerotrophic
2. Poor swamp forest	3	14	17	12	5	51	Downstream from 1; not adapted to strong water flow
3. Cedar string bog and fen	3	10	10	12	4	39	Better drainage than 2
4. Larch string bog and fen	3	9	9	12	4	37	Similar to 3; sheet flow
5. Black spruce feather moss forest	2	9	2	2	10	25	Gentle water flow on semiconvex template
6. Sphagnum bog	2	8	2	1	7	20	Isolated; little standing water
7. Sphagnum heath	2	6	2	2	5	17	Wet, soggy, and on convex template

Source: After Gosselink and Turner, 1978, and Heinselman, 1970.

continual supply of nutrients with the flooding river system conditions may be a second important factor in maintaining these high productivities.

Productivity was found to be low under both continually flooded conditions and drained conditions. S. L. Brown (1981) found that much of the variation in biomass productivity of cypress wetlands in Florida could be explained by the variation in nutrient inflow, as measured by phosphorus. Productivity is lowest there when nutrients are brought into the system solely by precipitation and is highest when large amounts of nutrients are passed through the wetlands by flooding rivers. Brown suggested that rather than there being a simple relationship between wetland productivity and hydrology, there is a more complex relationship among hydrology, nutrient inputs, and wetland productivity, decomposition, export, and nutrient cycling. Hydrology, then, also influences wetland productivity by being the main pathway through which nutrients are transported to many wetlands.

The influence of hydrologic conditions on freshwater marsh productivity is less certain. If peak biomass or similar measures are used as indicators of marsh productivity, studies can easily indicate a higher macrophyte productivity in sheltered, non-flowing marshes than in wetlands open to flowing conditions or coastal influences. For example, Robb (1989), as described by Mitsch (1992b), measured consistently higher macrophyte biomass in wetlands isolated from surface fluxes with artifical dikes than in wetlands open to coastal fluxes along Lake Erie (Table 4–8). Several explanations are possible: (1) the coastal fluxes may also be serving as a stress as well as a subsidy on the macrophytes; (2) the open marshes may be exporting a significant amount of their productivity; and (3) the diked wetlands have more predictable hydroperiods. A study of the influence of flow-through conditions on water column primary productivity of constructed marshes found that after two years of experimentation, productivity was higher in high-flow wetlands than in low-flow wetlands (Fig. 4–17). It appears that while the macrophyte productivity may take many years to respond to the difference in hydrology, the water column pro-

Table 4–8. Selected Macrophyte Measurements at Peak Biomass from Diked and Undiked Wetlands of Ohio's Coastal Lake Erie

Measure of Vegetation Structure	*Average ± std error*	
	Diked (impounded) Wetlands (n=6)	*Undiked (open to Lake Erie) Wetlands (n=4)*
Biomass g dry wt/m^2	897 ± 277	473 ± 149
# species/plot[a]	1.7 ± 0.3	1.4 ± 0.3
# stems/m^2	597 ± 211	241 ± 59

[a]Only species > 10% by weight per plot. Plots were 0.5 m^2 randomly placed in each wetland (3 to 6 per wetland).

Source: From Mitsch, 1992b, based on data from Robb, 1989

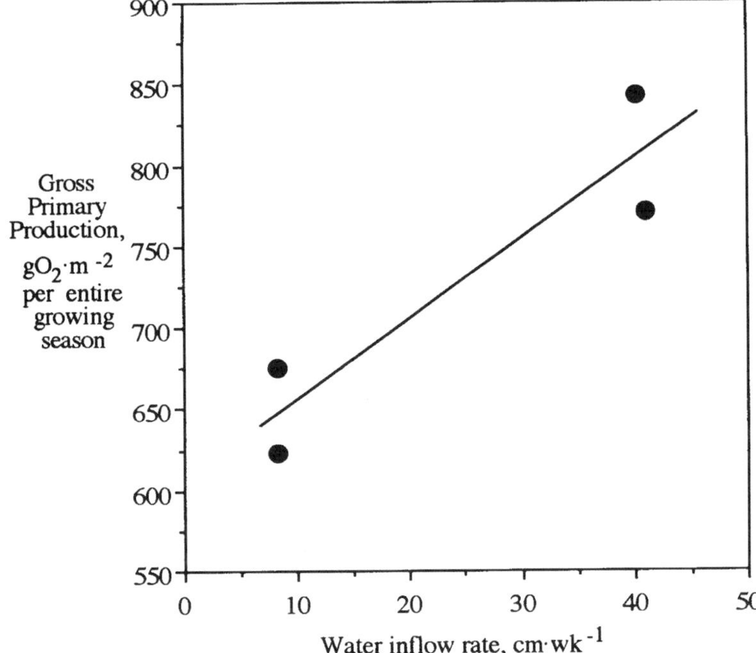

Figure 4–17. Aquatic gross primary productivity per growing season for two years versus average flow-through conditions of constructed freshwater marshes at Des Plaines River Wetland Demonstration Project in Northeastern Illinois. (*From Cronk and Mitsch, unpub. manuscript*)

ductivity, which is mainly due to attached and planktonic algae, responds relatively quickly to different hydrologic conditions.

Saltwater tidal wetlands subject to frequent tidal action are generally more productive than those that are only occasionally inundated. For example, Steever et al. (1976) showed a direct relationship between tidal range (as a measure of water flux) and end-of-season peak biomass of *Spartina alterniflora* (Fig. 4–18). They attributed the relationship to a nutrient subsidy and a flushing of toxic materials such as salt with vigorous tidal fluxes. Whigham et al. (1978) further suggested that freshwater tidal wetlands may be even more productive than saline tidal wetlands because they receive the energy and nutrient subsidy of tidal flushing while avoiding the stress of saline soils.

Despite the overwhelming evidence of the influence of hydrology on wetlands, some investigators have cautioned against always ascribing a direct linkage between hydrologic variables and wetland productivity. Richardson (1979) states that "a definitive statement about the influence of water levels on net primary productivity for all wetland types is impossible, since responses of individ-

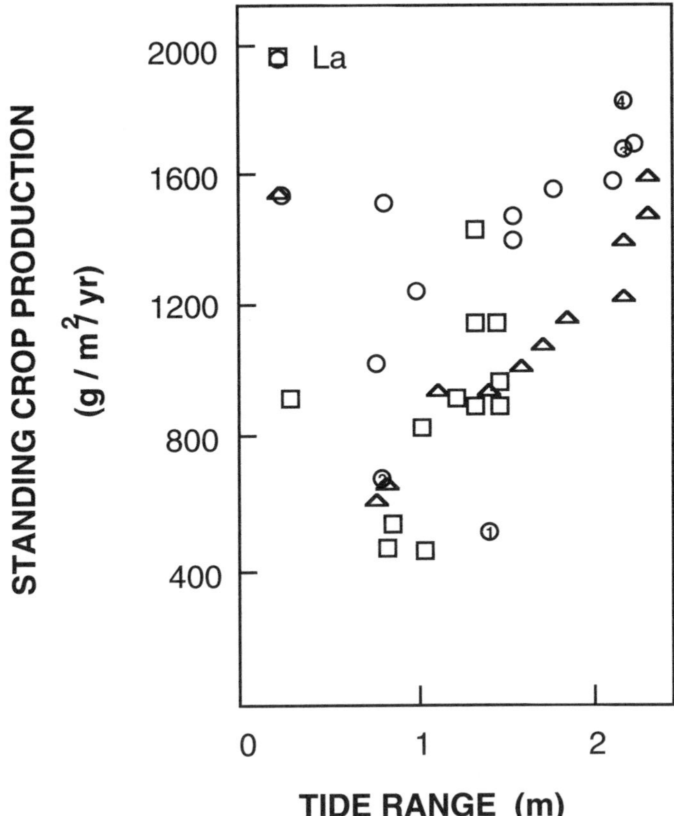

Figure 4–18. Production of *Spartina alterniflora* versus mean tidal range for several Atlantic coastal salt marshes. Different symbols indicate different data sources. La refers to Mississippi River delta marshes. (*After Steever et al., 1976*)

ual species to water fluctuations vary." Water level fluctuations, however, are not necessarily related to the volume of flow-through of water and the associated nutrients and allochthonous energy. Furthermore, although individual species vary in their responses to water levels and hydrology, ecosystem-level responses may be more consistent.

Organic Accumulation and Export

Wetlands can accumulate excess organic matter either as a result of increased primary productivity (as described above) or decreased decomposition and export. Notwithstanding the discrepancies from short-term litter decomposition studies, peat accumulates to some degree in all wetlands as a result of these processes. The effects of hydrology on decomposition pathways are even less clear than the effects on primary productivity discussed above. Brinson et al. (1981a) concluded that it cannot be assumed that increased frequency or duration of flooding will necessarily increase or decrease decomposition rates. They suggested, however, that alternating wet and dry conditions may lead to optimum litter decomposition rates, whereas completely anaerobic conditions caused by constant flooding are the least favorable conditions for decomposition.

Although litter decomposition rates have been measured in several wetlands, field results do not consistently supporting this view. Brinson (1977), in a study of an alluvial tupelo swamp in North Carolina, found that the decomposition of litter was most rapid in the river, slower on the wet swamp floor, and slowest on a dry levee. W. E. Odum and Heywood (1978) found that leaves of freshwater tidal marsh plants decomposed more rapidly when permanently submerged than when periodically or irregularly flooded. They suggest that this may be due to (1) better access to detritivores in the water, (2) a more constant physical environment for decomposer bacteria and fungi, (3) a greater availability of dissolved nutrients, and (4) a more suitable environment for leaching. Chamie and Richardson (1978), on the other hand, stated that "periodic or even constant flooding of a soil's surface, characteristic of wetlands, leads to an overall decrease in the activity of soil fauna" and causes slow anaerobic decomposition to dominate. Deghi et al., (1980), in a study of decomposition in cypress wetlands in Florida, found that the decomposition of cypress needles occurred more rapidly in wet areas than in dry ones but that there was no difference in decomposition rates between deep and shallow sites. Van der Valk et al. (1991) demonstrated at the Delta Marsh in south-central Manitoba that litter from several emergent plants decayed at a slightly faster rate in wetlands that were flooded approximately 1 m above normal water levels. They suggest that "when litter is not inundated [as was apparently the case in the normal water level sites], it rapidly dries out, and this adversely affects microbial populations."

The importance of hydrology for organic export is obvious. A generally higher rate of export is to be expected from wetlands that are open to the flow-through of water. Riparian wetlands often contribute large amounts of organic detritus to streams, including macro-detritus such as whole trees. For many years salt marshes and mangrove swamps were considered major exporters of their production (for example, 45 percent estimated by Teal (1962) for a salt marsh; 28 percent measured by Heald (1969) for a mangrove swamp), but the generality of this concept is not accepted by coastal ecologists (Nixon, 1980; see Chap. 5). Hydrologically isolated wetlands such as northern peatlands have much lower organic export. For example, Bazilevich and Tishkov (1982) found that only 6 percent of the net productivity of a fen in Russia was exported by surface and subsurface flows.

Nutrient Cycling

Nutrients are carried into wetlands by hydrologic inputs of precipitation, river flooding, tides, and surface and groundwater inflows. Outflows of nutrients are controlled primarily by the outflow of waters. These hydrologic/nutrient flows are also important determinants of wetland productivity and decomposition (see previous sections). Intrasystem nutrient cycling is generally, in turn, tied to pathways such as primary productivity and decomposition. When productivity and decomposition rates are high, as in flowing water or pulsing hydroperiod wetlands, nutrient cycling is rapid. When productivity and decomposition processes are slow, as in isolated ombrotrophic bogs, nutrient cycling is also slow.

The hydroperiod of a wetland has a significant effect on nutrient transformations and on the availability of nutrients to vegetation (see Chap. 5). Nitrogen availability is affected in wetlands by the reduced conditions that result from waterlogged soil. Typically, a narrow oxidized surface layer develops over the anaerobic zone in wetland soils, causing a combination of reactions in the nitrogen cycle—nitrification and denitrification—that may result in substantial losses of nitrogen to the atmosphere. Ammonium nitrogen often accumulates in wetland soils since the anaerobic environment favors the reduced ionic form over the nitrate common in agricultural soils.

The flooding of wetland soils by altering both the pH and the redox potential of the soil, influences the availability of other nutrients as well. The pH of both acid and alkaline soils tends to converge on a pH of 7 when they are flooded (see Chapter 5). The redox potential, a measure of the intensity of oxidation or reduction of a chemical or biological system, indicates the state of oxidation (and hence availability) of several nutrients. Phosphorus is known to be more soluble under anaerobic conditions. Several studies have documented higher concentrations of soluble phosphorus in poorly drained soils than in oxidized conditions (e.g., Redman and Patrick, 1965; Patrick and Khalid, 1974). This is partially caused by the hydrolysis and reduction of ferric and aluminum phosphates to more soluble compounds. The availability of major ions such as potassium and magnesium and several trace nutrients such as iron, manganese, and sulfur is also affected by hydrologic conditions in the wetlands (Gambrell and Patrick, 1978; Mohanty and Dash, 1982). Chemical transformations in wetlands are discussed in more detail in the next chapter.

WETLAND HYDROLOGY STUDIES

Measurement of Wetland Hydrology

It is curious that so little attention has been paid to hydrologic measurements in wetland studies, despite the importance of hydrology in ecosystem function. A

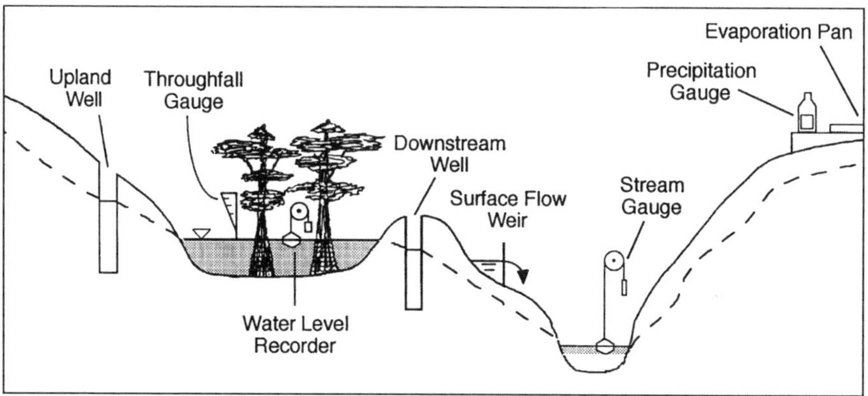

Figure 4–19. Placement of typical measuring equipment for monitoring a water budget of an alluvial wetland.

great deal of information can be obtained with only a modest investment in supplies and equipment. A diagram summarizing many of the hydrology measurements typical for developing a wetland's water budget is given in Figure 4–19. Water levels can be recorded continuously with a water level recorder or during site visits with a staff gauge. With records of water level, all of the following hydrologic parameters can be determined: hydroperiod, frequency of flooding, duration of flooding, and water depth (Gosselink and Turner, 1978). Water level recorders can also be used to determine the change in storage in a water budget, as in Equation 4.1.

Evapotranspiration measurements are more difficult to obtain, but there are several empirical relationships such as the Thornthwaite Equation that can use meteorological variables (Chow, 1964). Evaporation pans can also be used to estimate total evapotranspiration from wetlands, although pan coefficients are highly variable (Linsley and Franzini, 1979). Evapotranspiration of continuously flooded non-tidal wetlands can also be determined by monitoring the diurnal water level fluctuation as described in Figure 4–15.

Precipitation or throughfall or both can be measured by placing a statistically adequate number of rain gauges in random locations throughout the wetland or by utilizing weather station data. Surface runoff to wetlands can usually be determined as the increase in water level in the wetland during and immediately following a storm after throughfall and stemflow have been subtracted. Weirs can be constructed on more permanent streams to monitor surface water inputs and outputs.

Groundwater flows are usually the most difficult hydrologic flows to measure accurately. In some cases, a few shallow wells placed around a wetland will help indicate the direction of groundwater flow. Estimates of permeability are required to quantify the flows. In other cases, groundwater input or loss can be determined as the residual of the water budget, although this method has limited accuracy (Carter et al., 1979).

Hydrology and Wetland Classification

Hydrologic conditions are so important in defining wetlands that they are often used by scientists to classify these ecosystems. It is no coincidence that classification and mapping of wetlands based on biotic features (dominant vegetation) often matches the hydrologic conditions of the different wetlands very well. For example, peatlands have been classified according to whether they have water flow from surrounding mineral soils or if they are in flow-isolated basins. Salt marshes and salt marsh vegetation are defined and subdivided according to the frequency and depth of tidal inundation. Bottomland forests are zoned according to flooding frequency, and certain deep swamps are classified according to stillness or movement of water. Some of the classifications for particular wetland types are described in Chapters 8–14. Overall wetland classifications, which are based in whole or in part on hydrologic conditions, are described in Chapter 18.

Research Needs

There are several needs and shortcomings in wetland hydrology studies. Some of these, originally were listed by Carter et al. (1979), are still valid today:

1. the need for improving, refining, and perhaps simplifying existing techniques for hydrologic measurements;
2. the need for making accurate measurements of all the hydrologic inputs and outputs to representative wetland types and estimating the errors inherent in various measurement techniques;
3. the need to quantify the soil-water-vegetation relationships of wetlands and to improve our basic understanding of these relationships
4. the need to make in-depth, long-term studies of different wetland types under different environmental conditions; and
5. the need to continue developing models based on hydrologic data so that we can develop better analyses and predictive capability.

Wetland researchers and managers should recognize the importance of hydrologic studies and research to augment the more frequently studied biological components of wetlands. These two aspects are closely related.

Biogeochemistry of Wetlands

5

Wetlands have biogeochemical cycles with a combination of chemical transformations and chemical transport processes not shared by many other ecosystems. Wetland soils, also known as hydric soils, can become highly reduced when submerged relatively rapidly but usually have a narrow oxidized surface zone at their surface that allows for aerobic processes. Transformations of nitrogen, phosphorus, sulfur, iron, manganese, and carbon that occur within the anaerobic environment affect the availability of these minerals. Some also cause toxic conditions, while others such as denitrification and methanogenesis cause a loss of chemicals to the atmosphere. Many of the transformations are mediated by microbial populations that are adapted to the anaerobic environment. Chemicals are hydrologically transported to wetlands through precipitation, surface flow, groundwater, and tides. Wetlands dominated only by precipitation are generally nutrient poor, whereas there is a wide variability of concentrations of chemicals flowing into wetlands from the other three sources. Wetlands have been shown to be nitrogen and phosphorus sinks in several studies, although there is a growing consensus that not all wetlands are nutrient sinks, nor are the patterns consistent from season to season or year to year. Wetlands are often chemically coupled to adjacent ecosystems by the export of organic materials, although the direct effects on adjacent ecosystems have been difficult to quantify. Although wetlands are similar to terrestrial and aquatic ecosystems in that they can be high-nutrient or low-nutrient systems, there are several differences, particularly in the importance of sediment storage of nutrients and in the functioning of the vegetation in the cycle of different nutrients.

The transport and transformation of chemicals in ecosystems, known as *biogeochemical cycling,* involve a great number of interrelated physical, chemical, and biological processes. The unique and diverse hydrologic conditions in wetlands (discussed in the previous chapter) markedly influence biogeochemical processes. These processes result not only in changes in the chemical forms of materials but also in the spatial movement of materials within wetlands, as in water-sediment exchange and plant uptake (Atlas and Bartha, 1981), and with the surrounding ecosystems, as in organic exports. These processes, in turn, determine overall wetland productivity. The interrelationships among hydrology, biogeochemistry, and response of wetland biota were summarized in Figure 4–1.

The biogeochemistry of wetlands can be divided into (1) intrasystem cycling through various transformation processes and (2) the exchange of chemicals between a wetland and its surroundings. Although few transformation processes are unique to wetlands, the permanent or intermittent flooding of these ecosystems causes certain processes to be more dominant in wetlands than in either upland or deep aquatic ecosystems. For example, while *anaerobic* or oxygenless conditions are sometimes found in other ecosystems, they prevail in wetlands. The soils in wetlands are characterized by waterlogged conditions during part or all of the year, which produce reduced conditions that, in turn, have a marked influence on several biochemical transformations unique to anaerobic conditions.

This intrasystem cycling, along with hydrologic conditions, influences the degree to which chemicals are transported to or from wetlands. An ecosystem is considered biogeochemically *open* when there is an abundant exchange of materials with its surroundings. When there is little movement of materials across the ecosystem boundary, it is biogeochemically *closed.* Wetlands can be in either category. For example, wetlands such as bottomland forests and tidal salt marshes have a significant exchange of minerals with their surroundings through river flooding and tidal exchange, respectively. Other wetlands such as ombrotrophic bogs and cypress domes have little material except for gaseous matter that passes into or out of the ecosystem. These latter systems depend more on intrasystem cycling than on throughput for their chemical supplies.

WETLAND SOIL

Types and Definitions

Wetland soil is both the medium in which many of the wetland chemical transformations take place and the primary storage of available chemicals for most wetland plants. It is often described as a *hydric soil,* defined by the U.S. Soil Conservation Service (1987) as "a soil that is saturated, flooded, or ponded long enough during the growing season to develop anaerobic conditions in the upper part. Wetland soils are of two types: (1) mineral soils or (2) organic soils (also called *Histosols).* Nearly all soils have some organic material; but when a soil has less than 20 percent to 35 percent organic matter (on a dry weight basis), it is considered a mineral soil. The U.S. Soil Conservation Service (1975) defined organic soils and organic soil materials under either of two conditions of saturation:

1. are saturated with water for long periods or are artificially drained and, excluding live roots, (a) have 18 percent or more organic carbon if the mineral fraction is 60 percent or more clay, (b) have 12 percent or more organic carbon if the mineral fraction has no clay, or (c) have a proportional content of organic carbon between 12 and 18 percent if the clay content of the mineral fraction is between zero and 60 percent; or
2. are never saturated with water for more than a few days and have 20 percent or more organic carbon.

Any soil material not defined by the above is considered mineral soil material. Where mineral soils occur in wetlands such as in some freshwater marshes or riparian forests, they generally have a soil profile made up of horizons, or layers. The upper layer of wetland mineral soils is often organic peat composed of partially decayed plant materials.

Although the above definition of organic soil is applicable to many types of wetlands, particularly to northern peatlands (see Chap. 12), peat, a generic term for relatively undecomposed organic soil, is not usually that strictly defined. Clymo (1983), for example, reported that most peats contain less than 20 percent unburnable inorganic matter (and therefore usually contain more than 40 percent organic carbon) but that some soil scientists allow up to 35 percent unburnable inorganic matter, and commercial operations sometimes allow 55 percent.

Organic soils are different from mineral soils in several physiochemical features (Table 5-1):

1. Organic soils have lower bulk densities and higher water-holding capacities than do mineral soils. Bulk density, defined as the dry weight of soil material per unit volume, is generally 0.2 to 0.3 g/cm^3 when the organic soil is well decomposed (Brady, 1974), although sphagnum moss peatland soils have bulk densities as low as 0.04 g/cm^3. By contrast, mineral soils generally range between 1.0 and 2.0 g/cm^3. Bulk density is low in organic soils because of their high porosity, or percentage of pore spaces. Peat soils generally have at least 80 percent pore spaces and are thus 80 percent water by volume when flooded (Verry and Boelter, 1979). Mineral soils generally range from 45 percent to 55 percent total pore space regardless of the amount of clay or texture (Patrick, 1981).
2. Both mineral and organic soils have wide ranges of possible hydraulic conductivities; the latter depends on their degree of decomposition (Table 4–6).

Table 5–1. Comparison of Mineral and Organic Soils in Wetlands

	Mineral Soil	*Organic Soil*
Organic Content, percent	Less than 20 to 35	Greater than 20 to 35
Organic Carbon, percent	Less than 12 to 20	Greater than 12 to 20
pH	Usually circumneutral	Acid
Bulk Density	High	Low
Porosity	Low (45–55%)	High (80%)
Hydraulic Conductivity	High (except for clays)	Low to high [a]
Water Holding Capacity	Low	High
Nutrient Availability	Generally high	Often low
Cation Exchange Capacity	Low, dominated by major cations	High, dominated by hydrogen ion
Typical Wetland	Riparian forest, some marshes	Northern peatland

[a]See Chapter 4

Organic soils may hold more water than mineral soils, but, given the same hydraulic conditions, they do not necessarily allow water to pass through more rapidly.

3. Organic soils generally have more minerals tied up in organic forms unavailable to plants than do mineral soils. This follows from the fact that a greater percent of the soil material is organic. This does not mean, however, that there are more total nutrients in organic soils; very often the opposite is true in wetland soils. For example, organic soils can be extremely low in bioavailable phosphorus or iron content—enough to limit plant productivity.
4. Organic soils have a greater cation-exchange capacity, defined as the sum of exchangeable cations (positive ions) that a soil can hold. Figure 5–1 summarizes the general relationship between organic content and cation-exchange capacity of soils. Mineral soils have a cation-exchange capacity that is dominated by the major cations (Ca^{++}, Mg^{++}, K^{+}, and Na^{+}). As organic content increases, both the percentage and the amount of exchangeable hydrogen ions increase (Gorham, 1967). For sphagnum moss peat, the high cation capacity is caused by long-chain polymers of uronic acid (Clymo, 1983).

Organic Soil Origin and Decomposition

Organic soil is composed primarily of the remains of plants in various stages of decomposition and accumulates in wetlands as a result of anaerobic conditions created by standing water or poorly drained conditions. Two of the more important characteristics of organic soil, including soils commonly termed *peat* and *muck,* are the botanical origin of the organic material and the degree to which it is decomposed (Clymo, 1983). Several of the properties that have been discussed, including bulk density, cation-exchange capacity, hydraulic conductivi-

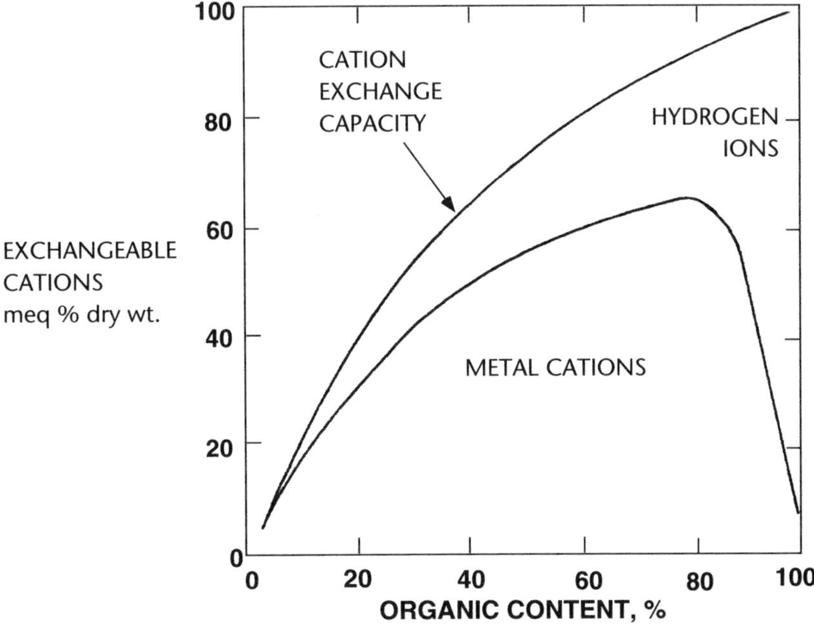

Figure 5–1. Relationship between cation exchange capacity and organic content for wetland soils. Note that at low organic content (mineral soils), the cation exchange capacity is saturated by metal cations; when the organic content is high, the exchange capacity is dominated by hydrogen ions. *(After Gorham, 1967)*

ty, and porosity, are often dependent on these characteristics. Therefore, it is often possible to predict the range of the physical properties of an organic soil if the origin and state of decomposition can be observed in the field or laboratory.

The botanical origin of the organic material can be (1) mosses, (2) herbaceous material, and (3) wood and leaf litter. For most northern peatlands, the moss is usually sphagnum, although several other moss species can dominate if the peatland is receiving inflows of mineral water. Organic soils can originate from herbaceous materials from grasses such as reed grass *(Phragmites)*, wild rice *(Zizania)*, and salt marsh cord grass *(Spartina)*, or from sedges such as *Carex* and *Cladium*. Organic soils can also be produced in freshwater marshes by plant fragments from a number of nongrass and nonsedge plants, including cattails *(Typha)* and water lilies *(Nymphaea)*. In forested wetlands the peat can be a result of woody detritus or leaf material or both. In northern peatlands the material can originate from birch *(Betula)*, pine *(Pinus)*, or tamarack *(Larix)*, and in southern deepwater swamps, the organic horizon can be composed of material from cypress *(Taxodium)* or water tupelo *(Nyssa)* trees.

The state of decomposition, or humification, of wetland soils is the second key character of organic peat (Clymo, 1983). As decomposition proceeds, albeit at a very slow rate in flooded conditions, the original plant structure is changed physically and chemically until the resulting material little resembles the parent material. As peat decomposes, bulk density increases, hydraulic conductivity decreases (see Chap. 4), and the quantity of larger (>1.5 mm) fiber particles decreases as the material becomes increasingly fragmented. Chemically, the amount of peat "wax," or material soluble in nonpolar solvents, and lignin increase with decomposition, whereas cellulose compounds and plant pigments decrease (Clymo, 1983). When some wetland plants such as salt marsh grasses die, the detritus rapidly loses a large percentage of its organic compounds through leaching (Turner, 1978; Teal, 1986). These readily soluble organic compounds are thought to be easily metabolized in adjacent aquatic systems (Gosselink, 1984). Details of decomposition rates for particular wetland types are given in Chapters 8 through 14.

Organic Soil Classification and Characteristics

Organic soils are classified into four groups, three of which are considered hydric soils:

1. *Saprists* (muck). Two-thirds or more of the material is decomposed, and less than one-third of plant fibers are identifiable.
2. *Fibrists* (peat). Less than one-third of material is decomposed, and more than two-thirds of plant fibers are identifiable.
3. *Hemists* (mucky peat or peaty muck). Conditions between saprist and fibrist soil.
4. *Folists*. Organic soils caused by excessive moisture (precipitation > evapotranspiration) that accumulate in tropical and boreal mountains; these soils are not classified as hydric soils, for saturated conditions are the exception rather than the rule.

Organic soil is generally dark in color, ranging from dark black soils characteristic of mucks such as those found in the Everglades in Florida to the dark brown color of partially decomposed peat from northern bogs.

Wetland Mineral Soil Characteristics

Mineral soils, when flooded for extended periods, develop certain characteristics that allow for their identification. One characteristic of many hydric mineral soils that are semipermanently or permanently flooded is the development of gray or sometimes greenish or blue-gray color as the result of a process known as *gleying*. This process, also known as *gleization*, is the result of the chemical reduction of iron (see Iron and Manganese Transformations below). Another characteristic of mineral soils that are seasonally flooded, characterized by alternate wetting and drying, is a process called *mottle formation.* Mottles that are orange/reddish-brown (because of iron) or dark reddish-brown/black (because of manganese) spots seen throughout an otherwise gray (gleyed) soil matrix suggest intermittently exposed soils with spots of iron and manganese oxides in an otherwise reduced environment. Mottles are relatively insoluble, enabling them to remain in soil long after it has been drained (Diers and Anderson, 1984). Because the development of gleys and mottles is mediated by microbiological processes, as described below, the rate at which they are formed depends on

1. the presence of sustained anaerobic conditions;
2. sufficient soil temperature (5°C is often considered "biological zero," below which much biological activity ceases or slows considerably); and
3. the presence of organic matter as a substrate for microbial activity.

If any of these three conditions is absent, gleying and mottle formation will not take place (Diers and Anderson, 1984).

In practice, the determination of whether a mineral soil is a hydric soil is a complicated process, but it is often done by determining soil color relative to a standard color chart (Fig. 5–2). Soils that contain low chromas (as indicated by the darker color chips in Fig. 5–2) of black, gray, or dark brown and red indicate hydric soils.

Another characteristic of some mineral wetland soils is the presence of an *oxidized rhizosphere* that results from the capacity of many hydrophytes to transport oxygen through above-ground stems and leaves to below-ground roots (Armstrong, 1964; Moorhead and Reddy, 1988; see Chap. 6). Excess oxygen, beyond the root's metabolic needs, diffuses from the roots to the surrounding soil matrix, forming deposits of oxidized iron along small roots. When a wetland soil is examined, these oxidized rhizosphere deposits can often be seen as thin traces through an otherwise gley matrix.

CHEMICAL TRANSFORMATIONS IN WETLAND SOILS

Oxygen and Redox Potential

When soils, whether mineral or organic, are inundated with water, anaerobic conditions usually result. When water fills the pore spaces, the rate at which oxygen can diffuse through the soil is drastically reduced. Diffusion of oxygen in an aqueous solution has been estimated at 10,000 times slower than oxygen diffusion through a porous medium such as drained soil (Greenwood, 1961; Gambrell and Patrick, 1978). This low diffusion leads relatively quickly to anaerobic, or reduced, conditions, with the time required for oxygen depletion

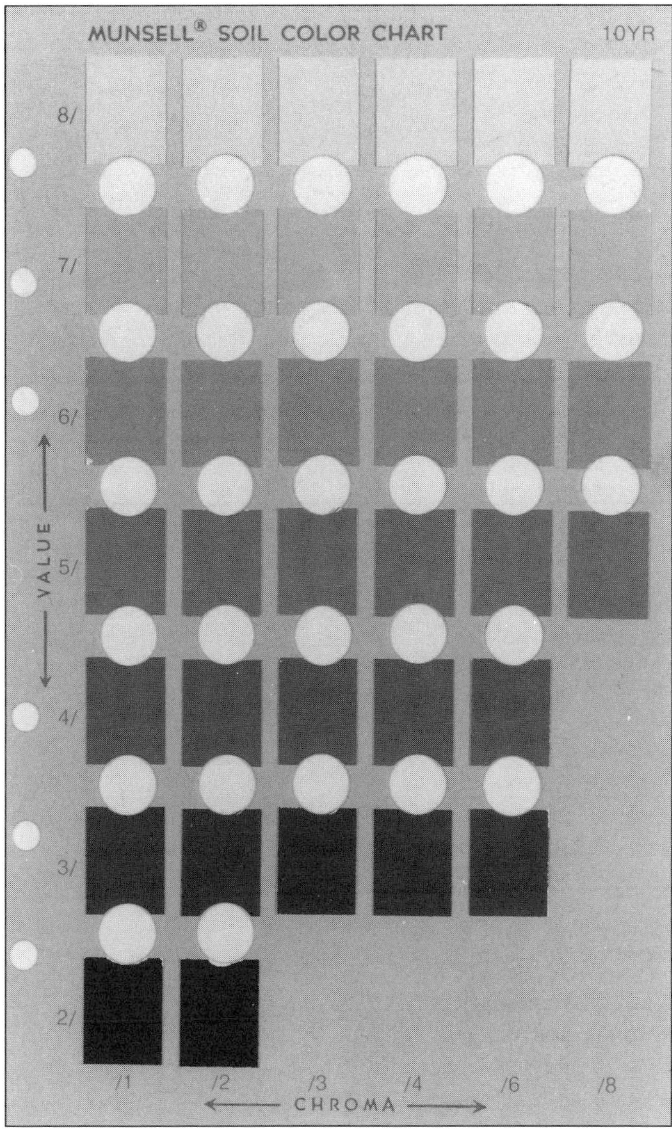

Figure 5–2. Hydric soils are sometimes identified by comparing the soil color with standard soil color charts such as the Munsell® Soil Color Chart shown here. The hue, given in the upper right-hand corner of the chart indicates the relation to standard spectral colors such as red (R) or yellow (Y). The value notation (vertical scale) indicates the soil lightness (darker with lower value) and the chroma (horizontal scale) indicates the color strength or purity, with grayer soils to the left. Chromas of 2 or less generally indicate hydric soils. *(Macbeth Division of Kollmorgen Instruments Company, 1992; reprinted with permission)*

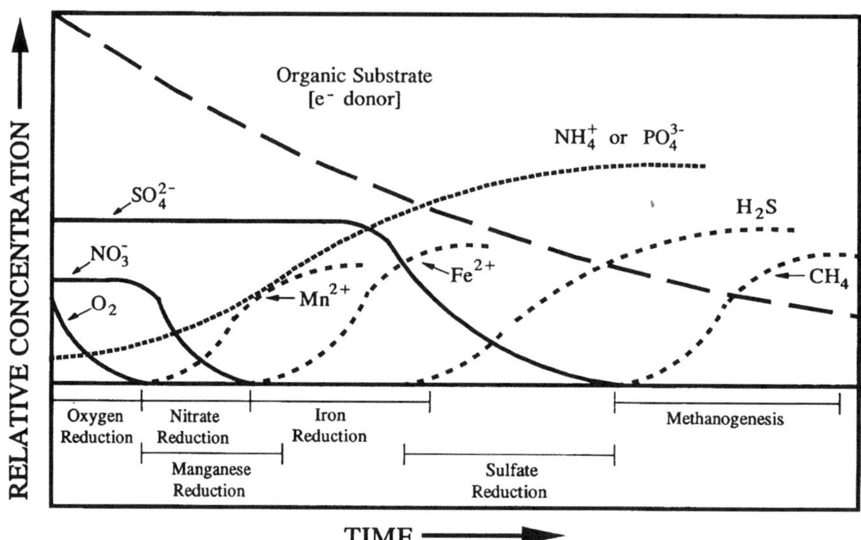

Figure 5–3. Sequence in time of transformations in soil after flooding, beginning with oxygen depletion and followed by nitrate and then sulfate reduction. Increases are seen in reduced manganese, reduced iron, hydrogen sulfide, and methane. Note the gradual decrease in organic substrate (electron donor) and increases in available ammonium and phosphate ions. Graph can also be interpreted as relative concentrations with depth in wetland soils. *(By K. R. Reddy, University of Florida, Gainesville; reprinted with permission)*

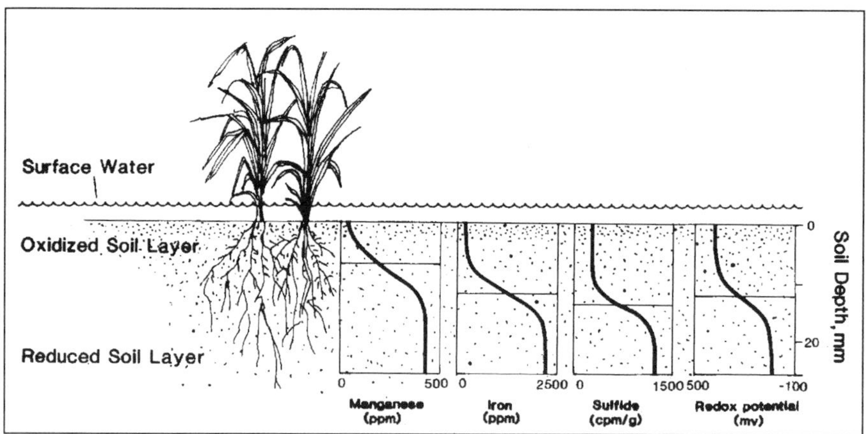

Figure 5–4. Characteristics of many wetland soils showing a shallow oxidized soil layer over a reduced soil layer, and soil profiles of sodium acetate-extractable manganese, ferrous iron, sulfide, and redox potential. *(After Patrick and Delaune, 1972, and Gambrell and Patrick, 1978)*

on the order of several hours to a few days after inundation begins (Fig. 5–3). The rate at which the oxygen is depleted depends on the ambient temperature, the availability of organic substrates for microbial respiration, and sometimes on chemical oxygen demand from reductants such as ferrous iron (Gambrell and Patrick, 1978). The resulting lack of oxygen prevents plants from carrying out normal aerobic root respiration and strongly affects the availability of plant nutrients and toxic materials in the soil. As a result, plants that grow in anaerobic soils generally have a number of specific adaptations to this environment (see Chap. 6).

It is not always true that oxygen is totally depleted from the soil water of wetlands. There is usually a thin layer of oxidized soil, sometimes only a few millimeters thick, at the surface of the soil at the soil-water interface (Fig. 5–4). The thickness of this oxidized layer (Gambrell and Patrick, 1978) is directly related to

1. the rate of oxygen transport across the atmosphere-surface water interface,
2. the small population of oxygen-consuming organisms present,
3. photosynthetic oxygen production by algae within the water column, and
4. surface mixing by convection currents and wind action.

Even though the deeper layers of the wetland soils remain reduced, this thin oxidized layer is often very important in the chemical transformations and nutrient cycling that occur in wetlands. Oxidized ions such as Fe^{+++}, Mn^{+4}, NO_3^-, and $SO_4^=$ are found in this microlayer, whereas the lower anaerobic soils are dominated by reduced forms such as ferrous and manganous salts, ammonia, and sulfides (Mohanty and Dash, 1982). Because of the presence of oxidized ferric iron (Fe^{+++}) in the oxidized layer, the soil often is a brown or brownish-red color. In contrast, the color of the reduced sediments, dominated by ferrous iron (Fe^{++}), often ranges from bluish-gray to greenish-gray because of gleying.

Redox potential, or oxidation-reduction potential, a measure of the electron pressure (or availability) in a solution, is often used to quantify further the degree of electrochemical reduction of wetland soils. *Oxidation* occurs not only during the uptake of oxygen but also when hydrogen is removed (e.g., $H_2S \longrightarrow S^{-2} + 2H^+$) or, more generally, when a chemical gives up an electron (e.g., $Fe^{++} \longrightarrow Fe^{+++} + e^-$). *Reduction* is the opposite process of giving up oxygen, gaining hydrogen (hydrogenation), or gaining an electron.

Redox potential can be measured in wetland soils and is a quantitative measure of the tendency of the soil to oxidize or reduce substances (Faulkner and Richardson, 1989). When based on a hydrogen scale, redox potential is referred to as E_H and is related in theory to the concentrations of oxidants {ox} and reductants {red} in a redox reaction by the *Nernst Equation:*

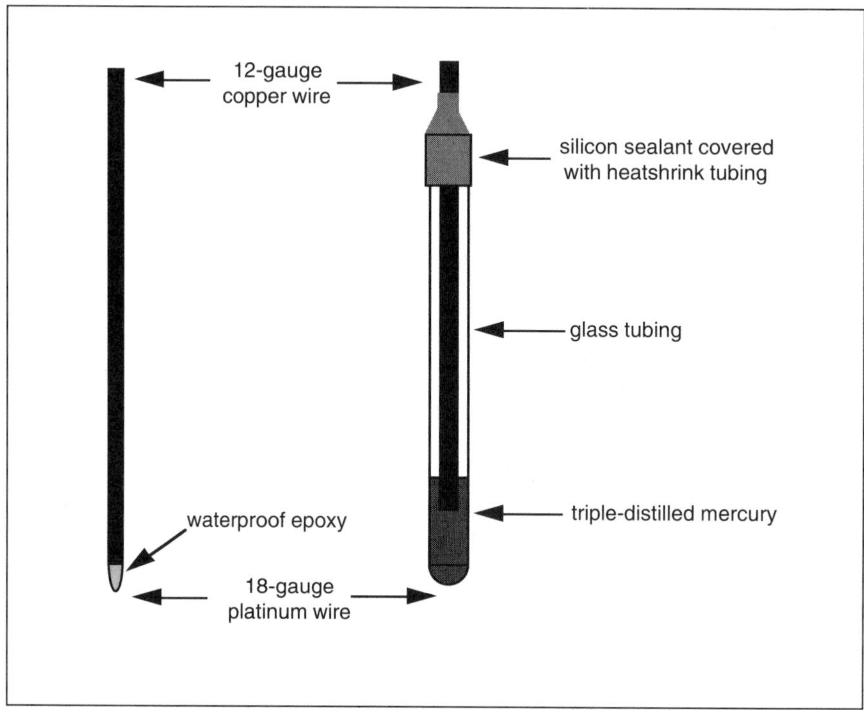

Figure 5–5. Two designs for constructing redox electrodes for measurement of redox potential in wetlands. *(From Faulkner et al., 1989; copyright © 1989 by Soil Science Society of America, reprinted with permission)*

$$E_H = E^o + 2.3\ [RT/nF]\ log\ [\{ox\}/\{red\}] \tag{5.1}$$

where

E^o = potential of reference, mv

R = gas constant = 81.987 cal deg^{-1} mole^{-1}

T = temperature, ^{o}K

n = number of moles of electrons transferred

F = Faraday constant = 23,061 cal/mole-volt

Redox potential can be measured by inserting an inert platinum electrode into the solution in question (Fig. 5–5). Electric potential in units of millivolts (mv) is measured relative to a hydrogen electrode ($H^+ + e \longrightarrow H$) or to a calomel reference electrode. As long as free dissolved oxygen is present in a solution, the redox potential varies little (in the range of +400 to +700 mv). However, it becomes a sensitive measure of the degree of reduction of wetland soils after oxygen disappears, ranging from +400 mv down to -400 mv (Gambrell and Patrick, 1978).

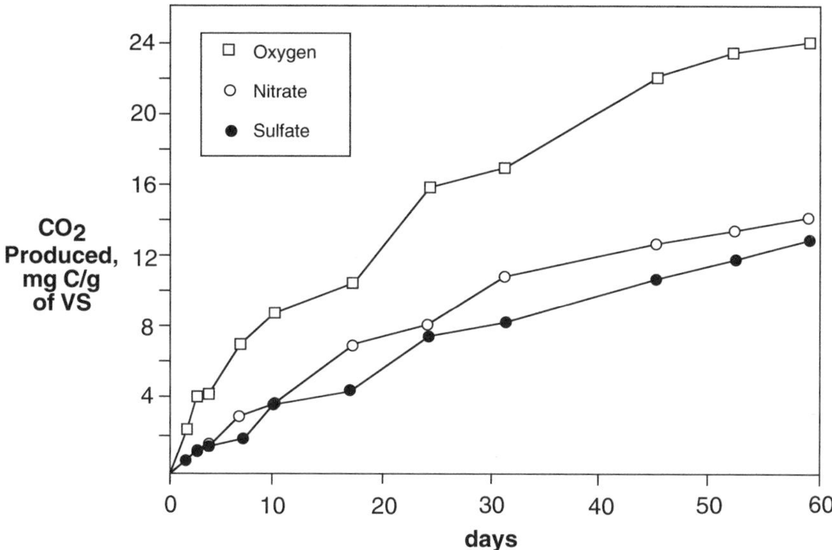

Figure 5–6. Decomposition of sediment organic carbon with oxygen, nitrate, and sulfate as terminal electron acceptor. The conditions were maintained at redox potentials of 312, 115, and -123 mv, respectively for oxygen, nitrate, and sulfate. *(From Reddy and Graetz, 1988; copyright © 1988 by Donald D. Hook, used with permission)*

As organic substrates in a waterlogged soil are oxidized (donate electrons), the redox potential drops as a sequence of reductions (electron gains) take place (Fig. 5–3). Because organic matter is one of the most reduced of substances, it can be oxidized when any number of terminal electron acceptors is available, including O_2, NO_3^-, Mn^{+4}, Fe^{+++}, or $SO_4^=$. Rates of organic decomposition are most rapid in the presence of oxygen and less so for electron acceptors such as nitrates and sulfates (Reddy et al., 1986; Reddy and Graetz, 1988; Fig. 5-6). The oxidation of organic substrate is described by the following equation, which illustrates the organic substrate as an electron (e^-) donor (Fig. 5–3).

$$[CH_2O]_n + nH_2O \longrightarrow nCO_2 + 4n\, e^- + 4n\, H^+ \tag{5.2}$$

Various chemical and biological transformations take place in a predictable sequence (Fig. 5–3), within predictable redox ranges to provide electron acceptors for this oxidation or decomposition (Table 5–2). The first and most common is through aerobic oxidation when oxygen itself is the terminal electron acceptor at a redox potential of between 400 and 600 mv.

$$O_2 + 4\, e^- + 4\, H^+ \longrightarrow 2\, H_2O \tag{5.3}$$

Table 5–2. Oxidized and Reduced Forms of Several Elements and Approximate Redox Potentials for Transformation

Element	*Oxidized Form*	*Reduced Form*	*Appoximate Redox Potential for Transformation, mv*
Nitrogen	NO_3^- (Nitrate)	N_2O, N_2, NH_4^+	250
Manganese	Mn^{+4} (Manganic)	Mn^{++} (Manganous)	225
Iron	Fe^{+++} (Ferric)	Fe^{++} (Ferrous)	120
Sulfur	$SO_4^=$ (Sulfate)	$S^=$ (Sulfide)	-75 to -150
Carbon	CO_2 (Carbon Dioxide)	CH_4 (Methane)	-250 to -350

One of the first reactions that occur in wetland soils after they become anaerobic (i.e., the dissolved oxygen is depleted) is the reduction of NO_3^- (nitrate) first to NO_2^- (nitrite) and ultimately to N_2O or N_2; nitrate becomes an electron acceptor at approximately 250 mv.

$$2\,NO_3^- + 10\,e- + 12\,H^+ \longrightarrow N_2 + 6\,H_2O \tag{5.4}$$

As the redox potential continues to decrease, manganese is transformed from manganic to manganous compounds at about 225 mv.

$$MnO_2 + 2\,e^- + 4\,H^+ \longrightarrow Mn^{++} + 2\,H_2O \tag{5.5}$$

Iron is transformed from ferric to ferrous form at about 120 mv, while sulfates are reduced to sulfides at -75 to -150 mv.

$$Fe(OH)_3 + e^- + 3\,H^+ \longrightarrow Fe^{++} + 3\,H_2O \tag{5.6}$$

$$SO_4^= + 8\,e^- + 9\,H^+ \longrightarrow HS^- + 4\,H_2O \tag{5.7}$$

Finally, in the most reduced conditions, the organic matter itself or carbon dioxide becomes the terminal electron acceptor at about -250 mv, producing low molecular weight organic compounds and methane gas, as, for example,

$$CO_2 + 8e^- + 8\,H^+ \longrightarrow CH_4 + 2\,H_2O \tag{5.8}$$

These redox potentials are not precise thresholds, for pH and temperature are also important factors in the rates of transformation. These major chemical transformations and others related to the nitrogen, iron, manganese, sulfur, and carbon cycles are discussed below.

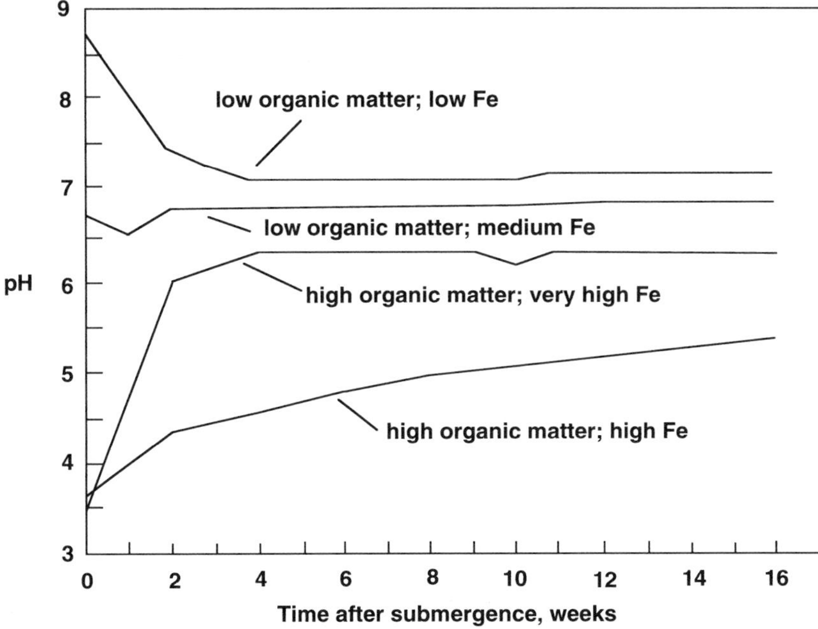

Figure 5–7. Changes in pH of soils of different organic and iron content after flooding. *(Ponnamperuma, 1972, as modified by Faulkner and Richardson, 1989; copyright © 1989, Lewis Publishers, Chelsea, MI, used with permission)*

pH

Soils and overlying waters of wetlands occur over a wide range of pH. Organic soils in wetlands are often acidic, particularly in peatlands in which there is little groundwater inflow. On the other hand, mineral soils often have more neutral or alkaline conditions. There are specific connections between redox potential and pH because the specific redox potential at which chemicals are stable in either reduced or oxidized states is pH dependent and can be shown on redox-pH stability diagrams (see, e.g., Stumm and Morgan, 1970). The general consequence of flooding previously drained soils is to cause alkaline soils to decrease in pH and acid soils to increase in pH (Fig. 5–7), the former stemming from the buildup of CO_2 and then carbonic acid, the latter stemming from the reduction of ferric iron hydroxides, as shown in Equation 5.6 (Ponnamperuma, 1972). This neutral pH is generally in the range of from 6.7 to 7.2. This convergence to neutrality is generally what can be expected to happen with soil pH when lands that had previously been drained become flooded, as occurs in the construction of wetlands. Fennessy (1991) found a convergence of soil pH toward neutrality for four constructed wetland basins in Illinois, three of which had alkaline soils and

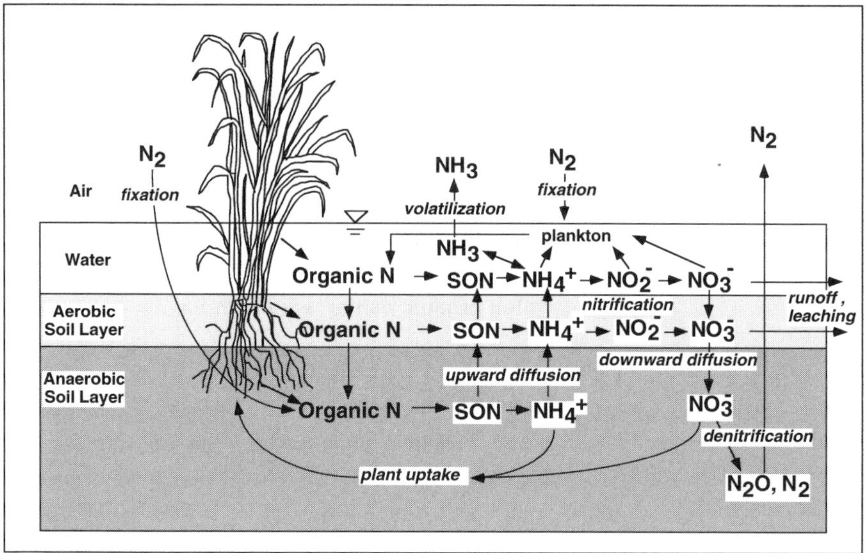

Figure 5–8. Nitrogen transformations in wetlands. SON indicates soluble organic nitrogen.

one that had acidic organic soils. For some organic soils high in iron content, submergence does not always increase pH (Fig. 5–7). Peat soils often remain acidic during submergence through the slow oxidation of sulfur compounds, producing sulfuric acid, and the production of humic acids and selective cation exchange by sphagnum mosses. This is discussed in more detail in Chapter 12.

Nitrogen Transformations

Nitrogen is often the most limiting nutrient in flooded soils, whether the flooded soils are in natural wetlands or on agricultural wetlands such as rice paddies (Gambrell and Patrick, 1978). Limitations of the element have been noted at least temporarily for salt marshes (e.g., Sullivan and Daiber, 1974; Valiela and Teal, 1974), freshwater inland marshes (Klopatek, 1978), and freshwater tidal marshes (Simpson et al., 1978). Nitrogen transformations in wetland soils involve several microbiological processes, some of which make the nutrient less available for plant uptake. The nitrogen transformations that dominate wetland soils are shown in Figure 5–8 and are summarized below. The ammonium ion is the primary form of mineralized nitrogen in most flooded wetland soils, although much of the nitrogen can be tied up in organic forms in highly organic soils. The presence of an oxidized zone over the anaerobic or reduced zone is critical for several of the pathways.

Nitrogen mineralization refers to "the biological transformation of organically combined nitrogen to ammonium nitrogen during organic matter degradation" (Gambrell and Patrick, 1978). This pathway occurs under both anaerobic and aerobic conditions and is often referred to as *ammonification.* Typical formulas for the mineralization of a simple organic nitrogen compound, urea, are given as

$$NH_2 \cdot CO \cdot NH_2 + H_2O \longrightarrow 2NH_3 + CO_2 \tag{5.9}$$

$$NH_3 + H_2O \longrightarrow NH_4^+ + OH^- \tag{5.10}$$

Once the ammonium ion (NH_4^+) is formed, it can take several possible pathways. It can be absorbed by plants through their root systems or by anaerobic microorganisms and converted back to organic matter. It can also be immobilized through ion exchange onto negatively charged soil particles. Because of the anaerobic conditions in wetland soils, ammonium would normally be restricted from further oxidation and would build up to excessive levels were it not for the thin oxidized layer at the surface of many wetland soils. The gradient between high concentrations of ammonium in the reduced soils and low concentrations in the oxidized layer causes an upward diffusion of ammonium, albeit very slowly, to the oxidized layer. This ammonium nitrogen then is oxidized through the process of *nitrification* in two steps by *Nitrosomonas* sp.:

$$2NH_4^+ + 3O_2 \longrightarrow 2NO_2^- + 2H_2O + 4H^+ + energy \tag{5.11}$$

and by *Nitrobacter* sp.:

$$2NO_2^- + O_2 \longrightarrow 2NO_3^- + energy \tag{5.12}$$

Nitrification can also occur in the oxidized rhizosphere of plants where adequate oxygen is often available to convert the ammonium-nitrogen to nitrate-nitrogen (Reddy and Graetz, 1988).

Nitrate (NO_3^-), as a negative ion rather than the positive ammonium ion, is not subject to immobilization by the negatively charged soil particles and is thus much more mobile in solution. If it is not assimilated immediately by plants or microbes *(assimilatory nitrate reduction)* or is lost through groundwater flow stemming from its rapid mobility, it will have the potential of going through *dissimilatory nitrogenous oxide reduction,* a term that refers to several pathways of nitrate reduction (Wiebe et al., 1981). The most prevalent are reduction to ammonia and *denitrification.* Denitrification, carried out by microorganisms in anaerobic conditions, with nitrate acting as a terminal electron acceptor, results in the loss of nitrogen as it is converted to gaseous nitrous oxide (N_2O) and molecular nitrogen (N_2):

$$C_6H_{12}O_6 + 4NO_3^- \longrightarrow 6CO_2 + 6H_2O + 2N_2 \tag{5.13}$$

Denitrification has been documented as a significant path of loss of nitrogen from salt marshes (e.g., Kaplan et al., 1979; Whitney et al., 1981) and rice cultures (e.g., Patrick and Reddy, 1976; Mohanty and Dash, 1982). As illustrated in Figure 5–8, the entire process occurs after (1) ammonium-nitrogen diffuses to the aerobic soil layer, (2) nitrification occurs, (3) nitrate-nitrogen diffuses back to the anaerobic layer, and (4) denitrification, as described in Equation 5.13, occurs. Because nitrate-diffusion rates in wetland soils are seven times faster than ammonium-diffusion rates (Reddy and Patrick, 1984; Reddy and Graetz, 1988), ammonium diffusion and subsequent nitrification limit the entire process of nitrogen loss. Denitrification is inhibited in acid soils and peat and is therefore thought to be of less consequence in northern peatlands (Etherington, 1983).

Nitrogen fixation results in the conversion of N_2 gas to organic nitrogen through the activity of certain organisms in the presence of the enzyme nitrogenase. It may be the source of significant nitrogen for some wetlands. Nitrogen fixation, which is carried out by certain aerobic and anaerobic bacteria and blue-green algae, is favored by low oxygen tensions because nitrogenase activity is inhibited by high oxygen (Etherington, 1983). In wetlands, nitrogen fixation can occur in overlying waters, in the aerobic soil layer, in the anaerobic soil layer, in the oxidized rhizosphere of the plants, and on the leaf and stem surfaces of plants (Reddy and Graetz, 1988). Bacterial nitrogen fixation can be carried out by nonsymbiotic bacteria, by symbiotic bacteria of the genus *Rhizobium,* or by the actinomycetes. Whitney et al. (1975, 1981), and Teal et al. (1979) documented the fact that bacterial fixation was the most significant pathway for nitrogen fixation in salt marsh soils. On the other hand, both nitrogen-fixing bacteria and nitrifying bacteria are virtually absent from the low pH peat of northern bogs (Moore and Bellamy, 1974). Cyanobacteria (blue-green algae), as nonsymbiotic nitrogen fixers, are also frequently found in waterlogged soils of wetlands and can contribute significant amounts of nitrogen. For example, Buresh et al. (1980a) reported a rate of 25 kg N ha^{-1} yr^{-1} for blue-green algae nitrogen fixation compared to only 0.1–0.5 kg N ha^{-1} yr^{-1} by heterotrophic bacteria in flooded soils in Louisiana. Nitrogen fixation by blue-green algae is also important in northern bogs and rice cultures, which are often too acidic to support large bacterial populations (Etherington, 1983).

Iron and Manganese Transformations

Below the reduction of nitrate on the redox potential scale comes the reduction of manganese and of iron (see Equations 5.5 and 5.6). Each element occurs in two oxidation states, as shown in Table 5–2. Iron and manganese are found in wetlands primarily in their reduced forms (ferrous and manganous, respectively), and both are more soluble and more readily available to organisms in those forms. Manganese is reduced slightly before iron on the redox scale, but otherwise it behaves similarly to iron. The direct involvement of bacteria in the reduction of MnO_2 (Equation 5.5) has been questioned by some, although several experiments have shown the generation of energy by the bacterial reduction of oxidized manganese (Laanbroek, 1990).

Iron can be oxidized from ferrous to the insoluble ferric form by chemosynthetic bacteria in the presence of oxygen:

$$4\ Fe^{++} + O_2(aq) + 4\ H^+ \longrightarrow 4\ Fe^{+++} + 2\ H_2O \qquad (5.14)$$

Although this reaction can occur nonbiologically at neutral or alkaline pH, microbial activity has been shown to accelerate ferrous iron oxidation by a factor of 10^6 in coal mine drainage water (Singer and Stumm, 1970). It is thought that a similar type of bacterial process exists for manganese.

Iron bacteria are thought to be responsible for the oxidation to insoluble ferric compounds of soluble ferrous iron that originated in anaerobic groundwaters in northern peatland areas. These "bog-iron" deposits form the basis of the ore that has been used in the iron and steel industry (Atlas and Bartha, 1981). Iron in its reduced ferrous form causes a gray-green coloration (gleying) of mineral soils ($Fe(OH)_2$) instead of its normal red or brown color in oxidized conditions ($Fe(OH)_3$). This gives a relatively easy field check on the oxidized and reduced layers in a mineral soil profile (see description above in Wetland Soils).

Iron and manganese in their reduced forms can reach toxic concentrations in wetland soils. Ferrous iron, diffusing to the surface of the roots of wetland plants, can be oxidized by oxygen leaking from root cells, immobilizing phosphorus and coating roots with an iron oxide, and causing a barrier to nutrient uptake (Gambrell and Patrick, 1978).

Sulfur Transformations

Sulfur occurs in several different states of oxidation in wetlands, and like nitrogen, it is transformed through several pathways that are mediated by microorganisms (Fig. 5–9). Although sulfur is rarely present in such low concentrations that it is limiting to plant or animal growth in wetlands, the hydrogen sulfide that is characteristic of anaerobic wetland sediments can be very toxic to plants and microbes, especially in saltwater wetlands where the concentration of sulfates is high. The release of sulfides when wetland sediments are disturbed causes the odor familiar to those who carry out research in wetlands—the smell of rotten eggs. In the redox scale, sulfur compounds are the next major electron acceptors after nitrates, iron, and manganese, with reduction occurring at about -75 to -150 mv on the redox scale (see Table 5–2). The most common oxidation states (valences) for sulfur in wetlands are:

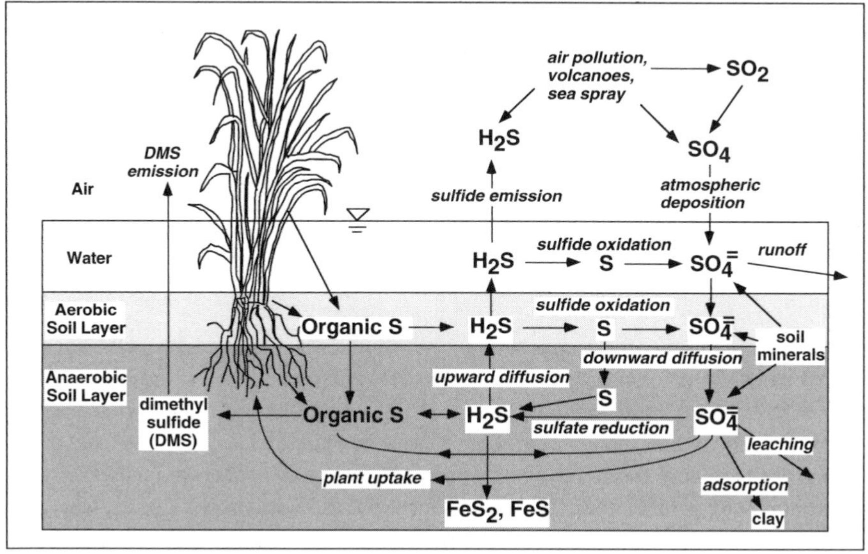

Figure 5–9. Sulfur transformations in wetlands.

Form	*Valence*
$S^=$ (sulfide)	-2
S (elemental sulfur)	0
S_2O_3 (thiosulfates)	+2
$SO_4^=$ (sulfates)	+6

Sulfate reduction can take place as *assimilatory sulfate reduction* in which certain sulfur-reducing obligate anaerobes such as *Desulfovibrio* bacteria utilize the sulfates as terminal electron acceptors in anaerobic respiration:

$$4H_2 + SO_4^= \longrightarrow H_2S + 2H_2O + 2OH^- \quad (5.15)$$

This sulfate reduction can occur over a wide range of pH, with highest rates prevalent near neutral pH.

There have been a few measurements of the rate at which hydrogen sulfide is produced in and released from wetlands, and those measurements have ranged from 0.004 to 2.6 gS m^{-2} yr^{-1} (Table 5–3). Although the data in Table 5–3 vary over a wide range, it can probably be safely generalized that salt water wetlands have higher rates of emission per unit area (Castro and Dierberg, 1987). Sulfur can also be released to the atmosphere as organic sulfur compounds (methyl and dimethyl sulfide), and this flux is thought by some to be as important as or more important than H_2S emissions from some wetlands (Faulkner and Richardson, 1989).

Sulfides are known to be highly toxic both to microbes and rooted higher plants. The negative effects of sulfides on higher plants have been described by

Table 5–3. Biogenic Emissions of H_2S from Wetlands

Wetland Type	*Sampling Location*	*Emission Rates, $gS\ m^{-2}\ yr^{-1}$*			*Source*
		Mean	*Min.*	*Max.*	
Freshwater					
Marsh	North Carolina	0.60	0.08	1.27	Aneja et al. (1981)
Marsh	Florida	0.08	0.05	0.11	Castro and Dierberg (1987)
Swamp	Ivory Coast	2.63[a]	—	—	Delmas et al. (1980)
Cypress Swamp	Florida	0.004	0.001	0.006	Castro and Dierberg (1987)
Cypress Swamp	North Carolina, New York, Georgia	—	0.001	0.16	Adams et al. (1981)
Saltwater					
Salt Marsh	North Carolina	0.15	0.04	0.41	Aneja et al. (1981)
Salt Marsh	Florida, Texas, North and South Carolina, Delaware, Louisiana		0.02	601.6	Adams et al. (1981)
Salt Marsh	New York	0.55	0.00	41.5	Hill et al. (1978)
Salt Marsh	Massachusetts	2.16	1.26	5.04	Steudler and Peterson (1984)
Salt Marsh (*Juncus*)	Florida	0.008	0.003	0.015	Castro and Dierberg (1987)
Mangrove Swamp	Florida	0.11	0.00	—	Castro and Dierberg (1987)

[a]Waterlogged soil only

Source: Castro and Dierberg, 1987

Ponnamperuma (1972) as attributable to a number of causes, including the following:

1. the direct toxicity of free sulfide as it comes in contact with plant roots;
2. the reduced availability of sulfur for plant growth because of its precipitation with trace metals;
3. the immobilization of zinc and copper by sulfide precipitation.

In wetland soils that contain high concentrations of ferrous iron (Fe^{++}), sulfides can combine with iron to form insoluble ferrous sulfides (FeS), thus reducing the toxicity of the free hydrogen sulfide (Gambrell and Patrick, 1978). Ferrous sulfide gives the black color characteristic of many anaerobic wetland soils; one of its common mineral forms is pyrite, FeS_2, the form of sulfur commonly found in coal deposits.

Sulfides can be oxidized by both chemoautotrophic and photosynthetic microorganisms to elemental sulfur and sulfates in the aerobic zones of some wetland soils. Certain species of *Thiobacillus* obtain energy from the oxidation of hydrogen sulfide to sulfur, whereas other species in this genus can further oxidize elemental sulfur to sulfate. These reactions are summarized as follows:

$$2H_2S + O_2 \longrightarrow 2S + 2H_2O + energy \tag{5.16}$$

and

$$2S + 3O_2 + 2H_2O \longrightarrow 2H_2SO_4 + energy \tag{5.17}$$

Photosynthetic bacteria such as the purple sulfur bacteria found on salt marshes and mud flats are capable of producing organic matter in the presence of light according to the following equation:

$$CO_2 + H_2S + light \longrightarrow CH_2O + S \tag{5.18}$$

This reaction uses hydrogen sulfide as an electron donor rather than the H_2O used in the more traditional photosynthesis equation with water serving as the electron donor. This reaction often takes place under anaerobic conditions where hydrogen sulfide is abundant but at the surface of sediments where sunlight is also available.

Carbon Transformations

Although biodegradation of organic matter by aerobic respiration is limited by the reduced conditions in wetland soils, several anaerobic processes can degrade

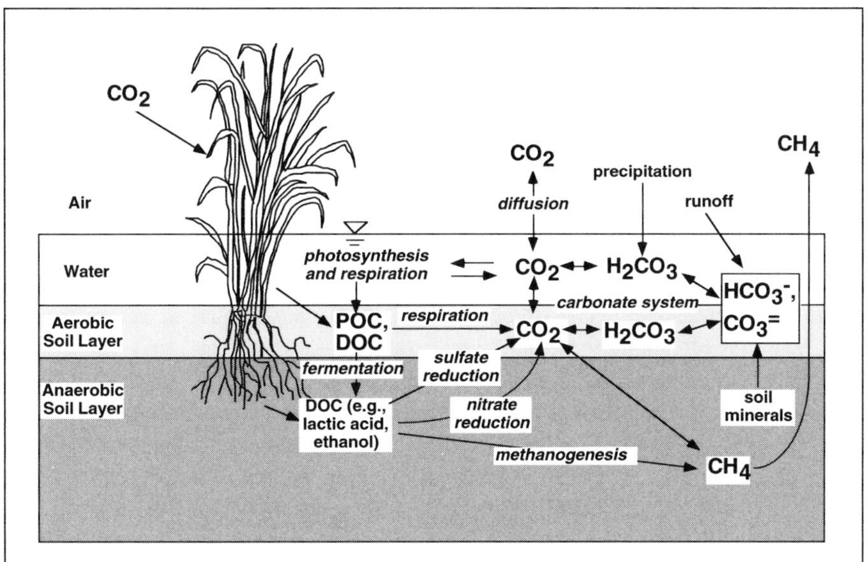

Figure 5–10. Carbon transformations in wetlands. POC indicates particulate organic carbon; DOC indicates dissolved organic carbon.

organic carbon. The major processes of carbon transformation under aerobic and anaerobic conditions are shown in Figure 5–10. The *fermentation* of organic matter, which occurs when organic matter itself is the terminal electron acceptor in anaerobic respiration by microorganisms, forms various low-molecular-weight acids and alcohols and CO_2 such as lactic acid

$$C_6H_{12}O_6 \text{—>} 2CH_3CH_2OCOOH \quad (5.19)$$
(lactic acid)

or ethanol

$$C_6H_{12}O_6 \text{—>} 2CH_3CH_2OH + 2CO_2 \quad (5.20)$$
(ethanol)

Fermentation can be carried out in wetland soils by either facultative or obligate anaerobes. Although *in situ* studies of fermentation in wetlands are rare, it is thought that "fermentation plays a central role in providing substrates for other anaerobes in sediments in waterlogged soils" (Wiebe et al., 1981). It represents one of the major ways in which high-molecular-weight carbohydrates are broken down to low-molecular-weight organic compounds, usually as dissolved organic carbon, which are, in turn, available to other microbes (Valiela, 1984).

Methanogenesis occurs when certain bacteria (methanogens) use CO_2 as an electron acceptor for the production of gaseous methane (CH_4), as shown in Equation 5.8, or alternatively, use a low-weight organic compound such as one from a methyl group:

$$CH_3COO^- + 4H_2 \longrightarrow 2CH_4 + 2H_2O \qquad (5.21)$$

Methane, which can be released to the atmosphere when sediments are disturbed, is often referred to as *swamp gas* or *marsh gas*. Methane production requires extremely reduced conditions, with a redox potential of between -250 and -350 mv, after other terminal electron acceptors (O_2, NO_3^-, and $SO_4^=$) have been reduced.

A comparison of methanogenesis between freshwater and marine environments has generally revealed that the rate of methane production is higher in the former apparently because of the lower amounts of sulfate in the water and sediments (Valiela, 1984). The rates of methanogenesis from both saltwater coastal wetlands and freshwater wetlands, however, have a considerable range (Table 5–4). A comparison of the rates of methane production from different studies is difficult because different methods are used and because the rates depend on both temperature and hydroperiod. Methanogenesis is seasonal in temperate-zone wetlands; Harriss et al. (1982) noted maximum methane production in a Virginia freshwater swamp in April–May, whereas Wiebe et al. (1981) found methane production to peak generally in late summer in a Georgia salt marsh. Furthermore, Harriss et al. (1982) measured a net uptake of methane by the wetland during a drought, when the wetland soil was exposed to the atmosphere. Researchers in boreal wetlands have found beaver ponds to have much higher methane flux rates than other wetland types (Naiman et al., 1991; Roulet et al., 1992) and neutral fens to have higher rates than acid fens and bogs (Crill et al., 1988).

Carbon-Sulfur Interactions

The sulfur cycle is important in some wetlands for the oxidation of organic carbon. This is particularly true in most coastal wetlands where sulfur is abundant. In general, methane is found at low concentrations in reduced soils when sulfate concentrations are high (Gambrell and Patrick, 1978; Valiela, 1984). Possible reasons for this phenomenon include (1) competition for substrates that occurs between sulfur and methane bacteria; (2) the inhibitory effects of sulfate or sulfide on methane bacteria; (3) a possible dependence of methane bacteria on products of sulfur-reducing bacteria; and (4) a stable redox potential that does not drop low enough to reduce CO_2 because of an ample supply of sulfate. More recent evidence suggests that methane may actually be oxidized to CO_2 by sulfate reducers (Valiela, 1984).

Table 5–4. Methane Production Rates for Various Freshwater and Saltwater Wetlands

Type of Wetland	*Rate of Methane Production* $mgC\ m^{-2}\ day^{-1}$	*Reference*
Freshwater Wetlands		
rice paddies, Spain	36–252	Seiler et al. (1984)
rice paddies, Italy	120–285	Schutz et al. (1989)
rice field, Texas	45–159	Sass et al. (1990)
Michigan swamp, average annual rate	110	Baker-Blocker et al. (1977)
Dismal Swamp, Virginia		Harriss et al. (1982)
minimum	1	
maximum	15	
Louisiana tidal freshwater marsh		C. J. Smith et al. (1982)
annual average	440	
Minnesota peatland		Crill et al. (1988)
forested bog	58	
acid fen	77	
forested fen	107	
open bog	221	
neutral fen	244	
Minnesota boreal wetlands		Naiman et al. (1991)
permanently wet areas, including beaver ponds		
annual average	22–30	
peak in summer	100–200	
occasionally inundated meadows and forests		
annual average	0.5–1.1	
peak	10	
Canadian boreal wetlands		Roulet et al. (1992)
conifer swamps	<8	
beaver ponds	30–90	
thicket swamps	0.1–88	
bogs	6–21	
Saltwater Wetlands		
Georgia salt marsh, average annual rate		Wiebe et al. (1981)
all *Spartina*	0.8	
intermediate *Spartina*	29	
short *Spartina*	109	
Louisiana salt marsh		Delaune et al. (1983a)
salt	12	
brackish	200	
Louisiana salt marsh	14	C. J. Smith et al. (1982)

Table 5–5. Carbon Dioxide Release, gC m^{-2} yr^{-1} from Mineralization of Organic Matter in a New England Salt Marsh and Wisconsin Freshwater Lake Sediments

Pathway	*Salt Marsh* $gC\ m^{-2}\ yr^{-1}$	*Lake Sediment* $gC\ m^{-2}\ yr^{-1}$
Aerobic Respiration	361	—
Nitrate Reduction	5	8[a]
Fermentation-Sulfate Reduction	432	61
Methanogenesis	6	254
Total	804	—

[a]Estimated from previous study
Sources: Ingvorsen and Brock, 1982; Howes et al., 1984, 1985

Sulfur-reducing bacteria require an organic substrate, generally of low molecular weight, as a source of energy in converting sulfate to sulfide (Equation 5.15). The process of fermentation described above can conveniently supply these necessary low molecular weight organic compounds such as lactate or ethanol (see Equations 5.19 and 5.20 and Fig. 5–10). Equations for sulfur reduction, also showing the oxidation of organic matter, are given as follows (from Valiela, 1984):

$$\underset{\text{(lactate)}}{2CH_3CHOHCOO^-} + SO_4^{=} + 3H^+$$

$$\longrightarrow 2CH_3COO^- + 2CO_2 + 2H_2O + HS^- \qquad (5.22)$$

and

$$\underset{\text{(acetate)}}{CH_3COO^-} + SO_4^{=} \longrightarrow 2CO_2 + 2H_2O + HS^- \qquad (5.23)$$

The importance of this fermentation–sulfur-reduction pathway in the oxidation of organic carbon to CO_2 in saltwater wetlands was demonstrated for a New England salt marsh by Howarth and Teal (1979, 1980), and Howes et al. (1984, 1985). Table 5–5 shows that fully 54 percent of the carbon dioxide evolution from the salt marsh was caused by the fermentation–sulfur-reduction pathway, with aerobic respiration accounting for another 45 percent. Only a small percent of carbon release (0.7 percent) was caused by methanogenesis. A similar study using the same methods in a salt marsh in Georgia yielded sulfate-reduction rates that were one-third of those measured in the New England salt marsh. This

Table 5–6. Major Types of Dissolved and Insoluble Phosphorus in Natural Waters

Phosphorus	*Soluble Forms*	*Insoluble Forms*
Inorganic	orthophophates ($H_2PO_4^-$, $HPO_4^=$, PO_4^{-3})	clay-phosphate complexes
	polyphosphates	
	ferric phosphate ($FeHPO_4^+$)	metal hydroxide-phosphates, e.g. vivianite $Fe_3(PO_4)_2$; variscite $Al\,(OH)_2H_2PO_4$
	calcium phosphate ($CaH_2PO_4^+$)	minerals, e.g., apatite ($Ca_{10}(OH)_2(PO_4)_6$)
Organic	dissolved organics, e.g., sugar phosphates, inositol phosphates, phospholipids, phosphoproteins	insoluble organic phosphorus bound in organic matter

Source: After Stumm and Morgan, 1970

difference was attributed by the authors to less underground organic productivity in the Georgia marsh (Howarth and Giblin, 1983).

Few if any studies have estimated the importance of sulfate reduction on the oxidation of organic carbon in freshwater wetlands. Several studies of freshwater lakes, however, suggest that methanogenesis results in a greater oxidation of organic carbon than does sulfate reduction. Smith and Klug (1981) found that 2.5 times more organic carbon was mineralized by methanogenesis than through sulfate reduction in a lake in Michigan. Ingvorsen and Brock (1982) estimated that four times more carbon was mineralized via methanogenesis than by sulfate reduction in anaerobic lake sediments in Lake Mendota, Wisconsin (Table 5–5). Estimates of the absolute values of these processes when compared with those for the salt marsh (Table 5–5) support the generalizations that oxidation of organic carbon by methane production is dominant in freshwater wetlands, whereas oxidation of organic carbon by sulfate reduction is dominant in saltwater wetlands (Capone and Kiene, 1988).

Phosphorus Transformations

Phosphorus is one of the most important chemicals in ecosystems, and wetlands are no exception. It has been described as a major limiting nutrient in northern bogs (Heilman, 1968), freshwater marshes (Klopatek, 1978; Mitsch and Reeder,

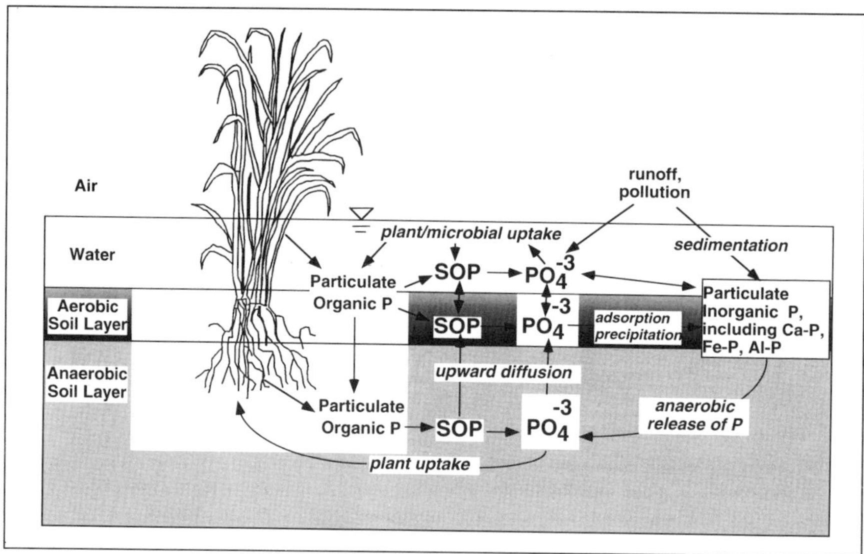

Figure 5–11. Phosphorus transformations in wetlands. SOP indicates soluble organic phosphorus.

1992), and southern deepwater swamps (Mitsch et al., 1979a; S. L. Brown, 1981). In other wetlands such as agricultural wetlands (Gambrell and Patrick, 1978) and salt marshes (Whitney et al., 1981), phosphorus is an important mineral, although it is not considered a limiting factor because of its relative abundance and biochemical stability. Phosphorus retention is considered one of the most important attributes of natural and constructed wetlands, particularly those that receive nonpoint source pollution or wastewater (see Chemical Mass Balances of Wetlands, this chapter, and Chapter 17).

Phosphorus occurs as soluble and insoluble complexes in both organic and inorganic forms in wetland soils (Table 5–6). It occurs in a sedimentary cycle (Figure 5–11) rather than in a gaseous cycle such as the nitrogen cycle. At any one time a major proportion of the phosphorus in wetlands is tied up in organic litter and peat and in inorganic sediments, with the former dominating peatlands and the latter dominating mineral soil wetlands. The principal inorganic form is orthophosphate, which includes the ions PO_4^{-3}, $HPO_4^{=}$, and $H_2PO_4^{-}$; the predominant form depends on pH. The analytical measure of biologically available orthophosphates is sometimes called *soluble reactive phosphorus,* although the equivalence between that term and orthophosphates is not exact. Dissolved organic phosphorus and insoluble forms of organic and inorganic phosphorus are generally not biologically available until they are transformed into soluble inorganic forms.

Although phosphorus is not directly altered by changes in redox potential as are nitrogen, iron, manganese, and sulfur, it is indirectly affected in soils and sediments by its association with several elements, especially iron, that are so altered (Mohanty and Dash, 1982). Phosphorus is rendered relatively unavailable to plants and microconsumers by

1. the precipitation of insoluble phosphates with ferric iron, calcium, and aluminum under aerobic conditions;
2. the adsorption of phosphate onto clay particles, organic peat, and ferric and aluminum hydroxides and oxides;
3. the binding of phosphorus in organic matter as a result of its incorporation into the living biomass of bacteria, algae, and vascular macrophytes.

The precipitation of the metal phosphates and the adsorption of phosphates onto ferric or aluminum hydroxides and oxides are believed to result from the same chemical forces, namely those involved in the forming of complex ions and salts. The adsorption of phosphates onto oxides and hydroxides of aluminum and iron and subsequent precipitation as insoluble ferric phosphates and aluminum phosphates can occur in acid soils. The precipitation of calcium phosphates dominates in pH greater than 7.0 (Lindsay, 1979; Stevenson, 1986). Richardson (1985) suggested that the best prediction of phosphorus sorption on a wetland soil is the amount of oxalate-extractable iron and aluminum in the soil.

The sorption of phosphorus onto clay particles is believed to involve both the chemical bonding of the negatively charged phosphates to positively charged edges of the clay and the substitution of phosphates for silicate in the clay matrix (Stumm and Morgan, 1970). This clay-phosphorus complex is particularly important for many wetlands, including riparian wetlands and coastal salt marshes, because a considerable portion of the phosphorus brought into these systems by flooding rivers and tides is brought in sorbed to clay particles. Thus phosphorus cycling in many mineral soil wetlands tends to follow sediment pathways of sedimentation and resuspension. Because most wetland macrophytes obtain their phosphorus from the soil, the sedimentation of phosphorus sorbed onto clay particles is an indirect way in which the phosphorus is made available to the biotic components of the wetland. In essence, the plants transform inorganic phosphorus to organic forms that are then stored in organic peat, mineralized by microbial activity, or exported from the wetland (Fig. 5–11).

When soils are flooded and conditions become anaerobic, several changes in the availability of phosphorus result. A well-documented phenomenon in the hypolimnion of lakes is the increase in soluble phosphorus when the hypolimnion and the sediment-water interface become anoxic (Mortimer, 1941–1942; Ruttner, 1963). In general, a similar phenomenon often occurs in wetlands on a compressed vertical scale. As ferric (Fe^{+++}) iron is reduced to more soluble ferrous (Fe^{++}) compounds, phosphorus that is in a specific ferric phosphate analytically known as reductant-soluble phosphorus (Gambrell and Patrick, 1978; Faulkner and Richardson, 1989) is released into solution. Other reactions that may be important in releasing phosphorus upon flooding are the hydrolysis of ferric and aluminum phosphates and the release of phosphorus sorbed to clays and hydrous oxides by the exchange of anions (Ponnamperuma, 1972). Phosphorus can also be released from insoluble salts when the pH is changed either by the production of organic acids or by the production of nitric and sulfuric acids by chemosynthetic bacteria (Atlas and Bartha, 1981). Phosphorus sorption onto clay particles, on the other hand, is highest under acidic to slightly acidic conditions (Stumm and Morgan, 1970).

CHEMICAL TRANSPORT INTO WETLANDS

The inputs of materials to wetlands occur through geologic, biologic, and hydrologic pathways typical of other ecosystems (Likens et al., 1977). The geologic input from weathering of parent rock, although poorly understood, may be important in some wetlands. Biologic inputs include photosynthetic uptake of carbon, nitrogen fixation, and biotic transport of materials by mobile animals such as birds. Except for gaseous exchanges such as carbon fixation in photosynthesis and nitrogen fixation, however, elemental inputs to wetlands are generally dominated by hydrologic inputs.

Precipitation

Table 5–7 describes the typical chemical characteristics of precipitation as measured in several locations in North America. The levels of chemicals in precipitation are variable but very dilute. Relatively higher concentrations of magnesium and sodium are associated with maritime influences, whereas high calcium indi-

Table 5–7. Chemical Characteristics of Bulk Precipitation (mg/l)

Chemical	*Georgia*[a]	*Newfoundland*[b]	*Wisconsin–Minnesota*[b]	*New Hampshire*[c]
Ca^{++}	0.17	0.8	1.0–1.2	0.13
Mg^{++}	0.05	—	—	0.04
Na^{+}	—	5.2	0.2–0.5	0.11
K^{+}	0.14	0.3	0.2	0.06
NO_3^-	0.26	—	—	1.47
NH_4^+	—	—	—	0.19
Cl^-	—	8.9	0.1	0.40
$SO_4^=$	—	2.2	1.4	2.6
P (soluble)	0.017	—	—	0.01

[a]Schlesinger, 1978
[b]Gorham, 1961, as summarized by Moore and Bellamy, 1974
[c]Likens et al., 1985

cates continental influences. Precipitation tends to contain contaminants at higher concentrations in short storms and when precipitation is infrequent.

Human influence on the chemicals in precipitation has been significant, particularly because of the burning of fossil fuels and subsequent increased concentrations of sulfates and nitrates in the atmosphere, the causes of the familiar acid rain (more properly called acid deposition). The data in Table 5–7 for the chemical characteristics of precipitation for New Hampshire are weighted averages over several years and do not reflect recent trends. Long-term monitoring at this site has shown that after several years of relatively high concentrations of sulfates in precipitation, there has been a steady decrease in the concentrations of sulfates and basic cations. Sulfate concentrations in precipitation decreased about 40 percent from about 3.1 mg/l in the late 1960s to 1.9 mg/l in the 1980s, probably caused by the decline in the emissions of sulfur dioxide in the northeastern United States over that period (Driscoll et al., 1989). Long-term trends of nitrates in precipitation, at least in eastern and parts of the midwestern United States, did not show similar decreases, because emissions from sources such as automobiles continued to increase over that period (R. A. Smith et al., 1987). Some wetlands such as northern peatlands or southern cypress domes are fed primarily by precipitation and therefore may be especially susceptible to anthropogenic inputs in precipitation especially where acidic deposition is severe (Gorham et al., 1984a; Gorham, 1987). In fact, well before there was a general concern for their effects on aquatic and terrestrial systems, the possible effects of anthropogenic inputs of acid precipitation to certain wetlands were discussed in biogeochemical studies of English peatlands by Gorham (1956, 1961).

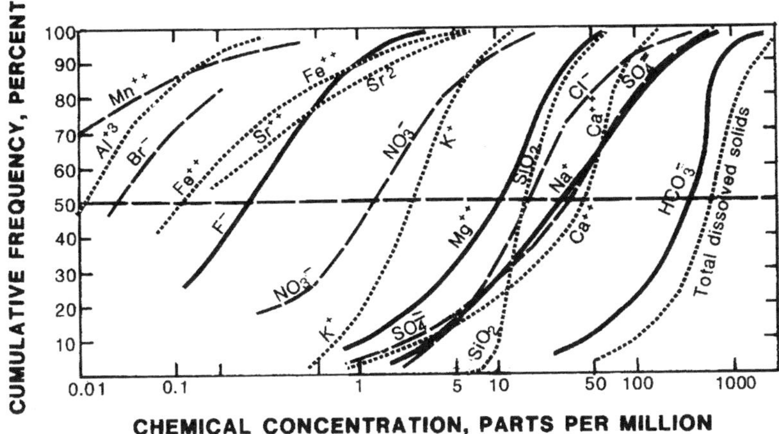

Figure 5–12. Cumulative frequency curves for concentrations of various dissolved minerals in surface waters. Horizontal dashed line indicates median concentration. *(After Stumm and Morgan, 1970, Davis and DeWiest, 1966)*

Table 5–8. Average Chemical Concentrations (mg/l) of Ocean Water and River Water

Chemical	Seawater[a]	"Average" River[b]
Na^+	10,773	6.3
Mg^{++}	1,294	4.1
Ca^{++}	412	15
K^+	399	2.3
Cl^-	19,344	7.8
$SO_4^=$	2,712	11.2
$HCO_3^-/CO_3^=$	142	58.4
B	4.5[c]	0.01[c]
F	1.4[c]	0.1[c]
Fe	<0.01[c]	0.7
SiO_2	<0.1–>10[c]	13.1[c]
N	0–0.5[c]	0.2[c]
P	0–0.07[c]	0.02[c]
Particulate Organic Carbon	0.01–1.0[c]	5–10[c]
Dissolved Organic Carbon	1–5[c]	10–20[c]

[a]Riley and Skirrow, 1975
[b]D. A. Livingston, 1963
[c]Burton and Liss, 1976

Streams, Rivers, and Groundwater

As precipitation reaches the ground in a watershed, it will infiltrate into the ground, pass back to the atmosphere through evapotranspiration, or flow on the surface as runoff. When enough runoff comes together, sometimes combined with groundwater flow, in channelized streamflow, its mineral content is different from that of the original precipitation.

There is not, however, a typical water quality for surface and subsurface flows. Figure 5–12 describes the cumulative frequency of the ionic composition of freshwater streams and rivers in the United States. The curves demonstrate the wide range over which these chemicals are found and the median values (50 percent line) of these ranges. The "average" concentration of dissolved materials in the world's rivers is given in Table 5–8. The variability in concentrations of chemicals in runoff and streamflow that enter wetlands is caused by several factors:

1. *Groundwater Influence.* The chemical characteristics of streams and rivers depend on the degree to which the water has previously come in contact with underground formations and on the types of minerals present in those forma-

tions. Soil and rock weathering, through dissolution and redox reactions, provides major dissolved ions to waters that enter the ground. The dissolved materials in surface water can range from a few milligrams per liter, found in precipitation, to 100 or even 1,000 milligrams per liter (mg/l). The ability of water to dissolve mineral rock depends, in part, on its nature as a weak carbonic acid. The rock being mineralized is also an important consideration. Minerals such as limestone and dolomite yield high levels of dissolved ions, and granite and sandstone formations are relatively resistant to dissolution.

2. *Climate.* Climate influences surface water quality through the balance of precipitation and evapotranspiration. Arid regions tend to have higher concentrations of salts in surface waters than do humid regions. Climate also has a considerable influence on the type and extent of vegetation on the land, and it therefore indirectly affects the physical, chemical, and biological characteristics of soils and the degree to which soils are eroded and transported to surface waters.
3. *Geographic Effects.* The amounts of dissolved and suspended materials that enter streams, rivers, and wetlands also depend on the size of the watershed, the steepness or slope of the landscape, soil texture, and the variety of topography (R. Lee, 1980). Surface waters that have high concentrations of suspended (insoluble) materials caused by erosion are often relatively low in dissolved substances. On the other hand, waters that have passed through groundwater systems often have high concentrations of dissolved materials and low levels of suspended materials. The presence of upstream wetlands also influences the quality of water entering downstream wetlands (see Coupling with Adjacent Ecosystems in this chapter).
4. *Streamflow/Ecosystem Effects.* The water quality of surface runoff, streams, and rivers varies seasonally. It is most common that there is an inverse correlation between streamflow and concentrations of dissolved materials (Hem, 1970; Dunne and Leopold, 1978). During wet periods and storm events, the water is contributed primarily by recent precipitation that becomes streamflow very quickly without coming into contact with soil and subsurface minerals. During low flow, some or much of the streamflow originates as groundwater and has higher concentrations of dissolved materials. This inverse relationship, however, is not always the case, as illustrated by the relationship found in several years of study at the Hubbard Brook, New Hampshire, watershed (Fig. 5–13). There the small watershed had remarkably similar concentrations of dissolved substances despite a wide range of streamflow, caused by the biotic and abiotic "regulation" of water quality in the small forested watershed and the stream itself (Bormann et al., 1969; Likens et al., 1985).

The relationship between particulate matter and streamflow is often found to be a nonlinear positive relationship (Fig. 5–13). Part of this relationship is based on the fact that streamflow from a small watershed is often low in the growing season, even for relatively heavy precipitation, because of high interception and evapotranspiration. On the other hand, precipitation in the nongrow-

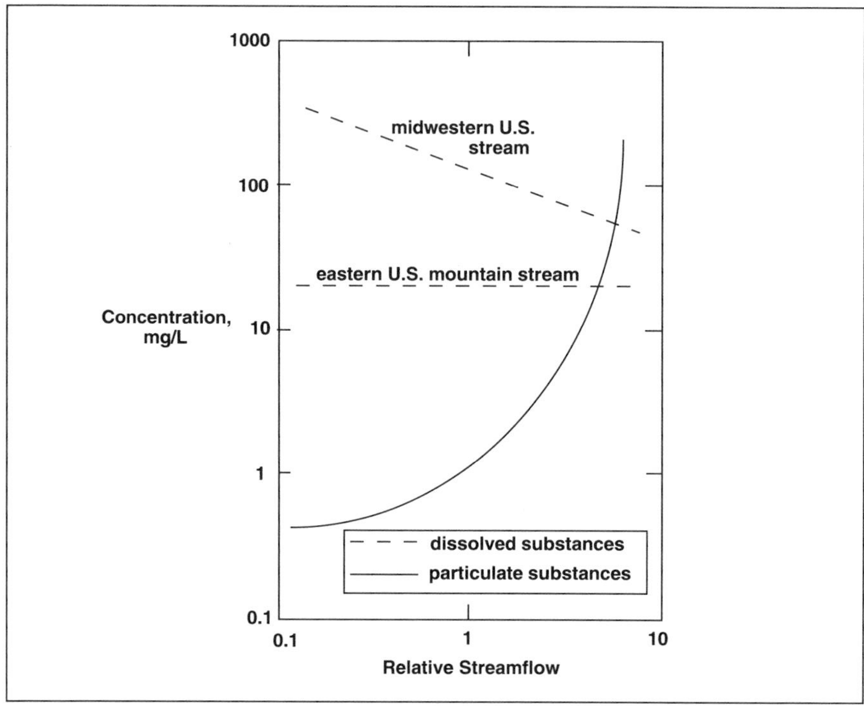

Figure 5–13. Generalized relationships between streamflow and concentrations of dissoved and particulate substances. Data for midwestern U.S. stream is generalization from several studies. Data for a small eastern U.S. watershed is Hubbard Brook, New Hampshire. *(Redrawn from Likens et al., 1985; copyright © 1985,by Springer-Verlag, reprinted with permission)*

ing season, often combined with saturated soils and little evapotranspiration, creates high streamflow and sediment transport. In one example, Omernik (1977) showed that for a large sample of streams throughout the United States, concentrations of nitrogen and phosphorus increased with discharge in disturbed watersheds because of increased erosion but decreased with streamflow in natural watersheds presumably because of reduced erosion and increased dilution.

As with all generalizations, the increase in particulate matter concentration with flow is not always the case. For example, in many streams in the midwestern United States, bioturbation stemming from active fish populations (e.g., common carp, *Cyprinus carpio*) and summer algal blooms can actually cause sediment concentrations to be higher in low-flow conditions typical of late summer.

5. *Human Effects.* Water that has been modified by humans through, for example, sewage effluent, urbanization, and runoff from farms often drastically alters the chemical composition of streamflow and groundwater that reach

wetlands. If drainage is from agricultural fields, higher concentrations of sediments and nutrients and some herbicides and pesticides might be expected. Urban and suburban drainage is often lower than that from farmland in those constituents, but it may have high concentrations of trace organics, oxygen-demanding substances, and some toxics.

In some cases, poor-quality waters have been discharged into wetlands for many years. Several studies of the influence of pollution sources on wetlands have documented the effects of municipal wastewater (Grant and Patrick, 1970; Boyt et al., 1977), coal mine drainage (Wieder and Lang, 1982; Mitsch et al., 1983a, 1983b, 1983c), highway construction (McLeese and Whiteside, 1977), stream channelization (Maki et al., 1980), and sulfate pollution (J. Richardson et al., 1983). Some of these alterations are described in more detail in Chapter 16.

Estuaries

Wetlands such as salt marshes and mangrove swamps are continually exchanging tidal waters with adjacent estuaries and other coastal waters. The quality of these waters differs considerably from that of the rivers described above. Although estuaries are places where rivers meet the sea, they are not simply places where seawater is diluted with freshwater (J. W. Day et al., 1989). Table 5–8 contrasts the chemical makeup of average river water with the average composition of seawater. The chemical characteristics of seawater are fairly constant worldwide compared with the relatively wide range of river water chemistry. Total salt concentrations typically range from 33 0/00 to 37 0/00. Although seawater contains almost every element that can go into solution, 99.6 percent of the salinity is accounted for by 11 ions. In addition to seawater dilution, estuarine waters can also involve chemical reactions when sea and river waters meet, including the dissolution of particulate substances, flocculation, chemical precipitation, biological assimilation and mineralization, and adsorption and absorption of chemicals on and into particles of clay, organic matter, and silt (J. H. Day, 1981). In most estuaries and coastal wetlands, biologically important chemicals such as nitrogen, phosphorus, silicon, and iron come from rivers, whereas other important chemicals such as sulfates and bicarbonates/carbonates come from ocean sources (J. W. Day et al., 1989).

CHEMICAL MASS BALANCES OF WETLANDS

A quantitative description of the inputs, outputs, and internal cycling of materials in an ecosystem is called an *ecosystem mass balance* (Whigham and Bayley, 1979; Nixon and Lee, 1986). If the material being measured is one of several elements such as phosphorus, nitrogen, or carbon that are essential for life, then the mass balance is called a *nutrient budget.* In wetlands, mass balances have been developed both to describe ecosystem function and to determine the importance of wetlands as sources, sinks, and transformers of chemicals. Extensive lit-

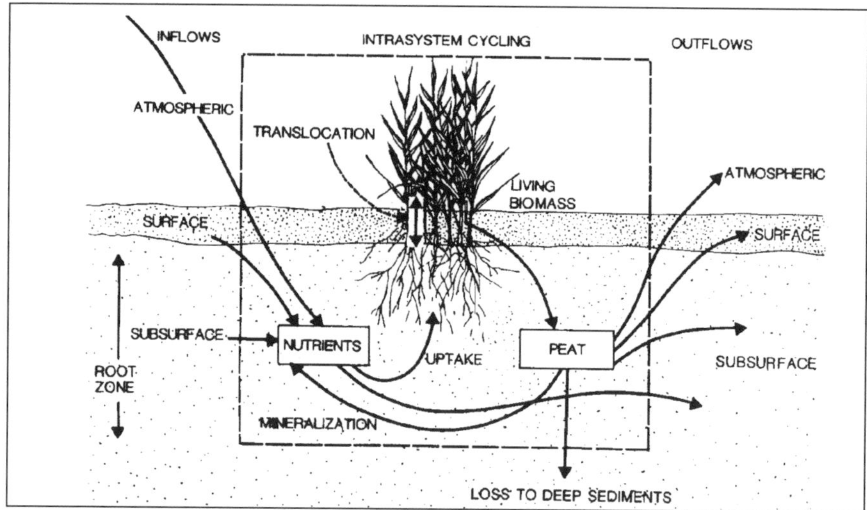

Figure 5–14. Generalized diagram of components of a wetland mass balance, including inflows, outflows, and intrasystem cycling.

erature reviews on the subject of the influence of wetlands on water quality, the net result of these mass balances, have been provided by Nixon and Lee (1986), Johnston et al. (1990), and Johnston (1991).

A general mass balance for a wetland, as shown in Figure 5–14, illustrates the major categories of pathways and storages that are important in accounting for materials passing into and out of wetlands. Nutrients or chemicals that are brought into the system are called *inputs* or *inflows.* For wetlands, these inputs are primarily through hydrologic (pathways described in Chapter 4) such as precipitation, surface and groundwater inflow, and tidal exchange. Biotic pathways of note that apply to the carbon and nitrogen budgets are the fixation of atmospheric carbon through photosynthesis and the capture of atmospheric nitrogen through nitrogen fixation.

Hydrologic *exports,* or *losses* or *outflows,* are by both surface water and groundwater unless the wetland is an isolated basin that has no outflow such as a northern ombrotrophic bog. The long-term burial of chemicals in the sediments is also considered a nutrient or chemical outflow, although the depth at which a chemical goes from internal cycling to permanent burial is an uncertain threshold. The depth of available chemicals is usually defined by the root zone of vegetation in the wetland. Biologically mediated exports to the atmosphere are also important in the nitrogen cycle (denitrification) and in the carbon cycle (respiratory loss of CO_2). The significance of other losses of elements to the atmosphere such as ammonia volatilization and methane and sulfide releases is not well understood, although they are potentially important pathways for individual wetlands as well as for the global cycling of minerals (see Chemical Transformations in Wetland Soils in this chapter).

Intrasystem cycling involves exchanges among various *pools,* or *standing*

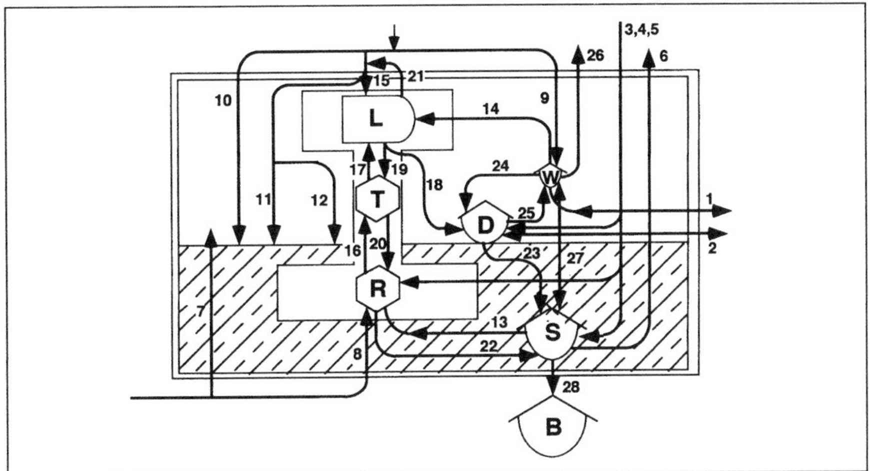

Figure 5–15. Model of major chemical storages and flows in wetlands. *Storages:* L = above-ground shoots or leaves; T = trunks and branches, perennial above-ground storage; R = roots and rhizomes; W = dissolved and suspended particulates in surface water; D = litter or detritus; S = near-surface sediments; B = deep sediments essentially removed from internal cycling.

Flows: 1 and 2 are exchanges of dissolved and particulate material with adjacent waters; 3-5 are nitrogen fixation in sediments, rhizosphere microflora, and litter; 6 is denitrification by sediments (N_2 and N_2O); 7 and 8 are groundwater inputs to roots and surface water; 9 is atmospheric deposition on water; 10 is on land; 11 and 12 are aqueous deposition from the canopy and its stemflow; 13 is uptake by roots; 14 is foliar uptake from surface water; 15 is uptake from rainfall; 16 and 17 are translocations from roots through trunks and stems to leaves; 18 is the production of litter; 19 and 20 are the readsorption of materials from leaves through trunks and stems to roots and rhizomes; 21 is leaching from leaves; 22 is death or sloughing of root material; 23 is incorporation of litter into sediments or peat; 24 is uptake by decomposing litter; 25 is release from decomposing litter; 26 is volatilization of ammonia; 27 is sediment-water exchange; 28 is long-term burial in sediments. *(Figure and caption from Nixon and Lee, 1986)*

stocks, of chemicals within a wetland. This cycling includes pathways such as litter production, remineralization, and various chemical transformations discussed earlier. The *translocation* of nutrients from the roots through the stems and leaves of vegetation is another important intrasystem process that results in the physical movement of chemicals within a wetland.

Figure 5–15 illustrates more details of the major pathways and storages that investigators should consider when developing mass balances of chemicals for wetlands. Major exchanges with the surroundings are shown as exchanges of particulate and dissolved material with adjacent bodies of water (Pathways 1 and 2), exchange through groundwater (Pathways 7 and 8), inputs from precipitation (Pathways 9 and 10), and burial in sediments (Pathway 28). Exchanges specific to a nitrogen mass balance, namely nitrogen fixation (Pathways 3, 4, 5), denitrification (Pathway 6), and ammonia volatilization (Pathway 26) are also shown in the diagram. A number of intrasystem pathways such as stem-flow (Pathway 12), root sloughing (Pathway 22), detrital-water exchanges (Pathways 24 and 25), and sediment water exchanges (Pathway 27) can be very important in determining the fate of chemicals in wetlands but are extremely difficult to measure. Few if any investigators have developed a complete mass balance for wetlands that includes measurement of all of the pathways shown in Figure 5–15, but the diagram remains a useful guide to those considering studies of wetlands as sources, sinks, or transformers of chemicals as described below.

The chemical balances that have been developed for various wetlands are extremely variable. A few generalizations emerge from these studies:

1. Wetlands serve as sources, sinks, or transformers of chemicals, depending on the wetland type, hydrologic conditions, and the length of time the wetland has been subjected to chemical loadings. When wetlands serve as sinks for certain chemicals, the long-term sustainability of that situation depends on the hydrologic and geomorphic conditions, the spatial and temporal distribution of chemicals in the wetland, and ecosystem succession. Wetlands can become saturated in certain chemicals after a number of years, particularly if loading rates are high (R. A. Kadlec, 1983; Knight et al., 1987).
2. Seasonal patterns of nutrient uptake and release are characteristic of many wetlands. In temperate climates, retention of certain chemicals such as nutrients is greatest in the growing season primarily because of higher microbial activity in the water column and sediments and secondarily because of greater macrophyte productivity.
3. Wetlands are frequently coupled to adjacent ecosystems through chemical exchanges that significantly affect both systems. Ecosystems upstream of wetlands are often significant sources of chemicals to wetlands, whereas downstream aquatic systems often benefit either from the ability of wetlands to retain certain chemicals or from the export of organic materials.
4. Contrary to popular opinion that all wetlands are highly productive, wetlands can be either highly productive ecosystems rich in nutrients or systems of low productivity caused by a scarce supply of nutrients.
5. Nutrient cycling in wetlands differs from both aquatic and terrestrial ecosystem cycling in temporal and spatial dimensions. For example, more nutrients are tied up in sediments and peat in wetlands than in most terrestrial systems, and deepwater aquatic systems have autotrophic activity more dependent on nutrients in the water column than in the sediments.
6. Anthropogenic changes have led to considerable changes in chemical cycling in many wetlands. Although wetlands are quite resilient to many chemical

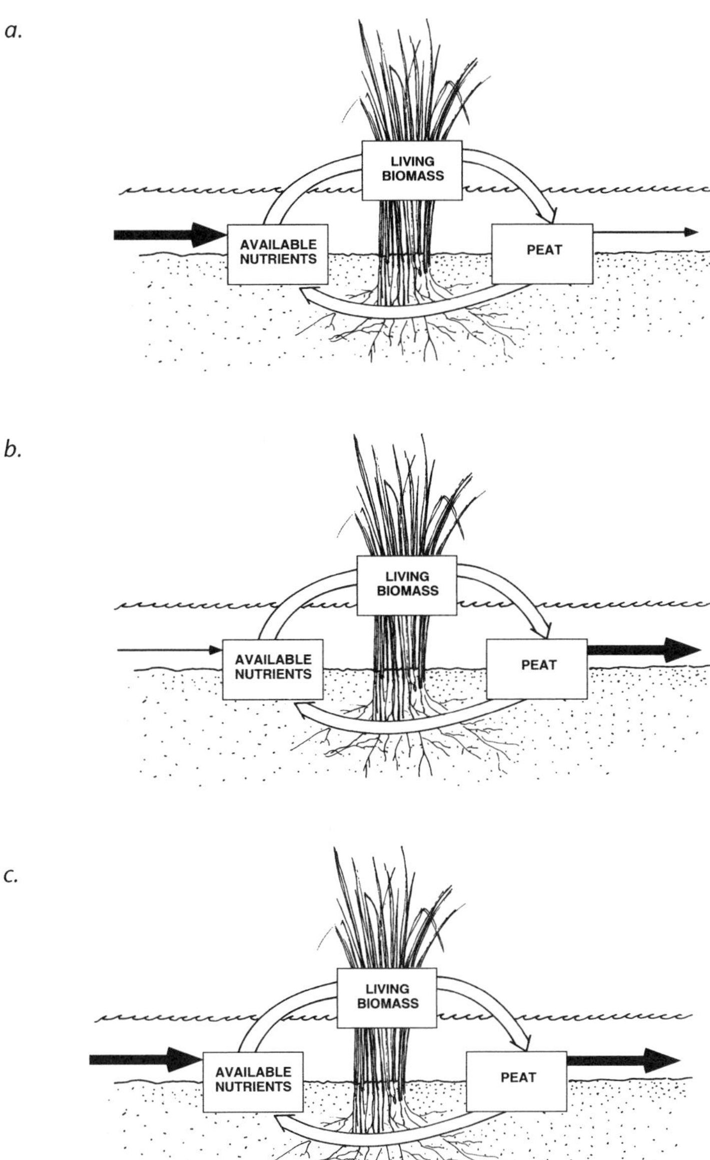

Figure 5–16. Diagrams of wetland as *a.* inorganic nutrient sink, *b.* source of total nutrients, and *c.* transformer of inorganic nutrients (inflow) to organic nutrients (outflow).

Table 5–9. Studies of Wetlands as Sources of Sinks of Nitrogen and Phosphorus

Type of Wetland, Location[a]	*Nutrient Sink*[b] *N*	*P*	*Reference*
Freshwater Marshes (tidal)			
Tinicum Marsh, Pa.	Yes	Yes	Grant and Patrick (1970)
Hamilton Marsh, N.J.	S	S	Simpson et al. (1978)
Freshwater Marshes (nontidal)			
4 marshes, Wis.	Yes	S	G. F. Lee et al. (1975)
Waterhyacinth marsh, Fla.[c]	Yes	S	Mitsch (1977); Vega and Ewel (1981)
Theresa Marsh, Wis.	S	S	Klopatek (1978)
Typha marsh, Wis.[e]	—	Yes	Fetter et al. (1978)
Marsh, Fla.[d, e]	—	Yes	Dolan et al. (1981)
Managed marsh, N.Y.	I	I	Peverly (1982)
Lake Erie coastal marsh, Ohio[c]	—	Yes	Mitsch and Reeder (1991, 1992); Klarer and Millie (1989)
Constructed marshes, No. Ill.	Yes	Yes	Sather (1992; Mitsch (1992a)
Northern Peatlands			
Arctic tundra, Alaska	Yes	—	Barsdate and Alexander (1975)
Forested peatland, Mich.	No	Yes	Richardson et al. (1978)
Thoreau's Bog, Mass.	Yes	—	Hemond (1980)
Black spruce bog, Minn.	Yes	Yes	Verry and Timmons (1982); Urban and Eisenreich (1988)
Thuja peatland, Mich.[e]	Yes	I	R. H. Kadlec (1983)

continued on next page

inputs, the capacity of wetlands to assimilate anthropogenic wastes from the atmosphere or hydrosphere is not limitless.

Wetlands as Sources, Sinks, or Transformers of Nutrients

There has been much discussion and research by wetland scientists about whether wetlands are nutrient *sources, sinks,* or *transformers.* A wetland is considered a sink if it has a net retention of an element or a specific form of that element (e.g., organic or inorganic), that is, if the inputs are greater than the outputs (Fig. 5–16a). If a wetland exports more of an element or material to a downstream or adjacent ecosystem than would occur without that wetland, it is considered a source (Fig. 5–16b). If a wetland transforms a chemical from, say, dissolved to particulate form but does not change the amount going into or out of the wetland, it is considered to be a transformer (Fig. 5–16c). Part of the interest in the source-sink-transformer question was stimulated by studies that hypothesized the importance of salt marshes as *sources* of particulate carbon for the adjacent estuaries and other studies that

Table 5–9 *(continued)*

Type of Wetland, Location[a]	*Nutrient Sink*[b] *N*	*P*	*Reference*
Forested Swamps			
Riverine cypress swamp, S.C.	Yes	Yes	Kitchens et al. (1975)
Mixed hardwood swamp, Fla.	Yes	Yes	Boyt et al. (1977)
Taxodium-Nyssa swamp, So. Ill.	—	Yes	Mitsch et al. (1979a)
Floodplain swamp, N.C.	—	I	Kuenzler et al. (1980)
Riparian forest, Ga.	Yes	Yes	Lowrance et al. (1984)
Fraxinus lakeside wetland, Wis.	Yes	Yes	Johnston et al. (1984)
Taxodium strand, Fla.	—	Yes	Nessel and Bayley (1984)
Taxodium pond[e]	Yes	Yes	Dierberg and Brezonik, (1984, 1985)
Swamp forests, La.[d]	Yes	I	Kemp and Day (1984); Kemp et al. (1985)
Riparian forest, Md.	Yes	Yes	Peterjohn and Correll (1984)
Floodplain forest, Fla.	S	S	Elder (1985)
Nyssa swamp, N.C.	Yes	No	Brinson et al. (1984)
Reedy Creek swamp, Fla.[e]	Yes	No	Knight et al. (1987)
Tidal Salt Marsh			
Delaware	—	No	Reimold and Daiber (1970)
Massachusetts[f]	Yes	Yes	Valiela et al. (1973)
Georgia	—	No	Gardner (1975)
Chesapeake Bay, Md.	No	No	Stevenson et al. (1977)
Flax Pond, N.Y.	S	S	Woodwell and Whitney (1977); Woodwell et al. (1979)
Great Sippewissett Marsh, Mass.	S	—	Valiela et al. (1978); Teal et al. (1979)
Sapelo Island, Ga.	Yes	S	Whitney et al. (1981)
Louisiana	Yes	Yes	Delaune et al. (1981); Delaune et al. (1983b)

[a]all results on at least annual budget unless otherwise noted
[b]S=seasonal sink; I=inconsistent results from multiple-year study
[c]9-month period
[d]10-month period
[e]receiving waste water
[f]sewage sludge on experimental plots

suggested the importance of wetlands as *sinks* for certain chemicals, particularly nitrogen and phosphorus. The two concepts of one wetland being a source and a sink for various materials are not mutually exclusive; a wetland can be a sink for an inorganic form of a nutrient and a source for an organic form of the same nutrient. The desire of wetland scientists to determine conclusively whether wetlands are sources or sinks of nutrients has often been hampered by the imprecise use of the words "source" and "sink" and by the inadequacy of the techniques used to measure the nutrient fluxes in a wetland nutrient budget.

Table 5–9 gives some examples of results of studies of wetlands as nitrogen and phosphorus sinks. There is no consensus on this question for wetlands in general, and, in fact, there is little agreement in the literature even about particular nutrients in specific wetland types. All that can be said with certainty is that many wetlands act as sinks for particular inorganic nutrients and many wetlands are sources of organic material to downstream or adjacent ecosystems.

Freshwater Marshes

One of the first studies that identified freshwater wetlands for their role as nutrient sinks was of Tinicum Marsh near Philadelphia (Grant and Patrick, 1970). The study, entitled "Tinicum Marsh as a Water Purifier," had among its goals to determine "the role of Tinicum Marsh wetlands in the reduction of nitrates and phosphates. . . ." That study found decreases in phosphorus (as PO_4^{-3}), nitrogen (as NO_3^- and NH_3), and organic materials (as biochemical oxygen demand, BOD), in water flowing from the adjacent river over the marsh. G. F. Lee et al. (1975) summarized the results of two research projects on the effects of freshwater marshes on water quality in Wisconsin and concluded that there were both beneficial (i.e., marsh acts as sink) and detrimental (i.e., marsh acts as source) effects of the marshes on water quality, but that "from an overall point of view, it appears that the beneficial effects outweigh the detrimental effects." Mitsch (1977) found that 49 percent of the total nitrogen and 11 percent of the total phosphorus were removed by a floating water hyacinth *(Eichhornia crassipes)* marsh receiving wastewater in north-central Florida, whereas Vega and Ewel (1981) found a 16 percent retention of phosphorus for the same system a few years later. The former study found a seasonal change in the loss of nitrate nitrogen, with greater uptake in the summer months. Studies by Klopatek (1978) in a Wisconsin riverine marsh and by Simpson et al. (1978) in a tidal freshwater wetland also showed the capacity for marsh wetlands to be at least a seasonal sink for inorganic forms of nitrogen and phosphorus. A two-year study of the potential of a managed marsh wetland in upper New York State to remove nutrients from agricultural drainage gave inconsistent results, with the wetland acting as a source of nitrogen and phosphorus in the first year and a net sink in the second year (Peverly, 1982).

Studies of a freshwater marsh along Lake Erie's shoreline have shown that the wetland is effective in ameliorating nutrient loading from an agricultural watershed to the lake and that the effectiveness depends on the amount of annual runoff and the level of the lake (Klarer and Millie, 1989; Mitsch and Reeder, 1991, 1992). Several studies, e.g., by Fetter et al. (1978) in Wisconsin and Dolan et al. (1981) in Florida, have demonstrated the capacity of natural marshes to be sinks for nutrients when wastewaters are added to wetlands (see Chapter 17). Reflecting interest in controlling nonpoint source pollution, several studies have investigated whether freshwater marshes can be sinks of nutrients and sediments when receiving nonpoint sources from both rural and urban areas (Hey et al., 1989; Olson, 1992; Mitsch, 1992a).

Bogs and Fens

There have been a number of studies of the nitrogen retention capacity of natural bogs and fens, for they generally have no or simple outflows and rely to a great extent on inputs from precipitation (Johnston, 1991). Although peatlands are generally anaerobic, denitrification has not generally been considered a major pathway for nitrogen loss in these systems, at least in studies in Alaska, Massachusetts, and Minnesota (respectively, Barsdate and Alexander, 1975; Hemond, 1980; Urban and Eisenreich, 1988). Studies by Kadlec and Tilton (1979) and Richardson and Marshall (1986) investigated the role of fens in Michigan in retaining nutrients. (The former study involved the addition of wastewater.) A multiple-year study by Kadlec (1983) demonstrated that a peatland in Michigan that received wastewater was consistently a sink for nitrogen (75–81 percent removal) but began to export phosphorus after several years of phosphorus retention.

Forested Swamps

The functioning of forested wetlands as nutrient sinks was suggested by Kitchens et al. (1975) in a preliminary winter-spring survey of a swamp forest-alluvial river swamp complex in South Carolina. These scientists found a significant reduction in phosphorus as the waters passed over the swamp. They assumed this to be the result of biological uptake by aquatic plant communities. Several studies in Florida were initiated to investigate the value of wetlands as nutrient sinks. These studies, from the Center for Wetlands at the University of Florida, included the purposeful disposal of high-nutrient wastewater in cypress domes (H. T. Odum et al., 1977; Ewel and Odum, 1984; Dierberg and Brezonik, 1984, 1985) and the long-term inadvertent disposal of wastewater by small communities in forested wetlands (Boyt et al., 1977; Nessel and Bayley, 1984). In all of these studies, the wetlands acted as sinks for nitrogen and phosphorus. In the cypress dome experiments, the nutrients were essentially retained in the water, sediments, and vegetation with little surface outflow. Boyt et al. (1977) described nutrient uptake that occurred in a mixed hardwood swamp that had received domestic sewage effluent for 20 years. They found that the total phosphorus and the total nitrogen in the outflow were reduced by 98 percent and 90 percent, respectively, compared with the inflow. In at least one contrasting study, a net export of phosphorus was seen in six out of seven years in a multi-year study of a forested wetland in Florida that received treated wastewater (Knight et al., 1987). The wetland, then, was serving as a source of phosphorus, probably the result of the entrainment of phosphorus that had accumulated in the forested wetland litter and soils.

Mitsch et al. (1979a) developed a nutrient budget for an alluvial river swamp in southern Illinois and found that ten times more phosphorus was deposited with sediments during river flooding (3.6 g-P m^{-2} yr^{-1}) than was returned from the swamp to the river during the rest of the year. The swamp was thus a sink for a significant amount of phosphorus and sediments during that particular year of flooding, although the percent of retention was low (3–4.5 percent) because a very large volume of water passed over the swamp during flooding conditions (Mitsch, 1992a). A study by Kuenzler et al. (1980) typified several chemical budget studies that have been developed for the Coastal Plain floodplain swamps of North Carolina. They found that 94 percent of the phosphorus transported to these wetlands was carried by surface water. They also found that there was a significant retention of phosphorus by the swamp, resulting in low concentrations of phosphorus downstream of the wetland.

Kemp and Day (1984) and Peterjohn and Correll (1984) described the fate of nutrients as they are carried into forested wetlands by agricultural runoff. The former study found that a Louisiana swamp forest acted primarily as a transformer system, removing inorganic forms of nitrogen and serving as a net source of organic nitrogen, phosphate, and organic phosphorus. The latter study in a riparian Maryland forest described the removal of nitrogen and phosphorus from runoff and groundwater as the runoff passed through approximately 50 m of riparian vegetation. Significant reductions of both nutrients from runoff were noted in the study. Additional studies are described in more detail in Chapters 13 and 14. A similar study of a floodplain forest in Georgia found a 14 percent retention and a 61 percent denitrification of nitrogen (for a total reduction of 75 percent of the incoming nitrogen) and a 30 percent retention of phosphorus (Lowrance et al., 1984). The Maryland and Georgia studies did not consider any river flooding in the calculations of their nutrient budgets.

The importance of floodplain wetlands to the speciation of nitrogen and phosphorus in a major river system was investigated by Elder (1985) on the Appalachicola River in northern Florida. That study found that there was little change in the total flows of total nitrogen and total phosphorus as the river flowed 170 km downstream. Nevertheless, the study found considerable transformations of these nutrients. There was a net import of ammonia and soluble reactive phosphorus and a net export of organic nitrogen in both dissolved and particulate forms. Elder argues that the floodplain wetlands, which are generally autotrophic, are responsible for a "net import of inorganic nutrients and a net export of organics" in the river system.

Salt Marshes

Salt marshes have the longest history of nutrient budget studies, have been the subject of the most comprehensive studies of nutrient dynamics, and probably have generated the most controversy about the source-sink question. A critical evaluation of 20 years of research on the role of salt marshes in nutrient cycles was presented by Nixon (1980) with the following general conclusion:

> On the basis of very little evidence, marshes have been widely regarded as strong terms (sources or sinks) in coastal marine nutrient cycles. The data we have available so far do not support this view. In general, marshes seem to act as nitrogen transformers, importing dissolved oxidized inorganic forms of nitrogen and exporting dissolved and particulate reduced forms. While the net exchanges are too small to influence the annual nitrogen budget of most coastal systems, it is possible that there may be a transient local importance attached to the marsh-estuarine nitrogen flux in some areas. Marshes are sinks for total phosphorus, but there appears to be a remobilization of phosphate in the sediments and a small net export of phosphate from the marsh.

Various studies that led to the above summary had depicted salt marshes as either sources or sinks of nutrients. The salt marsh was described as a source of phosphorus for the adjacent estuary in studies in Delaware (Reimold and Daiber, 1970; Reimold, 1972), Georgia (Gardner, 1975), and Maryland (Stevenson et al., 1977). Studies of Flax Pond Marsh in Long Island, New York (Woodwell and Whitney, 1977; Woodwell et al., 1979), found a net input of organic phosphorus and a net discharge of inorganic phosphate. Nitrogen budgets of the same marsh (Woodwell et al., 1979) and of the Great Sippewissett Marsh in Massachusetts (Valiela et al., 1978) indicated seasonal changes in nitrogen species but, within the accuracy of the measurements, a probable balance between inputs and outputs of nitrogen. Whitney et al. (1981), in a summary of many years of research on Sapelo Island, Georgia, described that marsh as generally balanced with regard to phosphorus and a net sink of nitrogen primarily because denitrification greatly exceeded nitrogen fixation. In some studies, e.g., Valiela et al. (1973), salt marshes have been shown to be nutrient sinks when fertilized with sewage sludge.

Seasonal Patterns of Uptake and Release

The fact that a wetland is a sink or a source of nutrients on a year-by-year basis suggests nothing about the seasonal differences in nutrient uptake and release. Van der Valk et al. (1979) described the general pattern of uptake of both nitrogen and phosphorus by wetlands as follows: During the growing season, there is a high rate of uptake of nutrients by emergent and submerged vegetation from the water and sediments. Increased microbiological immobilization of nutrients and uptake by algae and epiphytes also lead to a retention of inorganic forms of nitrogen and phosphorus. By the time the higher plants die, they have translocated a substantial portion of the nutrient material back to the roots and rhizomes. A substantial portion of the nutrients, however, are lost to the waters through litter fall and subsequent leaching. This generally leads to a net export of nutrients in the fall and early spring.

Several of the studies described above, particularly some on freshwater marshes (Lee et al., 1975; Klopatek, 1978; Simpson et al., 1978) and on salt marshes (Woodwell and Whitney, 1977; Valiela et al., 1978), documented the seasonal changes of nutrient export that occurred in these wetlands (Table 5–9). Lee et al. (1975) found that the marshes that they studied act as nutrient sinks during the summer and fall and as nutrient sources in the spring. This pattern had two potential benefits, according to the authors: (1) the problem of lake enrichment downstream of the marshes was decreased in the summer, the time of the most serious algal blooms in lakes, and (2) it might be economically and ecologically reasonable to treat waters for nutrient removal only during the periods of high flow, when the marsh is exporting nutrients, allowing the marsh to be the nutrient removal system when it is acting as a sink.

The seasonality of nutrient retention by freshwater marshes was also observed in another Wisconsin riverine marsh by Klopatek (1978), who found that the marsh acted as both a source and a sink of nutrients and that the pattern depended on the hydrologic conditions, the anaerobiosis in the sediments, and the activity of microbes and emergent macrophytes. Simpson et al. (1978) described the movement of inorganic nitrogen and phosphorus in a freshwater tidal marsh in New Jersey and concluded: "It appears almost all habitats of freshwater tidal marshes may be sinks for inorganic N and PO_4-P during the vascular plant growing season and that certain habitats may continually function as sinks." Woodwell and Whitney (1977) found that there was a seasonal shift from uptake of phosphate in the cold months to export of phosphate in the warm months in a New York salt marsh. This pattern is opposite to the seasonal pattern that would be expected if plant uptake were the dominant sink in the growing season.

Coupling with Adjacent Ecosystems

In the beginning of this chapter, wetlands were described as either being "open" or "closed" to hydrologic transport. For those that are open to export, the chemicals are often transformed from inorganic to organic forms and transported to downstream ecosystems. This connection can be from riparian wetlands to adjacent streams, rivers, and downstream ecosystems, from tidal salt marshes to the estuary, or from in-stream riverine marshes to the river itself. There is considerable evidence that watersheds that drain wetland regions export more organic material but retain more nutrients than do watersheds that do not have wetlands (Mulholland and Kuenzler, 1979; Brinson et al., 1981a; Elder, 1985; Johnston et al., 1990). For example, in Figure 5–17 the slope of the line for a swamp-draining watershed is much steeper than that for upland watersheds, indicating a much greater organic export for a given runoff. Nixon (1980) summarized total organic export data from

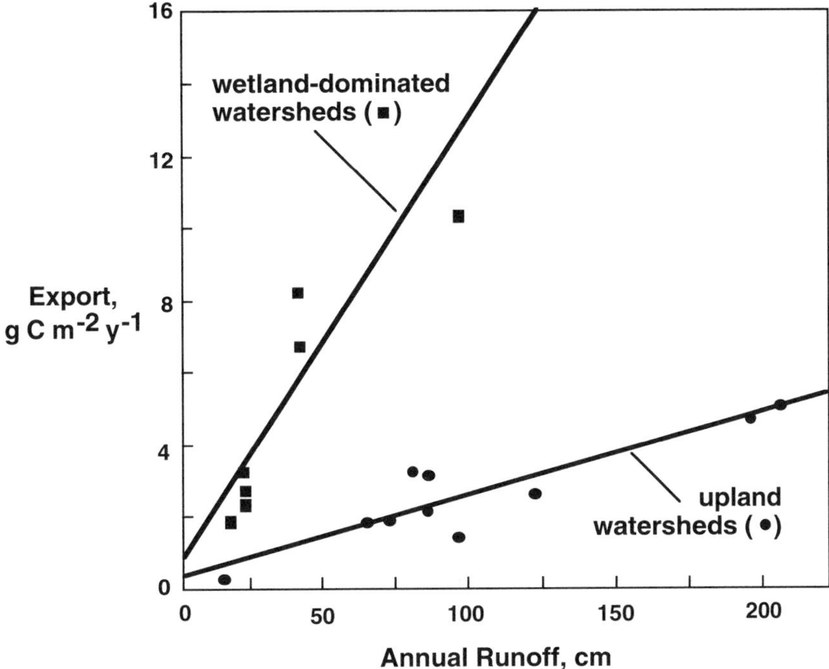

Figure 5–17. Organic carbon export from wetland-dominated watersheds compared with nonwetland watersheds. *(After Mulholland and Kunzler, 1979)*

several studies of salt marshes in the United States and suggested a general range of 100 to 200 g m^{-2} yr^{-1} carbon export from salt marshes to adjacent estuaries. This is an order of magnitude greater than the export of carbon from freshwater wetland watersheds as shown in Figure 5–17, indicating that coastal marshes, with their high productivity and frequent tidal exchange, may be more consistent exporters of organic matter than freshwater wetlands. The export indicated in Figure 5–17, however, is based on the entire watershed, some of which is not wetlands.

In an evaluation of the cumulative effects of wetlands on stream water quality in a multi-county region in Minnesota, Johnston et al. (1990) found that the extent of wetland in the watershed related, through a principal components analysis, only to lower conductivity, chloride, and lead. All are indicators of urban runoff and urban areas generally have fewer wetlands. They also found lower annual concentrations (and hence export) of inorganic suspended solids, fecal coliform, nitrates, and conductivity and flow-weighted ammonium-nitrogen and total phosphorus were related to wetland proximity. The authors suggested that "These findings do not necessarily mean that wetlands farther upstream from a sampling station are less important to water quality than proxi-

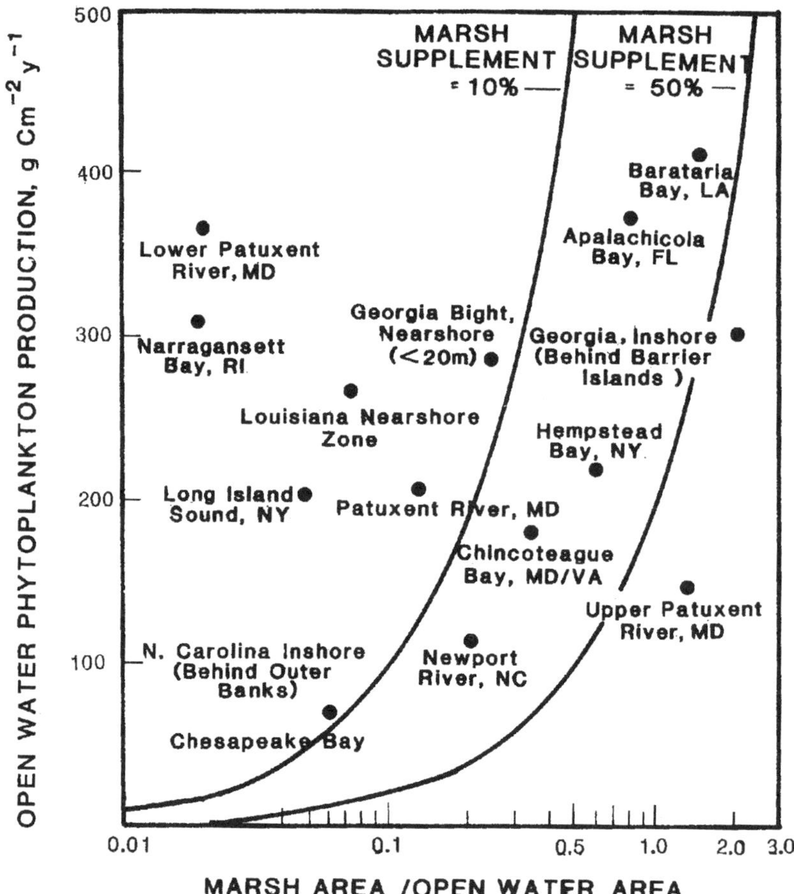

Figure 5–18. Relationship between estuarine primary productivity and relative area of surrounding salt marshes. *(After Nixon, 1980)*

mal wetlands; just that their effects on nutrients are not detectable very far downstream, or are offset by downstream inputs."

The effects of export on adjacent ecosystems have generally been difficult to quantify, although some attempts have been made to establish a cause-and-effect relationship. Figure 5–18 shows the general scatter of data that results when estuarine aquatic productivity is plotted versus the amount of adjacent salt marshes. The graph shows both highly productive estuaries with few adjacent marshes and estuaries of low productivity amid great expanses of marshes, but the general trend is for greater estuarine productivity with more coastal wetlands. Some investigators have presented evidence that certain estuarine organ-

Table 5–10. Characteristics of High-Nutrient (Eutrophic) and Low-Nutrient (Oligotrophic) Wetlands

Characteristic	*Low-Nutrient Wetland*	*High-Nutrient Wetland*
Inflows of nutrients	Mainly precipitation	Surface and ground water
Nutrient cycling	Tight closed cycles; adaptations such as carnivorous plants and nutrient translocations	Loose open cycle; few adaptations to shortages
Wetland as source or sink of nutrients	Neither	Either
Exporter of detritus	No	Usually
Net primary productivity	Low (100–500 g m^{-2} yr^{-1})	High (1,000–4,000 g m^{-2} yr^{-1})
Examples	Ombrotrophic bog; cypress dome	Floodplain wetland; many coastal marshes

isms benefit from the adjacent coastal wetland. W. E. Odum (1970) and Heald (1969; 1971) demonstrated the importance of mangrove export on commercial and sport fishing in south Florida. Turner (1977) correlated commercial shrimp harvesting and the type and area of adjacent coastal wetlands.

High- and Low-Nutrient Wetlands

There is a misconception in some ecological texts that all wetlands are high-nutrient, highly productive ecosystems. Although many types of wetlands such as tidal marshes and riparian forests have more than adequate supplies of nutrients and consequently are highly productive, there are many wetland types that have low supplies of nutrients. Table 5–10 lists some of the characteristics of high-nutrient (eutrophic) and low-nutrient (oligotrophic) wetlands. The terms *eutrophic* and *oligotrophic,* usually used to denote the trophic state of lakes and estuaries, are appropriate for wetlands despite the difference in structure. In fact, the use of the terms originated in the classification of peatlands (Hutchinson, 1973) and was later adapted for open bodies of water.

The intrasystem cycling of nutrients in wetlands depends on the availability of nutrients and the degree to which processes such as primary productivity and decomposition are controlled by the wetland environment. The availability of nitrogen and phosphorus is significantly altered by anoxic conditions as discussed earlier in this chapter, but these nutrients do not necessarily limit production or consumption. Ammonium nitrogen is high in most wetland soils, and soluble inorganic phosphorus is often in abundance. Intrasystem cycling thus is often limited by the effects that hydrologic conditions have on pathways such as primary productivity and decomposition, as discussed in Chapter 4. It is therefore possible to have wetlands with extremely rapid yet open nutrient cycling (e.g., some high-nutrient freshwater marshes) and wetlands with extremely slow nutrient cycling (e.g., low-nutrient ombrotrophic bogs).

Comparison with Terrestrial and Aquatic Systems

One of the major differences between wetlands and drier upland ecosystems is that more nutrients are tied up in organic deposits and are lost from ecosystem cycling as peat deposits or organic export. Because wetlands are more frequently open to nutrient fluxes than are upland ecosystems, they may not be as dependent on the recycling of nutrients; wetlands that are not open to these fluxes often have lower productivities and slower nutrient cycling than comparable upland ecosystems. There has also been a terrestrial bias in wetland research that has focused much of the attention of nutrient budget studies on the uptake and storage of nutrients in the vegetation rather than on extending it to include the soil and microbial processes (Johnston, 1991).

Wetlands are similar to deep aquatic systems in that most of the nutrients are often permanently tied up in sediments and peat. In most deep aquatic systems, the retention of nutrients in organic sediments is probably longer than in wetlands, although few such comparisons have been made. Wetlands, however, usually involve larger biotic storages of nutrients than do deep aquatic systems, which are primarily plankton dominated. Thus aquatic system nutrient cycling in the autotrophic zone of lakes and coastal waters is more rapid than it is in the autotrophic zone of most wetlands. Another obvious difference between wetlands and lakes or coastal waters is that most wetland plants obtain their nutrients from the sediments, whereas phytoplankton depend on nutrients dissolved in the water column. Wetland plants have often been described as "nutrient pumps" that bring nutrients from the anaerobic sediments to the aboveground strata. Phytoplankton in lakes and estuaries can be viewed as being "nutrient dumps" that take nutrients out of the aerobic zone and, through settling and death, deposit the nutrients in the anaerobic sediments. Thus the plants in these two environments can be viewed as having decidedly different functions in nutrient cycling.

Anthropogenic Effects

Human influences have caused significant changes in the chemical cycling in many wetlands. These changes have taken place as a result of land clearing and subsequent erosion, hydrological modifications such as stream channelization and dams, and pollution. Increased erosion in the uplands leads to increased deposition of sediments in the lowland wetlands such as forested swamps and coastal salt marshes. This increased accumulation of sediments can cause increased biochemical oxygen demand (BOD) and can alter the hydrologic regime of the wetlands in a relatively short time. Stream channelization and dams can lead to a change in the flooding frequency of many wetlands and thus alter the inputs of nutrients. Dams generally serve as nutrient traps, retaining materials that would otherwise nourish downstream wetlands. In some cases stream channelization has led to stream downcutting that ultimately drains wetlands.

Sources of pollution also have had localized effects on nutrient cycling in wetlands. Pollutants such as BOD, toxic materials, oils, trace organics, and metals have frequently been added purposefully or accidentally to wetlands from municipal industrial wastes and urban and rural runoff. The effects of toxic materials on wetland nutrient cycles are poorly understood, although wetlands were often used for the disposal of such materials. These effects are discussed in greater detail in Chapter 16.

Biological Adaptations to the Wetland Environment

6

The wetland environment is in many ways physiologically harsh. Major stresses are anoxia and the wide salinity and water fluctuations characteristic of an environment that is neither terrestrial nor aquatic. Adaptations to this environment have an energy cost, either because an organism's cells operate less efficiently (conformers) or because it expends energy to protect its cells from the external stress (regulators). At the cell level all organisms have similar adaptations, although primitive organisms (protists) appear to show more novelty. Adaptations of protists include the ability to respire anaerobically, to detoxify end products of anaerobic metabolism, to use reduced organic compounds in the sediment as energy sources, and to use mineral elements (N, Mn, Fe, S) in the sediment as alternative electron acceptors when oxygen is unavailable.

Higher plants and animals have a wider range of responses available to them because of the flexibility afforded by the development of organ systems and division of labor within the body, mobility, and complex life history strategies. One important adaptation in vascular plants is the development of pore space in the cortical tissues, which allows oxygen to diffuse from the aerial parts of the plant to the roots to supply root respiratory demands. Animals have developed structural or physiological adaptations to reduced oxygen availability such as specialized organs or organ systems, mechanisms to increase the oxygen gradient into the body, better means of circulation, more efficient respiratory pigment systems, and changed behavior patterns.

In general, in higher plants and animals, salt stresses have been met with specialized tissues or organs to regulate the internal salt concentration or to protect the rest of the body from the effects of salt (osmoregulators) or with increased metabolic and physiological tolerance to salt at high concentrations (osmoconformers).

Wetland environments are characterized by stresses that most organisms are ill equipped to handle. Aquatic organisms are not adapted to deal with the periodic drying that occurs in many wetlands. Terrestrial organisms are stressed by long periods of flooding. Because of the shallow water, temperature extremes on the wetland surface are greater than would ordinarily be expected in aquatic environments. But the most severe stress is probably the absence of oxygen in flooded wetland soils, which prevents organisms from respiring through normal metabolic pathways. In the absence of oxygen, the supply of nutrients available to plants is also modified, and concentrations of certain elements and organic compounds can reach toxic levels. In coastal wetlands, salt is an additional stress to which organisms must respond. It is not surprising that those plants and animals regularly found in wetlands have evolved functional mechanisms to deal with these stresses. Adaptations can be broadly classified as those that enable the organism to tolerate stress and those that enable it to regulate stress. *Tolerators* (also called *resisters*) have functional modifications that enable them to survive and often to function efficiently in the presence of stress. *Regulators* (alternatively called *avoiders*) actively avoid stress or modify it to minimize its effects. The specific mechanisms for either tolerating or regulating are many and varied. In general, bacteria show biochemical adaptations that are also characteristic of the range of cell-level adaptations found in more complex multicellular plants and animals. Vascular plants show both structural and physiological adaptations. Animals have the widest range of adaptations, not only through biochemical and structural means but also by using to advantage their mobility and their life-history patterns.

PROTISTS

Protists are one-celled organisms that have little mobility; therefore, the range of adaptations open to them is limited. Most adaptations of this group are metabolic. Because the metabolism of all living cells is similar, adaptations of this group are characteristic of cell-level adaptations in general, although some of the bacterial responses to anoxia are beyond anything found in higher organisms.

Anoxia

When an organic wetland soil is flooded, the oxygen available in the soil and in the water is rapidly depleted through metabolism by organisms that normally use oxygen as the terminal electron acceptor for oxidation of organic molecules.

The rate of diffusion of molecular oxygen through water is orders of magnitude slower than through air, and cannot supply the metabolic demand of soil organisms under most circumstances. When the demand exceeds the supply, dissolved oxygen is depleted, redox potential in the soil drops rapidly, and other ions (nitrate, manganese, iron, sulfate, and carbon dioxide) are progressively reduced (see Chap. 5 and Table 5–2).

Although some abiotic chemical reduction occurs in the soil, virtually all of these reductions are coupled to microbial respiration. When oxygen concentrations first become limiting, most cells, bacterial or otherwise, use internal organic compounds as electron acceptors. The glycolytic or fermentation pathway of sugar metabolism results in the anaerobic production of pyruvate, which is subsequently reduced to ethyl alcohol, lactic acid, or other reduced organic compounds, depending on the organism. A number of bacteria also have the ability to couple their oxidative-respiratory reactions to the reduction of inorganic ions (other than molecular oxygen) in the surrounding medium, using them as electron acceptors. Many bacterial species are facultative anaerobes, capable of switching from aerobic to anaerobic respiration. But others have become so specialized that they can grow only under anaerobic conditions and rely on specific electron acceptors other than oxygen in order to respire. *Desulfovibrio* is one such genus. It uses sulfate as its terminal electron acceptor, forming sulfides that give the marsh its characteristic rotten egg odor. Other microbially mediated chemical reactions in anoxic sediments were discussed in Chapter 5.

Most bacteria require organic energy sources. In contrast, nonphotosynthetic autotrophic bacteria are adapted to using energy of reduced inorganic compounds in wetland muds as an energy source for growth. The genus *Thiobacillus*, for example, captures the energy in the sulfide bonds formed by *Desulfovibrio* and in the process converts sulfide into elemental sulfur. Members of the genus *Nitrosomonas* oxidize ammonia to nitrite. *Siderocapsa* can capture the energy released in the oxidation of the ferrous ion to the ferric form (Nester et al., 1973). Thus adaptations to the anoxic environment of wetland soils enable bacteria not only to survive in it but sometimes to require it and even to obtain their energy from it.

Salt

In a freshwater aquatic or soil environment, the osmotic concentration of the cytoplasm in bacterial cells is higher than that of the surrounding medium. This enables the cells to develop turgor, that is, to absorb water until the turgor pressure of the cytoplasm is balanced by the resistance of their cell walls. In coastal wetlands, organisms must cope with high and variable external salt concentrations. The dangers of salts are twofold—osmotic and directly toxic. The immediate effect of an increase in salt concentration in a cell's environment is osmotic. If the osmotic potential surrounding the cell is higher than that of the cell cytoplasm, water is drawn out of the cell and the cytoplasm dehydrates. This is a rapid reaction that can occur in a matter of minutes and may be lethal to the cell. Even the "tightest" membranes leak salts passively so that in the absence of any active regulation by the cell, inorganic salts gradually diffuse into the cell. Although this absorption of inorganic ions such as Na^+ may relieve the osmotic gradient across the cell membrane, these ions at high concentrations in the cytoplasm are also toxic to most organisms, posing a second threat to survival.

Bacteria have adapted in a number of ways to cope with these twin problems of osmotic shock and toxicity. There is no evidence that cells are able to retain water against an osmotic gradient. Instead, in order to maintain its water potential, the internal osmotic concentration of salt-adapted cells is usually slightly higher than the external concentration. Indeed, the high specific gravities of halophiles—literally "salt-loving" organisms—can be accounted for only by the presence of inorganic salts at high concentrations (Ingram, 1957). Analyses of cell contents show, however, that the balance of specific ions is usually quite different from that of the external solution. For example, potassium is usually accumulated and sodium is usually diluted relative to external concentrations (Table 6–1). Active transport mechanisms that accumulate or excrete ions across cell membranes are universal features of all living cells, and although they depend on a cellular supply of biological energy, there is no evidence that large energy expenditures are needed to maintain the gradients shown in Table 6–1. They can be maintained in the cold, in cells apparently carrying out little metabolism (Ginsburg et al., 1971). This suggests that the potassium ions are loosely bound or complexed within the cytoplasm or that the cytoplasmic water has a more ordered structure than external water and such cytoplasmic ions as potassium and sodium are less free than in external solutions but still free enough to be osmotically and physiologically active (Kushner, 1978).

Although inorganic ions seem to make up the bulk of the osmotically active cell solutes in some halophilic bacteria, in others the internal salt concentration can be substantially lower than the external concentration. In these organisms the rest of the osmotic activity is supplied by organic compounds. For example, the halophilic green alga *Dunaliella virigus* contains large amounts of glycerol, the concentration varying with the external salt concentration; and certain salt-tolerant yeasts regulate internal osmotic concentration with polyols such as glycerol and arabitol. The enzymes of these organisms seem to be salt sensitive, and it has been suggested that the organic compounds act like "compatible solutes" that raise the osmotic pressure without interfering with enzymatic activity (Kushner, 1978).

The steric configuration of enzymes of salt-sensitive organisms can easily be modified by salt at high concentrations, and where enzymes are activated by specific ions, NaCl can interfere with activity. The enzymes of true halophiles, in contrast, are able to function normally in the presence of inorganic ions at high concentrations or even to require them.

Table 6–1. Internal Ionic Concentrations of Bacteria Growing with NaCl at Different Concentrations

	Ion Concentration (M)			
	External Medium		*Cell Cytoplasm*	
Bacterium	Na^+	K^+	Na^+	K^+
Vibrio costicola[a]	1.0	0.004	0.68	0.22
	0.6	0.008	0.50	0.52
	1.0	0.008	0.58	0.66
	1.6	0.008	1.09	0.59
	2.0	0.008	0.90	0.57
Paracoccus halodinitrificans	1.0	0.004	0.31	0.47
Pseudomonas 101	1.0	0.0055	0.90	0.71
	2.0	0.0055	1.15	0.89
	3.0	0.005	1.04	0.67
Marine pseudomonad B-16	0.3	—	0.12	0.37
Unidentified salt-tolerant rod	0.6	0.04	0.05	0.34
	4.4	0.04	0.62	0.58
Halobacterium cutirubrum[b]	3.33	0.05	0.80	5.32

[a]The difference between K^+ concentration in different studies may be related to whether the cultures were in a stationary phase or were growing exponentially.

[b]Several workers have reported similar or higher results for K^+ in *H. cutirubrum* or *H. halobium*.

Source: After Kushner, 1978

VASCULAR PLANTS

Emergent plants are structurally much more complex than protists. This has enabled them to develop a wider range of adaptations than bacteria to hypoxia and to salt, but at the same time on an evolutionary scale, it has probably eliminated some adaptations found in unicellular organisms such as the ability to use reduced inorganic compounds in the sediment as a source of energy.

Vascular emergent plants are sessile, but only their roots are in an anoxic or salty environment. Typically if the roots of a flood-sensitive upland plant are inundated, the oxygen supply to the roots rapidly decreases. This shuts down the aerobic metabolism of the roots, impairs the energy status of the cells, and reduces nearly all metabolically mediated activities such as cell extension and division and nutrient absorption. Even when cell metabolism shifts to anaerobic glycolysis, ATP production is reduced. Toxic metabolic endproducts of fermentation may accumulate, causing cytoplasmic acidosis and eventually death (Roberts, 1988). Hypoxia is soon followed by pathological changes in the mitochondrial structure, including swelling, the reduction of cristae number, and the development of a transparent matrix. The complete destruction of mitochondria and other organelles occurs within 24 hours (Vartapetian, 1988). Hypoxia changes the chemical environment of the root, increasing the availability of reduced minerals such as iron, manganese, and sulfur. These may accumulate to

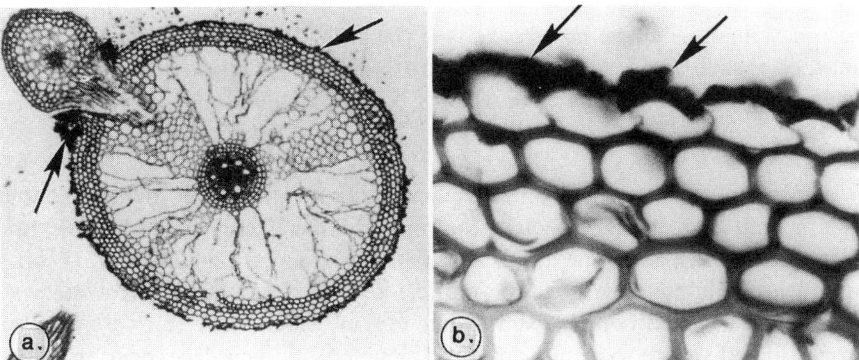

Figure 6–1. Light photomicrographs of *Spartina alterniflora* roots: *a.* cross section of a streamside root; arrows indicate the presence of red ferric deposits on the root epidermis. x192; *b.* streamside root cross section showing the presence of similar materials on the external walls of the epidermal cells. x1,143. Note the extensive pore space (aerenchyma) in the roots. *(From Mendelssohn and Postek, 1982; copyright © 1982 by the Botanical Society of America, reprinted with permission)*

toxic levels in the root (Ernst, 1990). In addition, if the sensitive plants are in a marine environment, they suffer osmotic shock and salt toxicity.

In contrast to flood-sensitive plants, flood-tolerant species (hydrophytes) possess a range of adaptations that enable them either to tolerate stresses or to avoid them.

Anoxia

Structural Adaptations

Aerenchyma. Virtually all wetland plants have elaborate structural mechanisms to avoid root anoxia. The primary plant strategy in response to flooding is the development of air spaces (*aerenchyma*) in roots and stems, which allow the diffusion of oxygen from the aerial portions of the plant into the roots (Fig. 6–1). In plants with well-developed aerenchyma, the root cells no longer depend on the diffusion of oxygen from the surrounding soil, the main source of root oxygen to terrestrial plants. Unlike the plant porosity of normal mesophytes, which is usually a low 2–7 percent of volume, up to 60 percent of the plant body of wetland species consists of pore space. Air spaces are formed by either cell separation during the maturation of the organs or by cell breakdown. They result in a honeycomb type of structure. Air spaces are not necessarily continuous throughout the stem and roots. The thin lateral cellular partitions within the aerenchyma, however, are not likely to impede internal gas diffusion significantly (Armstrong, 1975). Flood-tolerant wetland species, respond to flooding by increased root aerenchyma development (Burdick and Mendelssohn, 1990; Pezeshki et al., 1991). The same kind of cell lysis and air space development has been described in submerged stem tissue (Jackson, 1990). Roots of flood-tolerant species such as rice form aerenchyma even in aerated apical cells (Webb and Jackson, 1986), and it is not clear whether this "preadaptation" is ethylene mediated (Jackson, 1985).

Hormones. Recent research has implicated hormonal changes in the initiation of structural adaptations, especially the concentration of ethylene in hypoxic tissues. When ethylene production is stimulated by flooded conditions, it, in turn, stimulates cellulase activity in the cortical cells of a number of plant species, with the subsequent collapse and disintegration of cell walls (Kawase, 1981). Ironically, ethylene production from methionine is oxygen dependent; the conversion of its precursor, 1-aminocyclopropane-1-caraboxylic acid (ACC), to ethylene requires oxygen (Adams and Yang, 1977, 1979; Lurssen et al., 1979). It is not entirely clear, therefore, how ethylene accumulation in anaerobic environments occurs. One possibility is that partial oxygen depletion stimulates ethylene production (Jackson, 1982; Jackson et al., 1984), perhaps enhanced by carbon dioxide accumulation (Raskin and Kende, 1984). Another is that the slow diffusion rate of ethylene in water retards loss from the plant, resulting in its accumulation in water-logged tissue (Jackson, 1985). Not all the effect of ethylene occurs in the hypoxic tissues; there is strong evidence that ethylene precursors produced in hypoxic tissues diffuse to aerobic tissues, are converted into ethylene, and modify plant response in those aerobic zones (Fig. 6–2; Bradford and Yang, 1980; Jackson, 1988).

In addition to aerenchyma development, ethylene has been reported to stimulate the formation of *adventitious roots,* which develop in both flood-tolerant (e.g., willow) and flood-intolerant (e.g., tomato) plants just above the anaerobic zone when these plants are flooded (Wample and Reid, 1979; Jackson, 1985). Another response stimulated by submergence is rapid stem elongation in such aquatic and semiaquatic plants as the floating-heart (*Nymphoides peltata*) and rice (*Oryza sativa*) (Malone and Ridge, 1983; Raskin and Kende, 1983). All these morphological responses to flooding are mechanisms that increase the oxygen supply to the plant either by growth into aerobic environments or by enabling oxygen to penetrate more freely into the hypoxic zone.

Effectiveness of Aerenchyma. Because flooding interrupts the flow of oxygen from soil to root, a key question is the efficacy of adaptations to correct that interruption. The sufficiency of the oxygen supply to the roots depends on root permeability (i.e., how leaky the root is to oxygen, which can move out into the surrounding soil), the root respiration rate, the length of the diffusion pathway from the upper parts of the plant, and root porosity (the pore space volume). Models of gas diffusion show that, under most circumstances, root porosity is the overriding factor governing internal root oxygen concentration (Fig. 6–3). The effectiveness of aerenchyma in supplying oxygen to the roots has been

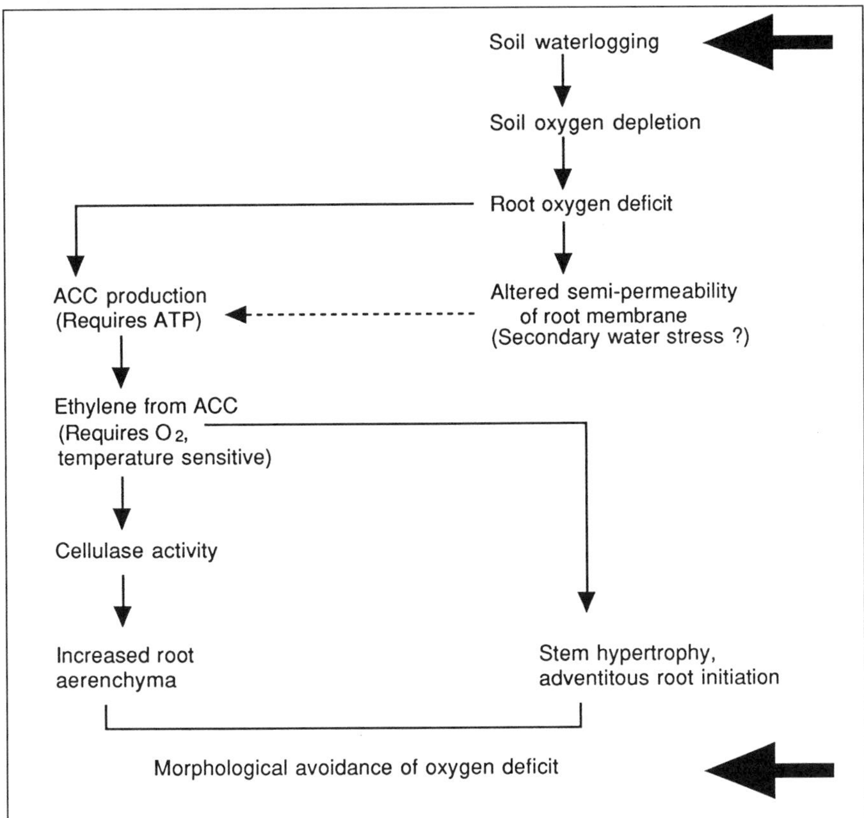

Figure 6–2. Schematic of the development of morphological acclimation characteristics in response to flooding stress. *(From McLeod et al., 1988; copyright © 1988 by Donal D. Hook, reprinted with permission)*

demonstrated in a number of plant species. For example, the root respiration of flood-tolerant *Senecio aquaticus* was only 50 percent inhibited by anaerobiosis, whereas that of *S. Jacobaea*, a flood-sensitive species, was almost completely inhibited. Greater root porosity in the tolerant species was the primary factor that contributed to the difference (Lambers et al., 1978). The most extensively studied flood-tolerant plant is rice. Rice plants grown under continuous flooding develop greater root porosity than unflooded plants, and this maintains the oxygen concentration in the root tissues. When deprived of oxygen, rice root mitochondria degraded in the same way as did flood-sensitive pumpkin plants, showing that the basis of resistance in flooded plants was by the avoidance of root anoxia, not by physiological changes in cell metabolism (Levitt, 1980). The water lily (*Nuphar luteum)* is a particularly interesting example of an adaptation of this type. Air moves into the internal gas spaces of young leaves on the water

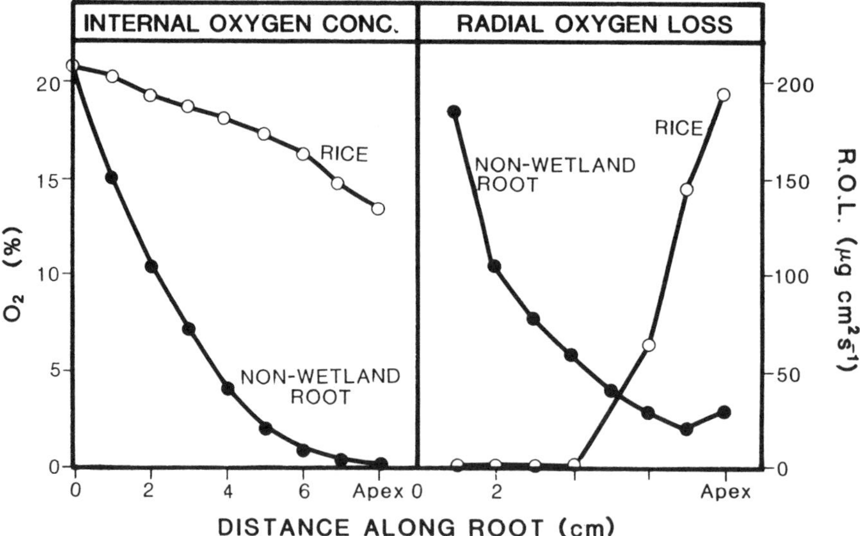

Figure 6–3. Electrical analog model predictions of internal oxygen balance and radial oxygen loss (R.O.L.) along a nonwetland adapted root and a wetland-adapted rice root. Root length = 8 cm; root radius = 0.05 cm; soil oxygen consumption = 4 x 10^{-5} cm^3 (O_2) cm^{-3} soil sec^{-1}. R.O.L. is the rate of diffusion of oxygen out of the root into the substrate. *(From Armstrong, 1978; copyright © 1978 by Technomics Publication Co., reprinted with permission)*

surface and is forced down through the aerenchyma of the stem into the roots by a slight pressure generated by the heating of the leaves. The older leaves lose their capacity to support pressure gradients, and so the return flow of gas from the roots is through the older leaves, which are rich in carbon dioxide from root respiration (Dacey, 1980).

Other Structural Adaptations. The species of woody trees that are successful (as opposed to tolerant) in the wetland habitat are few and include the mangroves (*Rhizophora* and *Avicennia*), cypress (*Taxodium*), tupelo (*Nyssa*), willow (*Salix*), and a few others. As with herbaceous species, an adequate ventilating system seems to be essential for their growth. Many trees and herbaceous species produce adventitious roots above the anoxic zone. These roots are able to function normally in an aerobic environment. The red mangrove (*Rhizophora* spp.) grows on arched prop roots in tropical and subtropical tidal swamps around, the world (see Chap. 10). These prop roots have numerous small pores, termed *lenticels* above the tide level, which terminate in long, spongy, air-filled, submerged roots. The oxygen concentration in these roots, embedded in anoxic mud, may remain as high as 15–18 percent continuously, but if the lenticels are blocked, this concentration can fall to 2 percent or less in two days (Scholander et al., 1955). Similarly, the black mangrove (*Avicennia* spp.) produces thousands of *pneumatophores* (air roots) about 20–30 cm high by 1 cm in diameter, spongy, and studded with lenticels. They protrude out of the mud from main roots and are exposed during low tides. The oxygen concentration of the submerged main roots has a tidal pulse, rising during low tide and falling during submergence, reflecting the cycle of emergence of the air roots (Scholander et al., 1955). The "knees" of bald cypress (*Taxodium distichum*) have also been thought to improve gas exchange to the root system (see Chap. 13).

Rhizosphere Oxygenation. **When hypoxia is moderate, the magnitude of** oxygen diffusion through many wetland plants into the roots is apparently large enough not only to supply the roots but also to diffuse out, oxidize the adjacent anoxic soil, and produce an oxidized rhizosphere (see also Chap. 7; Teal and Kanwisher, 1966; Howes et al., 1981; Laanbroek, 1990). Through scanning electron microscopy coupled with X-ray microanalysis, Mendelssohn and Postek (1982) showed that the brown deposits found around the roots of *Spartina alterniflora* are composed of iron and manganese deposits formed when root oxygen comes in contact with reduced soil ferrous ions (Fig. 6–1). It has been suggested (Armstrong, 1975) that oxygen diffusion from the roots is an important mechanism that moderates the toxic effects of soluble reduced ions such as manganese in anoxic soil. These ions tend to be reoxidized and precipitated in the rhizosphere, which effectively detoxifies them. In a similar vein, McKee et al. (1988) determined that soil redox potentials were higher and that pore water sulfide concentrations were three to five times lower in the presence of the aerial prop roots of the red mangrove or pneumatophores of the black mangrove than in nearby bare mud soils. This suggests the diffusion of oxygen from the mangrove roots into the soil. An interesting possibility is that the root systems of these flood-tolerant plants may modify sediment anoxia enough to allow the survival of nearby nontolerant plants (Ernst, 1990).

Water Uptake

Plants intolerent to anaerobic environments typically show decreased water uptake despite the abundance of water, probably as a response to an overall reduction of root metabolism. Decreased water uptake results in symptoms similar to those seen under drought conditions: closing of stomata, decreased carbon dioxide uptake, decreased transpiration, and wilting (Mendelssohn and Burdick, 1988). Stomatal closure is mediated by the hormone abscisic acid (Dorffling et al., 1980), the concentrations of which increase in leaf tissues as a result of root flooding. The increased concentration probably results from the reduced export of leaf abscisic acid to the flooded roots (Jackson, 1990). The adaptive advantage of these responses is probably the same as for drought-stricken plants—to minimize water loss and accompanying damage to the cytoplasm. The accompanying depression of the photosynthetic machinery is generally seen as an unavoidable corollary.

Nutrient Absorption

One of the earliest processes affected by hypoxia is the absorption by plants of nutrients from the substrate. The availability of many nutrients in the soil is modified by an anoxic environment. In general, flood-intolerant plant species lose the ability to control nutrient absorption because of the tissue energy deficit brought on by hypoxia. In contrast, nutrient uptake in most flood-tolerant species appears unchanged perhaps because the plant can maintain near-normal metabolism (Mendelssohn and Burdick, 1988). The major nutrients that have been studied are those most affected by soil anaerobiosis: nitrogen, phosphorus, iron, manganese and sulfur.

Nitrogen. In reduced soils nitrates are replaced by ammonium, although plants absorb the oxidized form (nitrate). Despite this change in the sediment supply of available nitrogen, most wetland plants that have been studied are able to maintain normal rates of nitrogen uptake. This ability is related to three factors: first, the possibility that ammonium is oxidized to nitrate in the rhizosphere through radial oxygen loss from the roots; second, the possibility that some wetland species are able to absorb the ammonium directly; and third, the ability of the plant to maintain metabolic activity sufficient to absorb nutrients, which is known to be an energy-requiring process (Morris and Dacey, 1984; Schat, 1984).

Phosphorus. The availability of phosphorus generally increases in waterlogged soils. It is, however, precipitated by iron (which also increases in availability). Generally, studies have shown reduced uptake by intolerant species probably because of the energy requirement, but no effect or enhanced uptake by flood-tolerant species.

Iron and Manganese. Because extremely small concentrations of iron and manganese are required by plants, they can reach toxic levels in many environments. Both elements are reduced and become much more available in flooded soils, where, because of their high concentrations, they escape every metabolic control and become concentrated in plant tissues (Ernst, 1990). Wetland plants tolerate these elements by means of several adaptations. First, the oxidized rhizosphere can precipitate and reduce the concentration that reaches the roots. Second, much of the minerals taken up into the tissues can be sequestered in cell vacuoles, in the shoot vascular tissue, or in scenescing tissues where they do not influence the metabolism of healthy cytoplasm. Third, many wetland plants appear to have a much higher than average metabolic tolerance for these ions (Ernst, 1990).

Sulfur. Sulfur as sulfide is toxic to plant tissues. The element is reduced to sulfide in anaerobic soils and accumulates to toxic concentrations in salt marshes (Goodman and Williams, 1961; Koch and Mendelssohn, 1989). Although sulfate uptake is metabolically controlled, sulfide can enter the plant without control and is found in elevated concentrations in many flood-adapted species under highly reduced conditions. In experiments with *Spartina alterniflora*, a salt marsh species, and *Panicum hemitomon*, a freshwater marsh species, Koch et al. (1990) reported that the activity of alcohol dehydrogenase (ADH), the enzyme that catalyzes the terminal step in alcohol fermentation, was significantly inhibited by hydrogen sulfide and that this inhibition may help to explain the physiological mechanism of sulfide phytotoxicity often seen in salt marshes. This suppression of an anaerobic pathway by high concentrations of sulfides is in addition to the known suppression of aerobic energy production that occurs in wetland plants (Allam and Hollis, 1972; Havill et al., 1985; Pearson and Havill, 1988). Sulfur tolerance in wetland plants varies widely probably because of the variety of detoxification mechanisms available. These include the oxidation of sulfide to sulfate; the accumulation of sulfate in the vacuole; the conversion to gaseous hydrogen sulfide, carbon disulfide, and dimethylsulfide and their subsequent diffusive loss; and the metabolic tolerance to elevated sulfide concentrations (Ernst, 1990).

Respiration

The presence of root air space is not a sufficient condition for the effective oxygen transport to the root. If oxygen leaks from the cortex into the surrounding sediments before it reaches the root tips where metabolic activity is greatest, it is of little value to the plant. Smits et al. (1990b) reported different patterns of oxygen leakage among several plant species; some showed oxygen leakage along the whole length of roots, and others (water lily, for example) leaked oxygen only from the 1 cm apex.

Mendelssohn and Postek (1982), found that the oxidized rhizosphere was only 1/50 as well developed around the roots of inland marsh plants as streamside ones, indicating that root oxygen availability was limited in inland sites. This limitation is probably related to the more anoxic conditions inland, placing greater demand on the ability of the root aerenchyma to deliver oxygen from aboveground.

Under conditions of oxygen deprivation, plant tissues respire anaerobically, as described for bacterial cells. In most plants, pyruvate, the end product of glycolysis, is decarboxylated to acetaldehyde, which is reduced to ethanol (Fig. 6–4). Both of these compounds are potentially toxic to root tissues. Flood-tolerant plants often have adaptations to minimize this toxicity. For example, under anaerobic conditions, *S. alterniflora* roots show much increased activity of alcohol dehydrogenase, the inducible enzyme that catalyzes the reduction of acetaldehyde to ethanol. The increase in the enzyme indicates a switch to anaerobic respiration, and it explains why acetaldehyde does not accumulate in the root tissue. Ethanol does not accumulate either, although its production is appar-

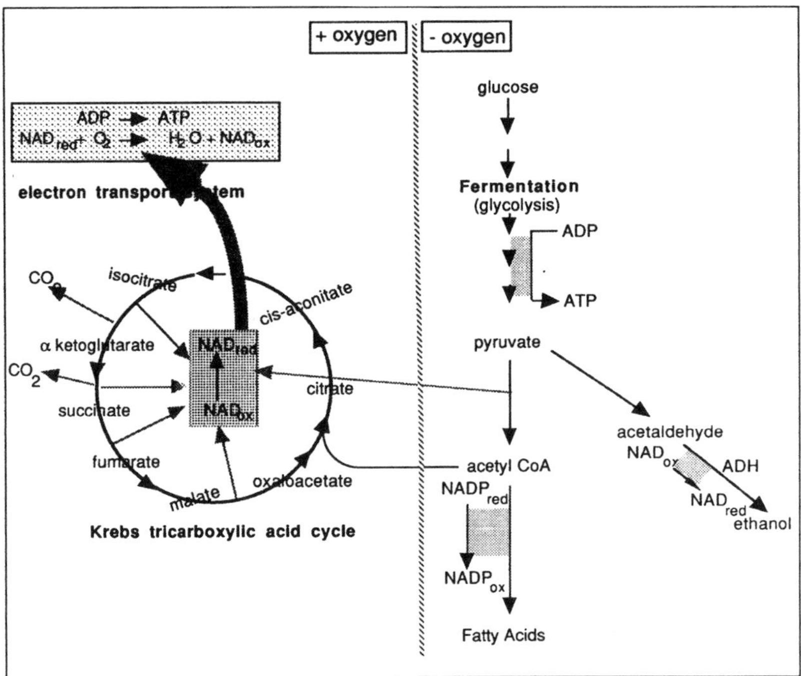

Figure 6–4. Schematic of metabolic respiration pathway in flood tolerant plants. ADH = alcohol dehydrogenase, NAD = nicotinamide adenine dinucleotide, and NADP = NAD phosphate; subscripts refer to oxidized (ox) and reduced (red) forms.

ently stimulated (Mendelssohn et al., 1982). It diffuses from rice roots during anaerobiosis (Bertani et al., 1980), thus preventing a toxic buildup, and it is probable that this occurs in *Spartina* also. Smits et al. (1990a) reported a positive correlation between the number of alcohol dehydrogenase isozymes and ethanol production in the roots of a number of aquatic macrophytes. They proposed that high enzyme polymorphism confers a selective advantage in hypoxic sediments. Similarly, in a flood-tolerant *Echinochloa* species, an extra ADH isozyme not present in aerobic soils was induced under hypoxia (Fox et al., 1988).

Another metabolic strategy reduces the production of alcohol by shifting the metabolism to accumulate nontoxic organic acids instead (Fig. 6–4). MacMannon and Crawford (1971) suggested that malic acid accumulation may be a characteristic feature of wetland species. In flood-intolerant species, excess malate is converted to pyruvate and then to acetaldehyde and ethanol. In contrast, they suggested that flood-tolerant species lack the malic enzyme and consequently malate accumulates. This idea was challenged by A. M. Smith and ap Rees (1979), who maintain that malate accumulation is not a mechanism by which wetland plants adapt to flooding. The complex metabolic response is still not entirely clear. Mendelssohn et al. (1981) reported that two different kinds of adaptation occur in *S. alterniflora*. Vigorous plants in the well-aerated streamside zone had low levels of ADH activity combined with high levels of adenosine triphosphate (ATP) and high adenylate energy charge (AEC) ratios, indicative of active aerobic metabolism. Farther inland the soil redox potential fell, root ADH remained low, malate accumulated, and ATP and the AEC ratio dropped. Still farther inland, when the soil redox potential fell below -200 mv, root cells suddenly showed high ADH activity accompanied by an increase of ATP and the AEC ratio to levels as high as those for streamside plants. The behavior of streamside plants is characteristic of plants in aerobic environments. The inland plant response can be interpreted as a shift to alcoholic fermentation. The accumulation of malate in the intermediate zone cannot easily be interpreted in part because malate is an intermediate in several metabolic pathways. The fermentation pathway provides a steady source of energy to the roots, but it is an inefficient pathway of metabolism. Under these anaerobic conditions, plants grow poorly probably because the carbohydrate energy source becomes exhausted and the flow of photosynthates from the leaves is interrupted (Schumacher and Smucker, 1985).

The metabolic problem encountered by plants deprived of oxygen is the loss of the electron acceptor that enables normal energy metabolism through ATP formation and use. The metabolic bottleneck in this process is often the electron-accepting coenzyme nicotinamide adenine dinucleotide (NAD), which is reduced in the oxidative steps of carbohydrate metabolism, and then reoxidized in the mitochondria by molecular oxygen to yield the biological energy currency ATP. In the absence of oxygen, reduced NAD accumulates and "jams" the metabolic system, blocking ATP generation (Fig. 6–4). As described above, in fermentation, acetaldehyde takes the place of oxygen, reoxidizing reduced NAD. NAD phosphate (NADP) serves a similar function in lipid synthesis. Reduced NADP is oxidized during the synthesis of fatty acids from acetyl coenzyme A and can be recycled in oxidative pentose phosphate metabolism to provide metabolic intermediates and oxidized NADP for more fatty acid synthesis. This cycle does not require oxygen. In some plant species it has been observed that fatty acids accumulate under hypoxic conditions and their lipid syntheses may even continue (Fox et al., 1988). This has obvious adaptive significance, especially for seed germination and growth under water or in other anaerobic environments.

Whole Plant Strategies

We have discussed a number of structural and biochemical mechanisms by which plants avoid or tolerate hypoxia. In addition, many species have evolved avoidance or escape strategies by life-history adaptations. The most common of these strategies are the timing of seed production in the nonflood season either by delayed or accelerated flowering (Blom et al., 1990); the production of bouyant seeds that float until they lodge on high, unflooded ground; and the germination of seeds while the fruit is still attached to the tree (vivipary), as in the red mangrove (see Chap. 10). In many riparian wetlands, flooding occurs primarily during the winter and early spring, when trees are dormant and much less susceptible to anoxia than they are during the active growing season.

Salt

At the cell level plants behave toward salt in much the same way as bacteria do, and their adaptive strategies are identical. Vascular plants, however, have also developed adaptations that take advantage of their structural complexity. These include barriers to prevent or control the entry of salts and organs specialized to excrete salts. In both cases, specialized cells bear most of the burden of the adaptation, allowing the remaining cells to function in a less hostile environment. It was generally thought that the air space of the root cortex is freely accessible from the rhizosphere; consequently, the endodermis formed the first real barrier to the upward movement of solutes from the soil. This was confirmed by the observation that the roots of plants in high salt environments often have much higher salt concentrations (and must also have higher salt tolerance) than the leaves do. X-ray microanalysis of the salt-resistant *Puccinellia peisonis* showed a decreasing concentration of sodium and an increasing concentration of potassium in the roots from the outer cortex through the endodermis to the stele (Stelzer and Lauchli, 1978). Both the inner cells of the cortex and the passage cells of the stele seemed to be barriers to sodium transport, whereas potassium moved through fairly freely. The selectivity for potassium (or the exclusion of sodium) was seen to be common in bacteria also (Table 6–1). In contrast to this picture, Moon et al. (1986) reported that in the mangrove *Avicennia marina,* the primary barrier to the passive salt incursion into the plant apoplast is at the root periderm and exodermis so that the root cortex is protected from high salt concentrations. Uptake through the symplasm is restricted mainly to the terminal third- and fourth-order roots, and the large lateral roots serve primarily as the means of vascular transport and support.

As a result of the filtering out of salt at the root apoplast, the sap of many halophytes is almost pure water. The mangroves *Rhizophora, Laguncularia,* and *Sonneratia* exclude salts almost completely. Their sap concentrations are only about 1–1.5 mg NaCl per ml (compared with about 35 mg/ml in seawater). *Avicennia* has a higher sap concentration of about 4–8 mg/ml (Scholander et al., 1966), or about 10 percent of external concentration (Moon et al., 1986). When fluid is released from the leaves of these species by pressure, it is almost pure distilled water. Thus both the root and the leaf cell membrane act like ultrafilters.

The leaf cytoplasm must have an osmotic potential higher than the osmotic potential of the sap in order to retain water, and in the mangroves, 50–70 percent of this osmotic potential is obtained from sodium and chloride ions. Most of the remainder is presumably organic. In *Batis*, a succulent halophyte, NaCl alone makes up 90 percent of the total osmotic concentration (Scholander et al., 1966).

Some plants that do not exclude salts at the root or are "leaky" to salt have secretory organs. The leaves of many salt marsh grasses, for example, characteristically are covered with crystalline salt particles excreted through specialized salt glands embedded in the leaf. These glands do not function passively; instead, they selectively remove certain ions from the vascular tissues of the leaf. In *Spartina*, for example, the excretion is enriched in sodium, relative to potassium. These two mechanisms, salt exclusion and salt secretion, protect the shoot and leaf cells of the plant from high concentrations of salt and maintain an optimum ionic balance between mono- and divalent cations and between sodium and potassium. At the same time the osmotic concentration of the cells of salt-tolerant plants must be maintained at a level high enough to allow the absorption of water from the root medium. Where the inorganic salt concentration is kept low, organic compounds make up the rest of the osmoticum in the cells.

Photosynthesis

One adaptation that many wetland plant species share with plants in other stressed environments, especially in drought-stressed environments, is the C_4 biochemical pathway of photosynthesis (formally called the Hatch-Slack-Kortschak pathway, after the discoverers). It gets its identity from the fact that the first product of CO_2 incorporation is a four-carbon compound, oxaloacetic acid. This pathway is outlined in simplified form in Figure 6–5, and C_4 plants are compared with C_3 plants in Table 6–2. (The first compound resulting from CO_2 incorporation in C_3 plants is a 3-carbon compound phosphoglyceric acid.) C_3 plants are much more common than C_4 plants. Although water is a universal feature of wetlands, plants in saline wetland habitats have much the same problem of water availability as plants in arid areas do. In both cases the water potential of the substrate is very low. In arid zones it is low because the soil is dry; in saline wetlands it is low because of the salt content. In wetlands, the water uptake is accompanied by a mass flow of dissolved salts to the roots; their absorption must be regulated, at an energy cost, by the plant. Therefore, in both environments, mechanisms that reduce water loss (transpiration) provide an adaptive advantage. For a plant to take up carbon dioxide for photosynthesis, its stomata must be open, and if they are open during the bright hours of the day, water loss is excessive. Plants that fix carbon by the C_4 pathway can use carbon dioxide more effectively than other plants. They are able to withdraw carbon dioxide from the atmosphere until its concentration falls below 20 ppm (as compared with 30–80 ppm for C_3 plants). This is achieved by using phosphoenolpyruvate (PEP), which has a high affinity for CO_2, as the carbon dioxide acceptor instead of the ribulose diphosphate acceptor of the conventional pathway. In addition, the

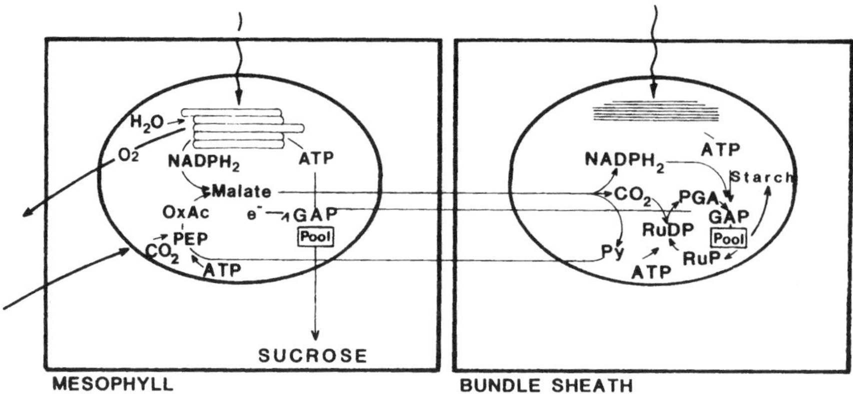

Figure 6–5. A much simplified diagram of CO_2 fixation via the Hatch-Slack-Kortschak pathway in C_4 plants. PEP = phosphoenol pyruvate; OxAc = oxaloacetate; PGA = 3-phosphoglyceric acid; GAP = 3-phosphoglyceraldehyde; RuP = ribulose-5-phosphate; Py = pyruvate. PGA is also produced by carboxylation of C_2 compounds that appear in the pool; the regeneration of PEP from PGA, in which water is given off, is not shown. *(From Larcher, 1991; copyright © 1991 by Springer-Verlag, reproduced with permission)*

malate formed by this carboxylation is nontoxic and can be stored in the cell until it can be decarboxylated and the released carbon dioxide is fixed through the normal C_3 pathway (Fig. 6–5). The production of PEP and the C_4 metabolism provide a possible pathway for recycling carbon dioxide from cell respiration. In addition, plants using the C_4 pathway of photosynthesis have low photorespiration rates and the ability to use efficiently even the most intense sunlight. These differences make C_4 plants more efficient than most C_3 plants both in their rates of carbon fixation and in the amount of water used per unit of carbon fixed (Table 6–2; Fitter and Hay, 1987). Finally, Armstrong (1975) suggested that water-conservation mechanisms in wetland plants have the additional function of reducing the rate at which soil toxins are drawn toward the root. This increases the probability of detoxifying them as they move through the oxidized rhizosphere.

Among the common wetland angiosperms that have been shown to photosynthesize through the C_4 pathway are *S. alterniflora, S. townsendii, S. foliosa, Cyperus rotundus, Echinochloa crus-galli, Panicum dichotomiflorum, P. virgatum, Paspalum distichum, Phragmites communis*, and *Sporobolus cryptandrus*. It is obvious that this adaptation is fairly common in the wetland environment. Table 6–3 compares some of the photosynthetic attributes of two salt marsh species, *S. alterniflora*, a C_4 plant, and *Juncus roemerianus*, a C_3 plant. The tall form of *S. alterniflora* grows along creek banks; the short form grows farther inland in poorly drained soils. Compared with *J. roemerianus*, *S. alterniflora* has a higher rate of photosynthesis, a lower CO_2 compensation concentration (the

Table 6–2. Comparison of Aspects of Photosynthesis of Herbaceous C_3 and C_4 Plants

Photosynthetic Characteristics	C_3	C_4
Initial CO_2 fixation enzyme	PEP carboxylase[a]	RuBP carboxylase[a]
Location of initial carboxylation	Mesophyll	bundle sheath
Theoretical energy requirement for net CO_2 fixation, CO_2:ATP:NADPH	1:3:2	1:5:2
CO_2 compensation concentration, ppm CO_2	30–70	0–10
Transpiration ratio, g H_2O transpired/g dry wt	450–950	250–350
Optimum day temperature for net CO_2 fixation, °C	15–25	30–47
Response of net photosynthesis to increasing light intensity	saturation at 1/4 to 1/2 full sunlight	Proportional to or saturation at full sunlight
Maximum rate of net photosynthesis, mg CO_2/dm^2 leaf surface/hr	15–40	40–80
Maximum growth rate, g m^{-2} day^{-1}	19.5	30.3
Dry matter production, g m^{-2} day^{-1}	2,200	3,860

[a]RuBP = Ribulose-bis-phosphate; PEP = Phosphoenolpyruvate
Source: Based on data from Black, 1973; Fitter and Hay, 1987

CO_2 concentration in the leaf cellular spaces when photosynthesis is reduced to zero), a lower respiration rate in the light, and a higher temperature optimum. Water use efficiency (Table 6–3) is an important index of the ability of plants to photosynthesize with a minimum water loss, especially in arid or saline environments, where available water is in scarce supply. *S. alterniflora* is almost twice as efficient in this respect as *J. roemerianus*.

One characteristic in which *J. roemerianus* does not fit the typical C_3 plant pattern is its photosynthetic response to light intensity (Fig. 6–6). Both species increase their photosynthetic rates as light intensity increases to full sunlight, although C_3 plants typically reach saturation at no more than one-half full intensity. The short form of *S. alterniflora* behaves in a way intermediate to the tall form and the C_3 plant. It is not known whether this reflects a switch from C_4 to C_3 metabolism under conditions of oxygen stress. Although the C_4 adaptations suggest some selective advantages for plants in wetlands, such advantages would not be enough to displace well-adapted C_3 species such as *J. roemerianus*.

ANIMALS

Animals are exposed to the same range of environmental conditions in wetlands as protists and plants, but because of their complexity, their adaptations are more

Table 6–3. Comparison of Photosynthetic Characteristics of Two Wetland Plant Species: *Spartina alterniflora*, a C_4 Plant; and *Juncus roemerianus*, a C_3 Plant

	Spartina alterniflora		
Photosynthetic Characteristics	*Tall form*	*Short form*	Juncus roemerianus
Maximum seasonal net photosynthetic rate, mg $CO_2/cm^2 \cdot sec$ (month)	90 (Sept.)	65 (July)	60 (March)
Photosynthetic light response (Fig. 6–6)	Nonsaturating	Saturating	Nonsaturating
CO_2 compensation concentration, mc CO_2/l	12	84	84
Photorespiration at 21% O_2, mg $CO_2/cm^2 \cdot sec$ (% of photosynthesis)	6.7 (11)	18.2 (40)	9.1 (54)
Temperature optimum (summer), °C	30–35	30–35	25
Water use efficiency, mg $CO_2/g\ H_2O$	15	12–15	8–9

Source: Based on data from Giurgevich and Dunn, 1978, 1979

varied. The adaptation may be as varied as a biochemical response at the cell level, a physiological response of the whole organism such as a modification of the circulatory system, or a behavioral response such as modified feeding habits. Furthermore, although it is convenient to discuss the specific response mechanisms to individual kinds of environmental stresses, in reality an organism must respond simultaneously to a complex of environmental factors, and it is the success of this integrated response that determines its fate. For example, one possible response to stress is avoidance by moving out of the stress zone. But in wetlands that might mean moving from an anoxic zone within the soil to the surface, where temperature extremes and dessication pose a different set of physiological problems. Thus the organism's successful adaptations are often compromises that enable it to live with several competing environmental demands.

Anoxia

At the cell level the metabolic responses of animals to anoxia are similar to those of bacteria. Evolutionary development, however, has put a premium on aerobic metabolism, and so the higher animals (in an evolutionary sense) tend to

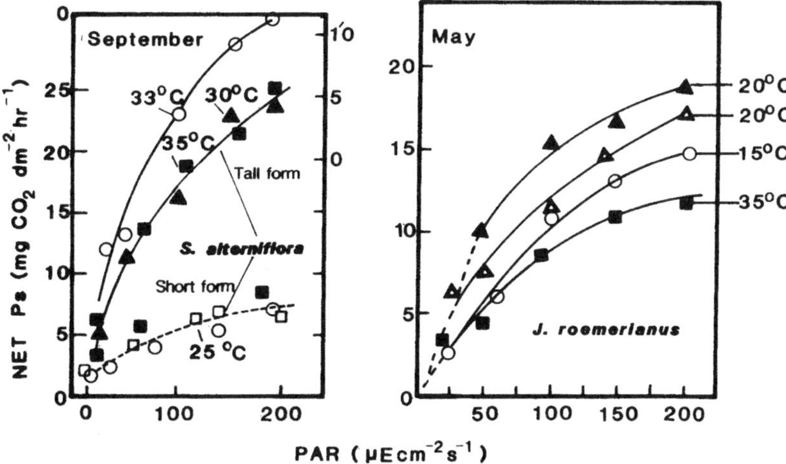

Figure 6–6. Light-photosynthesis curves for *Spartina alterniflora* and *Juncus roemerianus* at different ambient temperatures. PAR = photosynthetically active radiation. *(After Giurgevich and Dunn, 1978, 1979)*

have less ability to adapt to anaerobic conditions than primitive ones. The vertebrates and the more complex invertebrates are limited in anaerobic respiration to glycolysis or to the pentose monophosphate pathway whose dominant end product is lactate. In all higher animals, the internal cell environment is closely regulated. As a result, most adaptations are organism-level ones to maintain the internal environment. Vernberg and Vernberg (1972) list six major kinds of adaptations that marine organisms have evolved to control gaseous exchange:

1. development or modification of specialized regions of the body for gaseous exchange; for example, gills on fish and crustacea, parapodia on polychaetes;
2. mechanisms to improve the oxygen gradient across a diffusible membrane; for example, by moving to oxygen-rich environments or by moving water across the gills by ciliary action;
3. internal structural changes such as increased vascularization, a better circulation system, or a stronger pump (the heart);
4. modification of respiratory pigments to improve oxygen-carrying capacity;
5. behavioral patterns such as decreased locomotor activity or closing a shell during low oxygen stress;
6. physiological adaptations, including shifts in metabolic pathways and heart pumping rates

Examples of different kinds of adaptations are numerous. We give a few here to illustrate their diversity. Crabs inhabit a wide range of marine habitats. The number and the total volume of gills of crabs living on land are less per unit of body weight than those of aquatic species. In addition, the gills of some intertidal crabs have become highly sclerotized apparently to provide support so that the gill leaves do not stick together when the crab is out of water (Vernberg and Vemberg, 1972). Tube-dwelling amphipods apparently can function efficiently with low oxygen supplies. At saturated oxygen tensions, they exhibit an intermittent rhythm of ventilation. At low tide, when the oxygen in their burrows drops to very low levels, they ventilate continuously but do not hyperventilate as free-swimming amphipods do. Because of the resistance of their tubes to water flow, hyperventilation would be energetically expensive for tube-dwelling amphipodes (Vernberg, 1981).

Many marine animals associated with anoxic wetland soils have high concentrations of respiratory pigments or pigments that have unusually high affinities for oxygen or both. These include the nematode *(Enoplus communis)*, the Atlantic bloodworm (*Glycera dibranchiata*), the clam (*Mercenaria mercenaria*), and even the land crab (*Carooma quannumi*) (Vernberg and Vernberg, 1972; Vernberg and Coull, 1981).

Fiddler crabs (*Uca* spp.) illustrate the complex behavioral and physiological patterns to be found in the intertidal zone. These crabs are active during low tides, feeding daily when the marsh floor is exposed. (Incidentally, this pattern of activity is based on an innate lunar rhythm, not on a direct sensing of low water levels. When transported miles from the ocean, fiddlers continue to be active at the time that low tide would occur in their new location.) When the tide rises, they retreat to their burrows, where the oxygen concentration can become very low because fiddler crabs apparently do not pump water in their burrows. Not only are these species relatively resistant to anoxia, but also their critical oxygen tension (that is, the tension below which respiratory activity is reduced) is low, 0.01–0.03 atmospheres for inactive and 0.03–0.08 atmospheres for active crabs. They can continue to consume oxygen down to a level of 0.004 atmospheres (Vernberg and Vernberg, 1972). When oxygen levels get very low in the burrows, the crabs simply become inactive and consume very little oxygen. They may remain that way for several tidal cycles without harm.

Intertidal bivalves close their valves tightly or loosely when the tide recedes. Widdows et al. (1979) found that four different bivalves had lower respiration rates in air (valves closed) than in water. All could respire anaerobically, but the accumulation of end products of anaerobic respiration depended on how tightly their shells were closed and thus how much oxygen they received. The tolerance to anoxia may change during the life of an organism. The larvae of fiddler crabs, which are planktonic, are much more sensitive to low oxygen than the burrowing adults. An interesting but rather unusual adaptation is that of a gastrotrich (*Thiodasys sterreri*), which is reported to be able to use sulfide as an energy source under extreme anaerobiosis (Maguire and Boaden, 1975).

Salt

Like their responses to oxygen stress, the major mechanisms of adaptation by animals to salt involve control of the body's internal environment. Most simple marine animals are *osmoconformers,* that is, their internal cell environment follows closely the osmotic concentration of the external medium. But in animals that have greater body complexity, *osmoregulation,* that is, control of internal osmotic concentration, is the rule. This is particularly true of animals that inhabit the upper intertidal zone, where they are exposed to widely varying salinities and to prolonged periods of dessication. *Euryhaline* organisms can tolerate wide fluctuations in salinity. *Stenohaline* organisms, on the other hand, survive within fairly narrow osmotic limits. Most marsh organisms must be euryhaline, but they can be either osmoconformers or osmoregulators.

Figure 6–7 illustrates the imperfect osmoregulation found in penaeid shrimp. The hemolymph concentration of a perfect osmoconformer would follow the solid line of isotonicity. In contrast, a perfect osmoregulator would have a constant internal concentration, and would be illustrated by a horizontal line on the graph. The brown shrimp (*Penaeus aztecus*) is intermediate between these two positions. The internal environment varies much less than the external medium does, but it is not constant. At low external salt concentrations, the shrimp is hyperosmotic, indicating a water potential gradient into the organism, usually achieved by concentrating sodium and chloride ions. This probably requires the expenditure of less energy than the hypoosmotic regulation shown at high external salt concentrations. In this circumstance the water potential gradient is directed out of the animal, which means that dehydration would occur if the body covering were not to some extent impervious to water movement. Animals that possess the ability for hypoosmotic regulation must have some mechanism to lower the osmotic concentration of the body. This is accomplished through special regulatory organs and organ systems, chiefly renal organs (kidney, antennal glands, or more primitive nephridia), gills, salt-secretory, nasal, or rectal glands and the specialized excretory functions of the gut. These organs are able to move ions across cell membranes against the concentration gradient, concentrating them in some excretory product such as urine.

Figure 6–8 illustrates the differences in adaptation to salinity changes in species of crabs that inhabitat different environments. The aquatic species *Cancer* is an osmoconformer. Species that are submerged most of the time but are subject to wider osmotic fluctuations (*Hemigrapsus* and *Pachygrapsus*) are imperfect osmoregulators. The other species (including *Uca*, a common marsh crab) are from the high intertidal zone. They are excellent osmoregulators, possessing adaptations obviously useful in controlling the variable salinity and frequent dessication of their habitat. Regulation in these species is controlled both by differences in exoskeleton permeability and by specialized organs. The exoskeleton of terrestrial crabs is less permeable to water and salt than those of

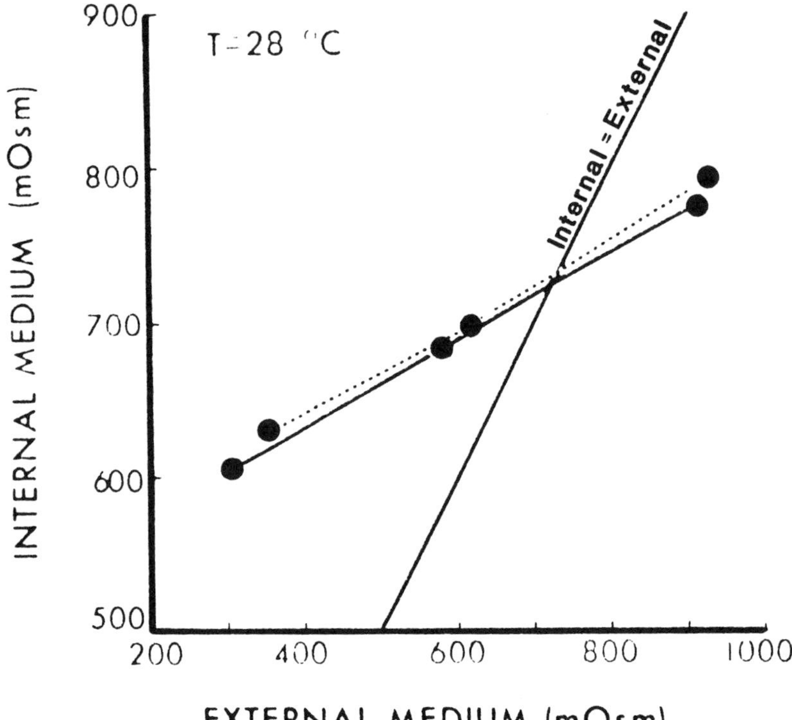

Figure 6–7. Mean hemolymph osmolality (mOsm) of 3.7 (.....) and 6.7g (______) penaeid shrimp *Penaeus aztecus* at 28° C. (*From Bishop et al., 1980*). The solid (internal = external) line indicates conditions expected of a perfect osmoconformer. A horizontal line would indicate perfect osmoregulation.

semiterrestrial species, which, in turn, are less permeable than subtidal crabs. The antennal glands seem to control the concentrations of specific ions in the hemolymph. Osmoregulation, however, is controlled by the gills and the posterior diverticulum of the alimentary canal (Vernberg and Vernberg, 1972). The complexity of the adaptations is illustrated by the permeability of the foregut of the land crab (*Geocarcinus lateralis*) to both water and salt. This permeability varies with time and environmental circumstance and is under neuroendocrine control (Mantel, 1968).

Other Adaptations

Reproduction

As might be expected, adaptations to specific habitats involve virtually every facet of an organism's existence. We have focused primarily on the immediate response of the individual to stresses of the wetland environment. But in terms of species survival, reproduction is equally important. In an evolutionary sense,

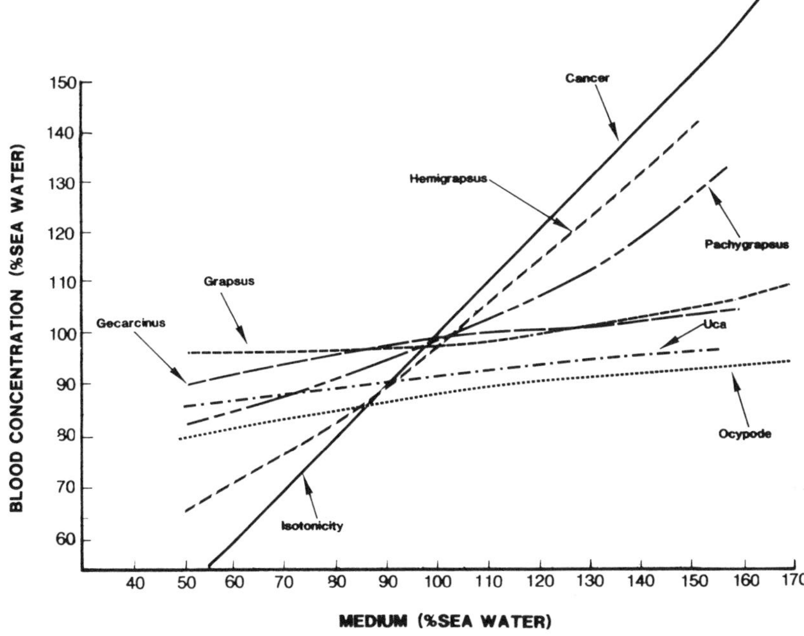

Figure 6–8. Comparative osmoregulation of crabs showing different degrees of adaptation to the marine environment. *Cancer* is an aquatic species; *Hemigrapsus* and *Pachygrapsus*, low intertidal zone; the other species are terrestrial or high intertidal zone inhabitants. *(From Gross, 1964; copyright © 1964 by the Biological Bulletin, reprinted with permission)*

a species strategy must be to produce reproductively active offspring at minimum energy cost. For infauna in the marsh sediment, where mobility is restricted, this is often accomplished by direct contact of organisms for fertilization and by direct growth of offspring through the elimination of larval stages or by larvae that remain in place. Epibenthic organisms in a fluctuating environment, in contrast, often produce great numbers of pelagic larvae that are widely distributed by currents and tides. Reproductive behavior is complex, and it is not always easy to see any adaptive significance in the responses that have been observed. For example, the subtidal clam *Rangia cuneata* requires a salinity shock (of about 5 ppt up or down) to release its gametes even though the female may be gravid more than half of the year. When salinities remain constant, for example, when an area is impounded, the clam eventually dies out. The intertidal crab *Sesarma cinereum* requires low estuarine salinity for larval development. The fourth zoeal stage, in particular, is sensitive to salinity, and best development occurs at 26.7 parts per thousand. In the succeeding megalops stage, however, it can withstand a wide range of salinities and temperatures (Costlow et al., 1960).

Feeding

As with reproductive adaptations, the broad range of animal feeding responses closely reflects their habitats. Adaptations of feeding appendages, for example, seem to be more closely related to feeding habits than to taxonomic relationships. Many organisms that exist in marsh sediments are adapted for the direct absorption of dissolved organic compounds from their environment. For example, infaunal polychaetes can supply a major portion of their energy requirements from the rich supply of dissolved amino acids present in their environment, but epifaunal species are unable to take advantage of amino acids at concentrations typical of their environment (Vernberg, 1981). Many mud-dwelling organisms have one or more adaptations to selective feeding on microscopic particles by means of pseudopods, cilia, mucus, and setae, or they may ingest substrate unselectively. Sikora (1977) suggested that the appendages of many macrobenthic organisms (shrimp, crabs) are adapted to feeding on microscopic meiobenthic organisms and that these latter organisms are major intermediaries in the marsh/estuary food chain.

Wetland Ecosystem Development

7

Wetland ecosystems have traditionally been considered transitional seres between open lakes and terrestrial forests. The accumulation of organic material from plant production was seen to build up the surface until it was no longer flooded and could support flood-tolerant terrestrial forest species (autogenic succession). Although there are well-documented successional sequences showing this line of development, there are also many examples that counter classical successional theory. An alternative hypothesis is that the vegetation found at a wetland site consists of species adapted to the particular environmental conditions of that site (allogenic succession). Observed zonation patterns, in this view, reflect underlying environmental gradients rather than autogenic successional patterns. Present evidence seems to lead to the conclusion that both allogenic and autogenic forces act to change wetland vegetation and that the idea of a regional terrestrial climax is inappropriate.

If one looks at ecosystem attributes as indices of ecosystem maturity, wetlands appear to be mature in some respects and young in others. Generally, productivity is high, some production is exported, and mineral cycles are open, all indications of young systems. On the other hand, most wetlands accumulate much structural biomass in peat, all wetlands are detrital systems, spatial heterogeneity is generally high, and life cycles are complex. These properties indicate maturity. Although the input of water and nutrients varies in different types of wetlands by as much as five orders of magnitude, ecosystem response in terms of productivity, biomass, and nutrient storage varies by only a factor of from 2 to 4. Wetland ecosystem response is controlled by stores of soil organic material that stabilize the flooding pattern and provide a steady source of nutrients to the plants, minimizing the impact of external supplies.

At landscape scales, observed patterns of wetlands, aquatic and upland habitats, reflect a complex and dynamic interaction of physical (allogenic) and biotic (autogenic) forces acting on the geomorphic template of the landscape.

WETLAND PLANT DEVELOPMENT

The beginning and subsequent development of a plant community is characterized by many events and conditions, including the availability of viable seeds or other propagules at a site, appropriate environmental conditions for germination and subsequent growth, and replacement by plants of the same or different species as site conditions change in response to both abiotic and biotic factors. The concept of succession, that is, the replacement of plant species in an orderly sequence of development, in particular, has exerted a strong influence on plant ecology throughout this century. In recent years this fairly narrow interpretation of development has been superseded by the demonstration that there are multiple pathways and controls that affect plant invasion and subsequent change.

The concept of plant succession has a long history. It was first clearly enunciated by Clements (1916) and applied to wetlands by the English ecologist W. H. Pearsall in 1920 and by an American, L. R. Wilson, in 1935. E. P. Odum (1969) adapted and extended the ideas of those early ecologists to include ecosystem properties such as productivity, respiration, and diversity. The classical use of the term *succession* involves three fundamental concepts: (1) vegetation occurs in recognizable and characteristic *communities*; (2) community change through time is brought about by the biota (that is, changes are *autogenic*); (3) changes are linear and *directed* toward a mature stable *climax* ecosystem (Odum, 1971). Using this definition of succession, all wetlands are regarded as transitional *seres* in a *hydrarch* successional sequence to a terrestrial forest climax.

Although the classical concept of succession has been a dominating paradigm of great importance in plant ecology, it is presently in disarray. As early as 1917, Gleason enunciated an *individualistic* hypothesis to explain the distribution of plant species. His ideas have developed into the *continuum* concept (Whittaker, 1967; McIntosh, 1980) that holds that the distribution of a species is governed by its response to its environment (*allogenic* succession). Because each species adapts differently, no two occupy exactly the same zone. The observed invasion/replacement sequence is also influenced by the chance occurrence of propagules at a site. The result is a continuum of overlapping sets of species, each responding to subtly different environmental cues. In this view, no communities exist in the sense used by Clements, and although ecosystems change, there is little evidence that this is directed or that it leads to a particular climax.

The Classical Idea of Succession

The Community Concept

The idea of the community is particularly strong in wetland literature. Historic names for different kinds of wetlands—marshes, swamps, carrs, fens, bogs, reedswamps—often used with the name of a dominant plant (sphagnum bog, leatherleaf bog, cypress swamp)—signify our recognition of distinctive associations of plants that are readily recognized and at least loosely comprise a community. One reason these associations are so clearly identified is that zonation patterns in wetlands often tend to be sharp, having abrupt boundaries that call attention to vegetation change and, by implication, the uniqueness of each zone.

Although wetland communities were historically identified qualitatively, the application of objective statistical clustering techniques supports the community idea, at least in some instances. For example, the classical syntaxonomical treatment of European *Spartina* communities (a semiquantitative analysis of vegetation stands based on dominants and observed similarities) resulted in the classification of these marshes into a number of subassociations (Beeftink and Gehu, 1973). A numerical classification of the same areas, based on similarity ratios and a statistical clustering technique, identified virtually the same subassociations (Kortekaas et al., 1976) (Fig. 7–1). It is important to notice that different degrees of clustering occur with these data. Three main groups are associated with the dominance of three different *Spartina* species: *S. maritima*, *S. alterniflora*, and *S. townsendii*. These three groups break down further into various subassociations. The decision about what level of similarity, if any, identifies a community is entirely arbitrary.

The identification of a community is also to some extent a conceptual issue that is confused by the scale of perception. Field techniques are adequate to describe the vegetation in an area and its variability. But its homogeneity—one index of community—may depend on size. For example, Louisiana coastal marshes have been classified into four zones, or communities, based on the dominant vegetation (Chabreck, 1972). If the size of the sampling area is large enough, any sample within one of these zones will always identify the same species. If smaller grids are used, however, differences appear within a zone. The intermediate marsh zone is dominated on a broad scale by *S. patens;* but aerial imagery shows patterns of vegetation within the zone, and intensive sampling and cluster analysis of the vegetation reveal at least five subassociations that are characteristic of intermediate marshes (Fig. 7–2). Is the intermediate marsh a community? Are the subassociations communities? Or is the community concept a pragmatic device to reduce the bewildering array of plants and possible habitats to a manageable number of groups within which there are reasonable similarities of ecological structure and function?

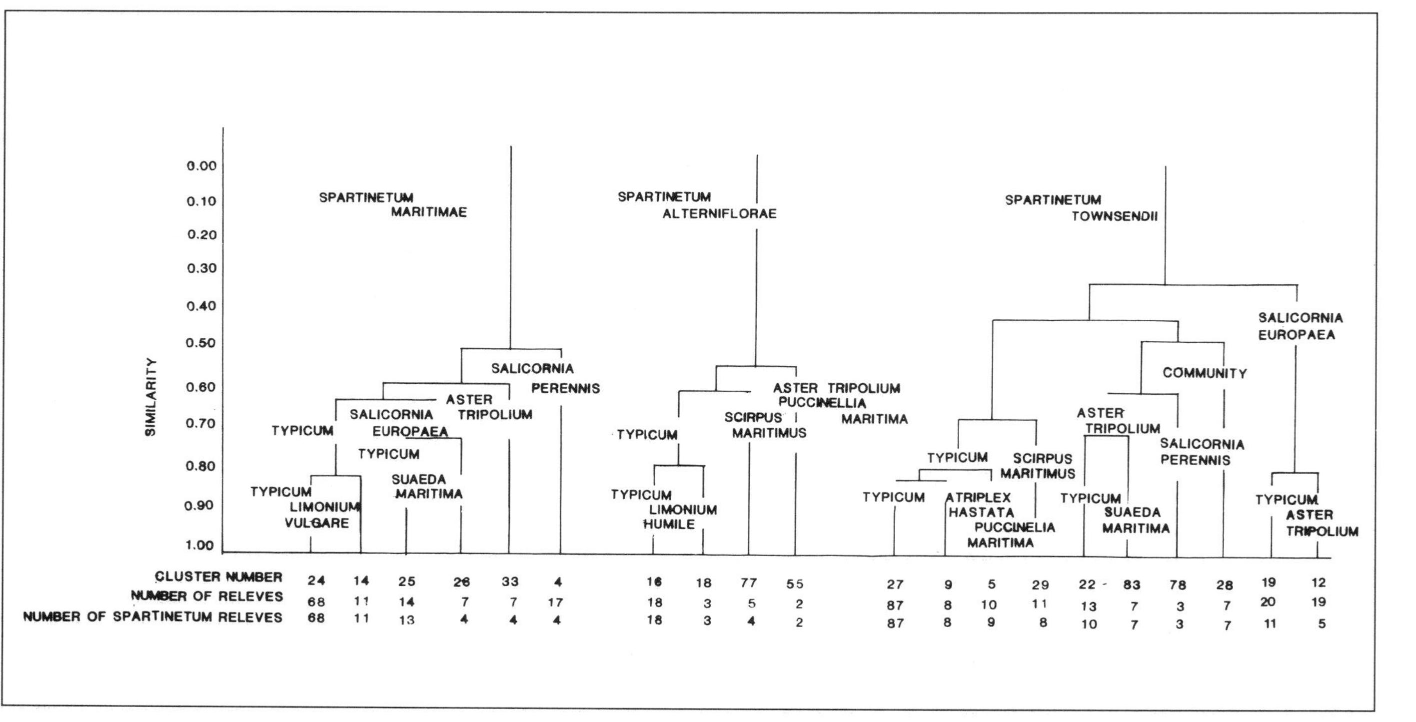

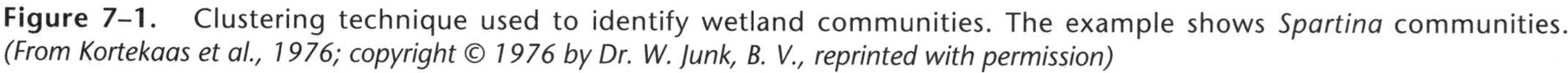

Figure 7–1. Clustering technique used to identify wetland communities. The example shows *Spartina* communities. *(From Kortekaas et al., 1976; copyright © 1976 by Dr. W. Junk, B. V., reprinted with permission)*

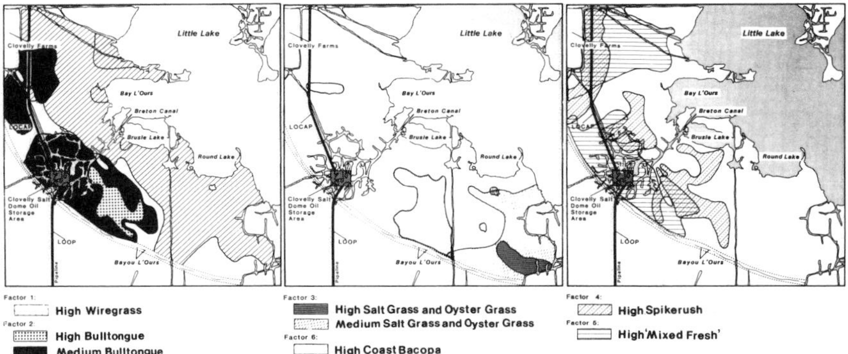

Figure 7–2. Classification of vegetation associations in an intermediate marsh in coastal Louisiana. The different associations were determined by a statistical factor analysis of species density. This figure shows the distribution of the first six factor associations. They were paired into three maps to facilitate comparison of contrasting community distributions. (High dominance represents a factor pattern weight of 0.65–0.99; medium of 0.25-0.64.) *(From Sasser et al., 1982)*

The Concept of Autogenic Succession

In the classical view of succession, wetlands are considered transient stages in the *hydrarch development* of a terrestrial forested climax community from a shallow lake (Fig. 7–3). In this view, lakes gradually fill in as organic material from dying plants accumulates and minerals are carried in from upslope. At first change is slow because the source of organic material is single-celled plankton. When the lake becomes shallow enough to support rooted aquatic plants, however, the pace of organic deposition increases. Eventually, the water becomes shallow enough to support emergent marsh vegetation, which continues to build a peat mat. Shrubs and small trees appear. They continue to transform the site to a terrestrial one not only by adding organic matter to the soil but also by drying it through enhanced evapotranspiration. Eventually, a climax terrestrial forest occupies the site (see, for example, W. S. Cooper, 1913). The important point to note in this description of hydrarch succession is that most of the change is brought about by the plant community itself as opposed to externally caused environmental changes.

How realistic is this concept of succession? It is certainly well documented that forests do occur on the sites of former lakes (Larsen, 1982), but the evidence that the successional sequence leading to these forests was autogenic is not clear. Because peat building is crucial to filling in a lake and its conversion to dry land, key questions involve the conditions for peat accumulation and the limits of that accumulation. Peat underlies many wetlands, often in beds 10 or more meters deep. Several scientists (McCaffrey, 1977; Delaune et al., 1983c) have shown that in coastal marshes it has accumulated and is still accumulating at rates varying from less than 1 to about 15 mm/yr. Most of this accumulation seems to be associ-

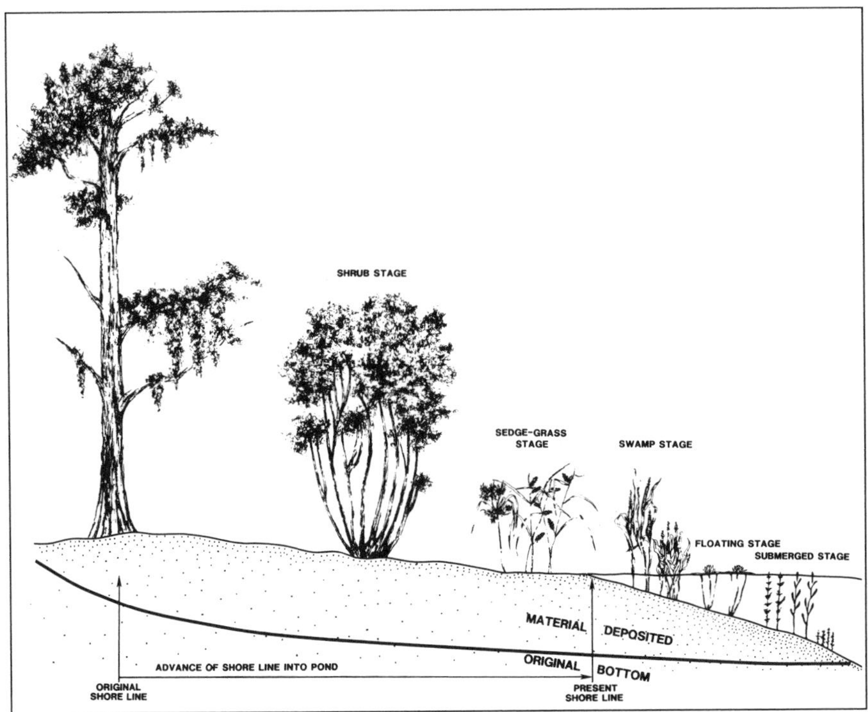

Figure 7–3. Diagram of classical hydrarch succession at the edge of a pond. *(After Wison and Loomis, 1967)*

ated with rising sea levels (or submerging land). By contrast, northern inland bogs accumulate peat at rates of from 0.2 to 2 mm/yr (see Chap. 12).

In general, accumulation occurs only in anoxic sediment. When organic peats are drained, they rapidly oxidize and subside, as farmers who cultivate drained marshes have discovered. As the wetland surface accretes and approaches the water surface or at least the upper limit of the saturated zone, peat accretion in excess of subsidence must cease. It is hard to see how this process can turn a wetland into a dry habitat that can support terrestrial vegetation unless there is a change in hydrologic conditions that lowers the water table. For example, Cushing (1963) used paleoecological techniques to show that most of the peatlands in the Lake Agassiz Plain (Minnesota and south-central Canada) formed during the mid-Holocene (beginning about 4,000 years ago) during a moist climatic period when surface water levels rose about 4 m (see Chap. 12).

To the extent that marsh vegetation traps inorganic sediments and thus enhances the rate of mineral deposition, mineral soil accretion may also be considered autogenic, but this would require flooding conditions to carry the mineral material into the marsh.

Linear, Directed Vegetation Changes

If one is convinced that identifiable communities exist and that they change because of autogenic processes, a further criterion is necessary to qualify the changes as successional; they must be linear and directed toward a stable climax. The scientific literature is replete with schematic diagrams showing the expected successional sequence from wetland to terrestrial forest (Fig. 7–4). Most of these are based on observed zonation patterns, assuming that these spatial patterns presage the temporal pathway of change. That this classical pattern does occur in some instances is clearly demonstrated in soil profiles. The relict remains of vegetation in soil profiles represent a temporal sequence, the oldest at the bottom of the profile. Figure 7–5 shows such a sequence, demonstrating a direct succession from a pioneer salt marsh to a tidal woodland.

In the case of the Gulf Coast wetland succession as outlined by Penfound and Hathaway (1938) (Fig. 7–4b), we now know that the sequence of change is from freshwater to saline vegetation, almost directly opposite to the direction of change they envisioned (Neill and Deegan, 1986). In the Gulf Coast, fresh marshes form as a result of rapid delta growth in active river mouths. They change to salt marshes and then to open bays because the river shifts its mouth to another location, resulting in salinization and subsidence in the abandoned delta lobe (Neill and Deegan, 1986).

Paleoecological analyses of peat beds shed considerable light on the concept of succession. Studies based on the macrofossil record were initiated as early as the seventeenth century (McIntosh, 1985) and were used by Clements (1916, 1924), as a basis for understanding succession. The discipline expanded rapidly after the development of the study of the pollen record in the 1920s. Although paleoecological studies reveal the kind of developmental sequence (succession) portrayed in Figure 7–5 (see also Futyma, 1988), the longer (older) sequences were taken by Clements to represent a record of shifts in the climax vegetation as a result of climate change (McIntosh, 1985). Fossil records, mostly from northern peat bogs, suggest two generalizations: (1) In some sites the present vegetation has existed for several thousands of years (Redfield, 1972); (2) Climatic change and glaciation had major impacts on plant species composition and distribution; generally bogs expanded during warm, wet periods and contracted during cool, drier periods, although the influence of local topographic, drainage and other site conditions often masked regional climatic shifts. (Dopson et al., 1986; Casparie and Streefkerk, 1992). For example, Fig. 7–6 shows the development of the moisture content of the peat-forming environment over the past 11,000 years from two sites in The Netherlands. Although these sites are only about 2 km apart, one from the central part of the Hunze Valley, and the other from the valley slope, there are such great differences between the two curves that a direct influence of climate on peat formation is difficult to discern (Casparie and Streefkerk, 1992). Pollen sequences, however, are generally consistent across Europe and North America, indicating response to similar global climate shifts (McIntosh, 1985).

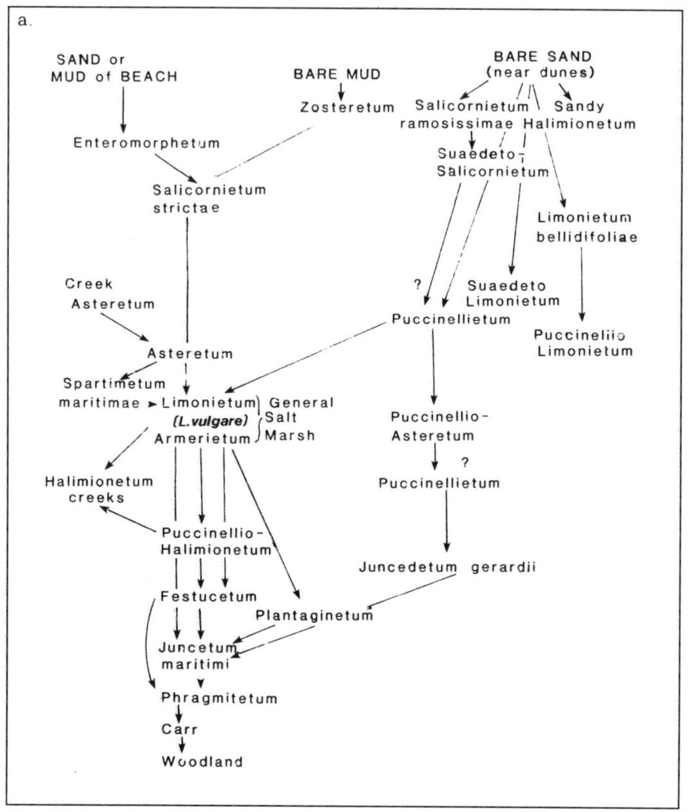
a.
SAND or
MUD of BEACH
BARE MUD
BARE SAND
(near dunes)
Zosteretum
Salicornietum
ramosissimae
Sandy
Halimionetum
Enteromorphetum
Suaedeto-
Salicornietum
Salicornietum
strictae
Limonietum
bellidifoliae
Creek
Asteretum
?
Suaedeto
Limonietum
Puccinellietum
Puccinellio
Limonietum
Asteretum
Spartinetum
maritimae
Limonietum
(L. vulgare)
Armerietum
General
Salt
Marsh
Puccinellio-
Asteretum
?
Halimionetum
creeks
Puccinellietum
Puccinellio-
Halimionetum
Juncedetum gerardii
Festucetum
Plantaginetum
Juncetum
maritimi
Phragmitetum
Carr
Woodland

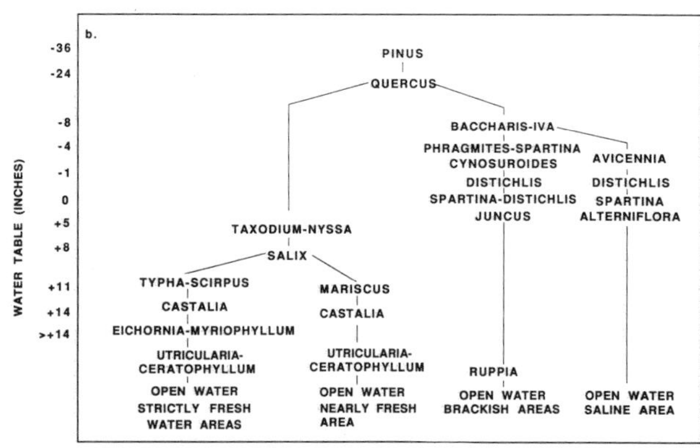
b.
WATER TABLE (INCHES)
-36
-24
-8
-4
-1
0
+5
+8
+11
+14
>+14
PINUS
QUERCUS
BACCHARIS-IVA
PHRAGMITES-SPARTINA
CYNOSUROIDES
AVICENNIA
DISTICHLIS
DISTICHLIS
SPARTINA-DISTICHLIS
SPARTINA
JUNCUS
ALTERNIFLORA
TAXODIUM-NYSSA
SALIX
TYPHA-SCIRPUS
MARISCUS
CASTALIA
CASTALIA
EICHORNIA-MYRIOPHYLLUM
UTRICULARIA-
CERATOPHYLLUM
UTRICULARIA-
CERATOPHYLLUM
RUPPIA
OPEN WATER
STRICTLY FRESH
WATER AREAS
OPEN WATER
NEARLY FRESH
AREA
OPEN WATER
BRACKISH AREAS
OPEN WATER
SALINE AREA

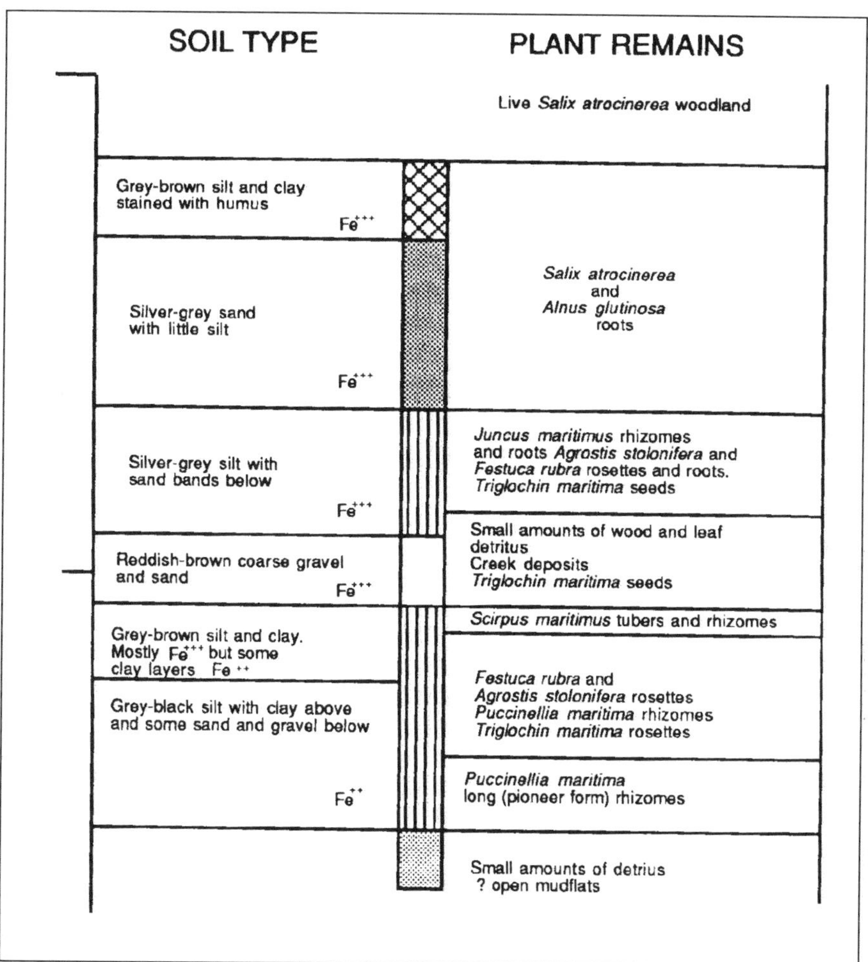

Figure 7–5. Soil profile 150 m landward of the present seaward limit of tidal woodland with evidence from plant remains of direct succession from pioneer salt marsh to tidal woodland at Fal estuary, Cornwall. *(From Ranwell, 1972; copyright © 1972 by D.S. Ranwell, reprinted with permission)*

Figure 7–4 (*opposite page*). Presumed successional relationships among coastal plant communities: *a.* England and *b.* northern Gulf Coast. (a. *after Chapman, 1960; b. from Penfound and Hathaway, 1938; copyright © 1938 by the Ecological Society of America, reprinted with permission*)

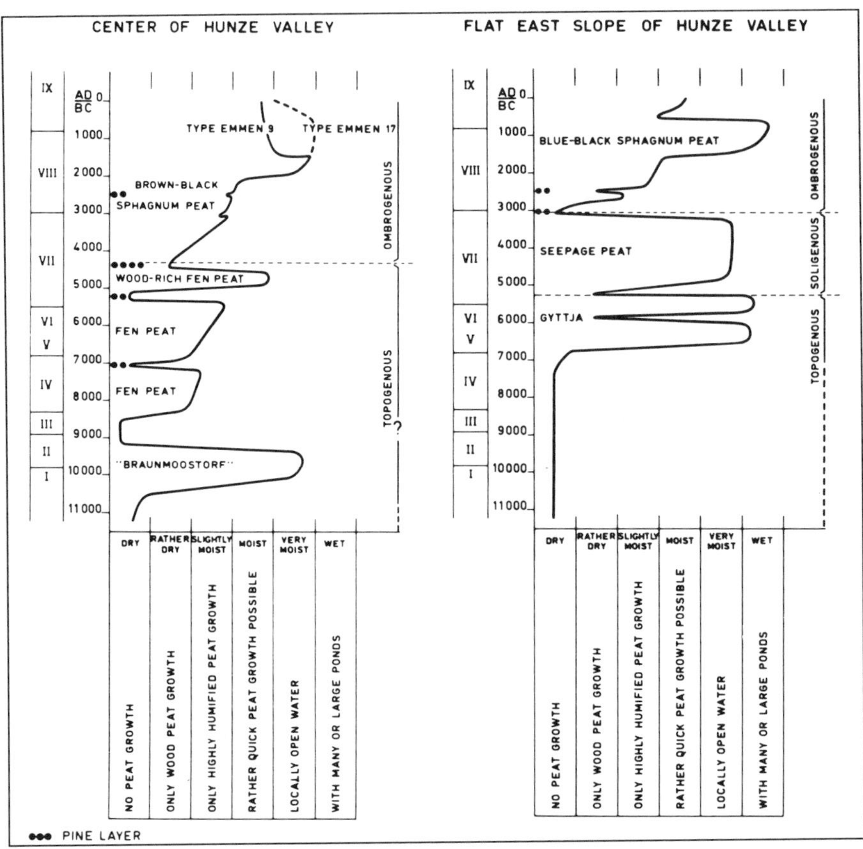

Figure 7–6. Peat sequence at two sites in the Hunze valley, The Netherlands, showing the degree of moisture plotted against time. (*From Casparie and Streefkerk, 1992; copyright © 1992 by Kluwer Academic Publishers, reprinted with permission*)

Paleoecological studies cast serious doubt on the concept of plant communities. For example, West (1964), as quoted in McIntosh (1985), wrote:

> We may conclude that our present plant communities have no long history in the Quaternary, but are merely temporary aggregations under given conditions of climate, other environmental factors, and historical factors.

The Continuum Idea

The presumed succession example illustrated in Figure 7–4b shows that zonation does not necessarily indicate succession. In fact, those who support the continuum idea maintain that zonation simply indicates an environmental gradient

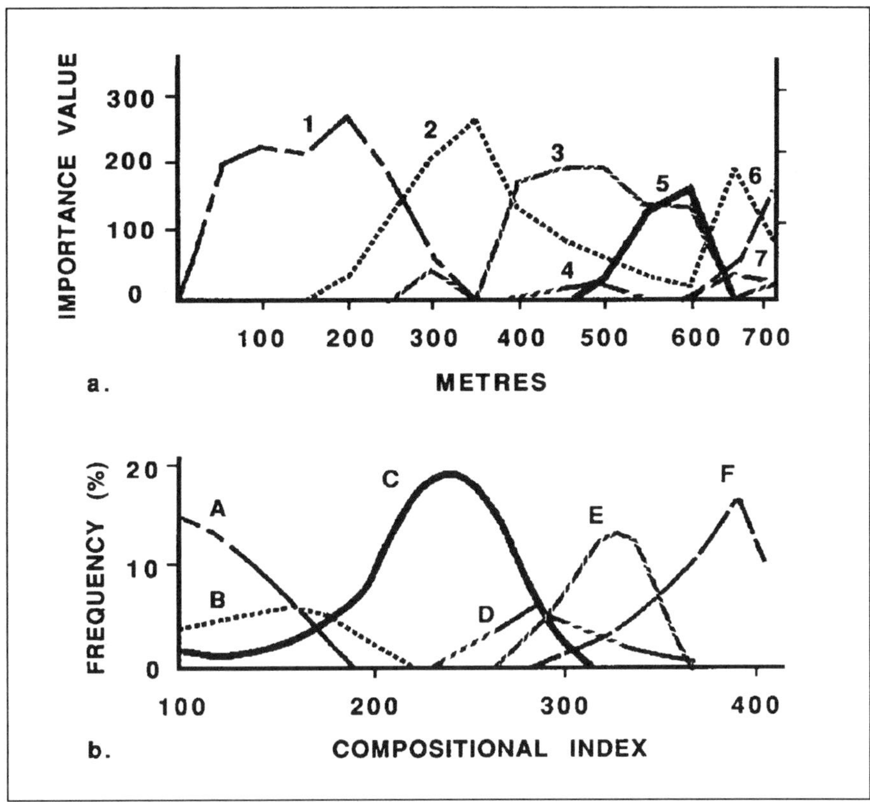

Figure 7–7. Examples of gradient analysis of wetlands: *a.* swamp forest where **the species are *a.* (1) *Larix laricina,* (2) *Tthuja occidentalis,* (3) *Ulmus americana,*** (4) *Fraxinus nigra,* (5) *Acer saccharinum,* (6) *Acer saccharum,* and (7) *Fraxinus americana* and *b.* submersed aquatic vegetation in Wisconsin lakes, where the species are (A) *Elatine minima, b. Potamogeton epihydrus,* (C) *Eleocharis acicularis,* (D) *Potamogeton praelongus,* (E) *Zosterella dubia,* and (F) *Myriophyllum exalbescens. (From van der Valk, 1982; copyright © 1982 by International Scientific Publications, reprinted with permission. Based on original data from Beschel and Webber, 1962, and Curtis, 1959)*

to which individual species are responding. The reason zonation is so sharp in many wetlands, they argue, is that environmental gradients are "ecologically" steep and groups of species have fairly similar tolerances that tend to group them on these gradients. Figure 7–1 can easily be interpreted to support this contention, for it shows that the similarity level between two groups of plants is never more than 0.85 (1.00 indicates identity) and may be as low as 0.50. Figure 7–7 shows the distribution of swamp trees and submersed aquatic vegetation along an ordination axis. Although the species overlap, the distribution of each seems to be distinct, leaving no reflection of a community. The idea that each species is found where the environment is optimal for it makes perfect sense to ecophysiologists and autecologists who interpret the success of a species in terms of its environmental adaptation.

One major difference between classical community ecologists and proponents of the continuum idea is the greater emphasis put on allogenic processes by the latter. In wetlands, abiotic environmental factors often seem to overwhelm biotic forces. Under these circumstances, the response of the vegetation is determined by these abiotic factors. Hydrologic conditions, for example, were described in Chapter 4 as having a particular significance for wetland structure and function. In coastal areas plants can do little to change the tidal pulse of water and salt. Tidal energy may be modified by vegetation as stems create friction that slows currents or as dead organic matter accumulates and changes the surface elevation. But these effects are limited by the overriding tides. These wetlands are often in dynamic equilibrium with the abiotic forces, an equilibrium that E. P. Odum (1971) called *pulse stability*. On the Louisiana coast, as mentioned above, the major abiotic force seems to be the high subsidence rate, which overrides any autogenic changes. There appear to be few, if any, examples of wetland ecosystems that became terrestrial without a concurrent allogenic lowering of the water level. Even the example given above (Fig. 7–5), documenting a classical change from salt marsh to woodland, resulted in a tidal woodland dominated by flood-tolerant trees, not a terrestrial ecosystem.

In the lower energy environment of a northern peatland, in contrast, hydrologic flows can be dramatically changed by biotic forces, resulting in distinctive patterned landscapes (Glaser and Wheeler, 1980; Glaser et al., 1981; Glaser, 1983c; Siegel, 1983; Foster et al., 1983; Rochefort et al., 1990; see Chap. 12). Thus changes in wetlands may be autogenic but are not necessarily directed toward a terrestrial climax. In fact, wetlands in dynamically stable environmental regimes seem to be extremely stable, contravening the central idea of succession. Walker (1970) found from pollen profiles that the successional sequence in northern peatlands was variable: There were reversals and skipped stages that may have been influenced by the dominant species first reaching a site (Fig. 7–8). A bog, not some type of terrestrial forest, was the most common end point in most of the sequences described.

Seed Banks

If, as the continuum idea suggests, plant communities are artifacts of human minds, then the development of plants on a site can be explained only in terms of the response of individual species to local conditions. One factor is historical—the previous history of the site—which determines what propagules are present for future invasion. This—the sediment *seed bank*—has been found to be extremely variable—both in space and time. Pederson and Smith (1988) made the following generalizations about marsh seed banks:

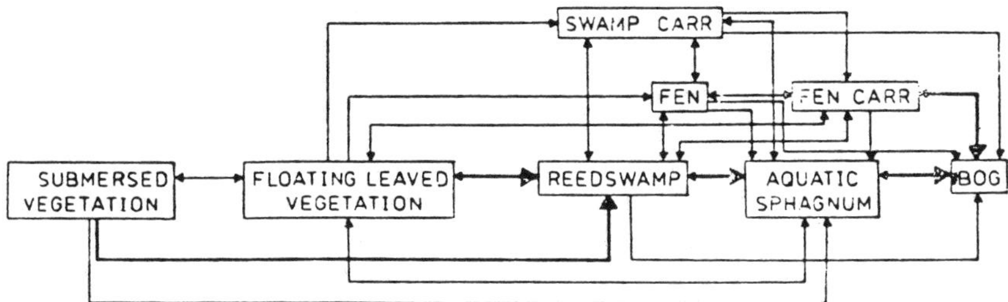

Figure 7–8. Successional sequences reconstructed from stratigraphic and palynological studies of post-glacial British peatlands. Thicker lines indicate the more common transitions. (*From van der Valk, 1982; copyright © 1982 by International Scientific Publications, reprinted with permission; After Walker, 1970*)

1. Freshwater marshes with drawdowns produce the greatest number of seeds. This is confirmed also by Siegley et al. (1988) and by Welling et al. (1988b).
2. Seed banks are dominated by the seeds of annual plants and flood-intolerant species. Areas that contain emergent plants have greater seed densities than mudflats. Perennials generally produce fewer seeds that have shorter viability than annuals. They are more likely to reproduce by asexual means such as rhizomes. Saline zones produce few seeds. The salt marsh is an example of a perennial-dominated system in which most reproduction is asexual.
3. Seed distribution decreases exponentially with the depth of the sediment.
4. Water is a major factor in seed banks. Seeds are concentrated along drift lines. The kinds of seeds produced depend on the flooding regime—by submergents when deep flooded, emergents when periodically flooded, and flood-intolerant annuals during drawdowns.

The germination of seedlings from a seed bank is similarly influenced by many factors that vary in space and time. Welling et al. (1988a) stated that differences in environmental conditions seem to have less impact on the distribution of seedlings along an elevation gradient than the distribution of seeds. Nevertheless, environmental factors such as flooding, temperature, soil chemistry, soil organic content, pathogens, nutrients (Gerritsen and Greening, 1989; Willis and Mitsch, unpub. manuscript) and allelopathy have been shown to influence recruitment. Water, in particular, is a critical variable because most seeds require moist but not flooded conditions for germination and early seedling growth. As a result of this restrictive moisture requirement, it is common to find even aged stands of trees at low elevations in riparian wetlands, reflecting seed germination during relatively uncommon years when water levels were unusually low during the spring and summer.

Postrecruitment processes play a major role in the distribution of adult plants at a site, leading to plant assemblages that cannot be predicted from seed bank alone (Welling et al., 1988a). Thus in coastal areas where the dominant plant species, *Spartina alterniflora*, occurs in large monotypic stands, it is often the pioneer species and remains dominant throughout the life of the marsh. In contrast, in tidal and nontidal freshwater marshes, the seed bank is much larger and richer, and the first species to invade a site may later be replaced by other species. For example, in one study primary succession on delta islands along the Louisiana coast is characterized by willows (*Salix* spp.) on the higher portions of the intertidal zone and arrowheads (*Sagittaria* spp.) on lower elevations (Shaffer et al., 1992). After 15 years, willows enriched with many additional understory species still dominated the high ground, whereas the arrowhead at intermediate elevations had been replaced by a rich mix of annual and perennial grasses and had died out at low elevations, leaving bare mudflats.

The Environmental Sieve Model

Detailed studies of seed banks, recruitment, and the subsequent growth of individual species lead to the conclusion that both allogenic and autogenic forces act to change wetland vegetation and that the idea of a regional terrestrial climax for wetlands is inappropriate. Van der Valk (1981) replaced the autogenic succession concept with a Gleasonian model (Fig. 7–9) in which the presence and the abundance of each species depend on its life history and its adaptation to the environment of a site. He classified all plant species into life history types based on potential life span, propagule longevity, and propagule establishment requirements. Each life history type has a unique set of characteristics and thus potential behavior in response to controlling environmental factors such as water-level changes. These environmental factors compose the "environmental sieve" in van der Valk's model. As the environment changes, so does the sieve and hence the species present. This is a useful conceptual model for understanding wetland change. For example, L. M. Smith and Kadlec (1985) tested the model's ability to predict species composition in a fresh marsh after a fire and were satisfied with the qualitative results.

The sieve model does not, however, explicitly recognize autogenic processes. These could easily be included with a feedback loop showing that the environmental sieve itself can be modified to some extent by the wetland vegetation present. An excellent example of this kind of feedback loop is discussed by Weller (1981). In midwestern prairie pothole marshes, the vegetation can be wiped out by a population explosion of herbivorous muskrats. New emergent vegetation cannot become established until a dry year exposes the soil. A typical succession of flood-tolerant grasses and annual broad-leaved plants follows until robust cattails outcompete them. This sets the stage for another muskrat explosion (see Chap. 11). This cycle is highly variable in both space and time,

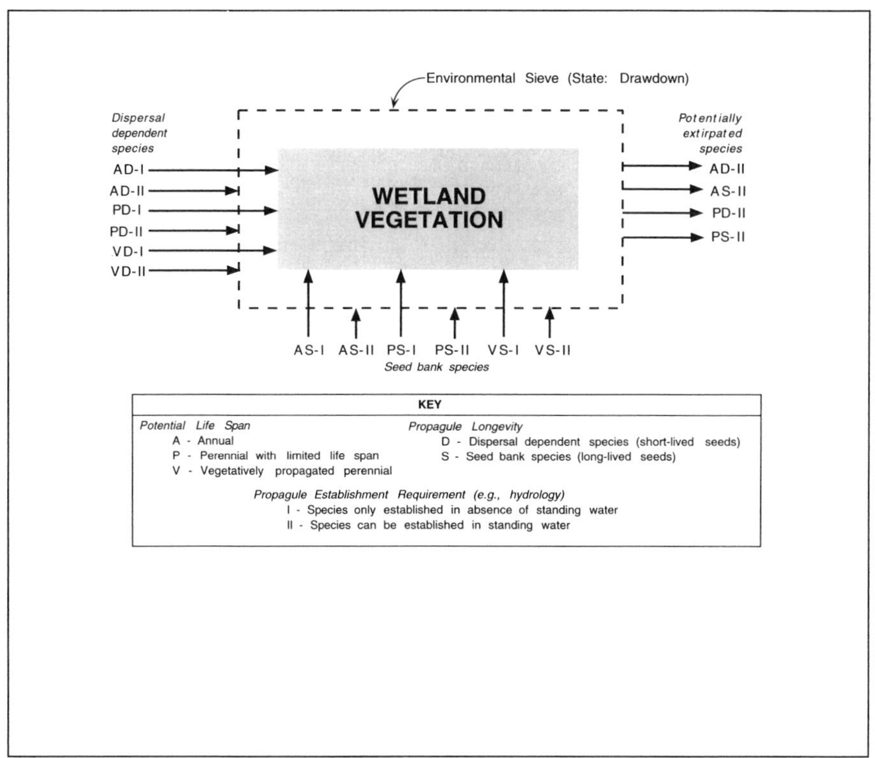

Figure 7–9. General model of Gleasonian wetland succession proposed by van der Valk. (*From van der Valk, 1981; copyright © 1981 by the Ecological Society of America, reprinted with permission*)

depending heavily on an abiotic wet-dry climatic cycle as well as the biotic vegetation/muskrat interaction.

Summarizing a discussion of wetland development Niering (1989) stated that

> Traditional successional concepts have limited usefulness when applied to wetland dynamics. Wetlands typically remain wet over time exhibiting a wetland aspect rather than succeeding to upland vegetation. Changes that occur may not necessarily be directional or orderly and are often not predictable on the long term. Fluctuating hydrologic conditions are the major factor controlling the vegetation pattern. The role of allogenic factors, including chance and coincidence, must be given new emphasis. Cyclic changes should be expected as water levels fluctuate. Catastrophic events such as floods and droughts also play a significant role in both modifying yet perpetuating these systems.

ECOSYSTEM-LEVEL PROCESSES

So far in this chapter we have discussed vegetational changes in wetlands. E. P. Odum (1969) described the maturation of ecosystems as a whole in an article entitled "The Strategy of Ecosystem Development." Immature ecosystems, in general, are characterized by high production to biomass (P:B) ratios; an excess of production over community respiration (P:R ratio > 1); simple, linear, grazing food chains; low species diversity; small organisms; simple life cycles; and open mineral cycles. In contrast, mature ecosystems such as old-growth forests tend to use all their production to maintain themselves and therefore have P:R ratios about equal to 1 and little if any net community production. Production may be lower than in immature systems, but the quality is better, that is, plant production tends to be high in fruits, flowers, tubers, and other materials that are rich in protein. Because of the large structural biomass of trees, the P:B ratio is small. Food chains are elaborate and detritus based, species diversity is high, the space is well organized into many different niches, organisms are larger than in immature systems, and life cycles tend to be long and complex. Nutrient cycles are closed; nutrients are efficiently stored and recycled within the ecosystem.

It is instructive to see how wetland ecosystems fit into this scheme of ecosystem development. Do their ecosystem-level characteristics fit the classical view that all wetlands are immature transitional seres? Or do they resemble the mature features of a terrestrial forest? For the wetland ecosystems covered in this book, Table 7–1 displays an evaluation of the system attributes discussed by Odum (1969). For comparison we have included a generalized developing (immature) and a mature ecosystem from Odum's article. The quantitative values are very rough because they represent means that reflect wide variation and were derived from incomplete data. Nevertheless, the table provides the following interesting insights:

1. Wetland ecosystems have properties of both immature and mature ecosystems. For example, nearly all of the nonforested wetlands have P:B ratios intermediate between developing and mature systems and have P:R ratios greater than one. Primary production tends to be very high compared with most terrestrial ecosystems. These attributes are characteristic of immature ecosystems. On the other hand, all of the ecosystems are detrital based, with complex food webs characteristic of mature systems.
2. Odum (1971) used live biomass as an index of structure or "information" within an ecosystem. Hence a forested ecosystem is more mature in this respect than a grassland. This relationship is reflected in the high P:B ratios (immature) of nonforested wetlands and the low P:B ratios (mature) of forested wetlands. In a real sense, however, peat is a structural element of wetlands because it is a primary autogenic factor modifying the flooding characteristic of a wetland site. If peat were included in biomass, herbaceous wetlands would have the high biomass and low P:B ratios characteristic of more

mature ecosystems. For example, a salt or fresh marsh has a live peak biomass of less than 2 kg/m^2. But the organic content of a meter depth of peat (peats are often many meters deep) beneath the surface is on the order of 45 kg/m^2. This is comparable to the aboveground biomass of the most dense wetland or terrestrial forest. As a structural attribute of a marsh, peat is an indication of a maturity far greater than the live biomass alone would signify.

3. Mineral cycles vary widely in wetlands, from extremely open riparian systems in which surface water (and nutrients) may be replaced thousands of times each year to bogs in which nutrients enter only in precipitation and are almost quantitatively retained. An open nutrient cycle is a juvenile characteristic of wetlands directly related to the large flux of water through these ecosystems. On the other hand, even in a system as open as a salt marsh that is flooded daily, about 80 percent of the nitrogen used by vegetation during a year is recycled from mineralized organic material (Delaune and Patrick, 1979).
4. Spatial heterogeneity is generally well organized in wetlands along allogenic gradients. The sharp, predictable zonation patterns and abundance of land-water interfaces are examples of this spatial organization. In forested wetlands, vertical heterogeneity is also well organized. This organization is an index of mature ecosystems. In most terrestrial ecosystems, however, the organization results from autogenic factors in ecosystem maturation. In wetlands most of the organization seems to result from allogenic processes, specifically hydrologic and salinity gradients created by slight elevation changes across a wetland. Thus the "maturity" of a wetland's spatial organization consists of a high level of adaptation to prevailing microhabitat differences.
5. Life cycles of wetland consumers are usually relatively short but are often exceedingly complex. The short cycle is characteristic of immature systems, although the complexity is a mature attribute. Once again, the complexity of the life cycles of many wetland animals seems to be as much an adaptation to the physical pattern of the environment as to the biotic forces. A number of animals use wetlands only seasonally or only during certain life stages. For example, many fish and shellfish species migrate to coastal wetlands to spawn or for use as a nursery. Waterfowl use northern wetlands to nest and southern wetlands to overwinter, migrating thousands of miles between the two areas each year.

Allogenic versus Autogenic Processes

Despite evidence of the role of chance in wetland ecosystem change and continuing discussions of the relative importance of allogenic versus autogenic processes, it seems clear that both autogenic and allogenic processes are important in both the pathway of development and the final characteristics of the

Table 7–1. Ecosystem Attributes of Wetlands Compared with Odum[a] Successional Attributes

	Community Energetics			Community Structure			
Ecosystem Type	*P:R Ratio*	*P:B Ratio*	*Net Primary Productivity gC m^{-2} day^{-1}*	*Food Chains*	*Total Organic Matter, kg/m^2*	*Species Diversity*	*Spatial Heterogeneity*
Developing[a]	<1 or >1	High (2–5)	High (~2–3)	Linear, grazing	Small (<2)	Low	Poorly organized
Mature[a]	1	Low (<0.1)	Low (~1)	Weblike, detritus	large (~20)	High	Well organized
Freshwater Wetlands							
Northern peatlands and bogs	>1	0.1[b]	0.8 (0.2–1.4)	Weblike, detritus	7.8 (1.2–16)[b]	Low	Well organized
Inland freshwater marshes	>1	1.2[c]	3.9 (0.7–8.2)[d]	Weblike, detritus	0.75–2.3	High	Well organized
Tidal freshwater marshes	>1	1.2[e]	1.9[e]	Weblike, detritus	1.1 (0.4–2.3)	Fairly low	Well organized
Swamp forests	1.3 (1.1–1.5)[f]	0.07 (0.015–0.09)[g]	1.2 (0.5–1.9)[g]	Weblike, detritus	22.6 (7.4–4.5)[g]	Fairly low	Well organized
Riparian forests	≥1	0.06[h]	1.4[h]	Weblike, detritus	17.4 (10–29)[h]	High	Well organized
Saltwater Wetlands							
Salt marshes	1.5[i]	2[i]	2.2 (0.45–5.7)	Weblike, detritus	1.1	Low	Well organized
Mangroves	1.9 (0.7–3.3)[j]	—	3 (0–7.5)[j]	Weblike, detritus	11 (1–29)[k]	Plants: low; animals: high	Well organized

Ecosystem Type	*Life History*		*Nutrient Cycles*		*Selection Pressure*	
	Organism Size	*Life Cycle*	*Mineral Cycles*	*Role of Detritus*	*Growth Form*	*Production*
Developing[a]	Small	Short, simple	Open	Unimportant	r	Quantity
Mature[a]	Large	Long, complex	Closed	Important	K	Quality
Freshwater Wetlands						
Northern peatlands and bogs	Small to large	Long	Closed	Important	K	Quality?
Inland freshwater marshes	Fairly small	Short, complex	Closed	Important	K?	Quality
Tidal freshwater marshes	Small	Short, complex	Open	Important	r?	Quantity
Swamp forests	Plants: large; animals: small	Long, simple Short	Open	Important	Plants: K Animals: r	Quantity
Riparian forests	Plants: large; animals: small to large	Long Short to long	Open to closed	Important	K	Quality
Saltwater Wetlands						
Salt marshes	Small	Short, complex	Open	Important	r	Quantity
Mangroves	Plants: large; animals: small	Long, simple Short, complex	Open	Important	Plants: K Animals: r?	Quantity

[a]Odum, 1969, 1971
[b]Table 12–6
[c]van der Valk and Davis (1978a)
[d]Table 11–5 (includes below-ground vegetation)
[e]Table 9–2
[f]Table 13–7
[g]Table 13–5
[h]Table 14–7
[i]Gosselink, 1984
[j]Table 10–4
[k]Table 10–2

Table 7–2. Comparison of a Young Fresh Marsh on the Louisiana Coast, with a Mature Salt Marsh and a Freshwater Floating Marsh

Ecosystem Attribute	*Young Fresh Marsh (Atchafalaya Delta)*	*Mature Salt Marsh (Barataria Basin)*	*Mature Freshwater Floating Marsh (Barataria Basin)*
Age, Year	10	1,000–1,500	1,000–1,500
Elevation, cm above local mean water level	–2-22	–0.1 – +0.1	+3
Dominant sediment	fine sand/silt	Clay	No inorganic
Organic content	low <10%	moderate 10–30%	high 80–100%
Flooding frequency, times/yr	–	260	0
(Process)	(River flow; winds)	(Tides)	(Floating marsh)
Salinity, %	<5	17.5	0
Total Sediment N, %	0.35–0.66	0.56	1.5–1.8
Total Sediment P,ppm	210–240	–	900
Net Plant Production	Moderate	High	Moderate
Allocation	roots	top and roots	roots

Source: W. B. Johnson et al. (1985); Rainey (1979); Sasser and Gosselink (1984); Sasser (1977)

mature wetland ecosystem. This is illustrated by an example from the wetlands of the Mississippi River Delta on the northern coast of the Gulf of Mexico. Typically a delta wetland originates as a fresh marsh following the formation of mud flats by river sediment deposition. Wetlands in the Atchafalaya River began to develop in 1973 after severe spring floods retreated, leaving behind new islands in Atchafalaya Bay, at the mouth of the river. At this point the major processes determining wetland development were clearly allogenic and were dominated by seasonal floods and associated sediment deposition. The plant species found in these wetlands—predominantly willows (*Salix*) and arrowheads (*Sagittaria*)—had very little influence on their environment. They existed because they were adapted to the extreme variations they experienced. At this stage all indices point to riverine control (Table 7–2). Elevations and flooding regimes are variable, reflecting the high-energy and sediment content of the river; salt concentrations are low; mineral sediment deposition overwhelms organic deposition; sediment nutrient stores are low; and plant production is low. The successful plants are fast-growing perennial trees such as willows that have fibrous roots that bind and hold the sediment and herbaceous plants that store reserves in perennial roots, where they are impervious to severe spring floods.

Typically a delta such as the Atchafalaya continues to build out onto the shallow ocean shelf for about 1,000 years until the river shifts its course to another, more efficient channel. When that happens, the fresh river water no longer holds back the ocean and the peripheral wetlands.become increasingly saline. The inner wetlands, however, are still fed primarily by fresh water from the abundant rainfall. At that point the further development of similar river-dominated marshes diverges. One track becomes a marsh whose driving environmental control is saltwater tides. The other remains fresh and in a low-energy regime, it dramatically modifies its own environment until it becomes a floating mat.

Table 7–2 contrasts these two mature systems with their youthful precursor. The salt marsh is flooded almost daily on the Gulf Coast, but the flood water energy is low and sediments are fine silts and clays. The marsh elevation range is rather small, stabilizing close to local mean high water. The marsh is maintained at this elevation in the intertidal zone by a combination of inorganic sediment carried in by tidal waters and organic materials grown in place. Only salt-tolerant species are found, but they are highly adapted to their environment. As a result, plant diversity is low but productivity is high. Much of the nutrient demand for this growth is met by mineralization (recycling) of organic material in the soil. This is an ecosystem in which the biota are adapted to the salt and tides. But they also modify their own environment chiefly through the concentration of organic debris in the soil that alters marsh elevation (and hence flooding) and stabilizes the nutrient supply.

In interior marshes that salt does not reach, the fullest expression of autogenic development occurs. With the substrates supply almost entirely cut off, the sediments become increasingly organic. As a result they become increasingly light until the whole mat becomes buoyant enough to float. When that occurs the earlier unpredictable flooding regime is replaced by a stable one in which the sediment is always saturated but the surface is never flooded. The stress of variable flooding is entirely eliminated and is replaced by another. Because the surface no longer floods, its major source of nutrients—waterborne sediments—is lost. Although new nutrients can be "wicked up" from the water under the mat, almost all of the nutrient demand of the plants is probably met by recycling from organic peat in the soil. Total productivity is quite high, but most of it is allocated to the root production necessary to maintain the floating mat.

Thus differing environmental conditions lead to the development of quite different wetland ecosystems from similar origins. Both mature systems appear to be stable and well adapted to their environments. One, the salt marsh, represents development in response to both allogenic and autogenic processes. The floating freshwater marsh appears to have modified its own environment much more strongly than the salt marsh. This example is one of many that could be cited. The chapters in Parts III and IV describe others.

The Strategy of Wetland Ecosystem Development

In the previous sections we showed that wetlands possess attributes of both immature and mature systems and that both allogenic and autogenic processes are important. In this section we suggest that in all wetland ecosystems there is a common theme: Development insulates the ecosystem from its environment. At the level of individual species, this occurs through genetic (structural and physiological) adaptations to anoxic sediments and salt (Chap. 6). At the ecosystem level it occurs primarily through peat production, which tends to stabilize the flooding regime and shifts the main source of nutrients to recycled material within the ecosystem.

The intensity of water flow over and through a wetland can be described by the water renewal-rate (*t–1*), the ratio of throughflow to the volume stored on the site (see Chap. 4). In wetlands, *t–1* varies by five orders of magnitude (Fig. 7–10), from about one per year in northern bogs to as much as 7,500 per year in low-lying riparian forests. The nutrient input (except nitrogen fixation) to a site follows closely the water-renewal rate because nutrients are carried to a site by water. The amount of nitrogen delivered to a wetland site, for example, also varies by five orders of magnitude, from less than 1 g m^{-2} yr^{-1} in a northern bog to perhaps 10 kg m^{-2} yr^{-1} in a riparian forest (Fig. 7–10; Table 7–3). Not all of this nitrogen is available to the plants in the ecosystem because in extreme cases it is flowing through much faster than it can be immobilized, but these figures indicate the potential nutrient supply to the ecosystem.

In spite of the extreme variability in these outside (allogenic) forces, wetland ecosystems are remarkably similar in many respects (Table 7–3). Total stored biomass, including peat to 1 m depth, varies from 40 to 60 kg/m^2—less than

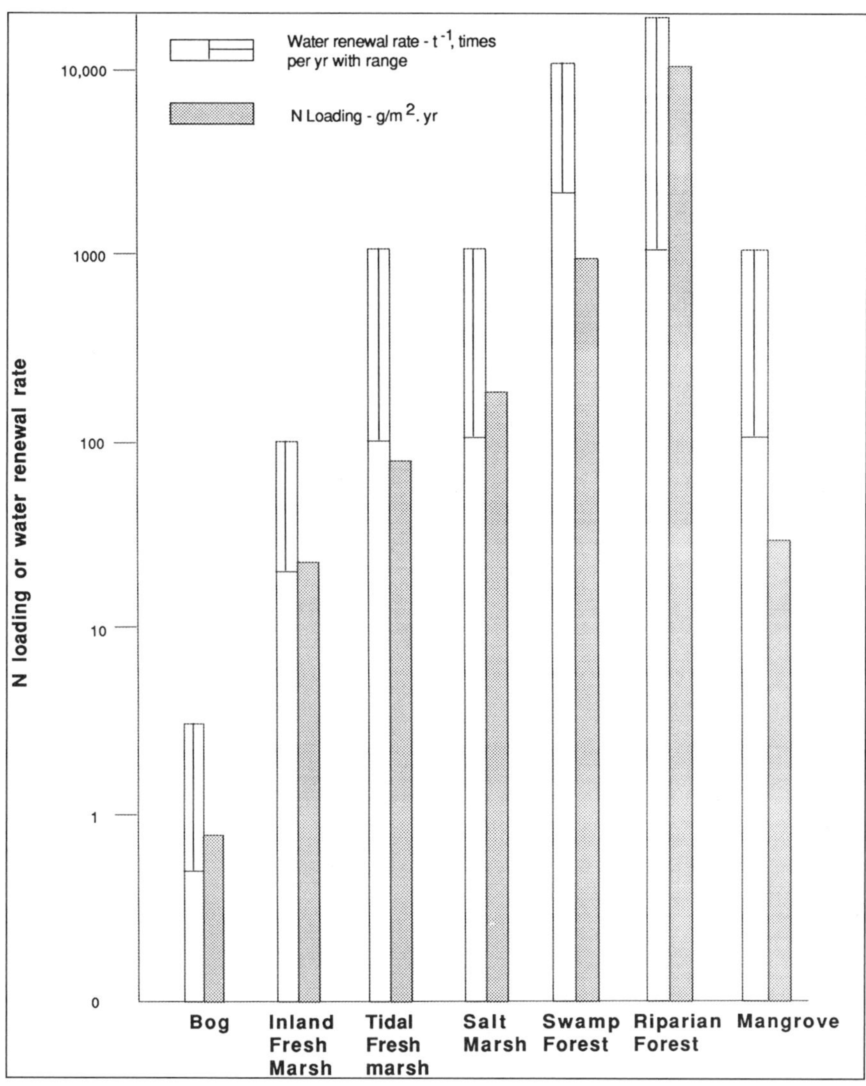

Figure 7–10. Renewal rates of water, and nitrogen loading of major wetland types.

twofold. Soil nitrogen similarly varies only about threefold, from about 500 to 1,500 g/m^2.

Net primary production, a key index of ecosystem function, varies by only a factor of about four. Mean values for different ecosystem types are usually in the range of 600 to 2,000 g m^{-2} yr^{-1} (Table 7–3). Thus although it has been shown in a number of studies of individual species (e.g., *Spartina alterniflora*; Steever et al. 1976) or ecosystems (e.g., cypress swamps; Conner et al., 1981) that produc-

Table 7–3. Comparison of Primary Productivity and Nitrogen Dynamics in Major Wetland Types[a]

	1	2	3	4	5	6	7
Wetland Type	*Net Primary Production, g m^{-2} yr^{-1} (range)*	*Total Biomass kg/m2*	*Soil Nitrogen, gN m^{-2} yr^{-1}*	*Nitrogen Loading, gN m^{-2} yr^{-1}*	*Plant Nitrogen Uptake gN m^{-2} yr^{-1}*	*Ratio N Throughput Soil Store, yr^{-1} (col 4/col 3)*	*Ratio N Throughput Uptake (col 4/col 5)*
Northern Bog	560 (153–1,943)	53	500	0.8	9	0.002	0.09
Inland Fresh Marsh	1980 (1,070–2,860)	46	1,600	22	48	0.01	0.46
Tidal Fresh Marsh	1,370 (780–2,100)	46	1,340	75	54	0.06	1.4
Salt Marsh	1,950 (330–3,700)	46	1,470	30–100	25	0.02–0.07	1.2–4.0
Swamp Forest	870 (390–1,780)	52	1,300	900	14	0.7	64
Riparian Forest	1,040 (750–1,370)	37	900	10,000	17	11	600
Mangrove	1,500 (0–4,700)	60	1,400	30	24	0.02	1.2
Range, All Wetlands	560–1,980)	37–60	500–1,470	0.8–10.000	9–54	0.0002–11	0.09–600

[a]Values are rough averages with large variability, based on data presented in chapters 8–14.

tion is directly proportional to the water renewal rate, when different wetland ecosystems that constitute greatly different water regimes are compared, the relationship breaks down. The apparent contradiction can be explained primarily by the role of stored nutrients, especially nitrogen, within the ecosystem. As the large store of organic nitrogen in the sediment (Table 7–3) mineralizes, it provides a steady source of inorganic nitrogen for plant growth. In most wetland ecosystems "new" nitrogen is not adequate or is barely adequate to supply the plants' demands but the demands are small in comparison to the amounts of stored nitrogen in the sediments (Table 7–3). As a result, most of the nitrogen demand is satisfied by recycling, even in systems as open as salt marshes. External nitrogen provides a subsidy to this basic supply. Therefore, growth is often apparently limited by the mineralization rate, which, in turn, is strongly temperature and hydroperiod dependent. Temperatures during the growing season are uniform enough to provide a similar nitrogen supply to plants in different wetland systems, except probably in northern bogs. There the low temperature and short growing season limit mineralization, and nutrient input is restricted. The combination of the two factors limits productivity.

Thus as wetland ecosystems develop, they become increasingly insulated from the variability of the environment by storing nutrients. Often the same process that stores nutrients, that is, peat accumulation, also reduces the variability of flooding, thus further stabilizing the system. The extreme example of the floating marsh was discussed earlier, but less extreme variations on the theme are common. The surface of marshes in general is built up by the deposition of peats and waterborne inorganic sediments. As the elevation increases, flooding becomes less frequent and sediment input decreases. In the absence of overriding factors, coastal wetland marshes in time reach a stable elevation somewhere around local mean high water. The surfaces of riparian wetlands similarly rise until they become only infrequently flooded. Northern bogs grow by peat deposition above the water table, stabilizing at an elevation that maintains saturated peat by capillarity. Prairie potholes may be exceptions to these generalizations. They appear to be periodically "reset" by a combination of herbivore activity and long-term precipitation cycles and achieve stability only in some cyclic sense.

LANDSCAPE PATTERNS IN WETLANDS

Many large wetland landscapes develop predictable and often complex patterns of aquatic, wetland, and terrestrial habitats or ecosystems. In high-energy environments these patterns appear to reflect abiotic forces but they are largely controlled by biotic processes in low-energy environments. At the high energy end of the spectrum, the microtopography and sediment characteristics of mature floodplains—complex mosaics of river channels, natural levees, back swamps, abandoned first and second terrace flats and upland ridges—reflect the flooding

Figure 7–11. Landscape patterns in wetlands: *(top)* the physically controlled pattern of tidal creeks in a Louisiana salt marsh; *(bottom)* a muskrat "eat-out" in a brackish marsh on the Louisiana coast. Note the high density of muskrat houses. *(Photograph by Robert Abernethy)*

pattern of the adjacent river (Gosselink et al., 1990a; see Chap. 14). The vegetation responds to the physical topography and sediments with typical zonation patterns. Salt marshes similarly develop a characteristic pattern of tidal creeks, creekside levees, and interior flats that determine the zonation pattern and vigor of the vegetation (Fig 7–11, top; see also Chap. 8)

At the low-energy end of the spectrum, the characteristic pattern of strings and flarks stretching for miles across northern peatlands appears to be primarily controlled by biotic processes. (Glaser, 1987; Rochefort et al., 1990; see Chap. 12). Similarly, in many freshwater marshes, herbivores can be major actors in the development of landscape patterns (Fig. 7–11, bottom). In actuality, both physical (climatic, topographic, hydrologic) and biotic (production rates, root binding, herbivory, peat-accumulation) processes combine in varying proportions and interact to produce observed wetland landscape patterns.

PART 3

COASTAL WETLAND ECOSYSTEMS

TIDAL SALT MARSH

Tidal Salt Marshes

8

The salt marsh, distributed worldwide along coastlines in mid- and high latitudes, is a complex ecosystem in dynamic balance with its surroundings. These marshes flourish wherever the accumulation of sediments is equal to or greater than the rate of land subsidence and where there is adequate protection from destructive waves and storms. The important physical and chemical variables that determine the structure and function of the salt marsh include tidal flooding frequency and duration, soil salinity, and nutrient limitation, particularly by nitrogen. The vegetation of the salt marsh, primarily salt-tolerant grasses and rushes, develops in identifiable zones in response to these and possibly other factors. Mud algae are also often an important component of the autotrophic community. The heterotrophic communities are dominated by detrital food chains, with the grazing food chain being much less significant and less diverse.

Salt marshes are among the most productive ecosystems of the world. Many measurements of the net primary productivity of salt marshes suggest that regional differences are related to available solar energy and to some extent to available nutrient imports by large rivers. Belowground production and anaerobic decomposition are major processes in the overall energy balance. Sulfur transformations in this environment assume much of the role of oxygen so that a major portion of the belowground flow of organic energy cycles through reduced sulfur compounds. The decomposition of dead vegetation in salt marshes at or near the surface enhances the protein content of the detritus for other marsh estuarine organisms. Nutrient cycle measurements in the marsh show few consistent patterns to indicate whether salt marshes are sources or sinks for dissolved nutrients.

Conceptual and simulation models of the salt marsh are well developed and have been significant aids to understanding their structure and function.

Beeftink (1977a) defined a salt marsh as a "natural or semi-natural halophytic grassland and dwarf brushwood on the alluvial sediments bordering saline water bodies whose water level fluctuates either tidally or non-tidally." Salt marshes, dominated by rooted vegetation that is alternately inundated and dewatered by the rise and fall of the tide, appear from afar to be vast fields of grass of a single species. In reality, salt marshes have a complex zonation and structure of plants, animals, and microbes, all tuned to the stresses of salinity fluctuations, alternate drying and submergence, and extreme daily and seasonal temperature variations. A maze of tidal creeks with plankton, fish, nutrients, and fluctuating water levels crisscrosses the marsh, forming conduits for energy and material exchange with the adjacent estuary. Studies of a number of different salt marshes have found them to be highly productive and to support the spawning and feeding habits of many marine organisms. Thus salt marshes and tropical mangrove swamps throughout the world form an important interface between terrestrial and marine habitats.

Salt marshes have been studied extensively. For many years the standard text on salt marshes was *Salt Marshes and Salt Deserts of the World* written by V. J. Chapman (1960). The ecosystem approach to wetlands was pioneered by scientists at the University of Georgia Marine Laboratory on Sapelo Island, Georgia, and early interest in this ecosystem was aroused by a symposium at that laboratory in 1958 (Ragotzkie et al., 1959) and continued by many researchers (see, e.g., Pomeroy and Wiegert, 1981). A less technical and excellent early description of salt marshes was written by Teal and Teal (1969). More recently a number of thorough reviews of regional coastal saline marshes were commissioned by the U.S. Fish and Wildlife Service: Northeast Atlantic coast marshes (Teal 1986), New England high salt marshes (Nixon, 1982), Southeast Atlantic Coast marshes (Wiegert and Freeman 1990), Northeast Gulf Coast irregularly flooded salt marshes (Stout, 1984), Gulf Coast deltaic marshes (Gosselink 1984), southern California coastal salt marshes (Zedler, 1982), and San Francisco Bay tidal marshes (Josselyn, 1983). Additional useful summaries include one on Florida coastal wetlands (Montague and Wiegert, 1990), one on New England salt marshes (Bertness, 1992), and small books by Chabreck (1988) and Allen and Pye (1992).

GEOGRAPHICAL EXTENT

Salt marshes are found in mid- and high latitudes along intertidal shores throughout the world. They are replaced by mangrove swamps along coastlines in tropical and subtropical regions (between 25°N and 25°S latitude; see Chap. 10). The distribution of salt marshes in North America is shown in Figure 8–1.

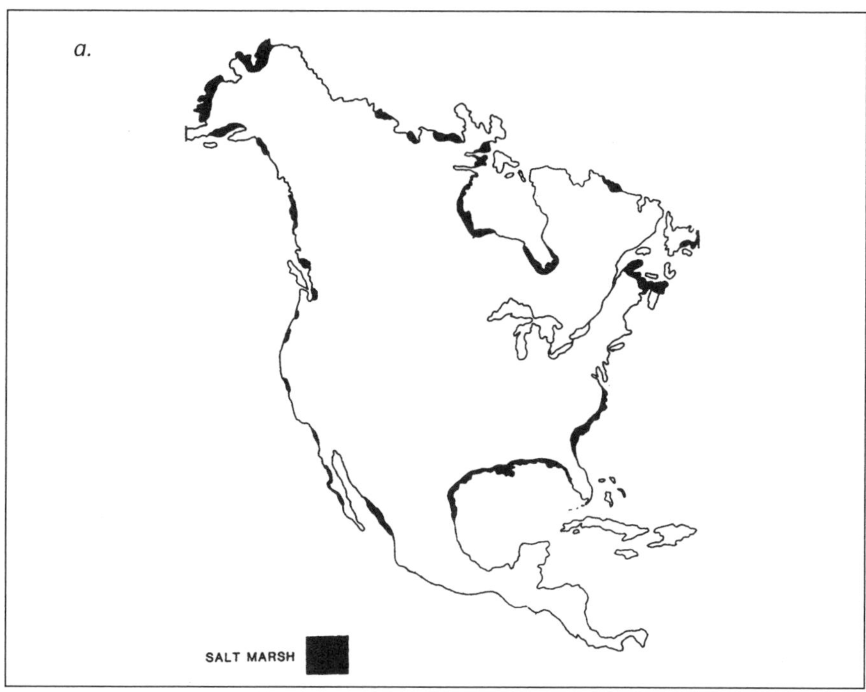

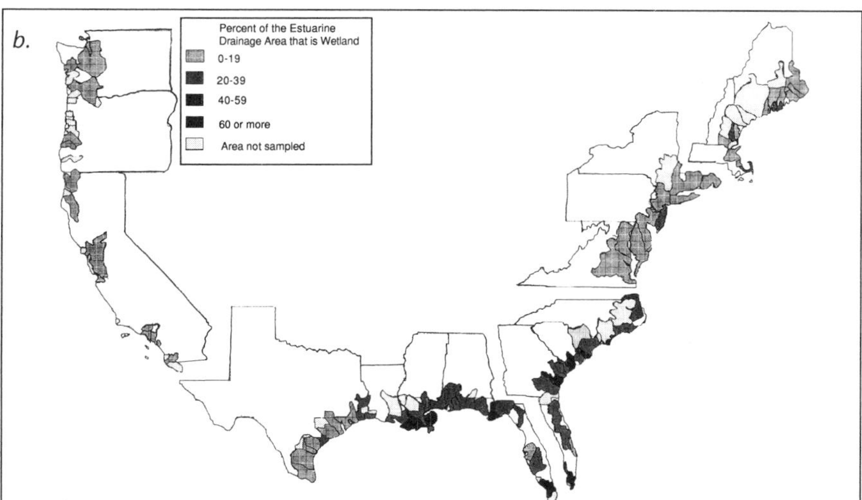

Figure 8–1. *a.* Distribution of salt marshes in North America. *b.* Distribution of wetlands in coastal drainage areas of the United States. (a. *after Chapman, 1977;* b. *from Field et al., 1991*)

Salt marshes can be narrow fringes on steep shorelines or expanses of several kilometers wide. They are found near river mouths, in bays, on protected coastal plains, and around protected lagoons. Different plant associations dominate different coastlines, but the ecological structure and function of salt marshes is similar around the world. Chapman (1960, 1974, 1975, 1976a, 1977) divided the world's salt marshes into nine geographical units. Those that apply to North America include the following:

1. *Arctic Marshes*. This group includes marshes of northern Canada, Alaska, Greenland, Iceland, northern Scandinavia, and Russia. Probably the largest extent of marshes in North America, as much as 300,000 km^2, occurs along the southern shore of the Hudson Bay. Those marshes, influenced by a positive water balance and numerous inflowing streams, can generally be characterized as brackish rather than saline (Ewing and Kershaw, 1986). Various species of the sedge *Carex* and the grass *Puccinellia phryganodes* often dominate. Parts of the southwestern coast of Alaska are dominated by species of *Salicornia* and *Suaeda*.
2. *Eastern North American Marshes*. These marshes, mostly dominated by *Spartina* (cordgrass) and *Juncus* (rush) species, are found along the East Coast of the United States and Canada and the Gulf Coast of the United States. This unit is further divided into three groups:

   a. Bay of Fundy group. River and tidal erosion is high in the soft rocks of this region, producing an abundance of reddish silt. The tidal range, as exemplified at the Bay of Fundy, is large, leading to a few marshes in protected areas and considerable depth of deposited sediments. *Puccinellia americana* dominates the lower marsh, and *Juncus balticus* is found on the highest levels.
   b. New England group. Marshes are built mainly on marine sediments and marsh peat, and there is little transport of sediment from the hard-rock uplands. These marshes range from Maine to New Jersey.
   c. Coastal Plain Group. These marshes extend southward from New Jersey along the southeastern coast of the United States to Texas along the Gulf of Mexico. Major rivers supply an abundance of silt from the recently elevated Coastal Plain. The tidal range is relatively small. The marshes are laced with tidal creeks. Mangrove swamps replace salt marshes along the southern tip of Florida. Because of the extensive delta marshes built by the Mississippi River, the Gulf Coast contains about 60 percent of the coastal salt and fresh marshes of the United States (Fig. 8–1; Field et al. 1991).

3. *Western North American Marshes*. Compared with the Arctic and the East Coast, salt marshes are far less developed along the western coasts of the United States and Canada because of the geomorphology of the coastline. A narrow belt of *Spartina foliosa* is often bordered by broad belts of *Salicornia* and *Suaeda*.

GEOMORPHOLOGY

The physical features of tides, sediments, freshwater inputs, and shoreline structure determine the development and extent of salt marsh wetlands within their geographical range. Coastal salt marshes are predominantly intertidal, that is, they are found in areas at least occasionally inundated by high tide but not flooded during low tide. A gentle, rather than steep, shoreline slope allows for tidal flooding and the stability of the vegetation. Adequate protection from wave and storm energy is also a physical requirement for the development of salt marshes. Sediments that build salt marshes originate from upland runoff, marine reworking of the coastal shelf sediments, or organic production within the marsh itself.

Although a number of different patterns of development can be identified, salt marshes can be classified broadly into those that were formed from reworked marine sediments on marine-dominated coasts and those that were formed in deltaic areas where the main source of sediment is riverine. The former type is typical of most of the North American coastline. Deltaic marshes develop mainly where large rivers debouch onto low-energy coasts, which in North America restricts them to the coasts of the South Atlantic and the Gulf of Mexico. The Mississippi River deltaic marshes are the major example of this type of development and are the most extensive coastal marshes in the United States.

Marine-Dominated Marsh Development

On marine-dominated coasts, salt marsh development requires sufficient shelter to ensure sedimentation and to prevent excessive erosion from wave action (Beeftink, 1977a). Some shoreline features that allow the development of salt marshes are shown in Figure 8-2. Marshes can develop at the mouths of estuaries where sediments are deposited by the river behind spits and bars that offer protection from waves and longshore currents. Chapman (1960) further described three situations in which salt marshes will develop:

1. Shelter of spits, offshore bars, and islands. Salt marshes will form along coastlines only where a bar, a neck of land (called a spit), or an island acts to trap sediment on its lee side and protects the marsh from the full forces of the open sea (Fig. 8–2b, d). The most extensive examples of this type of coastal salt marsh have developed behind outer barrier reefs along the Georgia and the Carolina coast. In their early development sedimentation is rapid, both vertically and laterally. The resulting marshes are low in the intertidal zone and are usually dominated by *Spartina alterniflora*. The drainage pattern is well developed, and there is a pronounced meandering and erosion of tributaries at the headward end. As the marsh matures, sedimentation is primarily vertical because lateral deposition is balanced by erosion at the seaward ends

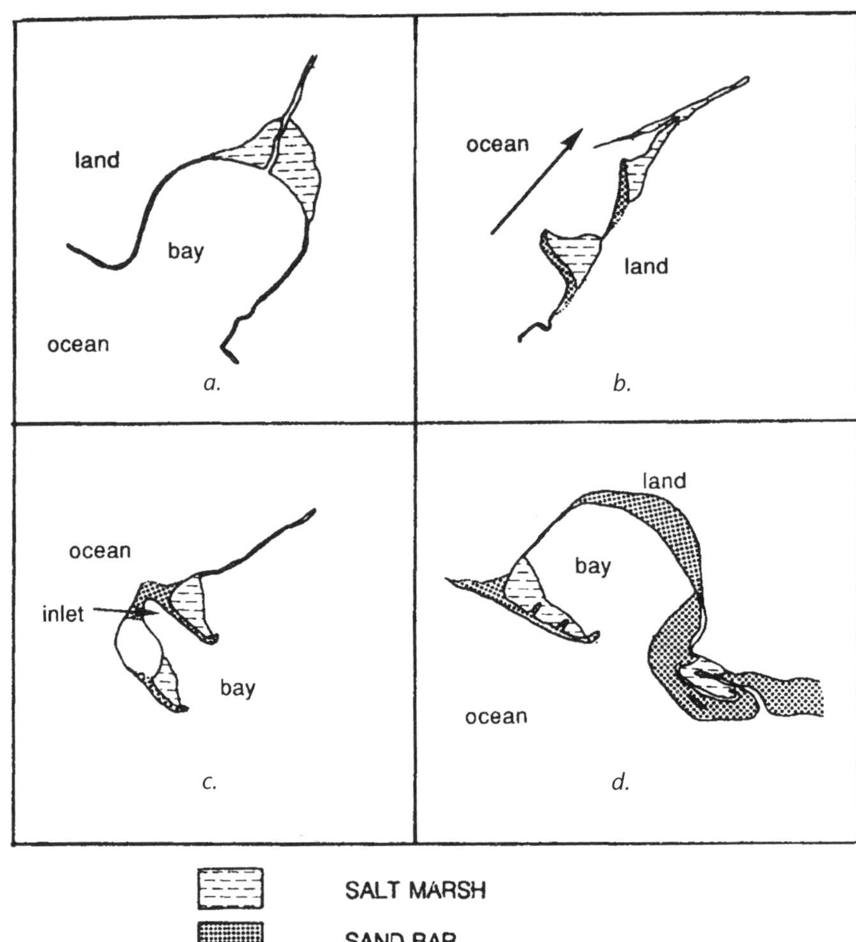

Figure 8–2. Diagram of typical shoreline features that allow for the development of salt marshes. *(After Chapman, 1960)*

of the marsh. As the elevation of the marsh approaches the highest excursion of the tide, sediment deposition slows and an equilibrium is attained between deposition and erosion. At maturity about half of the area is a monotypic low marsh stand of *S. alterniflora,* the rest a high marsh dominated by a mixture of short *S. alterniflora*, *Salicornia* spp., *Distichlis spicata,* and *Juncus roemerianus*. As the marsh continues to age, high marsh becomes increasingly dominant. The extensive drainage system is slowly filled so that surface flows dominate and tidal forces become increasingly weak. When this occurs in areas of excess rainfall, fresh marsh and terrestrial plant species start to

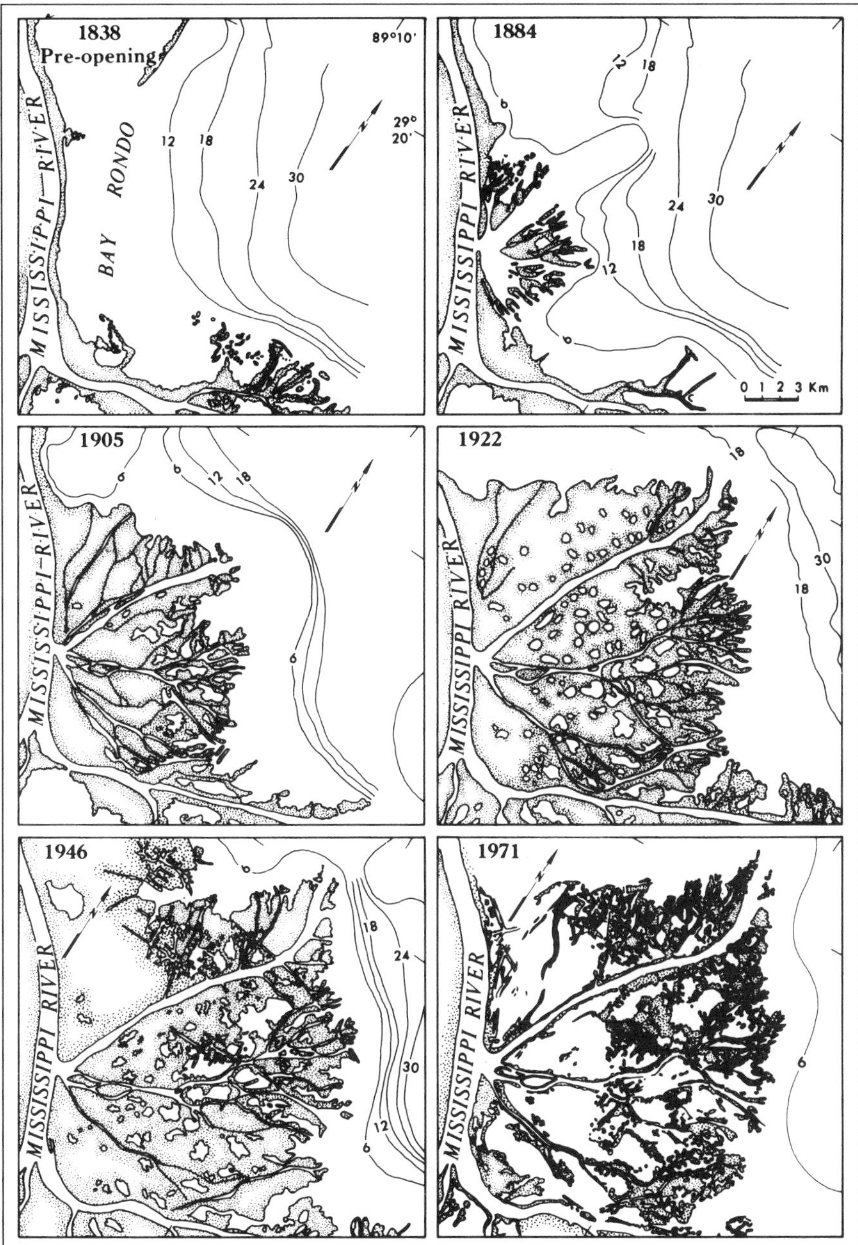

Figure 8–3. Deltaic marsh development of the Mississippi River. Historic maps illustrate the development and degradation of a crevasse splay at Cubits Gap Bay, 1838 to 1971. (*From Wells et al., 1982*)

invade (Frey and Basan, 1985; Wiegert and Freeman, 1990).

2. Protected bays. Several large bays such as the Chesapeake Bay, the Hudson Bay, the Bay of Fundy, and the San Francisco Bay are adequately protected from storms and waves so that they can support extensive peripheral salt marshes (Fig. 8–2a).
3. Some estuarine salt marshes have features of both marine and deltaic origins. They occur on the shores of estuaries where shallow water and low gradients lead to river sediment deposition in areas protected from destructive wave action (Fig. 8–2c). Tidal action must be strong enough to maintain salinities above about 5 ppt; otherwise, the salt marsh will be replaced by reeds, rushes, and other freshwater aquatic plants.

River-Dominated Marsh Development

Major rivers carrying large sediment loads build marshes into shallow estuaries or out onto the shallow continental shelf where the ocean is fairly quiet. In this situation, the first marshes developing on the newly deposited sediments are dominated by freshwater species. Typically, however, the river course shifts through geologic time, and the abandoned marshes, no longer supplied with fresh river water, become increasingly marine influenced. In the Mississippi River Delta, these marshes undergo a 5,000-year cycle of growth as fresh marshes, transition to salt marshes, and finally degradation back to open water under the influence of subsidence and marine transgression. During the last stage, the seaward edges of the marshes are reworked into barrier islands and spits in the same way as coastal marshes on the Atlantic Coast (Fig. 8–3).

MARSH DEVELOPMENT AND STABILITY

The long-term stability of a salt marsh is determined by the relative rates of two processes: sediment accretion on the marsh (including the production and the deposition of peat by growing plants), which causes it to expand outward and grow upward in the intertidal zone, and by coastal submergence caused by rising sea level and marsh surface subsidence. These two processes are to some extent self-regulating, for as a marsh subsides, it is inundated more frequently and thus receives more sediment and stores more peat (because the substrate is more anoxic and organic deposits degrade more slowly). Conversely, if a marsh accretes faster than it is submerging, it gradually rises out of the intertidal zone, is flooded less frequently, receives less sediment, and oxidizes more peat.

Local conditions have an overriding control over the balance achieved between accretion and submergence. The amount and type of sediment particles in the water column are determined by their source, whether coarse or fine, organic or inorganic, marine or riverine. The amplitude and frequency of the tide as well as the volume and seasonal variation of river flows onto the coast determine current speeds and hence the capacity to carry sediments. The morphology of the area determines the pattern of flow, erosion, and deposition.

Although physical processes probably dominate in coastal marshes, the effects of biota can also be significant. This is especially true when one considers the formation of peat, which is almost entirely caused by the *in situ* production of organic matter by marsh plants. In addition, the biota of the marsh control their physical environment in several other ways (O'Neil, 1949; Frey and Basan, 1985; Wiegert and Freeman, 1990): (1) Emergent grass dampens wind-generated waves, changing the sediment-transport capacity of flooding water compared to open water areas; (2) stems and leaves slow the water velocity, thus promoting sediment deposition; (3) changes caused by plants in the salinity of surrounding waters may influence the deposition of clays; (4) roots and rhizomes increase the stability of the sediment and its resistance to hydraulic erosion; (5) algal, bacterial, and diatom films help trap fine sediments; (6) colonial animals influence deposition and sediment structure, for example, oyster colonies directly modify the flow of water over them, and the dense concentration of fiddler crab burrows directly influences sediment permeability; (7) macroinvertebrates trap enormous quantities of suspended detritus, depositing it as feces or pseudofeces; and (8) grazing by waterfowl and mammals such as the nutria may completely denude an area, exposing it to tide and wind-driven erosive forces.

Gulf Coast of North America

A few examples will serve to illustrate the wide diversity of developmental paths in a salt marsh. The sea level has been quite stable throughout the world for the last 5,000 years. Along the northern Gulf Coast, however, submergence is currently rapid, mostly because of the subsidence of the surface by compaction of deltaic sediments and downwarping of the older Pleistocene surface. There, where tidal energy is low and Mississippi River sediments are no longer supplied to the coastal marshes (they are channeled directly into deep offshore waters), accretion is not keeping up with submergence and salt marshes are degrading rapidly. Delaune et al. (1983c) showed that the percent of open water in a Gulf coastal marsh was directly related to the rate of coastal submergence over an 85-year period. The average rate of coastal submergence was 1.2 cm/yr compared to a marsh accretion (sediment deposition) rate of 0.66 to 0.78 cm/yr. The accumulated aggradation deficit was closely related to wetland loss (Fig. 8–4). Deteriorating salt marshes along this coast receive most of their sediments from the reworking of marine and estuarine deposits during severe storms, especially hurricanes (Table 8–1) but stable marshes depend more on riverine input during spring floods (Baumann et al., 1984). Both inland and streamside salt marshes showed a net loss

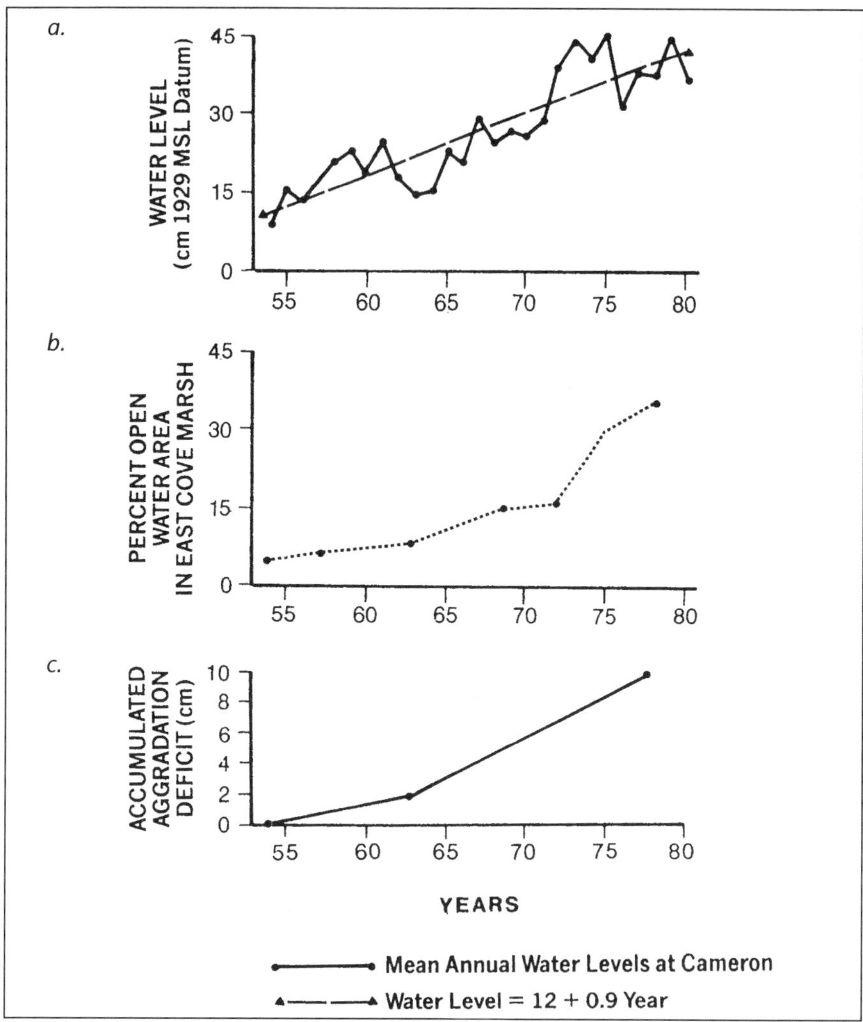

Figure 8–4. Relationships among *a.* apparent sea level rise, *b.* wetland conversion to open water, and *c.* a calculated aggradation deficit in a Gulf Coast salt marsh. *(From Delaune et al., 1983c; copyright © 1983 by the Society of Economic Paleontologists and Mineralogists, reprinted with permission)*

of elevation in this subsiding landscape, with a net gain only on streamside marshes when hurricane inputs are included (Table 8–1).

North Atlantic Coast

Along the North Atlantic Coast, the processes of accretion and submergence are apparently close to a dynamic equilibrium. There, the sea level has been rising at

Table 8–1. Rate of Sediment Accumulation of Salt Marshes

Vegetation Zone	Rate, cm/100 Years: Accretion	Subsidence	Net Accumulation	Age Years
Gulf Coast (*Louisiana*)[a]				
Barataria Bay (deteriorating)				
Spartina alterniflora-Spartina patens				
Streamside				
With hurricanes (4 yr)	150	123	27	
Without hurricanes (3 yr)	110	123	–13	
Inland				
With hurricanes	90	123	–33	
Without hurricanes	60	123	–63	
Four League Bay (stable)				
Streamside, no hurricanes	130	85	45	
Inland, no hurricanes	56	85	–29	
North Atlantic Coast (*Massachusetts*)[b]				
Spartina alterniflora	61	30	31	490
Spartina patens-Distichlis spicata	38	30	8	600
Juncus gerardi	32	30	2	1,200
Southern England (*Dorset*)[c]				
mudflats	340–390	10.50		
Spartina anglica zone	11–65	10.50		
Halimione zone	29–61	10.50		
high marsh	17–33	10.50		

[a]Data from Chapman (1960)
[b]Data from Baumann et al. (1984)
[c]Data from Gray (1992) and Allen (1992)

between only 1 and 3 mm/yr for the past several thousand years (Teal, 1986). Accretion in many marshes has been somewhat faster. Redfield (1965; 1972) gave a detailed description of the development of this kind of salt marsh in New England during the past 4,000 years in a general model of peat and sediment accumulation in the presence of a continually slowly rising sea level (Fig. 8–5). As the sea level rises ($HW_0 \longrightarrow HW_3$), the marsh extends inland and the upland is covered by marsh peat. At the same time, if sediments accumulate beyond the lower limit of marsh vegetation (called the thatch line) at a rate greater than the rise in sea level, the intertidal marsh migrates seaward to main-

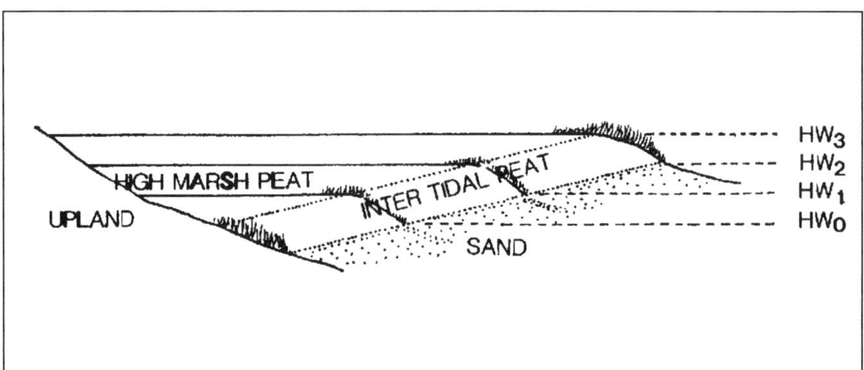

Figure 8–5. Development of a New England salt marsh. HW_0, HW_1, HW_2, and HW_3 refer to successive high water levels as sea level rises. *(From Redfield, 1965; copyright © 1965 by the American Association for the Advancement of Science, reprinted with permission)*

tain its critical elevation. The upper (high) marsh may also develop seaward over the old intertidal peat. The expansion of the marsh in this example occurs not primarily because of a large mineral sediment source and relatively high tidal energy compared to the Gulf Coast but because peat formation and aggradation exceed a submergence rate of only a fraction of the rate experienced in Louisiana (Table 8–1).

By examining old maps and by determining the depth of marsh peat and the sedimentation rate (which can be estimated from radioactive markers such as Pb-210 and Cs-137), scientists can measure the age of salt marshes. Such studies indicate that the oldest present-day salt marshes were formed during the last 3,000–4,000 years. In one study of a New England salt marsh (Table 8–1), it was found that the lower (seaward) marsh, dominated by *Spartina alterniflora*, accumulated sediments at a much greater rate than did the more inland upper marsh dominated by *Juncus gerardi*.

Subarctic North America

A third example of marsh development is that of the northern part of the North American continent, which is emerging slowly as the land rises at a rate of about 1 cm per year in response to the melting of the ice sheet that covered the land during the last ice age (Martini et al., 1980). As a result, in subarctic marshes such as those along the southern shore of Hudson Bay, the sea is retreating, shallow flats are being exposed and invaded by salt marsh species, and the whole marsh is expanding outward.

These examples show how delicately poised the salt marsh is geomorphically. A change in the rate of sea level rise or sedimentation of as little as a mil-

Table 8–2. Hydrologic Demarcation Between Lower Marsh and Upper Marsh in the Salt Marsh Ecosystem

	Submergences		
Marsh	*per day in daylight*	*per year*	*Maximum Period of Continuous Exposure, days*
Upper Marsh	<1	<360	≥10
Intertidal Marsh	>1.2	>360	≤9

Source: Data from Chapman (1960)

limeter or two per year can determine whether a marsh will expand, retreat, or degrade.

HYDROLOGY

Tidal energy represents a subsidy to the salt marsh that influences a wide range of physiographic, chemical, and biological processes, including sediment deposition and scouring, mineral and organic influx and efflux, flushing of toxins, and the control of sediment redox potential. These physical factors, in turn, influence the species that occur on the marsh and their productivity. The lower and upper limits of the marsh are generally set by the tide range. The lower limit is set by the depth and the duration of flooding and by the mechanical effects of waves, sediment availability, and erosional forces (Chapman, 1960). At least two or three days of continuous exposure is required during the seed-germination period for seedling establishment. The upland side of the salt marsh generally extends to the limit of flooding on extreme tides, normally between mean high water and extreme high water of spring tides (Beeftink, 1977a). Based on marsh elevation and flooding characteristics, the marsh is often divided into two zones, the upper marsh (or *high marsh*) and the lower marsh (or *intertidal marsh*). The upper marsh is flooded irregularly and has a minimum of at least ten days of continuous exposure to the atmosphere, whereas the lower marsh is flooded almost daily, and there are never more than nine continuous days of exposure (Table 8–2). This classification is simplistic, as will be clear from examples discussed in the Ecosystem Structure section. Nevertheless, it is a first approximation that has considerable practical value.

Tidal Creeks

A notable physiographic feature of salt marshes, especially low marshes, is the development of tidal creeks in the marsh itself. These creeks develop, as do

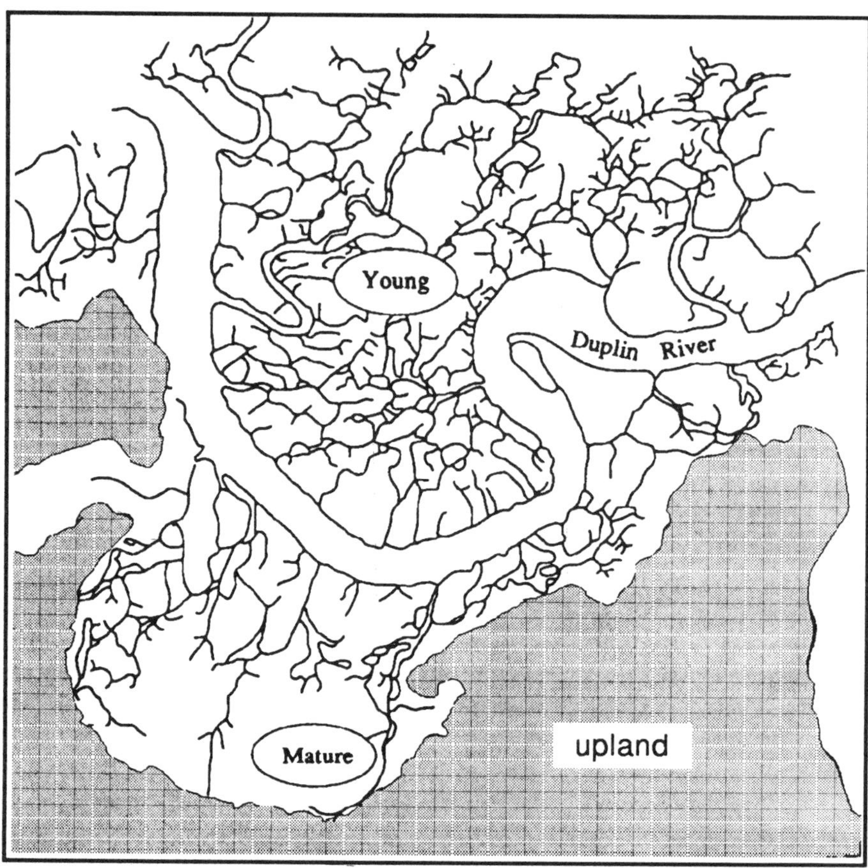

Figure 8–6. Drainage patterns of typical young and mature lagoonal *Spartina alterniflora* marshes in the Duplin River drainage, Doboy Sound, Georgia. (*From Wiegert and Freeman, 1990, after Wadsworth, 1979)*

rivers, "with minor irregularities sooner or later causing the water to be deflected into definite channels" (Chapman, 1960). Redfield (1965, 1972) suggested that these tidal creeks had already developed on sand flats before they were encroached upon by advancing intertidal peat. The creeks serve as important conduits for material and energy transfer between the marsh and its adjacent body of water. A tidal creek has salinity similar to that of the adjacent estuary or bay, and its water depth varies with tide fluctuations. Its microenvironments include different vegetation zones along its banks with aquatic food chains that are important to the adjacent estuaries. Because the flow in tidal channels is bidirectional, the channels tend to remain fairly stable, that is, they do not meander as much as streams that are subject to a unidirectional flow. As marshes mature and sediment deposition increases elevation, however, tidal creeks tend to fill in and their density decreases (Fig. 8–6).

Sediments

The sediment source and tidal current patterns determine the sediment characteristic of the marsh. Salt marsh sediments can come from river silt, from organic productivity in the marsh itself, or from reworked marine deposits. As a tidal creek rises out of its banks, water flowing over the marsh slows and drops its coarser grained sediment load near the stream edge, creating a slightly elevated streamside levee. Finer sediments drop out farther inland. This gives rise to a well-known "streamside" effect characterized by the greater productivity of grasses along tidal channels than inland, a result of the slightly larger nutrient input, higher elevation, and better drainage. The source of mineral sediment is not as important for the productivity of the marsh as elevation, drainage, and organic content, all of which are determined by local hydrologic factors.

Pannes

A distinctive feature of salt marshes is the occurrence of pannes (pans). The term *panne* is used to describe bare, exposed, or water-filled depressions in the marsh (Wiegert and Freeman, 1990), which may have different sources. In the higher reaches of the marsh, inundated by only the highest tides, pannes, also called *sand barrens* (Frey and Basan, 1985), appear to form where evaporation concentrates salts in the substrate, killing the rooted vegetation. These exposed barrens are often covered by thin films of blue-green algae. *Mud barrens* are naturally occurring depressions in the marsh that are intertidal and retain water even during low tide. These pannes are often barren of vascular vegetation or support submerged or floating vegetation because of the continuous standing water and the elevated salinities when evaporation is high. They are continually forming and filling due to shifting sediments and organic production. The vegetation that develops in a mud panne, for example, widgeon grass (*Ruppia* sp.), is tolerant of salt at high concentrations in the soil water. Relatively permanent ponds are formed on some high marshes and are infrequently flooded by tides (Redfield, 1972). Because of their shallow depth and their support of submerged vegetation, pannes are used heavily by migratory waterfowl. Pannes are a common feature of human intervention, occurring where free tidal movement has been blocked by roads or levees, where spoil deposits have elevated a site, or where soil excavation, for example, for highway construction, has occurred in a marsh.

CHEMISTRY

The development and zonation of vegetation in the salt marsh are influenced by several chemical factors. Three of the most important are the soil water salinity, which is linked with tidal flooding frequency; the availability of nutrients, par-

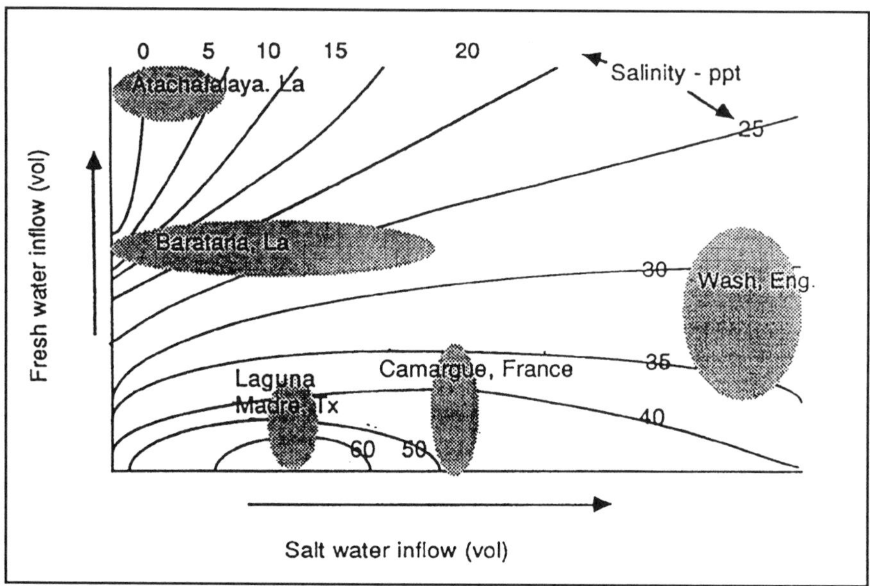

Figure 8–7. A conceptual diagram of the average salinities of salt marshes as related to the tidal range (salt water inflow) and fresh water supply (river inflow).

ticularly macronutrients such as nitrogen; and the degree of anaerobiosis, which controls the pathway of decomposition and nutrient availability.

Salinity

A dominant factor in the productivity and species selection of the salt marsh is the salinity of the overlying water and the soil water. The salinity in the marsh soil water depends on several factors (adapted from Morss, 1927, and Chapman, 1960):

1. *Frequency of Tidal Inundation.* The lower marsh soils that are flooded frequently tend to have a fairly constant salinity approximating that of the flooding seawater. On the other hand, the upper marsh that is only occasionally flooded experiences long periods of exposure that may lead to either higher or lower salt concentrations.
2. *Rainfall.* Frequent rainfall tends to leach the upper soil in the high marsh of its salts (Ranwell et al., 1964); frequent periods of drought, on the other hand, lead to higher salt concentrations in the soil.
3. *Tidal Creeks and Drainage Slopes.* The presence of tidal creeks and steep slopes that drain away saline water can lead to lower soil water salinity than that which would occur under poorly drained conditions.
4. *Soil Texture.* Silt and clay materials tend to reduce drainage rates and retain more salt than does sand.

5. *Vegetation.* The vegetation itself has an influence on soil salinity. Evaporation of water from the marsh surface is reduced by vegetation cover, but transpiration is increased. The net effect depends on the type of vegetation and the environmental setting. Salt marsh vegetation also changes the ion balance in soils when roots take up ions selectively from the surrounding soil solution (Smart and Barko, 1980).
6. *Depth to Water Table.* When groundwater is close to the surface, soil water salinity fluctuations are less.
7. *Fresh Water Inflow.* The inflow of fresh water in rivers, as overland flow, or in groundwater tends to dilute the salinity in both the salt marsh and the surrounding estuary. The early spring-flood periods along much of the eastern United States Coastal Plain lead to significant reductions in the salinity of downstream coastal marshes.
8. *Fossil Salt Deposits.* The presence of fossil salt deposits in the substrate can increase salt concentrations in the root zone, as occurs in Hudson and James Bays (Price and Woo, 1988).

The interaction of these factors on marsh salinity is illustrated by two examples. The first, as described in Figure 8–7, relates to the average salinity of marshes worldwide, as influenced by the relative size of the freshwater source (here shown as river inflow, although local rainfall contributes to the freshwater input) to the marine input as indicated by the range of the tide. In areas that experience a large tide range (e.g., the Wash, England), marshes tend to approximate the ambient marine water salinity even though rainfall may be significant. In coastal marshes adjacent to large rivers, on the other hand (e.g., the north coast of the Gulf of Mexico), fresh water dilutes marine sources and the marshes are brackish or even fresh. Extreme salinities are found in subtropical areas such as the Texas Gulf Coast where rivers and rainfall supply little fresh water and tides have a narrow range so that flushing is reduced. As a result, marine water is concentrated by evapotranspiration, often to double seawater strength or even higher.

A second example (Fig. 8–8) shows the development of a lateral salinity gradient as a function of the flooding frequency in an Atlantic Coast salt marsh. Near the adjacent tidal creek frequent tidal inundation keeps sediment salinity at or below sea strength. As the marsh elevation increases, the inundation frequency decreases and the finer sediments drain poorly. At the elevation shown in Fig. 8–8 as salt flats, infrequent spring tides bring in salt water that is concentrated by evaporation. Flushing is not frequent enough to remove these salts, and so they accumulate to lethal levels. Above this elevation tidal flooding is so infrequent that salt input is restricted, and flushing by rainwater is sufficient to prevent salt accumulation. In this way, the salt gradient set up by the interaction of marsh elevation, tides, and rain often controls the general zonation pattern of vegetation and its productivity. Within the salt marsh zone itself, however, all plants are salt tolerant, and it is misleading to account for plant zonation and

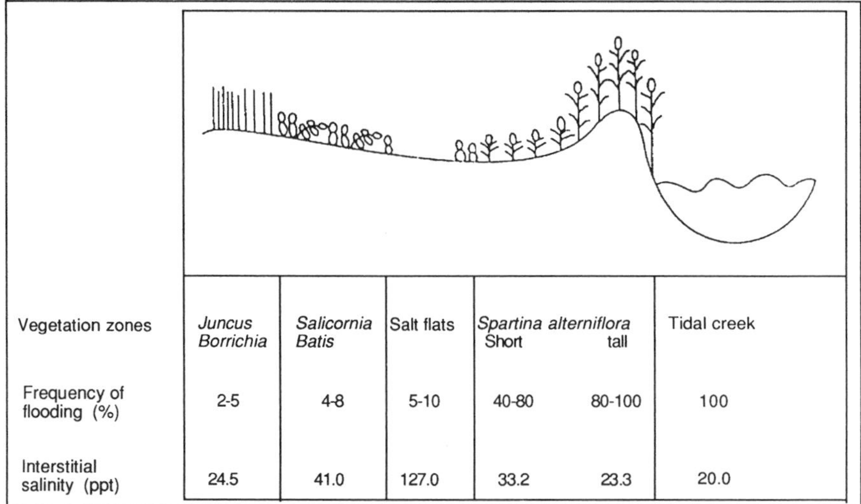

Vegetation zones	*Juncus Borrichia*	*Salicornia Batis*	Salt flats	*Spartina alterniflora* Short	tall	Tidal creek
Frequency of flooding (%)	2-5	4-8	5-10	40-80	80-100	100
Interstitial salinity (ppt)	24.5	41.0	127.0	33.2	23.3	20.0

Figure 8–8. The relation of a salt flat's interstitial salinity and its vegetation. *(From Wiegert and Freeman, 1990, after Antlfinger and Dunn, 1979)*

productivity on the basis of salinity alone. Salinity, after all, is the net result of many hydrodynamic factors, including tides, rainfall, freshwater inputs, and groundwater, as described above. When *Spartina* flourishes in the intertidal zone, it is responding also to tides that reduce the local salinity, remove toxic materials, supply nutrients, and modify soil anoxia. All of these factors collectively contribute to different productivities and different growth forms in the intertidal and high marshes.

Nutrients

Nitrogen

The availability of nutrients, particularly nitrogen and phosphorus, in the salt marsh soil is important for the productivity of the salt marsh ecosystem. Several wetland studies (e.g., those by Valiela and Teal, 1974, and Smart and Barko, 1980) have shown that salt marsh vegetation is primarily nitrogen limited. Figure 8–9 shows typical concentrations of total nitrogen and ammonium nitrogen in a salt marsh soil transect. A comparison of Figure 8–9a with Figure 8–9b shows that ammonium nitrogen, the primary form available to marsh vegetation, is only a small percentage (less than 1 percent) of the total nitrogen in the marsh soil. This is typical of organic wetland soils in general. Mendelssohn (1979), however, found ammonium to be the dominant form of available inorganic nitrogen in salt marsh interstitial water in North Carolina by one to two orders of magnitude over nitrate nitrogen. This results from the near anaerobic conditions usually present in the soil water, which precludes the buildup of nitrate nitrogen.

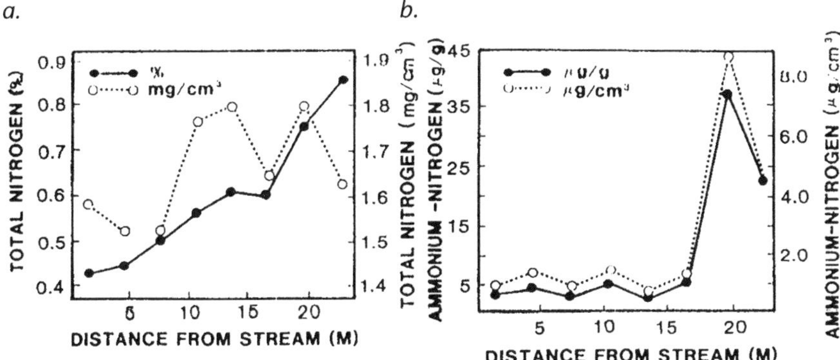

Figure 8–9. Variation in *a.* total nitrogen and *b.* extractable soil ammonium-nitrogen with distance inland from tidal stream in Louisiana salt marsh. *(From Buresh et al., 1980b; copyright © 1980 by the Estuarine Research Federation, reprinted with permission)*

Another interesting feature of Figure 8–9 is the increase in ammonium in the inland direction. Although, as stated above, salt marsh plant growth rates are generally limited by the supply of inorganic nitrogen, in this example ammonium concentrations are high precisely where growth is poorest. A number of scientists have observed this phenomenon and have tried to understand it (see Mendelssohn et al., 1982). It appears that other, nonnutrient, stresses related to the poor drainage and low redox potentials limit the ability of inland plants to assimilate ammonium nitrogen. Morris (1984) determined that under anaerobic conditions the rates of ammonium absorption by the roots of *Spartina alterniflora* and *S. patens* decreased by 60 percent and 40 percent, respectively. Absorption rates were further inhibited by high salinities in *S. patens*, but not in *S. alterniflora*. These results suggest that lowered ammonium absorption in the highly anaerobic inland marshes leads to accumulation in the substrate and that along the creek bank, available nitrogen is kept at low levels by the actively growing plants (Mendelssohn, 1979).

It is curious that although at least two studies have shown little relationship between soil nutrient concentrations (g/g dry wt. soil) and plant biomass (Broome et al., 1975a, b; Delaune et al., 1979), there tends to be a strong positive correlation between aboveground vegetation biomass and soil nutrient *density* (g/cm^3 wet soil; Table 8–3). Broome et al. (1975a) found that only soil-extractable phosphorus and zinc concentrations were correlated with plant biomass on a dry wieght basis; in the study by Delaune et al. (1979), only soil carbon concentration was related (negatively) to plant biomass. In contrast, when the nutrient content per unit volume of soil was examined, sodium, potassium, calcium, magnesium, total nitrogen, and extractable phosphorus were all positively correlated with aboveground plant biomass (Delaune et al., 1979).

Table 8–3. Concentration of Constituents of a Louisiana Salt Marsh Soil

Constituent	*Concentration, mg/g dry soil*	*Density, mg/cm³ soil volume*
Na^+	17	3.84[a]
K^+	1.753	0.40[b]
Ca^{++}	1.646	0.37[a]
Mg^{++}	3.864	0.87[a]
Total N	7.2	1.63[a]
Extractable P	0.123	0.028[b]
Organic C	12.0[a]	26.9

[a]Significantly correlated with above-ground plant biomass at the 0.05 level of significance.
[b]Significantly correlated with above-ground plant biomass at the 0.01 level of significance
Source: After Delaune et al. (1979)

Where bulk density varies widely from sample to sample, nutrient density on a volume basis gives a more consistent physiological characterization of the soil than concentration on a dry weight basis because it reflects the nutrients available to a plant relative to a volume of soil invaded by the plant's root system (Mehlich, 1972; Gosselink et al., 1984). Soils rich in minerals on a volume basis tend to be high in many plant nutrients and are more productive than mineral-poor soils; hence the positive correlations shown in Table 8–3. It is probable that the positive correlations stem as much from the buffering effect of mineral ions on the redox potential of the soil as from any direct nutritional impact.

Phosphorus

Phosphorus is a nutrient that often limits plant growth, but in salt marsh soils it accumulates in high concentrations and apparently does not limit growth. For example, the marsh sediments along the Georgia coast contain enough phosphorus to supply the marsh vegetation for several hundreds of years (Pomeroy et al., 1972).

Iron

Other plant nutrients have also been suggested as possibly limiting growth in salt marshes. D. A. Adams (1963) found that soluble iron concentrations in the marsh soil water were highest in the lower, more productive zones of the marsh. He also found in culture experiments that *S. alterniflora* deprived of iron became chlorotic or discolored due to lack of chlorophyll synthesis, leading him to suggest that iron may become limiting to plant growth. Subsequent studies (Haines and Dunn, 1976), however, have ruled this out. In fact, several micronutrients, including iron and manganese, are available in high concentrations in marsh soils because of the reducing conditions, and they are more likely to be in toxic concentrations than limiting. Iron, for example, is found in *Spartina* tissues at concentrations of about 10 times those in most crop plants.

Sulfur

Sulfur is an interesting marsh soil chemical because of its toxicity, acid-forming properties, and ability to store energy from organic sources. Seawater contains abundant sulfate. When this ion encounters the anoxic marsh soil, it is reduced by soil bacteria to sulfide, which, in turn, can form insoluble pyrites with iron. Hydrogen sulfide (which is responsible for the characteristic rotten egg odor of salt marsh sediments) is extremely toxic to plants (Hollis, 1967) and is probably responsible, at least in part, for the poor performance of inland marsh plants (Mendelssohn et al., 1982). In addition, when exposed to air, sulfides can be reoxidized to sulfate, forming sulfuric acid, with a resulting drop in soil pH. It has been suggested that local sediment drying (and oxidation) and the subsequent increase in acidity in the soil may account for some patchy death of plants in the marsh (Cooper, 1974). When salt marshes underlain with clays are drained for agricultural production, the sulfides in the resulting soils oxidize to sulfuric acid, and the soils become too acidic to support crop growth. This is the frequently observed "cat clay" phenomenon.

An intriguing property of sulfide is its ability to act as a storage compound for biologically fixed energy. In the anaerobic soil environment, bacteria reduce sulfate to sulfide by oxidizing organic compounds. They trap about 15 percent of the organic energy; the rest, or 75 percent, is transferred to the energy-rich sulfide radical. These sulfide compounds can later be reoxidized, and the stored energy can be used to fuel the growth of sulfur-oxidizing bacteria. An estimate of the importance of this pathway of energy flow in a New England salt marsh is given by Howarth and Teal (1979) and Howarth et al. (1983), who calculated that as much as 70 percent of the energy of net primary production flows through reduced inorganic sulfur compounds (see Chap. 5). Most of the stored sulfides are reoxidized on an annual basis by oxygen diffusing into the soil from marsh grass roots. A small percentage of the soluble sulfides, however, may be exported from the marsh in pore water. Other sulfides enter into the sediment food chain directly. For example, the colorless sulfur bacteria (*Beggiatoa* spp.) on marsh and intertidal flats oxidize soluble reduced sulfur compounds to elemental sulfur, storing the energy in bacterial biomass. The bacterial mats are easily suspended by waves and currents and are recycled into the water column (Grant and Bathmann, 1987).

ECOSYSTEM STRUCTURE

The salt marsh ecosystem has diverse biological components that include vegetation and animal and microbe communities in the marsh and plankton, inverte-

Table 8–4. Examples of Common Plant Species in Salt Marshes for Various Regions in North America

Location	Examples of Common Vegetation	
	Lower marsh	*Upper marsh*
Eastern North America		
New England	*Spartina alterniflora*	*Spartina patens* *Distichlis spicata* *Juncus gerardi* *Spartina alterniflora* (dwarf)
Coastal Plain	*Spartina alterniflora*	*Spartina patens* *Distichlis spicata* *Salicornia sp.* *Juncus roemerianus*
Bay of Fundy	*Spartina alterniflora*	*Spartina patens* *Limonium nashii* *Plantago oliganthos* *Puccinellia maritima* *Juncus gerardi*
Gulf of Mexico[a]		
North Florida/South Alabama & Mississippi	Dominant *Juncus roemerianus* Subdominant *Spartina patens* *Spartina alterniflora*	
Louisiana	Dominant *Spartina alterniflora* Subdominant *Spartina patens* *Distichlis spicata* *Juncus roemerianus*	
Arctic		
Northern Canada/Europe	*Puccinellia phryganodes*	*Carex subspathacea*
Western Alaska	*Puccinellia phryganodes*	*Puccinellia triflora* *Plantago maritima* *Triglochin sp.* *Plantago maritima*
Western North America		
Southern California	*Spartina foliosa*	*Salicornia pacifica* *Suaeda californica* *Batis maritima*

[a]Because of low tide range, low and high marsh distinctions are not clear.

Source: Many examples from Chapman, 1960, 1975

brates, and fish in the tidal creeks, pans, and estuaries. The discussion here will be limited to the biological structure of the marsh itself.

Vegetation

Salt marshes are dominated by halophytic flowering plants, often by one or a few species of grass. Some common plant species found in various regions of North America are given in Table 8–4.

The vegetation of the salt marsh has been divided into zones that are related to the upper and lower marshes described previously but that also reflect regional differences. Characteristic patterns found in the North American Atlantic Coast marshes are shown in Figure 8–10. Niering and Warren (1980) and Bertness (1992) described a generalized transect that shows how plant zonation in a New England salt marsh is determined by the physical and chemical features of the marsh (Fig. 8–10a). The intertidal zone or lower marsh next to the estuary, bay, or tidal creek is dominated by the tall form of *S. alterniflora* Loisel (cordgrass). In the high marsh, *S. alterniflora* gives way to extensive stands of *S. patens* (salt-meadow grass) mixed with *Distichlis spicata* (spike grass) and occasional patches of the shrub *Iva frutescens* (marsh elder) and various forbs. Beyond the *S. patens* zone and at normal high tide, *Juncus gerardi* forms pure stands. At the upper edge of a marsh inundated only by spring tides, two groups of species are common, depending on the local rainfall and temperature. Where rainfall exceeds evapotranspiration, salt-tolerant species give way to less tolerant species such as *Panicum virgatum, Phragmites australis, Limonium carolinianum* (sea lavender), *Aster* spp. (asters), and *Triglochin maritima* (arrow grass) (Niering and Warren, 1977). On the south-east New England Coast where evapotranspiration may exceed rainfall during the summer, salts can accumulate in these upper marshes, and salt-tolerant halophytes such as *Salicornia* spp. (saltwort) and *Batis maritima* flourish. Bare areas with salt efflorescence are common. Other features of New England salt marshes include well-flushed mosquito ditches lined with tall *S. alterniflora* and salt pannes containing short form *S. alterniflora* (Niering and Warren, 1980).

South of the Chesapeake Bay along the Atlantic Coast, salt marshes typical of the Coastal Plains appear. These marshes are similar in zonation to those in New England except that (1) tall *S. alterniflora* often forms only in very narrow bands along creeks, (2) the short form of *S. alterniflora* occurs more commonly in the wide middle zone, and (3) *Juncus roemerianus* replaces *J. gerardi* in the high marsh (Cooper, 1974; Montague and Wiegert, 1990). Frey and Basam (1985) described a typical marsh gradient in the barrier island area along the Georgia coast from a geological perspective (Fig. 8–10b). At maturity low-and high-marsh areas are approximately equal. The low marsh is almost entirely *S. alterniflora*, tall on the creek bank, and shorter behind the natural levee as elevation gradually increases in an inland direction. It may contain small vegetated

a. Southern New England

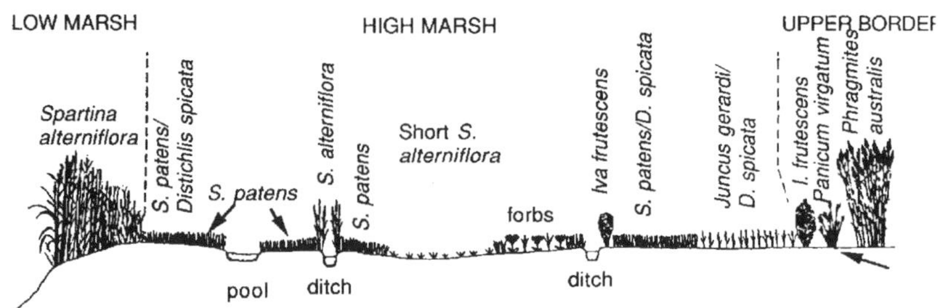

b. South Atlantic Coast

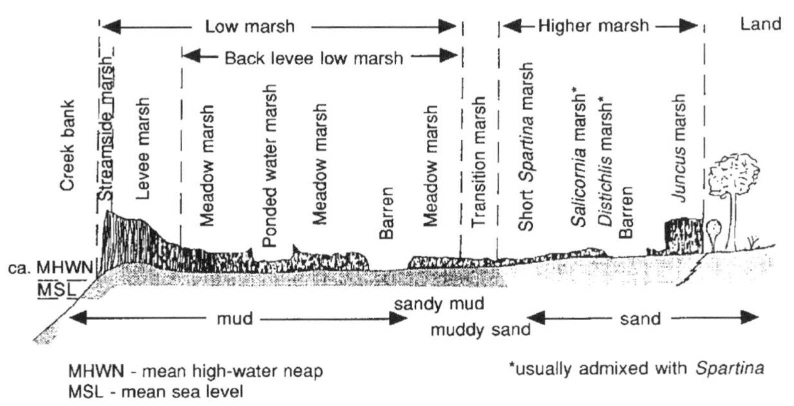

c. Northeastern Gulf of Mexico Coast

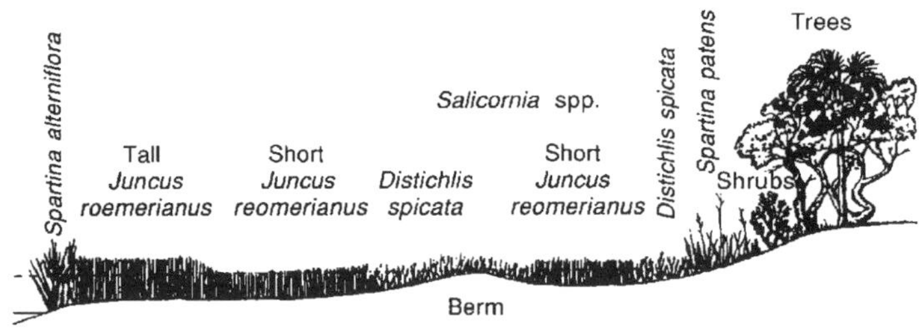

Figure 8–10. Zonation of vegetation in typical North American salt marshes. *a.* Southern New England*; b.* South Atlantic coast*; c.* Northeastern Gulf of Mexico. (a. *after Niering and Warren, 1980;* b. *after Wiegert and Freeman, 1990;* c. *from Montague and Wiegert, 1990, reprinted with permission)*

or unvegetated ponds and mud barrens. The high marsh is much more diverse, containing short *S. alterniflora* intermixed with associations of *Distichlis spicata*, *Juncus roemerianus*, and *Salicornia* spp.

Although *S. alterniflora* is the dominant salt marsh plant in most areas, along the Mississippi and northwest Florida coasts *J. roemerianus* is found in extensive monocultures (Fig. 8–10c). There is often a fringe of *S. alterniflora* along the seaward margin, followed, in an inland direction, by large areas of tall and short *J. roemerianus*. Mixtures of *S. patens* and *D. spicata* line the marsh on the landward edge, and *Salicornia* spp. can be found in small areas such as berms where salt accumulates.

Along the northern Gulf Coast, *S. patens* is the dominant species. It occurs in a broad zone inland of the more salt-tolerant *S. alterniflora*. For example, Chabreck (1972) estimated that *S. patens* dominated more than 200,000 hectares of coastal marsh in Louisiana.

Salt Marsh Plant Species

S. alterniflora is a stiff, leafy grass that can grow to 3 meters in height. It has two growth forms, tall and short, that are found in different parts of the marsh. The tall form (100 to 300 cm) generally occurs adjacent to tidal creeks and in the extreme low portions of the intertidal marsh. The short form (17 to 80 cm) is found away from the creeks in the upper marsh, generally in areas of greater salinity. Debate continues as to whether the differences in plant growth can be attributed to genetic or environmental causes. Anderson and Treshow (1980), in a review of the subject, concluded that the differences are both genetic, based on different ecotypes that evolved, and environmental, based on such factors as sediment anoxia and salinity stress.

The genus *Spartina* is native to North America, and only one native species, *Spartina maritima,* is found outside this continent—in Europe. *S. alterniflora,* however, was introduced to Britain and France about 1860 or 1870 and hybridized with *S. maritima*, forming a sterile species, *S. townsendii*. This sterile hybrid later produced a fertile amphidiploid (that is, the chromosome number of the original hybrid was doubled). *S. anglica*, the name given to this fertile hybrid, is extremely vigorous and is rapidly expanding along the European coast (Beeftink, 1977b). Since the 1960s, both *S. anglica* and *S. alterniflora* have been used extensively for reclaiming China's east coast along the China Sea (Chung, 1982, 1983, 1985, 1989). *Spartina*'s introduction there has led to increased arable land, the stabilization of the coastline, decreased soil salinity, and the production of animal fodder, green manure, and fuel (Chung, 1989).

Distichlis spicata (spike grass) is another common salt marsh plant that is widely distribution along coastal and inland saline wetlands in the United States. The plant is particularly tolerant of high soil salinity, and it may serve as a pioneer species in salt marsh development.

Mud Algae

Mud algal mats, dominated particularly by blue-green algae, diatoms, and green algae, are also present in the salt marsh. These communities are very diverse but are poorly known ecologically (Wiegert and Freeman, 1990). Algal species that are distributed throughout the world in salt marshes include the blue-greens *Lyngbya* and *Rivularia* and the green algae *Ulothrix, Rhizoclonium, Chaetomorpha, Ulva, Enteromorpha*, and *Monostroma* (Chapman, 1960; Ursin, 1972). In the Georgia salt marshes, hundreds of species of diatoms, dominated by *Cylindrotheca*, *Gyrosigma*, *Navicula*, and *Nitzschia*, make up 75–93 percent of the benthic algal biomass (Williams, 1962). In North and South Carolina, species composition is similar (Pomeroy et al., 1981; Hustedt, 1955). The species composition of the mud algae is dynamic. It is linked to the continual accretion or erosion of sediments. Species may completely change within a few weeks (Chapman, 1960). As with vascular plants, algae also tend to form zones in the marsh along tidal and substrate gradients. The algae, however, do not necessarily align in conformity with vascular vegetation zonation (Sullivan, 1978).

Consumers

Salt marshes, whose features are characteristic of both terrestrial and aquatic environments, provide a harsh environment for consumers. Salt is an additional stress with which they must contend. In addition, the variability of the environment through time is extreme. The dominant food source for marsh consumers is one plant species, *S. alterniflora,* that is limited in its nutritional value (Burkholder, 1956). Considering all these limitations, the number of consumers in the salt marsh is perhaps surprisingly diverse.

Salt marsh ecological studies have traditionally focused on plant productivity and ecosystem structure and function. Much less is known about faunal populations. The species composition and population size of resident species, however, appear remarkably similar from site to site, especially across the southeastern United States, where most of the tidal marshes are found (Montague and Wiegert, 1990). Excellent accounts of faunal populations are presented in Daiber (1977, 1982), Montague and Wiegert (1990), Montague et al. (1981), Davis and Gray (1966), and Pfeiffer and Wiegert (1981) and in the series of wetland community profiles published by the U.S. Fish and Wildlife Service (Nixon, 1982; Gosselink, 1984; Wiegert and Freeman, 1990; Zedler, 1982; Josselyn, 1983; Teal, 1986). Stout (1984) and Heard (1982) are useful sources for *Juncus*-dominated marshes.

It is convenient to classify consumers according to the type of marsh habitat they occupy, although the animals, especially in the higher tropic levels, move from one habitat to another. The marsh can be divided into three major habitats: an aerial habitat—the aboveground portions of the macrophytes, which is sel-

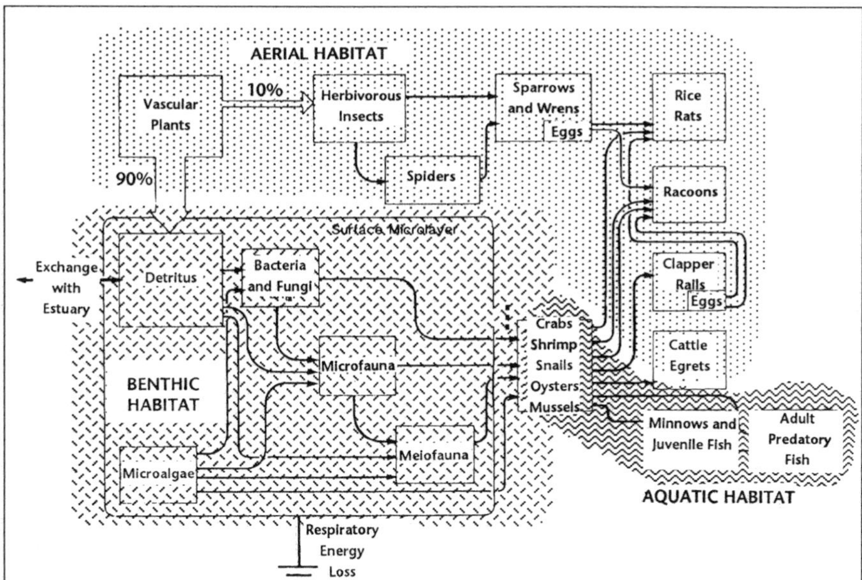

Figure 8–11. Salt marsh food web, showing the major consumer groups of the *aerial* habitat, *benthic* habitat, and *aquatic* habitat. (*Modified from Montague and Wiegert, 1990*)

dom flooded; a benthic habitat—the marsh surface and lower portions of the living plants; and an aquatic habitat—the marsh pools and creeks. Our modification of this classification presented by Montague and Wiegert (1990) is reflected in the following discussion.

Aerial Habitat

The aerial habitat is similar to a terrestrial environment and is dominated by insects and spiders that live in and on the plant leaves. This is the grazing portion of the salt marsh food web (Fig. 8–11). The most common leaf-chewing organisms are the arthropod *Orchelimum* (Smalley, 1960), the weevil *Lissorhoptrus*, and the squareback crab *Sesarma* (Pfeiffer and Wiegert, 1981). In addition, there are abundant sap-sucking insects (*Prokelisia marginata*, *Delphacodes detecta*) that ingest material translocated through the plant's vascular tissue or empty the contents of mesophyll cells. Some idea of the diversity of species occupying this habitat is shown in Table 8–5, which shows the number of herbivorous insect species associated with *S. alterniflora*, *S. patens*, and *S. foliosa* (a West Coast species). Numerous carnivorous insects are also found in this habitat. Pfeiffer and Wiegert (1981) list 81 species of spiders and insects in North Carolina, South Carolina, and Georgia *Spartina* marshes.

Finally, a number of birds, including the long-billed marsh wren (*Telmatodytes palustris griseus*) and the seaside sparrow (*Ammospiza maritima*),

Table 8–5. Comparisons of the Number of Herbivorous Species Occupying Various Insect Orders Associated with Three Intertidal Species of *Spartina*

	S. alterniflora	*S. patens*	*S. foliosa*
Orthoptera	6	10	0
Thysanoptera	3	8	5
Hemiptera	11	11	1
Homoptera	15	22	7
Coleoptera	19	18	10
Lepidoptera	5	—	2
Diptera	20	16	11
Hymenoptera	29	22	6
Total Herbivorous Insect Species	108	107	42

Source: After Pfeiffer and Wiegert, 1981; copyright © 1981 by Springer-Verlag, reprinted with permission.

feed or nest or both feed and nest in the marsh grasses. Wrens feed primarily on insects (Kale, 1965), and the sparrow apparently feeds on the marsh surface, eating worms, shrimp, small crabs, grasshoppers, flies, and spiders. The clapper rail (*Rallus longirostris*) is another permanent marsh resident, feeding primarily on cutworm moths and small crabs. Many nonpermanent insectivorous birds forage in the salt marsh periodically, entering from adjacent uplands habitats or migrating through. These include the sharp-tailed sparrow (*Ammospiza caudacuta*), swallows (*Iridoprocne bicolor*, *Hirundo rustica*, and *Stelgidopteryx ruficollis*), red-winged blackbirds (*Agelaius phoenicius*), and various gulls.

Benthic Habitat

Probably less than 10 percent of the aboveground primary production of the salt marsh is grazed by aerial consumers. Most plant biomass dies and decays on the marsh surface, and its energy is processed through the detrital pathway (Fig. 8–11). The primary consumers are microbial fungi and bacteria. These organisms, in turn, are preyed upon by meiofauna in the decaying grass, the surface microfilm of the marsh, and the decaying bases of plant shoots. Most of these microscopic organisms are protozoa, nematodes, harpacticoid copepods, annelids, rotifers, and larval stages of larger invertebrates (Fenchel, 1969). The larger invertebrates on the marsh surface are of two groups, foragers (deposit feeders) and filter feeders. In a general sense, they are considered aquatic because most have some kind of organ to filter oxygen out of water. Foragers include polychaetes, gastropod mollusks such as *Littorina irrorata* and *Melampus bidentatus*, and crustaceans such as *Uca* spp., the blue crab (*Callinectos sapidus*), and amphipods. These organisms browse on the sediment surface, ingesting algae, detritus, and meiofauna. The filter feeders such as the ribbed mussel (*Geukensia demissa*) and the oyster (*Crassostrea virginica*) filter particles out of the water column.

Table 8–6. Value of United States Landings of Commercial Fish and Shellfish Species, 1991

Species[a]	*Value (Millions of Dollars)*	*Percentage of Total*
Shrimp	510	15.5
Crabs	410	12.5
Salmon	360	10.9
Alaska Pollock	240	7.3
Lobsters	200	6.0
Cods	200	5.9
Scallops	160	4.9
Others (includes menhaden, flounders)	1,220	37.0

[a]Species that use coastal wetlands during some part of their life cycles are shrimp, crabs, salmon, and most "others" category, including menhaden and flounders.
Source: National Marine Fisheries Service, 1991 Fishery Statistics

Aquatic Habitat

Animals classified as aquatic overlap with those in the benthic habitat. For convenience we include in this group higher trophic levels (mostly vertebrates) and migratory organisms that are not permanent residents of the marsh. Few fish species are permanent residents of the marsh. Most feed along the marsh edges and in small, shallow marsh ponds and move up into the marsh on high tides. Werme (1981) found 30 percent of silverside (*Menidia menidia*) and mummichog (*Fundulus heteroclitus*) in a north Atlantic estuary up in the marsh at high tide. Ruebsamen (1972) reported that common fish in small salt marsh ponds in Louisiana are sheepshead minnow (*Cyprinodon variegatus*), diamond killifish (*Adinia xenica*), tidewater silversides (*Menidia beryllina*), gulf killifish (*Fundulus grandis*), and sailfin molly (*Poecilia latipinna*). Shrimp (*Penaeus* spp.) and blue crabs (*Callinectis sapidus*) are also common. Most other species use the marsh intermittently for shelter and for food but range widely. Many fish and shellfish spawn offshore or upstream and as juveniles migrate into the salt marsh, which offers an abundant food supply and shelter. They are concentrated along the edges of the marsh (Zimmerman et al., 1990). As subadults they migrate back into the estuary or offshore. This group of migratory organisms includes more than 90 percent of the commercially important fish and shellfish of the southeast Atlantic and Gulf coasts (Table 8–6).

Avian Fauna

Salt marshes support large populations of wading birds, including egrets, herons, willets, and even woodstorks and roseate spoonbills. Southern marshes also support vast populations of wintering waterfowl, including the mallard (*Anas platyrhynchos*), widgeon (*A. americana*), gadwall (*A. strepena*), redheads, (*Aythya americana*), and teals (*A. discors* and *A. crecca*). Black duck (*A. rubripes*) is a permanent resident in many marshes.

Mammals

Two mammals deserve attention because of their impact on salt marshes. The muskrat (*Ondatra zibethica*) is native to North America; the coypu or nutria (*Myocaster coypus*) is an exotic species introduced from South America. Both prefer fresh marshes but are also found in salt marshes. In Louisiana the muskrat appears to be displaced by nutria out of its preferred freshwater habitat into saline marshes. Both mammals are voracious herbivores that consume plant leaves and shoots during the growing season and dig up tubers during the winter. They destroy far more vegetation than they ingest and are responsible for "eat-outs" that degrade large areas of marsh (Fig. 7–11). These areas recover extremely slowly, especially in the subsiding environment of the northern Gulf Coast.

ECOSYSTEM FUNCTION

Major points that have been demonstrated in several studies about the functioning of salt marsh ecosystems include the following:

1. Annual gross and net primary productivity is high in much of the salt marsh—almost as high as in subsidized agriculture. This high productivity is a result of subsidies in the form of tides, nutrient import, and abundance of water that offset the stresses of salinity, widely fluctuating temperatures, and alternate flooding and drying.
2. The salt marsh is a major producer of detritus for both the salt marsh system and the adjacent estuary. In some cases, detrital material exported from the marsh is more important to the estuary than is the phytoplankton-based production in the estuary. Detritus export and the shelter found along marsh edges make salt marshes important as nursery areas for many commercially important fish and shellfish.
3. The grazing pathway is a minor energy flow in the salt marsh.
4. Leaves and stems of vegetation serve as surface areas for epiphytic algae and other epibiotic organisms. This enhances both the primary and the secondary productivity of the marsh.
5. Detrital decomposition, by breaking down and transforming indigestible plant cellulose, makes it available to consumers and is the major pathway of energy utilization in the salt marsh. Decomposers increase the protein content of the detritus and enhances its food value to consumers.
6. Salt marshes have been shown at times to be both sources and sinks of nutrients, particularly nitrogen.

These and other points will be discussed below.

TABLE 8–7. Estimates of Net Primary Production of *Spartina alterniflora* in North America (g m^{-2} yr^{-1})

Location	*Height Form*	*Above Ground*	*Below Ground*	*Below/Above*	*References*
Nova Scotia		514	720	1.4	Livingston and Patriquin (1981)
Massachusetts	tall	1320	3315	2.5	Valiela et al. (1976)
	short	420	3500	8.3	
Maine	short	705			Linthurst and Reimold (1978)
North Carolina	tall	1300	1360	1.0	Stroud (1976)
	short	330	420	1.3	
Georgia	tall	3700	2100	0.6	Stroud (1976)
	short	1300	2020	1.6	
Florida	tall	700			Kruczynski et al. (1978)
	short	130			
Alabama		2029	6218	3.1	Stout (1978)
Mississippi	tall	1964			de la Cruz (1974)
	short	1089			
Louisiana	tall	750–2600			Kirby and Gosselink (1976)
	tall	1473–2895			White et al. (1978)
	tall	1381			Hopkinson et al. (1980)
California (San Francisco Bay)	*S. foliosa*	274–1400			Josselyn (1983); Several studies–peak biomass

Primary Productivity

Tidal marshes are among the most productive ecosystems in the world, producing annually up to 80 metric tons per hectare of plant material (8000 g m^{-2} yr^{-1}) in the southern Coastal Plain of North America. The three major autotrophic units of the salt marsh include the marsh grasses, the mud algae, and the phytoplankton of the tidal creeks. Extensive studies of the net primary production of *Spartina alterniflora* have been conducted in salt marshes along the Atlantic and Gulf coasts of the United States. A comparison of many of the measured values of net above-ground and below-ground production is given in Table 8–7. Above-ground production shows a wide range, from as little as 330 g m^{-2} yr^{-1} to a high of 3700 g m^{-2} yr^{-1} in a Georgia study. Below-ground production is difficult to measure, and estimates vary widely from 220 to 2,500 g m^{-2} yr^{-1} (streamside) and 420 to 6,200 g m^{-2} yr^{-1} inland (Good et al., 1982).

Several major generalizations about primary productivity in salt marsh can be made:

Table 8–8. A Comparison of Annual Production Estimates (g m^{-2} yr^{-1}) of Different Salt Marsh Plant Species in the Same Region

Species	*Annual Net Aerial Primary Production*[a]		*Source*
Louisiana			
Distichlis spicata	1162–1291		White et al. (1978)
Juncus roemerianus	1806–1959		
Spartina alterniflora	1473–2895		
Spartina alterniflora	1342–1428		
Distichlis spicata	1967		Hopkinson et al. (1980)
Juncus roemerianus	3295		
Spartina alterniflora	1381		
Spartina cynosuroides	1134		
Spartina patens	4159		
Mississippi	*Above*	*Below*	
Juncus roemerianus	3078	7578	Stout (1978)
Spartina alterniflora	2029	6218	
Juncus roemerianus	1300		de la Cruz (1974)
Distichlis spicata	1072		
Spartina alterniflora	1473		
Spartina alterniflora	1342		

[a]In each study the methods are comparable.

Table 8–9. Summary of Annual Primary Production and Turnover Rates for *Spartina alterniflora*

	Primary Production, g dry wt m^{-2} yr^{-1}				
Methods	*Kaswadji et al. (1990)*[a]	*Kirby and Gosselink (1976)*[b]	*Hopkinson et al. (1980)*[c]	*Shew et al, (1981)*[d]	*Annual Turnover Rate (production to peak standing crop)*
Peak standing crop	831 ± 41	903	754	242	1.0
Milner-Hughes	831 ± 62	811	—	241	1.0
Smalley	1231 ± 252	1200	—	225	1.5
Weigert-Evans	1873 ± 147	1988	2658	1029	2.2
	(2733 ± 235)	—	—	(1038)	(3.3)
Lomnicki et al.	1437 ± 96	—	—	1028	1.7
	(2046 ± 125)	—	—	—	(2.5)

Numbers in parentheses are the results if negative production is counted as zero.

[a]Transect across a mostly inland site.

[b]Average of streamside and inland sites.

[c]Intermediate streamside to inland marsh site.

[d]Short marsh.

Source: Kaswadji et al., 1990

First, the above-ground productivity of *Spartina* is often higher along creek channels and in the low or intertidal marshes than in high marshes because of the increased exposure to tidal and freshwater flow. These conditions also produce the taller forms of *Spartina,* as discussed earlier.

Second, there is generally greater productivity in the southern Coastal Plain salt marshes than in those farther north because of the greater influx of solar energy and the longer growing season (R. E. Turner, 1976) but possibly also because of the nutrient-rich sediments carried by the rivers of that region. For example, White et al. (1978) argued that the productivity of salt marshes in Louisiana and Mississippi may be higher than Atlantic Coast marshes because of the higher nutrient import from the Mississippi River.

Third, the wide divergence in measurements of net primary productivity among sites is as much caused by the methods used for measurement as by true differences in productivity. Kirby and Gosselink (1976), White et al. (1978), and Hopkinson et al., (1980) discussed the variations obtainable with different methods used for calculating productivity. For example, estimates of *S. alterniflora* production, calculated from data from the same plots, varied from 1,470 to 2,900 g m^{-2} yr^{-1} in one study (White et al., 1978) and from 830 to 2,700 g m^{-2} yr^{-1} in another study (Tables 8–8, 8–9).

Fourth, below-ground production is sizable—often greater than aerial production (Table 8–7). Under unfavorable soil conditions, plants seem to put more of their energy into root production. Hence root:shoot ratios seem to be generally higher inland than at streamside locations.

Fifth, a comparison of different species (Table 8–8) shows that none comes out the winner in a productivity sweepstakes. Each is extremely productive in its preferred habitat.

The productivity of mud algae is less well understood, although there have been a few noteworthy studies, for example, Pomeroy (1959) in Georgia, Gallagher and Daiber (1974) in Delaware, Van Raalte and Valiela (1976) in Massachusetts, and Zedler (1980) in southern California. On the Atlantic Coast, the productivity of algae was one-third to one-fourth that of vascular plant productivity (see Table 8–10). Zedler (1980), however, found that algal net primary productivity in southern California was from 80 percent to 140 percent of vascular productivity. She hypothesized that the arid and hypersaline conditions of southern California favor algal growth over vascular plant growth. Several scientists have pointed out that algal biomass is more readily assimilated by animals and more nutritious than marsh grass. Hence it may be more important to the marsh food web than its relatively low productivity indicates (Wiegert and Freeman, 1990; Montague and Wiegert, 1990).

The Control of Primary Production

We have alluded to factors that control production in various places in this chapter. These factors have probably been studied more in salt marshes than in other

Table 8–10. Net Primary Productivity of Algae in Salt Marshes of North America

Location	Dominant Types of Algae	Net Primary Productivity, $g\ m^{-2}\ yr^{-1}$	NPP Algae / NPP Vascular Plants	References
Georgia	Pennate diatoms, green flagellates, blue-greens	324[a]	0.25	Pomeroy (1959) Teal (1962)
Delaware		160	0.33	Gallagher and Daiber (1974)
Massachusetts	Green filamentous	105	0.25	Van Raalte and Valiela (1976)
California	Blue-greens, filamentous greens, and diatoms	320-588	0.76-1.40	Zedler (1980)

[a]Assumes 1 g C = 2 g dry wt.

wetland systems. In a general sense plant production is a response to light (energy), water, nutrients, and toxins (negatively). Marsh plants grow in full sunlight, they appear to have a limitless water supply, and the sedimentary minerals on which they grow are generally rich in nutrients. From this one would expect uniformly high production rates. This is not true. Instead, the salt marsh environment sets limits on production that relate to all three parameters. Turner (1976) demonstrated that production generally declined with increasing latitude and related this phenomenon to the shorter length of the growing season and lower temperatures in the north. Beyond climatic differences, much of the observed variation in production is related to salt and to the toxic environment of an anaerobic soil.

Although water appears plentiful, the concentration of dissolved salt makes the salt marsh environment similar in many respects to a desert: The "normal" water gradient is from plant to substrate. In order to overcome the osmotic influence of salt, plants must expend energy to increase their internal osmotic concentration in order to take up water. As a result, numerous studies confirm that plant growth is progressively inhibited by increasing salt concentrations in the soil. This is true even for the salt-tolerant species of the salt marsh.

A second factor limiting production is the degree of anaerobiosis of the substrate (see Chap. 6). Vascular plants, even those that have developed adaptations to anaerobic conditions, grow best in aerobic soils. Many effects of anaerobiosis have been documented: reduced energy availability as the aerobic respiratory pathway is blocked, reduced nutrient uptake (Morris, 1984), the accumulation of toxic sulfides in the substrate (Koch and Mendelssohn, 1989), and changes in the availability of nutrients (Patrick and Delaune, 1972; see Chap. 6). Salt inhibition and oxygen depletion frequently occur together. *Spartina* grows shorter in the inland marsh because its drainage is poor and hence oxygen deficits are severe; salt also often accumulates in this environment.

Bertness (1992) put forth an environmental argument that *Spartina* essentially controls its productivity through negative feedback. Peat produced by the plant accumulates and compacts, leading to poor drainage in the upper marsh and hence lower productivity and shorter forms of *Spartina* there. As described by Bertness (1992), "As a strand of cord grass [*Spartina*] matures, it effectively destroys its own habitat." Along the creek banks, the soil has little peat due to rapid decomposition and export, and *Spartina* grows much better there.

One of the oldest paradigms of salt marsh ecology is the "tidal subsidy" (E. P. Odum, 1980). The hypothesis is that tides, by flushing salts and other toxins out of the marsh and by bringing in nutrients, stimulate marsh growth. Steever et al. (1976) verified this hypothesis by showing that the peak standing biomass of *S. alterniflora* along the New England coast was directly related to the tide range (see Chap. 4).

Finally, Morris et al. (1990) showed that the year-to-year variation in marsh production at a single site on the East Coast was correlated with the mean summer water level, which they equated with soil salinity (soil salinity was inversely correlated with the frequency of marsh flooding in this study).

These examples show that many differences in marsh production can be explained ultimately by their effects on water, nutrients, and toxins, although the pathways through which these three parameters are influenced may be subtle and tortuous.

Decomposition

Almost three-quarters of the detritus produced in the salt marsh ecosystem is broken down by bacteria and fungi (Teal, 1986). In his study of energy flow within the salt marsh environment, Teal (1962) estimated that 47 percent of the total net primary productivity was lost through respiration by microbes. Decomposition processes in salt marshes fragment dead leaves and stems into smaller sizes and upgrade the protein content by bacterial, fungal, and protozoal colonization of the substrate. E. P. Odum and de la Cruz (1967) found that as *Spartina* grass decomposed, the detritus increased in protein content from 10 percent to 24 percent on an ash-free basis (Fig. 8–12), that is, the carbon/nitrogen ratio decreased in the detritus. The authors concluded that "the bacteria-rich detritus is nutritionally a better food source for animals than is the *Spartina* tissue that forms the original base for most of the particulate matter."

A contrasting theory, however, has been proposed by Wiebe and Pomeroy (1972), who showed that detritus fragments are not colonized by large populations of bacteria and thus are not suitable as food for deposit feeders. Many of the nitrogenous compounds are probably resistant to decomposition and are not

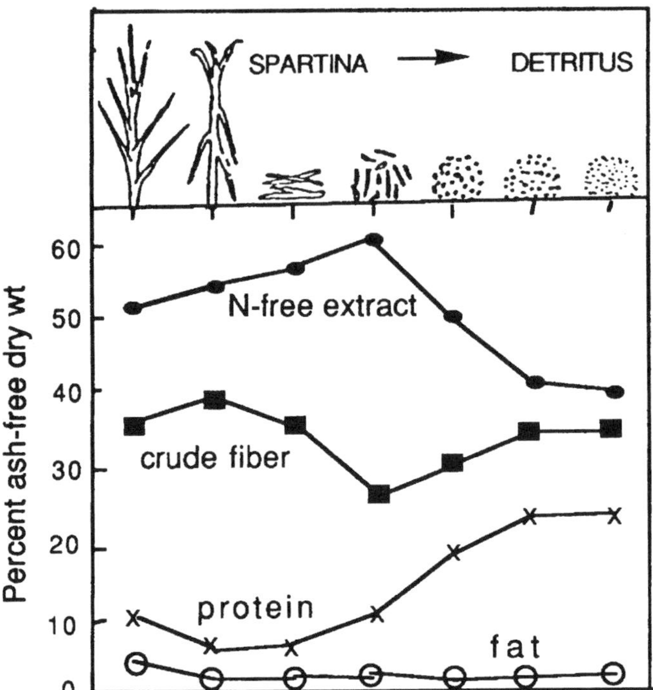

Figure 8–12. Nutritional composition of successive stages of decomposing *Spartina* grass, showing increasing protein content and decreasing carbohydrates. *(From E. P. Odum and de la Cruz, 1967; copyright © 1967 by the American Association for the Advancement of Science, reprinted with permission)*

readily available as food for detritivores (Teal, 1986). It appears that the detritus is grazed by nematodes and other microscopic benthic organisms that are, in turn, eaten by the large deposit feeders. Bacterial production on detritus is high, but the rapid turnover rate keeps the population low.

Several other investigators have looked at the dynamics of detrital decomposition in salt marshes (Burkholder and Bornside, 1957; Gosselink and Kirby, 1974; Pomeroy et al., 1977; White et al., 1978; Hackney and de la Cruz, 1980). Gosselink and Kirby (1974) found that a significant percentage (11–66 percent) of decomposed grass is converted to microbial biomass and that particle size is an important determinant of the rate of decomposition. Pomeroy et al. (1977) described the importance of soluble organic matter as well as particulate organic matter in the detrital food web of a salt marsh. They believe that this soluble organic matter (which may be as much as 25 percent of the initial dry weight of the dying grass) from both living and decomposing salt marsh vegetation is an important energy source for microorganisms in the marsh and the adjacent estuary. They also pointed out that anaerobic microorganisms such as nitrogen-reducing bacteria, sulfur-reducing bacteria, and methane-generating bacteria may be important consumers of organic substrate in the salt marsh, particularly below the surface. Because of the anoxic conditions, the decomposition of organic matter 20 cm or more beneath the marsh surface is slow. Still, Teal (1986) reported that less than 20 percent of the original litter remained after 2.5 years.

Organic Export

Several studies have demonstrated that a substantial portion (usually 20–45 percent) of the net primary productivity of a salt marsh is exported to adjacent estuaries. E. P. Odum and de la Cruz (1967) estimated that in a single tidal cycle there was a net export of about 140 kg and 25 kg of organic matter for spring and neap tides, respectively, from a 10 to 25 hectare salt marsh in Georgia. Wiegert et al., (1981) suggested that prorating this estimate over the watershed of the creek results in an export value of approximately 100 gC m^{-2} yr^{-1}. Other studies, however, found that the adjacent salt marsh is not the most important source of carbon for the estuary. Haines (1979) questioned the significance of wetland export based on stable carbon radioisotope data. Biggs and Flemer (1972) found that allochthonous material from uplands is the greatest source of organic carbon in the upper Chesapeake Bay; whereas Heinle and Flemer (1976) measured an annual flux of only 7.3 gC m^{-2} yr^{-1} of particulate carbon from a salt marsh in Maryland—less than 1 percent of the total marsh net production. These latter authors attributed the low export to relatively poor tidal exchange. Woodwell et al. (1977) also found a low net exchange of carbon between a salt marsh and the coastal waters of Long Island, New York. After a thorough review, Nixon (1980) concluded that the export of dissolved and particulate organic carbon was fairly universal but is often not large (see Chap. 5). A number of recent studies (Whiting et al., 1989; Wolaver and Spurrier, 1988; Chalmers et al., 1985) of carbon flux confirm that the magnitude is often small (as a proportion of primary production).

Probably the most extensive study of the dynamics of carbon export has been coupled with modeling studies conducted at the University of Georgia. As a result of iterative field investigations and model adjustments, a new picture emerged. The prediction of the model was that depending on the relationship between aerobic microbes and their consumers, the marsh could export considerable carbon or, conversely, act as a sink for carbon (Chalmers et al., 1985).

Energy Flow

Several studies have dealt with energy flow in parts of the salt marsh ecosystem, but only a few have considered the entire ecosystem. Most notable in this latter category is the study of a Georgia salt marsh by Teal (1962). Many of his values

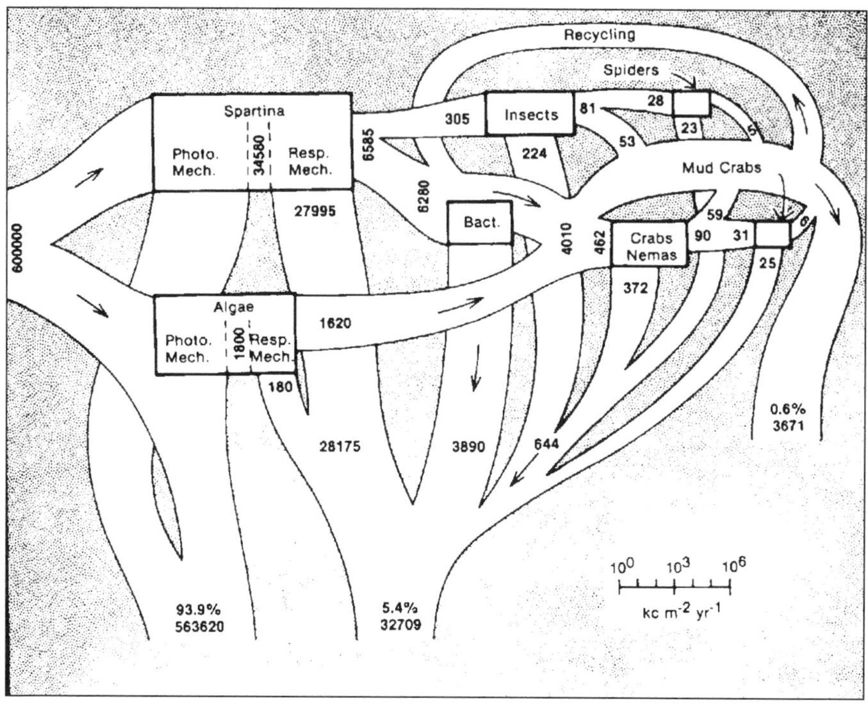

Figure 8–13. Energy flow diagram for a Georgia salt marsh. *(From Teal, 1962; copyright © 1962 by the Ecological Society of America, reprinted with permission)*

would be modified today, but the study remains a classic attempt to quantify energy fluxes in the salt marsh (Fig. 8–13). Gross primary productivity was calculated to be 6.1 percent of incident sunlight energy, verifying the observation that the salt marsh is one of the most productive ecosystems in the world. Only 1.4 percent of the incident light, however, was converted into organic material available to other organisms. Herbivorous insects, mostly plant hoppers (*Prokelisia*) and grasshoppers (*Orchelimum*), consumed only 4.6 percent of the *Spartina* net productivity. The rest of the *Spartina* net productivity and that of the mud algae passed through the detrital-algal food chain or was exported to the adjacent estuary. In the detrital food chain, primary and secondary consumers were dominated by only a few species such as the detritus-eating fiddler crab (*Uca*) and the carnivorous mud crab (*Eurytium*). An estimated 45 percent of the net production was exported from the marsh into the estuary in this study. As we saw in the previous section, this estimate may be unrealistically high. In a partial energy budget, Parsons and de la Cruz (1980) found that only 0.3 percent of the net primary productivity of a Mississippi *Juncus* marsh was ingested by three species of grasshoppers. They found that more than twice this amount of net primary productivity prematurely fell into the detrital mat because of grasshopper grazing.

Figure 8–14 shows the energy flow during summer and winter conditions in a salt marsh–estuary complex in New England (Nixon and Oviatt, 1973). The diagram emphasizes the importance of the marsh creeks and adjacent embayments in addition to the marsh itself. In this study, an estimated 23 percent of the net productivity of the salt marsh was exported to the embayment. Estimates of the productivity and consumption in the embayment itself let to the conclusion that the aquatic embayment is actually a heterotrophic ecosystem that depends on the import of organic matter from the autotrophic salt marsh.

Nutrient Budgets

Nutrients are carried into salt marshes by precipitation, surface water, ground water, and tidal exchange. Because many salt marshes appear to be net exporters of organic material (with incorporated nutrient elements), a nutrient budget would be expected to show that the marsh is a net sink for inorganic nutrients (to balance the nutrient budget). Recent studies have shown that this is not always the case.

One of the most ambitious studies of nutrient dynamics in the salt marsh ecosystem was carried out at the Great Sippewissett Marsh in Massachusetts (Valiela et al., 1978; Teal et al., 1979; Kaplan et al., 1979). A summary diagram of the nitrogen budget developed from this study is given in Figure 8–15. Valiela et al. (1978) estimated the amount of nitrogen in groundwater inputs, precipitation, and tidal exchanges in the marsh. Nitrogen from groundwater entering the marsh primarily as nitrate nitrogen (NO_3-N) and from tidal exchange represented the major fluxes of nitrogen to the salt marsh. Precipitation contributed significantly less nitrogen, mostly as NO_3-N and dissolved organic nitrogen (DON). Tidal exchange was found to result in a net export of nitrogen, mostly in the form of DON. Nitrogen fixation by bacteria was significant (Teal et al., 1979), and blue-green algal fixation was much less (Carpenter et al., 1978). Kaplan et al. (1979) found denitrification to be high in the salt marsh, particularly in muddy creek bottoms and in the short *Spartina* marsh. Figure 8–16 compares the denitrification loss in this salt marsh with the seasonal influx of groundwater and the loss of nitrates caused by tidal exchange.

Woodwell et al. (1979) described a budget of annual exchanges of inorganic nitrogen between a salt marsh complex and coastal waters in Long Island, New York (Fig. 8–17). They found a net export of ammonium nitrogen from the marsh during summer and fall (the marsh was a source) and a net import from coastal waters in winter and spring (the marsh was a sink). Nitrate was also imported in the winter and lost from the marsh in the summer. For the three inorganic nitrogen species studied (NH_4, NO_3, NO_2), there was a net export of about 10 kg N ha^{-1} yr^{-1}. As compared with the nitrogen uptake by the marsh vegetation, the loss from the marsh was only a fraction of intrasystem cycling. The export of nitrogen from salt marshes into Long Island Sound during the summer was about one-third of the amount of nitrogen that entered the sound from rivers during the same period.

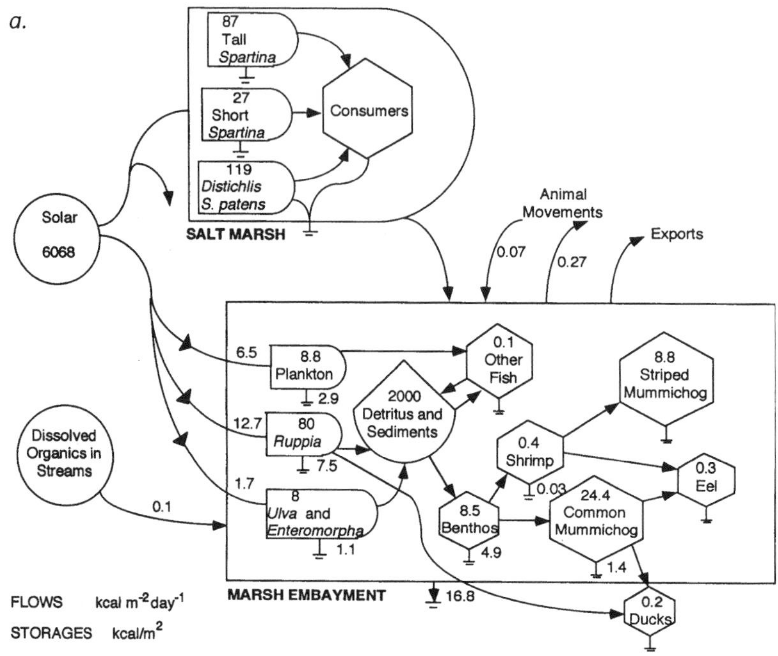
a.
87
Tall
Spartina
27
Short
Spartina
119
Distichlis
S. patens
Consumers
SALT MARSH
Solar
6068
Animal
Movements
0.07
0.27
Exports
6.5
8.8
Plankton
2.9
0.1
Other
Fish
8.8
Striped
Mummichog
2000
Detritus and
Sediments
12.7
80
Ruppia
7.5
Dissolved
Organics in
Streams
0.4
Shrimp
0.3
Eel
1.7
8
Ulva and
Enteromorpha
0.03
24.4
Common
Mummichog
0.1
8.5
Benthos
4.9
1.1
1.4
FLOWS kcal m-2 day-1
STORAGES kcal/m2
MARSH EMBAYMENT
16.8
0.2
Ducks

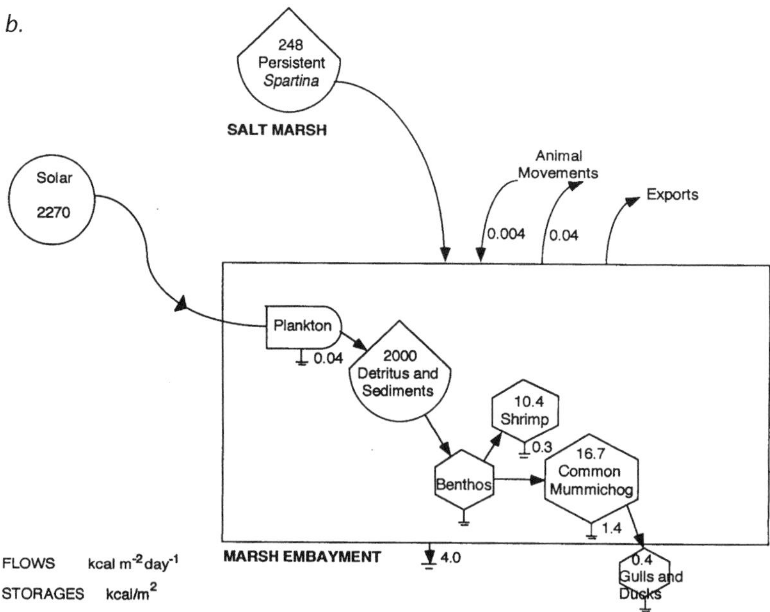
b.
248
Persistent
Spartina
SALT MARSH
Solar
2270
Animal
Movements
0.004
0.04
Exports
Plankton
0.04
2000
Detritus and
Sediments
10.4
Shrimp
0.3
Benthos
16.7
Common
Mummichog
1.4
FLOWS kcal m-2 day-1
STORAGES kcal/m2
MARSH EMBAYMENT
4.0
0.4
Gulls and
Ducks

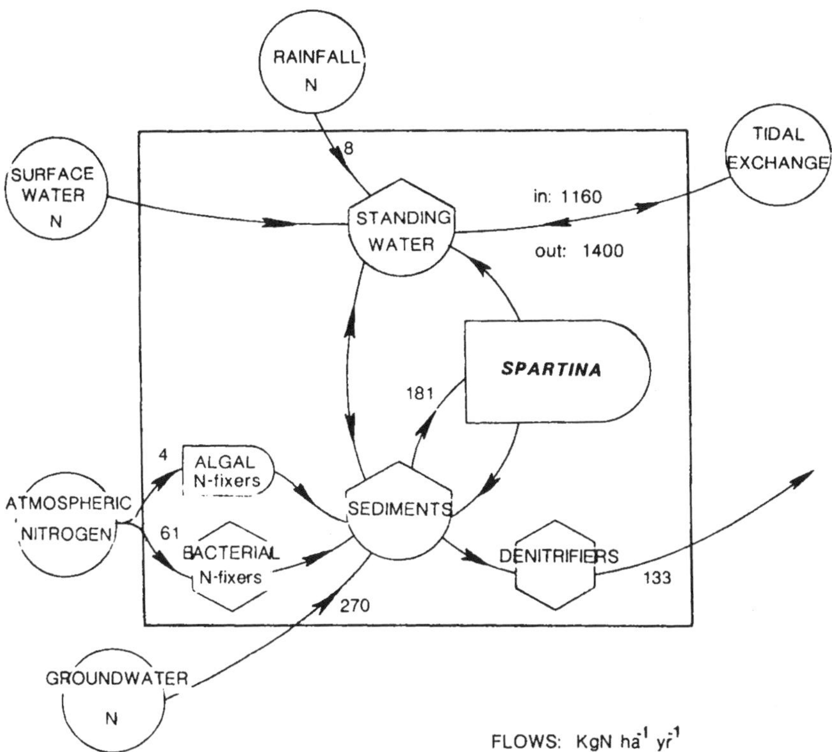

Figure 8–15. Nitrogen budget of Great Sippewissett Marsh, Massachusetts. *(Data from Valiela et al., 1978; Teal et al., 1979; Kaplan et al., 1979; and Valiela and Teal, 1979)*

The issue of whether there is a net uptake or a discharge of inorganic nutrients by the salt marsh remains unresolved. The study of the Great Sippewissett Marsh in Massachusetts indicated an approximate balance between inputs and outputs. The Long Island Marsh study found that the marsh was a source of nitrogen during the growing season and a sink in winter and spring. Aurand and Daiber (1973) found a net import of inorganic nitrogen into a Delaware salt marsh over a year, whereas Stevenson et al. (1977) found a net discharge of both nitrogen and phosphorus from a Chesapeake Bay salt marsh. Haines et al. (1977) concluded that compared to plant uptake, the leakage of nitrogen in Georgia salt marshes was small. Valiela et al. (1973), however, demonstrated that salt marsh-

Figure 8–14 (*opposite page*). Energy flow diagrams for salt-marsh-embayment complex in Rhode Island for composite *a.* summer and *b.* winter days. *(After Nixon and Oviatt, 1973)*

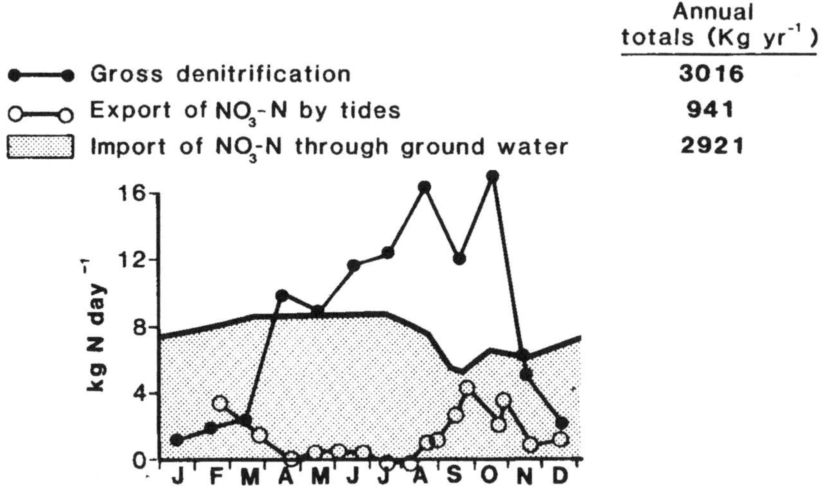

Figure 8–16. Seasonal exchanges of nitrate nitrogen in Great Sippewissett Marsh, Massachusetts. *(From Kaplan et al. 1979; copyright © 1979 by the American Society of Limnology and Oceanography, reprinted with permission)*

es can be nutrient sinks when high-nutrient waters pass through them or when high-nutrient waste water is applied to the marshes. Whiting et al. (1989) reported small imports of inorganic nitrogen and a small export of dissolved organic nitrogen. Wolaver and Spurrier (1988) found that both phosphate and total phosphorus were imported.

After a thorough review of previous studies, Nixon (1980) concluded that, in general, dissolved organic nitrogen and probably dissolved phosphorus were fairly universally exported and that nitrate and nitrite were generally imported, but that these fluxes were not very large components of the nutrient cycles of most estuaries. W. E. Odum et al. (1979) suggested that the balance between import and export depends on geophysical constraints, specifically the morphology of the basin, and on the magnitude of tidal and freshwater fluxes. They hypothesized that estuaries that contain constricted openings to the sea trap and recirculate materials, whereas wedge-shaped estuaries with large ocean exposures tend to export materials.

ECOSYSTEM MODELS

Several conceptual and mathematical models have been developed to describe the structure and function of the salt marsh ecosystem. The energy-flow diagrams of Teal (1962; Fig. 8–13) and Nixon and Oviatt (1973; Fig. 8–14) and the nitrogen budget of the Sippewisset Marsh (Fig. 8–15) represent conceptual mod-

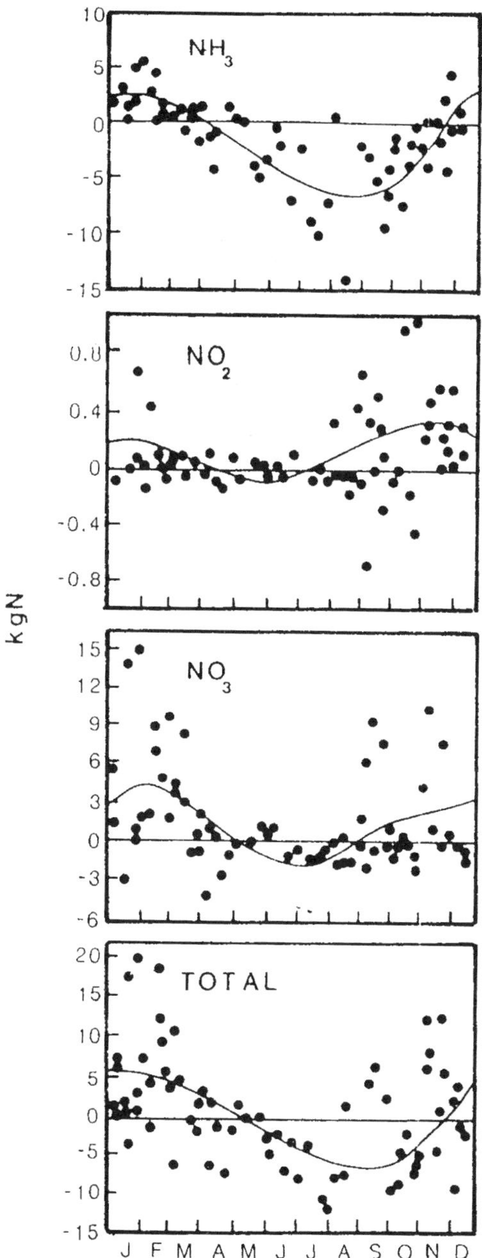

Figure 8–17. Net exchanges of inorganic nitrogen between Flax Pond salt marsh and adjacent estuary in New York. Negative numbers indicate export from marsh. *(From Woodwell et al., 1979; copyright © 1979 by the Ecological Society of America, reprinted with permission)*

els of the salt marsh and contain valuable information about the flow of energy and materials in the marsh. More important, however, these conceptual models have led to a number of simulation models of salt marshes, including those by Williams and Murdock (1972), Reimold (1974), Wiegert et al. (1975), Hopkinson and Day (1977), Zieman and Odum (1977), Wiegert and Wetzel (1979), Wiegert et al. (1981), Morris (1982), Chalmers et al. (1985), Morris and Bowden (1986), and Wiegert (1986).

Reimold (1974) expanded on the phosphorus model of a *Spartina* salt marsh developed by Pomeroy et al. (1972). The model has five phosphorus compartments, including water, sediments, *Spartina*, detritus, and detrital feeders. The model was used to simulate the effects of perturbations such as *Spartina* harvesting on the ecosystem and to help design subsequent field experiments. The simulations demonstrated that *Spartina* regrowth and the resulting phosphorus in the water depended on the time of year that the harvesting of the marsh grass took place.

Zieman and Odum (1977) developed a model of plant succession in the salt marsh and applied it to two sites along the Chesapeake Bay. The model was different from most ecosystem models. It used correlations to relate previously measured time-series data on plant growth to physical and biological characteristics of the marsh. The model was originally developed for use by the Army Corps of Engineers in determining where and how to dispose of dredged material in a manner that would promote the development of a viable salt marsh. The model considered four species of vascular plants, including two species of *Spartina*, and described their growth as a function of salinity, tide, light, pH, and temperature.

Hopkinson and Day (1977) combined a nitrogen budget model for a salt marsh complex in Louisiana with the carbon model of Day et al. (1973) and developed the simulation model shown in Figure 8–18. The model includes seven biological compartments and three nonliving nitrogen storages. The authors stated that nitrogen was the most important nutrient for modeling because it is generally the limiting nutrient in salt marshes. Forcing functions in the model include solar energy, water temperature, tidal fluctuations, nitrogen in rainfall and rivers, and the migration of aquatic fauna. The model simulations suggest that temperature may be the major parameter controlling salt marsh productivity and that the inclusion of sea-level fluctuations gives more realistic predictions.

Morris and Bowden (1986) developed a more detailed model of the sedimentation of exogenous and endogenous organic and inorganic matter, decomposition, above- and belowground biomass and production, and nitrogen and phosphorus mineralization in an Atlantic Coast salt marsh. They found that the model calculations of nitrogen and phosphorus export were sensitive to small changes in belowground production and the fraction of refractory organic matter, both parameters for which there is little good verification. The model demonstrates that mature marshes with deep sediments recycle proportionally more of the nutrients required for plant growth than do young marshes.

Wiegert and his associates (Wiegert et al., 1975; Wiegert et al., 1981; Wiegert and Wetzel, 1979; Chalmers et al., 1985; Wiegert, 1986) have put con-

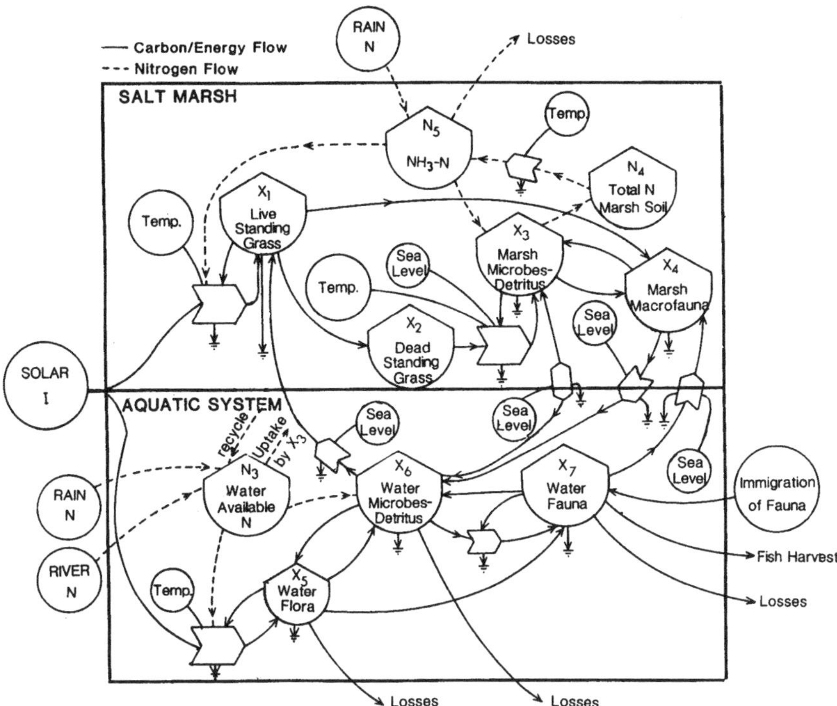

Figure 8–18. Simulation model of salt marsh in Louisiana. (*After Hopkinson and Day, 1977*)

siderable effort into the development of a simulation model for Sapelo Island, Georgia, salt marshes. The model, called MRSH1, has gone through several iterations; the general structure of version six of this model is shown in Figure 8–19. It has 14 major compartments and traces the major pathways of carbon in the ecosystem through *Spartina*, algae, grazers, decomposers, and several compartments of abiotic carbon storage. The model was originally constructed to answer three questions (Wiegert et al., 1981):

1. Is the Georgia salt marsh a potential source of carbon for the estuary, or is it a sink for carbon from offshore?
2. What organism groups are most responsible for the processing of carbon in the salt marsh?
3. What parameters are important (but poorly known) for the proper modeling of the salt marsh?

One of the most important revelations from this model-building process was the demonstration of the importance of the tidal export coefficient. It consistent-

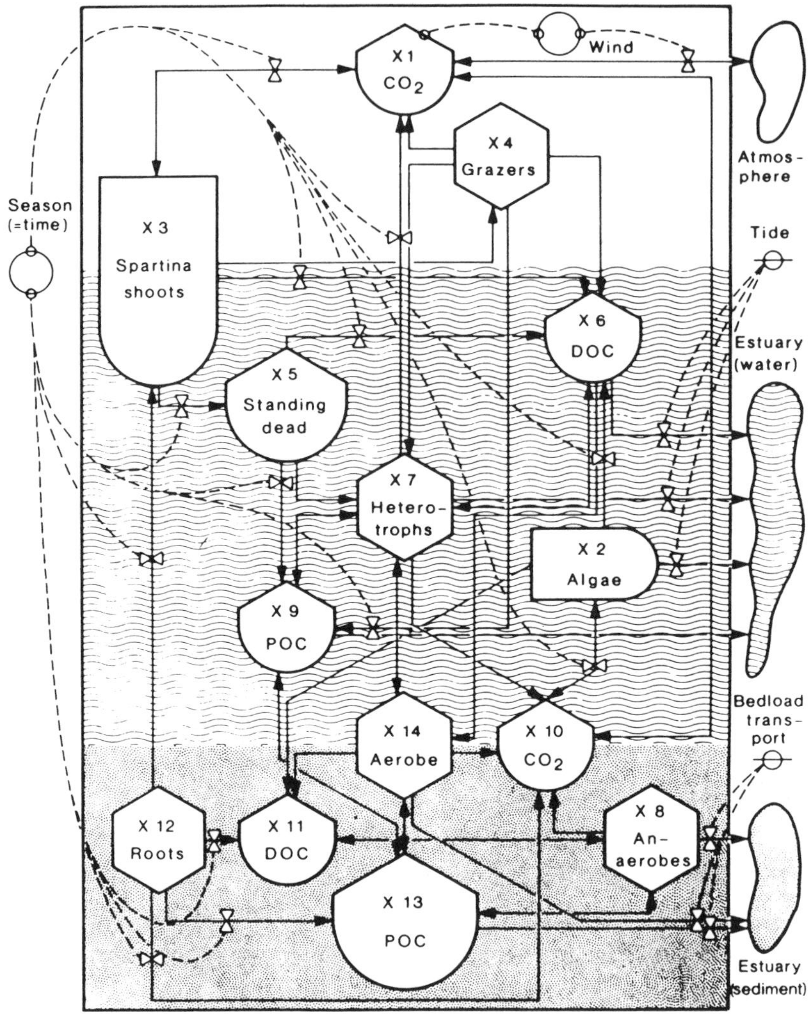

Figure 8–19. Simulation model of a Sapelo Island, Georgia salt marsh. (*From Wiegert et al., 1981; copyright © 1981 by Springer-Verlag, reprinted with permission*)

ly gave export values on the order of 1,000 gC m^{-2} yr^{-1}, well above the field measurements obtained by other researchers such as E. P. Odum and de la Cruz (1967). The model results led to intensive investigations conducted during the next five years of the seasonal variation in particulate and dissolved organic carbon concentrations in the river adjacent to the marsh (Chalmers et al., 1985). This resulted in a more sophisticated model and a revised hypothesis about the dynamics of carbon flux in which material is moved off the marsh in the guts of migratory feeding fish and birds or cycled from the marsh to the upper ends of tidal creeks by local rainfall at low tide and then redeposited on the marsh when tides rise. This is an excellent example of the use of a model as an intellectual tool to drive field research.

Sea Level Rise Coastal Models

Some of the most recent applications of ecological modeling to coastal marshes have been to investigate, on a regional scale, the impacts of future sea level rise due to global warming. A spatial cell-based simulation model developed by Park and colleagues (Park et al., 1991; J. K. Lee et al., 1991), named SLAMM (Sea Level Affecting Marshes Model), predicted that a 1 m sea level rise in the next century could lead to 26 to 82 percent loss of U.S. coastal wetlands, depending on the protection afforded. Most of the loss would occur in the Southeast, especially in Louisiana (Titus et al., 1991). J. K. Lee et al. (1991) predicted a 40 percent loss of wetlands of the Northeastern Florida coastline with a 1 m sea level rise, mostly as low salt marsh.

TIDAL FRESHWATER MARSH

Tidal Freshwater Marshes

9

Freshwater coastal wetlands are unique ecosystems that combine many features of salt marshes and freshwater inland marshes. They act in many ways like salt marshes, but the biota reflect the increased diversity made possible by the reduction of the salt stress found in salt marshes. Plant diversity is high, and more birds use these marshes than any other marsh type. Nutrient cycles are open. Nutrients are retained during the growing season but are lost during the winter. Because they are inland from saline parts of the estuary, they are often close to urban centers. This makes them more prone to human impact than coastal salt marshes.

Simpson et al. (1983) summarized some of the major uncertainties about these systems: "Although evidence suggests that freshwater tidal wetlands act seasonally as nutrient sinks, flux studies of one year or longer are lacking, and the question of whether freshwater tidal wetlands are sinks or sources of material to the estuary cannot now be resolved. Furthermore, the understanding of food chain relationships between freshwater tidal wetlands and the adjacent estuary is rudimentary. At best we can only guess at energy and material transfers between the wetlands and estuarine communities. Finally, there are few data on the short- and long-term effects of pollutants such as oils, pesticides, and heavy metals on species composition and community structure, although these wetlands serve as nursery grounds for commercially important fish and other wildlife."

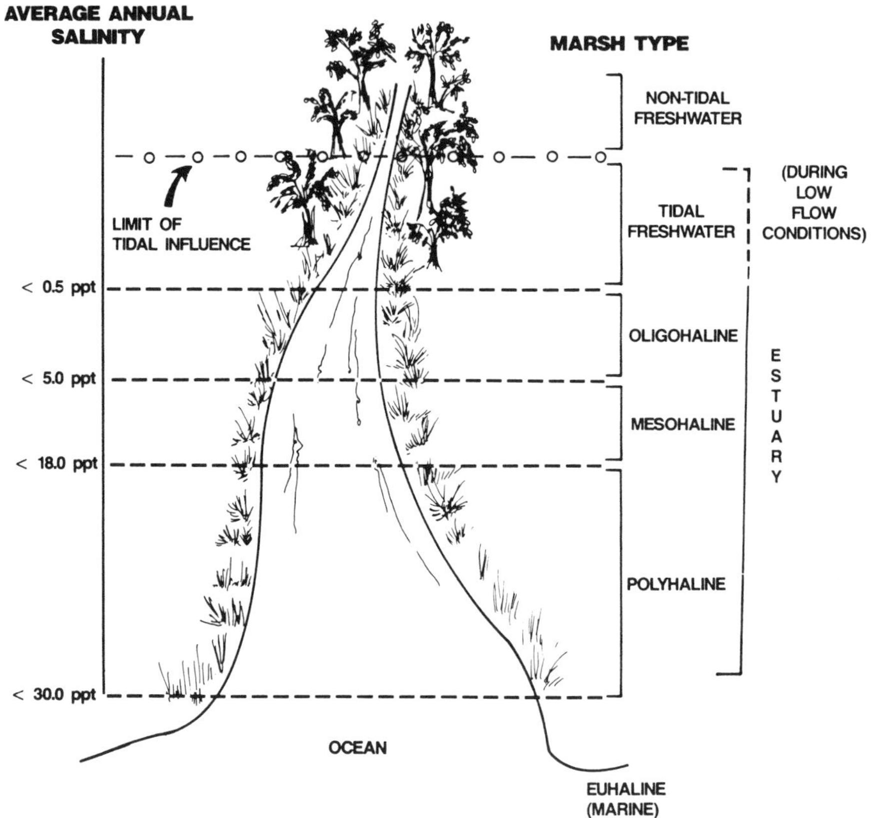

Figure 9–1. Coastal marshes lie on gradients of decreasing salinity from the ocean inland. Tidal freshwater marshes still experience tides but are above the salt boundary. Further inland, marshes experience neither salt nor tides. *(From W. E. Odum et al., 1984)*

This chapter is concerned with freshwater marshes that are close enough to coasts to experience significant tides but at the same time are above the reach of oceanic saltwater. This set of circumstances usually occurs where precipitation is high or fresh river water runs to the coast and where the morphology of the coast amplifies the tide as it moves inland. Tidal freshwater marshes are interesting because they receive the same "tidal subsidy" as coastal salt marshes but without the salt stress. One would expect, therefore, that these ecosystems might be very productive and also more diverse than their saltwater counterparts. As tides attenuate upstream the marshes assume more of the character of inland freshwater marshes (Chap. 11). The distinction between tidal and inland freshwater marshes is not clearcut because on the coast they form a continuum (Fig. 9–1). In this chapter we include the tidally dominated systems of the

Atlantic Coast and the extensive coastal freshwater marshes of the northern Gulf Coast even though the latter are influenced more by wind tides than by lunar tides.

Coastal freshwater marshes are not as well studied nor as well understood as other coastal wetlands, but interest in them is growing and several good reviews have recently been published (Good et al., 1978; Simpson et al., 1983; W. E. Odum et al., 1984; W. E. Odum, 1988). These are drawn on heavily in the following discussion.

GEOGRAPHICAL EXTENT

The physical conditions for tidal freshwater marsh development are adequate rainfall or river flow to maintain fresh conditions, a flat gradient from the ocean inland, and a significant tide range. These conditions occur predominantly along the middle and south Atlantic Coast in the United States. There a number of rivers bring fresh water to the coast, precipitation is moderate and fairly evenly spread throughout the year, and the broad coastal plain is flat and deep. Often the morphology of the system is such that tidal water is constricted as it moves inland, resulting in amplification of the tide range, typically to from 0.5 to 2 meters. Along the northern Gulf Coast, the tide range is small—less than 0.5 meters at the coast—and this range attenuates inland. Nevertheless, because the land slope is so small, freshwater marshes as far inland as 80 km experience some lunar tides, although these are overridden by wind tides and storm runoff.

W. E. Odum et al. (1984) estimated that Atlantic Coast tidal freshwater marshes cover about 164,000 hectares (Fig. 9–2). Most of these are along the middle Atlantic Coast, perhaps one-half in New Jersey. This area estimate does not include upstream nontidal marshes. The Gulf of Mexico coastal freshwater marshes are concentrated in Louisiana, where they cover about 468,000 hectares (Chabreck, 1972). Not all of these marshes are tidal.

GEOMORPHOLOGY AND HYDROLOGY

Coastal freshwater wetlands occur on many different kinds of substrates, but in spite of regional differences, their recent geological history is similar (W. E. Odum et al., 1984). Contemporary coastal marshes are recent (Holocene) in origin. They lie in river valleys that were cut during Pleistocene periods of lowered sea levels. When the sea level rose after the last glaciation (15,000 to 5,000 years B.P.), freshwater coastal marshes expanded rapidly as drowned river systems were inundated and filled with sediment. Except on the northern Gulf of Mexico coast, these marshes are probably still expanding because of the recent increased soil runoff associated with forest clearing, agriculture, and other human activities. On the Gulf Coast the rapid rate of subsidence is degrading fresh tidal marshes. A vertical section through a present-day tidal freshwater

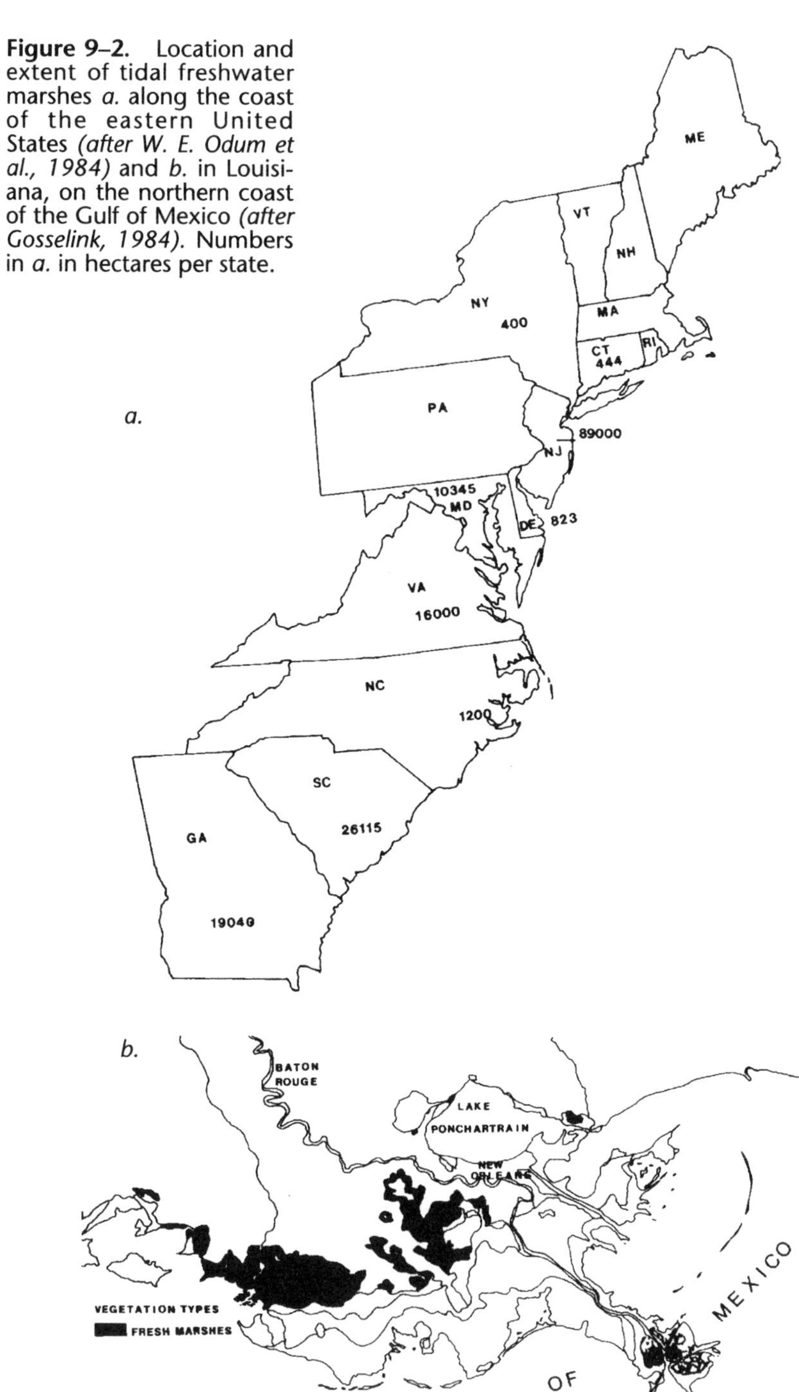

Figure 9–2. Location and extent of tidal freshwater marshes *a.* along the coast of the eastern United States *(after W. E. Odum et al., 1984)* and *b.* in Louisiana, on the northern coast of the Gulf of Mexico *(after Gosselink, 1984).* Numbers in *a.* in hectares per state.

marsh might show a sequence of sediments built on top of an eroded Pleistocene surface cut during a glacial period of lowered sea level. The sediments might include varying layers of riverine, estuarine, and marsh sediments, capped by recent tidal freshwater marsh sediments varying in thickness from less than 1 meter to more than 10 meters (W. E. Odum et al., 1984).

The vigorous tides and accompanying strong currents create a typical elevation gradient from tidal streams out into adjoining marshes. Odum et al. (1984) and Metzler and Rosza (1982) described similar cross-sectional profiles (Fig. 9–3a), with elevation increasing slowly from the stream edge to adjacent upland areas. The low marsh is usually geologically younger than the mature high marsh. A common feature, shown in Figure 9–3b, is a slight elevated levee along the margin of the creek, where overflowing water deposits much of its sediment load. Figure 9–3b shows the elevation decreasing away from the tidal channels, but Simpson et al. (1983) stated that this occurs primarily where the marshes have been impounded. High marshes are flooded to a depth of 30 cm for up to four hours on each tidal cycle; the low marshes are flooded deeper and may have standing water for 9–12 hours on each cycle. The elevation and flooding differences result in gradients of soil physical and chemical properties and in plant zonation patterns that are a consistent feature of freshwater tidal marshes.

Along the northern Gulf Coast the weak lunar tide is often minor compared to wind tides. Winds from the east and southeast, which dominate the weather patterns during the summer, blow water into the upper (inland) reaches of coastal estuaries. Conversely, during periods of north winds, common during the winter, water is blown out of the estuaries and abnormally low water periods result. These wind tides are responsible for water level changes of as much as a meter and sometimes more when heavy rains accompany summer storms. Thus the inundation pattern of Gulf Coast freshwater tidal marshes is not a regular pulse but an irregular pattern of flooding and drying, with longer duration than lunar tides.

Another difference between Gulf Coast and Atlantic Coast tidal freshwater marshes is their elevation profiles. Because they tend to be very wide—often tens of kilometers—Gulf coast marshes do not show a perceptible increase from low to high marsh. When the marshes occur at the edges of lakes, there is often a small natural levee caused by the sediment washover from the lake, but otherwise the elevation differences and vegetation patterns do not occur in distinct linear zones. Instead, associations of plant species occur in apparently random patches, and there is considerable intergrading around the edges.

SOIL AND WATER CHEMISTRY

Typical ranges of chemical parameters in both water and sediments of tidal freshwater marshes are shown in Table 9–1. Freshwater coastal marsh sediments are generally fairly organic, especially in the more mature high marsh. Along the

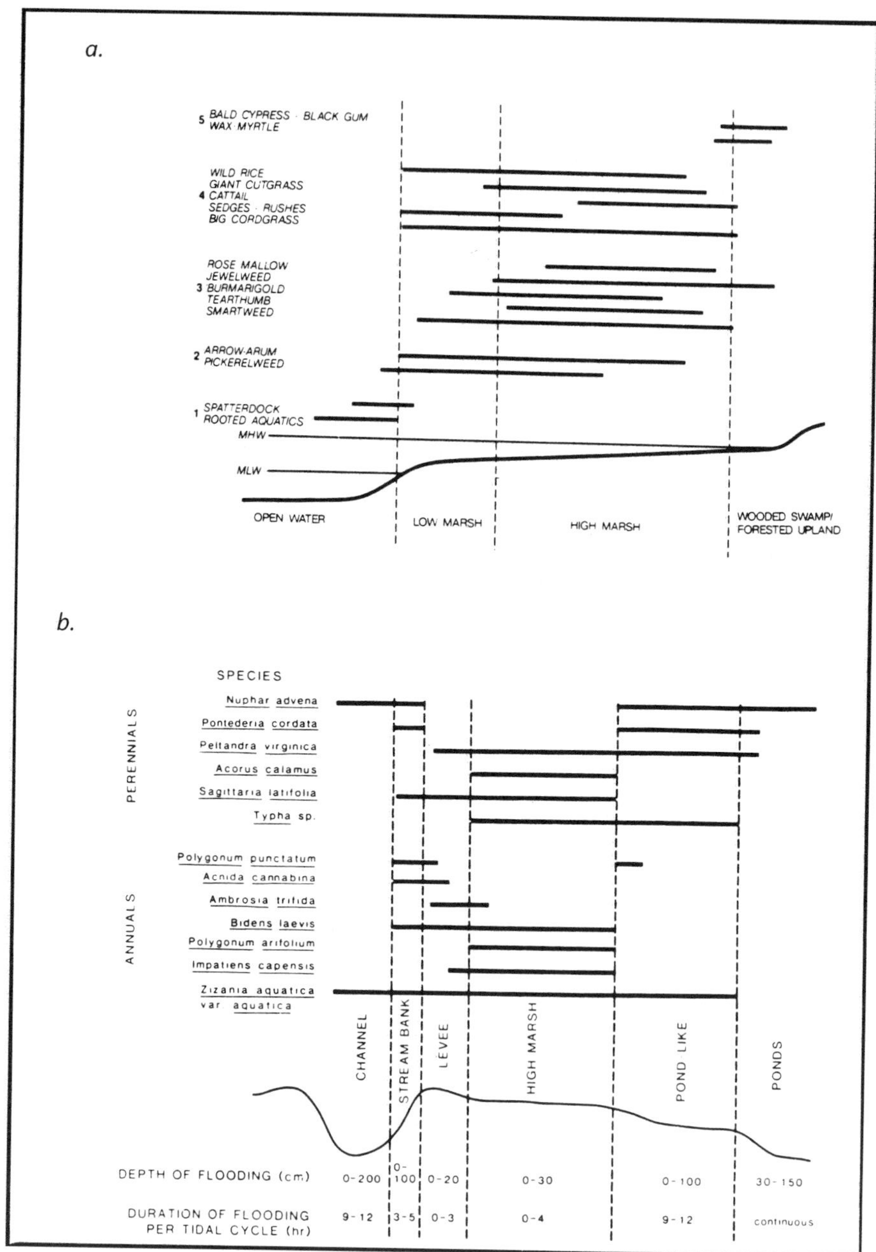

Figure 9–3. Diagrammatic cross sections across typical freshwater tidal marshes, showing elevation changes and typical vegetation. *(a. From W. E. Odum et al., 1984;* b. *from Simpson et al., 1983; copyright © 1983 by the American Institute of Biological Sciences, reprinted with permission)*

Table 9–1. Chemistry of Water and Sediments in Selected Coastal Freshwater Marshes

	Hamilton & Woodbury Creek Marsh, Delaware River[a]	*Herring Creek Marsh, James River, Va.*[b]	*Gulf Coastal Fresh Marsh, Barataria Bay, La.*[c]
Water			
Tide Range	3 m Semidiurnal	<1 m Semidiurnal	Very low, mostly wind tide
Dissolved oxygen, mg/1	4-13	7-12	2-8
Alkalinity, mg $CaCO_3$/1	–	0.39 (winter)	–
$NO_2 + NO_3 - N$, μg/1	40-300	500-1,600	40-370
$NH_3 - N$, μg/1	40-80	460-470	0-2,780
Total Kjeldahl nitrogen, μg/1	–	3,300-4,200	1,960
Reactive phosphorus, μg/1	5-20	30-40	35-340
Total phosphorus, μg/1	5-50	160-180	217
Sediment[d]			
Organic carbon, % of dry weight	14-40	20-50	6-68
Total Kjeldahl nitrogen, % of dry weight	1.03-1.64	1.5 ± 0.8 (S. D.)	1.5 – 1.8
Total phosphorus, % of dry weight	0.12-0.35	0.7	0.09
Cation exchange capacity, meq/100 g dry weight	–	40-67	–
pH	6.2-7.75	–	6.3

[a]Whigham et al., 1980
[b]D. D. Adams, 1978; Lunz et al., 1978; W. E. Odum et al., 1984
[c]Hatton , 1981; Chabreck, 1972; Ho and Schneider, 1976
[d]Top 20cm

Delaware River, they contain 14–40 percent organic material (Whigham and Simpson, 1975), in Virginia tidal freshwater marshes 20–70 percent (W. E. Odum, 1988). On the Gulf Coast they may be as low as 6.2 percent to as high as nearly pure organic peat (Chabreck, 1972; Sasser et al., 1991). The sediments are anaerobic except for a thin surface layer. This condition is reflected in the absence of nitrate in the sediment. Ammonium is present in the winter but is reduced to low levels in the summer by plant uptake. Total nitrogen levels are closely related to the organic content, for almost all of the sediment nitrogen is bound in organic form. Phosphorus is more variable (Table 9–1). The cation-exchange capacity (CEC) in James River marshes is 40–67 meq/100 g dry weight, which is high but typical for a highly organic soil (Wetzel and Powers, 1978). Soil acidity is generally close to neutral.

The water flooding the marsh varies in chemical composition according to the season and the source of water. Recorded concentrations of elements in the water have been high (Table 9–1). Since the detailed chemical studies reported for tidal freshwater wetlands have all been from more or less polluted sites, it is not clear how much the nutrient concentrations reflect the influence of the marsh as compared to upstream sources of materials.

ECOSYSTEM STRUCTURE

Vegetation

Macrophytes

Elevation differences across a freshwater tidal marsh correspond with different plant associations (Figure 9-3). These associations are not discrete enough to call communities, and the species involved change with latitude. Nevertheless, they are characteristic enough to allow some generalizations. On the Atlantic Coast, submerged vascular plants such as *Nuphar advena* (spatterdock), *Elodea* spp. (waterweed), *Potamogeton* spp. (pondweeds), and *Myriophyllum* spp. (water milfoil) grow in the streams and permanent ponds. The creek banks are scoured clean of vegetation each fall by the strong tidal currents, and they are dominated during the summer by annuals such as *Polygonum punctatum* (water smartweed), *Acnida cannabina* (water hemp), and *Bidens laevis* (bur marigold). The natural stream levee is often dominated by the *Ambrosia trifida* (giant ragweed). Behind this levee the low marsh is populated with broad-leaved monocotyledons such as *Peltandra virginica* (arrow arum), *Pontederia cordata* (pickerelweed), and *Sagittaria* spp. (arrowhead).

Typically the high marsh has a diverse population of annuals and perennials. W. E. Odum et al. (1984) called this the "mixed aquatic community type" in the Mid-Atlantic region. Leck and Graveline (1979) described a "mixed annual" association in New Jersey. In both cases, the area was dominated early in the season by perennials such as arrow arum. A diverse group of annuals—*Bidens laevis*, *Polygonum arifolium* (tear-thumb) and other smartweeds, *Pilea pumila* (clearweed), *Hibiscus coccineus* (rose mallow), *Acnida cannabina*, and others—assumed dominance later in the season. In addition to these associations, there are often almost pure stands of *Zizania aquatica* (wild rice), *Typha* spp. (cattail), *Zizaniopsis miliacea* (giant cutgrass), and *Spartina cynosuroides* (big cordgrass). In northern Gulf of Mexico coastal freshwater marshes, arrowheads (*Sagittaria* spp.) replace arrow arum (*Peltandra* spp.) and pickerelweed (*Pontedaria cordata*) at lower elevations. Elsewhere grasses often dominate the perennial species (especially *Panicum hemitomon* [maiden cane]). Many of the same annuals described above are mixed into the association.

The species composition of a tidal freshwater marsh does not appear to depend on the availability of seed in particular locations. Seeds of most species are found in almost all habitats, although the most abundant seed reserves are generally from species found in that marsh zone (Whigham and Simpson, 1975; Leck and Simpson, 1987). They differ, however, in their ability to germinate under the local field conditions and in seedling survival. Flooding is one of the main controlling physical factors. Many of the common plant species seem to germinate well even when submerged, for example, *Peltandra virginica* and *Typha latifolia*, whereas others such as *Impatiens capensis* (Leck 1979), *Cuscuta gronovii*, and *Polygonum arifolium* show reduced germination. (Leck and Graveline, 1979; Leck and Simpson 1987). Competitive factors also play a role: Arrow arum and cattail, for example, produce chemicals that inhibit the germination of seed (McNaughton, 1968; Bonasera et al., 1979); and shading by existing plants is apparently responsible for the inability of arrow arum plants to become established anywhere except along the marsh fringes (Whigham et al., 1979). Some species (*Impatiens capensis*, *Bidens laevis*, and *Polygonum arifolium*) are restricted to the high marsh because the seedlings are not tolerant of extended flooding (Simpson et al., 1983). Seed bank strategies differ in different zones of the marsh. Seed of most of the annuals in the high marsh germinate each spring so that there is little carryover in the soil. In contrast, perennials tend to maintain seed reserves. The seed of most species, however, appear to remain in the soil for a restricted period. In one study 31–56 percent of the seed were present only in surface samples, and 29–52 percent germinated only in sediment samples taken in early spring (Leek and Simpson, 1987). The complex interaction of all these factors has not been elucidated to the extent that it is possible to predict what species will be established where on the marsh.

Algae

In addition to vascular plants, phytoplankton and epibenthic algae abound in freshwater tidal marshes, but relatively little is known about them. In the Potomac River marshes, in one study, diatoms (Bacillariophytes) were the most common phytoplankton, with green algae (Chlorophytes) comprising about one-third of the population and blue-green algae (Cyanophytes) present in moderate numbers (Lippson et al., 1979). The same three phyla accounted for most of the epibenthic algae. Indeed, many of the algae in the water column are probably entrained by tidal currents off the bottom. In a study of New Jersey tidal freshwater marsh soil algae, Whigham et al. (1980) identified 84 species exclusive of diatoms. Growth was better on soil that was relatively mineral and coarse compared with growth on fine organic soils. Shading by emergent plants reduced algal populations in the summer months. Algal biomass is probably two to three orders of magnitude less than peak biomass of the vascular plants (Wetzel and Westlake, 1969), but the turnover rate is much more rapid.

Floating Marshes

An interesting variant in freshwater marshes is the floating marsh. Although it is usually found in nontidal areas, in the United States the largest area of floating marshes appears to be in the coastal Louisiana tidal marshes. Large expanses of floating marshes are also found in the Danube Delta (*Phragmites communis* marshes; Pallis, 1915), along the lower reaches of the Sud in Africa (*Papyrus* swamps; Beadle, 1974), in South America (floating meadows in lakes of the varzea; Junk, 1970), and Tasmania (floating islands in the Lagoon of Islands; Tyler, 1976). Floating marshes have also been reported in Germany, The Netherlands (Verhoeven, 1986), England (Wheeler, 1980), and North Dakota and Arkansas in the United States (Eisenlohr, 1972; Huffman and Lonard, 1983). In Louisiana floating marshes are usually floristically diverse but are dominated by ferns in spring and by *Panicum hemitomon* during summer and fall (Sasser and Gosselink, 1984). The marsh substrate is composed of a thick organic mat, entwined with living roots, that rises and falls (all year or seasonally) with the ambient water level (Swarzenski et al., 1991). This type of marsh is interesting in a successional sense because it appears to be an endpoint in development; it is freed from normal hydrologic fluctuation and sediment deposition. Hence in the absence of salinity intrusions, it appears to support a remarkably stable community (see also Chap. 7).

Consumers

Invertebrates

Coastal freshwater wetlands are used heavily by wildlife. The consumer food chain is predominantly detrital, and benthic invertebrates are an important link in the food web. Bacteria and protozoa decompose litter, gaining nourishment from the organic material. It appears unlikely that these microorganisms concentrate in large enough numbers to provide adequate food for macroinvertebrates (Wiebe and Pomeroy, 1972). Meiobenthic organisms, primarily nematodes, compose most of the living biomass of anaerobic sediments (J. P. Sikora et al., 1977). They probably crop the bacteria as they grow, packaging them in bite-sized portions for slightly larger macrobenthic deposit feeders. In coastal freshwater marshes, the microbenthos is composed primarily of amoebae (Thecamoebinids, a group of amoebae with theca, or tests). This is in sharp contrast to more saline marshes in which foraminifera predominate (Ellison and Nichols, 1976). The slightly larger macrobenthos is composed of amphipods, especially *Gammarus fasciatus*, oligochaete worms, freshwater snails, and insect larvae. Copepods and cladocerans are abundant in the tidal creeks. The Asiatic clam (*Corbicula fluminaea*), a species introduced earlier in the century, has spread through the coastal marshes of the southern states and as far north as the Potomac River (W. E. Odum et al., 1984). Caridean shrimp, particularly *Palaemonetes pugio*, are common, as are freshwater shrimp, *Macrobrachium* spp. The density and diversity of these benthic organisms are reported to be low compared with those in nontidal freshwater wetlands perhaps because of the lack of diverse bottom types in the tidal reaches of the estuary. No species are found exclusively in tidal freshwater systems. Instead, those found there appear to have a wide range (Diaz, 1977).

Nekton

Coastal freshwater marshes are important habitats for many nektonic species that use the area for spawning and year-round food and shelter and as a nursery zone and juvenile habitat. Fish of coastal freshwater marshes can be classified into four groups (Fig. 9–4). Most of them are freshwater species that spawn and complete their lives within freshwater areas. The three main families of these fish are cyprinids (minnows, shiners, carp), centrarchids (sunfishes, crappies, bass), and ictalurids (catfish). Juveniles of all species are most abundant in the shallows, often using submerged marsh vegetation for protection from predators. Predator species, the bluegill (*Lepomis macrochirus*), largemouth bass (*Micropterus salmoides*), sunfishes (*Lepomis* spp.), warmouth (*Lepomis gulosus*), and black crappie (*Pomoxis nigromaculatus*) are all important for sport fishing. Gars (*Lepisosteus* spp.), pickerels (*Esox* spp.), and bowfin (*Amios calya*) are other common predators often found in coastal freshwater marshes.

Some oligohaline or estuarine fish and shellfish that complete their entire life cycle in the estuary extend their range to include the freshwater marshes. Killifishes (*Fundulus* spp.) particularly the banded killifish (*F. diaphanus*) and the mummichog (*F. heteroclitus*) are abundant in schools in shallow freshwater marshes, where they feed opportunistically on whatever food is available. The bay anchovy (*Anchoa mitchilli*) and tidewater silverside (*Menidia beryllina*) are often abundant in freshwater areas also. The latter breeds in this habitat more than in saltwater areas. Juvenile hogchokers (*Trinectes maculatus*) and naked gobies (*Gobiosoma bosci*) use tidal freshwater areas as nursery grounds (W. E. Odum et al., 1984).

Anadromous fishes, which live as adults in the ocean, or semianadromous species, whose adults remain in the lower estuaries, pass through coastal freshwater marshes on their spawning runs to freshwater streams. For many of these species, the tidal freshwater areas are major nursery grounds for juveniles. Along the Atlantic Coast herrings (*Alosa* spp.) and shads (*Dorosoma* spp.) fit into this category. The young of all of them, except the hickory shad (*A. mediocris*), are found in peak abundance in tidal fresh waters, where they feed on small

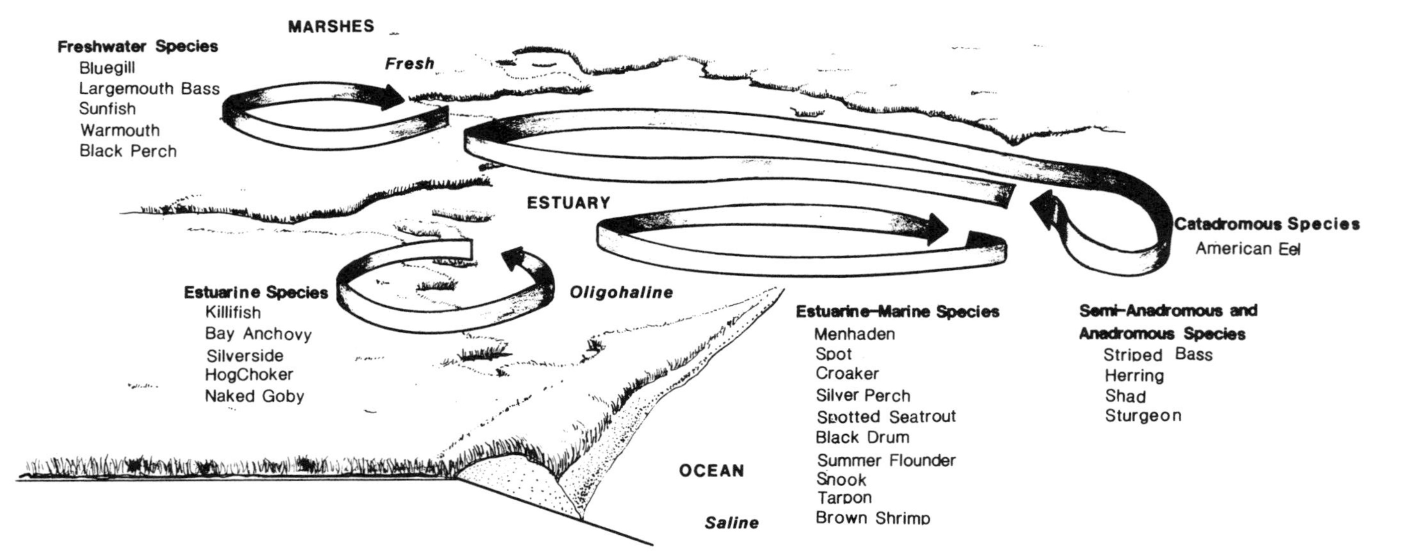

Figure 9–4. Fish and shellfish that use tidal freshwater marshes can be classified into four groups: freshwater, estuarine, anadromous and catadromous, and estuarine-marine.

invertebrates and, in turn, are an important forage fish for striped bass (*Morone saxatilis*), white perch (*Morone americana*), catfish (*Ictalurus* spp.), and others (W. E. Odum et al., 1984). As they mature late in the year, they migrate downstream to saline waters and offshore. Two species of sturgeon (*Acipenser brevirostrum* and *A. oxyrhynchus*) were important commercially in East Coast estuaries, but were seriously overfished and presently are rare. Both species spawn in nontidal and tidal fresh waters, and juveniles may spend several years there before migrating to the ocean (Brundage and Meadows, 1982).

The striped bass is perhaps the most familiar semianadromous fish of the mid-Atlantic Coast because of its importance in both commercial and sport fisheries. Approximately 90 percent of the striped bass on the East Coast are spawned in tributaries of the Chesapeake Bay system (Berggren and Lieberman, 1977). They spawn in spring in tidal fresh and oligohaline waters; juveniles remain in this habitat along marsh edges, moving gradually downstream to the lower estuary and nearshore zone as they mature. Because the critical period for survival of the young is the larval stage, conditions in the tidal fresh marsh area where these larvae congregate are important determinants of the strength of the year class (W. E. Odum et al., 1984).

The only catadromous fish species in Atlantic Coast estuaries is the American eel (*Anguilla rostrata*). It spends most of its life in fresh or brackish water, returning to the ocean to spawn in the region of the Sargasso Sea. Eels are common in tidal and nontidal coastal freshwater areas, in marsh creeks, and even in the marsh itself (Lippson et al., 1979).

The juveniles of a few species of fish that are marine spawners move into freshwater marshes, but most remain in the oligohaline reaches of the estuary. Species whose range extends into tidal freshwater marshes are menhaden (*Brevoortia tyrannus*), spot (*Leiostomus xanthurus*), croaker (*Micropogonias undulatus*), silver perch (*Bairdiella chrysoura*), spotted seatrout (*Cynoscion nebulosus*), black drum (*Pogonium cromis*), summer flounder (*Saralichthys dentatus*), snook (*Centropomus undecimalis*), and tarpon (*Megalops atlanticus*). Along the northern Gulf Coast, juvenile brown and white shrimp (*Penaeus* spp.) may also move into freshwater areas. These juveniles emigrate to deeper, more saline waters as temperatures drop in the fall.

Birds

Of all wetland habitats, coastal freshwater marshes may support the largest and most diverse populations of birds. W. E. Odum et al. (1984), working from a number of studies, compiled a list of 280 species of birds that have been reported from tidal freshwater marshes. They stated that although it is probably true that this environment supports the greatest bird diversity of all marshes, the lack of comparative quantitative data makes it difficult to test this hypothesis (W. E. Odum, 1988). Bird species include waterfowl (44 species); wading birds (15 species); rails and shorebirds (35 species); birds of prey (23 species); gulls, terns, kingfishers, and crows (20 species); arboreal birds (90 species); and ground and shrub birds (53 species). A major reason for the intense use of these marshes is the structural diversity of the vegetation provided by broad-leaved plants, tall grasses, shrubs, and interspersed ponds.

Dabbling ducks (family Anatidae) and Canada geese actively select tidal freshwater areas (R. E. Stewart, 1962) on their migratory flights from the north. They use the Atlantic Coast marshes in the late fall and early spring, flying farther south during the cold winter months. Most of these species winter in fresh coastal marshes of the northern Gulf, but some fly to South America. Their distribution in apparently similar marshes is variable; some marshes support dense populations, others few birds. The reasons for this spotty use are unclear. The birds feed in freshwater marshes on the abundant seeds of annual grasses and sedges and the rhizomes of perennial marsh plants and also in adjacent agricultural fields. They are opportunistic feeders, on the whole, ingesting from the available plant species. An analysis by Abernethy (1986) suggests that many species that frequent the fresh marsh early in the winter move down to the coastal salt marshes before beginning their northward migration in the spring. The reason for that behavior pattern is not known, but Abernethy speculated that the preferred foods of the freshwater marshes are depleted by early spring and the birds move into salt marshes that have not been previously grazed.

The wood duck (*Aix sponsa*) is the only species that nests regularly in coastal freshwater tidal marshes, although an occasional black duck (*Anas rubripes*) or mallard (*A. platyrynchos*) nest is found in the Atlantic Coast marshes.

Wading birds are common residents of coastal freshwater marshes. They are present year-round in Gulf Coast marshes but only during the summer along the Atlantic Coast. An exception is the great blue heron (*Ardia herodias*), which is seen throughout the winter in the north Atlantic states. Nesting colonies are common throughout the southern marshes, and some species (green herons [*Butorides striatus*] and bitterns [*Ixobrychus exilis* and *Botaurus lentiginosus*]), nest along the mid-Atlantic coast. They feed on fish and benthic invertebrates, often flying long distances each day from their nesting areas to fish.

Rails (*Rallus* spp.) and shorebirds, including the killdeer (*Charadrius vociferus*), sandpipers (Scolopacidae), and the American woodcock (*Philohela minor*), are common in coastal freshwater marshes. They feed on benthic macroinvertebrates and diverse seeds. Gulls (*Larus* spp.), terns (*Sterna* spp.), kingfishers (*Megaceryle alcyon*), and crows (*Corvus* spp.) are also common. Some are migratory; some are not. A number of birds of prey are seen hovering over freshwater marshes, including the common northern harrier (*Circus cyaneus*), the American kestrel (*Falco sparverius*), falcons (*Falco* spp.), eagles (*Haliaeetus leucocephalus*), ospreys (*Pandion haliaetus*), owls (Tytonidae), vultures (Cathartaidae), and the loggerhead shrike (*Lanius ludovicianus*). The number of these beautiful birds has been declining in recent years. Two of them, the southern bald eagle (*H. leaucocephalus*) and the peregrine falcon (*Falco peregrinus*), are listed as endangered.

Arboreal birds use the coastal freshwater marshes intensively during short periods of time on their annual migrations. Flocks of tens of thousands of swallows (*Hirundinidae*) have been reported over the upper Chesapeake freshwater marshes (R. E. Stewart and Robbins, 1958). Flycatchers (*Tyrannidae*) are also numerous. They often perch on trees bordering the marsh, darting out into the marsh from time to time to capture insects (W. E. Odum et al., 1984). Although coastal marshes may be used for only short periods of time by a migrating species, they may be important temporary habitats. For example, the northern Gulf coastal marshes are the first landfall for birds on their spring migration from South America. Often they reach this coast in an exhausted state and the availability of forested barrier islands for refuge and marshes for feeding is critical to their survival.

Sparrows and finches (Fringillidae), juncos (*Junco* spp.), blackbirds (Icteridae), wrens (Troglodytidae), and other ground and shrub birds are abundant residents of coastal freshwater marshes. W. E. Odum et al. (1984) indicated that ten species breed in mid-Atlantic Coast marshes, including the ring-necked pheasant (*Phasianus colchicus*), red-winged blackbird (*Agelaius phoeniceus*), American goldfinch (*Carduelis tristis*), rufous-sided towhee (*Pipilo erythrophthalmus*), and a number of sparrows. The most abundant are the red-winged blackbirds, dickcissels (*Spiza americana*), and bobolinks (*Dolichonyx oryzivorus*), which can move into and strip a wild rice marsh in a few days.

Amphibians and Reptiles

Although W. E. Odum et al. (1984) compiled a list of 102 species of amphibians and reptiles that frequent coastal freshwater marshes along the Atlantic Coast, many are poorly understood ecologically, especially with respect to their dependence on this type of habitat. None are specifically adapted for life in tidal freshwater marshes. Instead, they are able to tolerate the special conditions of this environment. River turtles, the most conspicuous members of this group, are abundant throughout the southeastern United States. Three species of water snakes (*Natrix*) are common. *Agkistrodon piscivorus* (the cottonmouth moccasin) is found south of the James River. And in the South, especially along the Gulf Coast, the American alligator's preferred habitat is the tidal freshwater marsh. These large reptiles used to be listed as threatened or endangered, but they have come back so strongly in most areas that they are presently harvested legally in Louisiana and Florida. They nest along the banks of coastal freshwater marshes, and the animal, identified by its high forehead and long snout, is a common sight gliding along the surface of marsh streams.

Mammals

The mammals most closely associated with coastal freshwater marshes are all able to get their total food requirements from the marsh, have fur coats that are more or less impervious to water, and are able to nest (or hibernate in northern areas) in the marsh (W. E. Odum et al., 1984). These include the otter (*Lutra canadensis*), muskrat (*Ondatra zibethicus*), nutria (*Myocastor coypus*), mink (*Mustela vison*), raccoon (*Procyon lotor*), marsh rabbit (*Silvilagus palustris*), and marsh rice rat (*Oryzomys palustris*). In addition, the opossum (*Didelphis virginiana*) and white-tailed deer (*Odocoileus virginianus*) are locally abundant. The nutria was introduced from South America some years ago and has spread throughout the Gulf Coast states and into Maryland, North Carolina, and Virginia. It is not likely to spread farther north because of its cold intolerance, but the south Atlantic marshes would seem to provide an ideal habitat. Nutria is more vigorous than the muskrat and has displaced it from the freshwater marshes in many parts of the northern Gulf. As a result, muskrat density is highest in oligohaline marshes. The muskrat, for some reason, is not found in coastal Georgia and South Carolina, nor in Florida, although it is abundant farther north along the Atlantic Coast. Muskrat, nutria, and beaver (*Castor canadensis*) can influence the development of a marsh. The first two species destroy large amounts of vegetation with their feeding habits (they prefer juicy rhizomes and uproot many plants when digging for them), their nest building, and their underground passages. Beavers have been observed in tidal freshwater marshes in Maryland and Virginia. Their influence on forested habitats is well known, but their impact on tidal freshwater marshes needs to be studied more closely (W. E. Odum et al. 1984).

ECOSYSTEM FUNCTION

Primary Production

Many production estimates have been made for freshwater coastal marshes. Productivity is generally high, usually falling in the range of 1,000 to 3,000 g m^{-2} yr^{-1} (Table 9–2). The large variability reported from different studies stems in part from a lack of standardization of measurement techniques, but real differences can be attributed to several factors:

1. The type of plant and its growth habit. Fresh coastal marshes, in contrast to saline ones, are floristically diverse, and productivity is determined, at least to some degree, by genetic factors that regulate the species' growth habits. Tall perennial grasses, for example, appear to be more productive than broad-leaved herbaceous species such as arrow arum and pickerelweed.

Table 9–2. Peak Standing Crop and Annual Production Estimates for Common Tidal Freshwater Vegetation Associations[a]

Vegetation Type	*Peak Standing Crop* g/m^2	*Annual Production* $g\ m^{-2}\ yr^{-1}$
Nuphar advena (spatterdock)	627	780
Peltandra virginica/Pontederia cordata (arrow arum/pickerelweed)	671	888
Zizania aquatica (wild rice)	1,218	1,578
Zizaniopsis miliacea (giant cutgrass)	1,039	2,048
Polygonum sp./*Leersia oryzoides* (smartweed/rice cutgrass)	1,207	–
Hibiscus coccineus (rose mallow)	1,141	869
Typha sp. (cattail)	1,215	1,420
Bidens spp. (bur marigold)	1,017	1,340
Acorus calamus (sweetflag)	857	1,071
Sagittaria latifolia (duck potato)	432	1,071
Amaranthus cannabinus (water hemp)	960	1,547
Ambrosia tirifida (giant ragweed)	1,205	1,205
Phragmites communis (common reed)	1,850	1,872
Spartina cynosuroides (big cordgrass)	2,311	–
Lythrum salicaria (spiked loosestrife)	1,616	2,100
Rosa palustris (swamp rose)	699	–

[a]Values are means of one to eight studies.

[b]Designation indicates the dominant species in the association.

Source: Summarized by W. E. Odum et al., 1984

2. Tidal energy. The stimulating effect of tides on production has been shown for salt marshes (see Chap. 8). Whigham and Simpson (1977) showed that the fresh marsh grass *Zizania aquatica* responded positively to tides, and the general trends in production shown by Brinson et al. (1981a) in nonforested freshwater wetlands support the idea that moving water generally stimulates production.
3. Other factors. Soil nutrients (Reader, 1978), grazing, parasites, and toxins are other factors that can limit production (de la Cruz, 1978).

The elevation gradient across a fresh coastal marsh and the resulting differences in vegetation and flooding patterns account for three broad zones of primary production. The low marsh bordering tidal creeks, dominated by broad-leaved perennials, is characterized by apparently low production rates. Biomass peaks early in the growing season. Turnover rates, however, are high, especially along the northern Gulf Coast (Visser, 1989), suggesting that annual production may be much higher than what can be determined from peak biomass. Much of the production is stored in belowground biomass (root:shoot >>1), but this biomass is mostly rhizomes rather than fibrous roots, decomposition is rapid, the litter is swept from the marsh almost as fast as it forms, the soil is bare in winter, and erosion rates are high.

The parts of the high marsh dominated by perennial grasses and other erect, tall species are characterized by the highest production rates of freshwater species, and root:shoot ratios are approximately 1. Because tidal energy is not as strong and the plant material is not so easily decomposed, litter accumulates on the soil surface, and little erosion occurs.

The high marsh mixed annual association typically reaches a large peak biomass late in the growing season. Most of the production is aboveground (root:shoot < 1), and litter accumulation is common; but in the absence of perennial roots, erosion rates might be expected to be greater than those where perennials dominate.

Decomposition

As with other marshes, little plant production is consumed directly by grazers. Although nutrias and muskrats are common in freshwater tidal marshes, herbivory is thought to account for the consumption of less than 10–40 percent of plant production (W. E. Odum, 1988). The remaining 60–90 percent becomes available to consumers through the detrital food chain. The vegetation is attacked by bacteria and fungi, aided by the fragmenting action of small invertebrates, and the bacteria-enriched decomposed broth feeds benthic invertebrates that, in turn, are prey to larger animals.

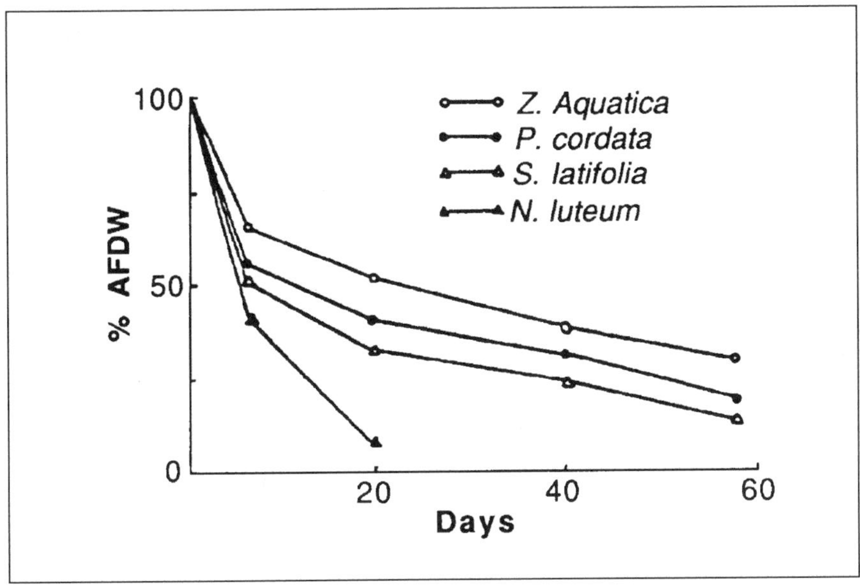

Figure 9–5. The rate of decay of leaves of *Zizania aquatica, Pontederia cordata, Sagittaria latifolia,* and *Nuphar luteum* as shown by the amount of material (ash free dry weight) remaining with time in submerged litterbags. Each data point represents four replicates. *(From W. E. Odum and Heywood, 1978; copyright © 1978 by Academic Press, reprinted with permission)*

Temperature is the major factor that controls litter decomposition; the higher the temperature, the higher the decay rate. The combined availability of oxygen and water is a second factor; plants submerged in anaerobic environments decompose slowly, as do exposed plants in dry environments. Optimum conditions for decomposition are found in a moist aerobic environment such as a regularly flooded tidal marsh. An important third factor in decomposition rates is the kind of plant tissue involved. In freshwater tidal marshes, two groups of plants can be identified (W. E. Odum et al., 1984). The broad-leaved perennials (*Pontederia cordata*, *Sagittaria latifolia*, and *Nuphar luteum* in Fig. 9–5) generally contain high leaf concentrations of nitrogen. Their detritus has high nutritional quality, as indicated by a low C:N ratio and their selection by detritus consumers over *Spartina* detritus (Dunn, 1978; Smock and Harlowe, 1983). In contrast, the high marsh grasses (*Zizania aquatica* in Fig. 9-5) are low in nitrogen (tissue concentration usually < 1 percent) and are composed primarily of long stems with much structural tissue that is resistant to decay. Their slow decay rates, combined with lower tidal energy on the high marsh, may explain why litter accumulates there and erosion rates are low.

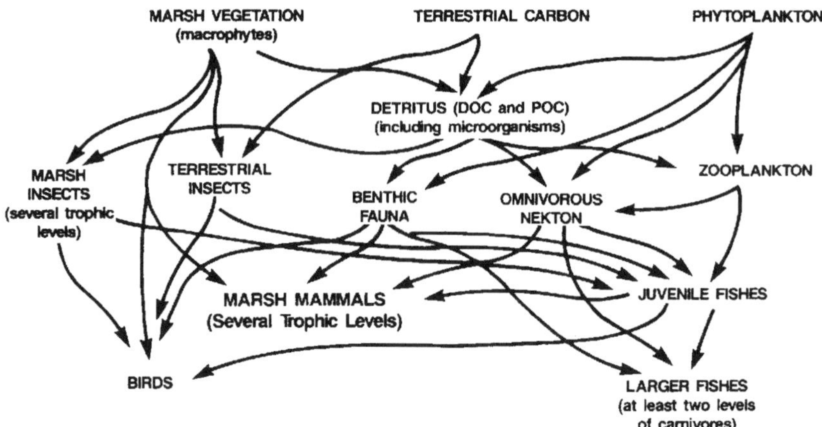

Figure 9–6. Energy flow diagram for a tidal freshwater marsh, showing the major groups of organisms and energy pathways. *(From W. E. Odum et al., 1984)*

Organic Export

Losses of organic carbon from marshes occur through respiration, flushing from the marsh surface, sequestering as peat below the root zone, conversion to methane that escapes as a gas, and export as biomass in the bodies of consumers that feed on the marsh. Only a handful of studies address carbon fluxes in freshwater tidal marshes, and so any conclusions must be considered tentative. W. E. Odum et al. (1984) hypothesized that young, low-lying marshes are subject to ice shearing of the vegetation during the winter and, because of the vigorous tidal action probably export significant quantities of both particulate and dissolved organic material. Older, higher marshes, which are less vigorously flushed, probably do not export much particulate organic matter but may export dissolved materials.

In anaerobic freshwater sediments, where, in contrast to salt marshes, little sulfur is available as an electron acceptor, carbon dioxide can be reduced to methane. Methanogenesis and fermentation are the predominant pathways of respiratory energy flow (Delaune et al. 1983a; W. E. Odum, 1988). The annual loss of carbon as methane from Gulf Coast freshwater marshes has been estimated as 160 g CH_4 m^{-2} yr^{-1} (C. J. Smith et al., 1982). In comparison, Lipschultz (1981) estimated a loss of only 10.7 g CH_4 m^{-2} yr^{-1} from a *Hibiscus*-dominated freshwater marsh in the Chesapeake Bay.

In marshes that are accreting vertically, organic matter is lost to deep sediments. There are few measurements of the magnitude of this loss in freshwater marshes along the Atlantic Coast, but in Gulf Coast freshwater marshes, where accretion is rapid (about 1 cm/yr), peat accumulation as organic carbon is 145–250 gC m^{-2} yr^{-1} (Hatton, 1981).

Energy Flow

Quantitative measurements of energy flow through the detrital food web in freshwater coastal wetlands are practically nonexistent. W. E. Odum et al. (1984) mentioned energy consumption values for some consumers, but these were generally calculated from the literature or were extrapolated from other kinds of marsh systems or both. This subject apparently is a fruitful area for research. Although it is not feasible to trace quantitatively the flow of organic energy in freshwater coastal marshes, W. E. Odum et al. (1984) presented a conceptual scheme that indicates the major functional groups and their interrelationships (Fig. 9–6). They identified three major sources of organic carbon. The largest is probably the vascular marsh vegetation, but organic material brought from upstream (terrestrial carbon) may be significant, especially on large rivers and where domestic sewage waters are present. Phytoplankton is a largely unknown quantity. Most of the organic energy flows through the detrital pool and is distributed to benthic fauna and deposit-feeding omnivorous nekton. These groups feed fish, mammals, and birds at higher trophic levels. The magnitude of the herbivore food chain, in comparison to the detritus one, is poorly understood. Insects are more abundant in fresh marshes than in salt marshes, but most do not appear to be herbivorous. Marsh mammals (O'Neil, 1949) and birds (T. J. Smith and Odum, 1981) apparently can "eat out" significant areas of vegetation, but direct herbivory is probably small in comparison to the flow of organic energy from destroyed vegetation into the detrital pool. The phytoplankton-zooplankton-juvenile fish food chain in fresh marshes is of interest because of its importance to humans. Zooplankton is an important dietary component for a variety of larval, postlarval, and juvenile fishes of commercial importance that are associated with tidal freshwater marshes (Van Engel and Joseph, 1968).

In addition to the flow of energy through living organisms, significant amounts of organic material are buried as peat or are used to reduce carbon dioxide to methane, which is lost to the atmosphere.

Nutrient Budgets

In general, nutrient cycling and nutrient budgets in coastal freshwater wetlands appear to be similar to salt marshes; they are fairly open systems that have the capacity to act as long-term sinks, sources, or transformers of nutrients. Most nutrient inputs are inorganic; these nutrients are transformed chemically or biologically to reduced or organic forms that appear as export products. Although marshes are leaky compared to forested ecosystems, they still recycle most of the nutrients used within the system; exports and imports are generally a relatively small percentage of the total material cycled.

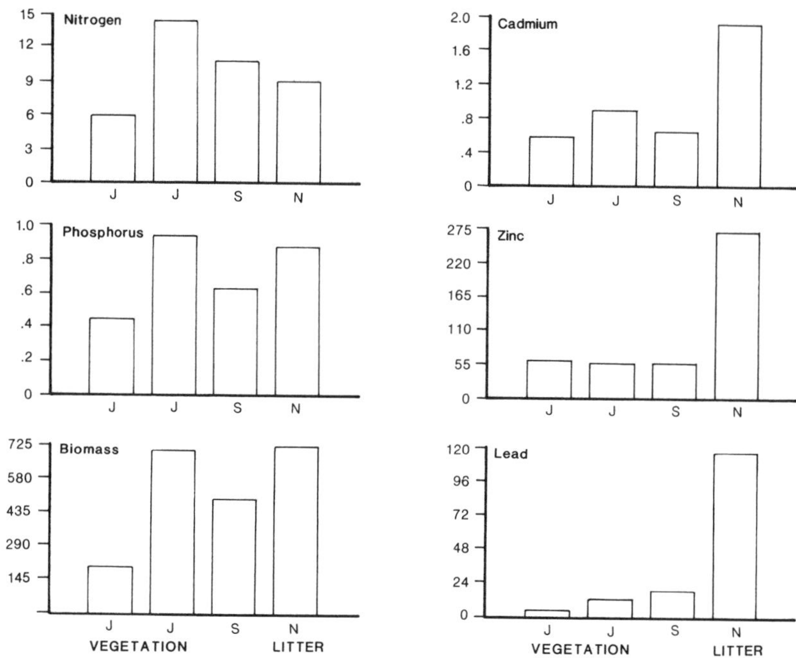

Figure 9–7. Biomass (g/m2), nutrient (g/m2), and heavy metal (mg/m2) standing stocks in the vegetation (June–September) and litter (November) in Woodbury Creek Marsh, New Jersey, for 1979. All values are means. *(Data from Simpson et al., 1983)*

Many marshes have been shown to be nutrient traps that purify the water flooding them. Whether this is true of freshwater coastal marshes depends on the age and ecological maturity of the marsh, the magnitude of upland runoff, anthropogenic effects such as sewage loading, and the magnitude of tidal action (Stevenson et al., 1977). Unfortunately, most nutrient studies of freshwater coastal marshes have been carried out in areas heavily influenced by nearby urban communities, and so it is difficult to know how representative they are. From a study of Woodbury Creek marsh in New Jersey (Simpson et al., 1983), tidal freshwater marshes appear to be net importers of nitrogen and phosphorus during the spring primarily because of the magnitude of upland runoff during this period. During spring and summer, nutrients were tied up in plant biomass. Since sediment concentrations do not seem to change much seasonally, the living biomass reflects seasonal storage (Fig. 9–7). On the low marsh only small amounts of nutrients were tied up in the sparse litter, but in the fall on the high marsh this was a major temporary store. After the vegetation died in the fall, there appeared to be a rapid net export of nutrients associated with the high decomposition rate and tidal flushing. During the year as a whole, the net balance of nitrogen and phosphorus indicated a net export. This may reflect the fact

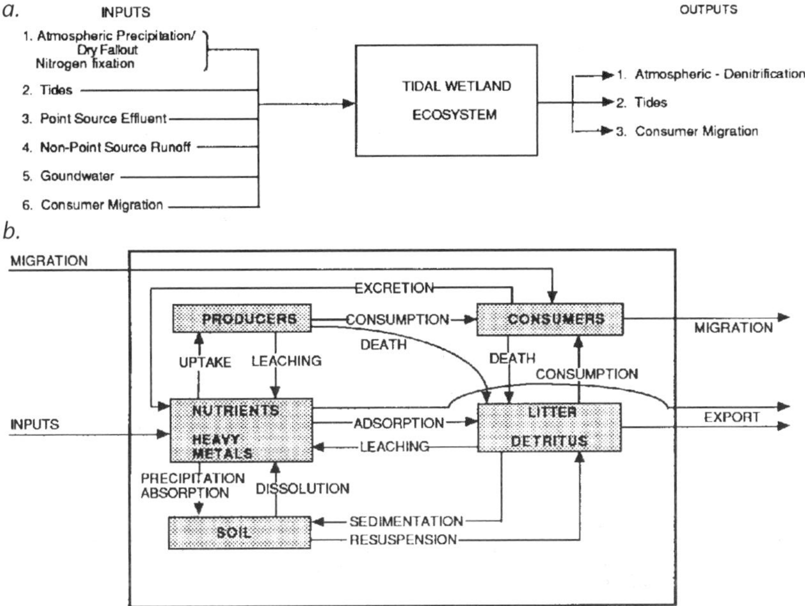

Figure 9–8. Model of tidal freshwater wetland showing: *a.* major inputs and outputs of materials, and *b.* major compartments and pathways in which nutrients and heavy metals are stored and move. *(From Simpson et al., 1983; copyright © 1983 by the American Institute of Biological Sciences, reprinted with permission)*

that the marshes studied were subject to unusually high loading with domestic sewage and were saturated with respect to those nutrients.

The accumulation of heavy metals in a high marsh in New Jersey was variable (Simpson et al., 1981; 1983). Cadmium, copper, lead, nickel, and zinc had accumulated in the litter at the end of the growing season in much higher concentrations than in the live vegetation (Fig. 9–7). Throughout the annual cycle, cadmium was always exported from the marsh, and nickel was imported in all months except June. Copper was imported during the growing season, and zinc and lead were imported primarily after it. In general, accumulation was greatest in the plant roots (Simpson et al., 1981).

ECOSYSTEM MODELS

There appear to be few conceptual models and fewer simulation models of tidal freshwater ecosystems. A coastal freshwater marsh model might combine the tidal forcing function of a salt marsh model (Zieman and Odum, 1977) with the diversity of species of an inland freshwater marsh. Such a model has not been produced. The trophic model of W. E. Odum et al. (1984, Fig. 9–6) and the general compart-

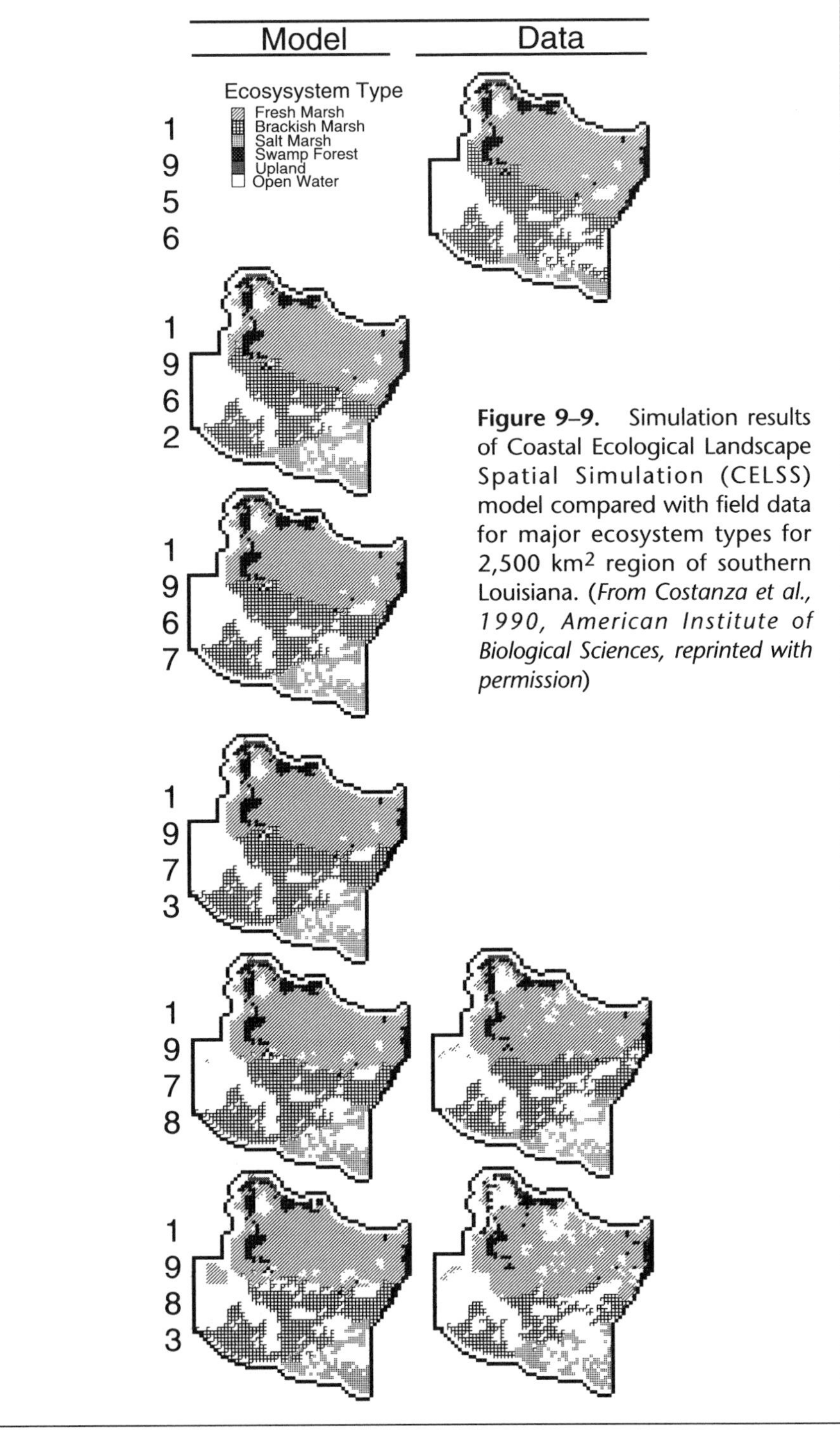

Figure 9–9. Simulation results of Coastal Ecological Landscape Spatial Simulation (CELSS) model compared with field data for major ecosystem types for 2,500 km^2 region of southern Louisiana. (*From Costanza et al., 1990, American Institute of Biological Sciences, reprinted with permission*)

mental model displayed by Simpson et al. (1983; Fig. 9–8) are useful conceptual aids to visualizing this ecosystem, but they do not incorporate important driving forces such as tides and nutrient loads except in the most general sense.

In an ambitious simulation modeling effort, scientists on the Gulf Coast modeled both the temporal and the spatial succession of coastal marshes, including tidal freshwater marshes, based on geomorphic and hydrologic changes that are occurring in the area. The model was used to predict the salinization of freshwater marshes and their erosion to open water. It also simulated the reversal of this effect by sediment carried by the Atchafalaya River, which is rapidly building a delta on the north coast of the Gulf of Mexico (Sklar et al., 1985, Costanza et al., 1988, 1990). That model, which simulated a 2500 km^2 region, is one of the first attempts to combine temporal and spatial modeling concepts for wetlands. Figure 9–9 illustrates the output for this Coastal Ecological Landscape Spatial Simulation (CELSS) model for a base-case run when model simulation results are compared to existing vegetation maps. Both the model and the field data illustrate the graduate "breakup" of freshwater marshes resulting from land subsidence, salt intrusion, and levee construction.

MANGROVE WETLAND

Mangrove Wetlands

10

The mangrove wetland replaces the salt marsh as the dominant coastal ecosystem in subtropical and tropical regions throughout the world. Mangrove wetlands are limited in the United States to the southern extremes of Florida (where there are approximately 2,700 km^2 of mangroves) and to Puerto Rico. The dominant plant species in mangrove wetlands are known for several adaptations to the saline wetland environment, including prop roots, pneumatophores, salt exclusion, salt excretion, and the production of viviparous seedlings. Mangrove wetlands have been classified according to their hydrodynamics and topography as fringe mangroves, riverine mangroves, basin mangroves, and dwarf or scrub mangroves. There is generally little understory. Mangrove wetlands have definite vegetation zonation patterns, although the importance of successional stages and physical conditions to this zonation has been much debated. It now appears that the zonation results from each species' optimal niche for productivity. Mangroves require a greater percentage of their energy for maintenance in high-salinity conditions. Massive diebacks of mangrove trees can occur when environmental change is rapid. Damage can be mitigated if there is freshwater or tidal flushing and exacerbated if the system is already under stress. The greater the hydrologic turnover (riverine > fringe > basin > scrub), the greater the productivity. The highest productivity occurs in riverine forests that are most open to both tidal action and inputs of nutrients from adjacent uplands. The least productive systems are dwarf mangroves that are found in nutrient-poor conditions, in hypersaline *soils, and at the northern extreme of the mangrove's range. The importance of both upland inflows and tidal exchange has been demonstrated with simulation models. Several complete energy budgets have been developed to described mangrove wetlands, and organic export studies and comparisons with estuarine productivities have verified the importance of these ecosystems to the secondary productivity of adjacent estuaries. Mangrove wetlands that are subject to increased tidal influence have the greatest percentage of their litterfall exported to adjacent waters, but herbivory in the wetlands is generally low.*

The coastal salt marsh of temperate middle and high latitudes gives way to its analog, the mangrove swamp, in tropical and subtropical regions of the world. The mangrove swamp is an association of halophytic trees, shrubs, and other plants growing in brackish to saline tidal waters of tropical and subtropical coastlines. This coastal, forested wetland (the wetland is called a *mangal* by some researchers) is infamous for its impenetrable maze of woody vegetation, its unconsolidated peat that seems to have no bottom, and its many adaptations to the double stresses of flooding and salinity. The word *mangrove* comes from the Portuguese word (*mangue*) for "tree" and the English word *grove* for "a stand of trees" (Dawes, 1981) and refers to both the dominant trees and to the entire plant community (W. E. Odum and McIvor, 1990).

Many myths have surrounded the mangrove swamp. It was described at one time or another in history as a haven for wild animals, a producer of fatal "mangrove root gas," and a wasteland of little or no value (Lugo and Snedaker, 1974). Researchers, however, have established the importance of mangrove swamps in exporting organic matter to adjacent coastal food chains, in providing physical stability to certain shorelines to prevent erosion, in protecting inland areas from severe damage during hurricanes and tidal waves, and in serving as sinks for nutrients and carbon. There is an extensive literature on the mangrove swamp on a worldwide basis—possibly more than 5,000 titles. This interest probably stems from the worldwide scope of these ecosystems and the many unique features that they possess. Much of the literature, however, concerns floristic and structural topics (excellent summaries by Chapman, 1976b, and Tomlinson, 1986), and major interest in the functional aspects of mangrove swamps has been expressed only since the early 1970s. Since that time, a significant literature on ecophysiology, primary productivity, stressors, food chains, and the detritus dynamics of mangrove ecosystems has been produced, but there has been surprisingly little work on nutrient cycling, mangrove restoration, valuation of mangrove resources, and responses of mangroves to sea level changes (Lugo, 1990a). Good summaries of mangrove ecosystem function are provided by Lugo and Snedaker (1974), W. E. Odum et al. (1982), Armentano (1990), and W. E. Odum and McIvor (1990).

GEOGRAPHICAL EXTENT

Mangrove swamps are found along tropical and subtropical coastlines throughout the world, usually between 25°N and 25°S latitude (Fig. 10–1a). Their limit in the Northern Hemisphere generally ranges from 24°N to 32°N latitude, depending on local climate and the southern limits of freezing weather. The frequency of frosts necessary to limit mangroves is not clearly established. For example, in the United States, mangrove wetlands are found primarily along the Atlantic and Gulf coasts of Florida up to 27–29° N latitude (Fig. 10–1b), north of which they are replaced by salt marshes (Kangas and Lugo, 1990). Schaeffer-Novelli et al. (1990) described mangroves as extending to 28°–30° south latitude along the Brazilian Coast. Chapman (1976b) suggested that three to four nights of a light frost are sufficient to kill even the hardiest mangrove species. Lugo and Patterson-Zucca (1977) showed that mangroves survived approximately five nonconsecutive days of frost in January 1977 in Sea Horse Key, Florida (latitude 29°N), but estimated that it would take 200 days for the forest to recover from cold damage. They also hypothesized that soil salinity stress could modify frost stress on mangroves, suggesting that the latitudinal limit of mangroves reflects a number of stresses rather than one factor. Kangas and Lugo (1990) suggested that the boundary between tropical mangroves and temperate salt marshes can be attributed to a combination of frost stress on mangroves and, in the absence of stress, a competitive advantage by mangrove vegetation over salt marsh grasses. They also suggested that the replacement of salt marsh grasses by mangrove trees "may be a special case of the more general phenomena of tree vegetation replacing herbaceous vegetation in successional sequences and along environmental gradients" such as the replacement of tundra by boreal forests along a temperature gradient and the replacement of grasslands by deciduous forests along moisture gradients.

Mangroves are divided into two groups—the Old World mangrove swamps and the New World and West African mangrove swamps. An estimated 68 species of mangroves exist, and their distribution is thought to be related to continental drift in the long term and possibly to transport by primitive humans in the short term (Chapman, 1976b). The distribution of these species, however, is uneven. The swamps are particularly dominant in the Indo–West Pacific region (part of the Old World group), where they contain the greatest diversity of species. There are 30–40 species of mangroves in that region, whereas there are only about 10 mangrove species in the Americas. It has often been argued, therefore, that the Indian-Malaysian region was the original center of distribution for the mangrove species (Chapman, 1976b). There is also a great deal of segregation between the mangrove vegetation found in the Old World region and the New World of the Americas and West Africa. Two of the primary genera of mangrove trees, *Rhizophora* and *Avicennia*, contain separate species in the Old and New Worlds, suggesting "that speciation is taking place independently in each region'' (Chapman, 1976b).

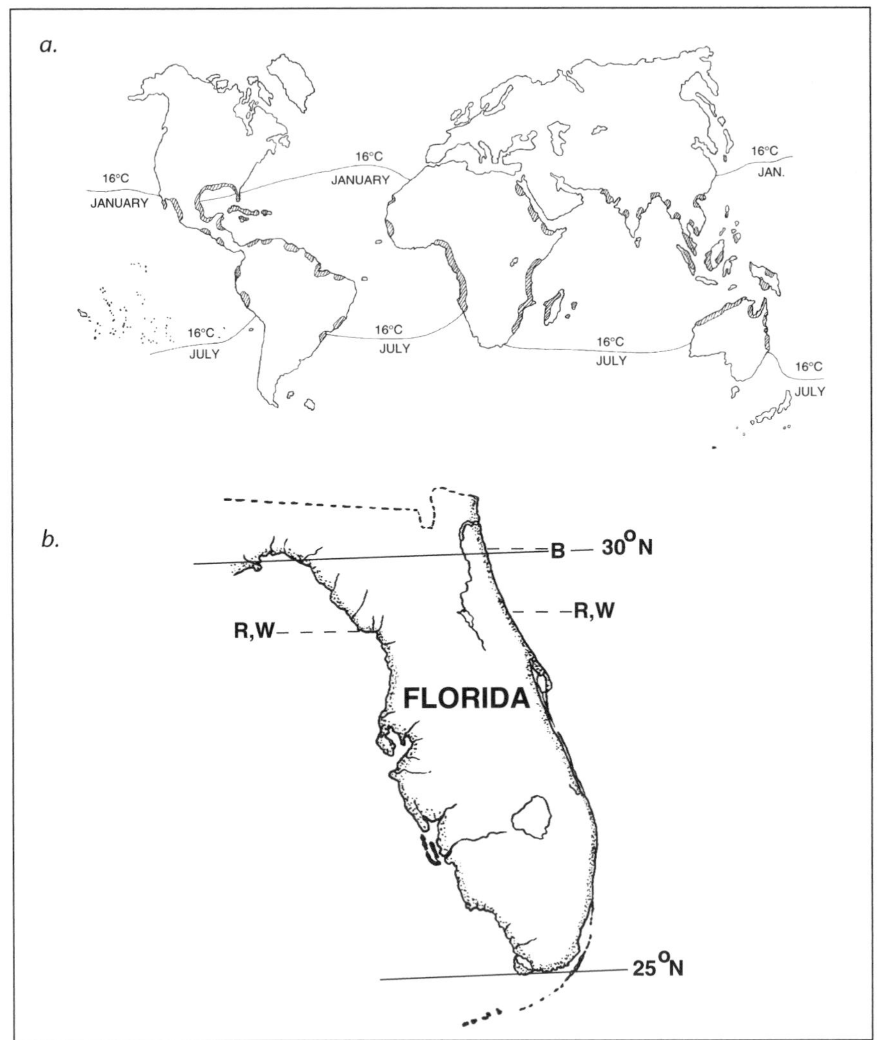

Figure 10–1. Distribution of mangrove wetlands *a.* in the world and *b.* on Florida coastline. R, B, and W indicate northern extent of red, black, and white-mangroves respectively along the Florida coastline. (a. *After Chapman, 1977;* b. *From W. E. Odum et al., 1982 and W. E. Odum and McIvor, 1990; by permission, University of Central Florida Press, Orlando*)

J. H. Davis (1940) described the Florida Coast as supporting more than 2,600 km^2 (1,000 square miles) of mangrove swamps, although Craighead (1971) revised the estimate down to 1,750 km^2 (675 square miles). Estimates from the National Wetlands Inventory, completed in 1982, are 2,730 km^2 (1,050 square miles) of mangroves in Florida (W. E. Odum and McIvor, 1990). The best development of mangroves in Florida is along the southwest coast, where the Everglades and the Big Cypress Swamp drain to the sea (see Fig. 10–1b). Mangroves extend up to 30 km (18 miles) inland along water courses on this coast. The area includes Florida's Ten Thousand Islands, one of the largest mangrove swamps in the world. Because of development pressure, a significant fraction of these original mangrove islands has been lost or altered. Patterson (1986) reported that there was a loss of 24 percent of mangroves on one of the most developed islands in this region, Marco Island, from 1952 to 1984. Mangroves are now protected in Florida, and it is illegal to remove them from the shoreline.

Mangrove swamps are also common farther north along Florida's coasts, north of Cape Canaveral on the Atlantic and to Cedar Key on the Gulf of Mexico, where mixtures of mangrove and salt marsh vegetation appear. One species of mangrove (*Avicennia germinans*) is found as far north as Louisiana and in the Laguna Madre of Texas, although the trees are more like shrubs at these extreme locations (Chapman, 1976b; 1977). Extensive mangrove swamps are also found throughout the Caribbean Islands, including Puerto Rico. Lugo (1988) estimated that there were originally 120 km^2 (46 square miles) of mangroves in Puerto Rico, although only half of those remained by 1975.

GEOMORPHOLOGY AND HYDROLOGY

There are several different types of mangrove wetlands, each having a unique set of topographic and hydrodynamic conditions. Like the coastal salt marsh, the mangrove swamp can develop only where there is adequate protection from wave action. Several physiographic settings favor the protection of mangrove swamps, including (1) protected shallow bays, (2) protected estuaries, (3) lagoons, (4) the leeward sides of peninsulas and islands, (5) protected seaways, (6) behind spits, and (7) behind offshore shell or shingle islands. Unvegetated coastal and barrier dunes usually develop where this protection does not exist, and mangroves can often be found behind the dunes (Chapman, 1976b).

In addition to the required physical protection from wave action, the range and duration of the flooding of tides exert a significant influence over the extent and functioning of the mangrove swamp. The tides constitute an important subsidy for the mangrove swamp, importing nutrients, aerating the soil water, and stabilizing soil salinity. Salt water is important to the mangroves in eliminating competition from freshwater species. The tides provide a subsidy for the movement and distribution of the seeds of several mangrove species (see Mangrove Adaptations, below). They also circulate the organic sediments in some fringe mangroves for the benefit of filter feeding organisms such as oysters, sponges, and barnacles and for deposit feeders such as snails and fiddler crabs (Kuenzler, 1974). Like salt marshes, mangrove swamps are intertidal, although a large tidal range is not necessary. Most mangrove wetlands are found in tidal ranges of from 0.5 m to 3 m and more. Mangrove tree species can also tolerate a wide range of inundation frequencies (Chapman, 1976b). *Rhizophora* spp., the red mangrove, is often found growing in continually flooded coastal waters below normal low tide. At the other extreme, mangroves can be found several kilometers inland along river banks where there is less tidal action. Lugo (1981) found that these inland mangroves depend on storm surges and "are not isolated from the sea but critically dependent on it as a source of fresh sea water."

The development of mangrove swamps is the result of topography, substrate, and freshwater hydrology as well as tidal action. A classification of mangrove wetland ecosystems according to their physical conditions was developed by Lugo and Snedaker (1974) and Lugo (in Wharton et al., 1976) and included six types. Cintrón et al. (1985) suggested a simplification of that classification system to four major types:

1. Fringe mangroves, including overwash islands
2. Riverine mangroves
3. Basin mangroves
4. Dwarf (or scrub) mangroves

The features of these types of mangrove wetlands are shown in Figure 10–2 and are discussed below.

Fringe Mangroves

Fringe mangrove wetlands are found along protected shorelines and along some canals, rivers, and lagoons (Fig. 10–2a). They are common along shorelines adjacent to land higher than mean high tide but are exposed to daily tides. In contrast to the overwash mangroves discussed in the next paragraph, fringe mangrove wetlands tend to accumulate organic debris because of the low-energy tides and the dense development of prop roots. Because the shoreline is open, these wetlands are often exposed to storms and strong winds that lead to the further accumulation of debris. Fringe mangroves are found on narrow berms along the coastline or in wide expanses along gently sloping beaches. If a berm is present, the mangroves may be isolated from freshwater runoff and would then have to depend completely on rainfall, the sea, and groundwater for their nutrient supply. These wetlands are found throughout south Florida along both coasts, and in Puerto Rico.

A special case of fringe mangroves are small islands and narrow extensions of larger landmasses (spits) that are "overwashed" on a daily basis during high tide. These are sometimes called overwash mangrove islands (Fig. 10–2b). The forests are dominated by the red mangrove (*Rhizophora mangle*) and a prop root system that obstructs the tidal flow and dissipates wave energy during periods of

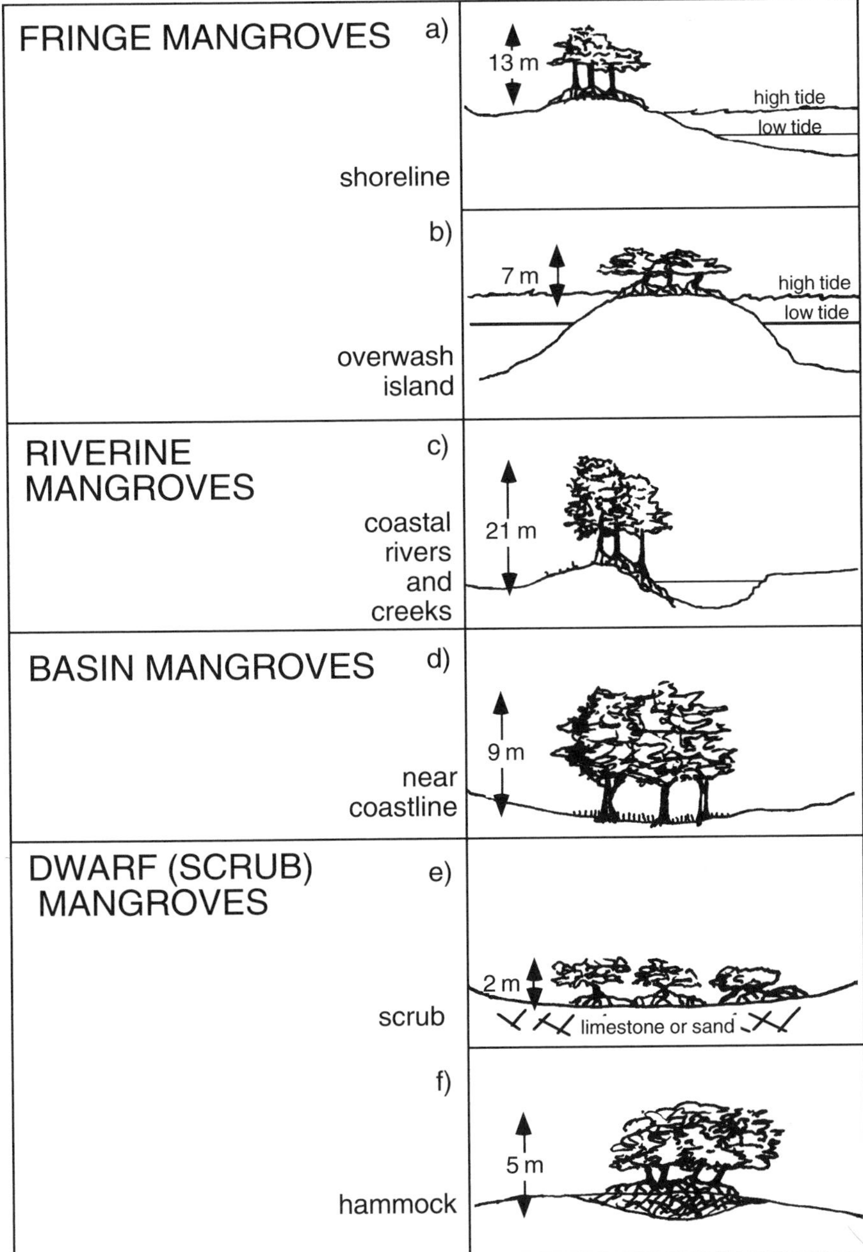

Figure 10–2. Classification of mangrove wetlands according to hydrodynamic conditions. General classification is (a and b) fringe mangroves, (c) riverine mangroves, (d) basin mangroves, and (e and f) scrub (dwarf) mangroves. (*After Wharton et al., 1976; Lugo, 1980, and Cintrón et al., 1985*)

heavy seas. Tidal velocities are high enough to wash away most of the loose debris and leaf litter into the adjacent bay. The islands often develop as concentric rings of tall mangroves around smaller mangroves and a permanent, usually hypersaline, pool of water. These wetlands are abundant in the Ten Thousand Islands region of Florida and the southern coast of Puerto Rico. They are particularly sensitive to the effects of ocean pollution.

Riverine Mangroves

Tall, productive riverine mangrove forests are found along the edges of coastal rivers and creeks, often several miles inland from the coast (Fig. 10–2c). These wetlands may be dry for a considerable time, although the water table is generally just below the surface. In Florida, freshwater input is greatest during the wet summer season, causing the highest water levels and the lowest salinity in the soils during that time. Riverine mangrove wetlands export a significant amount of organic matter because of their high productivity. These wetlands are affected by freshwater runoff from adjacent uplands and from water, sediments, and nutrients delivered by the adjacent river, and hence they can be significantly affected by upstream activity or stream alteration. The combination of adequate freshwater and high inputs of nutrients from both upland and estuarine sources causes these systems to be generally very productive, supporting large (16–26 m) mangrove trees. Salinity varies but is usually lower than that of the other mangrove types described here. The flushing of freshwater during wet seasons causes salts to be leached from the sediments (Cintrón et al., 1985).

Basin Mangroves

Basin mangrove wetlands occur in inland depressions, or basins, often behind fringe mangrove wetlands, and in drainage depressions where water is stagnant or slowly flowing (Fig. 10–2d). These basins are often isolated from all but the highest tides and yet remain flooded for long periods once tide water does flood them. Because of the stagnant conditions and less frequent flushing by tides, soils have high salinities and low redox potentials. These wetlands are often dominated by black mangroves (*Avicennia* spp.) and white mangroves (*Laguncularia* spp.), and the ground surface is often covered by pneumatophores from these trees.

Dwarf Mangroves

There are several examples of isolated, low-productivity mangrove wetlands that are usually limited in productivity because of the lack of nutrients or freshwater inflows. Dwarf mangrove wetlands are dominated by scattered, small (often less than 2 m tall) mangrove trees growing in an environment that is probably nutrient-poor (Fig. 10–2e). The nutrient-poor environment can be a sandy soil or limestone marl. Hypersaline conditions and cold at the northern extremes of the mangrove's range can also produce "scrub" or stressed mangrove trees, in riverine, fringe, or basin wetlands. True dwarf mangrove wetlands, however, are found in the coastal fringe of the Everglades and the Florida Keys and along the northeastern coast of Puerto Rico. Some of these wetlands in the Everglades are exposed to tides only during the spring tide or storm surges and are often flooded by freshwater runoff in the rainy season.

Hammock mangrove wetlands also occur as isolated, slightly raised tree islands in the coastal fringe of the Florida Everglades and have characteristics of both basin and scrub mangroves. They are slightly raised as a result of the buildup of peat in what was once a slight depression in the landscape (see Fig. 10–2f). The peat has accumulated from many years of mangrove productivity, actually raising the surface from 5 to 10 cm above the surrounding landscape. Because of this slightly raised level and the dominance by mangrove trees, these ecosystems look like the familiar tree islands or "hammocks" that are found throughout the Florida Everglades. They are different in that they are close enough to the coast to have saline soils and occasional tidal influences and thus can support only mangroves.

CHEMISTRY

Salinity

Mangrove swamps are found under conditions that provide a wide range of salinity. J. H. Davis (1940) summarized several major points about salinity in mangrove wetlands from his studies in Florida:

1. There is a wide annual variation in salinity in mangrove wetlands.
2. Saltwater is not necessary for the survival of any mangrove species but only gives mangroves a competitive advantage over salt-intolerant species.
3. Salinity is usually higher and fluctuates less in interstitial soil water than in the surface water of mangroves.
4. Saline conditions in the soil extend farther inland than normal high tide because of the slight relief, which prevents rapid leaching.

Figure 10–3 shows the wide spatial and temporal range of salinity, from constant seawater salinity of outer coast overwash and mangrove swamps to the brackish water in coastal rivers and canals, found in a region in Florida dominated by mangrove wetlands. Seasonal oscillations in salinity are a function of the

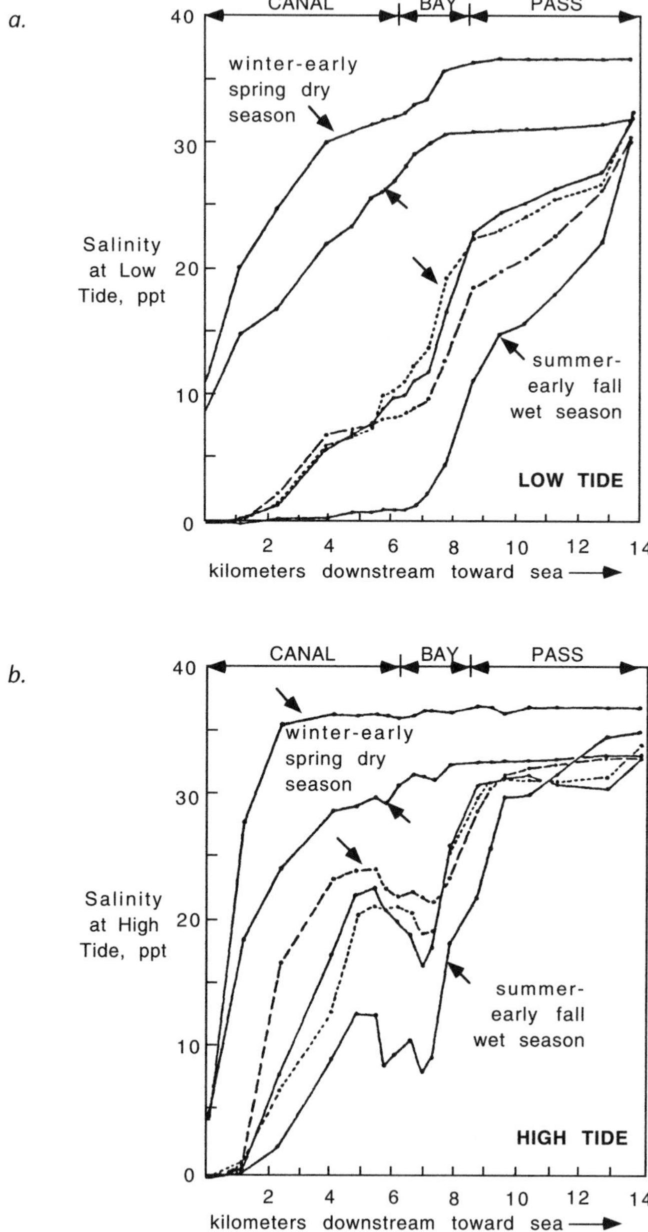

Figure 10–3. Variation in salinity from inland canal to open sea in a mangrove region of southwestern Florida during *a.* low tide, and *b.* high tide. Canal is channelized stream that flows into the bay. (*From Carter et al., 1973*)

height and duration of tides, the seasonality and intensity of rainfall, and the seasonality and amount of fresh water that enters the mangrove wetlands through rivers, creeks, and runoff. In Florida, as illustrated in Figure 10–3, summer wet season convective storms and associated freshwater flow in streams and rivers as well as an occasional hurricane in the late summer–early fall leads to the dilution of saltwater and the lowest salinity concentrations. Salinity is generally the highest in the dry season that occurs in the winter and early spring.

Soil salinity in mangrove ecosystems varies greatly from season to season and with mangrove type (Table 10–1). In riverine mangrove systems, the salinity is less than that of normal sea water because of the influx of fresh water. In basin mangroves, on the other hand, salinity can be well above that of sea water because of evaporative losses (>50 ppt). Boto and Wellington (1984) found soil salinity in mangroves in North Queensland, Australia, to range from an average of 30 to 50 ppt, generally above that of the overlying waters. The highest salinities were found where there was restriction of tidal exchange.

Dissolved Oxygen

Reduced conditions exist in most mangrove soils when they are flooded. The degree of reduction depends on the duration of flooding and the openness of the wetland to freshwater and tidal flows. Some oxygen transport to the rhizosphere occurs through the vegetation, although this overall contribution is localized and is probably small in the sediment oxygen balance. When creeks and surface runoff pass water through mangrove wetlands, the reduced conditions are not as severe because of the increased drainage and the continual importing of oxygenated waters (Chapman, 1976b).

Table 10–1. Soil Salinity Ranges for Major Mangrove Types

Hydrodynamic Type	*Soil Salinity, ppt*
Fringe mangroves	
Avicennia zone	59
Rhizophora zone	39
Riverine mangroves	10–20[a]
Basin mangroves	
Avicennia zone	>50
Laguncularia zone	low salinity
mixed forest zone	30–40

[a]Higher in dry season when less freshwater streamflow is available

Source: Cintrón et al., 1985

Mazda et al. (1990) reported on the variations in dissolved oxygen, salinity, and temperature in a Japanese mangrove that is connected by tides to the ocean through a coral reef but is occasionally isolated by a sandy sill. The oxygen content of the water that reaches the mangrove is the result of (1) the semidiurnal tidal cycle, which brings water to the mangrove at flood tide, (2) the diurnal pattern of dissolved oxygen from the productive coral reef, and (3) the presence or absence of the sandy sill that isolates the wetland (Fig. 10–4). The highest dissolved oxygen resulted when the flood tide occurred at the same time as peak dissolved oxygen in the coral reef; when the flood tide occurred in the early morning, the dissolved oxygen in the mangrove waters was less. When the mangrove was isolated by an increase in the sandy sill caused by a nearby typhoon (between August 3 and 4, as shown on Fig. 10–4b), the dissolved oxygen decreased rapidly to anoxic conditions within two days. Anoxia leads to decreased activity and the death of benthic algae in the swamp, and continues until the barrier beach is breached. These data show the importance of physical flushing by ocean water and off-shore productivity to mangrove oxygen supply.

Soil Acidity

Mangrove soils are often acidic, although in the presence of carbonate as is often the case in south Florida, the soil pore water can be close to neutrality. The soils are often highly reduced, with redox potentials from -100 to -400 mv. The highly reduced conditions and the subsequent accumulation of reduced sulfides in mangrove soils cause extremely acidic soils in many mangrove areas. Dent (1986, 1992) reported a measured accumulation of 10 kg-S per m^3 of sediment per 100 years in mangroves. When these soils are drained and aerated for conversion to agricultural land, the sulfides, generally stored as pyrites, oxidize to sulfuric acid (the phenomenon known as "cat clays," see Chap. 8) making traditional agriculture difficult (W. E. Odum and McIvor, 1990; Dent, 1992). Dent (1992) argued that the "dereclamation" of some previously "reclaimed" marginal coastal soils back to mangroves and salt marshes may be the best strategy for these acidic soils.

ECOSYSTEM STRUCTURE

Canopy Vegetation

As is evident in the coastal salt marshes, the stresses of waterlogged soils and salinity lead to a relatively simple flora in most mangrove wetlands particularly when compared to their upland neighboring ecosystem, the tropical rain forest (Cintrón et al., 1985). There are more than fifty species of mangroves throughout the world (Stewart and Popp, 1987), but fewer than ten species of man-

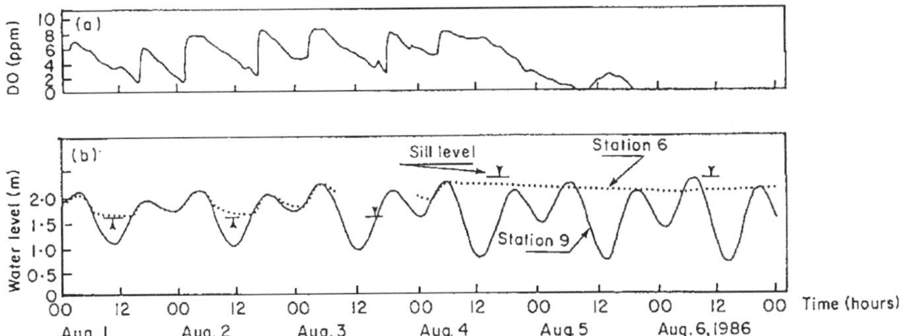

Figure 10–4. *a.* Dissolved oxygen in overlying water and *b.* water level in mangrove embayment on west coast of Amitori Bay, Japan. Station 6 refers to water level in the mangrove area, while Station 9 refers to the water level in the open ocean. Sill level indicates relative height of sand barrier that separated mangrove from ocean between August 3 and 4, 1986. (*From Mazda et al., 1990; copyright © 1990 by Academic Press Limited, London, reprinted by permission*)

groves are found in the New World region, and only three are dominant in the south Florida mangrove swamps—the red mangrove (*Rhizophora mangle* L.), the black mangrove (*Avicennia germinans* L., also named *A. nitida* Jacq.), and the white mangrove (*Laguncularia racemosa* L. Gaertn.). Buttonwood (*Conocarpus erecta* L.), although strictly not a mangrove, is occasionally found growing in association with mangroves or in the transition zone between the mangrove wetlands and the drier uplands. Each of the hydrologic types of mangrove wetlands described above is dominated by different associations of mangroves. Fringe mangrove wetlands are dominated by red mangroves (*Rhizophora*) that contain abundant and dense prop roots, particularly along the edges that face the open sea. Riverine mangrove wetlands are also numerically dominated by red mangroves, although they are straight-trunked and have relatively few, short prop roots (Lugo and Snedaker, 1974). Black (*Avicennia* spp.) and white (*Laguncularia* spp.) mangroves also frequently grow in these wetlands. Basin wetlands support all three species of mangroves, although black mangroves are the most common in basin swamps and hammock wetlands are mostly composed of red mangroves. Scrub mangrove wetlands are typically dominated by widely spaced, short (less than 2 m tall) red or black mangroves.

A comparison of the structural characteristics of the major hydrodynamic types of mangrove wetlands is provided in Table 10–2. These data were compiled from more than 100 mangrove research sites throughout the New World. Fringe mangroves generally have a greater density of large trees (> 10 cm dbh) compared to riverine and basin mangroves. The riverine wetlands, however, have the largest trees and hence a much greater basal area and tree height than

Table 10–2. Structural Characteristics of Canopy Vegetation for Major Mangrove Types [a]

Hydrodynamic Type	*Number of Tree Species*	*Number of Trees, #/ha*		*Basal Area, m^2/ha*		*Stand Height, m*	*Aboveground Biomass, kg/m^2*
		> 2.5 cm dbh	*>10 cm dbh*	*> 2.5 cm dbh*	*>10 cm dbh*		
Fringe mangroves	1.7 ± 0.1 (33)	4005 ± 642 (33)	852 ± 115 (31)	22.2 ± 1.5 (33)	14.6 ± 1.9 (31)	13.3 ± 2.6 (32)	0.8–15.9 (8)
Riverine mangroves	1.9 ± 0.1 (36)	1979 ± 209 (28)	661 ± 71 (32)	30.4 ± 3.5 (5)	32.6 ± 4.7 (32)	21.2 ± 4.8 (26)	1.6–28.7 (8)
Basin magroves	2.3 ± 0.1 (31)	3599 ± 400 (31)	573 ± 102 (21)	18.5 ± 1.6 (31)	10.6 ± 2.2 (21)	9.0 ± 0.7 (31)	—

[a]Data are based on mangrove sites in Florida, Mexico, Puerto Rico, Brazil, Costa Rica, Panama, and Ecuador. Values are average ± standard error (number of observations) except for aboveground biomass that is range (number of observations).

Source: Cintrón et al., 1985

do fringe or basin mangroves. The biomass of riverine mangroves is generally the highest, although data are generally difficult to compare because of different methods and sample sizes used in various observations. Cintrón et al. (1985) reported a range of aboveground biomass for the Florida mangroves of 9–17 kg/m^2 for riverine mangroves and from 0.8 to 15 kg/m^2 for fringe mangroves. Single measures of 0.8 kg/m^2 for a dwarf mangrove wetland and 9.8 kg/m^2 for a hammock mangrove (both in Florida) were also reported.

Understory Vegetation

One of the interesting aspects of certain mangrove swamps is the lack of conspicuous understory vegetation (Janzen, 1985; Corlett, 1986; Lugo, 1986) except for those in high rainfall regions of the world or in ecotones of low soil salinity (Lugo, 1986). Janzen (1985), with the agreement of Lugo (1986), hypothesized that "plants with low light resources cannot accumulate enough metabolites fast enough to meet the metabolic demands of the drain of the machinery and morphology of salt tolerance," but Lugo (1986) suggested that energy sources other than light (e.g., tidal flushing and freshwater inflow) and stresses other than salt tolerance (e.g., hydrogen sulfide, low oxygen, frost) are also important factors. He suggested a more general hypothesis: "Understory plants grow in those mangrove ecosystems where combinations of nutrients, light energy, soil oxygen, and freshwater meet the metabolic demands of drains caused by all environmental stressors converging on the site."

One understory genus that is found in many mangrove wetlands is the mangrove fern (*Acrostichum* spp.). There are three species of this fern found in the world, usually in mangrove regions of high rainfall or low soil salinity (Medina et al., 1990). One species studied in Puerto Rico, *A. aureum* L., was found to have a wide range of light tolerance, growing in full sun in a disturbed mangrove forest and in the understory of a white mangrove forest. It was hypothesized that the fern tolerates higher salinity in the shade because of lower evaporative demands and hence less salt accumulation in its leaf tissues in shady conditions (Medina et. al., 1990).

Plant Zonation and Succession

In trying to understand the vegetation of mangrove wetlands, most early researchers were concerned with describing plant zonation and successional patterns (Lugo and Snedaker, 1974). Some attempts were made to equate the plant zonation found in mangrove wetlands with successional seres, but Lugo (1980) warned that "zonation does not necessarily recapitulate succession because a zone may be a climax maintained by a steady or recurrent environmental condition." Davis (1940) is generally credited with the best early description of plant zonation in Florida mangrove swamps, especially in fringe and basin mangrove wetlands (Fig. 10–5). He hypothesized that the entire ecosystem was accumulating sediments and was therefore migrating seaward. Elaboration of the zonation pattern and theories for its occurrence were also provided by Egler (1952). Typically, *Rhizophora mangle* is found in the lowest zone, with seedlings and small trees sprouting even below the mean low tide in marl soils. Above the low tide level but well within the intertidal zone, full-grown *Rhizophora* with well-developed prop roots predominate. There tree height is approximately 10 m. Behind those red mangrove zones and the natural levee that often forms in fringe mangrove wetlands, basin mangrove wetlands, dominated by black mangroves (*Avicennia*) with numerous pneumatophores, are found. Flooding occurs only during high tides. Buttonwood (*Conocarpus erecta*) often forms a transition between the mangrove zones and upland ecosystems. Flooding occurs there only during spring tides or during storm surges, and soils are often brackish to saline (Chapman, 1976b).

The zonation of plants in mangrove wetlands led some researchers (e.g., J. H. Davis, 1940) to speculate that each zone is a step in an autogenic successional process that leads to freshwater wetlands and eventually to tropical upland forests or pine forests. Other researchers, led by Egler (1952), considered each zone to be controlled by its physical environment to the point that it is in a steady state or at least a state of arrested succession (allogenic succession). For example, with a rising sea level, the mangrove zones migrate inland; during periods of decreasing sea level, the mangrove zones move seaward. Egler thought that the impact of fire and hurricanes made conventional succession impossible in the mangroves of Florida. Another theory, advanced by Chapman (1976b), is that mangrove succession may be a combination of both autogenic and allogenic strategies or a "succession of successions." If that is the case, successional stages could be repeated a number of times before the next successional level is attained.

Lugo (1980) reviewed mangrove succession in light of Odum's criteria (1969; see Chap. 7) and found that except for mangroves on accreting coastlines, the traditional successional criteria do not apply. Succession in mangroves is primarily cyclic, and it exhibits patterns of stressed or "youthful" ecosystems, including slowed or arrested succession, low diversity, P/R greater than one, and open material cycles, even in mature stages. Lugo (1980) concluded that mangroves are

> true steady state systems in the sense that they are the optimal and self-maintaining ecosystems in low-energy tropical saline environments. In such a situation high rates of mortality, dispersal, germination, and growth are the necessary tools of survival. Unfortunately, these attributes could lead many to the identification of mangroves as successional systems.

It is no longer accepted dogma that mangrove wetlands are "land builders" that are gradually encroaching on the sea, as was suggested by Davis. In most cases, mangrove vegetation plays a passive role in the accumulation of sediments and the vegetation usually follows, not leads, the land building that is caused by current and tidal energies. It is only after the substrate has been established that the vegetation contributes to land building by slowing erosion and by increasing the rate of sediment accretion (Lugo, 1980). The mangrove's successional dynamics appear to involve a combination of (1) peat accumulation balanced by tidal export, fire, and hurricanes over years and decades, and (2) advancement or retreat of zones according to the fall or rise of sea level over centuries.

Ball (1980) has suggested another allogenic succession model for mangroves. In it interspecific competition predominates. She found that red mangroves did not grow in dry upland locations because they did not have a competitive advantage there but did dominate where salinity and intertidal water levels gave them a competitive advantage. In the same type of argument, Thibodeau and Nickerson (1986) suggested that red mangroves have a much lower ability to tolerate high sulfides typical of extremely reduced conditions than do black mangroves, and so red mangroves occur in regions that are frequently flushed by tides, whereas black mangroves are found in isolated basin settings where strongly reduced substrates containing high sulfides are found and pneumatophores can be of the greatest use (see Mangrove Adaptations below).

Mangrove Adaptations

Mangrove vegetation, particularly the dominant trees, has several adaptations that allow it to survive in an environment of high salinity, occasional harsh weather, and anoxic soil conditions. These physiological and morphological adaptations have been of interest to researchers and are among some of the most distinguishing features that the lay person notices when first viewing these wetlands. Some of the adaptations, as summarized by Kuenzler (1974) and Chapman (1976b) and shown in Figures 10–5 and 10–6, follow: (1) salinity control, (2) prop roots and pneumatophores, and (3) viviparous seedlings (see also Chap. 6).

Salinity Control

Mangroves are facultative halophytes, that is, they do not require salt water for growth but are able to tolerate high salinity and thus outcompete vascular plants that do not have this salt tolerance (W. E. Odum and McIvor, 1990). The ability of mangroves to live in saline soils depends on their ability to control the concentration of salt in their tissues. In this respect, mangroves are similar to other halophytes. Mangroves have the ability both to prevent salt from entering the plant at the roots (*salt exclusion*) and to excrete salt from the leaves (*salt secretion*). Salt exclusion at the roots is thought to be a result of reverse osmosis that causes the roots to absorb only freshwater from salt water. Root cell membranes

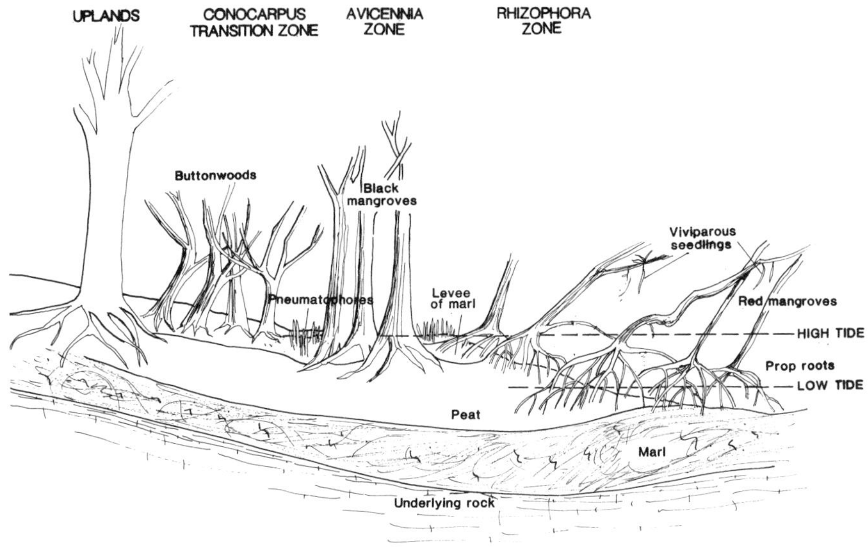

Figure 10–5. Zonation of Florida mangrove wetlands. Note adaptations of mangroves such as prop roots, viviparous seedlings, and pneumatophores (see Figure 10–6). *(After J. H. Davis, 1940)*

of mangroves species of *Rhizophora, Avicennia,* and *Laguncularia*, among others, may act as ultrafilters that exclude salt ions. Water is drawn into the root through the filtering membrane by the negative pressure in the xylem developed through transpiration at the leaves; this action counteracts the osmotic pressure caused by the solutions in the external root medium (Scholander et al., 1965; Scholander, 1968). There are also a number of mangrove species (e.g., *Avicennia* and *Laguncularia*) that have salt-secreting glands on the leaves to rid the plant of excess salt. The solutions that are secreted often have several percent of NaCl, and salt crystals can form on the leaves. Another possible way, still questioned as to its importance, in which mangroves discharge salt is through leaf fall. This leaf fall may be significant because mangroves produce essentially two crops of leaves per year (Chapman, 1976b).

Prop Roots and Pneumatophores

Some of the most notable features of most mangrove wetlands are the *prop roots* and *drop roots* of the red mangrove (*Rhizophora*; Fig. 10–6a) and numerous, small (usually 20 to 30 cm above the sediments, although they can be up to 1 m tall) pneumatophores of black mangroves (*Avicennia*) (Fig. 10–6b). The drop roots are special cases of the prop roots that extend from branches and other upper parts of the stem directly down to the ground, rooting only a few centimeters into the sediments. Oxygen enters the plant through small pores, called *lenticels*, that are found on both pneumatophores and on prop and drop roots. When lenticels are exposed to the atmosphere during low tide, oxygen is absorbed from the air and some of it is transported to and diffuses out of the roots through a system of aerenchyma tissues. This maintains an aerobic microlayer around the root system. When the prop roots or pneumatophores of mangroves are continuously flooded by stabilizing the water levels, those mangroves that have submerged pneumatophores or prop roots soon die (Macnae, 1963; J. H. Day, 1981).

In an interesting experiment to determine the importance of oxygen transport from aerial organs to the sediments, Thibodeau and Nickerson (1986) "capped" with plastic tubing the pneumatophores of *Avicennia germinans* in a fringe mangrove forest in the Bahamas. They observed a reduced soil-oxidation gradient surrounding the roots, indicating that the pneumatophores help the plant produce an oxidized rhizosphere. They also found that the greater the number of pneumatophores present in a given area, the more oxidized the soil. They described

Figure 10–6. Adaptations of mangroves including *a. (above)* prop and drop roots of red mangroves, and (*following pages*) *b.* pneumatophores of black mangroves, *c.* viviparous propagule hanging in red mangrove canopy, and *d.* red mangrove seedling germinated and rooted in sediments. (*Photo* a. *by M. T. Vogel, by permission; all others by W. J. Mitsch*)

Figure 10–6b.

the relationship as

$$E_H = -307 + 1.1\ pd \tag{10.1}$$

where

E_H = redox potential, mv

pd = pneumatophore density, no.per 0.25 m^2

Viviparous Seedlings

Red mangroves (and related genera in other parts of the world) have seeds that germinate while they are still in the parent tree; a long, cigar-shaped hypocotyl (viviparous seedling) develops while hanging from the tree (Fig. 10–6c). This is apparently an adaptation for seedling success where shallow anaerobic water and sediments would otherwise inhibit germination. The seedling (or propagule) eventually falls and often will root if it lands on sediments or will float and drift in currents and tides if it falls into the sea. After a time, if the floating seedling becomes stranded and the water is shallow enough, it will attach to the sediments and root (Fig. 10–6d). Often the seedling becomes heavier with time, rightens to a vertical position, and develops roots if the water is shallow. It is not well understood whether the contact with the sediments stimulates the root growth or if the soil contains some chemical compound that promotes root development (Chapman, 1976b). The value of the floating seedlings for mangrove dispersal and for invasion of newly exposed substrate is obvious. Rabinowitz (1978) described the obligate dispersal time (the time required during propagule dispersal for germination to be completed) to be 40 days for the red mangrove and 14 days for the black mangrove propagules. She also estimated that the red and black mangrove propagules could survive for 110 and 35 days, respectively. In contrast, J. H. Davis (1940) had found that red mangrove propagules could float for more that one year.

Figure 10–6c.

Figure 10–6d.

Consumers

W. E. Odum et al. (1982) found the following data from the literature describing faunal use of mangroves in Florida in terms of the number of species: 220 fish; 181 birds, including 18 wading birds, 29 water birds, 20 bird of prey, and 71 arboreal birds; 24 reptiles and amphibians; and 18 mammals. In general, a wide diversity of animals is found in mangrove wetlands; their distribution sometimes parallels the plant zonation described above. Many of the animals that are found in mangrove wetlands are filter feeders or detritivores, and the wetlands are just as important as a shelter for most of the resident animals as they are a source of food. Some of the important filter feeders found in Florida mangroves include barnacles (*Balanus eburneus*), coon oysters (*Ostrea frons*), and the eastern oyster (*Crassostrea virginica*). These organisms often attach themselves to the stems and prop roots of the mangroves within the intertidal zone, filtering organic matter from the water during high tide. Fiddler crabs (*Uca* spp.) are also abundant in mangrove wetlands, living on the prop roots and high ground during high water and burrowing in the sediments during low tide (Kuenzler, 1974). Many other invertebrates, including snails, sponges, flatworms, annelid worms, anemone, mussels, sea urchins, and tunicates, are found growing on roots and stems in and above the intertidal zone. Wading birds that are frequently found in mangroves include the wood stork (*Mycteria americana*), white ibis (*Eudocimus albus*), roseate spoonbill (*Ajaia ajaja*), cormorant (*Phalacrocorax* spp.), brown pelican (*Pelicanus occidentalis*), egrets, and herons (W. E. Odum and McIvor, 1990). Vertebrates that inhabit mangrove swamps include alligators, crocodiles, turtles, bears, wildcats, pumas, and rats (Kuenzler, 1974). Mangrove wetlands have been documented as important nursery areas and sources of food for sport and commercial fisheries in south Florida by Heald (1969; 1971) and W. E. Odum (1970) (see Organic Export discussed below). Important species include the spiny lobster (*Panulirus argus*), pink shrimp (*Penaeus duorarum*), mullet (*Mugil cephalus*), tarpon (*Megalops atlanticus*), snook (*Centropomus undecimalis*), and mangrove snapper (*Lutjanus apodus*).

ECOSYSTEM FUNCTION

Certain functions of mangroves such as gross and net primary productivity have been studied extensively, particularly in southern Florida. Other functional characteristics of these wetlands such as organic export and nutrient cycling have received much less quantitative work. Nevertheless, a picture of the dynamics of mangrove wetlands has emerged from several key studies. These studies have demonstrated the importance of the physical conditions of tides, salinity, and nutrients to these wetlands and have shown where natural and human-induced stresses have caused the most effect.

Primary Productivity

A wide range of productivity has been measured in mangrove wetlands due to the wide variety of hydrodynamic and chemical conditions encountered. Table 10–3 presents a balance of carbon flow in several fringe and basin mangrove swamps in Florida and Puerto Rico, and Table 10–4 summarizes daily productivity and litter production data for riverine, basin, and scrub mangrove wetlands from a number of field studies. Gross and net primary productivity is the highest in riverine mangrove wetlands, lower in basin mangrove wetlands, and the lowest in dwarf mangrove wetlands. Net primary productivity values given in Table 10–3 ranged from approximately 1,100 to 5,400 g m^{-2} yr^{-1} (assuming 1 g C = 2 g dry wt). J. W. Day et al. (1987) estimated the net aboveground primary productivity of a Mexican mangrove forest to range from 1607 g m^{-2} yr^{-1}

for a fringe mangrove to 2,458 g m^{-2} yr^{-1} for a riverine mangrove system. They attribute the higher productivity to the greater influence of nutrient loading and freshwater turnover at the riverine site. This range compares well with a study by S. Y. Lee (1990) of *Kandelia candel* mangroves in Hong Kong where mangrove plant net primary productivity averaged 1,950 to 2,440 g m^{-2} yr^{-1} depending on the method used for the productivity determination. The total net primary productivity, including all plants at the Hong Kong site, was approximately 4,400 g m^{-2} yr^{-1}; *Phragmites communis* contributed about 46 percent of the site productivity and macroalgae and phytoplankton another 3.2 percent. Organic production, as measured by litterfall, is summarized in Figure 10–7 for scrub, basin, fringe, and riverine mangrove systems, illustrating the same connection as

Table 10–3. Mass Balance of Carbon Flow (gC m^{-2} yr^{-1}) in Mangrove Forests in Florida and Puerto Rico

	Rookery Bay, Florida[a]		*Puerto*[b] *Rico*	*Fahkahatchee Bay, Florida*[c]		
	Fringe	*Basin*	*Fringe*	*Basin*	*Fringe*	*Fringe*
Gross Primary Productivity (GPP)						
Canopy	2055	3292	3004	3760	4307	5074
Algae	402	26	276			
Total	2457	3318	3280			
Respiration (plants)						
Leaves, stems	671	2022	1967	1172	1416	3084
Roots, above-ground	22	197	741	146	182	215
Roots, below-ground	?	?	?	?	?	?
Total	693	2219	2708	1318	1598	3299
Net Primary Production	1764	1099	572	2442	2709	1775
Growth		186	153			
Litterfall		318	237			
Respiration (heterotrophs)		197	135			
Respiration (total)		2416	2843			
Export		64	500			
Net Ecosystem Production (NEP)		838	-63			
Burial		?	?			
Growth		186	153			

[a]Lugo et al., 1975; Twilley, 1982, 1985; Twilley et al., 1986
[b]Golley et al., 1962
[c]Carter at al., 1973
Source: from Twilley, 1988

described above between hydrologic conditions and productivity. Litterfall is approximately 200 g m^{-2} yr^{-1} in scrub wetlands and 1200 g m^{-2} yr^{-1} in riverine swamps. The greater the hydrologic turnover is (riverine > fringe > basin > scrub), the greater the litter production is (Pool et al., 1975; Twilley, 1988). S. Y. Lee (1989) found litter productivity in his mangrove site in a highly managed tidal shrimp pond near Hong Kong to be about 1,100 g m^{-2} yr^{-1}, the identical litterfall measured by Flores-Verdugo et al. (1987) at a mangrove site on the Pacific shore of Mexico. The Hong Kong wetland could be considered to be similar to a fringe mangrove, for water changes in the pond every spring tide but remains in the pond for four to five days during neap tide periods, whereas the Mexican site was described as a mangrove in a lagoon that contains an ephemeral inlet that is open to the coastline for three to four months per year.

The important factors that control mangrove function in general and primary productivity in particular are: (1) tides and runoff; and (2) water chemistry (Carter et al., 1973; Lugo and Snedaker, 1974; Lugo, 1990b).

These factors are not mutually exclusive, for tides influence water chemistry and hence productivity by transporting oxygen to the root system, by removing the buildup of toxic materials and salt from the soil water, by controlling the rate of sediment accumulation or erosion, and by indirectly regenerating nutrients lost from the root zone. The chemical conditions that affect primary productivity are the soil water salinity and the concentration of major nutrients. Lugo and

Table 10–4. Primary Productivity, Respiration, and Litterfall Measurements for Three Types of Mangrove Wetlands

	Mangrove Wetland		
	Riverine	*Basin*	*Scrub*
Gross Primary Productivity[a] kcal m^{-2} day^{-1}	108	81	13
Total Respiration[a] kcal m^{-2} day^{-1}	51	56	18
Net Primary Productivity[a] kcal m^{-2} day^{-1}	57	25	0
Litter Production[b] kcal m^{-2} day^{-1}	14 ± 2	9.0 ± 0.4	1.5

[a] From several sites in Florida; based on CO_2 gas-exchange measurements; assumes 1 g organic matter = 4.5 kcal

[b]Average ± standard error from several sites in Florida and Puerto Rico; measured with litter traps; assumes 1 g organic matter = 4.5 kcal

Source: Based on data from Brown and Lugo, 1982

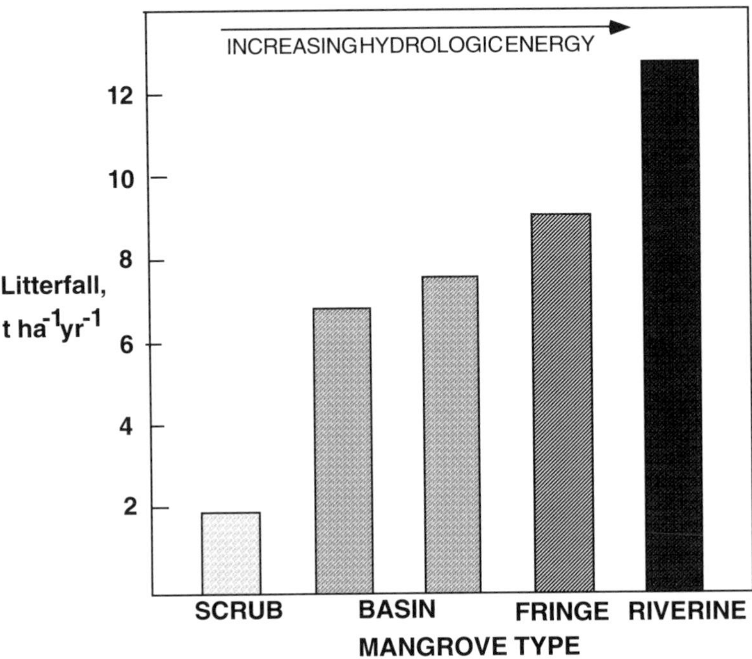

Figure 10–7. Litterfall rates in different hydrologic types of mangrove wetlands. The two values for basin mangroves are for monospecific and mixed forests. (*Twilley, 1988; copyright © by 1988 Springer-Verlag, New York, reprinted with permission*)

Snedaker (1974) have concluded that compared to mangrove wetlands such as dwarf mangroves that are isolated from the influence of daily tides, "environments flushed adequately and frequently by seawater and exposed to high nutrient concentrations are more favorable for mangrove ecosystem development; forests in these areas exhibit higher rates of net primary productivity."

The importance of chemical conditions to mangrove productivity is difficult to document because the chemical conditions include both stimulants (nutrients) and stressors (salinity). Both factors appear to be important for mangrove growth. Kuenzler (1974) suggested that even with low transpiration rates in mangroves as a result of high salinities, productivity can be high if nutrients are abundant. Carter et al. (1973) examined the productivity of mangrove canopy leaves along a gradient of fresh water to saline water in southwestern Florida. They found that mangroves increased both respiration and gross primary productivity with increased salinity but at different rates so that net primary productivity available for plant growth actually decreased with increased salinity (Fig. 10–8). Mangroves put more of their captured energy into growth rather than physiological maintenance when the water is low in salts. In salt water, more respiratory work is necessary to adapt physiologically to the saline conditions. Respiration increases as a metabolic cost of adapting to high salinities.

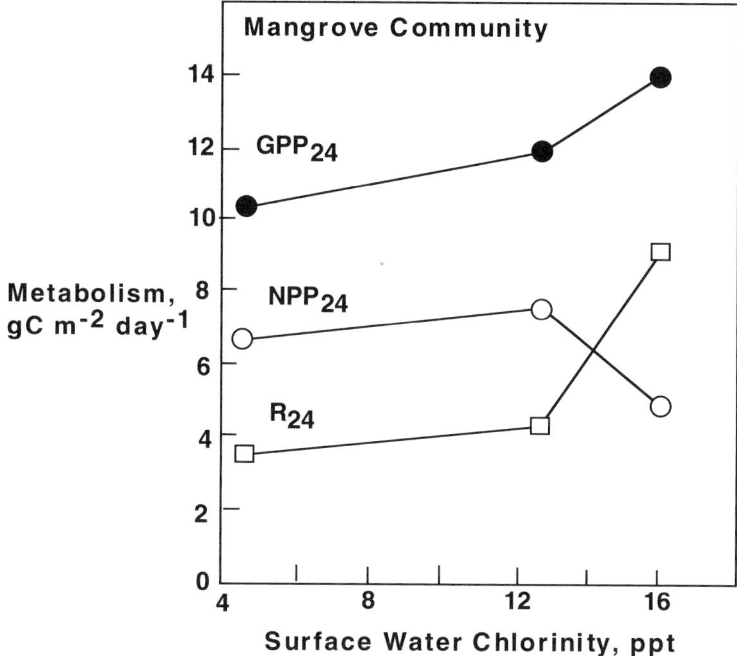

Figure 10–8. Changes in mangrove metabolism, as measured by carbon gas exchange with increased salinity (as measured by chlorine content) (*Lugo, 1990b as adapted from Carter et al. 1973*). GPP_{24} indicates gross primary productivity over 24 hours; NPP_{24} indicates net primary productivity over 24 hours; R_{24} indicates respiration over 24 hours. (*Copyright © 1990 by Elsevier Science Publishers, Amsterdam, reprinted with permission*)

It also became apparent in several of the mangrove productivity studies in Florida (Carter et al., 1973; Lugo et al., 1975) that a pattern of zonation of metabolism follows the zonation of species as described earlier. Lugo and Snedaker (1974) summarized the functional zonation in these studies as follows:

1. the gross primary productivity of the red mangrove decreased with increasing salinity;
2. the gross primary productivity of black and white mangroves increased with increasing salinity;
3. in areas of low salinities and under equal light conditions, the gross primary productivity of the red mangrove was four times that of the black mangrove;
4. in areas of intermediate salinity, the white mangrove had rates of gross primary productivity twice that of the red mangrove; and
5. in areas of high salinities, the white mangrove exhibited a gross primary productivity higher than that of the black mangrove, which, in turn, was higher than the red mangrove.

Thus the zonation of mangroves, as described by J. H. Davis (1940), Egler (1952), and others, has a functional basis for occurring. Species that are found growing out of their zone will have lower productivity than to those that are adapted to those conditions, and competition will eventually eliminate them from that zone.

Decomposition

The decomposition process in mangroves has been studied with litter bag measurements for a number of different plants (mainly red mangroves, *Rhizophora,* and black mangroves, *Avicennia*) and in a number of different mangrove types (Table 10–5). Decay rates range from 0.30 per year for red mangrove leaves in a basin mangrove wetland in Florida to 8.4 per year for red mangrove leaves in a fringe mangrove wetland. Lugo and Snedaker (1974) reported previously that decomposition is accelerated by moisture, with optimal decomposition occurring at about 50 percent moisture. Subsequent studies have shown that black mangrove leaves decay three times faster than do red mangrove leaves, a condition attributed to the higher C/N ratio in red mangrove leaves (Twilley, 1982). Mangrove leaves in a rapidly flushing environment also appear to decay more rapidly than those in a slowly flushing environment (Table 10–5). Because basin mangroves are generally dominated by black mangroves, it has been hypothesized that the faster decomposition of black mangrove leaves may lead to a greater export of dissolved organic material, rather than particulate organic matter, from basin mangroves when compared to riverine or fringe mangroves (Cintrón et al., 1985).

Mortality

Natural tree mortalities on large scales are frequently cited in the literature for mangrove wetlands (see summary in Jiménez et al., 1985). These mortalities are due to both natural causes, for example, hurricanes, droughts, and frosts, and human-influenced changes, for example, alteration of hydrology, dredging, and increased sedimentation (Lugo et al., 1981; Jiménez et al., 1985; Lugo, 1990a). Changes in the hydrologic conditions "either through alteration of regional hydrology or modification of the geomorphology of the mangrove basin" (Lugo, 1990a) are particularly important stresses on mangrove systems, often serving to exacerbate natural stresses. Stresses and massive mortalities are particularly noticeable in mangrove wetlands before or during a senescence stage when tree growth slows down and wide gaps in the canopy and lack of regeneration are typical (Jiménez et al., 1985). When massive diebacks of mangrove trees occur, a common denominator often appears to be rapid environmental change. The damage is mitigated if there is freshwater or tidal flushing or is exacerbated if the system is already under stress from factors such as excessive salinity, low

Table 10–5. Litter Degradation Rates in Mangrove Wetlands

Mangrove Type	*Location*	*Decay Rate, yr^{-1}*	*Half Life, days*
Fringe			
Rhizophora mangle	Florida	8.39	30
	Florida	2.55	99
	Florida	1.46	173
	Florida	1.10	231
	Florida	0.85	346
	Puerto Rico	2.55	99
	Mexico	1.42	178
	Brazil	1.23	206
Laguncularia racemosa	Brazil	3.08	82
Basin			
Rhizophora mangle	Florida	0.30	—
	Florida	1.22	231
	Florida	2.01	139
	Florida	1.28	231
	Florida	1.63	173
(Fast Flushing)	Florida	1.90	133
(Slow Flushing	Florida	0.97	260
Avicennia germinans			
(Fast flushing)	Florida	6.04	60
(Slow flushing)	Florida	3.89	65
Riverine	Mexico	2.33	99
	Colombia	2.81	87
Laguncularia racemosa	Mexico	1.71–4.7	54–147
Dwarf	Florida	1.46	173
	Florida	2.29	115
Hammock	Florida	2.29	115

Source: Cintrón et al., 1985; Flores-Verdugo et al., 1990

temperature, excessive siltation, or altered hydrology (Lugo et al., 1981; Jiménez et al., 1985). Lugo et al. (1981) and Jiménez et al. (1985) argued that even massive diebacks of mangroves reportedly due to biological causes such as the gall disease reported in the mangroves of Gambia (Teas and McEwan, 1982)—are manifestations of changes in riverine or tidal hydrology, not solely due to biological factors alone.

Export of Organic Material

Mangrove swamps are important exporters of organic material to the adjacent estuary. Heald (1971) estimated that about 50 percent of the aboveground productivity of a mangrove swamp in southwestern Florida was exported to the adjacent estuary as particulate organic matter. From 33 percent to 60 percent of the total particulate organic matter in the estuary came from *Rhizophora* (red mangrove) material. The production of organic matter was greater in the summer (the wet season in Florida) than other seasons although detrital levels in the swamp waters were greatest from November through February, which is the beginning of the dry season. Thirty percent of the yearly detrital export occurred during November. Heald also found that as the debris decomposed, its protein content increased. The apparent cause of this enrichment, also noted in salt marsh studies, is the increase of bacterial and fungal populations. Carter et al. (1973), basing their calculations on the work of Heald, estimated that at least 57 percent of the total energy base for Fahkahatchee Bay in the Florida Everglades came from mangrove forests. Brown and Lugo (1982) estimated an export of 58 gC m^{-2} yr^{-1} from a basin mangrove wetland in south Florida, with 55 percent of that export caused by tidal exchange and the remainder caused by freshwater runoff and seepage.

Flemming et al. (1990) demonstrated for a site in southeast Florida that the amount of mangrove detritus relative to seagrass (*Thalassia testudinum*) detritus decreased rapidly from 80 percent mangrove detritus along the stream, where the mangroves were found, to 10 percent mangrove detritus about 90 m beyond the mouth of the stream. Using a technique involving the measurement of $^{13}C/^{12}C$ ratios of organic matter, they found that mangroves were providing 37 percent and seagrasses 63 percent of the carbon to organisms in the bay, suggesting that mangrove detrital export is important to offshore water only in regions local to the wetlands and that it may be unimportant relative to other carbon sources farther offshore.

The estimates of organic carbon fluxes from mangroves establish the fact that many mangrove wetlands are important sources of detritus for adjacent aquatic systems whether the effect is far reaching or local. Studies by W. E. Odum (1970) and W. E. Odum and Heald (1972) established that this detritus is important to sport and commercial fisheries (Fig. 10–9). Through the examination of the stomach contents of more than 80 estuarine animals, Odum found that mangrove detritus, particularly from *Rhizophora*, is the primary food source in the estuary. The primary consumers of *Rhizophora* serve as prey to game fish such as tarpon, snook, sheepshead, spotted seatrout, red drum, jack, and jewfish. The primary consumers also used the mangrove estuarine waters during their early life stages as protection from predators and as a source of food (Lugo and Snedaker, 1974). It is reasonable to extrapolate from this and similar studies that the removal of mangrove wetlands would cause a significant decline in sport and commercial fisheries in adjacent open waters.

Twilley et al. (1986) summarized the comparison of leaf litter production and organic export for riverine, fringe, and basin mangrove systems (Fig. 10–10). They estimated the inundation depth per tide for these three mangrove systems as 3, 0.5, and 0.08 m, respectively, and reported that "as tidal influence decreases, the proportion of litterfall in each forest that is exported also decreases." Thus riverine mangrove systems export a majority of their organic litter (94 percent or 470 gC m^{-2} yr^{-1}), whereas basin mangroves export much less (21 percent, or 64 gC m^{-2} yr^{-1}), leaving the leaf litter to decompose or accumulate as peat. In a series of conceptual models that include the connection of mangrove ecosystems to their adjacent estuaries, Twilley (1988) illustrated that mangroves contributed an estimated 338 to 345 gC m^{-2} yr^{-1} to the estuary (per square meter of estuary), or from 83 to 86 percent of the allochthonous inputs, and from 39 to 52 percent of the entire fixed carbon pool available for secondary productivity in the estuary. These numbers illustrate the importance of the mangrove export to the overall secondary productivity of adjacent estuarine waters.

Flores-Verdugo et al. (1987) estimated that "a minimum of 74 percent, but probably closer to 90 percent, of mangrove litter fall reached the lagoon waters" at their semi-riverine Mexican mangrove site. By contrast, S. Y. Lee (1989)

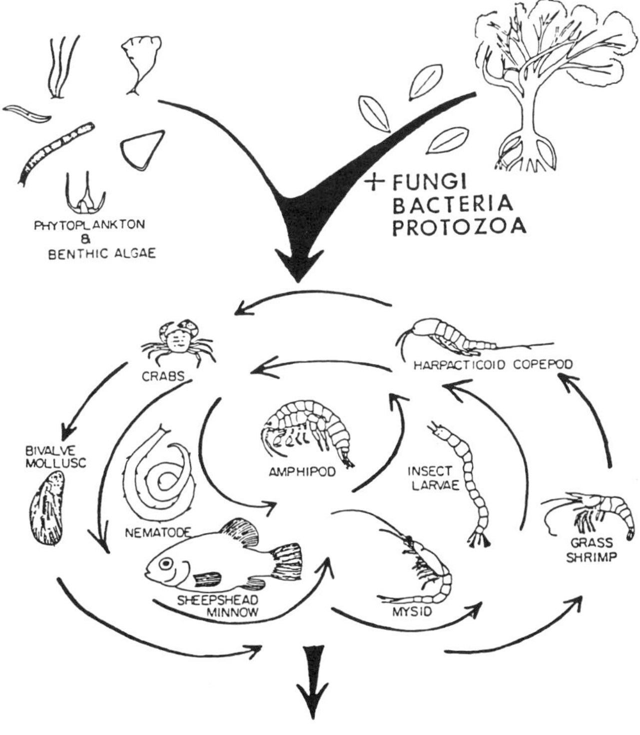

Figure 10–9. Detritus-based food web in south Florida estuary illustrating major contribution of mangrove detritus. (*From W. E. Odum, 1970*)

estimated that the export of litter from his mangrove site located above the mean water level was "unimportant" and that most of the litter production was consumed by crabs or microbial decomposition. S. Y. Lee (1991) found that the grazing of the living biomass in this mangrove was also low, with an estimated 2.8 to 3.5 percent of the net aboveground primary productivity consumed by herbivores.

The organic carbon export from mangroves includes not only particulate organic matter (POM), which is most often measured, but also dissolved organic matter (DOM), which has not received as much attention as POM in export studies. Snedaker (1990) argued that future research needs to focus on the role of DOM as (1) an alternative food source for animals in the lower parts of the food chain, (2) an energy source for heterotrophic microorganisms, and (3) a source of chemical cues for estuarine species.

Energy Flow

Golley et al. (1962) developed a synoptic energy budget of a Puerto Rican mangrove wetland for an average day in May. Of the total gross productivity

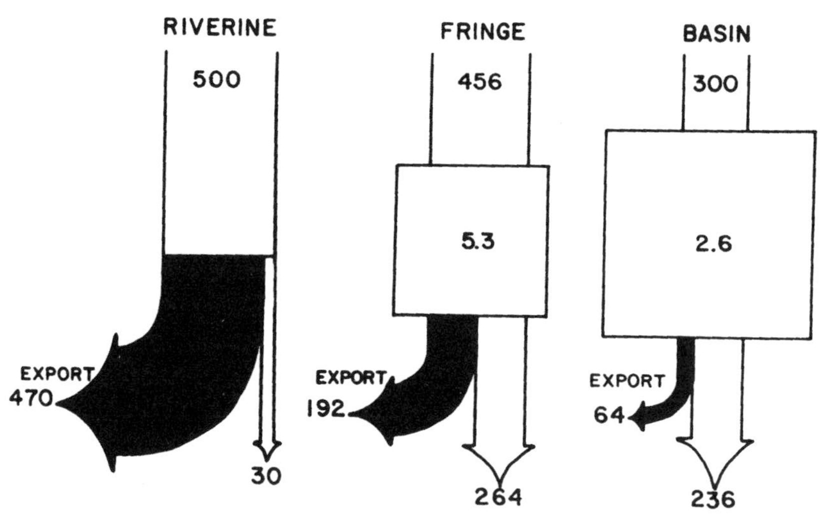

Figure 10–10. Summary of organic carbon inflows (litterfall), storages (litter standing crop), export to adjacent aquatic systems, and other losses (decomposition and peat production). Width of pathways and size of boxes proportional to flows and storages. Storages are in gC/m^2 while flows are in gC m^{-2} yr^{-1}. *(From Twilley et al., 1986; based on data from Boto and Bunt, 1981; Heald, 1969; and Pool et al., 1975; copyright © 1986 by the Ecological Society of America, reprinted with permission)*

of 82 kcal m^{-2} day^{-1}, a major portion is used by the plants themselves in respiration. The respiration of the prop roots of the red mangroves amounted to 20 kcal m^{-2} day^{-1}, whereas the export was about 14 kcal m^{-2} day^{-1} and the soil respiration was about 4kcal m^{-2} day^{-1}. Animal metabolism, estimated to be 0.8 kcal m^{-2} day^{-1}, made up a very minor part of the energy flow in this ecosystem. Lugo et al. (1975) found similar patterns of metabolism in mangrove swamps in southwest Florida, although the respiration of the red mangrove prop roots was much lower (0.6 kcal m^{-2} day^{-1}). The authors also found a higher production:respiration ratio than was found in the Puerto Rico study, convincing them of the importance of organic export from mangrove wetlands. A significant contribution to energy fixation by periphyton growing on red mangrove prop roots—a net productivity of 11 kcal m^{-2} day^{-1}—was also noted in the Florida study. Periphyton may have an important function of capturing and concentrating nutrients from incoming tidal waters for eventual use by the mangroves themselves.

Overall annual energy budgets (in terms of carbon) were summarized by Twilley (1988) for basin and fringe mangrove systems in Florida and compared with annualized data for the Golley et al. (1962) Puerto Rico study (Table 10–3). Gross primary productivity values, which ranged from 25,000 to 50,000 kcal m^{-2} yr^{-1} (assuming 1 gC = 10 kcal), establishes these systems as among the most productive in the world. In some of the wetlands, productivity due to epiphytic algae was significant, contributing as much as 16 percent of the gross productivity. Heterotrophic respiration was a relatively minor energy flow at about 1,300 to 2,000 kcal m^{-2} yr^{-1}, well below the autotrophic respiration. Net ecosystem productivity (NEP), which is the amount of energy (or organic carbon) remaining after respiration and export, was calculated to be a deficit for the Puerto Rico fringe wetland (i.e., the system was not accumulating carbon in the peat) compared to a substantial energy accumulation of 8,000 kcal m^{-2} yr in the basin mangrove in Florida's Rookery Bay. This comparison was summarized by Twilley (1988) as follows:

> These estimates of NEP between a fringe and a basin mangrove suggest that a large proportion of the NPP [net primary productivity] in the more inundated forests [fringe mangroves] is exported, while in the basin forests more of the net production is accumulated or utilized within the system. This supports the "open" versus "closed" concept of fringe and basin mangroves, respectively, as proposed by Lugo and Snedaker (1974) in relation to hydrologic energy.

ECOSYSTEM MODELS

Several qualitative and quantitative compartment models have been developed from research on the functional characteristics of mangroves in south Florida

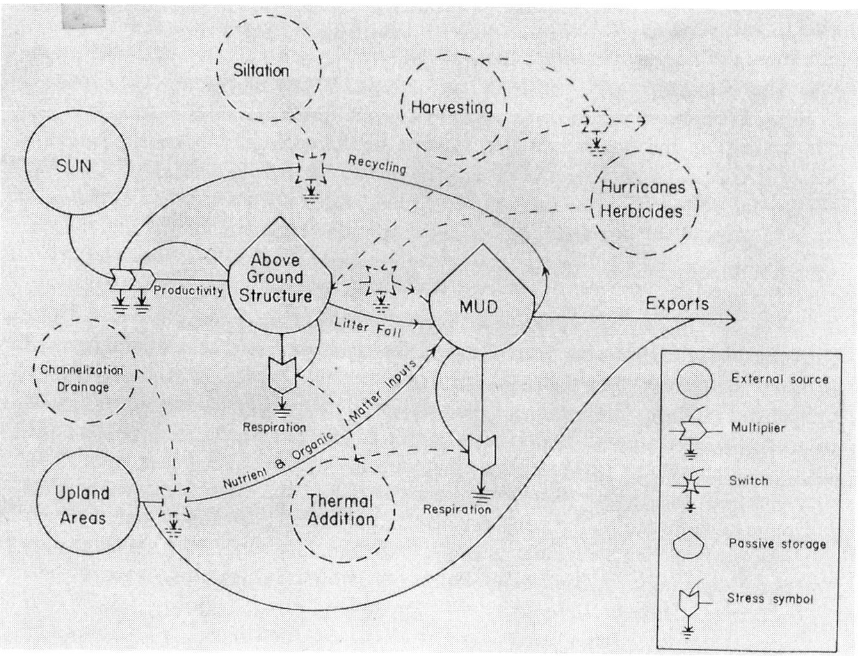

Figure 10–11. Conceptual and simulation models of mangrove wetlands showing *a.* (*above*) major stresses of siltation, harvesting, hurricanes, drainage, and thermal additions; and (*opposite page*) *b.* simple simulation model of mangrove productivity and nitrogen cycling; *c.* simulation model, including differential equations, that illustrates salinity and nutrient effects. (*Symbols are after H. T. Odum, 1983 and are summarized in a*). (a. *from Lugo and Snedaker, 1974; copyright © 1974 by Annual Review, Inc., reprinted with permission;* b. *and* c. *from H. T. Odum, 1983, based on data in Sell, 1977; copyright © 1983 by John Wiley and Sons, Inc., reprinted with permission*)

(Miller, 1972; Carter et al., 1973; Lugo and Snedaker, 1974; Lugo et al., 1976; Wharton et al., 1976; Sell, 1977; H. T. Odum et al., 1977b; Twilley, 1988). In a simple conceptual model of energy and material flows, wetland productivity was described as affected by activities such as channelization, drainage, siltation, harvesting, hurricanes, herbicides, and thermal additions (Fig. 10–11a). Simulations of similar models by Lugo et al. (1976) and Sell (1977), illustrated in Figure 10–11b, showed that the time required for the attainment of steady state of mangrove biomass is approximately the same as the average period between tropical hurricanes (approximately 20 to 24 years for Caribbean systems). This match suggests that mangroves may have adapted or evolved to go through one life cycle, on the average, between major tropical storms. On the average, a mangrove forest reaches maturity just as the next hurricane or typhoon hits. The models also demonstrated the importance of tidal exchange

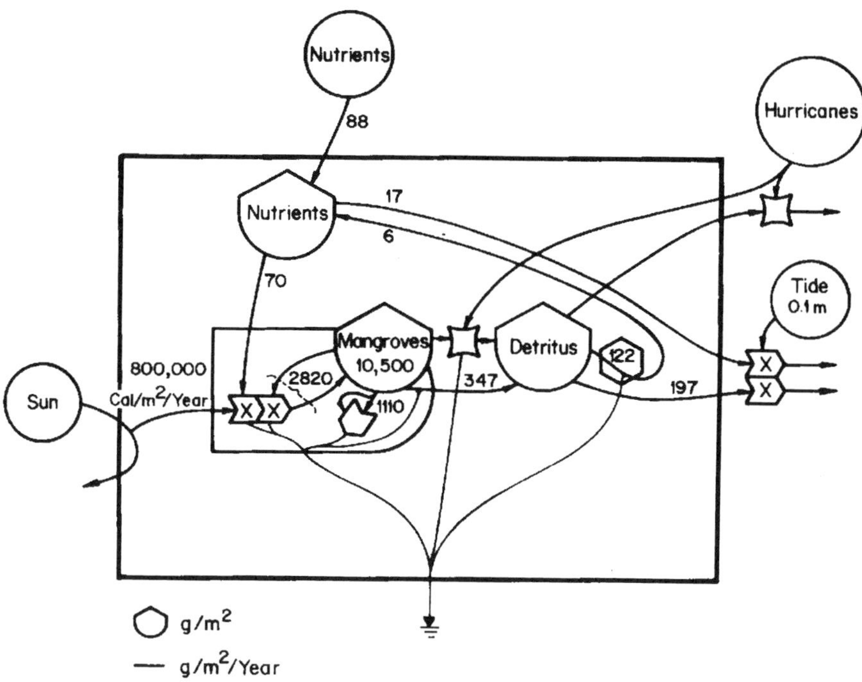

Figure 10–11b.

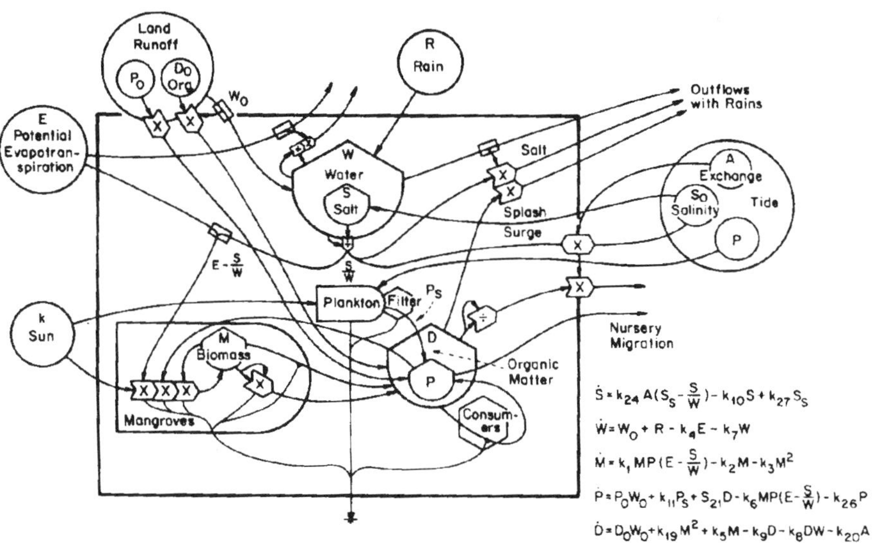

Figure 10–11c.

and terrestrial inputs of nutrients (Lugo et al., 1976). Figure 10–11c shows a more complex model that demonstrated the importance of freshwater sources of materials and tidal exchange. This model adds salinity and potential evapotranspiration as a variable and a forcing function, respectively—factors that affect mangrove productivity. Although some of these models were not verified with field data, the results pointed to the importance of freshwater inputs, tidal exchange, and salinity for mangrove wetland function and supported the theories put forward based on the field measurements.

A simulation model of the hydrology of a basin mangrove on Marco Island, Florida, was developed by Ritchie (1990) primarily as a submodel to a simulation model of the population dynamics of the black salt marsh mosquito (*Aedes taeniorhynchus*). The hydrology simulation model was calibrated and verified from field data. Water depths predicted by the model were highly correlated with measurements at two field sites.

PART 4

INLAND WETLAND ECOSYSTEMS

FRESHWATER MARSH

Freshwater Marshes

11

Freshwater inland marshes are perhaps the most diverse of the marsh types discussed in this book. The pothole marshes of the north-central United States and south-central Canada are individually small, occurring in the moraines of the last glaciation. Playas of north Texas and western New Mexico are similar in size but have different geological origins and are found in an arid climate. Coastal nontidal freshwater marshes and the Florida Everglades are other freshwater marshes.

Vegetation in freshwater marshes is characterized by graminoids such as the tall reeds Typha *and* Phragmites, *the grasses* Panicum *and* Cladium, *the sedges* Cyperus *and* Carex, *broad-leaved monocots such as* Sagittaria *spp., and floating aquatic plants such as* Nymphaea *and* Nelumbo. *Some inland marshes such as the prairie glacial marshes follow a five- to twenty-year cycle of (a) drought, when the marsh dries out and exposes large areas of mudflat on which dense seedling stands germinate; (b) reflooding after rain drowns out the annual seedlings but allows the perennials to spread rapidly and vigorously; (c) deterioration of the marshes, sometimes associated with concentrated muskrat activity; and (d) a lake stage, after most of the emergent vegetation has gone. This resets the cycle to stage (a). In contrast to bogs, inland marshes have high-pH substrates, high available soil calcium, medium or high loading rates for nutrients, high productivity, and high soil-microbial activity that leads to rapid decomposition, rapid recycling, and nitrogen fixation. Peat may or may not accumulate. Most of the primary productivity is routed thorough detrital pathways, but herbivory can be important, particularly by muskrats and geese.*

Consumers can have significant yet indirect effects on the detrital pathways as well. Inland marshes are valuable as wildlife islands in the middle of agricultural land and have been tested extensively as sites that can assimilate nutrients from human domestic wastes.

The wetlands discussed in this chapter are a diverse group. Nevertheless, they can be treated as a unit because of the fact that they are nontidal, freshwater systems dominated by grasses, sedges, and other freshwater emergent hydrophytes. Otherwise, they differ in their geological origins and in their driving hydrologic forces, and they vary in size from small marshes of less than a hectare in size to the immense sawgrass monocultures of the Florida Everglades. Terminology for wetlands, especially inland freshwater wetlands, is often confusing and contradictory. The major differences between European and North American words used to describe freshwater wetlands are illustrated in Table 3–1. In Europe, for example, the term *reedswamp* is often used to describe one type of freshwater marsh dominated by *Phragmites* spp. whereas in the United States, the word *swamp* usually refers to a forested wetland. Although the use of classifying terms connotes clear boundaries between different wetland types, in reality they form a continuum. The extremes of freshwater marshes are clearly different, but at the boundaries between two wetland types (e.g., marsh and bog), the distinction is not always clear.

As used in Europe, the term *fen,* one kind of marsh, refers to a peat forming wetland that receives nutrients from sources other than precipitation, usually through groundwater movement. Its peats are not acidic. Marshes (and reedswamps, as the terms is used in Europe) have mineral soils rather than peat and are generally eutrophic. American terminology has developed without much regard to whether the system is peat-forming. The fact is that most of the North American marshes, regardless of how far south they are and regardless of their geological origins, are peat forming. Based on the classification developed by the U.S. Fish and Wildlife Service (Cowardin et al., 1979), most of the marshes described in this chapter have been classified as palustrine, riverine, or lacustrine, persistent or nonpersistent, emergent wetlands.

For the purposes of this book we have divided inland nonforested wetlands into two groups. Bogs and fens are communities characterized by deep accumulation of peat. These peatlands are a boreal or high-altitude phenomenon generally associated with low temperatures and short growing seasons and are discussed in Chapter 12. All other nonforested, inland freshwater wetlands have been included in this chapter as freshwater marshes.

A number of good references on inland freshwater marshes exist. The reader is referred to Good et al. (1978), Gore (1983a), Hofstetter (1983), Prince and D'Itri (1985), Weller (1987), G. D. Adams (1988), van der Valk (1989),

Table 11–1. Areas of Freshwater Marshes in the United States

Description	*Area, hectares*
Inland fresh meadows	3,041,000
Inland shallow fresh marshes	1,606,000
Inland deep fresh marshes	941,000
Coastal shallow fresh marshes	896,000
Coastal deep fresh marshes	661,000

Source: Data from Shaw and Fredine, 1956

Kushlan (1990), and Gopal and Masing (1990) for additional reading on various aspects or types of freshwater marshes. In this chapter we refer extensively to studies of reed marshes in Czechoslovakia (Dykyjova and Kvet, 1978) because these International Biological Programme studies provide some of the best quantitative functional information on freshwater marshes.

GEOGRAPHICAL EXTENT AND GEOLOGICAL ORIGINS

Table 11–1 summarizes the extent of United States freshwater marshes covered in this chapter, classified essentially by depth (and duration) of flooding. The classification system reflects that study's primary interest in habitat value and waterfowl. These wetlands cover an estimated seven million hectares in the United States (Fig. 11–1). Countless other small marshes occur throughout the continent, many of which have not been inventoried. Several major groups of freshwater marshes, however, can be identified.

Prairie Potholes and Nebraska Sandhills

One of the dominant areas of freshwater marshes in the world is the prairie pothole region of North America (see also Chap. 3 and Figs. 3–11 and 3–12). Individual pothole marshes are usually small; they originated in millions of depressions formed by glacial action. They are found in greatest abundance in moraines of undulating glacial till, especially in a 780,000 km^2 region west of the Great Lakes in Minnesota, Iowa, and the Dakotas in the United States and in Alberta, Saskatchewan, and Manitoba in Canada (Fig. 3–11). It has been estimated that before the Europeans settled in this country there were 80,000 km^2 of these potholes in the region (Frayer et al., 1983; Leitch, 1989). They occur as far south as the southernmost advance of glaciers. These mid-latitude marshes are among the richest in the world because of the fertile soils and warm summer climate. Similar habitats farther north of the pothole region shown in Figure 3–11 are dominated by mosses and are properly considered peatlands. It has been esti-

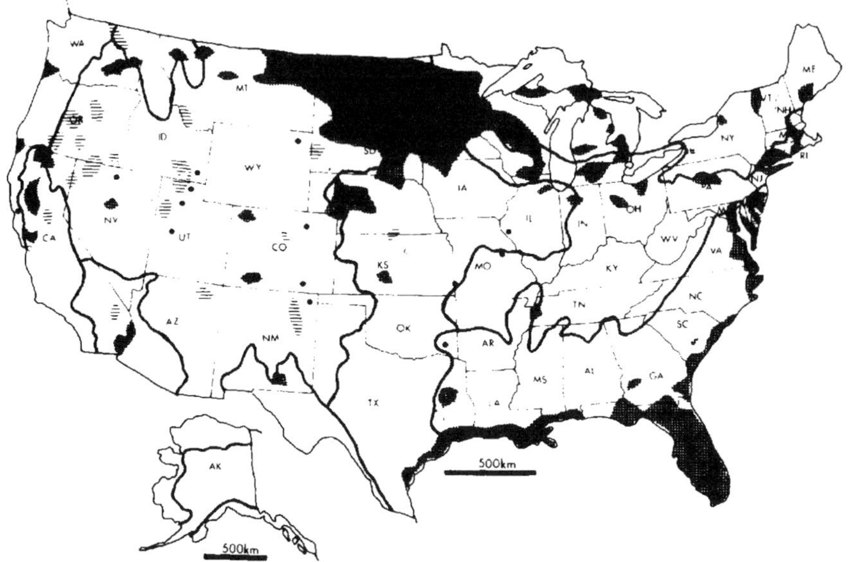

Figure 11–1. Distribution of major groups of inland marshes in the United States. Lines indicate major ecoregion divisions. Freshwater marshes are shown in dark cross-hatching; inland saline marshes are light hatched. (*From Hofstetter, 1983; copyright © 1983 by Elsevier Science Publishers, reprinted with permission*)

mated that much more than half of the prairie pothole marshes have been converted to other uses, particularly to agriculture (Leitch, 1989).

South of the prairie pothole region is the Nebraska sandhill region, a large (52,000 km^2) stabilized dune field. The region is now dominated by cattle ranches and irrigated crops. In lowlands between the dune formations there are about 5,600 km^2 of wetlands, mostly wet meadows whose water table is a few decimeters below the land surface (Novacek, 1989). The area is hydrologically significant not only because of the wetlands but also because the sandhills are important groundwater recharge and storage areas. The remaining wet meadows are thought to be threatened by declining water tables ostensibly caused by groundwater extraction for crop irrigation.

Near-Coast Marshes

In Chapter 9 tidally influenced freshwater marshes were discussed. Inland of tides and especially concentrated along the northern coast of the Gulf of Mexico and along the south Atlantic Coast of the United States, however, are large tracts of freshwater marshes that have vegetation and ecological functions similar in many ways to those of the prairie pothole marshes. One major difference is the relatively stable water level in the coastal systems that results from the influence of the adjacent ocean. These marshes originated in the same way as coastal salt marshes did, that is, they were formed as the sea gradually rose and inundated river valleys and the shallow coastal shelf after the last glacial period. Sediment and peat deposition associated with the inundation kept the marsh surface in the intertidal zone, and marshes spread inland when more land was flooded.

The Everglades

The largest single marsh system in the United States is in the Everglades of south Florida (Fig. 3–11; see also Chap. 3). It originally occupied an area of almost 10,000 km^2 extending in a strip up to 65 km wide and 170 km long from Lake Okeechobee to the brackish marshes and mangrove swamps of the southwest coast of Florida (Kushlan, 1991). Although some areas have been drained for agriculture, the remaining Everglades is still immense. An area of 4,200 km^2 has been preserved in a near-natural state, most of it as sawgrass (*Cladium jamaicense*) marshes. South Florida was built on a flat, low limestone formation, and so limestone outcrops appear throughout the Everglades. The average slope of the Everglades is only 2.8 cm/km so that fresh water from Lake Okeechobee flows slowly southward across the land as a broad sheet during the wet season. The freshwater head prevents salt intrusion from the Gulf of Mexico in the south. During the dry season, surface water dries up and is found only in the deepest sloughs and in some solution holes that serve as wildlife refuges (locally called *gator holes*) (Hofstetter, 1983). The Florida Everglades marshes are threatened by a combination of altered hydroperiods caused by human development, drainage for development, and polluted runoff from upstream agricultural activity. The water flow through this "river of grass" is heavily managed, particularly in the upper half of the original Everglades.

California Central Marshes

On the West Coast, in the valley between the Cascade mountain range and the Coastal Mountains in Washington, Oregon, and northern California, there used to be large tracts of marshland. Only a few pockets of marsh remain. The rest has been drained for agriculture.

Vernal Pools

In the western United States, particularly in California west of the Sierra Mountains, in the Central Valley, and on coastal terraces in the southwestern part of the state, shallow, intermittently flooded wet meadows called *vernal pools* dot the landscape (P. H. Zedler, 1987). These wetlands range from 50 m^2

to about 0.5 hectare in size and typically comprise less than 10 percent of a given landscape (A. Huffman, personal communication). They are wet in the cool winter and spring, when there is a bloom of plant life, invertebrates, and migratory waterfowl but they are generally dry during the warm summer months (Zedler, 1987). Vernal pools, although described as occurring throughout the country by some, are classically defined as seasonally flooded wetlands associated with a Mediterranean-type climate of wet winters and dry summers.

Great Lakes Marshes

Marshes occur along the shores of lakes, especially along the Laurentian Great Lakes of North America (Prince and D'Itri, 1985). They were formed originally in the deltas of rivers that flow into lakes and in protected shallow areas, often behind natural barrier beaches or levees thrown up by wave action on the shore or behind ancient beach ridges (Herdendorf, 1987). Since the shoreline was stabilized along the Great Lakes, many of the remaining Great Lakes marshes are now managed and protected from water level changes by artificial dikes (Herdendorf, 1987; Mitsch, 1989). There are an estimated 1,200 km^2 of wetlands around the Great Lakes in the United States (Table 11–2) and probably a larger area around the Great Lakes in Canada. Glooschenko and Grondin (1988) estimated that there are 9,000 km^2 of wetlands in southern Ontario alone, some of which are associated with Lakes Erie, Ontario, and St. Clair. Even though one wetland site on Lake Erie has been designated as a National Estuarine Research Reserve, questions have been raised about whether the wetlands and coastal rivers of the Laurentian Great Lakes should be called estuaries (Schubel and Pritchard, 1990).

Table 11–2. Extent of Coastal Freshwater Marshes Around the Laurentian Great Lakes Within the United States

Lake	*Shore Length, km*	*Number of Wetlands*	*Area of Wetlands, km^2*
Lake Superior	1598	348	267
Lake Michigan	2179	417	490
Lake Huron	832	177	249
Lake St. Clair	256	20	36
Lake Erie	666	96	83
Lake Ontario	598	312	84
TOTAL	6129	1370	1209

Source: Herdendorf, 1987

Riverine Marshes

Riverine wetlands are common throughout the continent. Wetlands are particularly extensive in the floodplain of the Mississippi River Valley and the many smaller rivers that empty into the south Atlantic. Although these wetlands are mostly forested (Chap. 14), marshes often border the forests or occupy pockets within them, sometimes in abandoned *oxbows*. At the headwaters of rivers beavers build wetlands by damming small streams. These small marshes often attract much wildlife, especially waterfowl.

Playas

Playas are an interesting group of marshes found on the high plain of northern Texas and eastern New Mexico. They are small basins that contain a clay or fine sandy loam hydric soil. Typically a playa has a watershed area of about 55 hectares and a wetland area of about 7 hectares. Because the climate of the high plain is arid, the size of the wetland is closely related to the size of the watershed that drains into it. An estimated 25,000 such basins occur on the high plain, but no complete inventory has been made. These wetlands are particularly important waterfowl habitats, and most of the ecological information available about them is related to habitat value. Virtually all playa watersheds are farmed. The water draining into the playa is therefore rich in fertilizers; furthermore, playas are often a source of irrigation water. Both irrigation and eutrophication have had strong effects on the size and quality of playa marshes (Guthery et al., 1982; Bolin and Guthery, 1982; Bolin et al., 1989).

HYDROLOGY

As with any other wetland, the flooding regime or hydroperiod of freshwater marshes determines their ecological character. The critical factors that determine the character of these wetlands are the presence of excess water and sources of water other than direct precipitation. Excess water occurs either when precipitation exceeds evapotranspiration or when the watershed draining into the marsh is large enough to provide adequate runoff or groundwater inflow. Along sea coasts, water levels tend to be stable over the long term because of the influence of the ocean (Fig. 11–2a). Water levels in inland marshes, in contrast, are much more controlled by the balance between precipitation and evapotranspiration, especially for marshes in small watersheds that are affected by restricted throughflow. Water levels of lacustrine marshes such as those found along the Laurentian Great Lakes are generally stable but are influenced by on the year-to-year variability of lake levels and whether the wetland is diked or open to the lake (Fig. 11–2b). Many marshes such as wet meadows, sedge meadows, vernal

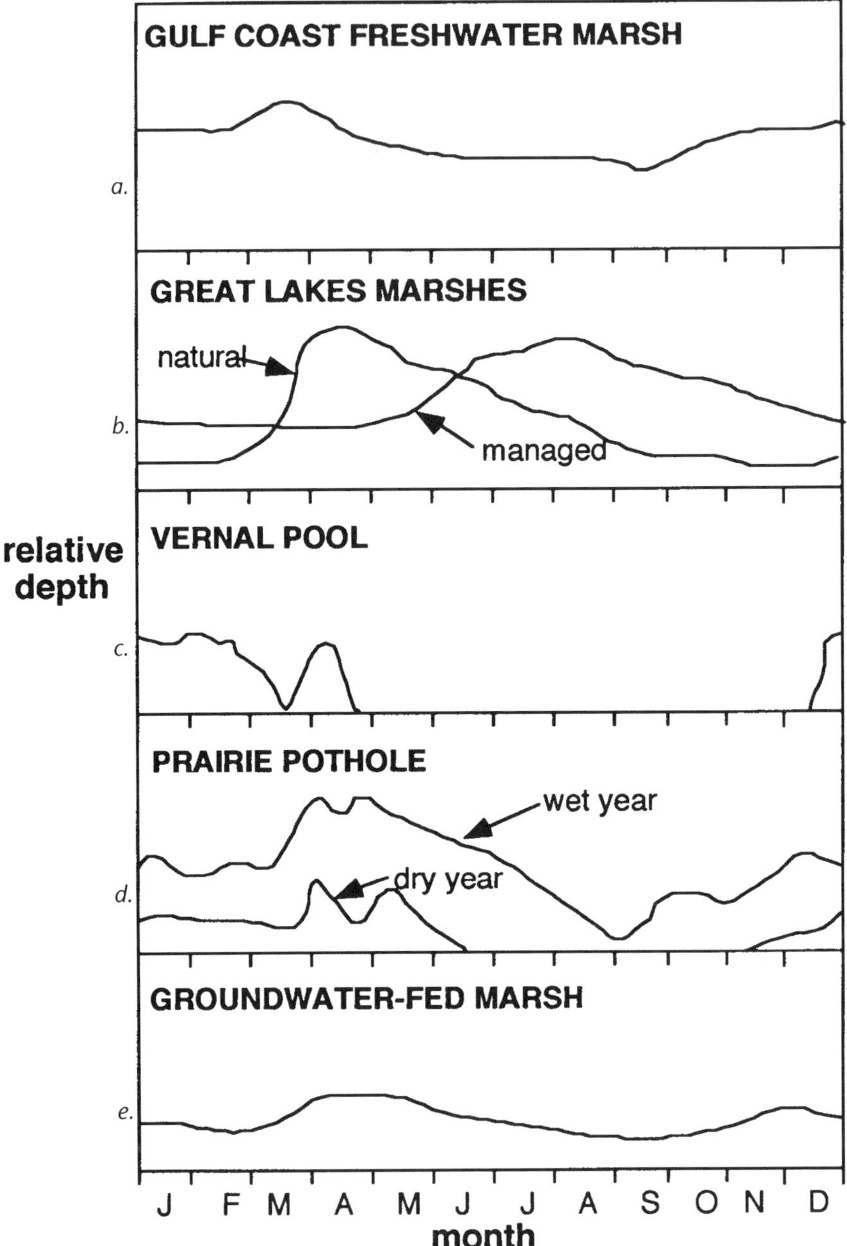

Figure 11–2. Typical hydroperiods for various freshwater marshes: *a.* coastal freshwater marsh in Louisiana; *b.* Great Lakes coastal marsh with and without dikes; *c.* vernal pool in California; *d.* prairie pothole; and *e.* groundwater-fed freshwater wetland.

pools, and even prairie potholes dry down seasonally (Fig. 11–2c, d), but the plant species found there reflect the hydric conditions that exist during most of the year.

The primary distinction between an ombrotrophic bog and a rheotrophic fen is the flow of water through the latter, which is the primary source of its nutrients. Marshes are similar to fens (some consider a fen a freshwater marsh; we defer the discussion of fens until Chapter 12), for they generally have a water source in addition to precipitation. For example, some marshes intercept groundwater supplies. Their water levels therefore reflect the local water table and contribute to it (Fig. 11–2e). Calculations of the seepage of water from marshes into the ground indicate that it occurs mostly at the margins (Millar, 1971). For this reason, seepage losses are greatest where the wetland shoreline is large relative to the water volume.

Although usually less than 20 percent of the water impounded in freshwater wetlands enters the groundwater supply, this volume may be an important contribution to the watershed (Weller, 1981). Other marshes collect surface water and entrained nutrients from watersheds that are large enough to maintain hydric conditions most of the time. For example, overflowing lakes supply water and nutrients to adjacent marshes, and riverine marshes are supplied by the rise and fall of the adjacent river. In cases in which flow from watersheds is seasonal, freshwater marshes can serve as biological and hydrologic "oases" during drought conditions (Mitsch and Reeder, 1992).

Because all of these water sources (except, in the short run, lakes) depend on precipitation, which is notoriously variable, the water regime of most inland marshes also varies in a way that is predictable only in a statistical sense. Stewart and Kantrud (1971) emphasized the duration of flooding in their classification of prairie pothole marshes. They classified wetlands as ephemeral, temporary, seasonally semipermanent, and permanent. In addition, they recognized that marshes may move through several of these classes over the span of a few years. Thus a marsh that would ordinarily be considered permanent might be in a "drawdown" phase that gives the appearance of an ephemeral marsh.

CHEMISTRY

The chemistry of inland marshes is best described in contrast to that of ombrotrophic bogs (Chap. 12) at one extreme and eutrophic tidal wetlands at the other (Chap. 9). Differences are related to the magnitude of nutrient and other chemical inputs and to the relative importance of groundwater inflow. Inland marshes generally are minerotrophic in contrast to northern bogs; that is, the inflowing water has a high specific conductivity resulting from the presence of dissolved cations. The peat is saturated with bases, and the pH, as a result, is close to neutral. Because nutrients are plentiful, productivity is higher than it is in bogs, bacteria are active in nitrogen fixation and litter decomposition, and turnover rates are high. The accumulation of organic matter results from high production rates, not from the inhibition of decomposition by low pH (as occurs in bogs).

Although inland wetlands are minerotrophic, they generally lack the high nutrient loading associated with tidal inundation of freshwater tidal marshes (Chap. 9). Flooding in inland marshes tends to be controlled by seasonal changes in local rainfall, subsequent runoff, and evapotranspiration; inundation of tidal areas occurs regularly, once or twice a day. In the latter case, even if tidewater nutrient levels are low, the large volumes of flooding water result in high loading rates. Because of these differences in surface flooding between inland and freshwater tidal marshes, groundwater flow is usually more important as a source of nutrients in inland marshes.

Table 11–3 lists nutrient concentrations reported for sediments in several inland freshwater marshes. The values vary widely depending on the substrate, the load-

Table 11–3. Parameters and Nutrient Concentrations for the Top 20 cm of Selected Inland Freshwater Marsh Soils

	Marsh Type			
	Riverine, Wisconsin[a]	*Coastal Floating Louisiana*[b]	*Lake Erie Marshes, Ohio*[c]	*Lacustrine, Czechoslovakia*[d]
Soil Parameters, top 20 cm				
pH	6.4–6.5	6–7	6.8–8.6 (natural) 5.6–10.5 (diked)	5.1
Organic matter, %	40.4–43.4	75	9–17 (natural) 11–31 (diked)	39
Total nitrogen, %	1.36–1.94	1.8–2.4	—	2.4
Total phosphorus, %	—	0.07–0.1	—	0.013
Available phosphorus, ppm	50–203	0.3[e]	26 ± 19[f]	130
Available potassium, ppm	98–230	—	662 ± 351[f]	550
Available calcium, ppm	5,700–12,700 (exchangeable)	—	5528 ± 2501[f]	5,140
Available magnesium, ppm	1,219–2,770	—	268 ± 185[f]	570

[a]Klopatek, 1978
[b]Sasser and Gosselink, 1984
[c]Mitsch, 1989, 1992b
[d]Dykyjova and Kvet, 1982, *Typha angustifolia* marsh
[e]PO_4 in interstitial water
[f]Average ± std. dev. for nine marshes

ing rate, and the nutrient uptake by the plants. The ion concentrations of freshwater marshes are high, and water is generally in a pH range of from 6 to 9. Organic matter can vary from a very high content (75 percent) as is found in floating freshwater marshes in coastal Louisiana to a low content (10–30 percent) in marshes fed by inorganic sediments from agricultural watersheds or open to organic export (Table 11–3). Concentrations of total (as distinguished from available) nutrients are reflections of the kinds of sediment in the marsh. Mineral sediments are often associated with high phosphorus content, for example, whereas total nitrogen is closely correlated to organic content. Dissolved inorganic nitrogen and phosphorus—the elements that most often limit plant growth—often vary seasonally from very low concentrations in the summer, when plants take them up as rapidly as

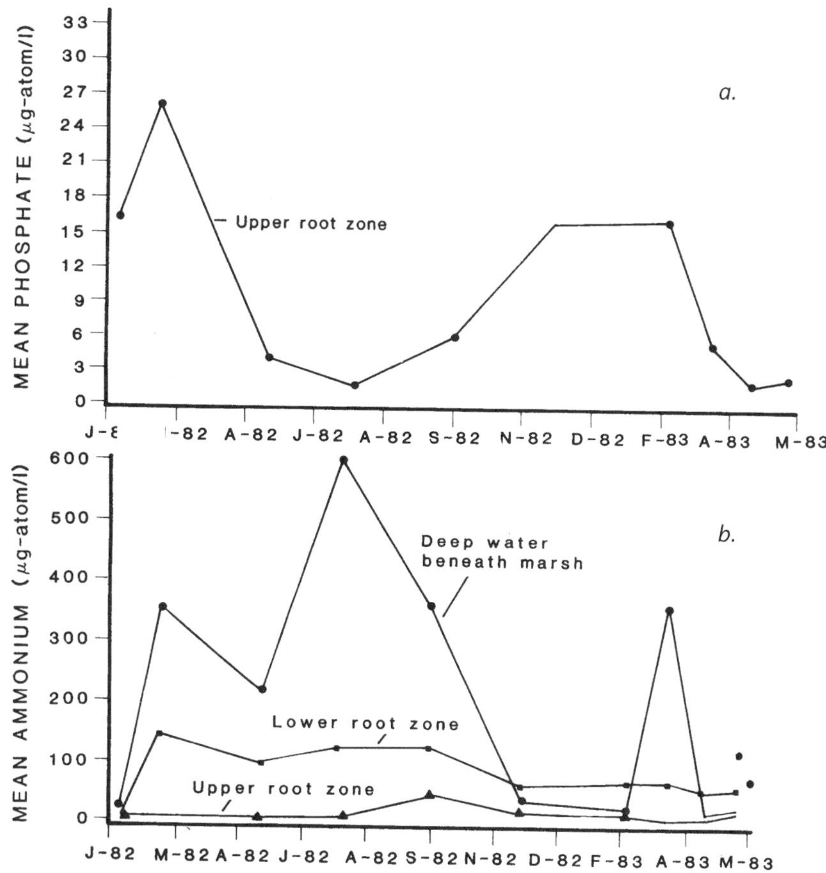

Figure 11–3. Concentration of *a.* phosphate and *b.* ammonium in interstitial water of freshwater marsh sediments in Louisiana. (*After Sasser et al., 1991*)

Table 11–4. Typical Dominant Emergent Vegetation in Different Freshwater Marshes

Marsh Type and Location	*Dominant Species*	*Reference*
Prairie glacial marsh, Iowa	*Typha latifolia* *Typha angustifolia* *Scirpus validus* *Scirpus fluviatilis* *Scirpus acutus* *Sparganium eurycarpum* *Carex* spp. *Sagittaria latifolia*	van der Valk and Davis (1978b); Weller (1981)
Riverine marsh, Wisconsin	*Typha latifolia* *Scirpus fluviatilis* *Carex lacustris* *Sparganium eurycarpum* *Phalaris arundinacea*	Klopatek (1978)
Lake Erie wetlands, Ohio	*Typha angustifolia*[a, b] *Cyperus erythrorhizos*[a, b] *Scirpus validus*[a] *Leersia oryzoides*[a] *Echinochloa Walteri*[a] *Ludwigia palustris*[a] *Phragmites australis*[a] *Scirpus acutus*[a] *Scirpus fluviatilis*[b] *Sparganium eurycarpum*[b] *Amaranthus tuberculatus*[b] *Nuphar advena*[b] *Nelumbo lutea*[b]	Robb (1989)

continued on next page

they become available, to high concentrations in the winter, when plants are dormant but mineralization continues in the soil.

In Figure 11–3a phosphate in a Louisiana floating marsh is shown to be high during the winter but low during the summer presumably because of plant uptake. Figure 11–3b shows the same phenomenon in a different way for ammonium (the dominant inorganic nitrogen form in anoxic soils). Here the NH_4^+ levels are high in deep waters underlying the marsh but decrease toward the surface, where plants remove it in the root zone. In Table 11–3 the high total soil nitrogen (1–2 percent) and phosphorus (0.01–0.1 percent) reflect an enormous reservoir of organic nutrients that can be mineralized and made available for plant use. Available inorganic nitrogen and phosphorus are one to three orders of magnitude lower. This available supply varies seasonally depending on both the rate of formation (mineralization) and uptake by the marsh plants. In most

Table 11–4 (continued)

Marsh Type and Location	*Dominant Species*	*Reference*
"Tule" marshes, California and Oregon	*Scirpus acutus* *Scirpus californicus* *Scirpus olneyi* *Scirpus validus* *Phragmites australis* *Cyperus* spp. *Juncus patens* *Typha latifolia*	Hofstetter (1983)
Floating freshwater coastal marsh, Louisiana	*Panicum hemitomon* *Thelypteris palustris* *Osmunda regalis* *Vigna luteola* *Polygonum sagittatum* *Sagittaria latifolia* *Decodon verticillatus*	Sasser and Gosselink (1984)
Everglades, Florida	*Cladium jamaicense* *Panicum hemitomon* *Rhynchospora* spp. *Eleocharis* spp. *Sagittaria latifolia* *Pontederia lanceolata* *Crinum americanum* *Hymenocallis* spp.	Hofstetter (1983)

[a]Diked marshes (impoundments)
[b]Undiked marshes

freshwater marshes the inflow probably contributes little to the concentration of available nutrients, but it does contribute significantly to the long-term pool of nutrients available to the vegetation.

ECOSYSTEM STRUCTURE

Vegetation

The vegetation of fresh inland marshes has been detailed in many studies. The dominant species vary from place to place, but the number of genera common to all locations in the temperate zone is quite remarkable. Table 11–4 lists dominant emergent species from sites that represent a wide range of different inland marshes. Common species include the graminoids and sedges *Phragmites australis (=P. communis)* (reed grass), *Typha* spp. (cattail), *Sparganium eurycarpum* (bur reed), *Zizania aquatica* (=*Z. palustris*; wild rice), *Panicum hemitomon*, *Cladium jamaicense*; and the sedges *Carex* spp., *Scirpus* spp. (bulrush), and *Eleocharis* spp. (spike rush). In addition, broad-leaved monocotyledons such as *Pontederia cordata* (pickerelweed) and *Sagittaria* spp. (arrowhead) are frequently found. Herbaceous dicotyledons are represented by a number of species, typical examples of which are *Ambrosia* spp. (ragweed) and *Polygonum* spp. (smartweed). Frequently represented also are such ferns as *Osmunda regalis* (royal fern) and *Thelypteris palustris* (marsh fern), and the horsetail, *Equisetum* spp. One of the most productive species in the world is the tropical sedge *Cyperus papyrus*, which flourishes on floating mats in southern Africa.

Marsh Vegetation Zonation

These typical plant species do not occur randomly mixed together in marshes. Each has its preferred habitat. Different species often occur in rough zones on slight gradients, especially flooding gradients. For example, Figure 11–4 illustrates the typical distribution of species along an elevation gradient in a prairie pothole marsh. Sedges (e.g., *Carex* spp., *Scirpus* spp.), rushes (*Juncus* spp.), and arrowheads (*Sagittaria* spp.) typically occupy the shallowly flooded edge of a pothole. Two species of cattail (*Typha latifolia* and *Typha angustifolia*) are common. The narrow-leaved species (*T. angustifolia*) is more flood tolerant than the broad-leaved cattail (*T. latifolia*) and may grow in water up to 1 m deep. The deepest zone of emergent plants is typically vegetated with hardstem bulrushes (*Scirpus acutus*). Beyond these emergents, floating-leaved and submersed vegetation will grow, the latter to depths dictated by light penetration. Typical floating-leaved aquatic hydrophytes include rhizomatous plants such as water lilies (*Nymphaea tuberosa* or *Nymphaea odorata*), water lotus (*Nelumbo lutea*), and

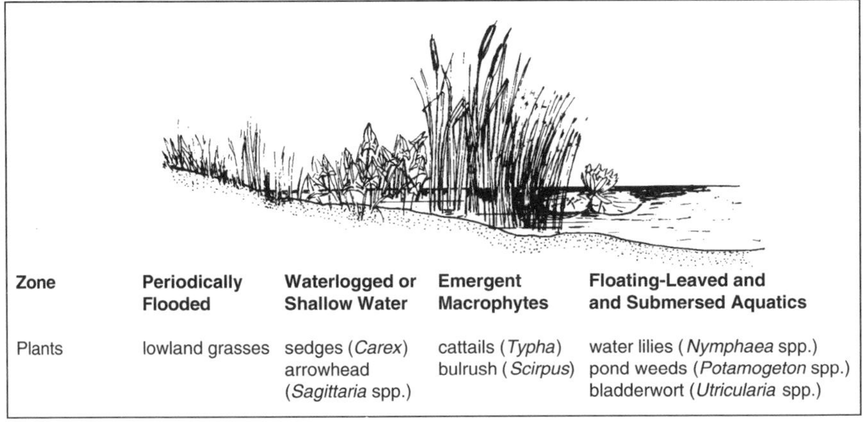

Figure 11–4. Cross-section through a freshwater marsh indicating plant zone according to water depth and typical plants found in each zone.

spatterdock (*Nuphar advena*) and stoloniferous plants such as water shield (*Brasenia Schreberi*) and smartweed (*Polygonum* spp.). Submersed hydrophytes include coontail (*Ceratophyllum demersum*), water millfoil (*Myriophyllum* spp.), pondweed (*Potamogeton* spp.), wild celery (*Vallisneria americana*), naiad (*Najas* spp.), bladderwort (*Utricularia* spp.), and waterweed (*Elodea canadensis*).

Although it is difficult to generalize about vegetation zonation in and adjacent to wetlands, there is some evidence that vegetation zonation depends on the size of the wetland. W. C. Johnson et al. (1987), using ordination techniques in some North Dakota prairie potholes, found that vegetation in small marshes formed essentially a continuum from upland edge to deep water, whereas vegetation had strong discontinuities, (i.e., sharp boundaries) between marsh vegetation and upland meadows in large marshes. This difference was attributed to greater wave action and ice scour that occur in large wetlands.

Floating Mats

A curious feature of many freshwater marshes is the development of a floating mat of vegetation (see Chemistry, above). This floating mat phenomenon has a wide geographical range: Floating mats of marsh vegetation have been observed from Louisiana (Sasser and Gosselink, 1984; see also Chap. 9) and to New Brunswick, Canada (Hogg and Wein, 1987, 1988a, b). In the Canadian study, the researchers concluded that the buoyancy of a floating *Typha* mat was caused by trapped gases (primarily nitrogen and methane) beneath the mat and the properties of the dominant vegetation itself (mainly *Typha* spp.). They determined that floating mats generally "grow from within" by accumulation of below-ground biomass rather than by accumulation of above-ground material at the mat surface. They also estimated that, without decomposition, the vertical growth of their floating *Typha* mat was about 6 cm/yr.

Marsh Cycles

A unique structural feature of prairie pothole marshes is the five-to-twenty-year cycle of dry marsh, regenerating marsh, degenerating marsh, and lake (Weller and Spatcher, 1965; van der Valk and Davis, 1978b) that is related to periodic droughts (Fig. 11–5). During drought years, standing water disappears. Buried seeds in the exposed mudflats germinate to grow a cover of annuals (*Bidens, Polygonum, Cyperus, Rumex*) and perennials (*Typha, Scirpus, Sparganium, Sagittaria*). When rainfall returns to normal, the mudflats are inundated. Annuals disappear, leaving only the perennial emergent species. Submersed species (*Potamogeton, Najas, Ceratophyllum, Myriophyllum, Chara*) also reappear. For the next year or more, during the regenerating stage, the emergent population increases in vigor and density. After a few years, however, these populations begin to decline. The reasons are poorly understood, but often

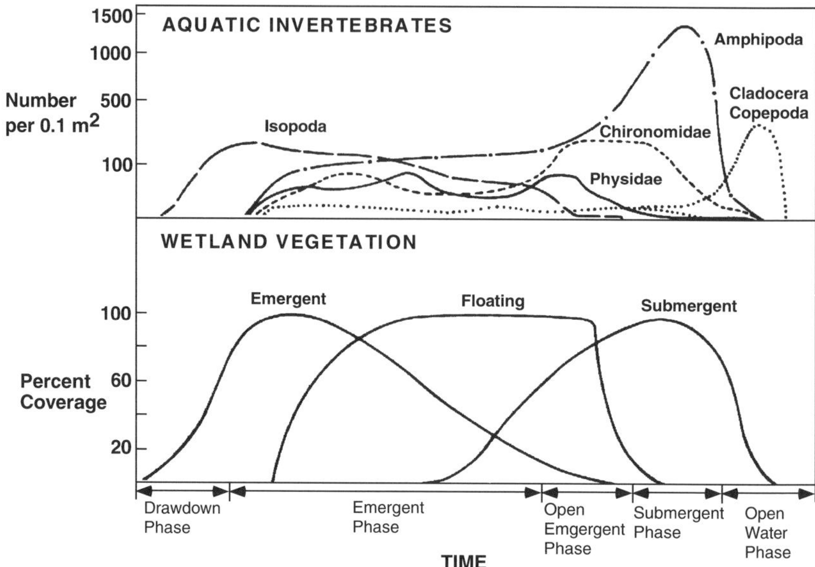

Figure 11–5. Drawdown cycle in prairie pothole freshwater marshes showing changes in vegetation and dominant aquatic macroinvertebrates. Time horizon is 5 to 20 years with drawdown cycle caused by periodic droughts. (*From Voigts, 1976; copyright © 1976 by American Midland Naturalist, reprinted with permission*)

muskrat populations explode in response to the vigorous vegetation growth. Their nest and trail building can decimate a marsh. Whatever the reason, in the final stage of the cycle, there is little emergent marsh; most of the area reverts to an open shallow lake or pond, setting the stage for the next drought cycle. The wildlife use of these wetlands follows the same cycle. The most intense use occurs when there is good interspersion of small ponds with submersed vegetation and emergent marshes with stands diverse in height, density, and potential food.

Seed Banks and Germination

Seed banks and fluctuating water levels interact in complicated ways to produce vegetation communities in freshwater marshes (see Chap. 7). As a general rule, seed germination is maximized under shallow water or damp soil conditions, after which many perennials can reproduce vegetatively into deeper water. Keddy and Reznicek (1986) illustrated their observation that fluctuating water levels along the Great Lakes allowed greater diversity of plant types and species in the coastal marshes and that these marshes sometimes have a density of buried seeds an order of magnitude greater than that of the more studied prairie marshes. The importance of other variables, particularly nutrients, for the germination and success of freshwater marsh plants is not as well understood. In one series of experiments, Gerritsen and Greening (1989) illustrated the importance of water levels and nutrient conditions for seed germination in seed banks from two marshes in Okefenokee Swamp in Georgia. Although water level was the important variable in determining which plants germinated, nitrogen was shown to limit marsh plant growth in drawdown conditions, and phosphorus limited the growth of a few species while they were inundated. In a series of experiments using natural (many species) and synthetic seed banks (two species only) under different hydrologic and nutrient conditions, Willis and Mitsch (unpubl. manuscript) found similar effects of the water level on germination but were unable to verify the importance of nutrient additions.

Other Factors

The particular species found in freshwater wetlands are also determined by many other environmental factors. Nutrient availability determines to a large degree whether a wetland site will support mosses or angiosperms (i.e., whether it is a bog or a marsh) and what the species diversity will be. It is not obvious for freshwater marshes that highly fertile wetlands are highly diverse. In fact, studies of freshwater marsh plant diversity published in the literature (e.g., Wheeler and Giller, 1982; Vermeer and Berendse, 1983; R. T. Day et al., 1988; D. R. J. Moore and Keddy, 1989; and Moore et al., 1989) suggest the opposite conclusion. For example, Moore et al. (1989) contrasted several fertile and infertile sites (as measured by the plant standing crop) in eastern Ontario and found the greatest species richness between 60–400 g/m^2 standing crop and much less richness at higher plant standing crops (>600 g/m^2) (Fig. 11–6). They also found rare species only at the infertile sites, suggesting that the conservation of infertile wetlands should be part of overall wetland management strategies.

Plant species also change with latitude, that is, as temperatures increase or decrease and the winters become more or less severe. Although the same genera may be found in the tropics and in the Arctic, the species are usually different, reflecting different adaptations to cold or heat. Because of their long isolation from one another, the flora of the North American continent differ at the species level from the European flora. Finally, soil salts, even in low concentrations, determine the species found on a site (McKee and Mendelssohn, 1989). Because many inland marshes are potholes that collect water that leaves only by evaporation, salts may become concentrated during periods of low precipitation, adversely affecting the growth of salt-intolerant species.

Alien Species

Alien plant species are often a part of the vegetation of freshwater marshes, particularly in areas that have been disturbed. Mitchell and Gopal (1991) discussed the invasion of plants such as *Eichhornia crassipes* (water hyacinth), *Salvinia molesta* (salvinia), and *Alternanthera philoxeroides* (alligator weed) in tropical

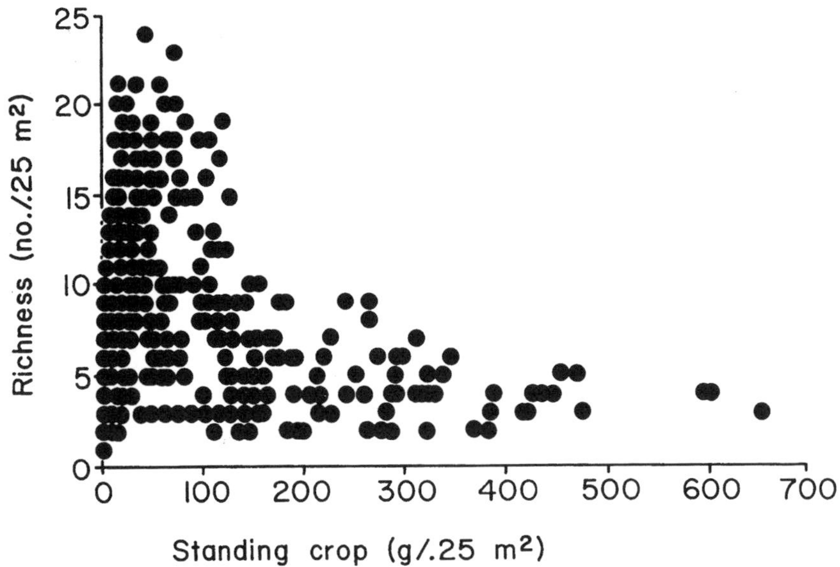

Figure 11–6. Species richness versus vegetation biomass in 0.25 m^2 quadrats from three wetland areas in Ontario, Quebec, and Nova Scotia. (*Moore et al., 1989; copyright © 1989 by Elsevier, reprinted with permission*)

and subtropical regions of the world. In the southeastern United States, *Eichhornia crassipes* (water hyacinth) is considered to be a nuisance plant because of its prolific growth rate. This free-floating aquatic plant can double the area that it covers in two weeks and has choked many waterways that have received high nutrient loads (Penfound and Earle, 1948). On the other hand, the plant has been praised for its ability to sequester nutrients and other chemicals from the water and has often been proposed as part of natural wetlands to purify wastewater (Mitsch, 1977; Ma and Yan, 1989). Although there are many theories about alien aquatic plants, there is some validity to the concept that disturbed ecosystems are most susceptible to biological invasions (Mitchell and Gopal, 1991).

It has been hypothesized that tropical regions are more susceptible to invasion than temperate regions because invading plants grow much more rapidly and are more noticeable in the tropics than in temperate latitudes (Mitchell and Gopal, 1991). In the freshwater marshes of the St. Lawrence and Hudson River valleys and in the Great Lakes region of North America, however, *Lythrum Salicaria* (purple loosestrife), a tall purple-flowered emergent hydrophyte, has spread at an alarming rate in this century, causing much concern to those who manage these marshes for wildlife (Stuckey, 1980; Balogh and Bookhout, 1989). The plant is aggressive in displacing native grasses, sedges, rushes, and even

Typha spp. Many freshwater marsh managers have implemented programs designed to control purple loosestrife with chemical and mechanical means. Other aquatic aliens such as the submersed *Hydrilla verticillata*, a plant native to Africa, Asia, and Australia, have invaded open, shallow water marshes in the United States (K. K. Steward, 1990) but rarely compete well with emergent vegetation.

Consumers

Perhaps one reason that small marshes of the prairie region and the western high plains harbor such a rich diversity of organisms and wildlife is that they are often natural islands in a sea of farmland. Cultivated land does not provide a diversity of either food or shelter, and many animals must retreat to the marshes, which have become the only natural habitats.

Decomposers

Like other wetland systems, inland marshes are detrital ecosystems. Unfortunately, we know very little about the many small benthic organisms that are the primary consumers in wetlands (inland marshes are no exception). It is probable that small decomposers such as nematodes and enchytraeids are relatively more important than larger decomposers in marshes compared to terrestrial woodlands.

Invertebrates

The most conspicuous invertebrates are the true flies (Diptera), which often make one's life miserable in the marsh. These include midges, mosquitoes, and crane flies. Many, especially in adult stages, are herbivorous; Cragg (1961) attributed about one-third of consumer respiration in a *Juncus* moor to herbivores, chiefly Diptera. But in the larval stages, many of the insects are benthic. For example, Weller (1981) states that midge larvae, which are called *bloodworms* because of their rich red color, "are found submerged in bottom soils and organic debris, serving as food for fish, frogs, and diving birds. When pupae surface and emerge as adults, they are exploited as well by surface-feeding birds and fish." Temporal cycles and spatial patterns of invertebrate species and concentrations reflect the natural seasonal cycle of insect growth and emergence superimposed on the vegetation cycles described above. Voigts (1976) described patterns of invertebrates in prairie potholes as they changed with hydrologic and vegetation conditions (Fig. 11–5). McLaughlin and Harris (1990) investigated insect emergence from diked and undiked marshes along Lake Michigan and found more insects, more insect biomass, and a greater number of taxa in the diked marsh and the greatest numbers and biomass in the sparsely vegetated zones of the wetlands rather than in open water or dense vegetation.

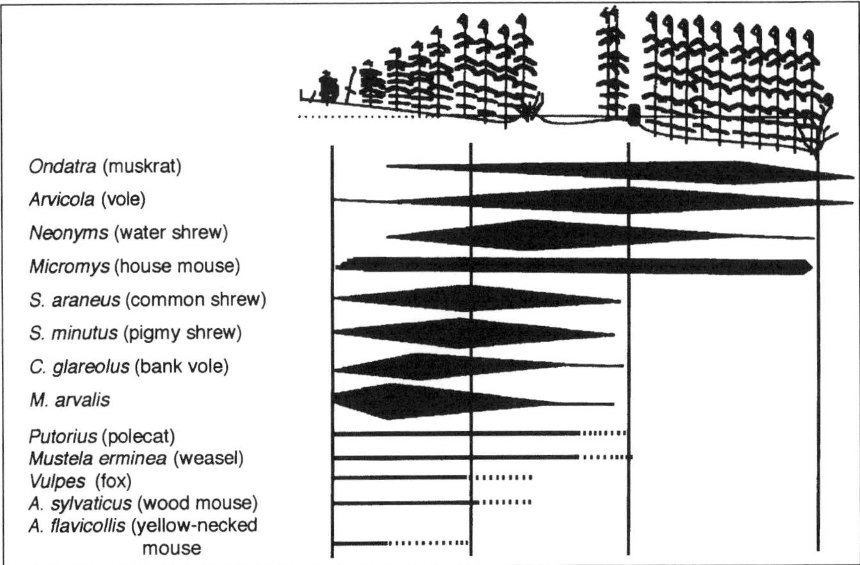

Figure 11–7. Distribution of small mammal populations in the littoral zone of a Czechoslovakian pond. The three zones are defined as terrestrial, limosal, and littoral from left to right. (*After Pelikan, 1978; copyright © 1978 by Springer-Verlag, reprinted with permission*)

Mammals

A number of mammals inhabit inland marshes. The most noticed probably is the muskrat (*Ondatra zibethicus*). This herbivore reproduces rapidly and can attain population densities that decimate the marsh, causing major changes in its character. Like plants, each mammalian species has preferred habitats. For example, Figure 11–7 shows the distribution of muskrats and other mammals on an elevation gradient in a Czechoslovak fishpond littoral marsh. Muskrats are found in the most aquatic areas, the water vole in overlapping but higher elevations and other voles found in the relatively terrestrial parts of the reed marsh. Most of the mammals are herbivorous.

Birds

Birds, particularly waterfowl, are also abundant in freshwater marshes. Many of these are herbivorous or omnivorous. Waterfowl are plentiful in almost all wetlands probably because of food richness and the diversity of habitats for nesting and resting. Waterfowl nest in northern freshwater marshes, winter in southern marshes, and rest in other marshes before resuming their migrations. Weller and Spatcher (1965) described how birds partition a typical marsh as they compete for food and nesting sites (Fig. 11–8). Different species distribute themselves

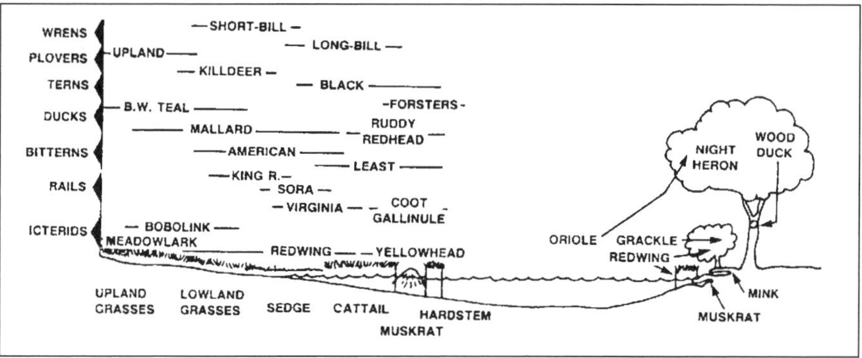

Figure 11–8. Typical distribution of birds across a freshwater marsh from open water edge across shallow water to upland grasses. Placement of muskrat and mink are also illustrated. (*From Weller and Spatcher, 1965*)

along an elevation gradient according to how well they are adapted to water. In northern marshes, the loon usually uses the deeper water of marsh ponds, which may hold fish populations. Grebes prefer marshy areas, especially during the nesting season. Some ducks (dabblers) nest in upland sites, feeding along the marsh-water interface and in shallow marsh ponds. Others (diving ducks) nest over water and fish by diving. For example, the black duck (*Anas rubripes*), one of the most popular ducks for naturalists and hunters alike, uses the emergent marsh as its preferred habitat (Frazer et al., 1990a, b). Geese and swans, along with canvasback ducks, are the major marsh herbivores. Wading birds usually nest colonially in wetlands and fish along the shallow ponds and streams. Rails live in the whole range of wetlands; many of them are solitary birds that are seldom seen. Songbirds and swallows are also abundant in and around marshes. They often nest or perch in adjacent uplands and fly into the marsh to feed.

Fish

One of the most difficult questions about which to generalize is whether freshwater marshes support much fish life. As a general rule, the deeper the water in the marsh and the more open the system is to large rivers or lakes, the more variety and abundance of fish that can be supported. The positive aspect of freshwater marshes as habitats and nurseries for fish was investigated by Derksen (1989) for a large Manitoba marsh complex and by Stephenson (1989) for Great Lakes marshes. Derksen found extensive use of marshes by northern pike (*Esox lucius*) with emigration from the marsh occurring primarily in the autumn. Stephenson investigated fish utilization for spawning and rearing of young in five marshes connected to Lake Ontario in Ontario, Canada. A total of 36 species of fish were collected, 23 to 27 species per marsh. The spawning adults of 23 species and

Table 11–5. Selected Primary Production Values for Inland Freshwater Marsh Species

Species	*Location*	*Net Primary Productivity* $g\ m^{-2}\ yr^{-1}$	*Reference*
Reeds and Grasses			
Glyceria maxima	Lake, Czechoslovakia	900–4,300[a]	Kvet and Husak (1978)
Phragmites communis	Lake, Czechoslovakia	1,000–6,000[a]	Kvet and Husak (1978)
P. communis	Denmark	1,400[a]	Anderson (1976)
Panicum hemitomon	Floating coastal marsh, Louisiana	1,700[b]	Sasser et al. (1982)
Schoenoplectus lacustris	Lake, Czechoslovakia	1,600–5,500[a]	Kvet and Husak (1978)
Sparganium eurycarpum	Prairie pothole, Iowa	1,066[b]	Vand der Valk and Davis (1978a)
Typha glauca	Prairie pothole, Iowa	2,297[b]	Van der Valk and Davis (1978a)
Typha latifolia	Oregon	2,040–2,210[a]	McNaughton (1966)
Typha sp.	Lakeside, Wisconsin	3,450[a]	Klopatek (1974)
Sedges and Rushes			
Carex atheroides	Prairie pothole, Iowa	2,858[b]	Van der Valk and Davis (1978a)
Carex lacustris	Sedge meadow, New York	1,078–1,741[a]	Bernard and Solsky (1977)
Juncus effusus	South Carolina	1,860[a]	Boyd (1971)
Scirpus fluviatilis	Prairie pothole, Iowa	943[a]	Van der Valk and Davis (1978a)
Broad-Leaved Monocots			
Acorus calamus	Lake, Czechoslovakia	500–1,100[a]	Kvet and Husak (1978)

[a]Above- and belowground vegetation
[b]Aboveground vegetation

young-of-the-year of 31 species were collected in the marshes, indicating the importance of these marshes for fish reproduction for Lake Ontario. Eighty-nine percent of the species encountered were using the marshes for reproduction (Stephenson, 1989).

Common carp (*Cyprinus carpio*) are able to withstand the dramatic fluctuations of water temperature and dissolved oxygen typical of shallow marshes and are thus abundant in many inland wetlands. They affect marsh vegetation by direct grazing, uprooting vegetation while searching for food (King and Hunt, 1967), and causing severe turbidity in the water column. For these reasons, carp are not considered desirable by many freshwater marsh managers.

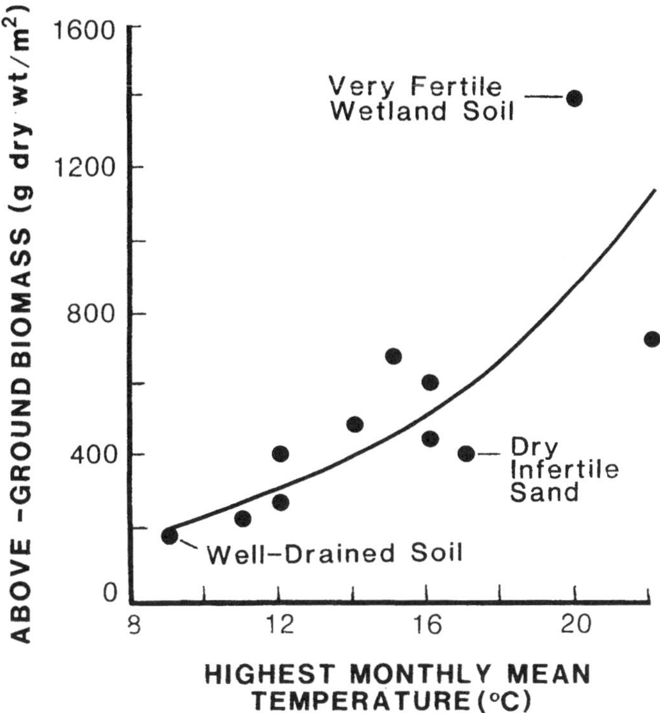

Figure 11–9. Relationship between highest mean monthly temperature and above-ground standing crop of various sedges in freshwater wetlands and uplands. Data points are for wetlands except where noted otherwise. (*From Gorham, 1974; copyright © 1974 by Blackwell Scientific Publications, Ltd., Oxford, reprinted with permission*)

ECOSYSTEM FUNCTION

Primary Production

The productivity of inland marshes has been reported in a number of studies (Table 11–5). Estimates are generally quite high, ranging upward from about 1,000 g m^{-2} yr^{-1}. Some of the best estimates, which take into account underground production as well as that above-ground, are from studies of fish ponds in Czechoslovakia. (These are small artificial lakes and bordering marshes used for fish culture.) These estimates, some indicating values of over 6,000 g m^{-2} yr^{-1}, are high compared with most of the North American work. The productivity is higher than even the productivity of intensively cultivated farm crops.

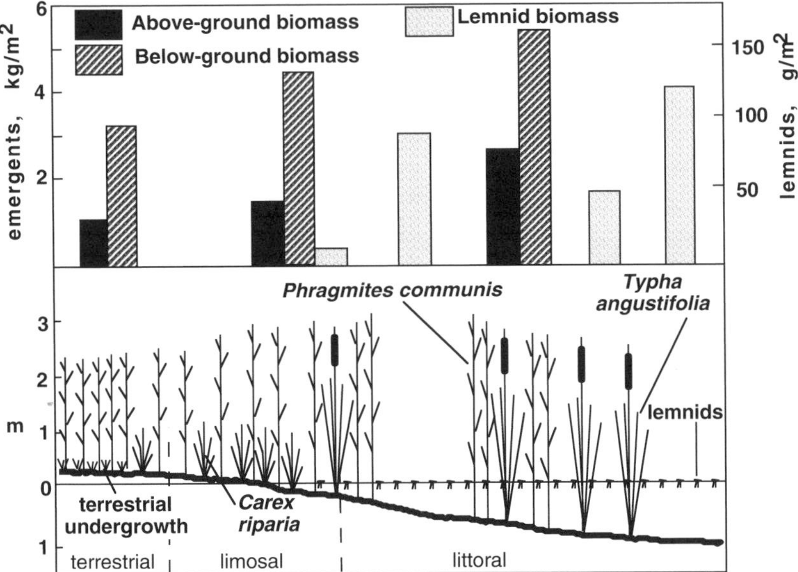

Figure 11–10 . Distribution of above-ground and below-ground biomass of emergent vegetation and lemnids across a reedbed (*Phragmites*) transect, showing relation to elevation and flooding. Three zones are as in Figure 11-7. (*From Kvet and Husak, 1978; copyright © 1978 by Springer-Verlag, redrawn with permission*)

Productivity variation is undoubtedly related to a number of factors. Innate genetic differences among species accounts for part of the variability. For example, in one study that used the same techniques of measurement (Kvet and Husak, 1978), *Typha angustifolia* production was determined to be about double that of *T. latifolia. T. angustifolia* however, is typically found in deeper water than those of the habitats of the other species, and so environmental factors were not identical for purposes of comparison. Regarding the environmental factors affecting production, Gorham (1974) established the close positive relationship between above-ground biomass and summer temperatures (Fig. 11–9).

The dynamics of underground growth are much less studied than those of above-ground growth (Bradbury and Grace, 1983; Hogg and Wein, 1987). In freshwater tidal wetlands (Chap. 9), relative root growth was shown to be related to the plant's life history. Annuals generally use small amounts of photosynthate to support root growth, whereas species with perennial roots and rhizomes often have root:shoot ratios well in excess of one. This relationship appears to hold true for inland marshes also. Perennial species in Iowa marshes (van der Valk and Davis, 1978a) had more below-ground than above-ground biomass, as did emergent macrophytes in Czechoslovak fish ponds (Fig. 11–10). It is interesting

Table 11–6. Comparison of the Root:Shoot Biomass Ratio, and Root Production to Biomass Ratio of the Principle Monospecific Reedswamp Communities of a Czechoslovakian Littoral Fish Pond

Species	*Root:Shoot Biomass Ratio*	*Root Production* / *Shoot Biomass*
Acorus calamus	1.8	0.5-0.9
Glyceria maxima	1.3-5.5	0.4
Phragmites communis	0.9-2.0	0.5-1.0
Schoenoplectus lacustris	2.3-3.9	1.0-1.2
Sparganium erectum	0.5-1.0	0.3
Typha angustifolia	0.9-1.2	0.8-0.9
Typha latifolia	0.4-0.6	0.4

Source: Based on data from Kvet and Husak (1978)

to note, however, that whereas biomass root:shoot ratios are usually greater than one, root-production-to-shoot biomass ratios are always less than one (Table 11–6). Since above-ground production is approximated by above-ground biomass, this latter ratio is an index of the allocation of resources by the plant, and it indicates that less than one-half of the photosynthate is translocated to the roots. The coexistence of large root biomass and relatively small root production suggests that the root system is generally longer lived (that is, it renews itself more slowly) than the shoot.

The emergent monocotyledons *Phragmites* and *Typha* have high photosynthetic efficiency (Fig. 11–11). For *Typha* it is the highest early in the growing season, gradually decreasing as the season progresses. *Phragmites*, in contrast, has a fairly constant efficiency rate throughout most of the growing season. The efficiencies of conversion of from 4 percent to 7 percent of photosynthetically active radiation is comparable to those calculated for intensively cultivated crops such as sugar beets, sugarcane, or corn.

Decomposition

With some notable exceptions, herbivory is considered fairly minor in inland marshes where most of the organic production decomposes before entering the detrital food chain. The decomposition process is much the same for all wetlands. Variations stem from the quality and resistance of the decomposing plant material, the temperature, the availability of inorganic nutrients to the microbial decomposers, and the flooding regime of the marsh. A conceptual diagram of decomposition in freshwater marshes that shows the complexity and interaction

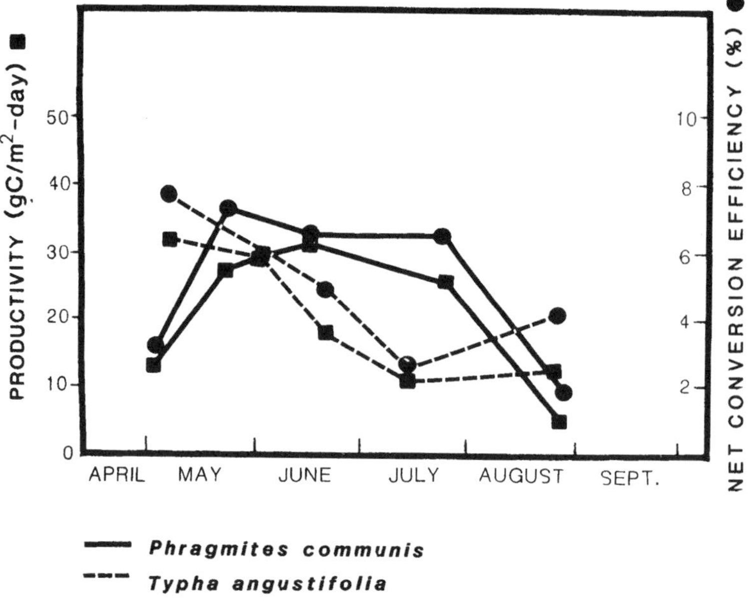

Figure 11–11. Productivity and net conversion efficiency of photosynthetically active radiation in stands of *Phragmites communis* and *Typha angustifolia.* (*After Dykyjova and Kvet, 1978*)

of the products of decomposition (Fig. 11–12) illustrates that the action of microbial organisms is not undertaken simply to incorporate plant organic matter into microbial cells. In addition, in the process, organic material is dissolved, nutrients are released to the substrate, other nutrients are absorbed by the microflora or adsorbed to fine organic particles, respiration oxidizes and releases organic carbon as carbon dioxide, ingested organic materials are released as repackaged fecal material, and dissolved organic matter may aggregate and flocculate into fine particles. The dynamics of these reactions in freshwater marshes are as poorly understood as are those in other wetlands. It is probably true, as it is in other detrital ecosystems, that the microbial decomposers are preyed on heavily by microscopic meiobenthic organisms, chiefly nematodes; the meiobenthic organisms are, in turn, a food source for larger macrobenthic organisms. Thus several links in the food chain may precede those that provide the commonly visible birds and other carnivores with their dinners.

Consumers play a significant role in detrital cycles. Most litter decomposition studies in freshwater marshes were done with senesced plant material during the winter, and low ($k = 0.002–0.007\ day^{-1}$) rates were generally measured (Table 11–7). But in a comparison of the decay of fresh biomass and senesced wetland plant leaves, Nelson et al. (1990a, b) found that samples of freshly harvested

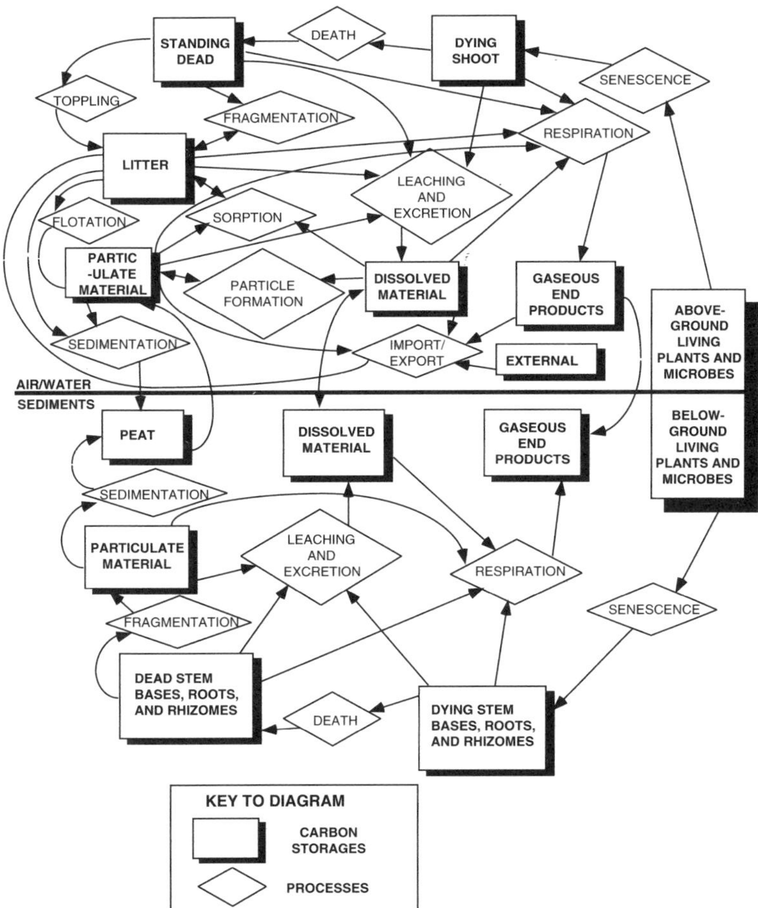

Figure 11–12. Conceptual model of decomposition in freshwater marshes. Boxes indicate storages of organic carbon; diamonds indicate processes. (*After Gallagher, 1978; copyright © 1978 by Academic Press, redrawn with permission*)

wetland plant material (*Typha glauca*) decomposed more than twice as fast (k = 0.024 day^{-1}) as naturally senesced material did (k = 0.011 day^{-1}). This comparison illustrates the more rapid decomposition that results when animals such as muskrats harvest live plant material. Muskrats also may play a positive role in the energy flow of a marsh as they harvest aquatic plants and standing detritus for their muskrat mounds. Wainscott et al. (1990) found in culturing experiments that litter from muskrat mounds supports substantially higher densities of microbes than litter from the marsh floor does. They suggested that these muskrat mounds may act like "compost piles" as they accelerate the decomposition and microbial growth than have become familiar to organic gardeners.

Table 11–7. Decomposition Rates of *Typha* spp. in Various Freshwater Marshes

Species	*Start Date*	*Duration,* day	*Decay Rate* (k), day^{-1}	*Reference*
Typha latifolia	November	180	0.0035	Boyd (1970)
T. angustifolia	October	626	0.0019	Mason and Bryant (1975)
T. glauca	Late autumn	525	0.0012	Davis and van der Valk (1978a)
T. glauca	Late autumn	330	0.0020	Davis and van der Valk (1978b)
T. latifolia	Late autumn	300	0.0104	Webster and Simmons (1978)
T. latifolia	September	348	0.0019	Puriveth (1980)
T. angustifolia	Early autumn	154	0.0047	Hill (1985)
T. glauca (green)	April	138	0.0240	Nelson et al. (1990a)
T. glauca (senesced)	April	138	0.0110	ibid.

Source: From Nelson et al., 1990a

Table 11–8. Partial Organic Budget for a Fish Pond Littoral Marsh in Czechoslovakia

	kcal m^{-2} yr^{-1}
Producers	
Gross production	4,500-27,000
Respiration	2,900-11,000
Net production	1,600-16,000
Consumers	
Decomposers (bacteria and fungi)	1760[a]
Decomposers (small invertebrates)	300[a]
Invertebrate macrofauna	—
Mammal consumption	232
Bird consumption	20[b]
Mammal production	5
Bird production	1

[a]Cragg, 1961
[b]Assuming production = 5% of consumption
Source: Based on data from Dykjova and Kvet (1978)

Consumption

Even though food chains begin in the detrital material of freshwater marshes, they develop into detailed webs that are still poorly understood. Benthic communities that feed on detritus form the basis of food for fish and waterfowl in the marshes. DeRoia and Bookhout (1989) found that chironomids made up 89 percent of the diet of blue-winged teal (*Anas discors*) and 99 percent of the diet of green-winged teal (*Anas crecca*) in a Great Lakes marsh. The direct grazing of freshwater marsh vegetation has occasionally been reported in the literature. Crayfish are often important consumers of macrophytes, particularly of submersed aquatic plants, in freshwater marshes. The red swamp crayfish (*Procambarus clarkii*) was shown to have effectively grazed on *Potamogeton pectinatus* in a freshwater marsh in California, where the plant decreased from 70 percent to 0 percent cover of the marsh while the crayfish population almost doubled (Feminella and Resh, 1989). The direct consumption of marsh plants by geese, muskrats, and other herbivores is common in some parts of the world. "Eatouts" causing large expanses of open water are the result of the inability of plants to survive after being clipped below the water surface by animals (Middleton, 1990).

Organic Export

Very little information is available about the export of organic energy from freshwater marshes. If other ecosystems are any guide, export is heavily governed by the flow of water across the marsh. Thus pothole marshes, which have small outflows, must export very little. Some dissolved organic materials may flow out in groundwater, but otherwise the primary loss is through living organisms that feed in the marshes and then move away. In contrast, lakeside and riverine marshes may export considerable organic material when periodically flushed.

Energy Flow

It is not possible to calculate a tight organic energy budget for inland marshes from the information that is now available. Several components, however, have been estimated for the littoral fish pond system in Czechoslovakia (Table 11–8). This allows at least some perspective to be developed about the major fluxes of energy. Net organic energy fixed by emergent plants ranges from 1,600 to 16,000 kcal m^{-2} yr^{-1}. (For the purpose of comparison, for a site producing 2,500 g m^{-2} yr^{-1} of biomass per year, the energetic equivalent is about 10,000 kcal m^{-2} yr^{-1}.) Most of this net production is lost through consumer respiration. An early study by Cragg (1961) suggested that microbial respiration in peat was about 1,760 kcal m^{-2} yr^{-1} in a *Juncus* moor. No other estimates appear to be available.

Invertebrates, especially the microinvertebrates, play an important role in sediments in the flow of organic energy through the ecosystem. Without doubt, they are important in fragmenting the litter so that it can be more readily attacked by bacteria and fungi (Fig. 11–12). These benthic organisms are also important intermediates in the transfer of energy to higher trophic levels. In Czechoslovak fish ponds, Dvorak (1978) calculated the average benthic macrofaunal biomass, composed mostly of mollusks and oligochaetes, to be 4,266 g/m^2. They were selectively fed on by fish. In quite a different marsh, a *Juncus* moor, Cragg (1961) estimated the respiration rate of the small soil invertebrates to be about 300 kcal m^{-2} yr^{-1}.

Pelikan (1978) calculated the energy flow through the mammals of a reedswamp ecosystem. The total energy consumption was 235 kcal m^{-2} yr^{-1}, mostly by herbivores that ingested 220 kcal m^{-2} yr^{-1}. Insectivores ingested 10, and carnivores ingested 1 kcal m^{-2} yr^{-1}. This amounted to about 0.55 percent of above-ground and 0.18 percent of below-ground plant production. Most of the assimilated energy was respired. The total mammal production was only 4.84 kcal m^{-2} yr^{-1}, which amounts to less than 1 g m^{-2} yr^{-1}. It seems evident that the indirect control of plant production by the muskrat is much more significant than the direct flow of energy through this group.

The same reedswamp ecosystem supported an estimated 83 nesting pairs of gulls (*Larus ridibundus*) and 20 pairs of other birds per hectare—about 13 passerines, 3 grebes, and the rest rails and ducks (Hudec and Stastny, 1978). The mean biomass was 44.4 kg/ha (fresh weight) for the gulls and 11.2 kg/ha for the remaining species. The production of eggs and young amounted to about 6,088 kcal/ha for gulls and 3,096 kcal/ha for the remaining species. If this production was considered to be 5 percent of the total consumption, the total annual flow of organic energy through birds would be about 20 $kcal/m^2$, or roughly 10 percent of the mammal contribution.

Collectively these estimates of energy flow through invertebrates, mammals, and birds account for less that 10 percent of net primary production. Most of the rest of the energy used for organic production must be dissipated by microbial respiration, but some organic production is stored as peat, reduced to methane, or exported to adjacent waters. Export has been particularly difficult to measure because so many of these marshes intercept the water table and may lose organic materials through groundwater flows.

Nutrient Budgets

A number of attempts have been made to calculate nutrient budgets for wetlands, but the results form no consistent picture because freshwater wetlands vary so widely in so many different ways. In a number of studies, freshwater

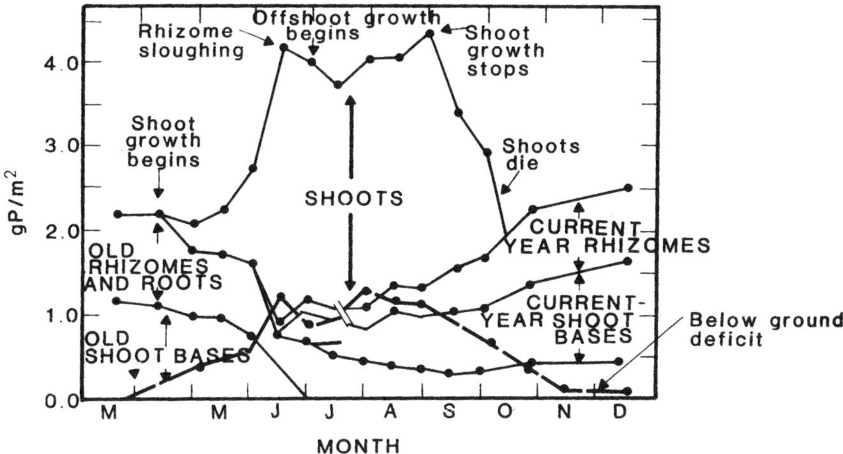

Figure 11–13. Seasonal stocks and below-ground deficit of phosphorus in *Typha latifolia* plant parts in a freshwater marsh on Lake Mendota, Wisconsin. (*After Prentki et al., 1978*)

wetlands have been evaluated as nutrient traps. These studies often emphasize input-output budgets. Because this subject has been discussed in some detail in Chapter 5, we emphasize in this chapter the role of vegetation in nutrient cycling.

Vegetation traps nutrients in biomass, but the storage of these nutrients is seasonally partitioned in above-ground and below-ground stocks. For example, the seasonal dynamics of phosphorus in *Typha latifolia* is described in Figure 11–13 (Prentki et al., 1978). Phosphorus stocks in the roots and rhizomes of *Typha* are mobilized into the shoots early in the growing season. Total stocks increase to more than 4 g P/m^2 during the summer. In the fall some phosphorus in the shoot is remobilized into the below-ground organs before the shoots die, but most of it is lost by leaching and in the litter. The calculated below-ground deficit is an indication of the magnitude of the phosphorus demand by the plant. It cannot be met by shifting internal supplies; it is largest during the period of active growth in the summer.

Shaver and Melillo (1984) illustrated the importance of nutrient availability for the efficiency of nutrient uptake of three freshwater marsh plants, *Carex lacustris, Calamagrostis canadensis,* and *Typha latifolia.* The efficiency of nutrient uptake, defined as the increase in plant nitrogen or phosphorus mass divided by the nitrogen or phosphorus mass available, decreased with increasing nutrient availability, suggesting that plant uptake becomes less important with higher nutrient inputs even though the nutrient content of the plants may increase. The higher concentrations of nutrients in the plants also cause higher concentrations in the litter, which, in turn, can stimulate microbial release. Thus as more nutrients become available to freshwater marshes, the marsh becomes more "leaky." Nutrients are lost from the system, and nutrient turnover in the vegetation increases.

In another set of measurements of nutrient conditions in below-ground organs of freshwater marsh macrophytes, C. S. Smith et al. (1988) measured the seasonal changes in several elements of above-ground and below-ground parts of *Typha latifolia* from a Wisconsin marsh. The ratio of the elements varied generally between 1:1 to 2:1 in above-ground: below-ground (A:B) stores, compared with a 2.2:1 ratio for A:B biomass because nutrients are more concentrated in root than in shoot tissues. They found that below-ground biomass concentrations of nitrogen, phosphorus, and potassium decreased significantly during the spring, supposedly because of shoot growth, whereas calcium, magnesium, manganese, sodium, and strontium showed little decrease in spring, suggesting that they are not limiting mineral reserves for spring growth.

Studies like those described above lead to several generalizations:

1. *The size of the plant stock of nutrients in freshwater marshes varies widely.* More nitrogen and phosphorus is retained in above-ground plant parts in mineral substrate wetlands (freshwater marshes) than in peat wetlands (Whigham and Bayley, 1979). The above-ground stock of nitrogen ranges from as low as 3 g/m^2 to as high as 29 g/m^2.
2. *The biologically inactivated stock of nutrients in plants is only a temporary storage that is released to flooding waters when the plant shoots die in the fall.* Where this occurs, the marsh may retain nutrients during the summer and release them in the winter.
3. *Nutrients retained in biomass usually account for only a small portion of nutrients that flow into marshes, and that percentage decreases with increased nutrient input.* For example, about 20 percent of the total nutrients that flowed into a Wisconsin marsh were detained in the marsh (Sloey et al., 1978).
4. *Marsh vegetation often acts as a nutrient pump, taking up nutrients from the soil, translocating them to the shoots, and releasing them on the marsh surface when the plant dies.* The effect of this pumping mechanism may be to mobilize nutrients that had been sequestered in the soil. This is clearly demonstrated in the nitrogen budget developed by Klopatek (1978) for a river bulrush (*Scirpus fluviatilis*) stand (Fig. 11–14). In this study 20.7 g N/m^2 was taken up annually from the sediment by the plants. More than 15 g N/m^2 was translocated to the shoots and lost through leaching and shoot senescence.
5. *Both nitrogen and phosphorus limitations are generally factors in freshwater marsh productivity, and their uptake rates are not independent of each other.* Despite a wide range of N:P ratios supplied in their experiments, Shaver and Melillo (1984) illustrated that their freshwater marsh plants accumulated nutrients at a much narrower range and that all plants had a similar "optimum" N/P ratio of approximately 8:1 by mass. Nevertheless, higher nitrogen concentrations in available solutions did increase the N:P ratios of the plants,

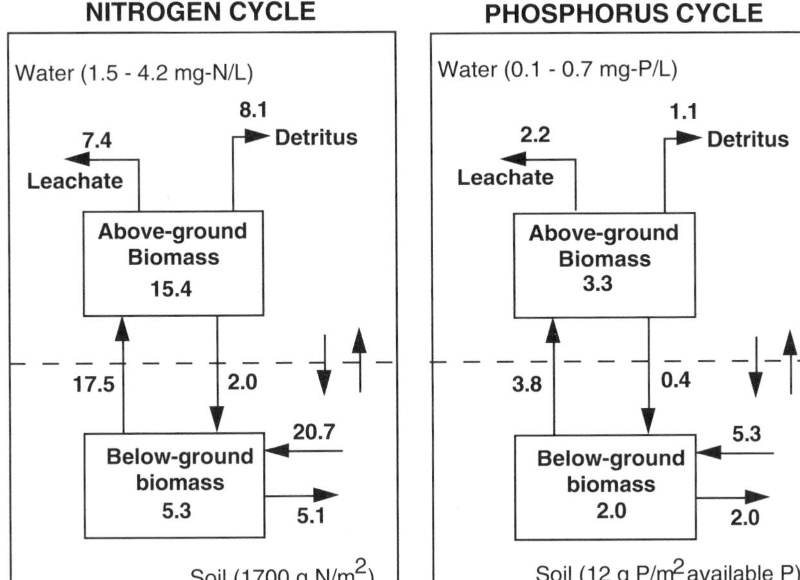

Figure 11–14. Flow of nitrogen and phosphorus through a river bulrush (*Scirpus fluviatilis*) stand in Wisconsin. Flows are in g m^{-2} yr^{-1} and storages are in g/m^2 of nitrogen and phosphorus respectively. (*From Klopatek, 1978; copyright © 1978 by Academic Press, redrawn with permission*)

and low N:P ratios in available solutions decreased the N:P ratios of the plants. Thus nitrogen and phosphorus uptake are not independent of each other. For a Manitoba *Scirpus acutus* marsh, Neill (1990c) found that neither nitrogen nor phosphorus increased net productivity when applied alone but that above-ground biomass nearly doubled when nitrogen and phosphorus were applied together. On the other hand, similar studies of a nearby marsh showed nitrogen limitation, indicating that differences in limiting factors are possible even in the same region (Neill, 1990c). In conditions in which water levels are more stable such as Louisiana's Gulf Coast, the addition of nitrogen fertilizer at a rate of 10 g NH_4^+-N/m^2 caused approximately a 100 percent increase in the growth of *Sagittaria lancifolia* (Delaune and Lindau, 1990).

6. *The role that nutrients play in plant productivity and species composition in freshwater marshes is influenced by hydrologic conditions.* For example, fluctuating water levels may affect nutrient limitations; when nitrogen and phosphorus fertilizers were added to prairie lacustrine marshes over two years, nitrogen additions stimulated the productivity of emergent macrophytes in the first year by increasing the *Scolochloa festucacea* biomass more

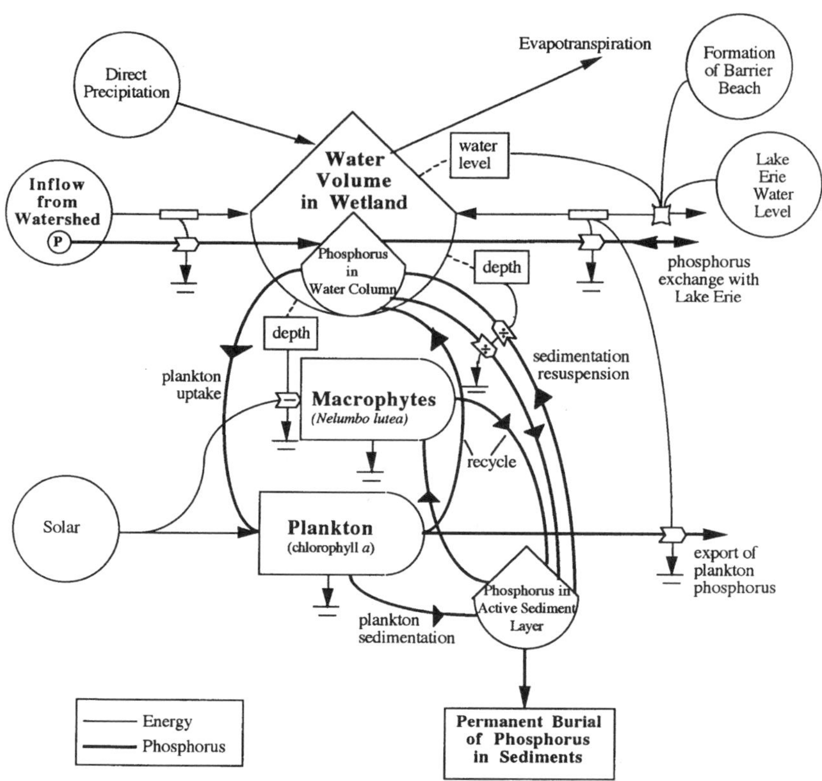

Figure 11–15. Diagram of simulation model for Lake Erie coastal freshwater wetland. Full description of this model is given in Mitsch and Reeder (1991). Simulations illustrated processes involved in the retention of phosphorus by this coastal wetland. (*Mitsch, 1992b; copyright © 1992 by International Association of Great Lakes Research, reprinted with permission*)

when the soil was flooded and the *Typha latifolia* biomass more in intermediate depths (0–20 cm) than in deep water (20–40 cm) or dry conditions (Neill, 1990a). This response to added nitrogen indicated nitrogen limitation in these marshes. In the second year of the experiment, nitrogen was not as limiting as it was in the first year, and some phosphorus limitation was noted. The species composition in these experiments changed when nitrogen was added, causing a decrease in the *Scolochloa festucacea* biomass and an increase in moist soil annuals. The phosphorus additions, however, had little effect on species composition (Neill, 1990b).

7. *The ability of freshwater marsh soils to retain nutrients varies widely.* Organic soils have large cation-exchange capacities. When these exchange

sites are saturated with hydrogen ions, as in nutrient-poor bogs, the possibility for retaining cations by displacing hydrogen ions may be large, whereas soils in which the exchange sites are already saturated with nutrients might have a limited additional capacity. In their Louisiana study of the outcome of ^{15}N labeled nitrogen, Delaune and Lindau (1990) found that very little nitrogen was lost from the soil organic pool because the labeled nitrogen was immobilized in the soil organic matter.

8. *In general, precipitation and dry fall account for less than 10 percent of plant nutrient demands in productive freshwater marshes.* Similarly, groundwater flows are usually small sources of nutrients. Surface inflow is usually a major source; it varies widely, sometimes providing many times the needs of the vegetation. Considering all of these variables, it is not surprising that each marsh seems to have its own unique budget.

ECOSYSTEM MODELS

Oliver and Legovic (1988) used a simulation model to evaluate the importance of nutrient enrichment in a freshwater marsh near a wading bird rookery in the Okefenokee Swamp in Georgia. The trophic level model illustrated the influence of the avian fauna on several aspects of the ecosystem, including detritus, macrophytes, and aquatic fauna. Benthic detritus increased 8.9 times, macrophytes 4.5 times, and fish 1.4 times the background levels as a result of the simulated influx of 8,000 wading birds into the marsh.

A simulation model was developed for a freshwater wetland along the coast of Lake Erie to investigate the roles of primary productivity, sedimentation, resuspension, and hydrology in the phosphorus-retention capability of that wetland (Mitsch and Reeder, 1991; Fig. 11–15). Open water plankton and aquatic beds dominated by *Nelumbo lutea* were the major producers in this wetland, and a barrier beach restricted outflow to certain seasons. In general, allocthonous and autothonous productivity resulted in a gross sedimentation of 13.3 mg-P m^{-2} day^{-1} in the model but a resuspension of 10.4 mg-P m^{-2} day^{-1}. The model was calibrated using data from a drought year (1988), for which a 17 percent retention of phosphorus was calculated. Predicted retention of phosphorus from 27 to 52 percent resulted from simulations of normal to high inflow (typical of normal and wet years), suggesting that the highest retention occurs during periods of high Lake Erie levels (Mitsch and Reeder, 1991). This model illustrated the importance of wetlands as buffers between agricultural watersheds and downstream aquatic ecosystems, but it suggests the importance of hydrologic conditions, both upstream and, in this case, downstream, for controlling the efficiency of this buffering capacity.

NORTHERN PEATLAND

Northern Peatlands

12

Bogs and fens are peatlands distributed primarily in the cool boreal zones of the world where excess moisture is abundant. Although many types of peatlands are identifiable, classification according to hydrologic and chemical conditions usually defines three types: (1) minerotrophic (true fens), (2) ombrotrophic (raised bogs), and (3) transition (poor fens). Ombrotrophic bogs are isolated from mineral-bearing groundwater and thus display lower pH, lower nutrients, lower minerals, and more dominance by mosses than minerotrophic fens do. Bogs can be formed in several ways, originating from either aquatic systems, as in flow-through succession or quaking bogs, or from terrestrial systems, as with blanket bogs. Bog acidity is caused by cation exchange with mosses, oxidation of sulfur compounds, and organic acids. In bogs the low nutrients and low pH lead to low primary productivity, slow decomposition, adaptive nutrient-cycling pathways, and peat accumulation. Several energy and nutrient budgets have been developed for peatlands, with the 1942 energy budget by Lindeman one of the most well-known. Relatively few mathematical models have been used for simulating the dynamics of northern peatlands.

As described in the introduction to Chapter 11, we have divided inland wetlands, for purposes of this book, into marshes and bogs. Bogs and to some extent fens belong to a major class of wetlands called *peatlands, moors,* or *mires* that occur as freshwater wetlands throughout much of the boreal zone of the world (Walter, 1973). *Bogs*, called *muskegs* in Canada, are acid peat deposits that generally contain a high water table, have no significant inflow or outflow streams, and support *acidophilic* (acid-loving) vegetation, particularly mosses. *Fens*, on the other hand, are open wetland systems that generally receive some drainage from surrounding mineral soils and are often covered by grasses, sedges, or reeds. They are in many respects transitional between marshes and bogs (see Table 3–1). As a successional stage in the development of bogs, fens are important and will be considered in that context here.

Bogs and fens have been studied and described on a worldwide basis more extensively than any other type of freshwater wetland. The older European peatland literature is particularly rich, establishing most of our basic understanding of these wetlands. Hundreds of North American studies, however, have since appeared in print. Peatlands have been studied because of their vast area in temperate climates, the economic importance of peat as a fuel and soil conditioner, and their unique biota and successional patterns. Bogs have intrigued and mystified many cultures for centuries because of such discoveries as the Iron Age "bog people" of Scandinavia, who were preserved intact for up to 2,000 years in the nondecomposing peat (Glob, 1969).

Because bogs and other peatlands are so ubiquitous in northern Europe and North America, many definitions and words, some unfortunate, that now describe wetlands in general originated from bog terminology; there is also considerable confusion in the use of terms such as *bog, fen, swamp, moor, muskeg, heath, mire, marsh, highmoor, lowmoor,* and *peatland* to describe these ecosystems (Heinselman, 1963; Stanek and Worley, 1983). The use of the words *peatlands* in general and *bogs* and *fens* in particular will be limited in this chapter to deep peat deposits, mostly of the cold, northern, forested regions of North America and Eurasia. Peat deposits also occur in warm temperate, subtropical, or tropical regions and we refer briefly to a major example of these, specifically the pocosins of the southeastern coastal plain.

As described by Curtis (1959) in Larsen (1982): "The bog . . . is a common feature of the glaciated landscapes of the entire northern hemisphere and has a remarkably uniform structure and composition throughout the circumboreal regions." Excellent references on bogs are Heinselman (1963; 1970), Moore and Bellamy (1974), Radforth and Brawner (1977), Clymo (1983), Gore (1983a), Moore (1984), C. W. Johnson (1985), Damman and French (1987), Glaser (1987), Crum (1988), and Verhoeven (1992). Godwin (1981) provides an interesting personal history of the study of bogs in the United Kingdom from a historical, archeological viewpoint.

GEOGRAPHICAL EXTENT

Bogs and fens are distributed in cold temperate climates of high humidity (generally resulting from maritime influences), mostly in the Northern Hemisphere, where precipitation exceeds evapotranspiration, leading to moisture accumulation.

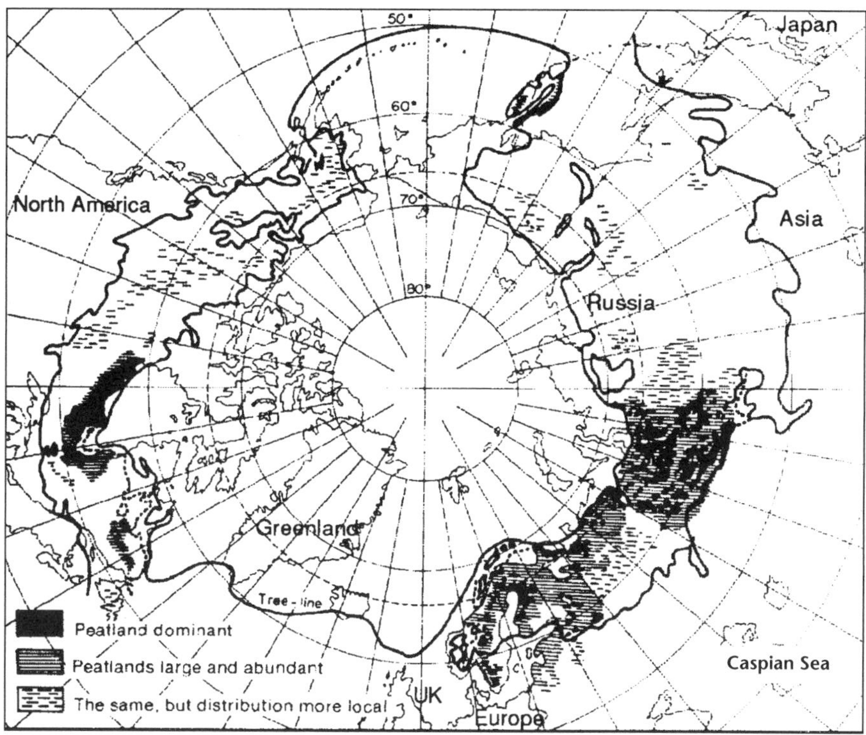

Figure 12–1. Peatland distribution in the northern hemisphere. The presence of peatlands is associated with the boreal lifezone and its subalpine equivalent in mountainous regions. (*After Terasmae, 1977*)

Extensive areas of bogs and fens occur in Scandinavia, eastern Europe, western Siberia, Alaska, Labrador, Canada, and the north-central United States (Fig. 12–1). The distribution of mires in North America is shown in Figure 12–2. Canada has approximately 110–130 million hectares of peatlands (Radforth, 1962; Zoltai, 1988) that give it the largest peat resources in the world. Their distribution approximates the boreal forest region delineated by the occurrence of black spruce. These peatlands dip into the United States in Minnesota and northern Michigan. Minnesota peatland distribution, depicted in Figure 12–3, was estimated by Soper (1919 in Glaser, 1987) to be about 2.7 million hectares. In the northeast United States, the zone of ombrogenous bogs dips into Maine. South of the Maine peatlands there tend to be small depressional systems that receive at least some nutrient inflow (Damman and French, 1987). In the United States, bogs usually develop in basins scoured out by the Pleistocene glaciers. One of the most studied bogs ecosystems in the United States is the Lake Agassiz region in northern Minnesota (Heinselman, 1963; 1970; 1975). West Coast bogs have been studied by Rigg (1925). Canadian bogs have been documented by Sjors (1961), Radforth, (1962), Reader and Stewart (1972), and Radforth and Brawner (1977).

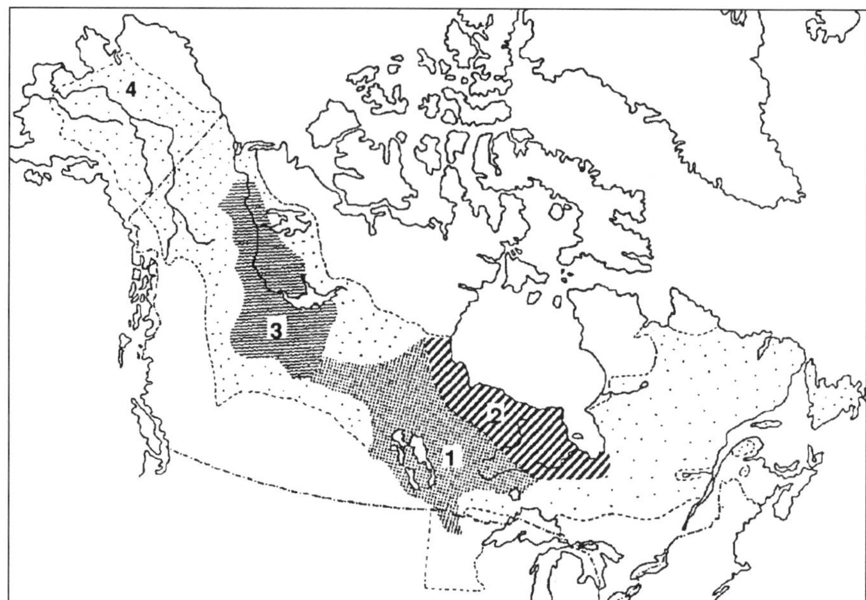

Figure 12–2. Major boreal peatlands of North America. Major peatland areas are: (1) Glacial Lake Agassiz region; (2) Hudson Bay lowlands; (3) Great Bear/Great Slave Lake region; and (4) the interior of Alaska. The lightly stippled area marks the boreal region of Alaska. (*From Glaser, 1987, as reported by Viereck and Little, 1972; Rowe, 1972; and Zoltai and Pollet, 1983*)

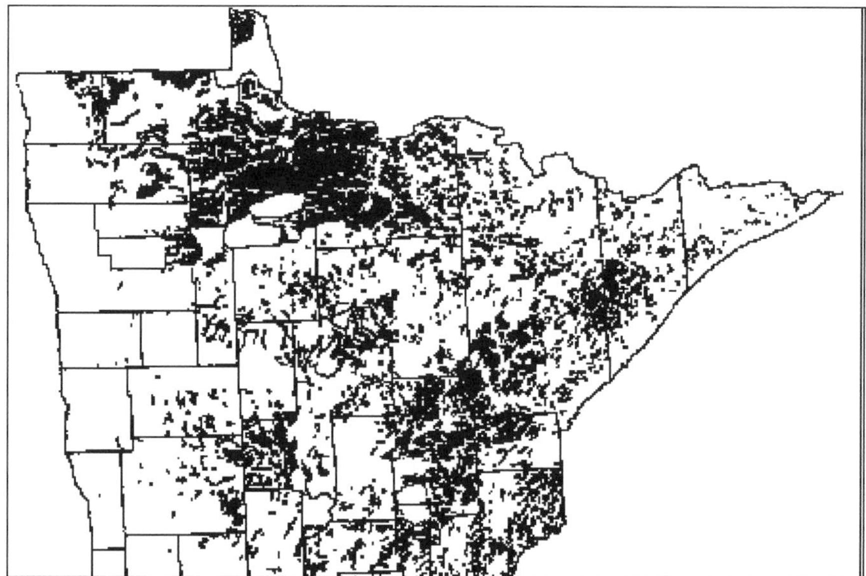

Figure 12–3. Peatlands distribution in northern MInnesota (shaded area). (*From Glaser, 1987, based on Minnesota Department of Natural Resources peatlands map, 1978*)

Although predominantly a northern phenomenon, peats can accumulate wherever drainage is impeded and anoxic conditions predominate regardless of temperature. Thus bogs are found as far south as northern Illinois and Indiana in the north-central United States and are also common in the unglaciated Appalachian Mountains in West Virginia (Wieder and Lang, 1983; Lynn and Karlin, 1985). The southern limit to bog species and hence to the bog wetland is thought to be determined by the intensity of solar radiation in the summer months when precipitation and humidity are otherwise adequate to support bogs farther south (Larsen, 1982, after Transeau, 1903). The Middle Atlantic coastal plain supports an expansive area of poorly drained peatlands, called *pocosins,* that are similar to more northern peatlands in that it is nutrient poor and dominated by evergreen and ericaceous woody vegetation (Sharitz and Gibbons, 1982). Pocosins once covered 1.2 million hectares, 70 percent of which were in North Carolina. Only about 370,000 hectares of the original 1.2 million hectares remained in 1980 (Richardson, 1983).

GEOMORPHOLOGY AND HYDROLOGY

Classification

Peatland classification schemes are intimately tied to hydrology, chemistry, and development (succession). Because the physical and biotic processes that form peatlands are complex and differ somewhat from region to region, many different classification systems have been proposed (e.g. Gore, 1983b; Moore, 1984; Wells and Zoltai, 1985). Moore (1984) described seven features on which classification schemes have been based: (1) floristics, (2) vegetation structure, (3) geomorphology (succession or development), (4) hydrology, (5) chemistry, (6) stratigraphy, and (7) peat characteristics. The last is used primarily for economic exploitation purposes. The other six are closely interrelated, leading to classification schemes that combine several natural features. Because most classifications are related to bog development, a discussion of this subject is appropriate background.

Peatland Development

Figure 12–4 illustrates the major processes, both physical and biotic, in peatland development. Two primary processes are the water balance and peat accumulation. A positive water balance, precipitation plus water inflow greater than evapotranspiration plus runoff ($P + S_i + G_i > ET + S_o + G_o$) is essential for peatland development and survival. The seasonal distribution of rains is important because peatlands require a humid environment year-around. For example, in dry, cold environments precipitation falls as snow, and peatlands will not develop. A second requirement for peatland development is a surplus of peat produc-

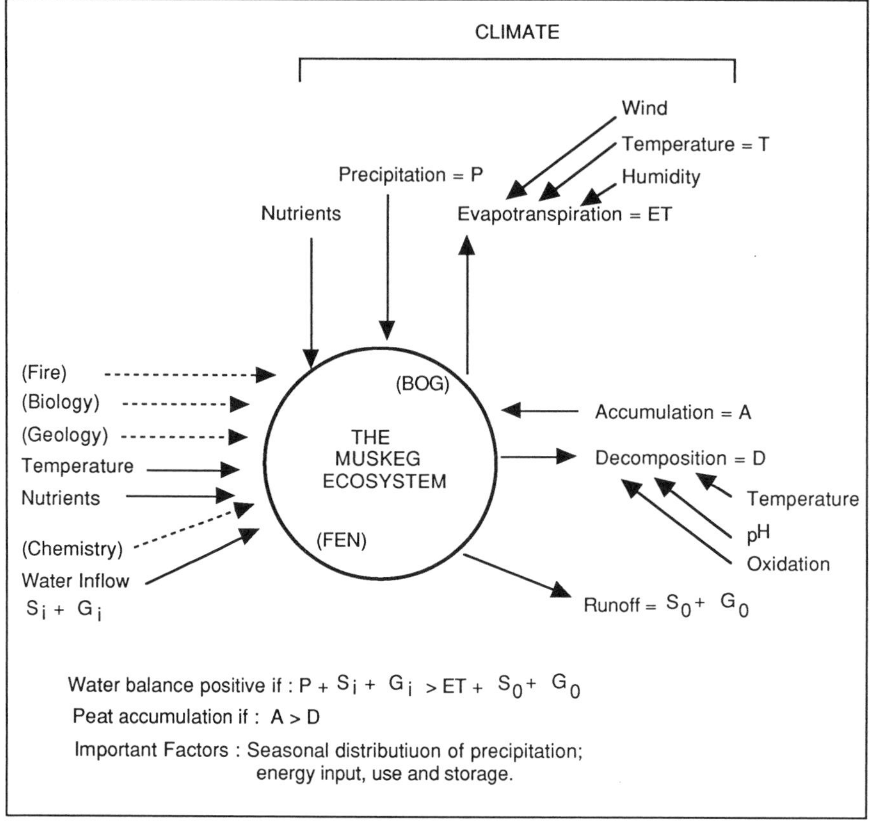

Figure 12–4. Diagram of hydrologic balance and peat accumulation of the Canadian muskeg ecosystem. (*After Terasmae, 1977*)

tion over decomposition. Although primary production is generally low in northern peatlands compared to other ecosystems, decomposition is even more depressed and so peat accumulates. This is a necessary condition for the development of ombrotrophic bogs (see description later in this section). The continued development of the ecosystem appears to be directly related to the amount of surplus water and peat (Terasmae, 1977). For example, in a cool, moist maritime climate, peatlands can develop over almost any substrate, even on hill slopes. In contrast, in warm climates where both evapotranspiration and decomposition are elevated, ombrotrophic peatlands seldom develop even when a precipitation surplus occurs.

Once formed, a bog is remarkably resistant to conditions that alter the water balance and peat accumulation. The perched water table, the water-holding capacity of the peat, and its low pH create a microclimate that is stable under fairly wide environmental fluctuation.

Given the conditions of water surplus and peat accumulation, bogs develop through variations of two processes: *terrestrialization* (the infilling of shallow lakes) or *paludification* (the blanketing of terrestrial ecosystems by overgrowth of bog vegetation). Three major bog-formation processes are commonly seen:

1. *Quaking bog succession.* This is the classical process of terrestrialization, as described in most introductory botany or limnology courses (Weber, 1908; Kratz and DeWitt, 1986). Bog development in some lake basins involves the filling in of the basin from the surface, creating a *quaking bog* (or *Schwingmoor* in German) (Fig. 12–5). Plant cover, only partially rooted in the basin bottom or floating like a raft, gradually develops from the edges toward the middle of the lake (Ruttner, 1963). A mat of reeds, sedges, grasses, and other herbaceous plants develops along the leading edge of a floating mat of peat that is soon consolidated and dominated by sphagnum and other bog flora. The mat has all of the characteristics of a raised bog. The older peat is often colonized by shrubs and then forest trees such as pine, tamarack, and spruce, which form uniform concentric rings around the advancing floating mat. These bogs develop only in small lakes that have little wave action; they receive their name from the quaking of the entire surface that can be caused by walking on the floating mat. After peat accumulates above the water table, isolating the sphagnum-dominated flora from their nutrient supply, the bog becomes increasingly nutrient poor (ombrotrophic). The development of a perched water table also isolates the bog from its groundwater and nutrient renewal. The result is a classic concentric, or excentric, raised ombrotrophic bog.
2. *Paludification.* A second pattern of bog evolution occurs when blanket bogs exceed basin boundaries and encroach on formerly dry land. This process of paludification can be brought about by climatic change, by geomorphological change, by beaver dams, by logging of forests, or by the natural advancement of a peatland (Moore and Bellamy, 1974). Often the lower layers of peat compress and become impermeable, causing a perched water table near the

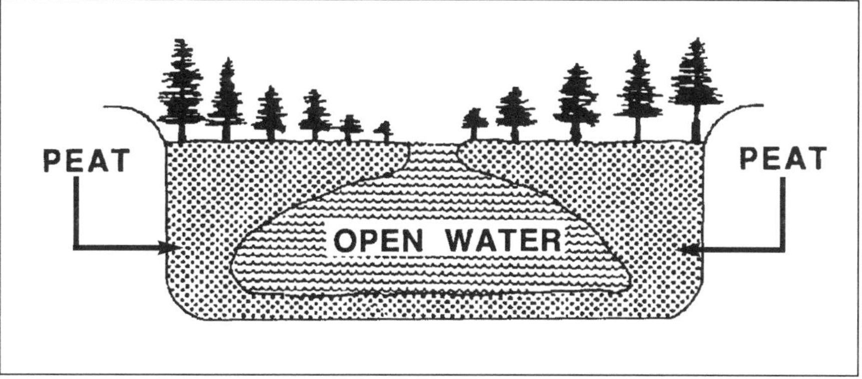

Figure 12–5. Typical profile of a quaking bog. (*After Moore and Bellamy, 1974*)

surface of what was formerly mineral soil. This causes wet and acid conditions that kill or stunt trees and allow only ombrotrophic bog species to exist (H. K. Smith, 1980). In certain regions the progression from forest to bog can take place in only a few generations of trees (Heilman, 1968).

3. *Flow-through succession.* Intermediate between terrestrialization and paludification is flow-through succession (also termed "*topogenous* development"), in which the development of peatland modifies the pattern of surface water flow. Moore and Bellamy (1974) described the development of a bog from a lake basin that originally had continuous inflow and outflow of surface and groundwater. The successional pattern is shown in Figure 12–6 in five stages. In the first stage, the inflow of sediments and the production of excess organic matter in the lake begin the buildup of material on the bottom of the lake. The growth of marsh vegetation continues the buildup of peat (stage 2), until the bottom rises above the water level and the flow of water is channelized around the peat. As the peat continues to build, the major inflow of water may be diverted (stage 3) and areas may develop that become inundated only during high rainfall (stage 4). In the final stage (stage 5), the bog remains above the groundwater level and becomes a true ombrotrophic bog. Figure 12–7 illustrates how a sedge peat bog can build up over 5,000 years to become a poor fen and a raised ombrotrophic bog.

The developmental processes described above lead to increasing isolation of bogs from surface and subsurface flows of both water and mineral nutrients. The degree of hydrologic isolation is the basis for a simple classification of mires (Moore and Bellamy, 1974):

1. *Minerotrophic peatlands.* These are true fens that receive water that has passed through mineral soil (Gorham, 1967). These peatlands generally have a high groundwater level and occupy a low point of relief in a basin. They are also referred to as *rheophilous* (flow-loving) in a classification by Kulczynski (1949) and *rheotrophic* by Moore and Bellamy (1974). Walter (1973) refers to these peatlands as *soligenous.*
2. *Ombrotrophic peatlands.* These are the true raised bogs that have developed peat layers higher than their surroundings and which receive nutrients and other minerals exclusively by precipitation. These are called *ombrophilous* (rain-loving) in Kulczynski's (1949) classification and *ombrogenous* by Walter (1973).
3. *Transition peatland.* These peatlands, often called poor fens, are intermediate between mineral-nourished (minerotrophic) and precipitation-dominated (ombrotrophic) peatlands. Another term used frequently for this class is *mesotrophic* (Moore and Bellamy, 1974).

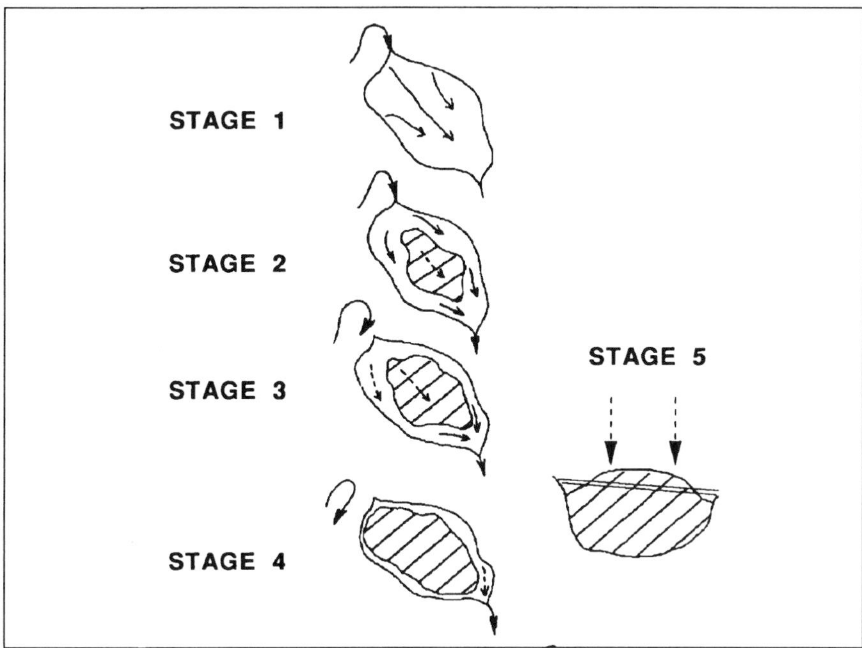

Figure 12–6. Flow-through succession of bog from lake basin. (*From Moore and Bellamy, 1974; copyright © 1974 by Springer-Verlag, reprinted with permission*)

Heinselman (1970) and Moore and Bellamy (1974), arguing for simplicity in the description of peatlands, stated that terms such as *soligenous* and *ombrogenous* actually refer to the hydrological and topographic origins of the peatlands and not to the mineral conditions of the inflowing water. These terms are not frequently used today.

Landscape Development

The developmental processes described above also determine large-scale patterns of landform development. In northern Europe, for example, bogs occur in relatively well-defined zones, as described by Walter (1973) and Moore and Bellamy (1974).

1. *Raised bogs.* These are peat deposits that fill entire basins, are raised above groundwater levels, and receive their major inputs of nutrients from precipitation (Fig. 12–8a, 12–8b). These bogs are primarily found in the boreal and northern deciduous biomes of northern Europe. Pine trees sometimes grow in these bogs in areas that have drier climates, although many are treeless. When a

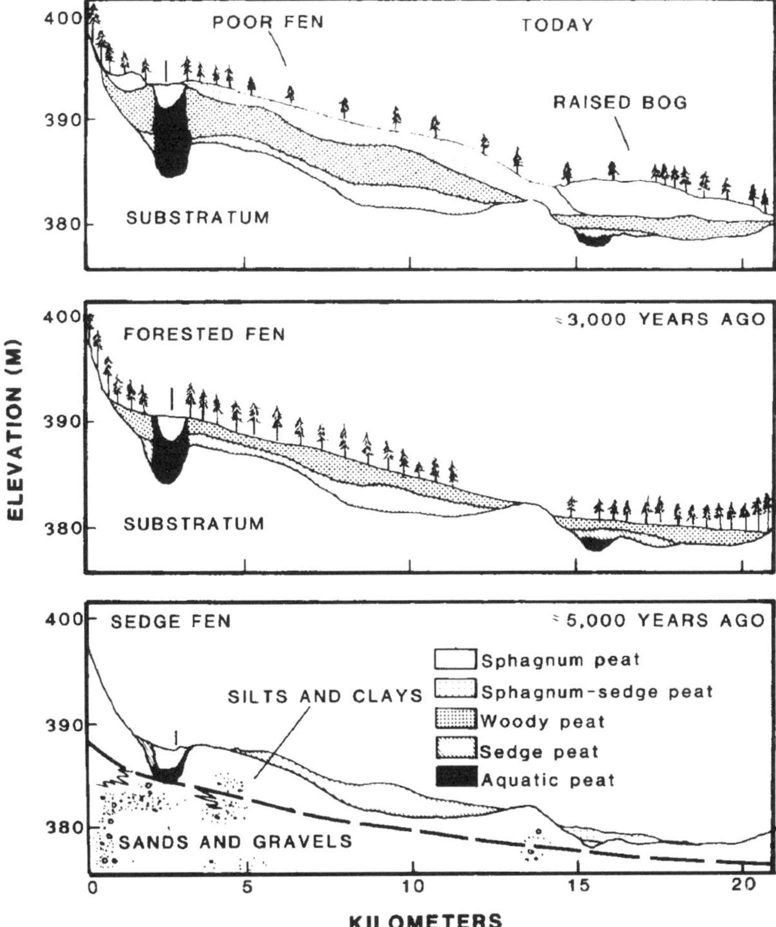

Figure 12–7. Peat accumulation over 5,000 years in a built-up peatland in northern Minnesota. Peat description based on plant remains. This vertical section runs southeast to northwest across the area depicted in Figure 12–15. (*From Boelter and Verry, 1977, based on data from Heinselman, 1963, 1970*)

concentric pattern of pools and peat communities forms around the most elevated part of the bog, the bog is called a *concentric domed* bog (Fig. 12–8a). Bogs that form from previously separate basins on sloping land and form elongated hummocks and pools aligned perpendicular to the slope are called *excentric raised* bogs (Fig. 12–8b). The former are found near the Baltic Sea, and the latter are found primarily in the North Karelian region of Finland.

2. *Aapa peatlands.* These wetlands (Fig. 12–8c), also called *string bogs* and *patterned fens*, are found primarily in Scandinavia and in the former Soviet

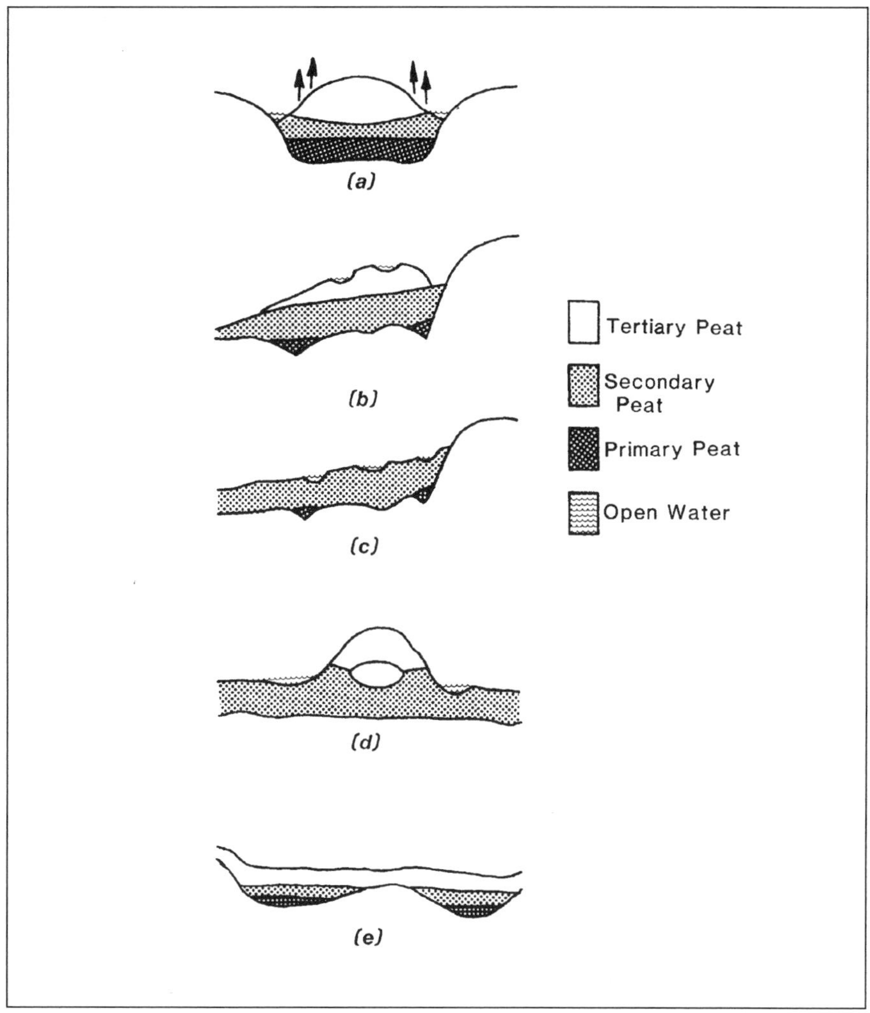

Figure 12–8. Diagrams of major peatland types, including *a.* raised bog (concentric), *b.* excentric raised bog, *c.* aapa peatland (string bogs and patterned fens), *d.* paalsa bog, and *e.* blanket bog. (*After Moore and Bellamy, 1974*)

Union north of the raised bog region. The dominant feature of these wetlands is the long, narrow alignment of the higher peat hummocks (strings) that form ridges perpendicular to the slope of the peatland and are separated by deep pools (*flarks* in Swedish). In appearance, they resemble a hillside of terraced rice fields (Walter, 1973).

3. *Paalsa bogs.* These bogs, found in the southern limit of the tundra biome, are large plateaus of peat (20 to 100 m in breadth and length and 3 m high) generally underlain by frozen peat and silt (Fig. 12–8d). The peat acts like an insulat-

ing blanket, actually keeping the ground ice from thawing and allowing the southernmost appearance of the discontinuous permafrost. In Canada as well, as much as 40 percent of the land area is influenced by cyrogenic factors. When peat overlies frozen sediments, it influences the pattern of the landscape. Many distinctive forms are similar to European aapa and palsa peatlands but are imbedded in a continuous peat-covered landscape (Gore, 1983a).

4. *Blanket bogs.* These wetlands (Fig. 12–8e) are common along the northwestern coast of Europe and throughout the British Isles. The favorable humid Atlantic climate allows the peat literally to "blanket" very large areas far from the site of the original peat accumulation. Peat in these areas can generally advance on slopes of up to 18 percent; extremes of 25 percent have been noted on slopes covered by blanket bogs in western Ireland (Moore and Bellamy, 1974). Thompson (1983) reported the distribution of patterned peatlands farther south in North America than had been reported previously.

Glaser (1983a, b; 1987), Siegel (1983), and Foster et al. (1983) described the formation of a patterned landscape in Northern Minnesota that is similar to aapa peatlands. It consists of bogs and water tracks. The latter have been further differentiated into *tree islands, strings,* and *flarks* (Fig. 12–9). Water flowing across peatlands is channeled by topographic features into definite zones, usually along the path of minerotrophic runoff. These water tracks are invaded by nutrient-demanding sedges such as *Carex lasiocarpa*. Their porous peat further channels the flow of water. Raised bogs develop wherever topographic obstructions divert the path of runoff. In these areas sphagnum grows and spreads outward until it is blocked by the main flows of the water tracks. The boundary between fen and bog is sharpened by the accumulation of dense, relatively impermeable sphagnum peat in the bog.

In the water track, a network of parallel, linear peat ridges (strings) and water-filled depressions (flarks) develops perpendicular to the direction of the water flow (Glaser et al., 1981; Glaser, 1983c). The pattern begins as a series of scattered pools on the down slope, wetter edge of the water track. These pools gradually coalesce into linear flarks. The peat accumulation in the adjacent strings and the increasing impermeability of decomposing peat in the flarks accentuate the pattern. Within the large water tracks, tree islands appear to be remnants of continuous swamp forests that were replaced by sedge lawns in the expanding water tracks (Glaser and Wheeler, 1980).

Stratigraphy

Accumulated deep peat deposits in northern peatlands have left a record of deposition thousands of years old, documenting the pathway of bog formation. This wealth of information has been exploited to study the history of climatic and associated geological and ecological change during the Quarternary epoch, that is, the geological period covering the waxing and waning of the great glaciations

Figure 12–9. Oblique aerial image of Cedarburg Bog in southeastern Wisonsin, showing a tree island in left center and a pattern of parallel linear peat ridges (strings) alternating with water-filled depressions (flarks) running diagonally across the lower half of the photograph from upper right to lower left. The strings stand out because they are vegetated with ericaceous shrubs and scrub trees over sphagnum moss, while the flarks are dominated by mosses and herbs. *(Photograph by Glen Guntenspergen)*

(Godwin, 1981; A. M. Davis, 1984; Solem, 1986). As an example, Figure 12–10 shows the terrestrialization of Cedar Bog Lake in south-central Minnesota, which has been reconstructed from the peat record (Glaser 1987, from data of Lindeman, 1941, and Cushing, 1963). About 8000 years B.P. (Fig. 12–10c) when the water surface was about 4 m below present levels, a sedge-dominated wetland spread over exposed mudflats on the sheltered southeast end of the lake basin. About 4000 years B.P., the warmer, drier period of the mid-Holocene ended and a cooler, moister period followed, during which most of the great peatlands of the Lake Agassiz plain formed. During that period, a cedar swamp grew across Cedar Bog lake (Fig. 12–10b) over the sedge marsh. Apparently that growth restricted the penetration of minerotrophic runoff from surrounding uplands, leading to the present ombrotrophic condition (Fig. 12–10a).

CHEMISTRY

Soil water chemistry is one of the most important factors in the development and structure of the bog ecosystem (Heinselman, 1970). Factors such as pH, mineral

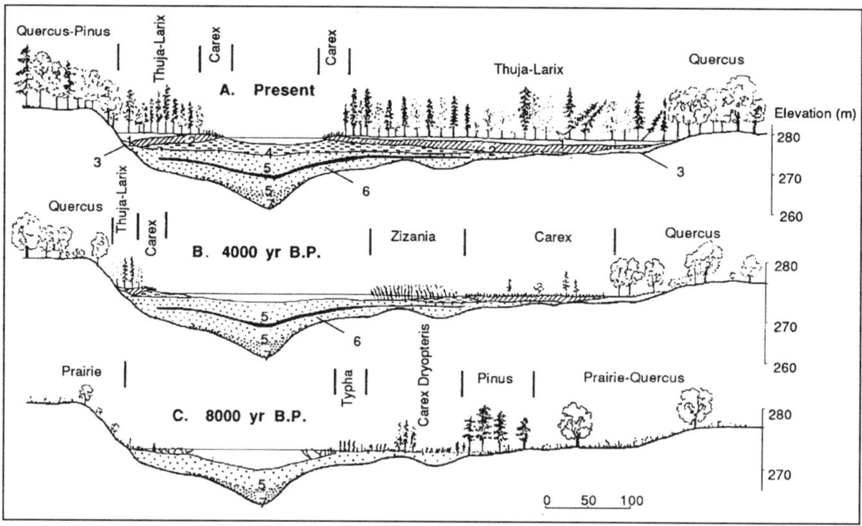

Figure 12–10. Terrestrialization at Cedar Bog Lake, south-central Minnesota. The three stages represent *a.* the present vegetation and lake level, *b.* the inferred vegetation and lake level 4,000 years ago during the mid-Holocene, and *c.* the inferred vegetation and stratigraphy during the early Holocene approximately 8,000 years ago. The Lake was 4 m lower during the early Holocene and the subsequent rise in the water table was accompanied by growth of the surrounding peatland. Sediment layers, defined by Lindeman (1941) are: (1) woody peat, (2) sedge peat, (3) sapropsammite, (4) coarse-detritus copropel, (5) marly copropel, (6) dark copropel, (7) sideritic copropel. Elevations are height above sea-level. (*From Glaser, 1987, after Cushing, 1963*)

concentration, available nutrients, and cation-exchange capacity influence the vegetation types and their productivity. Conversely, the plant communities influence the chemical properties of the soil water (Gorham, 1967). In few wetland types is this interdependence so apparent as in northern peatlands. Table 12-1 gives typical pH values and cation concentrations for different Northern Hemisphere peatlands, and Table 12–2 compares pH, mineral content, dissolved oxygen, and redox potential of bog waters with water in several other ecosystems in Wisconsin. The major features of peatland chemistry are discussed below.

Table 12–1. pH and Exchangeable Cations of Selected Peatlands

		Exchangeable Cations meq/100g				
Peatland	*pH*	*Ca^{++}*	*Mg^{++}*	*Na^{+}*	*K^{+}*	*Reference*
Western Europe						
(Mean values)						
Ombrotrophic bog	3.8	—	—	—	—	Moore &
Transition fens	4.1–4.8	—	—	—	—	Bellamy (1974)
Minerotrophic fens	5.6–7.5	—	—	—	—	
Alaska, U.S.A.						
Black spruce bog	3.6–4.0	15–16	5.3–6.4	—	3.8–4.2	Heilman (1968)
Minnesota, U.S.A.						
Alkaline peat (water soluble cations)						
Larix laricina stand	6.7–7.1	3.5–5.4	2.7–6.5	—	tr–3.1	Bares & Wali
Picea mariana stand	5.7–6.4	3.5–10.5	1.5–6	—	tr–3.2	(1979)
Sweden						
Marginal fen	5.2	19.5	—	0.43	0.49	Malmer & Sjors (1955)
Ireland						
Raised bog	4.7	2.0–4.0	—	—	0.64	Walsh & Barry (1958)
England						
Blanket bog	—	4.3	4.6	0.6	1.0	Gore & Allen (1956)
Michigan, U.S.A.						
Rich fen	5.1	52.4	7.7	1.2	0.45	Richardson et al. (1978)
Newfoundland, Can.						
Ombrotrophic bog	—	0.25 ± 0.2	0.46 ± 0.2	$3.89 \pm .1$	$0.10 \pm .02$	Damman
Minerotrophic bog	—	$2.8 \pm .36$	$1.11 \pm .06$	$5.76 \pm .26$	$0.22 \pm .03$	& French (1987)
England						
(English Lake District)						
Raised Bog	4.0–4.3	1.3–1.8	—	4.9–6.4	0.7–2.0	Gorham (1956)
Fen	6.1–7.6	2.3–17.5	—	3.6–6.0	0.17–1.75	

Table 12–2. Comparison of the Chemistry of Bog Water, Surface Stream, and Forest Soil Water in Wisconsin

	Depth to groundwater, cm	*pH*	*Hardness, mg $CaCO_3/l$*	*Specific conductivity, $\mu mho/cm$*	*Dissolved O_2, ppm*	*Redox Potential, mv*
Black spruce/tamarack bog	10	3.5	None	5.0	None	−364
Black spruce/tamarack bog	30.5	3.9	None	5.7	None	−335
Poor aspen stand	51	4.2	10	14.0	0.3	−305
White cedar/balsam fir bog	15	6.2	24	7.5	None	−42
White cedar/balsam fir bog	10	6.5	83	20.9	0.45	69
Lowland hardwoods	15	6.0	36	10.0	None	−81
Red pine/white pine upland	183	7.7	25	15.0	3.3	95
Creek draining cedar swamp	—	6.9	98	19.5	8.5	85
Creek from beaver dam	—	7.4	115	19.9	6.0	163
Lowland hardwoods (south)	61	8.2	107	28.0	0.4	161

Source: From Larsen (1982) after Wilde and Randall (1951)

Acidity and Exchangeable Cations

The pH of peatlands generally decreases as the organic content increases with the development from a minerotrophic fen to an ombrotrophic bog (Fig. 12–11). Fens are dominated by minerals from surrounding soils, whereas true bogs rely on a sparse supply of minerals from precipitation. Therefore, as a fen develops into a bog, the supply of metallic cations (Ca^{++}, Mg^{++}, Na^{+}, K^{+}) drops sharply (compare the concentration of cations in precipitation with concentrations in bogs and fens in Table 12–1). At the same time, as the organic content of the peat increases because of the slowing of the decomposition rate, the capacity of the soil to adsorb and exchange cations increases (see Chap. 5). These changes lead to the domination by hydrogen ions, and the pH falls sharply. Gorham (1967) found that bogs in the English Lake District had a pH range of 3.8–4.4 compared to noncalcareous fens, which had a pH range of 4.8–6.0. Pjavchenko (1982) assigned a pH range of 2.6–3.3 to oligotrophic (ombrotrophic) bogs and a range of 4.1–4.8 for mesotrophic (transitional) bogs; a pH greater than 4.8 defined a eutrophic (minerotrophic) fen.

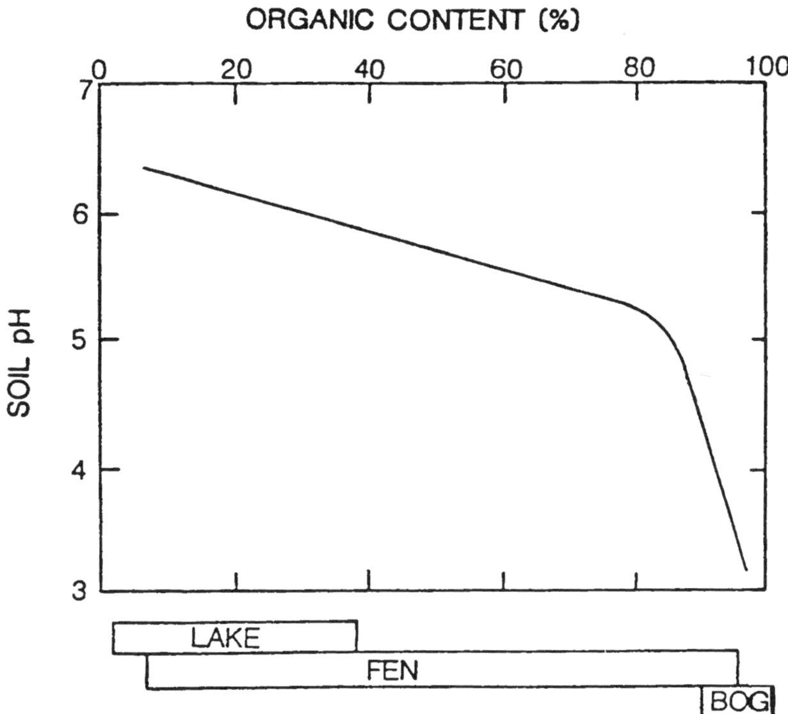

Figure 12–11. Soil pH as a function of organic content of peat soil. (*From Gorham, 1967*)

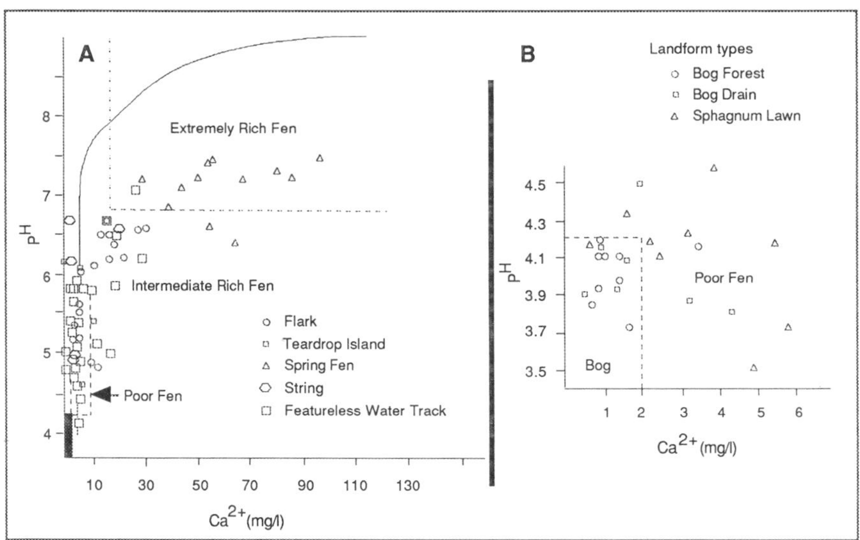

Figure 12–12. Water chemistry values for bog and fen samples in northern Minnesota. B. is an expansion of the bog section of A. (*From Glaser, 1987*)

Over the range of ombrotrophic bogs to rich fens, pH and the Ca^{++} ion concentration are reliably interrelated (Fig. 12–12). The rise in pH is related to the input of cations from a mineral source, either surface runoff or groundwater discharge. In addition, peatlands may receive a significant input of Ca^{++} from atmospheric deposition that is transported as dust from adjacent prairies (Glaser, 1987). The airborne source is apparently insufficient to account for calcium concentrations in rich fens, however. The theoretical curve relating pH and calcium (solid line in Fig. 12–12) pertains to solutions in equilibrium with the atmosphere with a Ca-balancing charge (Glaser, 1987). In rich fen samples, the pH is much lower than would have been predicted from the Ca concentration. These samples have apparently been oversaturated with calcite because of the complexing of Ca^{++} to organic compounds or the release of organic anions (Glaser, 1987). Although the data in Figure 12–12 indicate a continuum of pH and Ca values over the range of peatlands sampled, bogs are clearly different chemically from rich and intermediate fens. Glaser et al. (1990) emphasized the importance of soligenous and groundwater sources of minerals. As little as 10 percent of the water supply from groundwater may change the pH of a bog from 3.6 to 6.8, that is, from an ombrotrophic bog to a rich fen.

The causes of bog acidity are not entirely clear (Clymo and Hayward, 1982; Kilham, 1982; Sikora and Keeney, 1983; Gorham et al., 1984; Urban et al., 1985). Glaser (1987) cited five causes of low pH:

1. *Cation exchange by sphagnum.* Clymo (1963) and Clymo and Hayward (1982) concluded that this was the most important mechanism. Figure 12–13 shows a direct relationship between pH and exchangeable hydrogen on peat, presumably the result of the metabolic activity of the plants. Note that sphagnum peats have a higher exchangeable hydrogen and consequently a lower pH than sedge-dominated peats.
2. *Oxidation of sulfur compounds to sulfuric acid.* Organic sulfur reserves in peat, however, may be oxidized to acidic compounds.
3. *Acid atmospheric deposition.* Sulfur deposition is a significant source of acidity depending on the oxidation state of the sulfur and on the location of the bog. Acid sources in precipitation and dry deposition are usually small except close to sources of atmospheric pollution (Gorham, 1967; Brackke, 1976).
4. *Biological uptake of nutrient cations by plants.* Ions in the peat water are concentrated by evaporation and are differentially absorbed by the mosses. This affects acidity, for example, by the uptake of cations that are exchanged with plant hydrogen ions to maintain the charge balance.

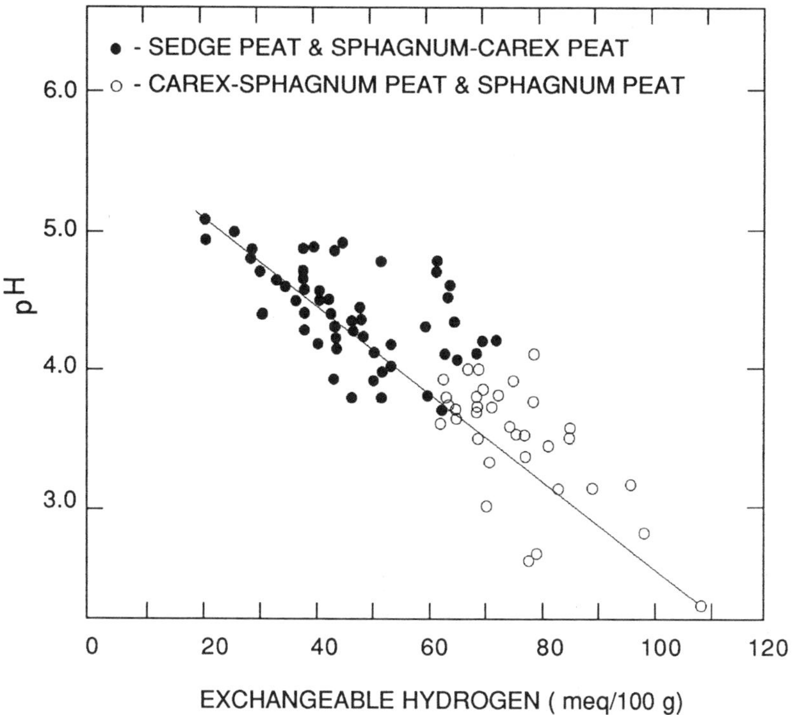

Figure 12–13. The relationship between pH and exchangeable hydrogen (*After Puustjarvi, 1957, from Walmsley, 1977; copyright © 1977 by the University of Toronto Press, reproduced with permission*)

5. *Buildup of organic acids by decomposition.* Gorham et al. (1984b) presented evidence supporting this source of bog acidity. Organic acids help buffer the system against the alkalinity of metallic cations brought in by rainfall and local runoff.

A detailed hydrogen budget constructed for a Minnesota bog complex implicated nutrient uptake as the major source of acidity (Urban et al., 1985; Table 12–3). About 15 percent of this represents ion exchange on cell walls of *Sphagnum*. Most of this acidity is neutralized by the release of cations during decomposition. Most of the rest of the acidity is generated by organic acid production from fulvic and other acids that result from the incomplete oxidation of organic matter (McKnight et al., 1985) and that buffer the pH of bogs throughout the world at a value of about 4. In addition to decomposition, the major source of alkalinity to neutralize the acids, the weathering of iron and aluminum and runoff are major processes.

Nutrients

Bogs are exceedingly deficient in available plant nutrients; fens that contain groundwater and surface water sources have considerably more nutrients (Moore and Bellamy, 1974). The paucity of nutrients in bogs leads to two significant results that are discussed later in this chapter: (1) the productivity of ombrotrophic bogs is lower than that of minerotrophic fens; (2) the characteris-

Table 12–3. Acidity Balance for a Minnesota Bog Complex

Sources	*Acidity (meq m^{-2} yr^{-1})*
Wet and dry deposition	-0.20 ± 10.7
Upland runoff	-44.3 ± 18.6
Nutrient uptake	827 ± 248
Organic acid production	263 ± 50
TOTAL	1044
Sinks	
Denitrification	12.2
Decomposition	784
Weathering	76
Outflow	142 ± 50
TOTAL	1044

Source: After Urban et al., 1985

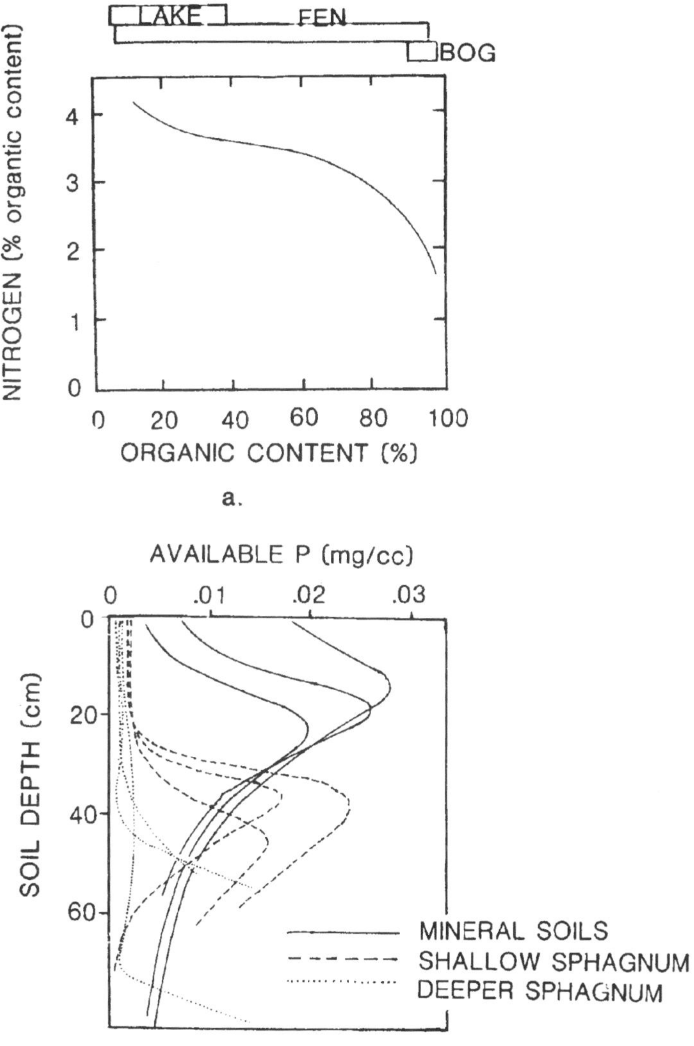

Figure 12–14. Nutrients in peatlands showing *a.* nitrogen content as a function of organic content, and *b.* available phosphorus as a function of depth for mineral soils and sphagnum soil. (a. *from Gorham, 1967;* b. *from Heilman, 1968; copyright © 1968 by the Ecological Society of America, reprinted with permission*)

tic plants, animals, and microbes have many special adaptations to the low-nutrient conditions.

The nutrient status of peatlands is controlled by a number of processes. Historically the major focus of research has been on surface and subsurface solute inflows, but the rate of peat accumulation and atmospheric inputs is also important (Hemond et al., 1987; Damman, 1990).

Figure 12–14a shows the decrease in nitrogen content of peat soils as the organic content increases. The nitrogen content is above 4 percent in minerotrophic fens but decreases to less than 2 percent in ombrotrophic bogs. The increased dominance of *Sphagnum* moss in bogs, generally with a nitrogen content of less than 1 percent (Gorham, 1967), contributes to this drop. Figure 12–14*b* shows the pattern of phosphorus levels with depth in ombrotrophic bogs and presents a comparison of that pattern with those in mineral soils. Nutrients are concentrated deeper in bogs than fens because of peat compression, the influence of past minerotrophic conditions, and a complex process of solute partitioning by freezing (Kadlec and Li, 1990). Much of the sequestered nutrients appears to be resistant to decomposition. In addition, the surface peat has an "insulating" effect, isolating surface plant life from the mineral water below (Gorham, 1967).

Some studies have attempted to find the ultimate limiting factor for bog primary productivity; this may be a complex and academic question because all available nutrients are in short supply and the growing season is short and cool. Most studies on the subject have shown that phosphorus and potassium are more important limiting factors than nitrogen. Goodman and Perkins (1968) found that potassium was the major limiting factor for the growth of *Eriophorum vaginatum* (cotton grass) in a bog in Wales, and Heilman (1968) found that levels of phosphorus and to a lesser extent potassium were deficient in black spruce (*Picea mariana*) foliage in a *Sphagnum* bog in Alaska. Bog formation in its latter stages is essentially limited to nutrients brought in by precipitation. Moore and Bellamy (1974) described the nutrient input from rainfall as adequate to support nitrogen accumulation in raised bogs in Denmark.

ECOSYSTEM STRUCTURE

Northern peatlands, particularly ombrotrophic bogs, support plants, animals, and microbes that have many adaptations to physical and chemical stresses. The organisms must deal with waterlogged conditions, acid waters, low nutrients, and extreme temperatures. The result is that specialized and unique flora and fauna have evolved in this wetland habitat.

Vegetation

Bogs can be simple sphagnum moss peatlands, sphagnum-sedge peatlands, sphagnum-shrub peatlands, bog forests, or any number or combination of acidophilic plants. Mosses, primarily those of the genus *Sphagnum,* are the most important peat-building plants in bogs throughout their geographical range. Mosses grow in cushionlike, spongy mats; water content is high, with the water, held by capillary action. *Sphagnum* grows shoots actively only in the surface layers (at a rate of about 1 to 10 cm annually); the lower layers die off and convert to peat (Walter, 1973).

A list of some of the major bog plants in North America is given in Table 12–4. *Sphagnum* often grows in association with cotton grass (*Eriophorum vaginatum*), various sedges (*Carex* spp.), and certain shrubs such as heather (*Calluna yelgaris*), leatherleaf (*Chamaedaphne calyculata*), cranberry and blueberry (*Vaccinium* spp.), and Labrador tea (*Ledum palustre*). Trees such as pine (*Pinus sylvestris*), crowberry (*Empetrum* spp.), spruce (*Picea* spp.), and tamarack (*Larix* spp.) are often found in bogs as stunted individuals that may be scarcely 1 m high yet several hundred years old (Ruttner, 1963; Malmer, 1975).

Heinselman (1970) described seven vegetation associations in the Lake Agassiz peatlands of northern Minnesota that are typical of many of those in North America (Fig. 12–15). These occur in an intricate mosaic across the landscape, reflecting the topography, chemistry, and previous history of the site that slopes from southeast to northwest, as shown in the cross section in Figure 12–7. The vegetation zones correspond closely to the underlying peat and to the present nutrient status of the site. The major zones are:

1. *Rich swamp forest.* These forested wetlands form narrow bands in very wet sites around the perimeter of peatlands. The canopy is dominated by northern red cedar (*Thuja occidentalis*), there are also some species of ash (*Fraxinus*

Table 12–4. Common Vegetation in Selected North American Peatlands

Peatland (Author)	*Ground Cover/ Herbaceous Plants*	*Shrubs*	*Trees*
Rich Fen (Michigan) (Richardson et al., 1978)	*Carex* spp. (sedge)	*Salix* sp. (willow) *Chamaedaphne calyculata* (leatherleaf) *Betula pumila* (bog birch)	—
Bog Swamp Forest (Minnesota) (Heinselman, 1970)	*Sphagnum* spp.	*C. calyculata* (leatherleaf) *B. pumila* (bog birch)	*Larix* (tamarack)
Bog Swamp (Minnesota)[a] (Heinselman, 1970)	*Sphagnum* spp.	*C. calyculata* (leatherleaf) *Kalmia* spp. (laurel) *Ledum palustre* (Labrador tea)	*Picea mariana* (black spruce)

[a]Widespread in North America

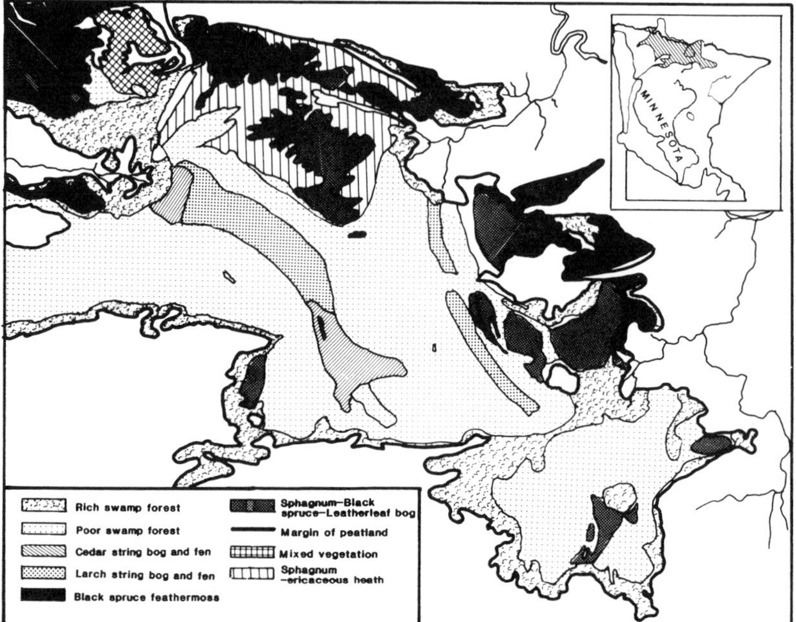

Figure 12–15. Distribution of seven vegetation associations in Lake Agassiz peatlands in northern Minnesota. (*After Heinselman, 1970*)

spp.), tamarack, and spruce. A shrub layer of alder, *Alnus rugosa*, is often present, as are hummocks of sphagnum moss.

2. *Poor swamp forest.* These swamps, occurring down slope of the rich swamp forests, are nutrient-poor ecosystems and are the most common peatland type in the Lake Agassiz region. Tamarack is usually the dominant canopy tree, with bog birch (*Betula pumila*) and leatherleaf in the understory and sphagnum forming 0.3–to 0.6-m-high hummocks.
3. *Cedar string bog and fen complex.* This is similar to type 2 except that tree-sedge fens alternate with cedar (*Thuja occidentalis*) on the bog ridges (strings) and treeless sedge (mostly *Carex*) in hollows (flarks) between the ridges.
4. *Larch string bog and fen.* In this type of string bog, similar to types 2 and 3, tamarack (*Larix*) dominates the bog ridges.
5. *Black spruce–feathermoss forest.* This type is a mature spruce forest that also contains a carpet of feather moss (*Pleurozium*) and other mosses. The trees are tall, dense, and even aged. This peatland occurs near the margins of ombrotrophic bogs and generally does not have standing water.
6. *Sphagnum–black spruce–leatherleaf bog forest.* This is a widespread wetland type in northern North America. Stunted black spruce is the only tree, and there is a heavy shrub layer of leatherleaf, laurel (*Kalmia* spp.), and Labrador

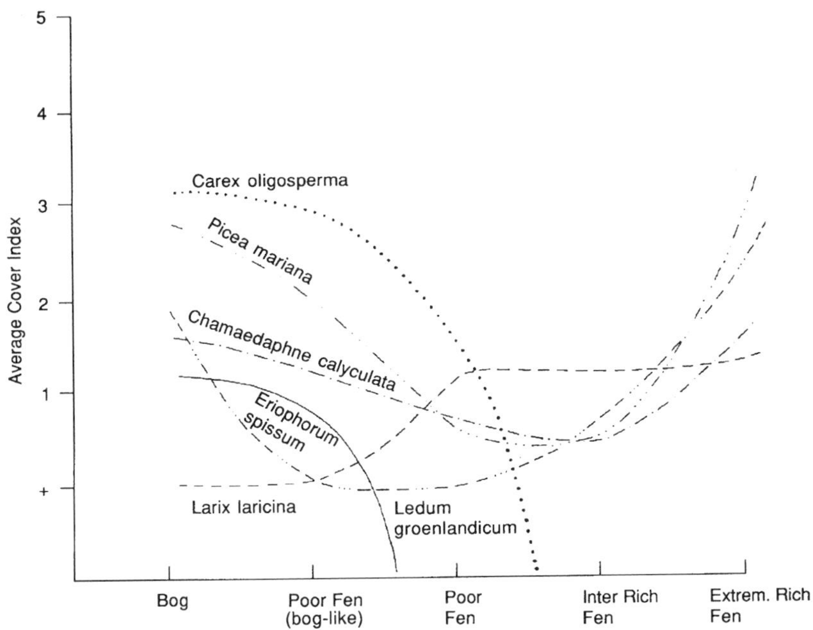

Figure 12–16. Gradient analysis of the major vascular plants that occur in both bogs and fens in Minnesota. Ordinate from Braun-Blanquet cover index. Ranges in water chemistry are bog (pH<4.2; Ca<2 mg/l), poor fen (bog-like) (pH 4.1–4.6; Ca 1.5–5.5 mg/l), poor fen (pH 4.1–5.8, Ca <10 mg/l), intermediate rich fen (pH 5.8–6.7; Ca 10–32 mg/l), extremely rich fen (pH>6.7; Ca>30 mg/l). *(From Glaser, 1987)*

tea growing in large "pillows" of sphagnum between spruce patches. This association is found in convex relief and is isolated from mineral-bearing water.

7. *Sphagnum–leatherleaf–Kalmia–spruce heath.* A continuous blanket of sphagnum moss is the most conspicuous feature; a low shrub layer and stunted trees (usually black spruce) are present in 5–10 percent of the area. Types 6 and 7, as shown in the lower end of the Figure 12–7 transect, occur on a raised bog.

In the Moore and Bellamy (1974) classification presented earlier in the chapter, zones 1 through 4 would be classified as minerotrophic, zone 5 as transitional, zone 6 as semi-ombrotrophic, and zone 7 as ombrotrophic.

Although *Sphagnum* species are the characteristic peat-forming ground cover of bogs, as sedges are of poor fens, there is a considerable overlap of species along the chemical gradient from nutrient poor to rich. For example, Figure 12–16 shows a direct gradient analysis of vascular plants found in both bogs and fens in northern Minnesota (Glaser, 1987). The sedges *Carex oligosperma* and

Eriophorum spissum decrease in cover abundance with the nutrient enrichment of the peat, whereas tamarack (*Larix laricina*) increases in abundance. Black spruce and the ericaceous shrubs Labrador tea and leatherleaf, however, show dual peaks, indicating that their distribution is not controlled by water chemistry but by another gradient such as water level.

Pocosin Vegetation

In contrast to the more northern peatlands, the woody vegetation of pocosins is dominated by evergreen trees and shrubs. Sharitz and Gibbons (1982) summarized the major associations of the North Carolina Green Swamp as reported by Kologiski (1977). Two broad community classes were identified, and their presence was related to fire frequency, soil type, and hydroperiod. Each has several distinct associations (Fig. 12–17). A *Pine-Ericalean* (pine and heath shrub) community develops on deep organic soils with long hydroperiods and frequent fire. Three associations were identified within this community: 1) pond pine (*Pinus serotina*) canopy with titi (*Cyrilla racemiflora*) and zenobia (*Zenobia pulverulenta*) shrubs; (2) pond pine and loblolly bay canopy with fetterbush; and (3) pond pine canopy with titi and fetterbush shrubs.

A *Conifer-Hardwood* community type is found on shallow organic soils with slightly shorter hydroperiods. Two associations in this group are (1) pond pine canopy with titi, fetterbush (*Lyonia lucida*), red maple (*Acer rubrum*), and black gum shrubs (*Nyssa sylvatica*); and (2) pond pine and pond cypress (*Taxodium ascendens*) canopy with red maple, titi, fetterbush, and black gum shrubs.

Peat buildup in pocosins has a major effect on hydrology and nutrients. Fire is a recurring influence that has been proposed as a major control of plant succession. This is illustrated in Figure 12–17, which shows the relationships among pocosin communities and other peripheral ones. Pocosins are found on the deepest peats that are always saturated at depth. Roots do not penetrate these deep peats into the underlying mineral soils. As a result, nutrients are limited and growth is stunted (Otte, 1981; Sharitz and Gibbons, 1982). Shallow peat burns allow the regeneration of stunted pocosins, whereas deeper burns would lead to a nonpocosin community.

Vegetation Adaptations

The vegetation in bogs and peatlands both controls and is adapted to its physical and chemical environment (Walter, 1973; Moore and Bellamy, 1974). Some of the conditions for which adaptations are necessary in peatlands are discussed below.

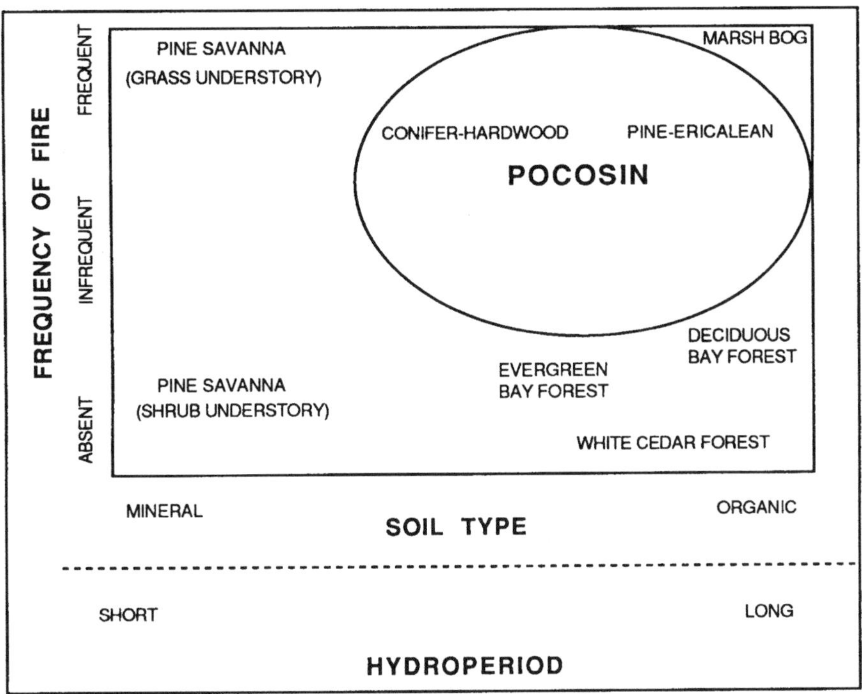

Figure 12–17. Proposed relationships among vegetation types, hydroperiod, and fire in pocosin habitats. (*From Sharitz and Gibbons, 1982, based on Kologiski, 1977*)

Waterlogging

Many bog plants, in common with wetland vegetation in general, have anatomical and morphological adaptation to waterlogged anaerobic environments. These include (1) the development of large intercellular spaces (aerenchyma or lacunae) for oxygen supply, (2) reduced oxygen consumption, and (3) oxygen leakage from the roots to produce a locally aerobic root environment. These adaptations were discussed in Chapter 6. Sphagnum, conversely, is morphologically adapted to maintain waterlogging. The compact growth habit, overlapping leaves, and rolled branch leaves form a wick that draws up water and holds it by capillarity. These adaptations enable sphagnum to hold water up to 15–23 times its dry weight (Vitt et al., 1975b).

Acidification of the External Interstitial Water

Sphagnum has the unique ability to acidify its environment, probably through the production of organic acids, especially polygalacturonic acids located on the cell walls (Clymo and Hayward, 1982). The galacturonic acid residues in the cell walls increase the cation-exchange capacity to double that of other bryophytes (Vitt et al., 1975a). The adaptive significance of this peculiarity of sphagnum is unclear. The acid environment retards bacterial action and hence decomposition, enabling peat accumulation despite low primary production rates. It has been suggested that the high cation exchange capacity also enables the plant to maintain a higher and more stable pH and cation concentration in the living cells than in the surrounding water (Glaser, 1987).

Nutrient Deficiency

Many bog plants have adaptations to the low nutrient supply that enable them to conserve and accumulate nutrients. Some bog plants, notably cotton grass (*Eriophorum* spp.), translocate nutrients back to perennating organs prior to litterfall in the autumn to provide nutrient reserves for the following year's growth and seedling establishment. The roots of other bog plants penetrate deep into peat zones to bring nutrients to the surface. Bog litter has been demonstrated to release potassium and phosphorus, often the most limiting nutrients, more rapidly than other nutrients, an adaptation that keeps these nutrients in the upper layers of peat (Moore and Bellamy, 1974).

Another adaptation to nitrogen deficiency in bogs is the ability of carnivorous plants to trap and digest insects. This special feature is seen in several unique insectivorous bog plants, including the pitcher plant (*Sarracenia* spp., Fig. 12–18) and sundew (*Drosera* spp.). Some bog plants also carry out symbiotic

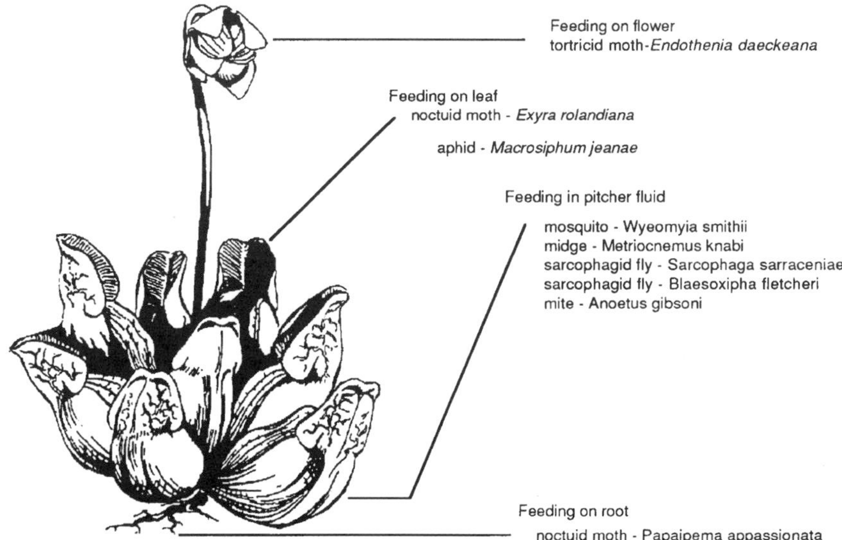

Figure 12–18. Invertebrate associates of the pitcher plant (*Sarracenia purpurea*). (*From Damman and French, 1987*)

nitrogen fixation. The bog myrtle (*Myrica gale*) and the alder develop root nodules characteristic of nitrogen fixers and have been shown to fix atmospheric nitrogen in bog environments.

Overgrowth by Peat Mosses

Many flowering plants are faced with the additional problem of being overgrown by peat mosses as the mosses grow in depth and in area covered. Adapting plants must raise their shoot bases by elongating their rhizomes or by developing adventitious roots. Trees such as pine, birch, and spruce are often severely stunted because of the moss growth and poor substrate; they grow better on bogs where the vertical growth of moss has ceased.

Consumers

Mammals

The populations of animals in bogs are generally low because of the low productivity and the unpalatability of bog vegetation. Animal density is closely related to the structural diversity of the peatland vegetation. For example, forested peatlands tend to support the greatest number of small mammals species, especially close to upland habitats (Stockwell and Hunter, 1985). Table 12–5 lists small mammals associated with peatlands in northern Minnesota. Large mammals tend to roam over larger landscapes and are thus not specific to individual peatland types. In northern Minnesota and New England, moose (*Alces alces*) are frequently found in small peatlands. White-tailed deer (*Odocoileus virginianus*) browse heavily in white cedar bogs in winter. Black bear (*Ursus americanus*) use peatlands for escape cover and for food. The woodland caribou (*Rangifer tarandus*) was the largest mammal that was largely restricted to peatlands, but it disappeared from Minnesota in 1936 (Nelson, 1947) probably as a result of hunting pressure. Smaller mammals closely associated with peatlands are beaver (*Castor canadensis*), lynx (*Lynx canadensis*), fishers (*Martes pennanti*), and snowshoe hares (*Lepus americanus*). Beaver is a fairly recent import into Minnesota peatlands. It moved in along drainage ditches, seldom penetrating deep into large peatlands, but it has had a significant effect on peatland flooding in Northern Minnesota (Naiman et al., 1991). Wet forests are becoming the only habitats where wide-ranging mammals such as the black bear, otter, and mink are found (Sharitz and Gibbons, 1982; L. D. Harris, 1989). This is not so much because peatlands are obligate habitats but because the clearing of upland forests has forced the remaining population into the remaining large tracts of forested wetlands.

Amphibians and Reptiles

Glaser (1987) reported only seven species of amphibians and four reptiles in northern Minnesota peatlands. Acid waters below pH 5 appear to be the major limiting factor in their ability to colonize bogs.

Table 12–5. Presence of Small Mammals in General Peatland Types in Northern Minnesota

Species	*Fen*	*Swamp Thicket*	*Swamp Forest*	*Forested Bog*	*Open Bog*	*Adjacent Upland*
Masked shrew (*Sorex cinereus*)	4	4	4	4	4	4
Water shrew (*Sorex palustris*)	2				2	
Arctic Shrew (*Sorex articus*)	4	4	1–4		1	1
Pygmy shrew (*Sorex hoyi*)	2–4	3	2–3	3	2	3
Short-tailed shrew (*Blarina brevicauda*)	2–4	4	3–4	2	1	4
Star-nosed mole (*Condylura cristata*)		2	0–4			
Eastern chipmunk (*Tamias striatus*)	0–1		0–1			4
Least chipmunk (*Eutamias minimus*)	0–1		0–2			3
Franklin ground squirrel (*Spermophilus franklinii*)	0–1				1	1
Red squirrel (*Tamiasciurus hudsonicus*)	0–1	1	4	4		4
Northern flying squirrel (*Glaucomys sabrinus*)			0–2			3
Deer mouse (*Peromyscus maniculatus*)			3–4	1	1	4
White-footed mouse (*Peromyscus leucopus*)	1	2	2–3			4
Southern red-backed vole (*Clethrionomys gapperi*)	4	4	4	4	4	4
Heather vole (*Phenacomys intermedius*)*						
Meadow vole (*Microtus pennsylvanicus*)	2–4	4	1–3	1	4	1
Southern bog lemming (*Synaptomys cooperi*)			0–4	0–4	2	
Northern bog lemming (*Synaptomys borealis*)	0–1				2	
Meadow jumping mouse (*Zapus hudsonius*)	2	3	0–3		1	3
Least weasel (*Mustela nivalis*)*						

Key:
4—characteristic 2—occasional 0 or Blank—not found
3—frequent 1—occurred *—reported to occur in peatlands

Source: Glaser, 1987; data from Nordquist and Birney, 1980, and Minnesota Department of Natural Resources, 1984

Birds

Many bird species are seen in peatlands during different times of the year. For example, Warner and Wells (1980) reported 70 species during the breeding season. Most of these are also common on upland sites, but a few depend on peatlands for survival (Glaser, 1987). These include the greater sandhill crane (*Grus canadensis*), great gray owl (*Strix nebulosa*), short-eared owl (*Asio flammeus*),

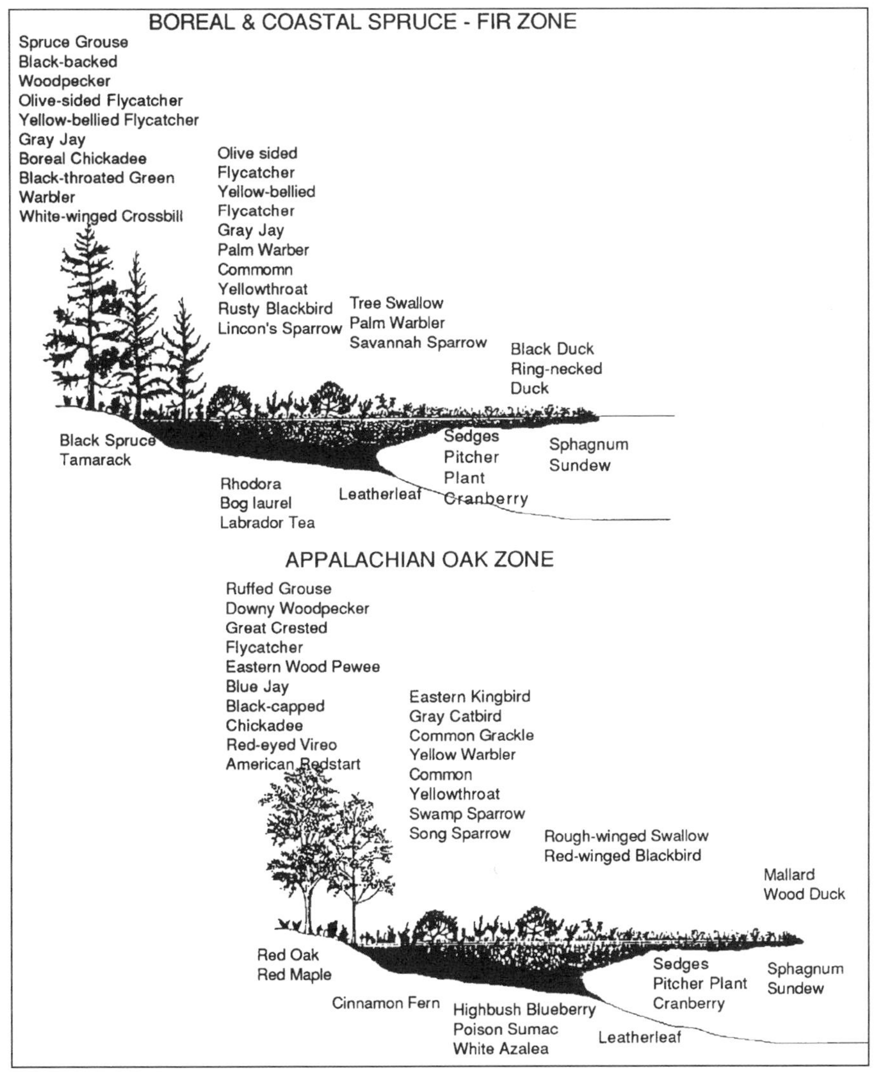

Figure 12–19. Comparison of bird distribution, typical of a lake-border bog in the northern and southern part of northeast United States. (*From Damman and French, 1987*)

sora rail (*Porzana carolina*), and sharp-tailed sparrow (*Ammospiza caudacuta*). In New England, Damman and French (1987) described the distribution of typical bird species in different habitat types of a lake-border bog (Fig. 12–19). As one moves from the Canadian border south, the species change, but the new species have analogous positions along the gradient.

Invertebrates in Pitcher Plants

Pitcher plants are interesting for the unique fauna associated with different parts of the plant. Pitcher plants (*Sarracenia* spp.) are obligate host to more invertebrate species than any other bog plant (Rymal and Folkerts, 1982; Fig. 12–18). In the water-filled pitcher are found a mosquito, a midge, two sarcophagid flies, and a mite. An aphid and three moths feed exclusively on the tissue. Other insects are associated with other parts of the plant.

ECOSYSTEM FUNCTION

The dynamics of the bog ecosystem reflect the realities of the harsh physical environment and the scarcity of mineral nutrients. These conditions result in several major features of bogs:

1. Bogs are systems of low primary productivity; often sphagnum mosses dominate and other vegetation is stunted in growth.
2. Bogs are peat producers whose rates of accumulation are controlled by a combination of complex hydrologic, chemical, and topographic factors. This peat contains a great store of nutrients, most of it below the rooting zone and thus unavailable to plants.
3. Bogs have developed several unique pathways to obtain, conserve, and recycle nutrients. The amount of nutrients in living biomass is small. Cycling is slow because of the nutrient deficiency of the litter and the waterlogging of the substrate. It is more active when peat production stagnates and when bogs receive increased nutrient inputs.

Primary Productivity

Major organic inputs to bog systems come from the primary production of the vascular plants, liverworts, mosses, and lichens. Little is known about algal productivity. Allochthonous organic carbon inputs can be significant in small bogs, but their importance decreases with the size of the peatland. Among vascular plants, ericaceous shrubs and sedges are the most important primary producers, and much of this production is below ground (Dennis and Johnson, 1970; Svensson and Rosswall, 1980). Mosses, especially sphagnum, account for one-third to one-half of the total production (Forrest and Smith 1975; Grigal, 1985).

Bogs and fens are less productive than most other wetland types and are generally less productive than the climatic terrestrial ecosystems in their region. Moore and Bellamy (1974) described the productivity of a forested sphagnum bog as about half that of a coniferous forest and a little more than a third that of a deciduous forest. Table 12–6 summarizes many measurements and ranges of

Table 12–6. Biomass and Net Primary Productivity of Selected Northern Bogs and Other Peatlands

Location	*Type of Peatland*	*Living Biomass, g dry wt/m²*	*Net Primary Productivity, g dry wt m⁻² yr⁻¹*	*Reference*
Europe				
Western Europe	General nonwooded raised bog	1,200	400–500	Malmer (1975)
Western Europe	Forested raised bog	3,700	340	Moore & Bellamy (1974)
U.S.S.R.	Eutrophic (minerotrophic) forest bog	9.700–11,000	400	Pjavchenko (1982)
	Mesotrophic (transition) forest bog	4,500–8,900	350	Pjavchenko (1982)
	Oligotrophic (ombrotrophic) forest bog	2,200–3,600	260	Pjavchenko (1982)
U.S.S.R.	Mesotrophic (transition) *Pinus-Sphagnum* bog	8,500	393	Bazilevich & Tishkov (1982)
England	Blanket bog (*Calluna-Eriophorum-Sphagnum*)		659 ± 53[a]	Forrest & Smith (1975)
England	Blanket bog		635	Heal et al. (1975)
Ireland	Blanket bog		316	Doyle (1973)
North America				
Michigan	Rich fen (*Chamaedaphne-Betula*)		341[b]	Richardson et al. (1976)
Minnesota	Forested peatland	15,941	1,014[b]	Reiners (1972)
	Fen forest	9,808	651[b]	Reiners (1972)
	Forested parched bog	10,070 (aboveground)	360	Grigal et al. (1985)
	Forested raised bog	3.100 (aboveground)	300	Grigal et al. (1985)
Manitoba	Peatland bog	—	1,943	Reader & Stewart (1972)
General	Northern bog marshes (Does not include bog forests or ombrotrophc bogs)	Aboveground	101–1,026[c]	Reader (1978)
		Belowground	141–513[d]	Reader (1978)

[a]Mean ± std. deviation for seven sites
[b]Aboveground only
[c]Range for nine bog marshes
[d]Range for five bog marshes

peatland biomass and productivity. According to Pjavchenko (1982), forested peatlands produce a range of from 260 to 400 g organic matter m^{-2} yr^{-1}, with the low value that of an ombrotrophic bog and the high value that of a minerotrophic fen. Malmer (1975) cited a typical range of from 400 to 500 g m^{-2} yr^{-1} for nonforested, raised (ombrotrophic) bogs in western Europe. In contrast, Lieth (1975) estimated the net primary productivity in the boreal forest to average 500 g m^{-2} yr^{-1} and that in the temperate forest to average 1,000 g m^{-2} yr^{-1}. The estimate for boreal forests probably includes bog forests as well as upland forests.

The measurement of the growth or primary productivity of sphagnum mosses presents special problems not encountered in productivity measurements of other plants (Clymo, 1970; Moore and Bellamy, 1974). The upper stems of the plant elongate, and the lower portions gradually die off, become litter, and eventually form peat. It is difficult to measure the sloughing off of dead material to litter. It is equally hard to measure the biomass of the plant at any one time because it is difficult to separate the living and dead material of the peat. The following two methods for measuring sphagnum growth give comparable results: (1) the use of "innate" time markers such as certain anatomical or morphological features of the moss, and (2) the direct measurement of changes in weight. Growth rates for sphagnum determined by these two techniques generally fall in the range of from 300 to 800 g m^{-2} yr^{-1} (Moore and Bellamy, 1974). Table 12–7 summarizes production data from a number of sources on three species of sphagnum (Rochefort et al., 1990), arranged by decreasing latitude. Although Damman (1979) and Wieder and Lang (1983) suggested that annual production should increase with decreasing latitude, only *S. magellanicum* showed such a trend. Evidently local and regional factors are more important than latitude. For example, production varied from year to year by a factor of 3 in one four-year study (Rochefort et al., 1990).

Decomposition and Peat Formation

The accumulation of peat in bogs is determined by the production of litter (from primary production) and the destruction of organic matter (decomposition). As with primary production, the rate of decomposition in peat bogs is generally low because of (1) waterlogged conditions, (2) low temperatures, and (3) acid conditions (Moore and Bellamy, 1974). Besides leading to peat accumulation, slow decomposition leads to slower nutrient recycling in an already nutrient-limited system.

A pattern of sphagnum decomposition with depth in a bog in southern England is shown in Figure 12–20. The decomposition rate is highest near the surface, where aerobic conditions exist. By 20 centimeters depth, the rate is about one-fifth of that at the surface. This is caused by anaerobic conditions, as illustrated by the sulfide-production curve (also shown in Fig. 12–20). Clymo (1965) attributed the bulk of the organic decomposition that does occur in peat

Table 12–7. Comparison of Selected Data on the Production of *Sphagnum* Species in Order of Decreasing Latitude

Species[a]	*Growth (mm/yr)*	*Production* ($g\ m^{-2}\ yr^{-1}$)	*Latitude (N)*	*Location*	*Mean annual Precipitation (mm)*	*Mean annual Temperature (°C)*	*Source*
fus	1.4–3.2	70	68°22'	N Sweden	600	2.9	Rosswall et al. (1975)
fus	—	250	63°09'	S Finland	532	3.5	Silvola and Hanski (1979)
fus	—	220–290	63°09'	S Finland	532	3.5	K. Tolonen, in Rochefort et al. (1990)
fus	7–16	195	60°62'	S Finland	632	4–4.8	Pakarinen (1978)
mag	9.5	70	59°50'	S Norway	1250	5.9	Pedersen (1975)
ang	14.7	500	—	—	—	—	
fus	9.8	90	56°05'	S Sweden	800	7.9	Damman (1978)
mag	7.8	100	—	—	—	—	
mag	10–18	50–100	55°09'	England	1270	9.3	S. B. Chapman (1965)
ang	28–34	110–240	54°46'	England	1980	7.4	Clymo and Reddaway (1971)
mag	14–15	230	54°46'	England	1980	7.4	Forrest and Smith (1975)
ang	—	240–330	—	—	—	—	
ang	38–43	110–440	54°46'	England	1980	7.4	Clymo (1970)
fus	6–7	75–83	54°43'	Quebec	791	4.9	Bartsch and Moore (1985)
ang	4–17	19–127	—	—	—	—	T. R. Moore (1989)
fus	—	270	54°28'	England	1375	7.4	Bellamy and Rieley (1967)
fus	30	424–801	54°20'	N Germany	714	8.4	Overbeck and Happach (1957)
mag	35–51	252–794	—	—	—	—	
ang	120–160	488–1656	—	—	—	—	
fus	—	50	49°53'	S Manitoba	517	2.5	Reader and Stewart (1971)
fus	17–24	240	49°52'	NE Ontario	858	0.8	Pakarinen and Gorham (1983)
fus	7–31	69–303	49°40'	NW Ontario	714	2.6	Rochefort et al. (1990)
mag	11–34	52–240	—	—	—	—	
ang	20–39	97–198	—	—	—	—	
mag	62	540	39°07'	West Virginia	1330	7.9	Wieder and Lang (1983)

[a]fus—*Sphagnum fuscum*; mag—*S. magellanicum*; ang—*S. angustifolium*

Source: Rochefort et al., 1990

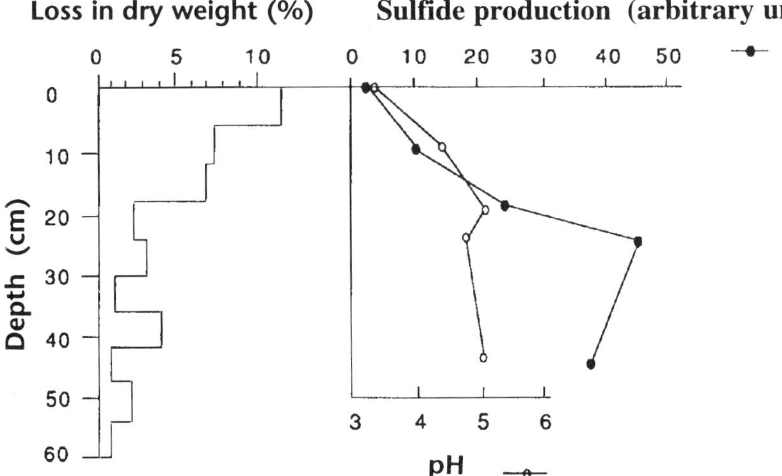

Figure 12-20. Patterns of sphagnum decomposition, sulfide production, and pH with depth in an English bog. Decomposition is during a 103-day period. (*After Clymo, 1965*)

bogs to microorganisms, although the total numbers of bacteria in these wetland soils are much fewer than in aerated soils (Moore and Bellamy, 1974). As pH decreases, the fungal component of the decomposer food web becomes more important relative to bacterial populations.

Chamie and Richardson (1978) described rates of decomposition of plant material from a rich fen in central Michigan. The decay rates and half-lives of various plant materials from that study are summarized on Table 12–8. Weight losses for leaves of sedge, willow, and birch were about 26 percent–36 percent after one year; leatherleaf leaves and stems decomposed more slowly, showing a 6–16 percent loss in one year. This decomposition rate is one-half to one-quarter the rates found either in more aerobic environments or at lower latitudes.

There has been considerable speculation about factors that give rise to patterned peatlands in Canada and the northern United States. The pattern of strings and flarks or hummocks and hollows, for example, appears to be related to differential rates of peat accumulation. Rochefort et al. (1990) determined that differential accumulation in a poor fen system in northwestern Ontario, Canada, was caused more by differences in peat decomposition rates than by differences in primary production rates. They found that even though the production rates of sphagnum in hummocks were generally about equal to rates in hollows or even lower than in minerotrophic hollows, hummock species had slower decomposition rates than hollow species. As a result, peat accumulated faster on hummocks than in hollows, and hummocks may be expanding at the expense of hollows.

The vertical accumulation rate of peat in bogs and fens is generally thought to

Table 12–8. Decomposition Rates for Various Plant Parts from Michigan Peatland[a]

Peatland Species	*Decay Coefficient (k), yr*$^{-1}$	*Half-life, yr*
Carex spp.		
Leaves	0.45	1.6
Salix spp.		
Leaves	0.46	1.5
Small stems	0.25	2.8
Large stems	0.18	3.8
Betula pumila (bog birch)		
Leaves	0.46	1.5
Small stems	0.15	4.7
Large stems	0.08	8.4
Chamaedaphne (leatherleaf)		
Leaves	0.17	4.1
Small stems	0.09	7.5
Large stems	0.09	7.5

[a]Assumes decomposition rate $Q/Q_o = e^{-kt}$; litterbags were used and placed in wetland.
Source: From Chamie and Richardson, 1978

be between 20 and 80 cm per 1,000 years in European bogs (Moore and Bellamy, 1974), although Cameron (1970) gave a range of from 100 to 200 cm per 1,000 years for North American bogs and Nichols (1983) reported an accumulation rate for peat of 150–200 cm/1,000 yr in warm, highly productive sites. Malmer (1975) described a vertical growth rate of 50–100 cm/1,000 yr as typical for western Europe. Assuming an average density of peat of 50 mg/ml, this rate is equivalent to a peat accumulation rate of 25–50 g m^{-2} yr^{-1}. This range compares reasonably well with the accumulation rate of 86 g m^{-2} yr^{-1} measured for a transition forested bog in the Soviet Union by Bazilevich and Tishkov (1982) (see Energy Flow section). In comparison, Hemond (1980) estimated a rapid accumulation rate of 430 cm/1,000 yr, or 180 g m^{-2} yr^{-1} for Thoreau's Bog, Massachusetts.

Durno (1961) compared rates of peat growth in vertical sections of peat in England and related them to climatic periods (Table 12–9). The growth of peat was rapid (110 cm/1,000 yr) during the Boreal Period, but it slowed down considerably in the wetter Atlantic climate (14–36 cm/1,000 yr). This was not expected, except that at the time it happened, the peatland was transforming itself from a minerotrophic fen to an ombrotrophic bog; the slower peat accumulation was thus probably based on low productivity during the bog stage. The surface layer, produced in the cool, moist Sub-Atlantic Period, had a higher rate of peat accumulation (48–96 cm/1,000 yr), but Durno (1961) suggested that this layer might subsequently be subjected to compression.

Table 12–9. Rates of Peat Growth (Depth) for Four Bogs in Southern Pennines Region, U.K., for Different Climatic Periods

Climatic Period	*Age of Peat, years*	*Growth Rate, cm/1,000 yr*
Sub-Atlantic	2,500	48-96
Sub-Boreal	5,000	12-48
Atlantic	7,200	14-36
Boreal	8,900	111

Source: Durno, 1961

Energy Flow

Bog ecosystems have energy-flow patterns similar to those in many other wetlands although the magnitudes of the flows are reduced. Low temperatures, flooding, and the chemical conditions of anaerobic sediments limit primary production (input) and slow decomposition (output). These patterns are different from those of most aquatic or terrestrial ecosystems found in the same north temperate region. The net result is that inputs, though low, generally exceed outputs and there is a buildup of peat.

Although few detailed studies of energy flow have been developed for bog ecosystems, one of the earliest energy budgets for any ecosystem was determined in a now classical study by Lindeman (1942) of Cedar Bog Lake, a small bog in northern Minnesota (Fig. 12–21). Although this energy budget is crude, the main features have stood the test of time. Very little of the incoming radiation (<0.1 percent) is captured in photosynthesis. The two largest flows of organic energy are to respiration (26 percent) and to storage as peat (70 percent). Energy flow in the simplified food web is primarily to herbivores (13 percent), and about 3.5 percent goes to decomposers. As the following two more recent budget measurements show, the peat-storage term is exceedingly high, and decomposition losses are probably underestimated.

Heal et al. (1975) described the energy flows and storage of an English blanket bog. Net primary productivity (635 g m^{-2} yr^{-1}) primarily stemmed from sphagnum mosses (300 g m^{-2} yr^{-1}). In contrast to Lindeman's results, only 1 percent of the productivity was consumed in the grazing food web, primarily by psyllid and tipulid flies. The primary energy flow was through the detrital food web, where about 85 percent was decomposed by microflora and 10 percent was accumulated as peat beneath the water table. This system showed a relatively low amount of peat buildup during the period of measurement.

Bazilivich and Tishkov (1982) presented a detailed breakdown of energy flow through a mesotrophic (transition) bog in the European region of the former Soviet Union. The bog is a sphagnum-pine community containing shrubs such as blueberry. The total energy stored in the bog was estimated to be in excess of 137 kg dry organic matter/m^2, with dead organic matter (to a depth of 0.6 m of peat) accounting for 94 percent of the storage. Living biomass was 8.5

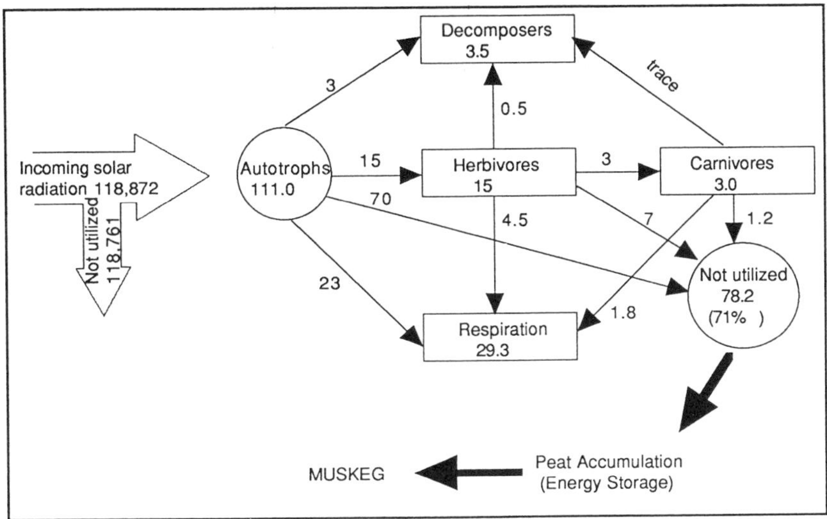

Figure 12–21. Diagram of the energy flow in Cedar Bog Lake, Minnesota. Stocks are in g • cal cm^{-2}, flows in g • cal cm^{-2} yr^{-1}. (*After Lindeman, 1942*)

kg/m^2, or about 6 percent of the organic storage. Gross primary productivity was 987 g m^{-2} yr^{-1}, with about 60 percent consumed by plant respiration, leaving 393 g m^{-2} yr^{-1} as net primary productivity (Table 12–10). The distribution of the net primary production came from trees (39 percent), algae (28 percent), shrubs (21 percent), mosses and lichens (9 percent), and grasses (3 percent). The net annual primary production, combined with other abiotic and biotic energy flows (42 g m^{-2} yr^{-1}), was primarily consumed by decomposers (284 g m^{-2} yr^{-1}). Much less was consumed in grazing food webs (7 g m^{-2} yr^{-1}). The decomposition of detritus and peat for this bog ecosystem is illustrated in Figure 12–22. Note the net accumulation of 86 g m^{-2} yr^{-1} of peat. Losses other than biotic decomposition, which accounted for most of the loss of organic matter, were chemical oxidation (15 g m^{-2} yr^{-1}) and surface and subsurface flows (25 g m^{-2} yr^{-1}).

Nutrient Budgets

Complete nutrient budgets for peatlands, particularly for ombrotrophic bogs, are rare in the literature. Crisp (1966) developed an overall nutrient budget for nitrogen, phosphorus, sodium, potassium, and calcium for a blanket bog-dominated watershed in the Pennines, England (Table 12-11). The only input considered was precipitation, which has relatively high concentrations of dissolved materials because of the proximity of the bog to the sea. The author found that the outputs of nutrients greatly exceeded the inputs from precipitation. This finding was partially the result of the omission of any estimations of

Table 12–10. Balance of Energy Inputs and Outputs of Mesotrophic Bog in the Soviet Union

Description of Energy Flow	*Energy Flow, g m^{-2} yr^{-1}*	*Percent of Total Input*
Inputs		
Gross primary productivity	987	95.9
Precipitation and runoff	34	3.3
Subsurface flow	6	0.6
Biotic input from other ecosystems	2	0.2
Total inputs	1,029	100.0
Outputs		
Plant respiration	594	57.7
Phytophage respiration	6	0.6
Carnivore respiration	1	0.1
Decomposer respiration	284	27.6
Consumption from other ecosystems	4	0.4
Abiotic oxidation	15	1.5
Subsurface outflows	5.5	0.5
Surface outflows	19.5	1.9
Total outputs	929	90.3
Accumulation		
Peat formation	86	8.3
Growth retention	14	1.4
Total accumulation	100	9.7

Source: Bazilevich and Tishkov, 1982

input caused by the weathering of the parent rock in the watershed. The budget does illustrate that the erosion of peat results in a major loss of nitrogen, exceeding the input of nitrogen by precipitation. The erosional losses of phosphorus, calcium, and potassium are 50 percent or more of the input by precipitation. Major outputs of calcium, sodium, and potassium in dissolved form illustrate that this budget is incomplete without the inclusion of weathering inputs. Sheep grazing and harvesting and the stream drift of organisms were insignificant losses from the system.

Richardson et al. (1978) presented some nutrient budgets of a fen in central Michigan (Fig. 12–23). The peat layer, measured to only 20 cm, represented more than 97 percent of nutrient storage in the fen. The uptake of nutrients by plants and litterfall were generally very low, lower, for example, than in the blanket bog previously described. This is reflected in extremely low productivity (represented as plant uptake in Fig. 12–23). Many investigators (e.g., Malmer,

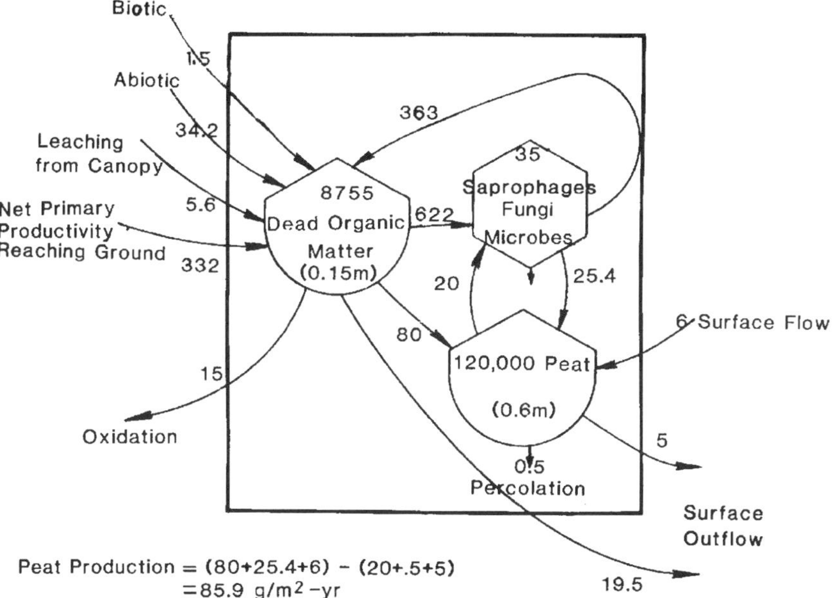

Figure 12–22. Detritus-peat production in mesotrophic bog in the former USSR. Flows are in g organic matter m^{-2} yr^{-1}; storages are in g organic matter m^{-2}. (*Based on data from Bazilevich and Tishkov, 1982*)

1962; Small, 1972; Tilton, 1977) found that nutrient content, and hence uptake rate, is lower in ombrotrophic bogs than in fens. The results of this Michigan study are therefore not representative of a typical fen. The budgets, however, show a general peatland phenomenon: The available nutrients are a small percentage of the total nutrients stored in the peat. Nitrogen fixation is not considered in the nitrogen budgets shown in Figure 12–23 and Table 12–11, but Schwintzer (1983) measured about 0.5 kg ha^{-1} yr^{-1} nonsymbiotic N-fixation in a weakly minerotrophic peatland. This is a small but significant additional input to the nitrogen budget (e.g., about 10 percent of bulk precipitation in Fig. 12–23).

Hemond (1980) summarized biogeochemical processes in a small floating mat sphagnum bog (Thoreau's Bog) in New England (Fig. 12–24). The bog is ombrotrophic and evaporative concentration determines the interstitial water concentrations of inert minerals such as chloride and sodium. Metal ions are accumulated from precipitation primarily by ion exchange or through active uptake by sphagnum in the case of potassium. Their retention depends on the exchange affinity of different metals. For example, lead is almost quantitatively retained, whereas most potassium is leached. The largest source of acidity for this bog is acid rain. Acidity is counteracted by biological processes in the bog, notably sulfate reduction and nitrate assimilation. The results for Thoreau's Bog contrast sharply with those of Crisp (1966) discussed above (Table 12–11).

Table 12–11. Nutrient Budget for Blanket Bog Watershed in Pennines Region of England

	Nutrient Flow, kg ha^{-1} yr^{-1}				
Flow	*Na*	*K*	*Ca*	*P*	*N*
Input by precipitation	25.54	3.07	8.98	0.69	8.20
Output					
Sale of sheep	0.002	0.005	0.019	0.012	0.053
Dissolved matter in stream	45.27	8.97	53.81	0.39	2.94
Drift of fauna in stream	0.001	0.005	0.001	0.005	0.057
Erosion of peat in stream	0.28	2.06	4.83	0.45	14.63
Total Output	45.44	11.04	58.66	0.86	17.68
Net Loss	20.01	7.97	49.68	0.15	9.48

Source: Crisp, 1966, as presented by Moore and Bellamy, 1974

Unlike the Pennines, where soil weathering leads to a net export of nutrients, Thoreau's Bog is a net sink.

Nitrogen budgets for Thoreau's Bog (Hemond, 1983) and for a perched raised bog complex (Urban, 1983) make an interesting comparison (Fig. 12–25). Although total nitrogen input is comparable in both systems, the Minnesota perched bog system catches some runoff from surrounding uplands, whereas nitrogen fixation is the largest source of biologically active nitrogen in Thoreau's Bog. Otherwise, the budgets are remarkably similar despite the differences in bog type and location. Both accumulate nitrogen in peat and lose a significant portion through runoff. Denitrification is an uncertain term. Peatlands appear to have the capacity for denitrification (Hemond, 1983), but the magnitude *in vivo* is uncertain.

ECOSYSTEM MODELS

Relatively few ecosystem models have been developed for northern peatlands, despite the abundance of research on those ecosystems. Conceptual models such as those by Heal et al. (1975) for an English blanket bog and Bazilivich and Tishkov (1982) for a transition bog in the Soviet Union have been useful for understanding peatland dynamics.

Logofet and Alexandrov (1982, 1984) simulated carbon and nitrogen flux through a simple multi-compartment model of a dwarf shrub-*Sphagnum* mesotrophic bog in the Soviet Union. Their simulation suggested that the assimilated organic matter is almost entirely lost through plant (62 percent) and

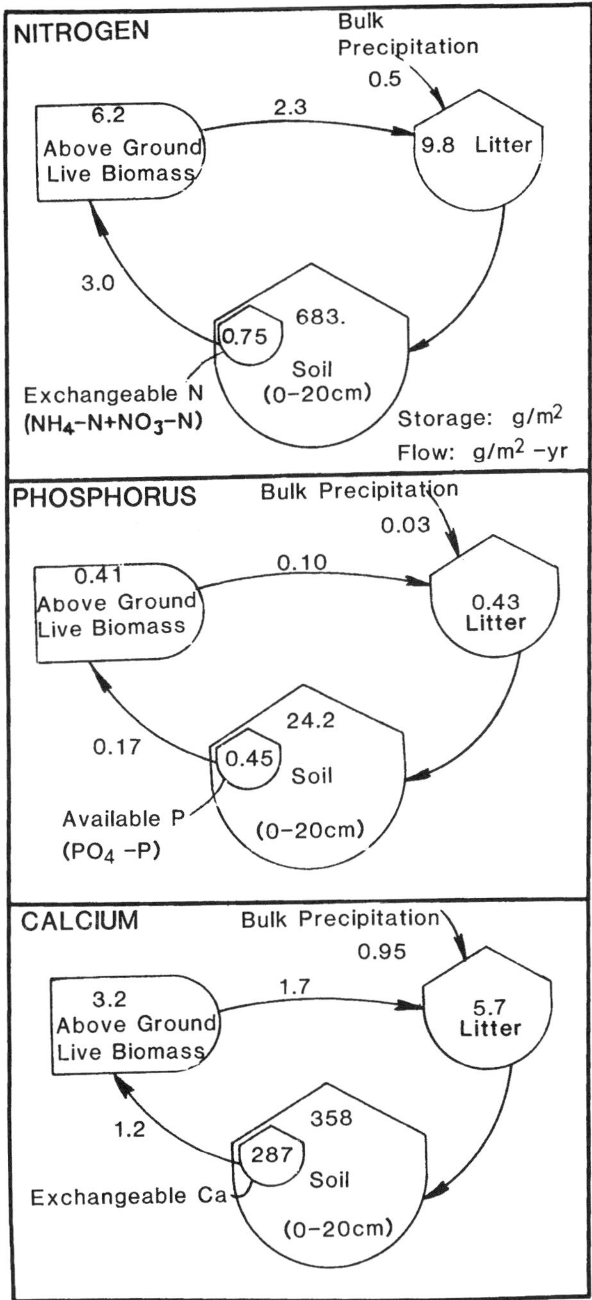

Figure 12–23. Nutrient budgets for nitrogen, phosphorus, and calcium for a leatherleaf and bog birch fen in Michigan. (*After Richardson et al., 1978*)

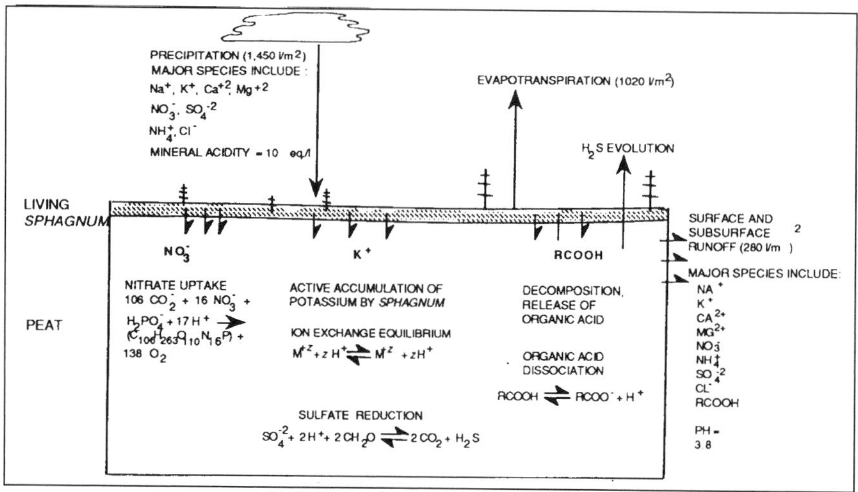

Figure 12–24. Diagram of a bog ecosystem, showing the major inputs and outputs and indicating chemical reactions within the bog ecosystem that are biogeochemically significant. The water balance is drawn on an annual basis. (*From Hemond, 1980; copyright © 1980 by the Ecological Society of America, reprinted with permission*)

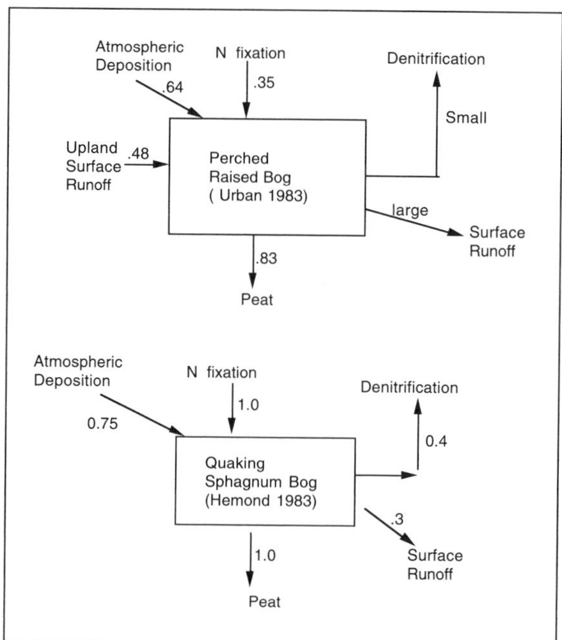

Figure 12–25. Nitrogen budgets for two northern ombrotrophic bog ecosystems. (*Data from Urban, 1983; and Hemond, 1983*)

microorganism (26 percent) respiration. About 7.4 percent of assimilated organic matter and 20 percent of allochthonous material were deposited as peat. Organic material had a total residence time of about 26 years in the system (8.5 years in plant material and 17 in other compartments). Those slow turnover rates are in sharp contrast with rates in most temperate wetlands.

There has been a fair amount of modeling of northern tundra (see Mitsch et al., 1982, for details) including peatlands, as part of the International Biological Programme's Tundra Biome studies (Miller and Tieszen, 1972; Miller et al., 1975, 1976). The ABISKO models (Bunnell and Dowding, 1974; Bunnell and Scoullar, 1975) emphasized carbon fluxes among plant, animal, and abiotic compartments in tundra ecosystems. Models have also been developed to describe the effects of oil spills on tundra ecosystems (McKay et al., 1975; Dauffenbach et al., 1981).

SOUTHERN DEEPWATER SWAMP

Southern Deepwater Swamps

13

*Southern deepwater swamps, primarily bald cypress–tupelo and pond cypress–black gum ecosystems, are freshwater systems with standing water throughout most or all of the year. Found throughout the Coastal Plain of the southeastern United States, they can occur under nutrient-poor conditions (cypress domes and dwarf cypress swamps) or under nutrient-rich conditions (lake-edge swamps, cypress strands, and alluvial river swamps). Two varieties of cypress are found in these systems: bald cypress (*Taxodium distichum*) in nutrient-rich systems and pond cypress (*Taxodium distichum *var.* nutans*) in nutrient-poor systems. Cypress and tupelo have developed several unique adaptations to the wetland environment, including knees and wide buttresses. Cypress seeds require a drawdown period for germination, and seedlings grow rapidly to get out of the water and are dispersed widely in flowing water systems. Swamp primary productivity is closely tied to hydrologic conditions. The highest productivity is in a pulsing hydroperiod wetland that receives high inputs of nutrients and lower productivity occurs in either drained or continuously flooded swamps. Consumption is primarily through detrital pathways, and decomposition depends on flooding, type of material, and average annual temperature. Cypress swamps have been shown to be nutrient sinks, particularly in studies of phosphorus budgets, and they have been investigated for their value as nutrient sinks when wastewater is applied. Modeling has been used to investigate deepwater swamp regions, particularly for the effects of logging, hydrologic modifications, and wastewater additions.*

One of the most enchanting of all the wetland environments in the United States is the southern freshwater forested swamp. The enchantment comes from the stately and venerable cypress trees, the fronds of Spanish moss hanging from the branches, the stillness of the tea-colored water below, and the quietness that is pierced only by the sound of a passing heron or a croaking tree frog. A more forbidding view of a forested swamp is engendered by the presence of mosquitoes, cottonmouth moccasins, floating logs, and submerged obstructions and the ease with which one can get lost in an environment that looks the same in every direction. This chapter is about the freshwater forested swamps of the southeastern United States, a group of some of the most mysterious wetlands in the world. Although most of these wetlands have been altered as a result of drainage attempts or logging, many persist today to remind us of the vast swamps that once were found in much greater abundance throughout the southeastern United States.

The freshwater deep swamp has been defined by Penfound (1952) as having "fresh water, woody communities with water throughout most or all of the growing season." In the southeastern United States, the cypress (*Taxodium* sp.) and tupelo/gum (*Nyssa* sp.) swamps are the major deepwater forested wetlands and are characterized by bald cypress–water tupelo and pond cypress–black gum communities with permanent or near-permanent standing water. These include cypress domes, which are often flooded with 0.3 m of water for most of the year, and alluvial cypress swamps, which may be permanently flooded with 1 m or more of water. Forested wetlands on floodplains that receive only seasonal pulses of flooding are discussed in Chapter 14. The reader is referred to an excellent tome called *Cypress Swamps* (Ewel and Odum, 1984) for more details about southern deepwater swamps. Dennis (1988) wrote an informative book describing, with photos as well as text, most of the remnant swamp forests in the southeastern United States. The book *Forested Wetlands* (Lugo et al., 1990), although not specifically limited to systems covered in this chapter, has much information on the functional aspects of southern deepwater swamps.

GEOGRAPHICAL EXTENT

The distribution of bald cypress and pond cypress (Fig. 13–1), both deciduous conifers, corresponds closely to the geographical extent of deepwater swamps discussed in this chapter. Bald cypress (*Taxodium distichum* [L.] Rich.) swamps are found as far north as southern Illinois in the Mississippi River floodplain and southern New Jersey along the Atlantic Coastal Plain. Pond cypress (*Taxodium distichum* var. *nutans* [Ait.] Sweet), described variously as either a different species or a subspecies of bald cypress, has a more limited range than bald cypress and is found primarily in Florida and southern Georgia; it is not present along the Mississippi River floodplain except in southeastern Louisiana. One of the main features that distinguishes pond from bald cypress is the leaf structure. Bald cypress has needles that spread from a twig in a flat plane, whereas pond

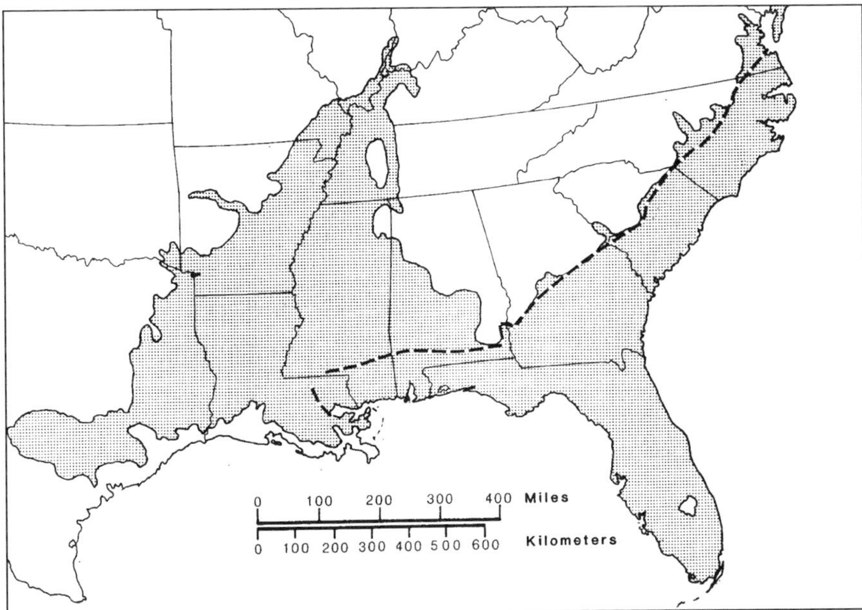

Figure 13–1. Distribution of bald cypress (*Taxodium distichum*) in southeastern United States. Dotted line indicates northern extent of pond cypress (*Taxodium distichum* var *nutans*). (*After Little, 1971, Map 84–E*)

cypress needles are appressed to the twig (C. A. Brown, 1984; Ewel and Odum, 1984). Both species are intolerant of salt and are found only in freshwater areas. Pond cypress is limited to sites that are poor in nutrients and are relatively isolated from the effects of river flooding or large inflows of nutrients. Another species indicative of the deepwater swamp is the water tupelo (*Nyssa aquatica* L.), which has a range similar to that of bald cypress along the Atlantic Coastal Plain and the Mississippi River, although it is generally absent from Florida except for the western peninsula. Water tupelo occurs in pure stands or is mixed with bald cypress in floodplain swamps.

Big Cypress Swamp

One of the most extensive areas of cypress swamps in the United States is an area contiguous to south Florida's Everglades. Called the Big Cypress Swamp, it was named not for the size of its trees but because of its large areal expanse (5,000 square kilometers). This area includes dwarf cypress, cypress domes, and cypress strands (see Geomorphology and Hydrology, below) interspersed with marshes and includes both subspecies of *Taxodium*. Much of the area is now preserved as part of the Big Cypress National Preserve, but extensive logging had already removed most of the large trees in areas such as in the Fakahatchee

Strand, where second growth strand forests are recovering (Littlehales and Niering, 1991). One of the most beautiful wetlands in Big Cypress Swamp is Corkscrew Swamp, a cypress strand that has been referred to as "An Emerald Kingdom" and is now a rookery for the rare wood stork (*Mycteria americana*).

Okefenokee Swamp

This wetland is actually a mosaic of many wetland types, including some pond cypress and black gum (*Nyssa sylvatica* var. *biflora*) swamps (see Schlesinger, 1978, and Cohen et al., 1984; see also Chapter 3). The Okefenokee Swamp is located mostly in Georgia at the Florida–Georgia border and forms the headwaters of the St. Mary River that flows to the Atlantic Ocean and the Suwannee River that flows to the Gulf of Mexico. Its name means "Land of Trembling Earth" because of the many floating peatlands. The cypress swamps in the Okefenokee have been subjected to logging, natural fires, and attempts at drainage, but they persisted and are mostly contained in the Okefenokee National Wildlife Refuge, established in 1937, and the Okefenokee Wilderness, established in 1974 (Thomas, 1976).

GEOMORPHOLOGY AND HYDROLOGY

Southern cypress-tupelo swamps occur under a variety of geologic and hydrologic conditions, ranging from the extremely nutrient-poor dwarf cypress communities of southern Florida to rich floodplain swamps along many tributaries of the lower Mississippi River Basin. A useful classification of deepwater swamps according to geological and hydrological conditions was developed by Wharton et al. (1976). That system includes the following types:

1. Still water cypress domes
2. Dwarf cypress swamps
3. Lake-edge swamps
4. Slow-flowing cypress strands
5. Alluvial river swamps

The physical features and flow conditions of these wetlands are summarized in Figure 13–2 and are described below.

Still Water Cypress Domes

Cypress domes (sometimes called cypress ponds or cypress heads) are poorly drained to permanently wet depressions dominated by pond cypress. They are generally small in size, usually 1 to 10 hectares, and are numerous in the upland

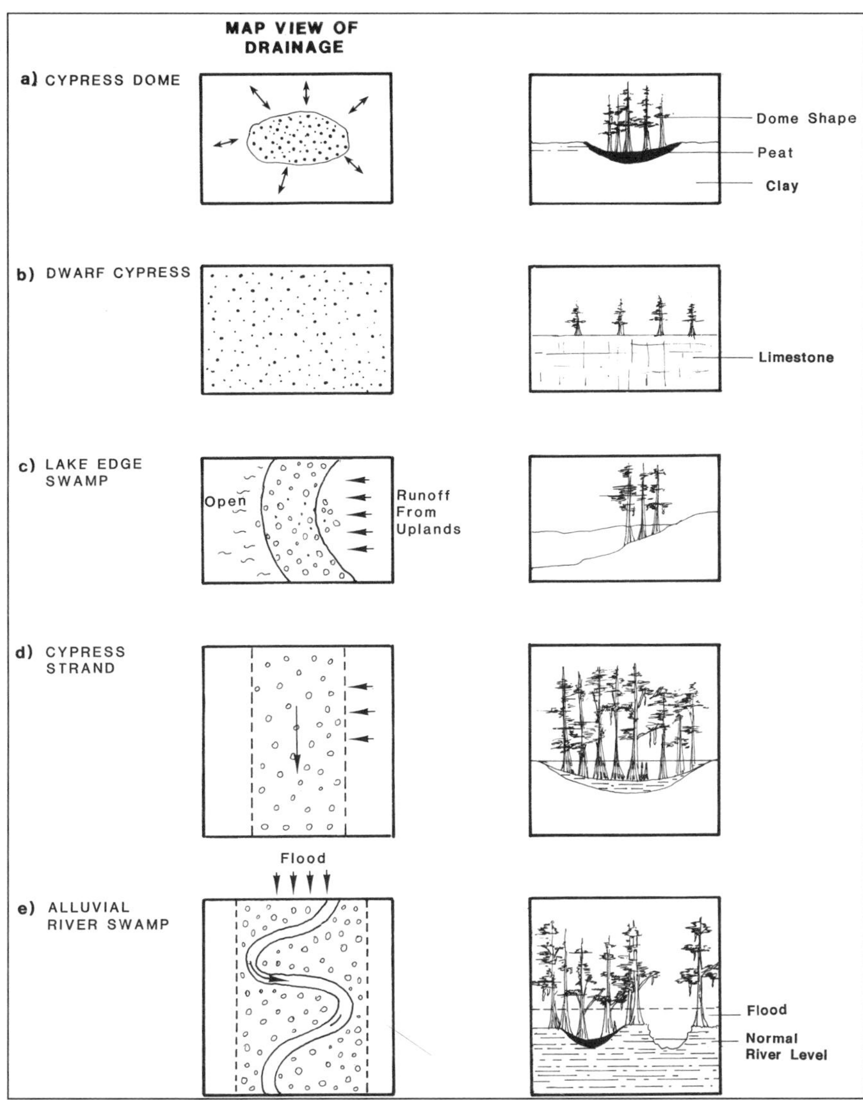

Figure 13–2. General profile and drainage pattern of major types of deepwater wetlands showing *a.* cypress dome; *b.* dwaft cypress; *c.* lake-edge swamp; *d.* cypress strand; and *e.* alluvial river swamp. (*After H. T. Odum, 1982*)

pine flatwoods of Florida and southern Georgia. Cypress domes are found in both sandy and clay soils and usually have several centimeters of organic matter that has accumulated in the wetland depression. These wetlands are called *domes* because of their appearance when viewed from the side: The larger trees are in the middle, and smaller trees are toward the edges (Fig. 13–2a). This phenomenon, it has been suggested, is caused by deeper peat deposits in the middle of the dome, fire that is more frequent around the edges of the dome, or a gradual increase in the water level that causes the dome to "grow" from the center outward (Vernon, 1947; Kurz and Wagner, 1953). A definite reason for this profile has not been determined, nor do all domes display the characteristic shape.

A typical hydroperiod for a Florida cypress dome is shown in Figure 4–3. The wet season is in the summer, and dry periods occur in the fall and spring. An example of a water budget for a Florida cypress dome is given in Figure 13–3a. The standing water in cypress domes is often dominated by rainfall and surface inflow, and there is little or no groundwater inflow. The cypress domes are sometimes underlain by an impermeable clay layer and sometimes by a *hardpan*, a layer of consolidated and relatively impermeable material. Both layers impede downward drainage, although there is often some loss of groundwater to the surrounding upland as it radiates outward from the dome rather than vertically (Heimburg, 1984). The major loss of water from the cypress dome, as shown in Figure 13–3a, is evapotranspiration. Radial groundwater loss is rapid during the dry season but relatively slow during the wet summers, when water levels surrounding the cypress dome are also high (Heimburg, 1984).

Dwarf Cypress Swamps

There are major areas in southwestern Florida, primarily in the Big Cypress Swamp and the Everglades, where pond cypress is the dominant tree but it grows stunted and scattered in a herbaceous understory marsh (Fig. 13–2b). The trees generally do not grow more than 6 to 7 m high and are more typically 3 m in height. These wetlands are called dwarf cypress or pigmy cypress swamps. The poor growing conditions are primarily caused by the lack of suitable substrate overlying the bedrock limestone that is found in outcrops throughout the region. The hydroperiod includes a relatively short period of flooding as compared with other deepwater swamps, and fire often occurs. The cypress, however, are rarely killed by fire because of the lack of litter accumulation and buildup of fuel. Although distinct from dwarf cypress swamps, individual small cypress trees are also found in scattered locations throughout the Florida Everglades. These trees are often grouped in clusters that have the appearance of small cypress domes.

Lake-Edge Swamps

Bald cypress swamps also are found as margins around many lakes and isolated sloughs in the Southeast, ranging from Florida to southern Illinois (Fig. 13–2c). Tupelo and water-tolerant hardwoods such as ash (*Fraxinus* spp.) often grow in association with the bald cypress. A seasonally fluctuating water level is characteristic of these systems and is necessary for seedling survival. The trees in these sys-

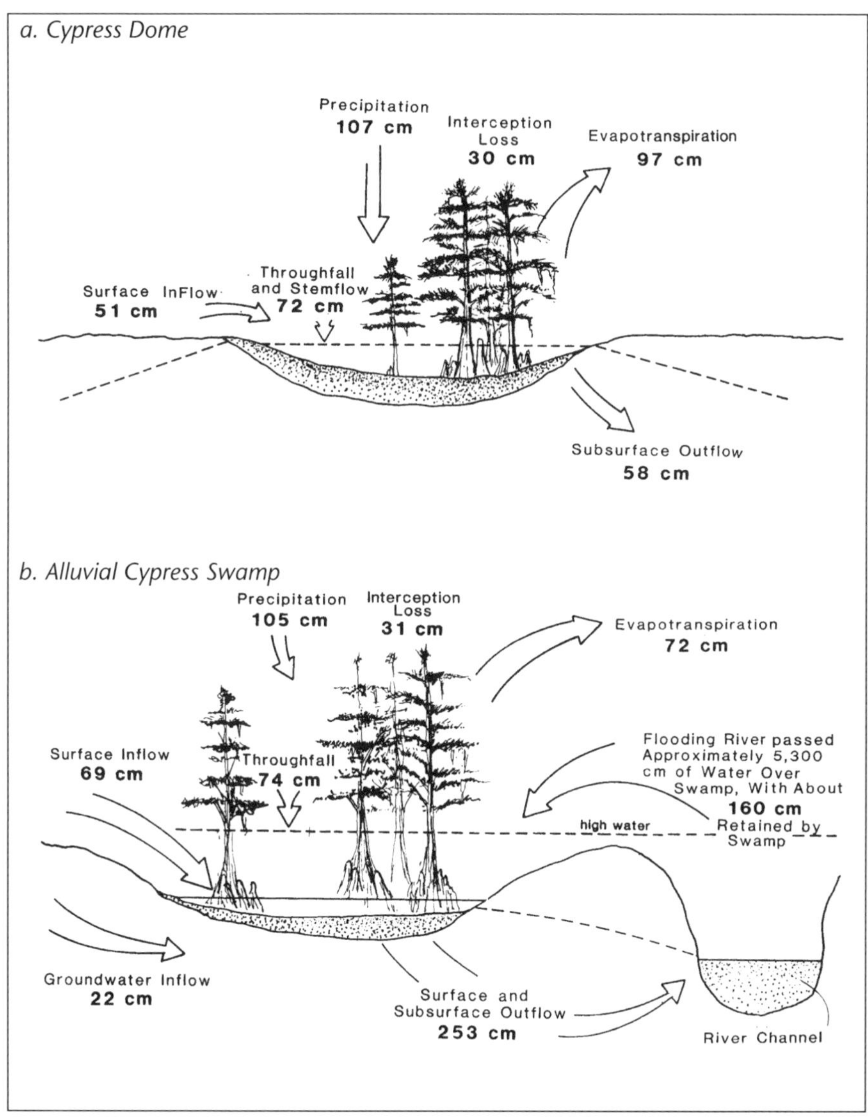

Figure 13–3. Annual water budgets for deepwater swamps for *a.* Florida cypress dome and *b.* southern Illinois cypress-tupelo alluvial swamp. (*a. Based on data from Heimburg, 1984; b. based on data from Mitsch et al., 1979a*).

tems receive nutrients from the lake as well as from upland runoff. The lake-edge swamp has been described as a filter that receives overland flow from the uplands and allows sediments to settle out and chemicals to adsorb onto the sediments before the water discharges into the open lake (Wharton et al., 1976). The importance of this filtering function, however, has not been adequately investigated.

Slow-Flowing Cypress Strands

A cypress strand is "a diffuse freshwater stream flowing through a shallow forested depression on a gently sloping plain" (Wharton et al., 1976). Cypress strands (Fig. 13–2d) are found primarily in south Florida, where the topography is slight, and rivers are replaced by slow-flowing strands with little erosive power. The substrate is primarily sand, and there is some mixture of limestone and remnants of shell beds. Peat deposits are shallow on higher ground and deeper in the depressions. The hydroperiod has a seasonal wet and dry cycle. The deeper peat deposits usually retain moisture even in extremely dry conditions. Much is known about south Florida strands from studies of Fakahatchee Strand by Carter et al. (1973) and Corkscrew Swamp by Duever et al. (1984).

Alluvial River Swamps

The broad alluvial floodplains of southeastern rivers and creeks support a vast array of forested wetlands, some of which are permanently flooded deepwater swamps (Fig. 13–2e). (Those forested wetlands that are temporarily flooded by streams and rivers are described in Chapter 14.) Alluvial river swamps, usually dominated by bald cypress or water tupelo or both, are confined to permanently flooded depressions on floodplains such as abandoned river channels (*oxbows*) or elongated swamps that usually parallel the river (*sloughs*). These alluvial wetlands sometimes are called *backswamps*, a name that distinguishes them from the drier surrounding bottomlands and indicates their hydrologic isolation from the river except during the flooding season. The water budget for an alluvial bald cypress-water tupelo swamp is shown in Figure 13–3b. The backswamps are noted for a seasonal pulse of flooding that brings in water and nutrient-rich sediments. Alluvial river swamps are continuously or almost continuously flooded. The hydrologic inflows are dominated by runoff from the surrounding uplands and by overflow from the flooding rivers.

CHEMISTRY

Wide ranges of acidity, dissolved substances, and nutrients are found in the waters of deepwater swamps (Table 13–1). Several facts should be noted from this wide range of water quality:

1. Deepwater swamps do not necessarily have acid water.
2. Nutrient conditions vary from nutrient-poor conditions in rainwater-fed swamps to nutrient-rich conditions in alluvial river swamps.
3. An alluvial river swamp often has water quality very different from that of the adjacent river.

Table 13–1. Water Chemistry in Selected Deepwater Swamps[a]

	North-Central Florida Cypress Dome[b]	*Lousiana Bayou Swamp*[c]	*Louisiana Atchafalaya Basin*[d]	*Southern Illinois Cypress-Tupelo Swamp*[e]	*Western Kentucky Cypress Slough*[f]
pH	4.51 ± 0.36 (51)	—	6.4–8.4	5.8–6.5 (4)	6.6–7.2 (8)
Conductivity, mhos/cm	60 ± 17 (41)	—	—	51–240 (9)	360–550 (8)
Alkalinity, mg $CaCo_3$/1	1.8 ± 222.9 (13)	—	38–179	12–84 (9)	—
Na^+, mg/1	4.95 ± 1.60 (38)	—	—	0.7–7.8 (7)	26.5–43.6 (4)
K^+, mg/1	0.34 ± 0.24 (38)	—	—	1.0–7.0 (7)	3.1–4.0 (4)
Mg^{++}, mg/1	1.37 ± 0.59 (39)	—	—	1.0–4.3 (7)	7–52 (4)
Ca^{++}, mg/1	2.87 ± 0.99 (39)	—	—	2.3–10.6 (7)	—
$SO_4^{=}$, mg/1	2.6 ± 2.7 (25)	—	—	0.5–4 (4)	9.2–38.6 (9)
Dissolved oxygen, mg/1	2.0 ± 1.8 (21)	—	1.8–9.9	0.9–4.0 (5)	5.8–10.9 (8)
Turbidity, NTU	2.8 ± 8.7 (34)	—	—	23–690 (8)	0.9–29 (9)
Color, mg Pt/1	456 ± 162 (43)	—	—	—	—
Total organic C, mg/1	40 ± 13 (32)	—	3.3–31.6	—	—
NO_3-N, mg N/1	0.08 ± 0.19 (63)	0.01–0.13 (24)	0.03–1.19	<.01 (6)	—
NH_3-N, mg N/1	0.14 ± 0.19 (63)	0.01–0.62 (24)	0.02–1.71	0.10–4.1 (6)	—
Total N, mg N/1	1.6 ± 1.3 (62)	0.58–1.82 (23)	0.47–9.70	0.60–4.7 (6)	—
ortho-P, mg P/1	0.07 ± 0.11 (61)	0.05–0.44 (24)	0.01–0.24	0.06–0.28 (9)	0.02–0.03 (33)
Total P, mg P/1	0.18 ± 0.38 (63)	0.15–0.66 (24)	0.08–0.56	0.17–0.47 (9)	0.08–0.20 (33)

[a]Numbers given as range (number of samples) except as otherwise noted
[b]Dierberg and Brezonik, 1984; average ± standard deviation
[c]Kemp and Day, 1984; range for three swamp sites
[d]Hern et al., 1980
[e]Mitsch et al., 1977
[f]Hill, 1983, and Mitsch et al., 1991; range for average of three sites

pH

Many deepwater wetlands, particularly alluvial river swamps, are "open" to river flooding and other inputs of neutral and generally well-mineralized waters. The pH of many alluvial swamps in the Southeast is 6 to 7, and there are high concentrations of dissolved ions (Table 13–1). Cypress domes, on the other hand, are fed primarily by rainwater and have acidic waters, usually in the pH range of 3.5 to 5.0, caused by humic acids produced within the swamp. The colloidal humic substances also contribute to the tea-colored or "blackwater" appearance of many acidic cypress domes.

Nutrients and Dissolved Ions

The buffering capacity of the water in cypress domes is low. There is little or no alkalinity and low concentrations of dissolved ions, and nutrients; on the other hand, swamps open to surface and groundwater inputs are generally rich in alkalinity, dissolved ions and nutrients. For example, conductivity ranges from only 60 μmhos/cm in cypress domes to 200 to 400 μmhos/cm in alluvial cypress swamps (Table 13–1). Cypress domes and dwarf cypress swamps are low in nutrients because of their relative hydrologic isolation. For example, average phosphorus levels of only 50 to 240 μg-P/l in cypress domes and 10 μg-P/l in a scrub cypress swamp were observed in Florida by S. L. Brown (1981). Mitsch (1984) also noted the range of total phosphorus to be 50 to 160 μg-P/l in the central pond of a similar cypress dome, with most of the phosphorus in inorganic form, whereas Dierberg and Brezonik (1984) found an average of 180 μg-P/l for 5 years of sampling in the same cypress dome (Table 13–1). On the other hand, phosphorus levels are often considerably higher in alluvial river swamps, particularly during flooding from the adjacent river. Kemp and Day (1984) reported phosphorus concentrations as high as 660 μg-P/l in Louisiana swamps, and Mitsch et al. (1977) reported concentrations as high as 470 μg-P/l in an alluvial cypress swamp in southern Illinois.

Swamp-River Comparisons

The isolation of an alluvial river swamp from its nearby river for most of the year often leads to remarkable differences in the water chemistry of the swamp and the river. Denitrification and sulfate reduction are dominant in the stagnant swamp but are less prevalent in the flowing river. Furthermore, dissolved ions in the backswamp are often lower in concentration than the same ions in the river because of the dilute nature of dissolved ions in the river when it is flooding the backswamp and also possibly because of nutrient uptake by vegetation. This physiochemical isolation of alluvial river swamps was noted in studies in Louisiana and southern Illinois. Water chemistry in Louisiana backswamps in the Atchafalaya Basin is distinct from that of the adjacent rivers and streams

Table 13–2. Water Chemistry Averages and Ranges for Alluvial Bald Cypress-Tupelo Swamp and Adjacent River in Southern Illinois

	Swamp Water			River Water		
Parameter	*n*[a]	$\overline{X}$[b]	*Flood*[c]	*n*[a]	$\overline{X}$[b]	*Flood*[c]
Dissolved oxygen, mg/1	5	2.2	—	8	8.9	—
pH	4	6.1	—	11	7.3	—
Alkalinity, mg $CaCO_3$/1	9	31	12	27	99	32-52
Ortho-Phosphate, mg P/1	9	0.16	0.16	27	0.15	0.05-0.14
Total soluble P, mg P/1	9	0.19	—	25	0.20	—
Total P, mg P/1	10	0.47	1.81	28	0.53	0.72-2.12
NO_2^-, mg N/1	6	< 0.01	< 0.01	18	0.02	< 0.01-0.06
NO_3^-, mg N/1	6	< 0.01	< 0.01	18	0.09	< 0.01-0.09
NH_4^+, mg N/1	6	1.00	0.18	15	0.27	0.08-0.24
Total Kjeldahl nitrogen, mg N/1	6	1.64	0.60	18	0.97	0.5-1.0
Ca^{++}, mg/1	7	6.6	4.1	21	27.6	3.7-6.0
Mg^{++}, mg/1	7	2.4	2.5	21	5.7	2.4-2.7
Na^+, mg/1	7	3.2	2.7	21	15.5	2.4-3.1
K^+, mg/1	7	3.3	—	21	5.9	3.0-3.2
$SO_4^=$, mg/1	4	2.7	7.0	15	15.9	8.5-13.1

[a]Number of separate sample dates at all stations
[b]Underlined values are significantly lower than river values on same sampling dates (paired t-test = 0.05)
[c]Average concentrations during flooding
Source: Based on data from Mitsch, 1979, and Dorge et al., 1984

except in the flooding season, when the waters of the entire region are well mixed (Beck, 1977; Hern et al., 1980). In southern Illinois several dissolved substances were significantly lower in the standing water of a riparian cypress-tupelo swamp than in the adjacent river (Table 13–2). The swamp had lower values of calcium, magnesium, sodium, potassium, sulfate, and nitrate than the river did, despite the fact that the swamp was flooded annually (Mitsch, 1979; Dorge et al., 1984).

ECOSYSTEM STRUCTURE

Canopy Vegetation

Southern deepwater swamps, particularly cypress wetlands, have unique plant communities that either depend on or adapt to the almost continuously wet envi-

Table 13–3. Distinction Between Bald Cypress and Pond Cypress Swamps

Characteristic	*Bald Cypress Swamp*	*Pond Cypress Swamp*
Dominant Cypress	*Taxodium distichum*	*Taxodium distichum* var. *nutans*
Dominant Tupelo or Gum (when present)	*Nyssa aquatica* (water tupelo)	*Nyssa sylvatica* var. *biflora* (black gum)
Tree Physiology	Large, old trees, high growth rate, usually abundance of knees and spreading buttresses	Smaller, younger trees, low growth rate, some knees and buttresses but not as pronounced
Location	Alluvial floodplains of Coastal Plain, particularly along Atlantic seaboard, Gulf seaboard, and Mississippi embayment	"Uplands" of Coastal Plain, particularly in Florida and southern Georgia
Chemical Status	Neutral to slightly acid, high in dissolved ions, usually high in suspended sediments and rich in nutrients	Low pH, poorly buffered, low in dissolved ions, poor in nutrients
Annual Flooding from River	Yes	No
Types of Deepwater Swamps	Alluvial river swamp, cypress strand, lake-edge swamp	Cypress dome, dwarf cypress swamp

ronment. The dominant canopy vegetation found in alluvial river swamps of southeastern United States includes bald cypress (*Taxodium distichum*) and water tupelo (*Nyssa aquatica*). The trees are often found growing in association in the same swamp, although pure stands of either bald cypress or water tupelo are also frequent in the southeastern United States. Many of the pure tupelo stands are thought to be the result of the selective logging of bald cypress (Penfound, 1952). Another vegetation association, the pond cypress–black gum (*Taxodium distichum* var. *nutans–Nyssa sylvatica* var. *biflora* [Walt.] Sarg.) swamp, is more commonly found on the uplands of the southeastern Coastal Plain, usually in areas of poor sandy soils without alluvial flooding. These same conditions are usually found in cypress domes. Some of the distinctions between bald cypress and pond cypress swamps are summarized in Table 13–3.

When deepwater swamps have been drained or when their dry period is extended dramatically, they can be invaded by pine (*Pinus* spp.) or hardwood species. In north-central Florida, a cypress-pine association indicates a drained cypress dome

(Mitsch and Ewel, 1979). Hardwoods that characteristically are found in cypress domes include swamp red bay (*Persea palustris*) and sweet bay (*Magnolia virginiana*) (S. L. Brown, 1981). In the lake-edge and alluvial river swamps, several species of ash (*Fraxinus* sp.) and maple (*Acer* sp.) often grow as subdominants with the cypress or tupelo or both. In the deep South, Spanish moss (*Tillandsia usneoides*) is found in abundance as an epiphyte on the stems and branches of the canopy trees. Schlesinger (1978) found that Spanish moss has a relatively high biomass and productivity as compared with epiphytic communities in many other temperate forests. A listing of some of the dominant canopy and understory vegetation in various cypress swamps is given in Table 13–4.

Fire

Fire is generally infrequent in southern deepwater swamps because of the standing water, but it can be a significant ecological factor during droughts or in swamps that have been artificially drained. For example, from 1970 to 1977, there were four fires in the Big Cypress National Preserve in southern Florida. An average of 500 hectares burned per fire (Duever et al., 1986, as cited in Ewel, 1990). It appears that fire is rare in most alluvial river swamps but can be more frequent in cypress domes or dwarf cypress swamps—as frequent as several times per century (Ewel, 1990; Fig. 13–4). Ewel and Mitsch (1978) investigated the effects of fire on a cypress dome in northern Florida and found that fire had a "cleansing" effect on the dome, selectively killing almost all of the pines and hardwoods and yet killing relatively few pond cypress; this suggests a possible advantage of fire to some shallow cypress ecosystems in eliminating competition that is less water tolerant.

Tree Adaptations

Knees

Several unique features about the trees in *Taxodium-Nyssa* swamps appear to be adaptations to the almost continuously flooded conditions. Cypress (bald and pond), water tupelo, and black gum are among a number of wetland plants that produce pneumatophores (see Chap. 6). In deepwater swamps these organs extend from the root system to well above the average water level (Fig. 13–5). On cypress, these "knees" are conical and typically less than 1 m in height, although some cypress knees are as tall as 3 to 4 m (Hook and Scholtens, 1978). Knees are much more prominent on cypress than on tupelo. Hall and Penfound (1939) discovered that cypress had more knees (three per tree) than tupelo did (one per tree) in a cypress-tupelo swamp in Louisiana. Pneumatophores on black gum in cypress domes are actually arching or "kinked" roots that approximate the appearance of cypress knees.

Table 13–4. Dominant or Abundant Vegetation of Deepwater Swamps of the Southeastern United States

	Cypress Dome[a]	*Alluvial River Swamp*[b]
Location	North-central Florida	Louisiana
Dominant Canopy Trees	*Taxodium distichum* var. *nutans* (pond cypress) *Nyssa sylvatica* var. *biflora* (swamp black gum)	*Taxodium distichum* (bald cypress) *Nyssa aquatica* (water tupelo)
Sub-dominant Trees	*Pinus elliottii* (slash pine) *Persea palustris* (swamp red bay) *Magnolia virginiana* (sweet bay)	*Acer rubrum* var. *drummondii* (Drummond red maple) *Fraxinus tomentosa* (pumpkin ash)
Shrubs	*Lyonia lucida* (fetterbush) *Myrica cerifera* (wax myrtle) *Acer rubrum* (red maple) *Cephalanthus occidentalis* (buttonbush) *Itea virginica* (Virginia willow)	*Cephalanthus occidentalis* (buttonbush) *Celtis laevigata* (hackberry) *Salix nigra* (black willow)
Herbs and Aquatic Vegetation	*Woodwardia virginica* (Virginia chain fern) *Saururus cernuus* (lizard's tail) *Lachnanthes caroliniana* (red root)	*Lemna minor* (duckweed) *Spirodela polyrhiza* (duckweed) *Riccia* sp. *Limnobium Spongia* (common frog's bit)

[a]After Monk and Brown, 1965; S. L. Brown, 1981; Marois and Ewel, 1983
[b]After Conner and Day, 1976

continued on next page

Table 13–4. continued

	Alluvial River Swamp[c]	*Scrub Cypress*[d]
Location	Southern Illinois	Southern Florida
Dominant Canopy Trees	*Taxodium distichum* (bald cypress) *Nyssa aquatica* (water tupelo)	*Taxodium distichum* var. *nutans* (pond cypress)
Sub-dominant Trees	—	*Pinus elliottii* (slash pine)
Shrubs	*Cephalanthus occidentalis* (buttonbush) *Itea virginica* (Virginia willow) *Rosa palustris* (swamp rose)	*Myrica cerifera* (wax myrtle) *Ilex cassine* *Stylingia sylvatica*
Herbs and Aquatic Vegetation	*Azolla mexicana* (water fern) *Spirodela polyrhiza* (duckweed)	*Panicum hemitomon*

[c]After Anderson and White, 1970; Mitsch et al., 1979a
[d]After S. L. Brown, 1981

continued on next page

Table 13–4. continued

	Still water Cypress Swamp[e]	*Cypress Strand*[f]
Location	Okefenokee Swamp, Georgia	Southwestern Florida
Dominant Canopy Trees	*Taxodium distichum* var. *nutans* (pond cypress)	*Taxodium distichum* (bald cypress)
Sub-dominant Trees	*Ilex cassine* *Nyssa sylvatica* var. *biflora* (swamp black gum)	*Taxodium distichum* var. *nutans* (pond cypress) *Salix caroliniana* (willow) *Annona glabra* (pond apple) *Acer rubrum* (red maple) *Sabal palmetto* (cabbage palm)
Shrubs	*Lyonia lucida* (fetterbush) *Itea virginica* (Virginia willow) *Leucothoe racemosa* *Clethra alnifolia*	*Myrica cerifera* (wax myrtle) *Persea borbonia* (red bay) *Hippocratea volubilis (liana)* *Toxicodendron radicans* (poison ivy)
Herbs and Aquatic Vegetation	*Eriocaulon compressum* (pipewort)	*Blechnum serrulatum* *Nephrolepsis exaltata* (Boston fern) *Thelypteris kunthii* (swamp fern) *Chloris neglecta* (finger grass) *Andropogon virginicus* (broom sedge) *Ludwigia repens*

[e]After Schlesinger, 1978
[f]After Carter et al., 1973

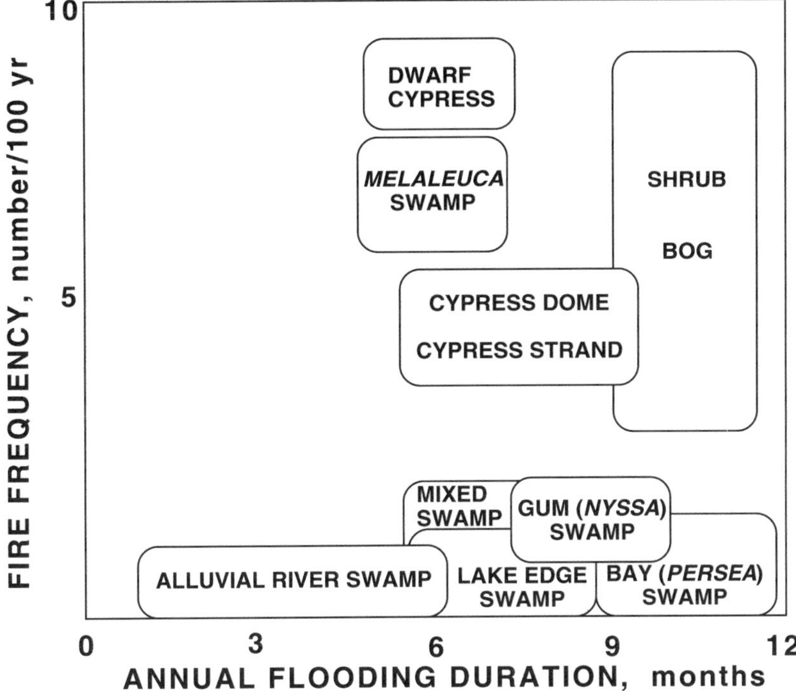

Figure 13–4. General relationship between hydroperiod and fire frequency for forested swamps in Florida. Diagram indicates names of deepwater swamps as defined in Figure 13–2. (*From Ewel, 1990; copyright © 1990 University of Central Florida Press, Orlando, redrawn with permission*)

The functions of the knees have been speculated about for the last century. It was thought that the knee might be an adaptation for anchoring the tree because of the appearance of a secondary root system beneath the knee that is similar to and smaller than the main root system of the tree (Mattoon, 1915; C. A. Brown, 1984). Observations of swamp and upland damage in South Carolina following tropical hurricane Hugo in 1990 showed that cypress trees often remained standing while hardwoods and pines did not, supporting the tree-anchoring theory for cypress root, knee, and buttress systems (K. Ewel, personal communication).

Other discussions of cypress knee function have centered on their possible functioning as sites of gas exchange for the root systems. Penfound (1952) argued that cypress knees are often absent where they are most needed—in deep water—and that the wood of the cypress knee is not aerenchymous; that is, there are no intercellular gas spaces capable of transporting oxygen to the root system. Kramer et al. (1952) concluded that the knees did not provide aeration to the rest of the trees. However, gas exchange does occur at the knees. Carbon dioxide evolved at rates from 4 to 86 g C/m^2 of tissue per day from cypress knees in

Figure 13–5. Among the several features of vegetation in cypress swamps are *a.* (*above*) cypress knees, and *b.* (*opposite page*) buttresses and large size of cypress trees.

Florida cypress domes (Cowles, 1975). S. L. Brown (1981) estimated that gas evolution from knees accounted for 0.04–0.12 gC m^{-2} day^{-1} of the respiration in a cypress dome, and 0.23 gC m^{-2} day^{-1} in an alluvial river swamp. This accounted for 0.3–0.9 percent of the total tree respiration but 5–15 percent of the estimated woody tissue (stems and knees) respiration. The fact that CO_2 is exchanged at the knee, however, does not prove that oxygen transport is taking place there or that the CO_2 evolved resulted from the oxidation of anaerobically produced organic compounds in the root system.

Buttresses

Taxodium and *Nyssa* species often produce swollen bases or buttresses when they grow in flooded conditions (Fig. 13–5b). The swelling can occur from less than 1 m above the soil to several meters, depending on the hydroperiod of the wetland. Swelling generally occurs along the part of the tree that is flooded at least seasonally, although the duration and frequency of the flooding necessary to cause the swelling are unknown. One theory described the height of the buttress as a response to aeration: The greatest swelling occurs where there is a continual wetting and soaking of the tree trunk but where the trunk is also above the normal water level (Kurz and Demaree, 1934). The value of the buttress swelling to ecosystem survivability is unknown; it may simply be a relict response that is of little use to the plant.

Figure 13–5b.

Seed Germination and Dispersal

The seeds of swamp trees, including cypress and tupelo, require oxygen for germination and flowing water for their dispersal. It has been demonstrated that cypress seeds and seedlings require very moist but not flooded soil for germination and survival (Mattoon, 1915; DuBarry, 1963). Occasional drawdowns, if only at relatively infrequent intervals, are therefore necessary for persistence of the trees in these deepwater swamps unless floating mats develop (see below). Otherwise, continuous flooding will ultimately lead to an open-water pond, although individual cypress trees may survive for a century or more.

The dispersal and survival of seeds in southern deepwater swamps depend on hydrologic conditions. Schneider and Sharitz (1986) found a relatively low number of viable seeds in a seed bank study of a cypress-tupelo swamp in South Carolina: an average of 127 seeds/m^2 were found for woody species (88 percent as cypress or tupelo) in the swamp compared to a seed density of 233 seeds/m^2 from an adjacent bottomland hardwood forest. They speculated that the continual flooding in the cypress-tupelo swamp leads to reduced seed viability. Huenneke and Sharitz (1986) and Schneider and Sharitz (1988) elaborated further on the importance of *hydrochory* or seed dispersal by water in these swamps. Hydrologic conditions, particularly scouring by flooding waters, are important factors in determining the composition, dispersal, and survival of seeds in riverine settings. Seeds are transported relatively long distances; Schneider and Sharitz (1988) reported that a distance of 1.8 km was necessary in their study to trap 90 percent of seeds from a given tree. They found that flowing water distributes seeds nonrandomly: The highest seed densities accumulate near obstructions such as logs, tree stumps, cypress knees, and tree stems and lowest seed densities in open water areas. They also found higher densities of seeds near the edge of the swamps (415 seeds/m^2) than near the center (175 seeds/m^2). Bald cypress and tupelo seeds are produced mainly in the fall and winter, which include the periods of both lowest (October) and highest (March) streamflows, giving the fallen seeds the widest possible range of hydrologic conditions. Schneider and Sharitz (1988) stated that "seed dispersal processes of many wetland species are sensitively linked to the timing and magnitude of hydrologic events" and concluded that this is probably the case for cypress-tupelo swamps.

Seedling Survival

After germination, cypress seedlings survive by rapid vertical growth to keep above the rising waters (Mattoon, 1916; Demaree, 1932; Huffman and Lonard, 1983). Conner and Flynn (1989) illustrated that when they are planted in flooded, intermittently flooded, and unflooded plots cypress seedlings showed high survival in the first two conditions (> 70 percent survival after three years) but much less survival (40 percent) in the unflooded site. Therefore, an optimum environment for seedling germination and survival is a short drawdown period for germination followed by intermittent to continuous shallow flooding. Conner and Flynn (1989) also noted the problem of seedling survival in Louisiana because of herbivory, particularly by the nutria (*Myocastor coypus*).

Longevity

Bald cypress trees may live for centuries and achieve great sizes. Virgin stands of cypress were typically from 400 to 600 years old (Mattoon, 1915). One individual cypress in Corkscrew Swamp in southwestern Florida was noted to be about 700 years old (Sprunt, 1961). Most of the virgin stands of cypress in this country were logged during the last century, however, and few individuals over 200 years old remain. Mature bald cypress are typically 30–40 m high and 1–1.5 m in diameter. Anderson and White (1970) reported a very large cypress tree in a cypress-tupelo swamp in southern Illinois that measured 2.1 m in diameter. C. A. Brown (1984) summarized several reports that documented bald cypress as large as 3.6–5.1 m in diameter. Bald cypress wood itself was used extensively for building materials in the Southeast, for the construction of southern antebellum homes and for box construction. It has been called the wood eternal by foresters because of its resistance to decay, and was used almost exclusively in external applications.

Understory Vegetation

The abundance of understory vegetation in cypress-tupelo swamps depends on the amount of light that penetrates the tree canopy. Many mature swamps appear as quiet, dark cathedrals of tree trunks devoid of any understory vegetation. Even when enough light is available for understory vegetation, it is difficult to generalize about its composition (Table 13–4). There can be a dominance of woody shrubs, of herbaceous vegetation, or of both. Fetterbush (*Lyonia lucida*), wax myrtle *(Myrica cerifera*), and Virginia willow (*Itea virginica*) are common as shrubs and small trees in nutrient-poor cypress domes. Understory species in higher nutrient river swamps include buttonbush (*Cephalanthus occidentalis*) and Virginia willow. Some continually flooded cypress swamps that have high concentrations of dissolved nutrients in the water develop dense mats of duckweed (e.g., *Lemna* spp., *Spirodela* spp., or *Azolla* spp.) on the water surface during most of the year. An experiment in enriching cypress domes with high-nutrient wastewater caused a thick mat of duckweed to develop in what was otherwise a nutrient-poor environment (H. T. Odum et al., 1977a).

Floating logs and old tree stumps often provide substrate for understory vegetation to attach and to flourish (Dennis and Batson, 1974). In some swamps, free or attached floating mats develop in much the same way that they develop in freshwater marshes (see Chaps. 9 and 11). Huffman and Lonard (1983) described floating mats from 0.5 m^2 to 0.5 hectares in a 1,200 hectare swamp in Arkansas. The mats were dominated by water willow (*Decodon* spp.) in association with Southern wild rice (*Zizania miliacea*) and water pennywort (*Hydrocotyle verticillata*) as other important species in the mats. The floating mats had the following successional pattern: (1) a pioneer stage as particles accumulate on partially decaying logs and water willow takes root in the mat; (2) a waterwillow–herbaceous stage as the mat, now floating, is invaded by a variety of marsh plants; (3) a water-willow—herb stage when additional woody plants, including bald cypress, become established; and (4) a tree stage after the submergence of the floating mat when only bald cypress can survive. Thus the mats are an important mechanism for *Taxodium* seed germination and survival, especially when drawdowns are infrequent (Huffman and Lonard, 1983).

Consumers

Invertebrates

Invertebrate communities, particularly benthic macroinvertebrates, have been analyzed in several cypress-tupelo swamps. A wide diversity and high numbers of invertebrates have been found in permanently flooded swamps. Species include crayfish, clams, oligochaete worms, snails, freshwater shrimp, midges, amphipods, and various immature insects. The organisms are highly dependent, either directly or indirectly, on the abundant detritus found in these systems. Beck (1977) found that a cypress-tupelo swamp in the Louisiana Atchafalaya Basin had a higher number of organisms (3,768 individuals/m^2) than the bayous (3,292/m^2), lakes (1,840/m^2), canals (1,593/m^2), and rivers (327/m^2). These high densities were attributed to the abundance of detritus and to the pulse of spring flooding. Sklar and Conner (1979), working in contiguous environments in Louisiana, found higher numbers of benthic organisms in a cypress-tupelo swamp (7,500 individuals/m^2) than were in a nearby impounded swamp (3,000/m^2) or in a swamp managed as a crayfish farm. Their study suggests that the natural swamp hydroperiod results in the highest numbers of benthic invertebrates.

Ziser (1978) and Bryan et al. (1976) surveyed the benthic fauna of alluvial river swamps. Oligochaetes and midges (Chironomidae), both of which can tolerate low dissolved oxygen conditions, and amphipods such as *Hyalella azteca*, which occur in abundance amid aquatic plants such as duckweed, usually dominate. In nutrient-poor cypress domes, the benthic fauna are dominated by Chironomidae, although crayfish, isopods, and other Diptera are also found there. Stresses stemming from low dissolved oxygen and periodic drawdowns are reasons for the low diversity and numbers in these domes.

The production of wood in deepwater swamps results in an abundance of substrate for invertebrates to colonize, although few studies have documented the importance of this substrate in swamps for invertebrates. Thorp et al. (1985) found that suspended *Nyssa* logs had three times as many invertebrates and twice as many taxa when they were placed in a swamp-influent stream than in the swamp itself or by its outflow stream. The swamp inflow had the highest number of mayflies (Ephemeroptera), stoneflys (Plecoptera), midges (Chironomids), and caddis flies (Trichoptera), whereas Oligochaetes were greatest in the swamp itself, supposedly because of anoxic, stagnant conditions.

Fish

Fish are both temporary and permanent residents of alluvial river swamps in the Southeast (Wharton et al., 1976; 1981). Several studies have noted the value of sloughs and backswamps for fish and shellfish spawning and feeding during the flooding season (e.g., Lambou, 1965, 1990; R. Patrick et al., 1967; Bryan et al., 1976; Wharton et al., 1981). The deepwater swamp often serves as a reservoir for fish when flooding ceases, although the backwaters are less than optimum for aquatic life because of fluctuating water levels and occasional low dissolved oxygen levels. Some fish such as bowfin (*Amia* sp.), gar (*Lepisosteus* sp.), and certain top minnows (e.g., *Fundulus* spp. and *Gambusia affinis*) are better adapted to periodic anoxia through their ability to utilize atmospheric oxygen. Often several species of forage minnows dominate cypress strands and alluvial river swamps, where most larger fish are temporary residents of the wetlands (Clark, 1979). Fish are sparse to nonexistent in the shallow cypress domes, except for the mosquito fish (*Gambusia affinis*).

Reptiles and Amphibians

Reptiles and amphibians are prevalent in cypress swamps because of their ability to adapt to fluctuating water levels. Nine or ten species of frogs are common in many southeastern cypress-gum swamps (Clark, 1979). Two of the most interesting reptiles in southeastern deepwater swamps are the American alligator and the cottonmouth moccasin. The alligator ranges from North Carolina through Louisiana, where alluvial cypress swamps and cypress strands often serve as suitable habitats. The cottonmouth moccasin (*Agkistrodon piscivorus*), a poisonous water snake that has a white inner mouth, is found throughout much of the range of cypress wetlands and is the topic of many a "snake story" of those who have been in these swamps. Other water snakes, particularly several species of *Natrix*, however, are often more important in terms of numbers and biomass and are often mistakenly identified as cottonmouth moccasins. The snakes feed primarily on frogs, small fish, salamanders, and crayfish (Clark, 1979).

ECOSYSTEM FUNCTION

Several generalizations about the ecosystem function of the southern deepwater swamp will be discussed in this section:

1. Swamp productivity is closely tied to its hydrologic regime.
2. Nutrient inflows, often coupled with hydrologic conditions, are major sources of influence on swamp productivity.
3. Swamps can be nutrient sinks whether the nutrients are a natural source or are artificially applied.
4. Decomposition is affected by the water regime and the degree of anaerobiosis.

Primary Productivity

There is a wide range of productivity in the cypress swamps of the southeastern United States. Brinson et al. (1981a), Conner and Day (1982), and Lugo et al. (1990) have compiled summaries of much of the data on deepwater swamp productivity. The studies show that primary productivity depends on hydrologic and nutrient conditions (Table 13–5) in a diversity of deepwater swamps: Florida cypress strands (Carter et al., 1973; Burns, 1978), Louisiana cypress-tupelo swamps (Conner and Day, 1976; Day et al. 1977; Conner et al. 1981), Florida cypress swamps (Mitsch and Ewel, 1979; S. L. Brown, 1981; Marois and Ewel, 1983), an Illinois alluvial river swamp (Mitsch, 1979; Mitsch et al. 1979a; Dorge et al., 1984), a Georgia pond cypress swamp (Schlesinger, 1978), a North Carolina alluvial swamp (Brinson et al., 1980), the Great Dismal Swamp of Virginia (Megonigal and Day, 1988; Powell and Day, 1991), and western Kentucky swamps (Mitsch et al., 1991). In almost all of these studies only

Table 13–5. Biomass and Net Primary Productivity of Deepwater Swamps in Southeastern United States

Location/Forest Type	*Tree Standing Biomass, kg/m²*	*Litter fall, g m⁻² yr⁻¹*	*Stem growth, g m⁻² yr⁻¹*	*Above-ground NPP,[a] g m⁻² yr⁻¹*	*Reference*
Louisiana					
Bottomland hardwood	16.5[b]	574	800	1,374	Conner and Day (1976)
Cypress-tupelo	37.5[b]	620	500	1,120	ibid.
Impounded managed Swamp	32.8[b c]	550	1,230	1,780	Conner et al. (1981)
Impounded stagnant Swamp	15.9[b c]	330	560	890	ibid.
Tupelo stand	36.2[b]	379	—	—	Conner and Day (1982)
Cypress stand	27.8[b]	562	—	—	ibid.
North Carolina					
Tupelo swamp	—	609–677	—	—	Brinson (1977)
Floodplain swamp	26.7[d]	523	585	1,108	Mulholland (1979)
Virginia					
Cedar swamp	22.0[b]	758	441	1,097[j]	Dabel and Day (1977); Gomez and Day (1982); Megonigal and Day (1988)
Maple gum swamp	19.6[b]	659	450	1,050[j]	ibid.
Cypress swamp	34.5[b]	678	557	1,176[j]	ibid.
Mixed hardwood swamp	19.5[b]	652	249	831[j]	ibid.
Georgia					
Nutrient-poor cypress swamp	30.7[e]	328	353	681	Schlesinger (1978)
Illinois					
Cypress tupelo	45[d]	348	330	678	Mitsch (1979); Dorge et al. (1984)

Table 13–5 continued

Location/Forest Type	*Tree Standing Biomass, kg/m²*	*Litter fall, g m⁻² yr⁻¹*	*Stem growth, g m⁻² yr⁻¹*	*Above-ground g m⁻² yr⁻¹*	*Reference*
Kentucky					
Cypress-ash slough	31.2	136	498	634	Taylor (1985); Mitsch et al. (1991)
Cypress swamp	10.2	253	271	524	ibid.
Stagnant cypress swamp	9.4	63	142	205	ibid.
Florida					
Cypress-tupelo (6 sites)	19 ± 4.7^f	—	289 ± 58^f	760[g]	Mitsch and Ewel (1979)
Cypress-hardwood (4 sites)	15.4 ± 2.9^f	—	336 ± 76^f	950[g]	ibid.
Pure cypress stand (4 sites)	9.5 ± 2.6^f	—	154 ± 55^f	—	ibid
Cypress-pine (7 sites)	10.1 ± 2.1^f	—	117 ± 27^f	—	ibid.
Floodplain swamp	32.5	521	1,086	1,607	S. L. Brown (1978)
Natural dome[h]	21.2	518	451	969	ibid.
Sewage dome[i]	13.3	546	716	1,262	ibid.
Scrub cypress	7.4	250	—	—	S. L. Brown and Lugo (1982)
Drained strand	8.9	120	267	387	Carter et al. (1973)
Undrained strand	17.1	485	373	858	ibid.
Large cypress strand	60.8	700	196	896	Duever et al. (1984)
Small tree cypress strand	24.0	724	818	1,542	ibid.
Sewage enriched cypress strand	28.6	650	640	1,290	Nessel (1978b)

[a]NPP = net primary productivity = litterfall + stem growth
[b]Trees defined as > 2.54 cm DBH (diameter at breast height)
[c]Cypress, tupelo, ash only
[d]Trees defined as >10 cm DBH
[e]Trees defined as > 4 cm DBH
[f]Average ± std error for cypress only
[g]Estimated
[h]Average of five natural domes
[i]Average of three domes; domes were receiving high nutrient wastewater
[j]Above-ground NPP is less than sum of litterfall plus stem growth because some stem growth is measured as litterfall

above-ground productivity was estimated. Powell and Day (1991) made direct measurements of below-ground productivity and found that it was highest in a mixed hardwood swamp (989 g m^{-2} yr^{-1}) and much lower in a more frequently flooded cedar swamp (366 g m^{-2} yr^{-1-}), a cypress swamp (308 g m^{-2} yr^{-1}) and a maplegum swamp (59 g m^{-2} yr^{-1}). These results suggest that the allocation of carbon to the root system, as a percent of total net primary productivity, decreases with increased flooding.

Ewel and Wickenheiser (1988) investigated the importance of the size of cypress domes and the relative location of trees in the dome on the basal area growth of pond cypress. They found that trees grew slowest at the edges and fastest near the center of the domes, a possible mechanism for the shape of the domes (see Geomorphology and Hydrology, above). They found no significant differences in tree growth among small, medium, and large cypress domes.

Figure 13–6 shows two similar relationships that have been suggested to explain the importance of hydrology on deepwater swamp productivity. In a study of cypress tree productivity in Florida, Mitsch and Ewel (1979) concluded that growth was low in pure stands of cypress (characterized by deep standing water) and in cypress-pine associations (characterized by dry conditions). The productivity of cypress was high in cypress-tupelo associations that were characterized by moderately wet conditions and in cypress-hardwood associations that have fluctuating water levels characteristic of alluvial river swamps (Fig. 13–6a). Conner and Day (1982) suggested a similar relationship between swamp productivity and hydrologic conditions and produced a similar curve, including some data points (Fig. 13–6b). Both of these curves suggest that the highest productivity results in systems that are neither very wet nor too dry but that have seasonal hydrologic pulsing.

S. L. Brown (1981) and Brinson et al. (1981a) found a similar relationship of productivity to hydrology. They analyzed productivity data from several forested wetlands, mostly southern deepwater swamps, and reported that productivity was highest in flowing water swamps, less in sluggish or slow-flowing swamps, and lowest in stillwater swamps. Flowing water swamps include alluvial river swamps where water and nutrients are periodically fed to the wetland from a flooding stream, whereas sluggish or slow-flowing swamps receive significant inputs from groundwater and surface runoff but not from river or stream flooding. Stillwater swamps receive their major inputs of water and nutrients from rainfall and are thus poorly nourished. S. L. Brown (1981) emphasized the importance of nutrient inflows as well as hydrologic conditions to productivity in cypress swamps (Fig. 13–7a). She pointed out that a stillwater cypress dome that received a low rate of nutrient in flow increased in productivity when it was enriched with high-nutrient wastewater. Nessel et al. (1982) and Lemlich and Ewel (1984) also demonstrated an increased growth of individual *Taxodium* trees in cypress domes when high-nutrient wastewater was applied.

a.

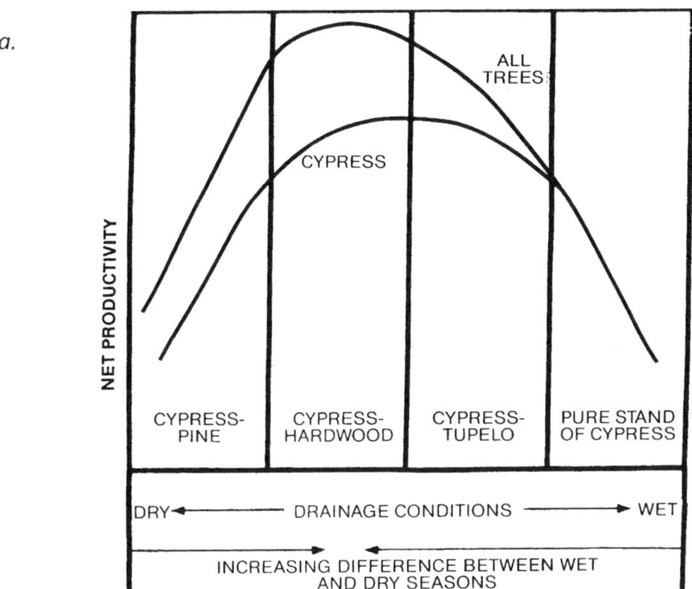

b.

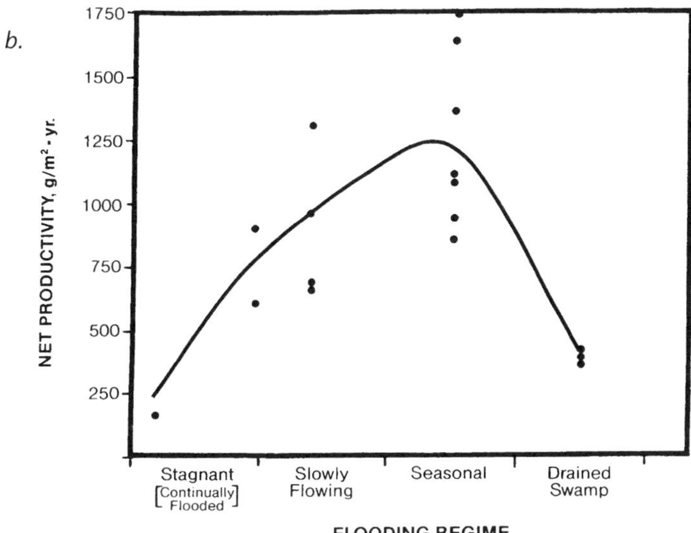

Figure 13–6. The relationship between productivity and hydrologic conditions for forested cypress swamps. (a. *from Mitsch and Ewel; copyright © by American Midland Naturalist, reprinted with permission;* b. *from Conner and Day, 1982; reprinted with permission of author*)

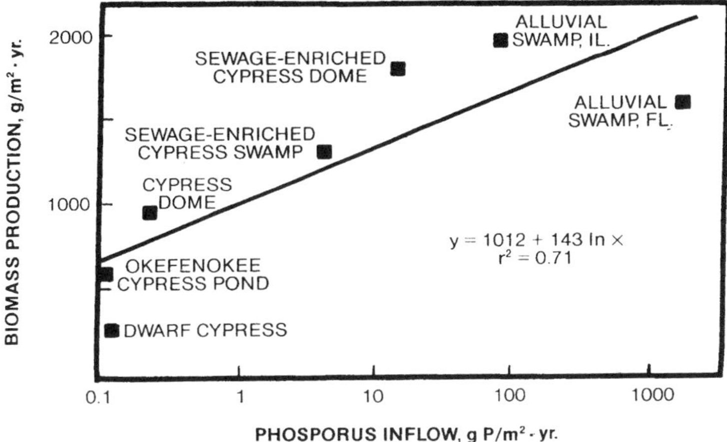

Figure 13–7. Relationships between hydrologic conditions and tree productivity in cypress swamps: *a. (above)* cypress swamp biomass productivity as a function of phosphorus inflow, *b. (opposite page)* increase in basal area of bald cypress trees in southern Illinois alluvial swamp as a function of river discharge for five-year periods. Data points indicate mean; bar indicates one standard error. (a. *after S. L. Brown, 1981;* b. *from Mitsch et al., 1979; copyright © 1979 by Ecological Society of America, reprinted with permission*)

The importance of flooding to the productivity of alluvial river swamps is further illustrated in Figure 13–7b, where the basal area growth of bald cypress in an alluvial river swamp in southern Illinois was strongly correlated with the annual discharge of the adjacent river. This graph suggests that higher tree productivity in this wetland occurred in years when the swamp was flooded more frequently than the average or for longer durations by the nutrient-rich river. Similar correlations were also obtained when other independent variables that indicate degree of flooding were used. Data points for tree growth and flooding for 1927–1936 and 1967–1976 were not included in the regression analysis due to extensive logging activity and the invasion of the pond by beavers during those periods (Mitsch et al., 1979a). Hydrologic inflows and nutrient inflows are coupled in most natural deepwater swamps, and Figures 13–7a and 13–7b reflect the same phenomenon.

The influence of the alteration of the hydroperiod on the productivity of cypress swamps is apparent from the above studies. When flooding of an alluvial river swamp is decreased, lower productivity may result. When still water swamps are drained, productivity of the swamp species will eventually decrease as water-intolerant species invade. Impounding wetlands, whether by artificial levees or beaver dams, leads to deeper and more continuous flooding and may decrease productivity.

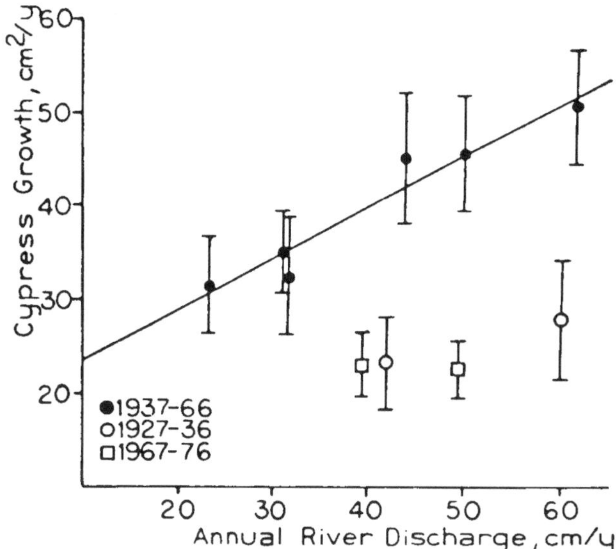

Figure 13–7b.

Decomposition

Biological utilization of organic matter is primarily through detrital pathways in deepwater swamps, although decomposition is impeded by the anaerobic conditions usually found in the sediments. The first-order decay coefficients for litter in a variety of southern deepwater swamps are summarized in Table 13–6. Brinson (1977) found that decomposition of *Nyssa aquatica* leaves and twigs in a North Carolina riparian tupelo swamp was greatest in the wettest site, whereas Duever et al. (1984) found that decomposition in a southern Florida cypress strand region was slowest in an area where no flooding occurred and more rapid in areas that were flooded from 50 percent to 61 percent of the time. Deghi et al. (1980) found no difference between decomposition of pond cypress needles in deep and shallow sites in Florida cypress domes, although decomposition was generally more rapid in wet sites than in dry sites.

F. P. Day (1982) and Yates (Yates and Day, 1983) investigated the decomposition of several types of forested wetland leaves in several forested wetland communities in the Great Dismal Swamp in Virginia (Table 13–6). Decomposition was generally slower there than in the Florida and North Carolina alluvial swamps. Variations in decomposition were seen from site to site. There was a faster breakdown of litter in seasonally flooded sites than in a mixed hardwood site that was not inundated. The data suggested, however, that the type of litter was often as important as flooding in differentiating among sites. Water tupelo leaves had the highest decomposition rates, and cedar and bald cypress had slower rates (F. P. Day, 1982). There was also quite a difference in decomposition rates measured in 1978–1979 (F. P. Day, 1982) and in 1980 (Yates and Day, 1983). The slower decomposition in the cypress and maple gum communities in 1980 was attributed to a drought during the summer and autumn and an abbreviated period of inundation in the spring (Yates and Day, 1983).

The decomposition of tree roots in forested wetlands has also been investigated for cypress swamps. F. P. Day et al. (1989) studied *Taxodium* root decomposition in experimental mesocosms and found little difference in mass loss between a continuously flooded and a periodically flooded mesocosm, although the periodically flooded roots generally had higher concentrations of nutrients (nitrogen and phosphorus) than the continuously flooded mesocosm did. The decomposition of root material was extremely slow after the initial rapid leaching losses in the first 6 months, and inhibited decay was correlated with abiotic variables such as redox potential, low oxygen concentrations, and pH. Tupacz and Day (1990) estimated the rate of root decomposition with litter bags and sediment cores in a continuation of the above-referenced Great Dismal Swamp studies, and found slowest decay on sites with the longest soil saturation and greater soil depths. Decay rates, expressed as linear rather than exponential, ranged from 0.48 to 1.00 mg g^{-1} day^{-1} in the litter bags. Roots of *Quercus* rather than *Taxodium* or *Nyssa* were most resistant to decay.

Generally, decomposition of both leaves and roots seem to be maximum in wet but not permanently flooded sites. The rates also increase with increases in average ground temperature and depend strongly on the quality (species, type of litter or roots) of the decomposing material.

Organic Export

Little organic matter is exported from stillwater and slow-flowing swamps such as cypress domes, dwarf cypress swamps, and cypress strands. Export from lake-edge swamps and alluvial river swamps, however, can be significant. Organic export is higher from watersheds that contain significant deepwater swamps than from those that do not (Mulholland and Kuenzler, 1979) (see Fig. 5–17). J. W. Day et al. (1977) found a high rate of 10.4 gC m^{-2} yr^{-1} exported as total organic carbon from a swamp forest in Louisiana. Further discussions of organic export from wetland-dominated watersheds are contained in Chapters 5 and 14.

Energy Flow

The energy flow of deepwater swamps is dominated by primary productivity of the canopy trees. Energy consumption is primarily accomplished by detrital decomposition. Significant differences exist, however, between the energy-flow patterns in low-nutrient swamps such as dwarf cypress swamps and cypress

Table 13–6. Decay Coefficients of Litter in Deepwater Swamps

	Material	*Site Description*	*Decay Coefficient* (k, yr^{-1})	*Reference*
Florida				
Cypress strand	Site leaves	On forest floor	0.86–1.39[b]	Burns (1978)
	Site leaves	On debris pile	0.69–0.75[b]	ibid.
	Site leaves	Dry site	0.47	Duever et al. (1984)
	Site leaves	Flooded 50% of time	0.23	ibid.
	Site leaves	Flooded 61% of time	0.30	ibid.
Cypress dome	Pond cypress leaves	Flooded 100% of time	1.21–1.40	Deghi et al. (1980)
	Pond cypress leaves	Dry site	0.50–0.69	ibid.
Scrub cypress	Site leaves		0.25	M. T. Brown et al. (1984)
North Carolina	Water tupelo leaves		1.89	Brinson (1977)
	Water tupelo twigs		0.28	ibid.
Virginia				
Cypress community	Bald cypress leaves		0.33	F. P. Day (1982)
	Mixed litter (1978–1979)		0.51–0.59	ibid.
	Mixed litter (1980)		0.28	Yates and Day (1983)
Maple gum communities	Tupelo leaves		0.65	F. P. Day (1982)
	Maple leaves		0.47	ibid.
	Mixed litter (1978–1979)		0.51–0.67	ibid.
	Mixed litter (1980)		0.29	Yates and Day (1983)
Cedar community	Cedar leaves		0.34	F. P. Day (1982)
	Mixed litter (1978–1979)		0.35–0.43	ibid.
	Mixed litter (1980)		0.49	Yates and Day (1983)

[a]Decay coefficient (k) based on exponential decay: $y = y_o e^{-kt}$
where y = final biomass
y_o = initial biomass
t = time in years

Source: Based partially on Brinson et al., 1981a

[b]Range due to different mesh size used in litterbags

domes, and high-nutrient swamps such as alluvial cypress swamps (Table 13–7). All of the cypress wetlands are autotrophic—productivity exceeds respiration. Gross primary productivity, net primary productivity, and net ecosystem productivity are highest in the alluvial river swamp that receives high-nutrient inflows. Buildup and/or export of organic matter are characteristic of all of these deepwater swamps but are most characteristic of the alluvial swamp. There are few allochthonous inputs of energy to the low-nutrient wetlands, and energy flow at the primary producer level is relatively low. The alluvial cypress-tupelo swamp depends more on allochthonous inputs of nutrients and energy, particularly from runoff and river flooding. In alluvial deepwater swamps, productivity by aquatic plants is often high, whereas aquatic productivity in cypress domes is usually low.

Sedimentation

One of the mechanisms by which nutrients enter many cypress swamps is through the process of sediment deposition during flooding conditions. Hupp and Morris (1990) and Hupp and Bazemore (1993) developed a dendrogeomorphic approach to determining sedimentation rates in forested wetlands at the Black Swamp on the Cache River in eastern Arkansas and two river sites in western Tennessee. They found that sedimentation rates averaged 0.25 cm/yr with a maximum of 0.60 cm/yr in cypress-tupelo swamps, in contrast to a sedimentation rate of 0.10 cm/yr when cypress and tupelo trees were absent. A distinct decrease in sedimentation correlated with an increase in land elevation in both studies (Fig. 13–8), suggesting that there is a tendency of low-elevation alluvial swamps to "fill in" with sediments. The study also showed that sedimentation rates have increased significantly at the Arkansas site from 0.01 cm/yr prior to 1945 to an average of 0.28 cm/yr over the past 19 years. Hupp and

Table 13–7. Estimated Energy Flow (kcal m^{-2} day^{-1}) in Selected Florida Cypress Swamps[a]

Parameter	*Dwarf Cypress Swamp*	*Cypress Dome*	*Alluvial River Swamp*
Gross Primary Productivity[b]	27	115	233
Plant Respiration[c]	18	98	205
Net Primary Productivity	9	17	28
Soil or Water Respiration	7	13	18
Net Ecosystem Productivity	2	4	10

[a]Assumes 1 gC = 10 kcal
[b]Assumes GPP = net daytime photosynthesis + nighttime leaf respiration
[c]Plant respiration = 2 (nighttime leaf respiration) + stem respiration + knee respiration
Source: After S. L. Brown, 1981

Bazemore (1993) suggested that channelized streams deposit fewer sediments into adjacent wetlands than unchannelized streams, an explanation for the higher sediment loads generally seen in channelized streams during flooding in contrast to natural channel streams.

Aust et al. (1991) described an experiment in which sedimentation was measured in a southwestern Alabama cypress-tupelo swamp that had been subjected to three types of land disturbance. They found that areas logged by helicopter and skidder trapped more sediments (0.12–0.22 cm/flood season) than the natural control area (0.11 cm/flood season), leading to an interesting speculation that logging improved the wetland's ability to retain sediments from the flooding river. Their study was apparently done in a season of relatively low flood conditions, as ^{137}Cs analysis indicated a mean sedimentation rate of 0.76 cm/yr during the preceding 35 years.

Nutrient Budgets

Nutrient budgets of deepwater swamps vary from "open" alluvial river swamps that receive and export large quantities of materials to "closed" cypress domes that are mostly isolated from their surroundings (Table 13–8). The importance of nutrient inflows to productivity is demonstrated in Figure 13–7.

Several nutrient budgets that illustrate intrasystem cycling as well as inputs and outputs have been developed for southern deepwater swamps. Yarbro

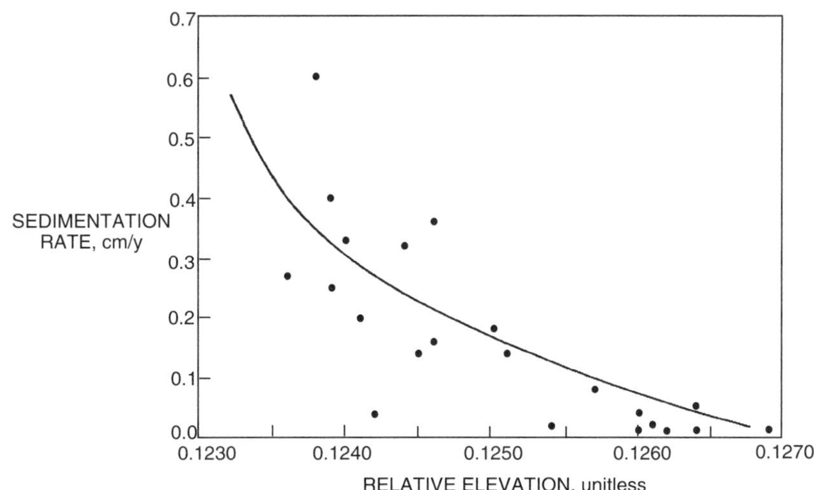

Figure 13–8. Sedimentation rate, primarily from flooding river, in Black Swamp, Arkansas, versus relative elevation of sampling site. Data suggest relatively higher sedimentation rates in lower wetland areas. (*From Hupp and Morris, 1990; copyright © 1990 by Society of Wetland Scientists, reprinted with permission*)

(1979; 1983) found that an alluvial swamp in North Carolina retained 0.3–0.7 gP m^{-2} yr^{-1} over two years of measurement. Intrasystem cycling through the vegetation accounted for 0.5 gP m^{-2} yr^{-1}, but only a fraction was retained in the woody biomass. Another phosphorus budget for an alluvial cypress-tupelo swamp in southern Illinois further demonstrated that swamps may serve as nutrient sinks (Fig. 13–9). A spring flood that spilled over from the adjacent river to the swamp contributed large amounts of nutrient-rich sediments. Estimated input attributed to sedimentation was 3.6 gP m^{-2} yr^{-1}, 26 times the contribution by throughfall. The flood was estimated to have passed 80 gP m^{-2} yr^{-1} over the swamp; the swamp retained about 4.5 percent of this flux. The total outflow of phosphorus from the swamp to the river for the remainder of the year was 0.34 gP m^{-2} yr^{-1}, which is a high rate of loss per unit area as compared with nonwetland ecosystems (Mitsch, 1979). The loss however, was only one-tenth of the input from the river to the swamp during flooding. As with the North Carolina phosphorus budget, only a small portion of the excess phosphorus from the flood was stored in the growth of the vegetation, although cycling was rapid. In both wetlands the major sink for phosphorus was the sediments that physically retained the deposited phosphorus or stored it in organic matter contributed by the highly productive vegetation.

The nutrient dynamics of deepwater swamps have also been investigated from the aspect that these swamps serve as "sinks" or ultimate deposits for nutrients. As described in Chapter 5, a number of studies have demonstrated that deepwater swamps can be nutrient sinks whether the nutrients are from a natural source or are artificially added. In South Carolina, Kitchens et al. (1975) found a 50 percent reduction in phosphorus as overflow waters passed from the river through the backwaters. In a similar study in Louisiana, J. W. Day et al. (1977) found that nitrogen was reduced by 48 percent and phosphorus decreased by 45 percent as water passed through a lake-swamp

Table 13–8. Phosphorus Inputs to Deepwater Swamps gP m^{-2} yr^{-1}

Swamp	*Rainfall*	*Surface Inflow*	*Sediments from River Flooding*	*Reference*
Florida				
Dwarf cypress	0.11	—[a]	0	S. L. Brown (1981)
Cypress dome	0.09	0.12	0	*ibid.*
Alluvial river swamp	—[a]	—[a]	3.1	*ibid.*
Southern Illinois				
Alluvial river swamp	0.11	0.1	3.6	Mitsch et al. (1979a)
North Carolina				
Alluvial tupelo swamp	0.02–0.04	0.1–1.2	0.2	Yarbro (1983)

[a]Not measured

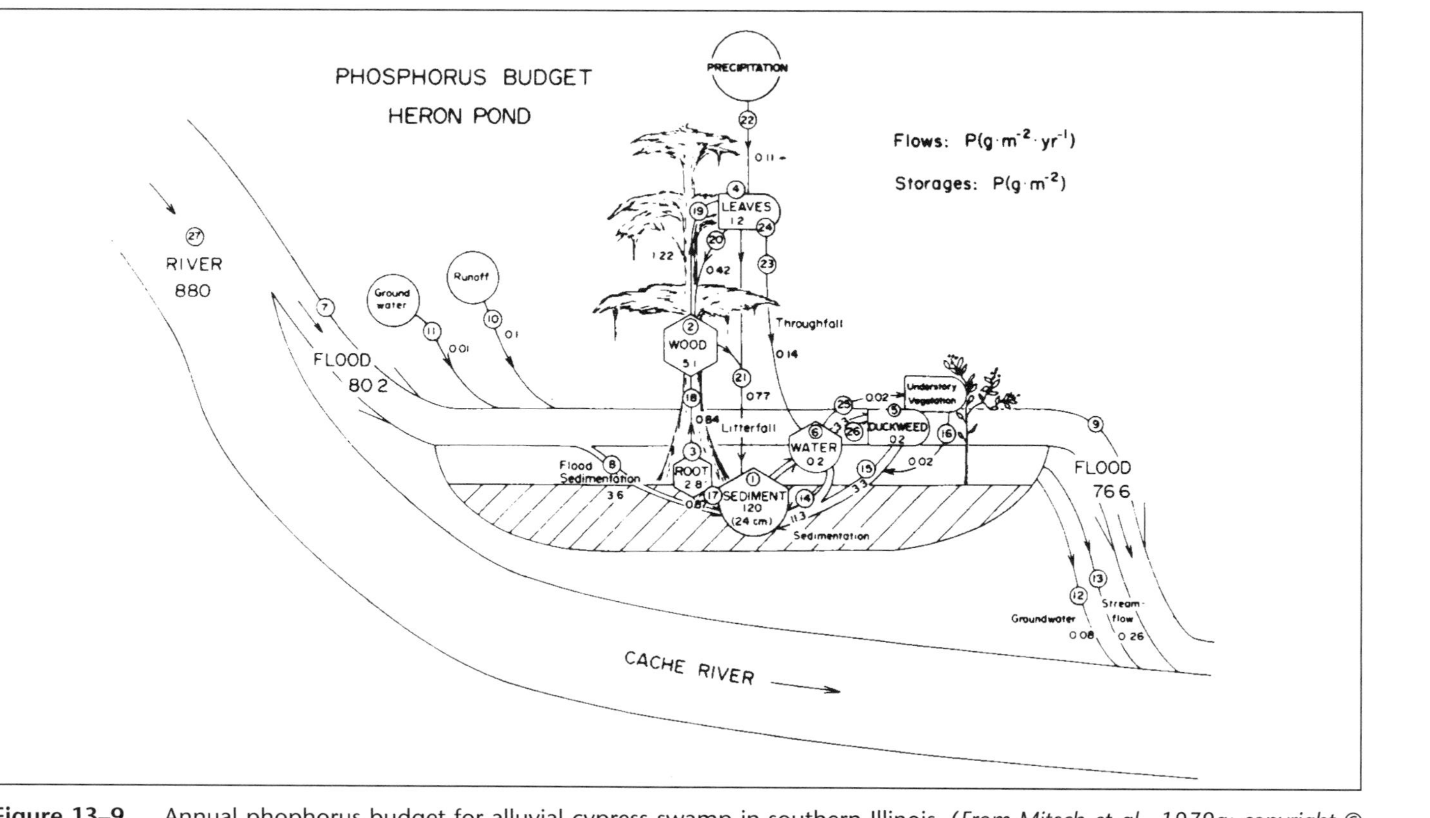

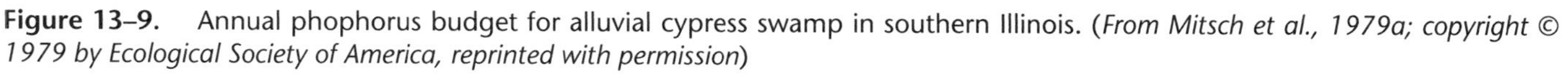

Figure 13–9. Annual phophorus budget for alluvial cypress swamp in southern Illinois. (*From Mitsch et al., 1979a; copyright © 1979 by Ecological Society of America, reprinted with permission*)

complex of Barataria Bay to the lower estuary. They attributed this decrease in nutrients to sediment interactions, including nitrate storage/denitrification and phosphorus adsorption to the clay sediments.

Dierberg and Brezonik (1983a, b; 1984; 1985), as part of the cypress dome recycling studies in Florida described by Ewel and Odum (1978; 1979; 1984), showed a high retention of nutrients by experimental cypress domes that received treated sewage at moderate loading rates of 2.5 cm/week. Dierberg and Brezonik (1985) investigated the nutrient-retention capacity of cypress swamp sediments in laboratory experiments to verify that the soil was retaining the nutrients as the field studies had shown. They found that the sediments retained more than 90 percent of the nitrate-nitrogen and phosphorus, whereas only 45 to 66 percent of the ammonium-nitrogen. These studies illustrated the long-term capacity of cypress dome sediments for retaining nutrients from wastewater.

Nessel (1978a, b) and Nessel and Bayley (1984) described the nutrient-retention capacity of a Florida cypress strand that had been enriched for 50 years with partially treated wastewater and found that these systems continued to be effective sinks for some nutrients. DeBusk and Reddy (1987) investigated the ammonium-nitrogen retention of sediment cores from that same Florida site using ^{15}N tracers and found that 21 days after application, 0.5 to 2.3 percent of the applied ^{15}N was in the floodwater and 11.4 to 17.3 percent in the sediment. This meant that from 82.2 to 86.3 percent of the ^{15}N was lost from the sediment-water system supposedly because of nitrification-denitrification and ammonia volatilization. Even the sediments that had received partially treated sewage for 50 years did not show reduced nitrogen-removal efficiency. The authors suggest that in these systems, the effectiveness of the nutrient removal is limited by phosphorus rather than nitrogen.

ECOSYSTEM MODELS

Hydrology Models

Several models have been developed of the hydrologic cycle of deepwater swamps or of regions that contain southern forested wetlands. Littlejohn (1977) demonstrated the importance of wetland conservation for groundwater protection with a water table model of the Gordon River region of southwestern Florida. Wiemhoff (1977) used a simple water model of an alluvial cypress-tupelo swamp in southern Illinois to demonstrate the water-storage potential of the wetland after river flooding. S. L. Brown (1978; 1984) developed a hydrology model for the Green Swamp in central Florida showing that the drainage of wetlands led to lower aquifers, drier streams, and greater floods. Water available to the region was reduced by 16 percent in the model when 40 percent of the wetlands were drained. In an application of a water-transport model already developed by the U.S. Environmental Protection Agency, Hopkinson and Day (1980a, b) simulated hydrology and nutrient transport in a Louisiana swamp-bayou complex. Those simulations demonstrated that canals with dredged material along their banks channeled flows directly from uplands into estuarine lakes and bayous reducing sheet flow across the adjacent wetlands. They suggested that removal of the canals would again allow runoff to pass through the swamps, thus decreasing upstream flooding and optimizing conditions for the swamp forests, the bayous, and the downstream estuary.

Ecosystem Dynamics

Several conceptual and simulation models have been developed to study the ecosystem dynamics of deepwater swamps. A review of some of these models is given in Mitsch et al. (1982, 1988), Mitsch (1983, 1988), and H. T. Odum (1982, 1984). One such simulation model of a Florida cypress swamp was used to investigate the management of cypress domes in Florida (Fig. 13–10). The model predicted long-term (100-years) effects of drainage, harvesting, fire, and nutrient disposal. Simulations showed that when water levels were lowered and trees were logged at the same time, the cypress swamp did not recover and was replaced by shrub vegetation. The model also suggested that if tree harvesting occurred without drainage, the cypress would recover because of the absence of fire. In a model that investigated understory productivity in cypress domes, Mitsch (1984) demonstrated annual patterns of aquatic productivity that peaked in the spring when maximum solar radiation is available through the deciduous cypress canopy. M. T. Brown (1988) developed a model that demonstrated the annual patterns of hydrology and nutrients in forested swamps typical of central and northern Florida. The model illustrated the impact of surface water diversion, lowering groundwater, and additions of advanced wastewater treatment effluents on the hydroperiod and nutrient dynamics of the swamps.

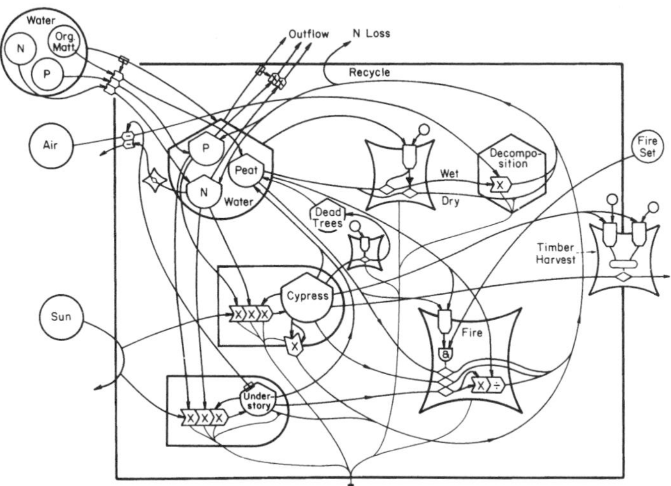

Figure 13–10. Simulation model of Florida cypress dome ecosystem. (*From H. T. Odum, 1983; copyright © 1983 by John Wiley and Sons, Inc., reprinted with permission*)

RIPARIAN WETLAND

Riparian Wetlands

14

Riparian wetlands, ecosystems in which soils and soil moisture are influenced by the adjacent stream or river, are unique because of their linear form along rivers and streams and because they process large fluxes of energy and materials from upstream systems. Major expanses of riparian ecosystems, called bottomland hardwood forests, are found in the southeastern United States, although many have been drained and cleared for agriculture. Less extensive but ecologically critical riparian systems in the western United States, especially in the arid Southwest, have nearly all been greatly modified as a result of water management practices, logging, and use by livestock. Riparian zones respond to structures and processes operating at continental scales (climate, geology); river system (intra-riparian) scales (elevation, timing, magnitude, duration of flooding, stream valley landform dynamics); and local trans-riparian scales (slope and moisture gradients, sediment sorting, biotic processes). Riparian zones span a wide range of environments and processes, the common thread being the linkage between riparian zone, river, and adjacent upland. Riparian systems of the American West often occur on low-order streams and are characterized by extreme and variable fluvial conditions and geomorphic responses. Southeastern riparian systems, which are the most extensive wetlands in the United States, in contrast, are generally low-lying, flat, extensive floodplains, with strong seasonal hydrologic pulses and well developed soils.

Flooding of the riparian zone affects soil chemistry by producing anaerobic conditions, importing and removing organic matter, and replenishing mineral nutrients. The plant communities of riparian ecosystems, which form in response to continental and watershed gradients, trans-riparian moisture, valley form, and soil gradients, are generally productive and diverse. Evidence shows that there is not always the expected positive relationship between periodic flooding and productivity. The riparian zone is valuable for many animals that seek its refuge, diversity of habitat, and abundant water or that use it as a corridor for migration. This is particularly true in arid regions, where riparian zones may support the only vegetation within many kilometers. The ecosystem functions of these systems are poorly understood, except that primary productivity is generally higher than that in adjacent uplands from the same region, and that the systems are open to large fluxes of energy and nutrients. The riparian ecosystem acts as a nutrient sink for lateral runoff from uplands but as a nutrient transformer for upstream-downstream flows. Few energy/nutrient models have been developed to describe these systems, although tree growth models have been applied to bottomland hardwood forests with some success.

The riparian zone of a river, stream, or other body of water is the land adjacent to that body of water that is, at least periodically, influenced by flooding. E. P. Odum (1981) described the riparian zone as "an interface between man's most vital resource, namely, water, and his living space, the land." A National Symposium on Strategies for Protection and Management of Floodplain Wetlands and Other Riparian Ecosystems (R. R. Johnson and McCormick, 1979) developed a definition of riparian ecosystems:

> Riparian ecosystems are ecosystems with a high water table because of proximity to an aquatic ecosystem or subsurface water. Riparian ecosystems usually occur as an ecotone between aquatic and upland ecosystems but have distinct vegetation and soil characteristics. Aridity, topographic relief, and presence of depositional soils most strongly influence the extent of high water tables and associated riparian ecosystems. These ecosystems are most commonly recognized as bottomland hardwood and floodplain forests in the eastern and central U.S. and as bosque or streambank vegetation in the west. Riparian ecosystems are uniquely characterized by the combination of high species diversity, high species densities and high productivity. Continuous interactions occur between riparian, aquatic, and upland terrestrial ecosystems through exchanges of energy, nutrients, and species.

Minshall et al. (1989) incorporated the U.S. Fish and Wildlife Service wetland definition (Cowardin et al., 1979) into their definition of riparian zones of the western states Great Basin region: "Land inclusive of hydrophytes and/or with soil that is saturated by ground water for at least part of the growing season within the rooting depth of potential native vegetation."

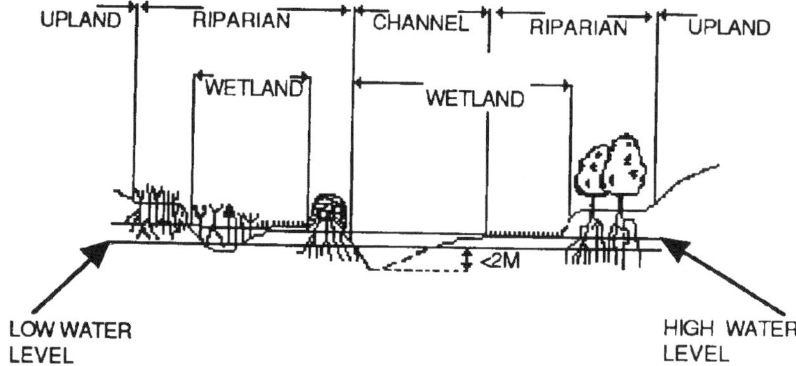

Figure 14–1. The relationship of riparian to wetland habitat. (*From Minshall et al., 1989*)

This definition includes both "wetlands" as defined by Cowardin et al. (1979) and more mesic adjacent lands called *transitional* ecosystems by Kovalchik (1987). The latter are subirrigated sites such as inactive floodplains, terraces, toe-slopes, and meadows that have seasonally high water tables that recede to below the rooting zone in late summer. Figure 14–1 compares riparian with wetland habitats. The latter are a subset of riparian habitats, based on the more restrictive wetland definition of water dependency.

Gregory et al. (1991) emphasized landscape pattern in their characterization of the riparian zone:

> Riparian zones are the interfaces between terrestrial and aquatic ecosystems. As ecotones, they encompass sharp gradients of environmental factors, ecological processes and plant communities. Riparian zones are not easily delineated but are composed of mosaics of landforms, communities, and environments within the larger landscape.

In general terms, riparian ecosystems are found wherever streams or rivers at least occasionally cause flooding beyond their channel confines or where new sites for vegetation establishment and growth are created by channel meandering (e.g., point bars) (Bradley and Smith, 1986). In arid regions, riparian vegetation may be found along or in ephemeral streams as well as on the floodplains of perennial streams. In most nonarid regions, floodplains and hence riparian zones tend to appear first along a stream "where the flow in the channel changes from ephemeral to perennial—that is, where groundwater enters the channel in sufficient quantity to sustain flow through nonstorm periods" (Leopold et al., 1964).

Riparian ecosystems can be broad alluvial valleys several tens of kilometers wide in the southern United States or narrow strips of streambank vegetation in the arid western United States. Brinson et al. (1981b) described the "abundance of water and rich alluvial soils" as factors that make riparian ecosystems different from upland ecosystems. They listed three major features that separate riparian ecosystems from other ecosystem types:

1. Riparian ecosystems generally have a linear form as a consequence of their proximity to rivers and streams.
2. Energy and material from the surrounding landscape converge and pass through riparian ecosystems in much greater amounts than those of any other wetland ecosystem; that is, riparian systems are open systems.
3. Riparian ecosystems are functionally connected to upstream and downstream ecosystems and are laterally connected to upslope (upland) and downslope (aquatic) ecosystems.

In the United States, the most extensive riparian ecosystems are the mesic bottomland hardwood forest ecosystems of the Southeast, including the broad, flat floodplain of the Mississippi River as well as the floodplains of numerous smaller rivers emptying into the Gulf of Mexico and the Atlantic Ocean. These southeast ecosystems are part of a flooding continuum that includes deepwater swamps (discussed in the previous chapter). These flat, low-lying floodplains are quite different from the steep, narrow, continuously changing riparian zones in the Northwest mountains and from the floodplains of ephemeral rivers of the arid Southwest.

Several proceedings and reports have been published about riparian wetlands, including the proceedings of national symposia on floodplain wetlands (R. R. Johnson and McCormick, 1979; Abell, 1984; R. R. Johnson et al., 1985; Hook and Lea, 1989), the proceedings of a series of workshops on southern bottomland hardwood forests (Clark and Benforado, 1981; Gosselink et al., 1990b), and general reviews on bottomland hardwood forests and riparian ecosystems (Brinson et al., 1981b; Wharton et al., 1982; Minshall et al., 1989; Faber et al., 1989; Sharitz and Mitsch, 1993).

TYPES AND GEOGRAPHICAL DISTRIBUTION

Table 14–1 gives the distribution of land in riparian forests in the contiguous United States based on riparian forest vegetation type groups from the U.S. Forest Service surveys. Abernethy and Turner (1987) estimated a total 1985 riparian forest area of about 22.9 million hectares, excluding California, Arizona, New Mexico, Hawaii, and Alaska. Forested wetland area in Arizona, California, and New Mexico is about 360,000 hectares (Brinson et al., 1981b), and 12 million hectares are estimated in Alaska. Between 1940 and 1980 the national loss rate was 0.27 percent per year, or about 2.8 million hectares. Most of this loss occurred in the south-central and southeastern United States, which has 58 percent of the total United States forested wetlands. The Forest Service

Table 14–1. Wetland Forest Area in the United States, 1980, and Rates of Change 1940–1980

Region	*Area, 1940 x 1000 ha*	*Area, 1980 x 1000 ha*	*Cumulative change 1940–1980, x 1000 ha/yr*	*% change per year*
North Central	6,900	5,000	-47.6	-0.69
North East	1,650	3,000	+33.8	+2.0
Pacific Coast (excluding California)	730	1,250	+12.9	+1.8
Rocky Mountain (excluding Arizona and New Mexico)	730	375	-9.0	-1.2
South Central	9,000	7,100	-47	-0.52
South East	6,600	6,100	-11.7	-0.18
TOTAL	25,600	22,800	-68.6	-0.27

Source: Abernethy and Turner, 1987

surveys do not include many Western United States riparian strips because minimum surveyed size is 38 m wide or 4 hectares in area. Thus the Pacific Coast and Rocky Mountain estimates are probably low. These figures differ slightly from those given in Chapter 3 because they were compiled by a different methodology. The trends, however, are the same from both sources.

Bottomland Hardwood Forests

Bottomland hardwood forests are one of the dominant types of riparian ecosystems in the United States. Historically the term "bottomland hardwood forest," or "bottomland hardwoods," has been used to describe the vast forests that occur on river floodplains of the southeastern United States. Huffman and Forsythe (1981) suggested that the term applies to floodplain forests of the eastern and central United States as well. Their definition of a bottomland hardwood forest includes the following:

1. The habitat is inundated or saturated by surface or groundwater periodically during the growing season.
2. The soils within the root zone become saturated periodically during the growing season.

3. The prevalent woody plant species associated with a given habitat have demonstrated the ability, because of morphological and/or physiological adaptation(s), to survive, achieve maturity, and reproduce in a habitat where the soils within the root zone may become anaerobic for various periods during the growing season.

Bottomland hardwood forests are particularly notable wetlands because of the large areas that they cover in the southeastern United States (Fig. 14–2) and because of the rapid rate at which they are being converted to other uses such as agriculture and human settlements (see also Chap. 3). This ecosystem is particularly prevalent in the lower Mississippi River alluvial valley as far north as southern Illinois and western Kentucky (Taylor et al., 1990) and along many streams that drain into the Atlantic Ocean on the south Atlantic Coastal Plain. The Nature Conservancy (1992) estimated that before European settlement this region supported about 21 million hectares of riparian forests; about 4.9 million hectares remained in 1991 (Fig. 3–9). The loss of riparian wetlands has been accompanied by fragmentation: Whereas single forest blocks used to cover hundreds of thou-

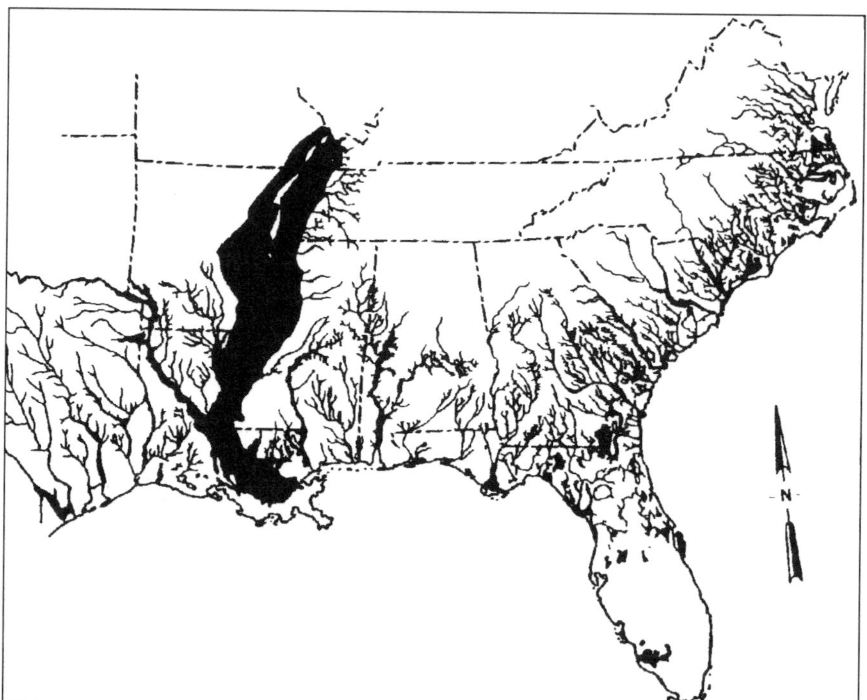

Figure 14–2. Extent of bottomland hardwood forests of southern United States. (*After Putnam et al., 1960*)

sands of hectares, there now remain isolated fragments, most less than 100 hectares in size, surrounded by agricultural fields (Gosselink et al., 1990c). The Atlantic Coastal Plain from Maryland to Florida is another area of dense riparian forests lining the many rivers that flow into the ocean. These forests have been logged, as have most in this country, but many are otherwise fairly intact.

Western Riparian Wetlands

Abernethy and Turner (1987) estimated a total 1980 forested area of 375,000 hectares in the Rocky Mountain states (excluding Arizona and New Mexico) and 1.25 million hectares in the Pacific Coast states of Washington and Oregon. Arizona was reported to contain about 113,000 hectares of riparian ecosystems (Babcock, 1968), New Mexico probably less. California has been reported to have lost more than 91 percent of its original wetland area (Dahl, 1990), including most of its riparian wetlands (Faber et al., 1989).

In contrast to the broad, flat, expansive southeastern riparian forests, western riparian zones tend to be narrow, linear features of the landscape, often lining streams with steep gradients and narrow floodplains. Along southeastern high-order rivers the contrast in elevation and vegetation between bottomland and upland is often subtle and the gradients is gradual, whereas in the West gradients are usually sharp and the visual distinctions are usually clear (Fig 14–3).

Western riparian zones have been extensively modified by human activity. Conversion to housing or agriculture is widespread. Damage from grazing animals is almost ubiquitous. In an area where vegetation is generally limited by the lack of water, riparian vegetation and the availability of water inevitably draw and concentrate cattle. Grazing along these primarily low-order streams results in increased erosion and channel downcutting (Kauffman and Kreuger, 1984). Higher-order streams have been modified for water use. They have been dammed; water withdrawals have depleted instream flows; the timing and size of floods and low flows have been modified; vegetation has been removed to reduce transpiration losses. In some areas essentially all water has been withdrawn and perennial streams have become ephemeral below a dam. In other cases, return flows from agricultural use enhance stream flow and have led to expanded and more stable riparian zones (W. C. Johnson et al., 1976; Williams and Wolman, 1984; Bradley and Smith, 1986; Faber et al., 1989; Szaro, 1989; Knopf and Scott, 1990; Rood and Mahoney, 1990).

In the Northwest, the major impact on streams and riparian zones has occurred because of logging. Beginning in the early 1800s streams were used to float logs from the forest to estuarine holding pens. Because many streams were too small to move logs efficiently, they were dammed, banks were cleared, and were otherwise modified to enhance logging. Major log drives still occur today in the rivers of the northwestern United States, British Columbia, and Alaska (Sedell and Duval, 1985).

Figure 14–3. A view of Red Meadow Creek, Montana, about 7,000 feet elevation. Typical linked series of wetlands from high elevation seeps fed by melting snow, through shrub-dominated steep avalanche wetland chutes, to extensive wet meadows dominated by *Carex* and *Juncus*. (*Photograph by L. C. Lee*)

Figure 14–4. Riparian zone along a stream in arid Wyoming about 4,800 feet elevation. Note distinct pattern of trees (*Populus* spp., *Craetagus* spp., *Alnus incana, Salix* spp.) in an otherwise treeless landscape. Adjacent farmland is irrigated. (*Photograph by L. C. Lee*)

GEOMORPHOLOGY AND HYDROLOGY

One way to describe these different systems is in terms of landform and process gradients that result in changing continua of riparian zones. Throughout this chapter we refer to processes on three major gradients that are nested in both space and time (Kovalchik, 1987; Minshall et al., 1989):

1. A continental gradient that includes the effects of east-west and latitudinal climatic gradients acting at the hydrologic basin level;
2. An intra-riparian continuum (R. R. Johnson and Lowe, 1985), reflecting changes in elevation, stream gradient, fluvial processes, and sediments along the length of a stream system (Frissell et al., 1986; Baker, 1989; Gregory et al., 1991);
3. A lateral trans-riparian gradient (R. R. Johnson and Lowe, 1985) across the riparian zone. This local topographic gradient reflects stream valley cross-sectional form and determines the local moisture regime and soil development.

Continental and Regional Scales

Climate

Climate—related to latitude, elevation, and continental weather patterns—determines precipitation patterns and the presence or absence of a significant spring

Table 14-2. A Comparison of Physical, Geomorphic, and

Attribute	*SOUTHWEST* *Upstream*	*Downstream*
Example	Rio San Pedro	Colorado River
Climate	arid	arid
Net precipitation	ET>>P	ET>>P
Landscape pattern	linear	braided
Stream gradient	steep	flat
Floodplain width	narrow	broad
Valley shape	V	alluvial fan
Fluvial type	very flashy	flashy
Fluvial stability (annual peak flood to mean flow)	10–100	
Geomorphic system stability	very low disequilibrium	low
Landscape contrast	very high	high
Vegetation	alder, poplar	cottonwood, willow, ash, walnut, screw bean, salt cedar
Human impacts	water diversion, dams, grazing, clearing for water retention	

thaw. In the eastern and south-central United States, the climate is mesic. Precipitation exceeds evapotranspiration on an annual basis, and even during late summer, moisture deficits are not usually severe. Mountains are nonexistent or not as high as in the West. As a result, streams are usually perennial from source to mouth (Table 14–2). In a westward direction moisture decreases; in the American West, evapotranspiration exceeds precipitation. In addition, along a latitudinal gradient in the Mid- and Southwest, precipitation declines so that in the southwestern United States in general evapotranspiration is much greater than precipitation (precipitation is <20 to 50 cm, and evapotranspiration is >100

Ecological Features of Western and Eastern United States Riparian Ecosystems

NORTHWEST		*SOUTHEAST*	
Upstream	*Downstream*	*Upstream*	*Downstream*
	N. Fork Flathead River		Mississippi River alluvial valley
semi-arid	semi-arid	mesic	mesic
$ET \geq P$	$ET > P$	$P > ET$	$P > ET$
linear	linear	linear	broad floodplain
steep	moderate/flat	steep/moderate	flat, meandering
narrow	moderate	moderate	very broad
V	alluvial fan	U	sinuous depositional flats
very flashy	flashy	flashy	seasonally pulsed
			2–5
10–100		5–10	
very low disequilibrium	fairly stable	moderately stable	stable (1000+ yr)
moderate	high	fairly low	low
willow, poplar, birch, hemlock	willow, sycamore, walnut, cottonwood	sycamore, oak, hackberry, silver maple, river birch	cypress, tupelo, many oaks spp., willow, hickory, loblolly pine
logging, water diversion, dams		logging, farming, flood control levees and draining	

cm). The picture is complicated by the rainshadow effect of the Rocky Mountain range, yielding major climatic contrasts between the moist, maritime northwestern coastal range and the arid eastern slope of the Rocky Mountains.

Evapotranspiration, which is governed primarily by air temperature, can control water levels even in mesic regions (Sharitz and Gibbons, 1982). In arid regions it may be the single most important influence on riparian zones. Hence western riparian zones are much more variable than the mesic eastern part of the United States. As examples of the foregoing, the delta of the Colorado River on the Gulf of California to its confluence with the Gila River is (or was before dam construction) perennial with hydro-riparian vegetation; upstream one encounters the meso-riparian vegetation of the intermittent Rio San Pedro. The intermittent headwater tributaries of the San Pedro are dry, desert, xero-riparian habitats (R. R. Johnson and Lowe, 1985). In the montane northwestern states, where high altitudes moderate evapotranspiration and increase precipitation, a mesic to semiarid climate gives rise to lush meso-riparian vegetation.

Watershed

In addition to continental factors—precipitation and temperature—the character of a watershed exerts a strong influence on river flooding. The local flooding regime of a riparian zone is determined by the elevation of the watershed (which, in turn, is related to precipitation and evapotranspiration), by the size of the watershed, and by its slope.

Drainage Area. Discharge volume and flooding duration are particularly related to drainage area of the stream basin upstream of the floodplain. For two basins in Arkansas, Bedinger (1981) found that sites that had larger upstream drainage areas had correspondingly longer times in standing water because small watersheds have rapid runoff and sharp flood peaks, whereas large watersheds have flood peaks that are less sudden and longer lasting (Table 14–3). Typically, midwestern and eastern U.S. riparian ecosystems on small watersheds or low-order streams are flooded for several days to several weeks during the spring thaw, whereas the flooding of bottomland forests on the high-order Mississippi River lasts for long periods of up to several months. In the West, snows store precipitation over large areas of the watershed, releasing it during the short spring period. This results in great floods that are not correlated with local precipitation.

Channel Slope. A stream that has a steep slope or gradient floods less frequently but with sharper flood peaks than a stream that contains a gentle slope. This is illustrated in Figure 14–5, which shows the recurrence interval of floods. In the steeply sloped California rivers, the flow of water tends to be extreme and to be associated with storms. For example, a flood 50 times the mean annual

Table 14–3. Relationship of Drainage Basin Size to Duration of Flooding of Bottomland Forests in Arkansas

Drainage Basin Area, mi^2	*Flooding Duration, Percent of time*
300	5-7
500-700	10-18
Several tens of thousands	18-40

Source: Based on data from Bedinger, 1981

flood can be expected every 20 years on the Santa Anna River, and a 50-year flood on the Gila River is 280 times the mean annual discharge (Graf, 1988). In comparison, a 20-year flood on eastern coastal plain rivers is only about 1.5 times the average annual flood (see also Chap. 4).

Intra-Riparian Scale

Within the constraints of the continental or regional environment, the riparian zone changes along the course of a river system from headwaters to its mouth. Vannote et al. (1980) described the structure and function of aquatic communities along a river system. The *intra-riparian continuum* (R. R. Johnson and Lowe, 1985) refers to the same continuum in the riparian corridor along a river's course—the characteristics of communities along the river are determined by the geologic setting and by the geomorphic, hydrologic, and other physical processes that provide the physical environment within which a riparian community develops.

Most rivers have three major geomorphic zones (Fig. 14–6): erosion, storage and transport, and sediment deposition. These zones have been variously characterized (Brakenridge, 1988; Graf, 1988; R. R. Harris, 1988; Minshall et al., 1989; and Faber et al., 1989).

1. The *zone of erosion* is in the headwaters and upper reaches of low-order streams. This zone is at high altitudes, at least in many areas of the West. The

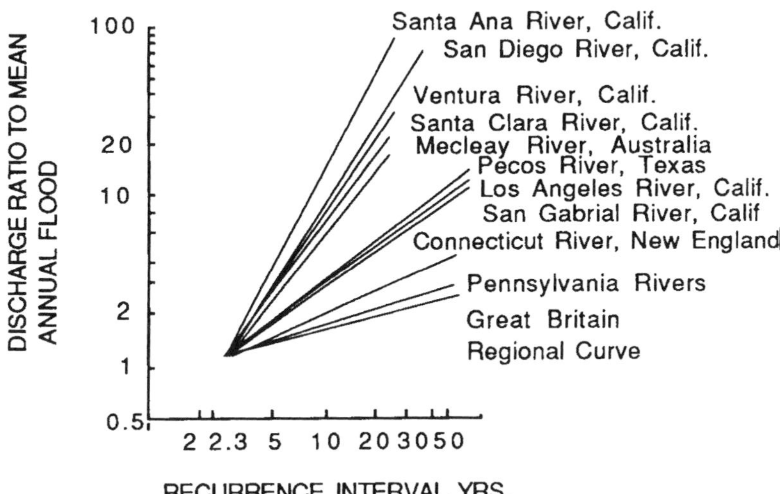

Figure 14–5. Relationship between mean annual flood (which has a recurrence interval of 2.33 years) and recurrence intervals for several rivers. (*From Faber et al., 1989*)

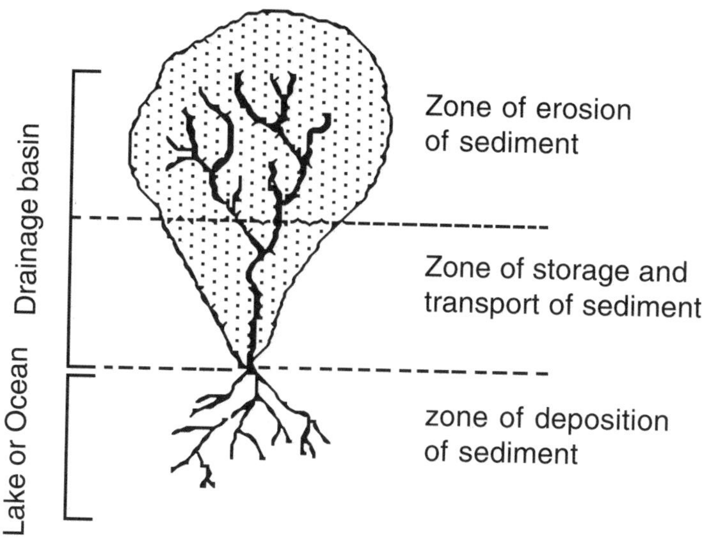

Figure 14–6. Three major geomorphic zones in the fluvial system. (*From Faber et al., 1989*)

stream course tends to be steep and straight, and the valleys are often V-shaped because they are scoured. The steep banks have narrow riparian zones. Flood frequency and duration vary widely, depending on precipitation. Depending on local geology, these headwater areas may contain fairly extensive, flat meadows (fens) that may accrete considerable organic peat at high altitudes.

2. The *zone of storage and transport* occurs below the zone of erosion. These mid-order streams are primarily conduits for sediment, nutrients, and water. They tend to be fairly steep, and their straight V- or U-shaped channels, with some coarse sediment deposition, form a narrow floodplain. These sediments are often scoured during high-energy floods. Flooding is variable and depends on the size of the watershed, the gradient, and local precipitation.
3. The *zone of deposition* is characteristic of high-order, low-gradient streams. Sediment deposition is much greater than erosion and transport, and valley slopes are gentle. These two factors lead to the development of broad floodplains and sinuous and meandering stream channels. Sediments grade from coarse at the channel to fine at the periphery of the floodplain. The flooding of the riparian zone tends to be seasonal, characterized by one or a few long-duration spring floods. At their downstream ends, rivers typically debauch into flat, broad valleys, where the channels become braided and the flow is often unconfined. These depositional rivers are characteristic of the southeastern Coastal Plain, running into coastal estuaries. In the American West, evapotran-

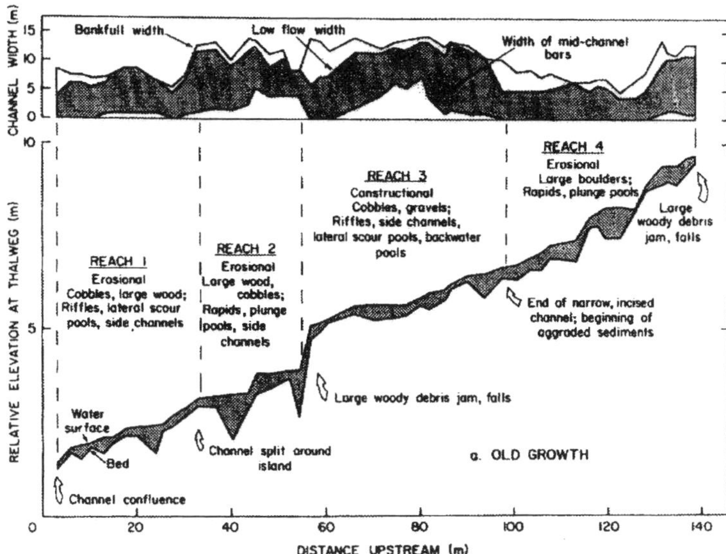

Figure 14–7. Variation in slope and channel width in an old-growth stream in the Oregon Cascades. (*From Frissell, 1986; copyright © 1986 by Springer-Verlag, reprinted with permission*)

spiration is the dominant process that determines the volume of the flow of these braided streams. As water evaporates and flow decreases, soil salts may accumulate.

Western Riparian Systems

Frissel (1986) diagramed a fourth-order stream flowing through an old-growth forest in the Oregon Cascades (Figure 14–7). Most of the reaches are erosional, interspersed with depositional (constructional) segments. In cross-section, similar streams assume a number of shapes related to geomorphic processes. In the eastern Sierra Nevadas, glaciated bedrock valleys at elevations greater than 2,500 m tend to have incised stream channels and narrow floodplains, moderate gradients, and limited sediment deposits (R. R. Harris, 1988; Fig. 14–8). (This type of valley occurs above any water diversions and therefore is not influenced by them.) U-shaped valleys in glacial till occur at slightly lower elevations (2,000–2,500 m). The floodplain gradient is moderate, and the floodplain width is the widest of all types classified. V-shaped valleys incised in glacial till span the widest elevation range and are almost always associated with diversions that reduce the mean annual streamflow below the diversion to less than 18 percent of the natural flow. These four channel types probably occur predominantly in erosional and transport sections of the stream (Fig. 14–6).

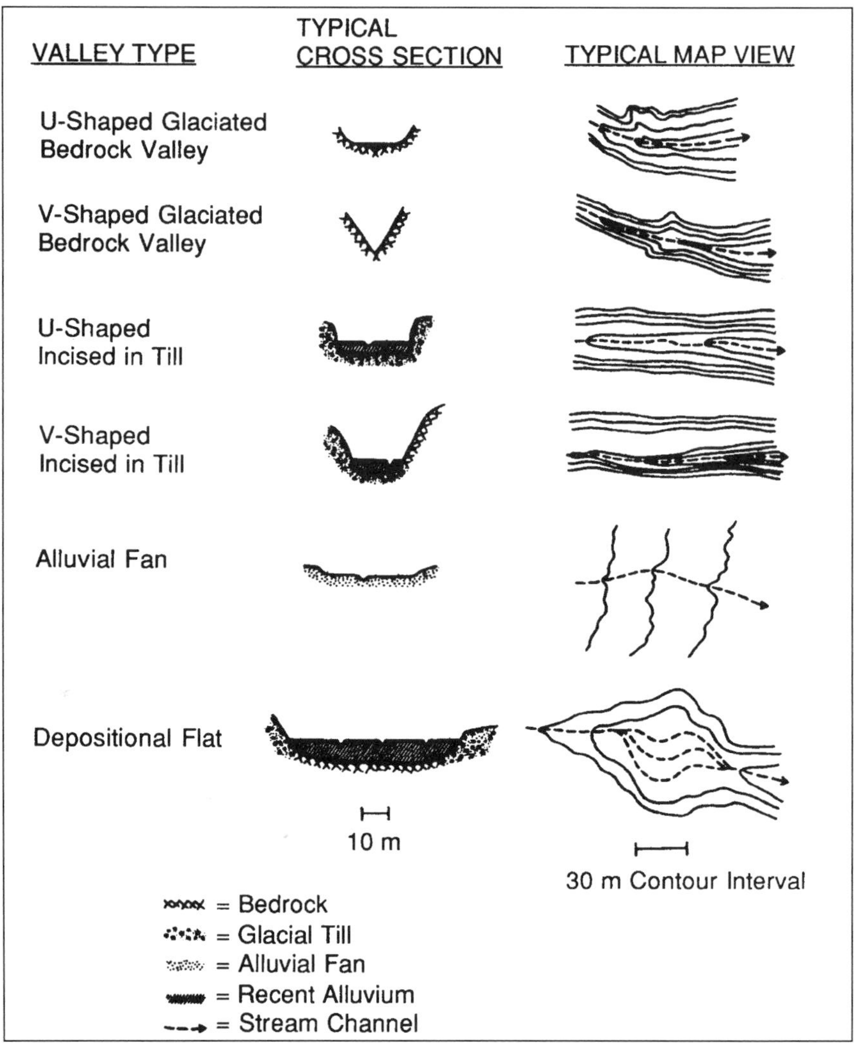

Figure 14–8. Classification of valley types in an eastern Sierra Nevada stream. (*From R. R. Harris, 1988; copyright © 1988 by Springer-Verlag, reprinted with permission*)

Depositional stream reaches bordered by alluvial fans are at elevations of less than 2,000 m, all containing braided channels (and all below diversions). Depositional flats are characterized by gentle gradients, broad floodplains, meandering channels, and high groundwater tables.

Structurally the temporal and spatial stability of these western rivers is fundamentally different from those of rivers in other regions of the United States. Because of the dramatic differences in peak floods compared to mean flows, the temporally unpredictable nature of flooding, and the coarseness of most sedimentary material in this region, arid-region channels are time-dependent systems that seldom reach any kind of equilibrium (Graf, 1988). When a large flood causes drastic changes in channel morphology, the channel is no longer in equilibrium with low flows. During an extended period without floods, the channel adjusts by infilling, development of vegetation, and other changes. As a result it is unable to accommodate subsequent floods, which again change its morphology, beginning the cycle of change again. The changes in morphology vary along the length of the channel, depending on local differences in slope, geology, the presence of logs, and other factors. A large flood can transform a narrow, meandering channel to a wide, shallow braided channel that results in the complete elimination of the floodplain. This occurred in the Gila River (Burkham, 1972) and the Cimarron River (Schumm and Lichty, 1963). The recovery of a meandering channel resulted from channel narrowing and increased bank cohesiveness after vegetation was established and began to trap silt and clay. A stream segment may have both braided and meandering segments or the geomorphic types may be nested with a low-flow meander channel in a high-flood braided network (Graf, 1988; Brakenridge, 1988).

Eastern Riparian Systems

Most of the extensive forests of the Mississippi River floodplain are characterized as zones of deposition. A typical broad floodplain in Eastern North America contains several major features (Leopold et al., 1964; Brinson et al., 1981b; Bedinger, 1981; Fig. 14–9).

1. The river channel meanders through the area, transporting, eroding, and depositing alluvial sediments.
2. Natural levees adjacent to the channel are composed of coarser materials that are deposited when floods flow over the channel banks. Natural levees, sloping sharply toward the river and more gently away from the flood plain, are often the highest elevation on the floodplain.
3. Meander scrolls are depressions and ridges on the convex side of bends in the river. They are formed as the stream migrates laterally across the floodplain. This type of terrain is often referred to as ridge and swale topography.
4. Oxbows, or oxbow lakes, are bodies of permanently standing water that result from the cutoff of meanders. Deepwater swamps (Chap. 13) or freshwater marshes (Chap. 11) often develop in oxbows in the southern United States.
5. Point bars are areas of sedimentation on the convex sides of river curves. As sediments are deposited on the point bar, the meander curve of the river tends to increase in radius and migrate downstream. Eventually, the point bar begins to support vegetation that stabilizes it as part of the floodplain.
6. Sloughs are areas of dead water that form in meander scrolls and along valley walls. Deepwater swamps can also form in the permanently flooded sloughs.

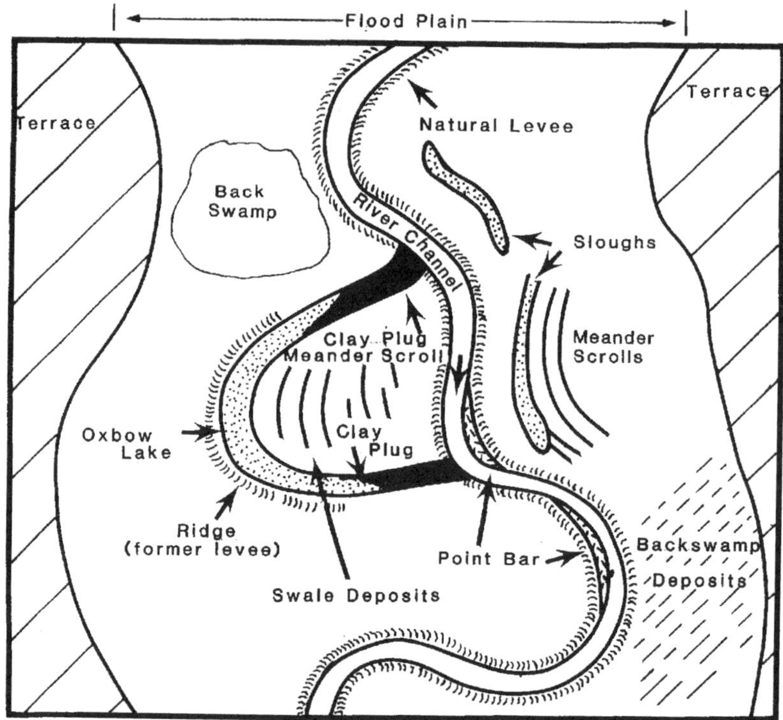

Figure 14–9. Major geomorphologic features of a southeastern U.S. floodplain. (*After Leopold et al., 1964, and Brinson et al., 1981*)

7. Backswamps are deposits of fine sediments that occur between the natural levee and the valley wall or terrace.
8. Terraces are "abandoned floodplains" that may have been formed by the river's alluvial deposits but are not hydrologically connected to the present river.

As is true of the western U.S. riparian systems, the importance of the river to the floodplain and the floodplain to the river in the eastern U.S. cannot be overemphasized. If either is altered, the other will change over time because floodplains and their rivers are in a continuous dynamic balance between the building and the removal of structure. In the long term, floodplains result from a combination of the deposition of alluvial materials (aggradation) and the downcutting of surface geology (degradation).

Two major aggradation processes are thought to be responsible for the formation of most floodplains: deposition on the inside curves of rivers (point bars) and deposition from overbank flooding. "As a river moves laterally, sediment is deposited within or below the level of the bankfull stage on the point bar, while at overflow stages the sediment is deposited on both the point bar and over the adjacent flood plain" (Leopold et al., 1964). The resulting floodplain is made up of alluvial sediments (or alluvium) that can range from 10 to 80 m thick. In the lower reaches of the Mississippi River, the alluvium, derived from the river over many thousands of years, generally progresses from gravel or coarse sand at the bottom to fine-grained material at the surface (Bedinger, 1981).

Degradation (downcutting) of floodplains occurs when the supply of sediments is less than the outflow of sediments, a condition that could be caused naturally with a shift in climate or synthetically with the construction of an upstream dam. There are few long-term data to substantiate the first cause, but a considerable number of "before-and-after" studies have verified stream degradation downstream of dams attributed to the trapping of sediments (Meade and Parker, 1985). In the absence of geologic uplifting, rivers tend to degrade slowly, and "downcutting is slow enough that lateral swinging of the channel can usually make the valley wider than the channel itself" (Leopold et al., 1964). The process is thus difficult to observe over short periods; both aggradation and degradation can only be inferred from the study of floodplain stratigraphy.

The formation of a riparian floodplain and terrace is shown in sequences A to B and C to E in Figure 14–10. When degradation occurs but some of the original floodplain is not downcut, that "abandoned" floodplain is called a terrace. Although it may be composed of alluvial fill, it is not considered part of the active floodplain. Aggradation and degradation can alternate over time, as shown in the sequence C to E in Figure 14–10. A third case, a dynamic steady state, can exist if aggradation resulting from the input of sediments from upstream is balanced by the degradation or downcutting of the stream (Brinson et al., 1981b). Figure 14–10 demonstrates that the same surface geometry can result from two dissimilar sequences of aggradation and degradation.

Trans-Riparian Scale

At a local site on a stream the riparian vegetation is determined by the cross-sectional morphology, including braiding of the stream, width of the floodplain, soil type, and elevation and moisture gradients. These are all determined in part by larger scale (continental, basin, stream system) processes that are modified by local biotic and physical processes.

Western Riparian Systems

Typical stream valley cross-sections of the western United States, as described by R. R. Harris (1988), were discussed earlier in this chapter and diagramed in Figure 14–8. These geomorphic valley types are predictably associated with distinctive vegetation patterns (Table 14–4; see the Ecosystem Structure section).

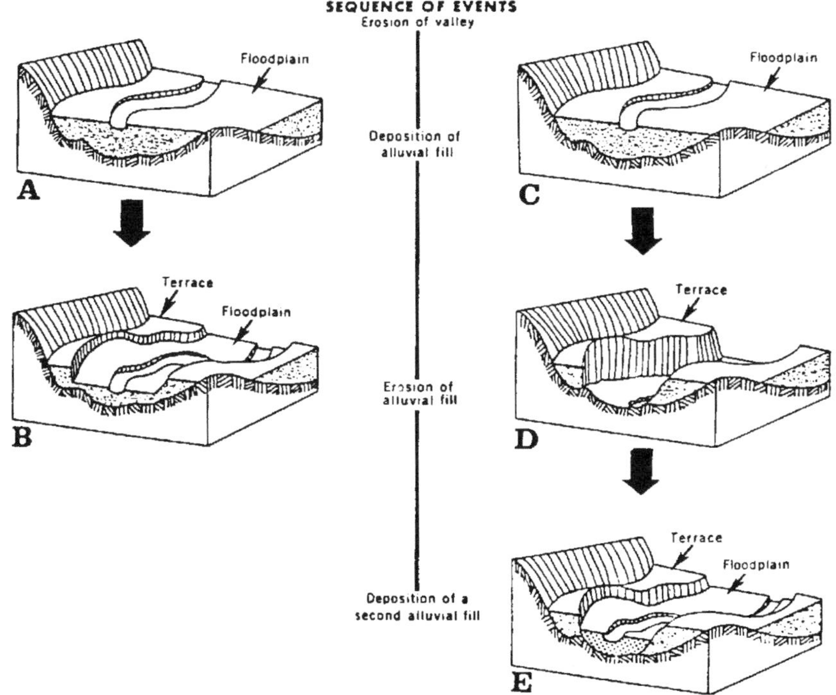

Figure 14–10. Two sequences in the development of floodplains and river terraces. (*From Brinson et al., 1981b, after Leopold et al., 1964*)

R. R. Harris (1988) and other researchers have used the riparian soil moisture regime, in part, to explain the plant associations. R. R. Johnson and Lowe (1985) showed a schematic diagram of the relationship of soil moisture to riparian communities across the floodplain of a typical western U.S. stream (Fig. 14–11). The relationship, however, is probably seldom as simple as that outlined in Figure 14–11. L. C. Lee et al. (1985), for example, reported that although a floodplain moisture gradient existed in a Montana riparian zone, the gradient did not limit the distribution of floodplain plant communities. In another study, Baker (1989) found that the riparian vegetation in western Colorado sites was determined by the macroscale factors of elevation and the size of the drainage basin as well as by the microscale (trans-riparian) variable of channel width.

Eastern Riparian Systems

The extensive bottomland hardwood forests of the southeastern United States often show clear zonation patterns related to elevation and flood frequency. This pattern is the basis for a classification developed at a workshop on bottomland hardwood wetlands held at Lake Lanier, Georgia, in 1980 (Figure 14–12). The fol-

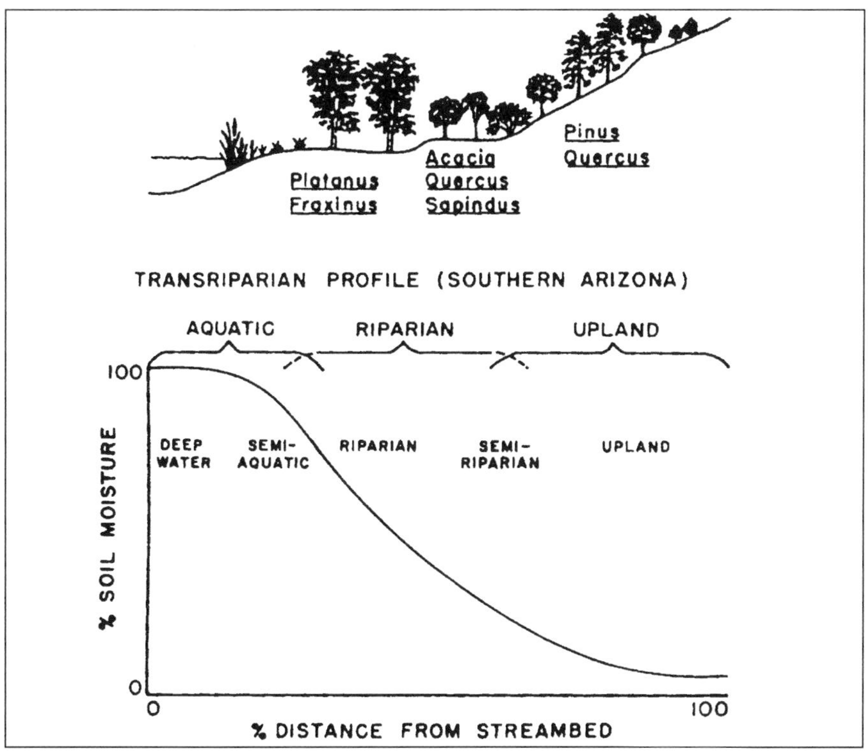

Figure 14–11. Schematic of the trans-riparian continuum on the moisture gradient from aquatic through riparian into the adjacent upland community. (*From R. R. Johnson and Lowe, 1985*)

lowing water regimes were described (Cowardin et al., 1979; Larson et al., 1981):

Zone II—Intermittently exposed. Surface water is present throughout the year except in years of extreme drought. The probability of annual flooding is nearly 100 percent, and vegetation exists in saturated or flooded soil for the entire growing season.

Zone III—Semipermanently flooded. Surface water or soil saturation persists for a major portion of the growing season in most years. Flooding frequency ranges from 51 to 100 years per 100 years. Flooding duration typically exceeds 25 percent of the growing season.

Zone IV—Seasonally flooded. Surface water or saturated soil is present for extended periods, especially early in the growing season, but is absent by the end of the season in most years. Flooding frequency ranges from 51 to 100 years per 100 years, and flooding duration is typically 12.5 to 25 percent of the growing season.

Table 14–4. Vegetation Characteristics of Different Valley Types in the Eastern Sierra Nevada

				Vegetation Type Associations		
Valley Type[a]	*Mean Riparian Zone Width,*[b] *m*	*Mean Species Richness*[c]	*Associated Vegetation Types*	*Substrate*	*Cross section Type*	*Diversion Status*
U-shaped			*Pinus contorta*–meadow	Sand, cobble	Incised	Undiverted
bedrock	7.0	8.0	*Salix* spp.–*Glycerea strista*	Gravel	Braided	Undiverted
U-shaped till			*Salix* spp.–*Cornus stolonifera*	Cobble	One-sided	NS
	27.0	13.0	*Populus tremuloides*–*Salix* spp.	NS	One-or two-sided	Undiverted
V-shaped till	21.0	9.0	***Betula occidentalis*–*Salix* spp.**	NS	NS	NS
Alluvial fan	27.0	9.0	*Chrysothamnus nauseosus*–*Artemisa tridutula*	Gravel	Braided	Diverted
			Populus trichocarpa–*Rosa woodsii*	Gravel	NS	Diverted
All			*Populus tremuloides*–*Populus trichocarpa*	Boulderc	NS	NS

[a]See Fig. 14–8 for description of valley types
[b]Average width of vegetation by valley type
[c]Average number of species per plot by valley type
Source: R. R. Harris, 1988

Zone V—Temporarily flooded. Surface water is present or soil is saturated for brief periods during the growing season, but the water table usually lies well below the soil surface for most of the season. A typical frequency of flooding is 11 to 50 years out of 100. Typical duration is 2–12.5 percent of the growing season.

Zone VI—Intermittently flooded. Soil inundation or saturation rarely occurs, but surface water may be present for variable periods without detectable seasonal periodicity. Flood frequency typically ranges from 1 to 10 years per 100 years. Total duration of flood events is typically less than 2 percent of the growing season.

It should not be assumed that these flooding conditions or zones occur in sequence from the river's edge to the uplands, as suggested in Figure 14–12 because the floodplain level does not necessarily rise uniformly from the river. In this classification of flooding, riparian ecosystems occur in zones II through VI. Zone I is the aquatic system of the stream bed; zone II contains deepwater swamp ecosystems more typical of the alluvial river swamp discussed in Chap. 13; zones II and III are typically considered wetland by most wetland scientists, but there is much debate over the inclusion of zones IV and V in the wetland definition when dealing with wetland management issues. As described by Larson et al. (1981):

> Within Zones II and III, water is an overriding environmental factor and is present on a nearly permanent basis, or at least for a major portion of the growing season. Within Zones IV and V, water is a signifi-

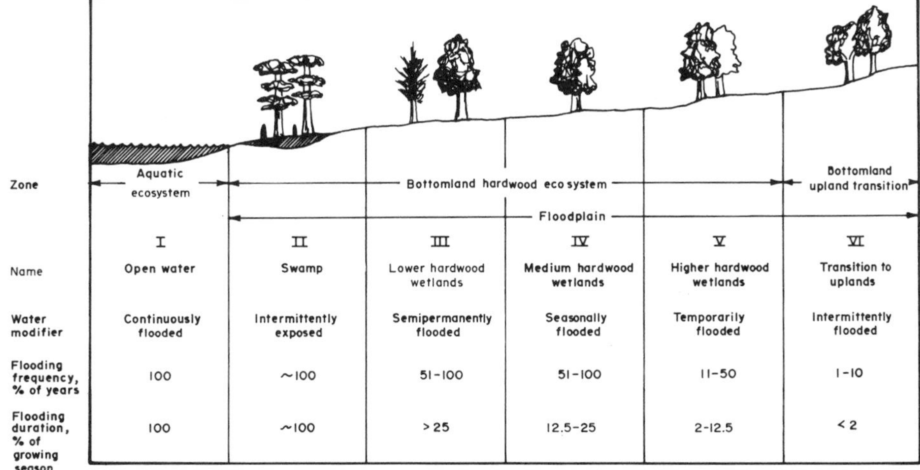

Figure 14–12. Zonal classification of southeastern United States bottomland forest wetlands showing average hydrologic conditions. (*After Clark and Benforado, 1981*)

cant determinant but its periodicity and duration indicate that other environmental factors such as nutrient status, competition, and soil texture are also determinants.

Zone VI ecosystems are sometimes not considered wetlands but are thought to constitute a zone of transition to upland ecosystems.

Although a description of zones across these extensive floodplains is useful in a descriptive sense, it has dubious value as a management tool because it ignores the strong interactions among the zones and of the whole riparian area with rivers and uplands. The zones occur because a gradient exists across the floodplain. The gradient itself presupposes flows of materials and energy to maintain it. The coupling is functionally tight (Brinson et al., 1981b; Gosselink et al, 1990b), for it extends to large landscapes in which river, floodplain, and uplands form both longitudinal and transverse continua that are are important to the integrity of the whole river system (Vannote et al., 1980; Forman and Godron, 1986; M. G. Turner, 1989; Gosselink et al., 1990c).

CHEMISTRY AND SOILS

Soil Structure

The physiochemical characteristics of the soils of riparian ecosystems are different from those of either upland ecosystems or permanently flooded swamps. In erosional and transport systems, sediments are coarse and soils poorly developed. In general, riparian soils on floodplains and low river terraces of the American Southwest are youthful and poorly developed. Outcrops of bedrock occur frequently.

Presumably, all riparian soils have an aquic moisture regime, but some riparian plant communities occur on nonaquic soils. They are probably remnant communities established on sites that used to be wetter than they are now (Padgett et al., 1989). Aquic soils must be saturated with water long enough to result in anaerobic conditions at least a few days a year, causing distinct mottles or gleyed soils.

In erosional and transport segments of streams, riparian soils are characterized by large fragments, for example, 10 to 23 percent >2 mm, 65–71 percent sand and remaining silt and clay, in aquic soils of the upper Gila River Basin (Brock, 1985). Riverwash consists of 63 percent coarse material, 35 percent sand, 2 percent silt, and 2 percent clay. Most soils in the mountainous region of Utah developed on alluvium, but colluvial and residual soils occur where underground flows are shallow, where there is late snowmelt, or where groundwater surfaces (Padgett et al., 1989). Organic soils may occur, primarily at high elevations where temperatures are cold and water is abundant.

In these dynamic systems the substrate can change rapidly and dramatically. For example, at one location on the San Francisco River, major areas of well-drained sandy soil 1.5 to 2 m deep were replaced by riverwash from a single severe flood (Brock, 1985).

Depositional systems in xeric areas such as the Great Basin region of Nevada, and western Utah riparian soils on alluvial fans, can be desert pavement that contains a smooth wind-formed surface of tightly fitting pebbles. Lime- and silica-cemented hardpans are common. The lower basin of these types of river typically has deep loam or silt clay loam soils that are saline or sodic (Minshall et al., 1989). Depositional southeastern U.S. bottomland forests develop deep alluvial fine-grained soils. Some of the most important soil properties are aeration, organic and clay content, and nutrient content. All of these characteristics are influenced by the flooding and subsequent dewatering of these ecosystems; the characteristics of the soil, in turn, greatly affect the structure and function of the plant communities that are found in the riparian ecosystem. Table 14–5 presents the typical physiochemical characteristics of southeastern floodplain soils by zone, including oxygenation, organic matter content, and mineral nutrients.

Soil Oxygen

Soil oxygen is one of the most important (yet changeable) characteristics of bottomland soils. Anaerobic conditions are created rapidly when the floodplain is flooded, sometimes in a period as short as a few days (see Fig. 5–3). When the floodplain is dewatered, aerobic conditions quickly return. Soil aeration is important for rooted vegetation. Most rooted plants are unable to function normally under extended periods of anoxia, although some plants have special adaptations that enable them to survive extended periods of little soil aeration (see Chap. 6). Soil aeration, defined as the capacity of a soil to transmit oxygen from the atmosphere to the root zone, is dramatically curtailed by flooding water, but it is also affected by several other characteristics of the soil, including texture, amount of organic matter, permeability to water flow, elevation of groundwater, and degree of compaction. Soils whose clay content is high and whose pore size is small impede drainage more than sandy or loamy soils do, thereby increasing the likelihood that they will be poorly aerated. The high organic matter content of bottomland soils can both increase and deplete soil oxygen. Organic matter usually improves the soil structure and thus increases the aeration of clay soils. On the other hand, decomposing organic matter creates an oxygen demand of its own. Soil aeration also depends on how close the groundwater level is to the soil surface. Finally, when bottomland soils are compacted, the air-filled pores may decrease dramatically, thereby decreasing soil aeration (W. H. Patrick, 1981).

These generalizations apply primarily to the deep, well-developed, fine-grained alluvial soils of the Southeast. In the American West, the coarse-textured immature riparian soils generally drain more rapidly and are better aer-

Table 14–5. Physiochemical Characteristics of Floodplain Soils by Zones as in Figure 14–12

	Zone				
Characteristic	*II*	*III*	*IV*	*V*	*VI*
Soil Texture	Dominated by silty clays or sands	Dominated by dense clays	Clays dominate surface; some coarser fractions (sands) increase with depth	Clay and sandy loams dominate; sandy soils frequent	Sands to clay
Sand:Silt:Clay (% composition)					
Blackwater	69:20:12	—	74:14:12	—	—
Alluvial	29:23:48	34:22:44	34:20:45	71:16:14	—
Organic Matter, %					
Blackwater	18.0	—	7.9	—	—
Alluvial	4.5	3.4	2.8	3.8	—
Oxygenation	Moving water aerobic; stagnant water anaerobic	Anaerobic for portions of the year	Alternating anaerobic and aerobic conditions	Alternating: mostly aerobic, occasionally anaerobic	Aerobic year-round
Soil Color	Gray to olive gray with greenish gray, bluish gray, and grayish green mottles	Gray with olive mottles	Dominantly gray on blackwater floodplains and reddish on alluvial with brownish gray and grayish brown mottles	Dominantly gray or grayish brown with brown, yellowish brown, and reddish brown mottles	Dominantly red, brown, reddish brown, yellow, yellowish red, and yellowish brown, with a wide range of mottle colors

pH					
Blackwater	5.0	—	5.1	—	—
Alluvial	5.0	5.3	5.5	5.6	—
Phosphorus (ppm)					
Blackwater	11.2	—	9.8	—	—
Alluvial	9.1	6.3	8.1	4.8	—
Calcium (ppm)					
Blackwater	607	—	346	—	—
Alluvial	1,079	752	669	186	—
Magnesium (ppm)					
Blackwater	98	—	36	—	—
Alluvial	154	140	145	39	—
Sodium (ppm)					
Blackwater	46	—	31	—	—
Alluvial	94	31	28	23	—
Potassium (ppm)					
Blackwater	48	—	29	—	—
Alluvial	51	28	32	20	—

Source: After Wharton et al., 1982

ated. They often show no mottling or other signs of prolonged saturation (Padgett et al., 1989). Conversely, saline or sodic soils may have sharp cemented horizons that prevent drainage.

Organic Matter

The organic content of bottomland soils is usually intermediate (2–5 percent) compared with the highly organic peats (20–60 percent) at one extreme and upland soils (0.4–1.5 percent) at the other (Wharton et al., 1982). The alternating aerobic and anaerobic conditions apparently slow down but do not eliminate decomposition. In addition, the high clay content provides a degree of protection against decomposition of litter and other organic matter on the floodplain. Furthermore, bottomland forests are generally more productive than upland forests, and therefore more organic litter is produced. Wharton et al. (1982) suggested that a 5 percent organic content of soils is a good indicator of periodically flooded areas as opposed to the more permanently flooded blackwater swamps of the southeastern United States.

The organic content of riparian soils depends on a number of processes, including primary production, allochthonous inputs, decomposition rates, and erosion. Organic content is low in the immature soils of western riparian systems where coarse-grained substrates and good aeration exist. The input of large woody organic matter and other allochthonous organic carbon, and the senescense of roots may increase organic content. In high-altitude zones that experience poor drainage, organic peats may develop.

Nutrients

The bottomland hardwood wetlands of the southeastern United States generally have ample available nutrients because of several processes. The high clay content of the soils results in high concentrations of nutrients such as phosphorus, which has a higher affinity for clay particles than for sand or silt particles. The high organic content results in higher concentrations of nitrogen than would be found in upland soils of low organic content (W. H. Patrick, 1981). The bottomland soils are also rich in nutrients because of continual replenishment during flooding. Because much of the sediment deposited by rivers consists of fine-grained clays and silts, nutrients such as phosphorus are likely to be deposited in greater amounts than if the material were coarse grained. Mitsch et al. (1979a), for example, found that clay-rich sediments that flooded an alluvial swamp in southern Illinois had phosphorus contents of 8.0 to 9.8 mg/g dry weight.

Anoxic conditions during flooding have several other effects on nutrient availability. Flooding causes soils to be in a highly reduced oxidation state and often causes a shift in pH, thereby increasing mobilization of certain minerals such as phosphorus, nitrogen, magnesium, sulfur, iron, manganese, boron, copper, and zinc (see Chap. 5). This can lead to both greater availability of certain nutrients and also to an accumulation of potentially toxic compounds in the soil. The low oxygen levels also cause a shift in the redox state of several nutrients to more reduced states, possibly making it more difficult for the plants to assimilate certain important elements such as nitrogen (Wharton et al., 1982).

Denitrification may be prevalent in soils that have adjacent oxidized and reduced zones, as is the case in periodically flooded riparian soils. Phosphorus is more soluble in flooded soils than in dry soils, but whether a shift from reduced to oxidized conditions and back again makes phosphorus more or less available has not been determined. Nevertheless, the periodic wetting and drying of riparian soils are important in the release of nutrients from leaf litter. The generally high concentrations of nutrients and the relatively good soil texture during dry periods suggest that the major limiting factor in riparian ecosystems is the physical stress of inadequate root oxygen during flooding rather than the inadequate supply of any mineral.

In riparian systems of the American West, nutrient levels are probably more closely related to the flux of nutrients than to soil stocks, except in high-altitude organic fens. The alder (*Alnus tenuifolia*) is a common nitrogen-fixing tree of the western riparian zones. In the Pacific Northwest concentrations of available nitrogen are higher in alder-dominated stands than in coniferous riparian zones, reflecting this capability (Goldman, 1961). In arid zones, nutrients and salts accumulate through evapotranspiration, sometimes to toxic levels, where streams carry water and dissolved salts to the riparian zone.

ECOSYSTEM STRUCTURE

Hierarchical Organization and Landscape Pattern

Throughout this chapter we have emphasized the role that scale plays in the organization and functioning of riparian ecosystems. Specifically, riparian structures and processes are determined by continental/regional climatic and geologic processes, intra-riparian processes along the length of a stream system, and trans-riparian phenomena across a stream and its riparian zone. These scales are closely interrelated. The plants growing at a point on a western stream may be confined geographically to the west by continental barriers to diffusion; they are adapted to the elevation of the site (intra-riparian scale), and they respond to the moisture content of the soil, which is related to trans-riparian elevation and flooding phenomena.

The spatial scales of structures are related to the time scales of events. Thus in southeastern bottomland systems sediment deposition and erosion occurring over thousands of years produce major geomorphic features—soil type, point bars, ridge and swale topography, floodplains. Vegetation communities develop over hundreds of years as a function of soil type, topography, and hydrology. River flow varies in irregular wet and dry cycles of tens of years. Seedling establishment is often closely tied to these cycles.

The annual cycle is the dominant time scale, reflecting the seasonality of temperature and hydrology, driving plant production, decomposition, and secondary biotic production. At shorter time scales, chemical processes and biota respond to short-term flooding and drying at microscales and minute-to-hour time scales (Gosselink et al., 1990c).

These types of hierarchy were clearly described for aquatic systems by Vannote et al. (1980), Frissell et al. (1986), and others, and applied to riparian zones by R. R. Johnson and Lowe (1985), Baker (1989), Gosselink et al. (1990c), and Gregory et al. (1991). Figure 14–13 shows a representation of this spatial and temporal hierarchy in the organization of stream-valley forms and riparian biota.

The vegetation of riparian ecosystems develops in response to processes at all three scales. Although local flooding or moisture regime often appears to be the immediate cause of specific vegetation associations, that regime itself reflects both micro- and macroscale processes.

Gregory et al. (1991) described these landscape scale interactions in riparian systems (Fig. 14–14):

> The narrow, ribbon-like networks of streams and rivers intricately dissect the landscape, accentuating the interaction between aquatic and surrounding terrestrial ecosystems. . . . The linear nature of lotic ecosystems enhances the importance of riparian zones in landscape ecology. River valleys connect montane headwaters with lowland terrains, providing avenues for the transfer of water, nutrients, sediments, particulate organic matter, and organisms. . . . Nutrients, sediments, and organic matter move laterally and are deposited onto floodplains, as well as being transported off the land into the stream. River valleys are important routes for the dispersal of plants and animals . . . and provide corridors for migratory species.

Vegetation

Southeastern Bottomland Forests

The vegetation of southeastern riparian ecosystems is dominated by diverse trees that are adapted to the wide variety of environmental conditions on the floodplain. The most important local environmental condition is the hydroperiod, which determines the "moisture gradient," or—as Wharton et al. (1982) prefer—the "anaerobic gradient," which varies in time and space across the floodplain. The plant species found along this gradient are related to the bottomland zones described above (Clark and Benforado, 1981). The vegetation in zone II is adapted to the wettest part of the floodplain that experiences almost continuous flooding. These *Taxodium-Nyssa* swamps were discussed in Chapter 13.

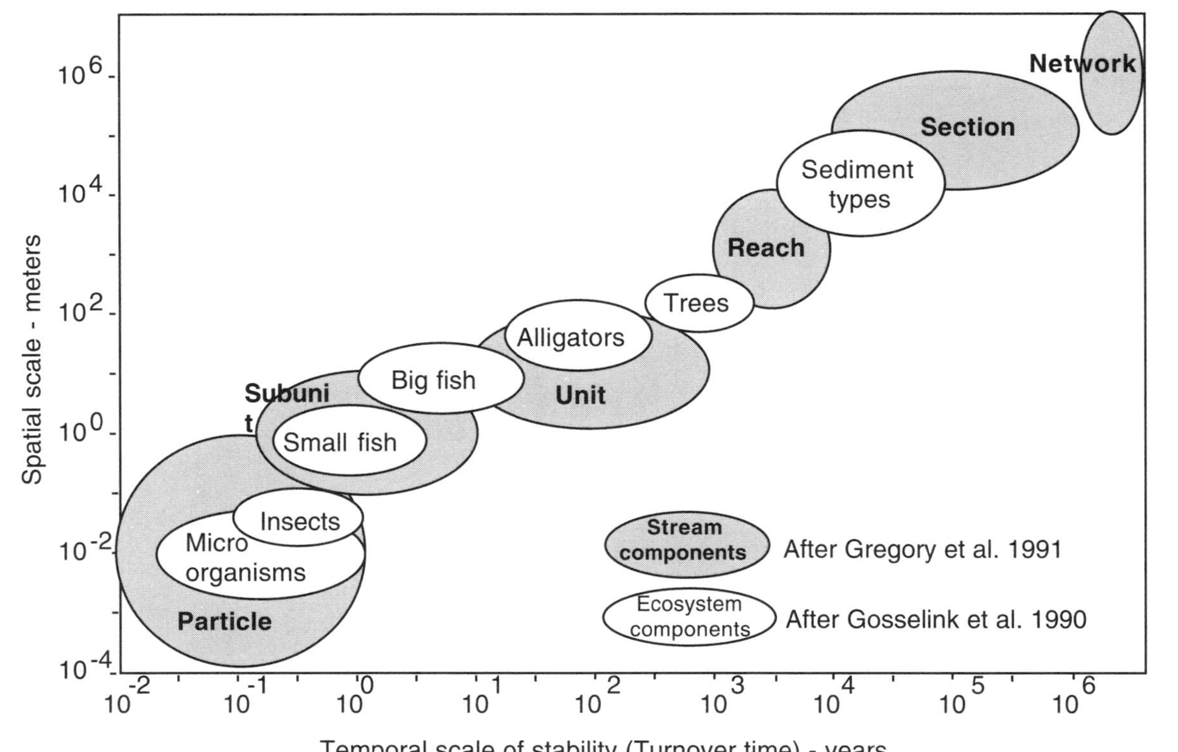

Figure 14–13. Temporal and spatial scales of hierarchical organization of valley landforms (*after Gregory et al., 1991*) and biota (*after Gosselink et al., 1990c*). The turnover time is a measure of the longevity or persistence of the features.

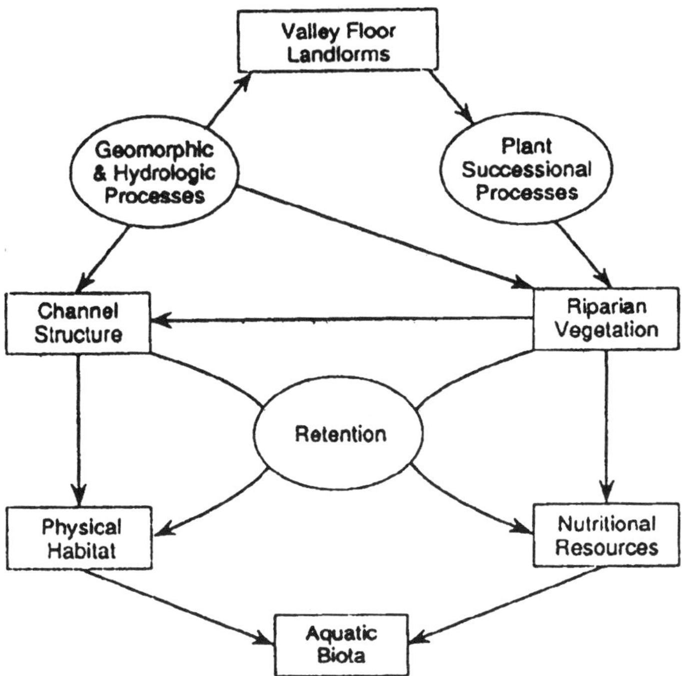

Figure 14–14. Relationships among geomorphic processes, terrestrial plant succession, and aquatic ecosystems in riparian zones. Directions of arrows indicate predominant influences of geomorphic and biological components (rectangles) and physical and ecological processes (circles). (*From Gregory et al., 1991; copyright © 1991 by the American Institute of Biological Sciences, reprinted with permission*)

Upslope from the deep swamps the soils are semipermanently inundated or saturated and support an association of black willow (*Salix nigra*), silver maple (*Acer saccharinum*), and sometimes cottonwood (*Populus deltoides*) in the pioneer stage (zone III). A more common association in this zone includes overcup oak (*Quercus lyrata*) and water hickory (*Carya aquatica*), which often occur in relatively small depressions on floodplains. Also found in this zone are green ash (*Fraxinus pennsylvanica*), red maple (*Acer rubrum*), and river birch (*Betula nigra*). New point bars that form in the river channel often are colonized by monospecific stands of willow, silver maple, river birch, or cottonwood.

Higher still on the floodplain (zone IV), in seasonally flooded or saturated bottomlands inundated one to two months during the growing season, are found an even wider array of hardwood trees. Common species include laurel oak (*Quercus laurifolia*), green ash (*Fraxinus pennsylvanica*), American elm (*Ulmus americana*), and sweetgum (*Liquidambar styraciflua*) as well as hackberry

(*Celtis laevigata*), red maple (*Quercus rubra*), willow oak (*Quercus phellos*), and sycamore (*Platanus occidentalis*). Pioneer successional communities in this zone can consist of monotypic stands of river birch or cottonwood. Notice that the zones are not exclusive; species often occur in two or more zones.

Temporarily flooded bottomlands at the highest elevations of the floodplain (zone V) are flooded for less than a week to about a month during each growing season. Several oaks, tolerant of occasionally wet soils, appear here. These include swamp chestnut oak (*Quercus michauxii*), cherrybark oak (*Quercus falcata* var. *pagodifolia*), and water oak (*Quercus nigra*). Hickories (*Carya* spp.) are often present in this zone in associations with the oaks. Two pines, spruce pine (*Pinus glabra*) and loblolly pine (*Pinus taeda*), occur at the edges of this zone in many bottomlands.

Plant zonation is not linear topographically, nor is it vegetationally discrete. Figure 14–15 is a cross section of the microtopography of an alluvial floodplain in the southeastern United States. The complex microrelief does not show a smooth change from one zone to the next. The levee next to the stream, in fact, is often one of the most diverse parts of the floodplain because of fluctuations in its elevation (Wharton et al., 1982).

Similarly, plant associations have diffuse edges. For example, Gemborys and Hodgkins (1971) showed the overlap of dominant plants species along a moisture gradient in a southwestern Alabama riparian zone (Fig. 14–16), and Robertson (1987) used ordination techniques to show that the distribution of plant species formed a continuum along a complex flooding-soil texture (drainage-aeration) gradient in a southern Illinois upland swamp.

Arid and Semiarid Riparian Forests

The riparian forests of the semiarid grasslands and arid western United States differ from those found in the humid eastern and southern United States. The natural upland ecosystems of this region are grasslands, deserts, or other nonforested ecosystems, and so the riparian zone is a conspicuous feature of the landscape (Brinson et al., 1981b; Johnson and Lowe, 1985). The riparian zone in arid regions is also narrow and sharply defined in contrast with the wide alluvial valleys of the southeastern United States. As stated by R. R. Johnson (1979), "when compared to the drier surrounding uplands, these riparian wetlands with their lush vegetation are attractive oases to wildlife and humans alike."

There are significant differences between the flora of riparian ecosystems in the eastern and western United States. One notable feature is the general absence of oak (*Quercus* spp.) in the West (Brinson et al., 1981b). Also, few species are common to both regions because of the differences in climate, although species of cottonwood (*Populus* spp.) and willow (*Salix* spp.) are found across the United States. Keammerer et al. (1975) and W. C. Johnson et al. (1976) described the

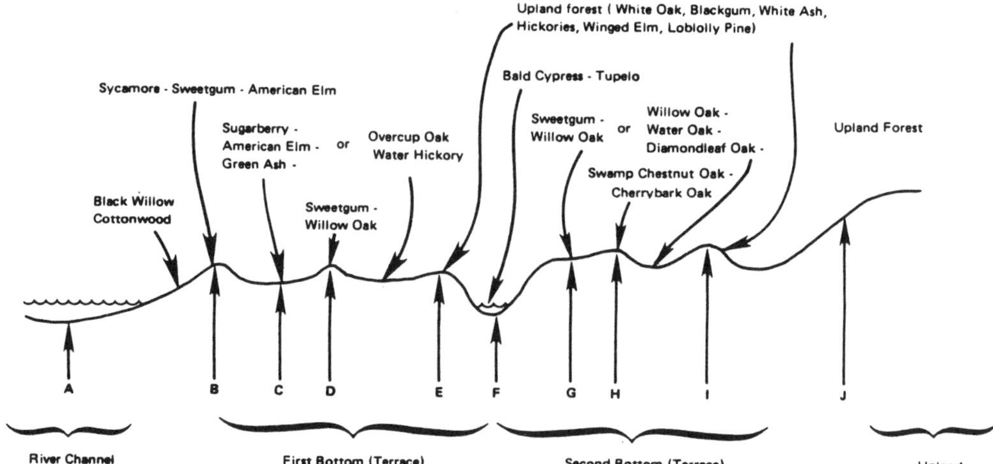

Figure 14–15. General relationship between vegetation associations and floodplain microtopography in southeastern floodplains. A=river channel; B=natural levee; C=backswamp or first terrace; D=low first terrace; E=high first terrace ridge; F=oxbow; G=second terrace flats; H=low second terrace ridge; I=high second terrace ridge; J= upland. (*From Wharton et al., 1982*)

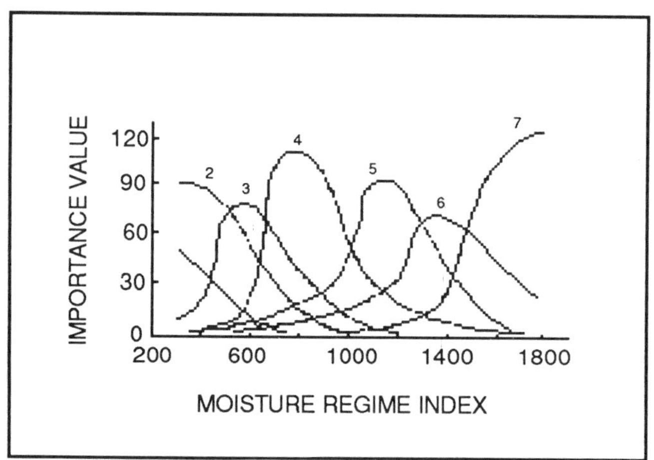

Figure 14–16. Importance value curves for the seven dominant species in a small stream bottomland in the Alabama coastal plain versus soil mixture index. 1. *Cornus florida*; 2. *Pinus palustris*; 3. *Quercus nigra*; 4. *Liquidambar styraciflua*; 5. *Nyssa sylvatica v. sylvatica*; 6. *Magnolia virginiana*; 7. *Nyssa sylvatica v. biflora*. (*From Gemborys and Hodgkins, 1971; copyright © 1971 by the Ecological Society of America, reprinted with permission*)

riparian forest along the Missouri River in North Dakota, where common species included cottonwood, willow, and green ash in the lower terraces and American elm, box elder (*Acer nugundo*), and bur oak (*Quercus macrocarpa*) at the higher elevations (Fig. 14–17).

Southwestern riparian zones commonly support cottonwood (*Populus fremontii*), willow (*Salix* spp.), sycamore (*Platanus wrightii*), ash (*Fraxinus pennsylvanica velutina*), walnut (*Juglans major*), and at higher elevations, alder (*Alnus tenuifolia*). Western riparian vegetation is often characterized by mesquite (*Prosopis* spp.) and catclaw acacia (*Acacia gregii*) or even by the cactus saguaro (*Cereus gigantea*) (R. R. Johnson and Lowe, 1985). The Rio Grande in Texas and New Mexico has a continuum of riparian vegetation, from a domination by screwbean (*Prosopis pubescens*) in the xeric south to associations of Fremont cottonwood, Goodding willow (*Salix gooddingii*), Russian olive (*Elaeagnus angustifolia*), and salt cedar (*Tamarix chinensis*) in the north. The last two species were introduced in the last 50 years and have changed the character and successional characteristics of many riparian woodlands in the arid west (Brinson et al., 1981b).

Figure 14–18 shows the elevational distribution and overlap of dominant overstory and undergrowth riparian plant species in Utah and southeastern Idaho (Padgett et al., 1989). Understory species produce a better partition of the environment than overstory, presumably because the latter have a broader ecological amplitude. Among the overstory trees different *Salix* species occur with changes in altitude; *S. exigua* is found at the lowest elevations, and *S. glauca* is found in alpine areas.

Several studies have related riparian vegetation associations in the western United States to physical site factors. R. R. Harris (1988) found a close relationship between eight vegetation types identified by ordination analysis and six valley forms in the eastern Sierra Nevada (Table 14–4). Similarly, Szaro (1989) found a strong relationship between site factors and plant associations for riparian forests in Arizona and New Mexico. He classified 24 forest and scrub community types based on cluster analysis and found that elevation was the most important determinant of vegetation at a site. Elevation, stream bearing, and stream gradient together explained 84 percent of the variation in the data. Baker (1989) found elevation and valley width to be closely correlated with species composition. Finally, L. C. Lee (1983) determined that for riparian vegetation community types in a Pacific Northwest river system, the floristic gradient determined by discriminant analysis was significantly related to a similar gradient of soil development. He interpreted the vegetation data to represent a canopy-structure gradient from nonforested wetlands to a nonwetland forested association.

Riparian systems in arid areas support many plant species unique to these environments. With the rapid modification of almost every stream for water management and the invasion of a number of exotic plant species, some native plants are becoming increasingly rare. For example, Faber et al. (1989) list nine rare and endangered riparian plant species in southern California.

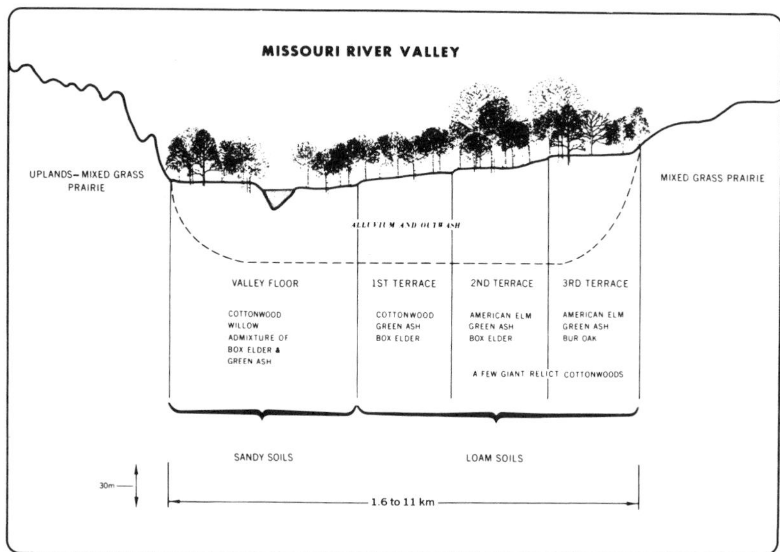

Figure 14–17. Cross section of Missouri River in semiarid North Dakota, showing major riparian tree species. (*From Brinson et al., 1981b, after Keammerer et al., 1975*)

Consumers

The riparian ecosystem provides a valuable and diverse habitat for many animal species. Some studies (e.g., Blem and Blem, 1975) have documented that forested floodplains are generally more populated with wildlife than adjacent uplands are. Because riparian ecosystems are at the interface between aquatic and terrestrial systems, they are a classic example of the ecological principle of the "edge effect" (E. P. Odum, 1979a). The diversity and abundance of species tend to be greatest at the ecotone, or "edge" between two distinct ecosystems such as a river and uplands. Brinson et al. (1981b) described four ecological attributes that are important to the animals of the riparian ecosystem:

1. *Predominance of woody plant communities.* This is particularly important in regions where the forested riparian zone is the only wooded region remaining, as in heavily farmed regions, or where the riparian forest was the only wooded area to begin with, as is the case in the arid West. Trees and shrubs not only provide protection, roosting areas, and favorable microclimates for many species, but they also provide standing dead trees and "snags" in streams that represent habitat value for both terrestrial and aquatic animals. The riparian vegetation also shades the stream, stabilizes the stream bank with tree roots, and produces leaf litter, all of which support a greater variety of aquatic life in the stream.
2. *Presence of surface water and abundant soil moisture.* The stream or river that passes through a riparian ecosystem is an important source of water, especially in the Far West, and of food for consumers such as waterfowl and fish-eating birds;

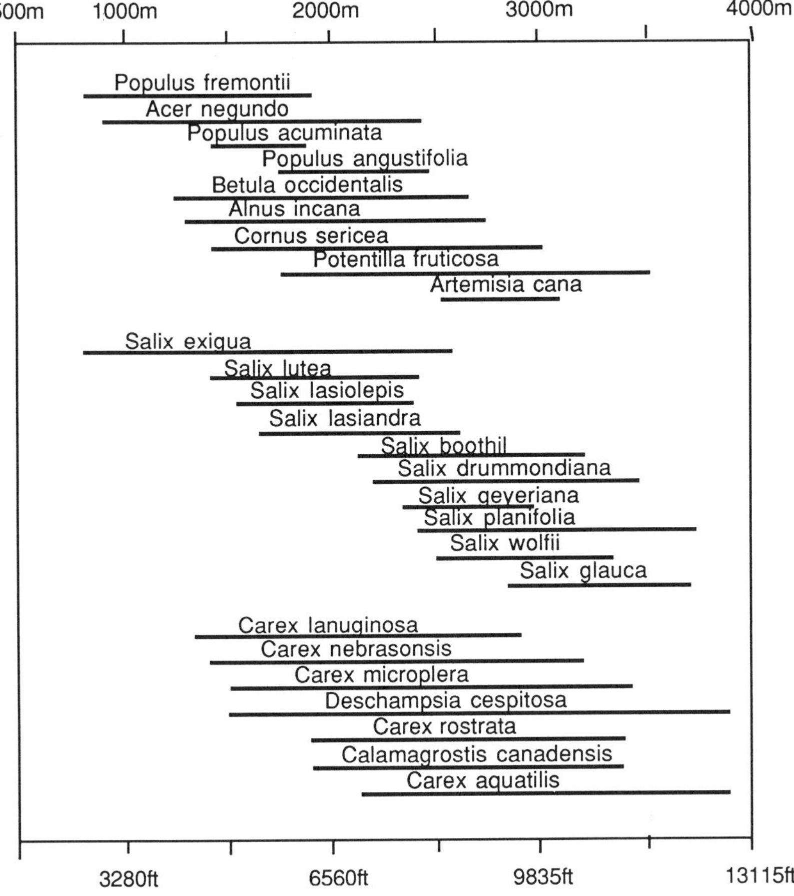

Figure 14–18. Elevational distribution of some dominant overstory and undergrowth riparian plant species in Utah. (*From Padgett et al., 1989, developed from data published by Welsh et al., 1987*)

an area of protection and travel for beavers and muskrats; and a reproduction haven for amphibians. The floodplain is also important for many aquatic species, particularly as a breeding and feeding area for fish during the flood season.

3. *Diversity and interspersion of habitat features.* The riparian zones in both eastern and western North America form an array of diverse habitats, from permanently flooded swamps to organic fens and infrequently flooded forests. This diversity, coupled with the edge effect discussed above, provides for a great abundance of wildlife.
4. *Corridors for dispersal and migration.* The linear nature of bottomland and other riparian ecosystems along rivers provides protective pathways for animals such as birds, deer, elk, and small mammals to migrate among habitats. Fish

migration and dispersal may also depend on maintaining riparian ecosystems along the stream or river.

Eastern Riparian Systems

A comprehensive description of the fauna that inhabit floodplain wetlands is not possible here because of the great number of birds, reptiles, amphibians, fish, mammals, and invertebrates that use the floodplain environment. Several good reviews on the fauna of bottomland and riparian wetlands are given by Fredrickson (1979), Wharton et al. (1981; 1982), Brinson et al. (1981b), Minshall et al. (1989), and Faber et al. (1989). The surveys that have been made in these areas show a rich diversity of animals. The animals, like the plants, are accustomed to certain zones in the floodplain, although animal, plant, and geomorphic zones do not always coincide (Wharton et al., 1981). Fredrickson (1979) described the distribution of amphibians, reptiles, mammals, and birds across southeastern floodplains, as shown in Figure 14–19. Few generalizations are possible. Several animals such as the beaver, river otter, snakes of the genus *Natrix,* the cottonmouth, and several frogs are present near the water. Deer, foxes, squirrels, certain species of mice, the copperhead, and the canebrake rattlesnake exist at the uplands edge. Most of the food chains are detrital, based on the organic production of the vegetation. Other characteristics of the fauna, associated with the sequence of flooding and dewatering, are high mobility, arboreal (tree-inhabiting) abilities, swimming ability, and the ability to survive inundation (Wharton et al., 1981). Most of the species in Figure 14–19 have one or more of these traits.

Welcomme (1979), Risotto and Turner (1985), and R. E. Turner (1988a, 1988b) reported a strong relationship between fishery yields and floodplains of lowland rivers because fish spawn and feed in floodplains during flood stages on the river (Lambou, 1990; Hall and Lambou, 1990) and because the productivity of large, lowland rivers depends on the exchange of nutrients with the floodplains (Junk et al., 1989).

Western Riparian Systems

Arid western riparian areas are particularly diverse and under enormous pressure from human development. Faber et al. (1989) summarized the fauna of the southern California riparian systems as follows: Insect fauna are estimated at 27,000–28,000 species that play an important role in the riparian community as both predators and prey. Fish populations are limited since most of the streams are intermittent. They are disappearing rapidly because of habitat destruction. Amphibians are present around undisturbed mountain streams. Eighty-eight species of breeding birds are strictly riparian and another twenty-three use riparian habitats extensively. Many species of nonbreeding birds also use riparian zones for food and rest during migration. The group most affected by riparian habitat loss consists of seventy-six species of passerine birds, of which fifty-nine

a.

RIVER OTTER
Lutra canadensis
MINK
Mustela vison
WHITE-TAILED DEER
Odocoileus virginianus
BEAVER
Castor canadensis
GRAY FOX
Urocyon cinereoargenteus
RACCOON
Procyon lotor
CONGO EEL
Amphiuma sp
GRAY SQUIRREL
Sciurus carolinensis
SIREN
Siren sp
BOBCAT
Lynx rufus
GREEN TREEFROG
Hyla cinerea
COTTONMOUTH
Agkistrodon piscivorus
GRAY TREEFROG
Hyla versicolor
BROAD-BANDED WATER SNAKE
Natrix fasciata
GOLDEN MOUSE
Peromyscus nuttalli
SWAMP RABBIT
Sylvilagus aquaticus
ALLIGATOR
Alligator mississippiensis
CANEBRAKE RATTLESNAKE
Crotalus horridus
BLACK RAT SNAKE
Elaphe obsoleta
BULLFROG
Rana catesbeiana
COPPERHEAD
Agkistrodon contortrix

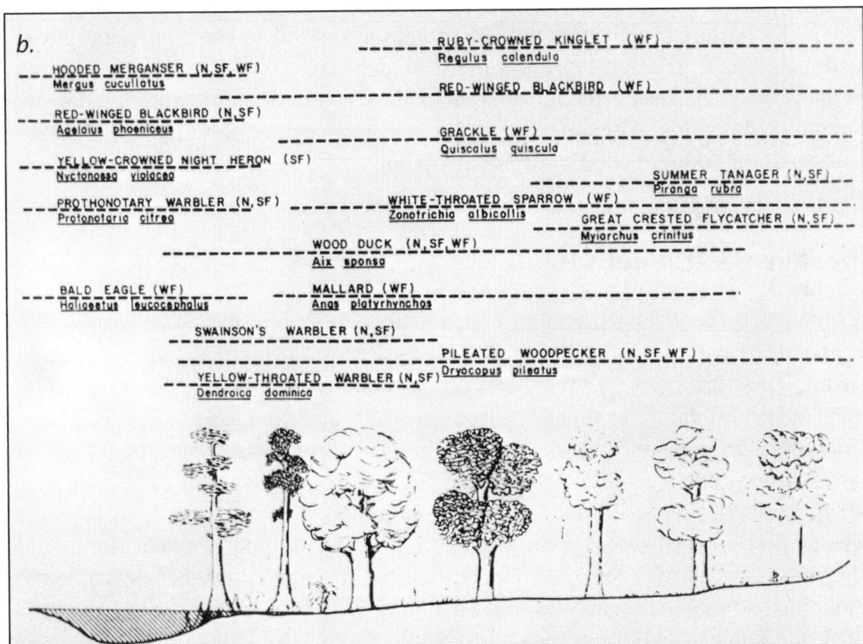

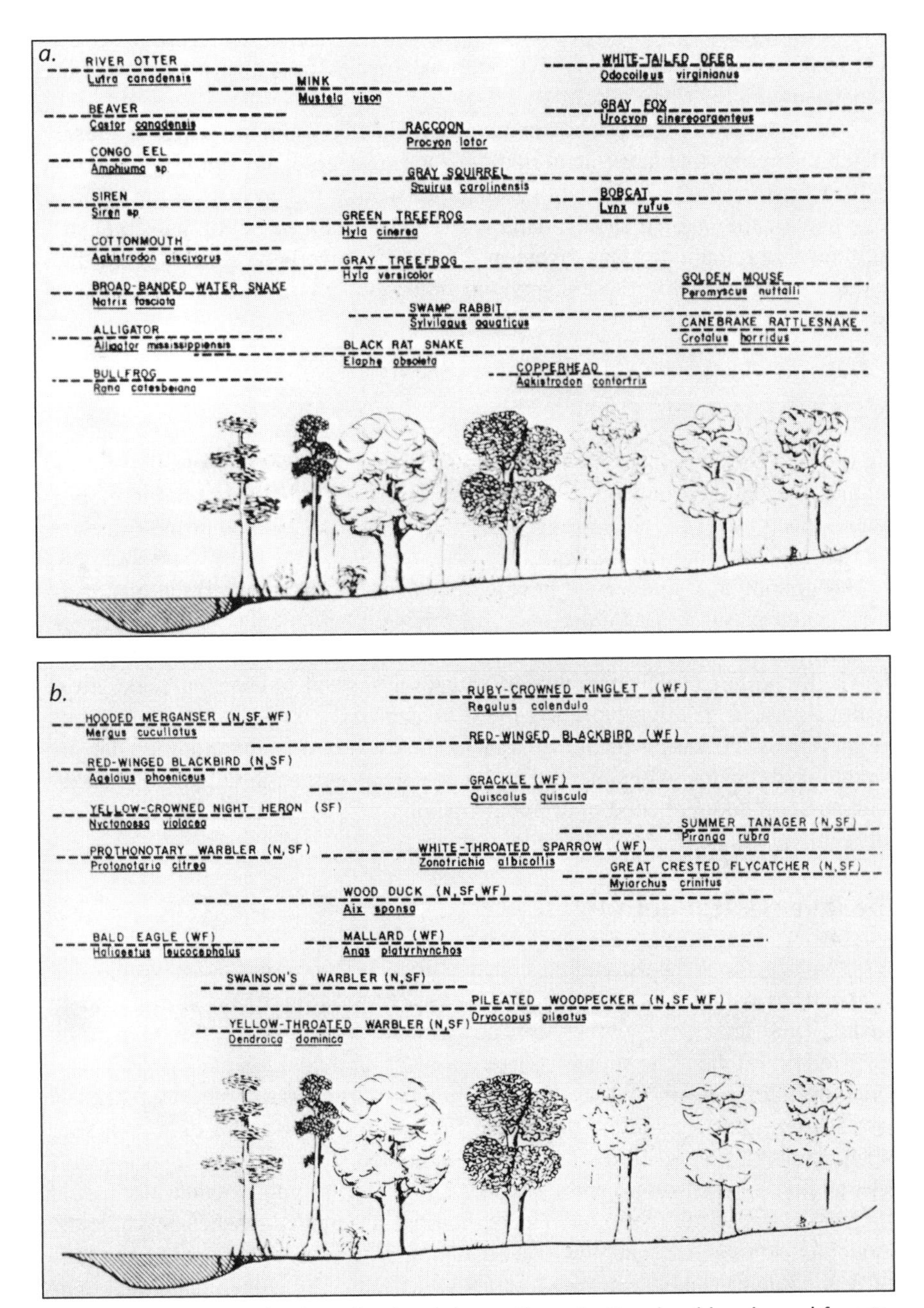

Figure 14–19. Distribution of animals in southern bottomland hardwood forests; *a.* amphibians, reptiles, and mammals, and *b.* birds in relation to nesting (N), summer foraging (SF), and winter foraging (WF). (*From Frederickson, 1979; copyright © 1979 by the American Water Resources Association, reprinted with permission*)

nest in the riparian zone. Forty-five mammal species in southern California are associated with the riparian habitat.

In northwestern United States and western Canada, salmon success is closely related to streams that have intact riparian zones (Sedell et al., 1980, 1982; Sedell and Luchessa, 1982). Snags and other large, woody material from the riparian zone historically created shoals, dammed sloughs, and caused log jams, enhancing the salmon habitat. Many rivers in this region have been isolated from their floodplains by clearing these snags and improving the channel for navigation (Sedell and Froggatt, 1984).

ECOSYSTEM FUNCTION

Riparian ecosystems have many functional characteristics that result from the unique physical environment. It is generally recognized that they are highly productive because of the convergence of energy and materials that pass through riparian wetlands in great amounts.

Throughout this chapter we have emphasized the close interrelationship of stream and riparian zone. Figure 14–20 illustrates schematically the role of riparian vegetation in the stream community. The interaction is much broader than that shown. McArthur (1989) enumerated aquatic-terrestrial linkages at scales from basin catchment to microbiota. At the drainage-basin level, the frequency and duration of the flooding of the riparian zone are functions of drainage density (length of channel/area of watershed). The flooding regime, in turn, controls sediment aeration and reduction, nutrient availability and the mineralization rate, and nutrient quality and flux to the adjacent stream.

The River Continuum

According to the river continuum concept (RCC) (Cummins, 1974; Vannote et al., 1980; Minshall et al., 1983, 1985), most organic matter is introduced to streams from terrestrial sources in headwater areas. The production/respiration (P/R) ratio is <1 (i.e., the stream is heterotrophic), and invertebrate shredders and collectors dominate the fauna. Organic matter is reduced in size as it travels downstream. Nutrient spirals (the processing and reprocessing of organic matter and minerals) are long. In midreaches of unbraided streams, more light is available, and phytoplankton prosper. The P/R ratio is >1. Organic matter input from upstream is fine; filter feeders dominate the flora. In braided reaches or where the floodplain is broad, however, the bank habitat is a major source of snags and logs that lead to debris dams that slow water flow and increase stream habitat diversity. The increased input of riparian coarse debris increases food diversity and increases heterotrophy. The P/R ratio is less than 1 and nutrient spirals tend to be tight and short (nutrients are rapidly processed). Terrestrial inputs are probably more nutritious than the fine reworked material from upstream. Even bacteria

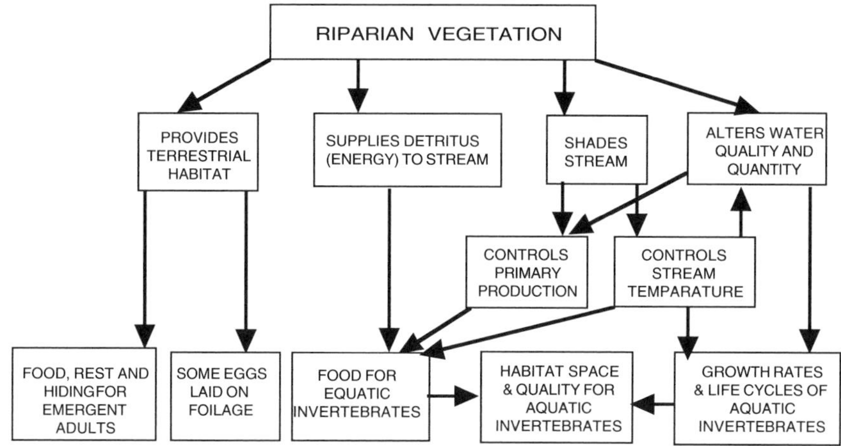

Figure 14–20. Some relationships between riparian vegetation and stream aquatic communities. (*After Knight and Bottorff, 1984*)

respond to the concentrations and sources of organic matter. The allelic frequency of metabolic enzymes is correlated with the habitat (McArthur, 1989). This suggests that bacterial flora and genetic selection pressures in a specific habitat (i.e., a floodplain-stream reach) are functions of linkages and interactions between stream and floodplain, including timing, quantity, quality, and source of organic material inputs.

Flood Pulse Concept

Junk et al. (1989) discussed the generality of a flood pulse concept (FPC) for floodplain/large river systems based on their experience in both temperate and tropical regions of the world. They dispute the river continuum concept as a generalizable theory because (1) most of the theory was developed from experience on low-order temperate streams, and (2) the concept is mostly restricted to habitats that are permanent and lotic. In essence, they believe that there are "biases and inadequacies of limnological paradigms when applied to floodplain systems." In the FPC, the pulsing of the river discharge is the major force controlling biota in river floodplains, and lateral exchange between the floodplain and river channel and nutrient cycling within the floodplain "have more direct impact on biota than nutrient spiralling discussed in the RCC."

Primary Productivity

Riparian wetlands are generally more productive than adjacent upland ecosystems because of their unique hydrologic conditions. The periodic flooding (flood pulses) usually contributes to higher productivities in at least three ways (Brinson et al., 1981b):

1. Flooding provides an adequate water supply for the vegetation. This is particularly important in arid riparian wetlands.
2. Nutrients are supplied and favorable alteration of soil chemistry results from the periodic overbank flooding. These alterations include nitrification, sulfate reduction, and nutrient mineralization.
3. The flowing water offers a more oxygenated root zone than if the water were stagnant. The periodic "flushing'' also carries away many waste products of soil and root metabolism such as carbon dioxide and methane.

Table 14–6 presents measurements of biomass and net primary productivity for several mid-eastern United States forested riparian ecosystems. In general, floodplain wetlands that have an annual unaltered cycle of wet and dry periods have an above-ground net biomass production (litterfall + stem growth) greater than 1,000 g m^{-2} yr^{-1}. This level of productivity is generally higher than that of forested wetlands that are permanently flooded or have sluggish flow, a point described in detail in Chapters 4 and 13. It is usually the case that both permanently flooded zones and rarely flooded zones are less productive than those zones that alternate between wet and dry conditions more frequently.

While it is generally believed that the pulsing floods on riparian wetlands necessarily lead to their high productivity (see Chap. 4 and Chap. 13), two studies of bottomland forested wetlands in Illinois have cast some doubt on this hypothesis. S. L. Brown and Peterson (1983) compared the above-ground productivity of two forested bottomlands in Illinois, one still water and the other flowing water. They found the productivity to be 50 percent higher in the still-water site than in the flowing-water site, suggesting that not all riparian wetlands are productive as a result of seasonal flooding. Mitsch and Rust (1984) investigated a 60-year record of annual tree-ring growth and the frequency of flooding in a forested riparian wetland in northeastern Illinois and found a general lack of correlation between measures of flooding and tree growth. They suggested a more complicated interplay, with floods having influences that are both positive (nutrient and water replenishment) and negative (anaerobic root zone and ice damage). Although neither of these studies conclusively eliminates the flood pulse–high productivity hypothesis, they did show that the relationship is more complicated than originally thought.

There have been few studies of the primary productivity of western U.S. riparian wetlands. Because these systems span a broader range of fluvial energy than southeastern floodplain forests (for which most productivity data are available), it would be interesting to determine whether the hydrologic energy-productivity relationship discussed above holds up in the western U.S. riparian ecosystems, or whether the energies of many montane streams are beyond subsidy levels, acting as stresses and reducing primary production.

Table 14–6. Selected Measurements of Biomass and Net Primary Productivity of Riparian Forested Wetlands

Forest Type	Stem Denisty no./ha	Basal Area, m^2/ha	Biomass kg/m^2 Above-ground	Biomass kg/m^2 Below-ground	Litter fall, $g\ m^{-2}\ yr^{-1}$	Stem growth, $g\ m^{-2}\ yr^{-1}$	Aboveground NPP,[a] $g\ m^{-2}\ yr^{-1}$	Reference
Louisiana								
Bottomland hardwood	1,710	24.3	16.5	—	574	800	1,374	Conner and Day (1976)
Minnesota								
Forested fen	3,348	25.1	10.0	—	412	334	746	Reiners (1972)
Illinois								
Floodplain forest	—	—	29.0	—	—	—	1,250	F. L. Johnson and Bell (1978)
Transition forest	—	—	14.2	—	—	—	800	F. L. Johnson and Bell (1978)
Floodplain forest	423	32.1	—	—	491	177	668	S. L. Brown and Peterson (1983)
Kentucky								
Floodplain forests	990 370	42.0 17.7	30.3 18.4	— —	420 468	914 812	1,334 1,280	Taylor (1985)

[a]NPP = net primary productivity = litterfall + stem growth

Source: After Brinson et al., 1981b

Decomposition

The decomposition of organic matter in riparian ecosystems is undoubtedly related to the intensity and duration of flooding, although consistent relationships have not been found. Brinson (1977) found in an alluvial swamp forest in North Carolina that the decomposition of litter was most rapid in the wettest site and slowest on the dry levee. Duever et al. (1984), however, found that decomposition was greater in dry sites than in those flooded for durations ranging from 16–50 percent of the time in Corkscrew Swamp in southwestern Florida. McArthur and Marzolf (1987) reported that the dissolved nutrient concentration of decomposing leaves depended on both the species and their location on the floodplain. The relationship was not straightforward between leachate concentrations and precipitation or dissolved organic carbon. The rate of decomposition in riparian wetlands is probably the greatest in areas that are generally aerobic but are supplied with adequate moisture. The decomposition is probably slightly slower at sites that are continually dry and lacking in moisture. Decomposition of litter in permanently anaerobic wetlands is probably the slowest (Brinson et al., 1981a). The research of McArthur (1989) and his associates mentioned earlier on the correlation of bacterial allele frequency with specific habitats illustrates the extreme complexity of the decomposition process.

Organic Export

There is considerable evidence that watersheds that drain areas of wetlands export more organic carbon than watersheds that do not have wetlands (see Fig. 5–17). The organic carbon is in both dissolved and particulate forms, although the particulate fraction is generally a small percentage of the total carbon in most rivers (Brinson et al., 1981b). The large organic carbon export from watersheds dominated by riparian ecosystems is attributed to the following factors:

1. A large surface area of litter, detritus, and organic soil that is exposed to the river water during flooding;
2. Rapid leaching of soluble organic carbon from some riparian soils and litter exposed to flooding;
3. The long time during which water is in contact with the floodplain, allowing for significant passive leaching;
4. River movement over floodplains during flooding that can physically erode and transport particulate organic carbon from the floodplain.

Both particulate and dissolved carbon have been shown to be valuable for downstream ecosystems, particularly for lacustrine and marine ecosystems. There is evidence that the particulate fraction is important as a source of energy for shredders and for filter-feeding organisms (Vannote et al., 1980). The dissolved organic carbon probably is most valuable as a source of carbon for microorganisms, which, in turn, convert it to particulate form (Brinson et al., 1981a, b).

Sedell and his coworkers (Sedell et al., 1980; Sedell and Froggatt, 1984) have documented the contribution of large woody material from the riparian zone not only as an energy source to the adjacent stream but as an important source of structure that modifies fluvial processes and enhances the interaction of the stream with its riparian zone.

Energy Flow

Studies of the overall energy flow of riparian wetlands are lacking, although there are several studies related to primary productivity. A useful energy-flow diagram of the riparian ecosystem is shown in Figure 14–21. The primary sources of energy are solar energy, wind, water flow, sediments, and nutrients. These flow through the riparian zone both longitudinally along the river course and laterally across the riparian zone. The major stresses include hydrological, geomorphological, physiological, and biomass removal. The stress effects, shown to affect different pathways in the riparian ecosystem, are greater if they influence secondary

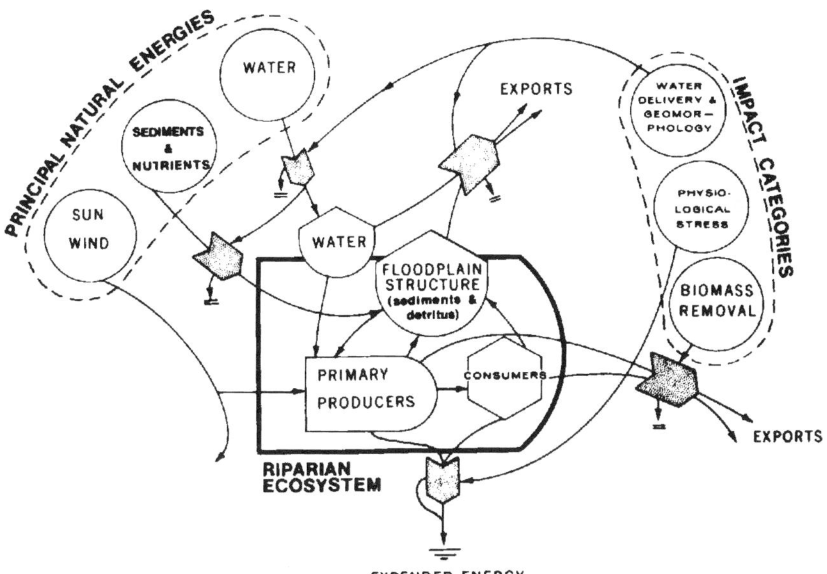

Figure 14–21 Model of energy flow of a bottomland hardwood ecosystem showing natural energies and major impacts. (*From Cairns et al., 1981; copyright © 1981 by Elsevier, reprinted with permission, after S. L. Brown et al., 1979*)

or tertiary energy flows (e.g., hydrologic modification) than if they affect the ecosystem within its structure (e.g., biomass removal) (Cairns et al., 1981).

It is possible to make only broad generalizations about riparian ecosystem energy flow. The food webs that develop in these ecosystems begin with the production of detritus, although a great complexity and diversity of animals develop in the food webs. A unique feature of riparian wetlands is that the detrital production supports both aquatic and terrestrial communities (Vannote et al., 1980; Brinson et al., 1981b; Gregory et al., 1991). The wetlands also receive and transport to downstream ecosystems a large amount of detrital energy. Most of what is known about energy flow in riparian ecosystems is limited to measures of primary productivity.

Nutrient Cycling

Several important points should be made about nutrient cycling in riparian ecosystems:

1. Riparian ecosystems have "open" nutrient cycles that are dominated by the flooding stream or river, runoff from upslope forests, or both, depending on stream order and season.
2. Riparian forests exert significant biotic control over the intrasystem cycling of nutrients, and seasonal patterns of growth and decay often match available nutrients.
3. Contact of water with sediments on the forest floor leads to several important nutrient transformations. Riparian wetlands can serve as effective sinks for nutrients that enter as lateral (trans-riparian) runoff and groundwater flow.
4. When the entire river system that flows through a watershed dominated by riparian wetlands is investigated, the riparian wetlands often appear to be nutrient transformers that change a net import of inorganic forms of nutrients to a net export of organic forms.

The nutrient cycles of riparian wetlands are open cycles characterized by large inputs and outputs caused by flooding. Cycling is also significantly affected by the many chemical transformations that occur when the soil is saturated or under water. Brinson et al. (1981b) described the cycling of nitrogen in a cypress-tupelo stream-floodplain complex. In winter, flooding contributes dissolved and particulate nitrogen that is not taken up by the canopy trees because of their winter dormancy. The nitrogen is retained, however, by filamentous algae on the forest floor and through immobilization by detritivores. In spring, the nitrogen is released by decomposition as the waters warm and as the filamentous algae are shaded by the developing canopy. The vegetation canopy begins to develop, increasing plant nitrogen uptake and lowering water levels through evapotranspiration. When the sediments are exposed to the atmosphere

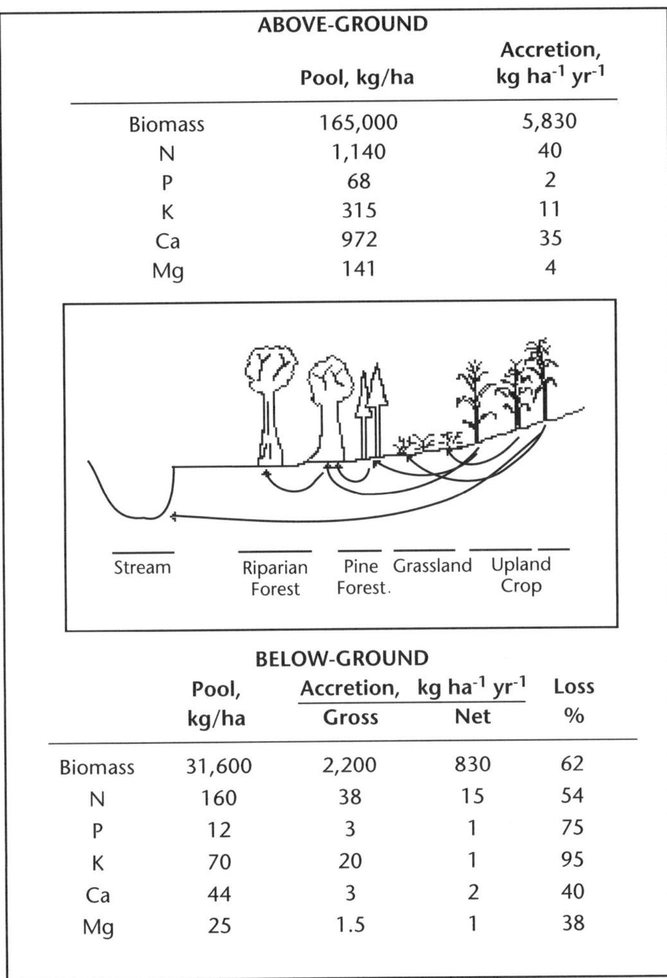

ABOVE-GROUND

	Pool, kg/ha	Accretion, kg ha^{-1} yr^{-1}
Biomass	165,000	5,830
N	1,140	40
P	68	2
K	315	11
Ca	972	35
Mg	141	4

BELOW-GROUND

	Pool, kg/ha	Accretion, kg ha^{-1} yr^{-1} Gross	Accretion, kg ha^{-1} yr^{-1} Net	Loss %
Biomass	31,600	2,200	830	62
N	160	38	15	54
P	12	3	1	75
K	70	20	1	95
Ca	44	3	2	40
Mg	25	1.5	1	38

Figure 14–22. Above- and below-ground vegetation nutrient pools and accretion rates for N, P, K, Ca, and Mg in a riparian zone along a southeastern United States stream. (*From Fail et al., 1989*)

as summer progresses, ammonification and nitrification are stimulated further, making the nitrogen more available for plant uptake. Nitrification, in turn, produces nitrates that are lost through denitrification when they are exposed to anaerobic conditions caused by subsequent flooding.

In some western riparian systems dominated by alder, soil-nitrogen levels are enhanced by nitrogen fixation, and these enhanced concentrations influence the adjacent stream (Goldman, 1961). Fail et al. (1989) measured the stocks and the annual accretion of tissue nutrients in a riparian forest in south Georgia (Fig. 14–22). In this forest most of the tissue nutrients were tied up in above-ground live biomass. The soil stores (as distinguished from below-ground tissue stores) were not measured.

Longitudinal Nutrient Exchange

Inflows and outflows of nutrients in riparian systems occur along two gradients, longitudinal, that is, along the stream course, and transverse from uplands across the riparian zone to the stream. At the stream-system level, Elder and Mattraw (1982) and Elder (1985) investigated the flux and speciation of nitrogen and phosphorus in the Apalachicola River in northern Florida, a river that is dominated by forested riparian ecosystems along much of its length. The authors of this study concluded that the floodplain forests were nutrient transformers rather then nutrient sinks, when the total export of nutrients by the watershed was compared to the upstream inputs; that is, the inputs and outputs were similar for both total nitrogen and total phosphorus but the forms of the nutrients were different. Compared to the influx, the stream discharge contained net increases in particulate organic nitrogen, dissolved organic nitrogen, particulate phosphorus and dissolved phosphorus, and net decreases in dissolved inorganic phosphorus and soluble reactive phosphorus (Fig. 14–23). This means that the riparian forests were net importers of inorganic forms of nutrients and net exporters of organic forms. These transformations, the authors argued, are important to secondary productivity in downstream estuaries.

Despite this study, it is clear that depositional riparian systems do trap considerable amounts of nutrients from upstream. The inflow of upstream nutrients is a major pathway in the local nutrient cycle. Phillips (1989) calculated that in ten study basins representative of most of the different types of riparian systems, 23 to 93 percent of upland sediment that reached the stream was subsequently stored in riparian wetlands. Table 14–7 illustrates the range of measurements of phosphorus input to riverine forests. Although these data are only approximations of the phosphorus contribution stemming from overbank and trans-riparian nutrient fluxes flooding, it is worth noting that they are usually several orders of magnitude greater than phosphorus inflows from precipitation and are at least on the same order of magnitude as intrasystem cycling flows. The relative importance of these inflows for alluvial river swamps was shown in Chapter 13.

Transverse Nutrient Exchange

Other studies have emphasized the role of riparian wetlands in trans-riparian nutrient fluxes. Thus riparian ecosystems have been measured as effective "filters" for nutrient materials that enter through lateral runoff and groundwater (Correll, 1986). Peterjohn and Correll (1984) found that a 50-meter-wide riparian forest in an agricultural watershed near the Chesapeake Bay in Maryland removed an estimated 89 percent of the nitrogen and 80 percent of the phosphorus that entered it from upland runoff, groundwater, and bulk precipitation.

Table 14–7. Rates of Phosphorus Retention in Riparian Forested Wetlands

Location	*Sedimentation Rate*	*kg* ha^{-1} yr^{-1}	*Source*
Cache River, southern Illinois	3.6 gP m^{-2} yr^{-1} contributed by flood as sedimentation for flood of 1.13-yr recurrence interval	36	Mitsch et al. (1979a)
Prairie Creek, north-central Florida	3.25 gP m^{-2} yr^{-1} as sedimentation from river overflow	32.5	S. L. Brown (1978)
Creeping Swamp, Coastal Plain, North Carolina	0.17 gP m^{-2} yr^{-1} sedimentation on floodplain floor from river overflow	1.72	Yarbro (1979)
Creeping Swamp, Coastal Plain, North Carolina	0.315–0.730 gP m^{-2} yr^{-1} based input-output budget of floodplain (most was filterable reactive phosphorus)	3.15–7.30	Yarbro (1979)
Kankakee River, northwestern Illinois	1.36 gP m^{-2} yr^{-1} contributed by unusually large spring flood lasting 62–80 days	13.6	Mitsch et al. (1979b)

Source: From Brinson et al., 1981b

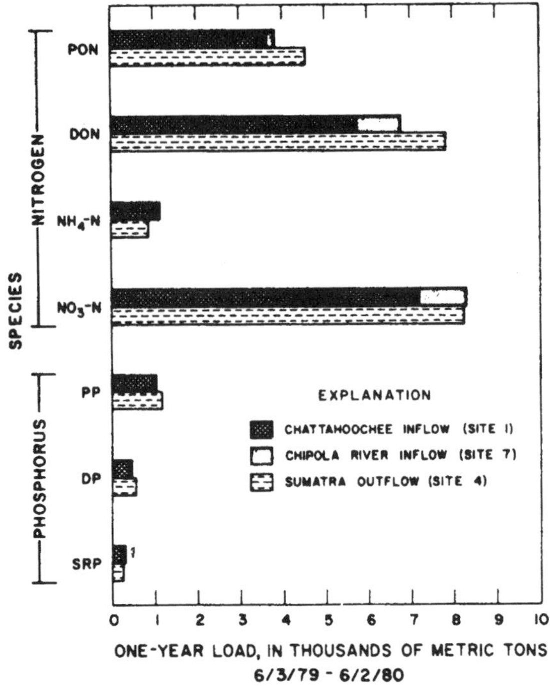

Figure 14–23. Inflows and outflows of various forms of nitrogen and phosphorus for Apalachicola River in northern Florida. (*From Elder, 1985*)

The study estimated that there was a net removal of 11 kg ha^{-1} yr^{-1} of particulate organic nitrogen, 0.83 kg ha^{-1} yr^{-1} of dissolved ammonium nitrogen, 47.2 kg ha^{-1} yr^{-1} of nitrate nitrogen, and 3.0 kg ha^{-1} yr^{-1} of particulate phosphorus. This study is only one of many (e.g., Schlosser and Karr, 1981a, b; C. A. Johnston et al., 1984; Lowrance et al., 1984; Jacobs and Gilliam, 1985; Cooper et al., 1986, 1987; Kuenzler and Craig, 1986; Jordan et al., 1986; Whigham et al., 1986; Cooper and Gilliam, 1987; Fail et al., 1989) that have demonstrated the way in which riparian ecosystems can be effective in removing as well as modifying nutrients and sediments from agricultural runoff before it reaches a stream or river. For example, Jacobs and Gilliam (1985) suggested that riparian buffer strips less than 16m in width in North Carolina were effective in denitrifying a substantial portion of nitrates in agricultural subsurface salvage water before it reached the adjacent stream.

ECOSYSTEM MODELS

Few energy-nutrient models have been developed for bottomland hardwood forests in particular or riparian wetlands in general. The scarcity of specific energy-nutrient simulation models can be attributed to the difficulty in quantifying the relationships between stream flooding and ecosystem productivity as much as any cause. The plant composition or vegetation change in a riparian site have been modeled in several ways: (1) direct specification, (2) environmental suitability, and (3) transition simulation (Auble, 1991). Direct specification involves a description of the current vegetation and a set of relations that specify the vegetation resulting from various actions. This kind of model is most useful for management decisions involving changes that are abrupt, for example, clearing for cropland and reservoir filling. This approach may not involve any mathematical characterization at all; it may simply represent a conceptual approach to vegetation change (Auble, 1991).

Environmental suitability models add environmental relations to the description of vegetation in a site. When the characterization of vegetation in various sites is linked to local environmental conditions, changes in the environment can be used to predict vegetation change. Environmental conditions need not be instantaneous; instead, they may represent long-term averages, extreme events, or other similar parameters. For example, Stromberg and Patten (1990) used site-specific regressives between discharge and stem radial growth increment of cottonwood to establish a flow standard based on average tree growth. Most simulation models include some aspects of the environmental-suitability approach since the relationship between individual species or species groups and the environment is a key aspect of most ecological models.

Transition simulation applies the relationship of vegetation and the environment to a set of incremental changes; that is, an incremental change in one or more environmental variables leads to a change in vegetation. The new vegetation condition then becomes the starting point for the next incremental change (Auble, 1991). This is the most flexible and commonly used modeling approach in ecology and has been applied to a large class of riparian systems. For example, C. W. Johnson (in press) developed a compartmental simulation model to examine changes in stand dynamics and composition of riparian forests, resulting from altered flows and meandering rate in the Missouri River, downstream from a large dam.

Transition models based on individual species growth and their responses to the environment include a suite of models based on SWAMP, a tree-growth simulation model of southern wetland forests. This model was developed by Phipps (1979) and applied to a bottomland forest in Arkansas and later to a forested wetland in Virginia (Phipps and Applegate, 1983). Unlike energy-nutrient models, this model simulates the growth of individual trees in the forest, summing the growth of all trees in a plot to determine plot dynamics. The model includes subroutines, including GROW, which "grows" trees on the plot according to a parabolic growth form; KILL, which determines the survival probabilities of trees and occasionally "kills" trees; and CUT, which enables the modeler to remove trees, as in lumbering or insect damage. Another important subroutine,

WATER, describes the influence of water-level fluctuations on tree growth. This subroutine assumes that tree growth will be suboptimal during the peak May–June growth period if water is either too high or too low. The effect of water levels on tree productivity is hypothesized to be a parabolic function as follows:

$$H = 1 - 0.05511(T - W)^{2} \qquad (14.1)$$

where H = growth factor related to the water table, T = water table depth of sample plot, and W = optimum water table depth.

The model begins with all tree species greater than 3 cm in diameter on a 20 m x 20 m plot and "grows" them on a year-by-year basis, generating results that depend on hydrologic conditions such as flooding frequency, depth to the water table, and other factors, including shading and simulated lumbering. Typical results from runs of the model based on data from the White River in Arkansas indicate the importance of altered hydroperiod and lumbering for the structure of the bottomland forest. Phipps (1979) acknowledges that "though in its present form the model is heavily dependent upon hypothetical relationships, it is felt that it can ultimately be of value to predict wetland vegetation change subsequent to hydrologic change."

FORMS (Tharp, 1978) is a similar kind of simulation model of a Mississippi River floodplain forest. FORFLO, a model based on SWAMP, was applied to evaluate the impact of an altered hydrologic regime on succession in a bottomland forest in South Carolina (Pearlstein et al., 1985; Brody and Pendleton, 1987).

Mitsch (1988) developed a simple simulation model that illustrated the simultaneous effects of river flooding and nutrient dynamics on forested wetland productivity. Model simulations illustrated that productivity (net biomass accumulation) ranged from highest to lowest in the following order of hydrologic conditions in agreement with the hydrologic control theory of Brinson et al. (1981a) (see Chap. 4):

pulsing > flowing > stagnant

Mitsch (1988) summarized these findings by stating that "The pulsing system had the highest productivity because of high nutrients and because the pulsing hydrology allowed the water level to decrease between floods to levels closer to the optimum requirements of the vegetation." This model also illustrated the flood pulse concept (FPC) of floodplain systems as proposed by Junk et al. (1989).

Auble et al. (in press) developed a simulation model to predict vegetation change as related to hydroperiod on the Gunnison River in Colorado. They field-calibrated the relationship between hydroperiod (flow-exceedance curves) and plant cover type, using gradient analysis. Predictions of changed flows (flow-exceedance results in changed hydroperiods) led to predictions of vegetation change in plots along the river valley. This approach has obvious utility for predicting the impact of river management alternatives that change hydrologic parameters.

PART 5

MANAGEMENT OF WETLANDS

Values and Valuation of Wetlands 15

Wetlands provide many services and commodities to humanity. At the population level, wetland-dependent fish, shellfish, fur animals, waterfowl, and timber provide important and valuable harvests and millions of days of recreational fishing and hunting. At the ecosystem level, wetlands moderate the effects of floods, improve water quality, and have aesthetic and heritage value. They also contribute to the stability of global levels of available nitrogen, atmospheric sulfur, carbon dioxide, and methane.

The valuation of these services and commodities for purposes of wetland management is complicated by the difficulty of comparing by some common denominator the various values of wetlands against human economic systems; by the conflict between a private owner's interest in the wetlands and the values that accrue to the public at large; and by the need to consider the value of a wetland as a part of an integrated landscape. Valuation techniques include the nonmonetized scaling and weighting approaches for comparing different wetlands or different management options for the same wetland, and common denominator approaches that reduce the various values to some common term such as dollars or embodied energy. These common denominator methodologies can include willingess to pay, opportunity costs, replacement value, and energy analysis. None of these approaches is without problems, and no universal agreement about their use has been reached.

The term *value* imposes an anthropocentric orientation on a discussion of wetlands. The term is often used in an ecological sense to refer to functional processes, as, for example, when we speak of the "value" of primary production in providing the food energy that drives the ecosystem. But in ordinary parlance, the word connotes something worthy, desirable, or useful to humans. The reasons that wetlands are often legally protected have to do with their value to society, not with the abstruse ecological processes that occur in wetlands; this is the sense in which the word *value* is used in this chapter. Perceived values arise from the functional ecological processes described in previous chapters but are determined also by human perceptions, the location of a particular wetland, the human population pressures on it, and the extent of the resource. Regional wetlands are integral parts of larger landscapes—drainage basins, estuaries. Their functions and their values to people in these landscapes depend on both their extent and their location. Thus the value of a forested wetland varies. If it lies along a river, it probably plays a greater functional role in stream water quality and downstream flooding than if it were isolated from the stream. If situated at the headwaters of a stream, a wetland would function in ways different from those of a wetland located near the stream's mouth. The fauna it supports depend on the size of the wetland relative to the home range of the animal. Thus to some extent each wetland is ecologically unique. This complicates the measurement of its "value." References for this subject are found in several literature reviews and annotated bibliographies (e.g., Reppert et al., 1979; Larson, 1982; Adamus, 1983; Sather and Smith, 1984; Shabman and Batie, 1988; Costanza et al., 1989; Leitch and Ekstrom, 1989; Reeder and Mitsch, 1990; Folke, 1991) and in a series of reports of the "Wetlands Are Not Wastelands" joint project of Wildlife Habitat Canada, Environment Canada, and Canadian Wildlife Service (see Manning et al., 1990).

WETLAND VALUES

Wetland values can conveniently be considered from the perspective of three hierarchical levels—population, ecosystem, and global.

Populations

The easiest values of wetlands to identify are the populations that depend on wetland habitats for their survival.

Animals Harvested for Pelts

Fur-bearing mammals and the alligator are harvested for their pelts (Fig. 15–1). In contrast to most other commercially important wetland species, these animals typically have a limited range and spend their lives within a short distance of their birthplaces. The most abundant are the muskrats (*Ondatra zibethicus*), with about 10 million pelts harvested each year (Fig. 15–1b) out of a total of 12 million for all species (Fig. 15–1a). Muskrats (Fig. 15–2, top) are found in wetlands throughout the United States except, strangely, the south Atlantic Coast. They

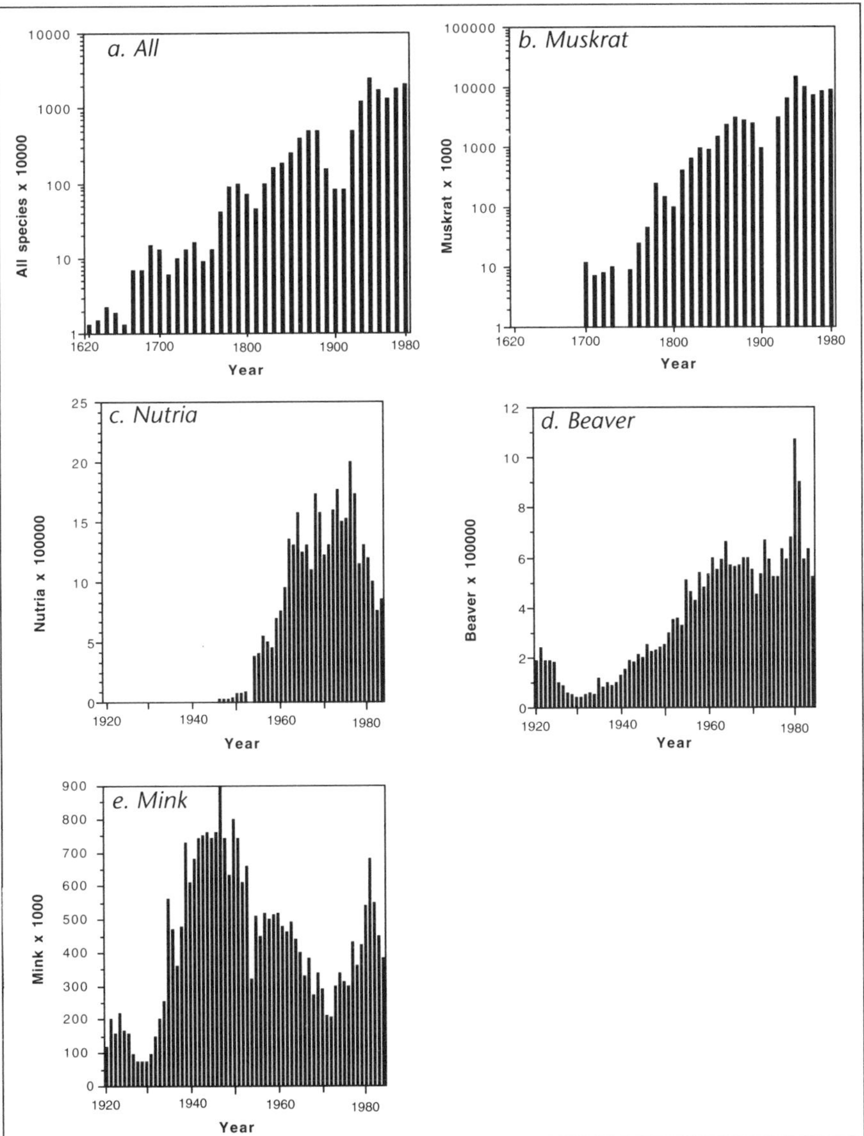

Figure 15–1. Fur pelt harvests for *a.* all species; *b.* muskrats by decade in North America; *c.* nutria in United States, post-1934; *d.* beaver and *e.* mink in Canada and United States, post-1934. *(Data from Novak et al., 1987)*

prefer fresh inland marshes, but along the northern Gulf Coast are more abundant in brackish marshes. About 50 percent of the nation's harvest is from the Midwest and 25 percent from along the northern Gulf of Mexico, mostly Louisiana (Chabreck, 1979).

The nutria (*Myocastor coypus*) is the next most abundant species. It is very much like a muskrat but is larger and more vigorous (Figs. 15-1c; 15-2; Kinler et al., 1987). This species was imported from South America to Louisiana and escaped from captivity in 1938, spreading rapidly through the state's coastal marshes. It is now abundant in freshwater swamps and in coastal marshes from which it may have displaced muskrats to more brackish locations. Ninety-seven percent of the U.S. nutria harvest occurs in Louisiana. In order of decreasing abundance (but not dollar value), other harvested fur animals are beaver (Fig. 15–1d), mink (Fig. 15–1e), and otter. Raccoons are also taken commercially and in the north-central states are second only to muskrats. Beavers are associated with forested wetlands, especially in the Midwest. Minnesota harvests 27 percent of the nation's catch.

The alligator represents a dramatic success story of a return from the edge of extinction to a healthy population (at least in the central Gulf states) that is now harvested under close regulation in Florida and Louisiana. The species was threatened by severe hunting pressure, not by habitat loss. When that pressure was removed, its numbers increased rapidly. Alligators are abundant in fresh and slightly brackish lakes and streams and build nests in adjacent marshes and swamps. The Louisiana harvest in 1979 was worth about $1.7 million (Office of Technology Assessment, 1984). In Louisiana, the industry was worth over $16 million in 1992, for both wild and farm-raised animals (G. Linscombe, La. Dept. Wildlife and Fisheries, personal communication).

Waterfowl and Other Birds

Eighty percent of America's breeding bird population and more than 50 percent of the 800 species of protected migratory birds rely on wetlands (Wharton et al., 1982). Wetlands, which are probably known best for their waterfowl abundance, also support a large and valuable recreational hunting industry. We use the term "industry" because hunters spend large sums of money in the local economy (estimated at $58 million for the Mississippi flyway alone) for guns, ammunition, hunting clothes, travel to hunting spots, food, and lodging. Most of the birds hunted are hatched in marshes in the far North, sometimes above the Arctic Circle, but are shot during their winter migrations to the southern United States and Central America. There are exceptions—the wood duck (*Aix sponsa*) breeds locally throughout the continent—but the generalization holds for most species. Different groups of geese and ducks have different habitat preferences, and these preferences change with the maturity of the duck and the season. Therefore, a broad diversity of wetland habitat types is important for waterfowl success. The freshwater prairie potholes of North America are the primary breeding place for waterfowl. There an estimated 50 to 80 percent of the continent's main game species are produced (Batt et al., 1989). Wood ducks prefer forested wetlands, however. During the winter, diving ducks (*Aythya* spp. and

Figure 15–2. Two animals that are frequently harvested for their pelts include *(top)* the muskrat (*Ondatra zibethicus*), and *(bottom)* the nutria (*Myocastor coypus*). (*Photo of muskrat by Al Staffan, Ohio Department of Natural Resources, reprinted with permission; nutria, copyright © by Greg Linscombe, Louisiana Department of Wildlife and Fisheries, reprinted with permission)*

Oxyura spp.) are found in brackish marshes, preferably adjacent to fairly deep ponds and lakes. Dabbling ducks (*Anas* spp.) prefer freshwater marshes and often graze heavily in adjacent rice fields and in very shallow marsh ponds. Gadwalls (*Anas strepera*) like shallow ponds with submerged vegetation.

The waterfowl value of wetlands such as the prairie pothole region of North America (see Chaps. 3 and 11) is unmistakable. The loss of wetlands in this region, which is estimated to be much more than half the original pothole area, has been suggested as a major factor in the decline in the nesting success of duck populations in North America (Sellers, 1973; Higgins, 1977; Cowardin et al., 1985; Harris, 1988; Batt et al., 1989; Swanson and Duebbert, 1989). In one study, Batt et al. (1989) examined waterfowl census data for the prairie pothole region over the 30-year period 1955 to 1985 and found that there was an average of almost 22 million waterfowl (dabbling and diving ducks) in the region, dominated by the mallard (average 3.7 million). There was a great deal of fluctuation in population size, however, from year to year. When these fluctuations were compared to the number of potholes flooded in May of each year, there was a clear positive correlation (Fig. 15–3), indicating the importance of wetland hydrology in the breeding success of waterfowl.

Since the mid-1970s there has been an overall decline in waterfowl populations in North America (Fig. 15–4). The trend is paralleled by population data for individual species, including mallards, widgeon, blue-winged teal, northern

Table 15-1. Population Estimates of the Ten Most Common Species of Breeding Ducks and Four Species of Geese, 1991, with Percent Change from the 1955–91 Average

Species	*Population (in thousands)*	*Change in 1991 (from 1955–91 Average)*
Mallard (*Anas platyrhynchos*)	5353 ± 188	-27
Gadwall (*Anas strepera*)	1573 ± 94	+22
American wigeon (*Anas americana*)	2328 ± 135	-14
Green-winged teal (*Anas crecca*)	1601 ± 88	-4
Blue-winged teal (*Anas discors*)	3779 ± 245	-10
Northern shoveler (*Anas clypeata*)	1663 ± 84	-8
Northern pintail (*Anas acuta*)	1794 ± 199	-62
Redhead (*Aythya americana*)	437 ± 37	-26
Canvasback (*Aythya valisneria*)	463 ± 57	-16
Scaup (*Aythya spp.*)	5247 ± 333	-7
Canada goose (*Branta canadensis*)	3750	
Snow goose (*Chen caerulescens*)	2440	
White-fronted geese (*Anser albifrons*)	492	
Brant (*Branta bernicla*)	275	

Source: Data from Bortner et al., 1991. Duck surveys on summer breeding grounds, goose surveys during fall and winter.

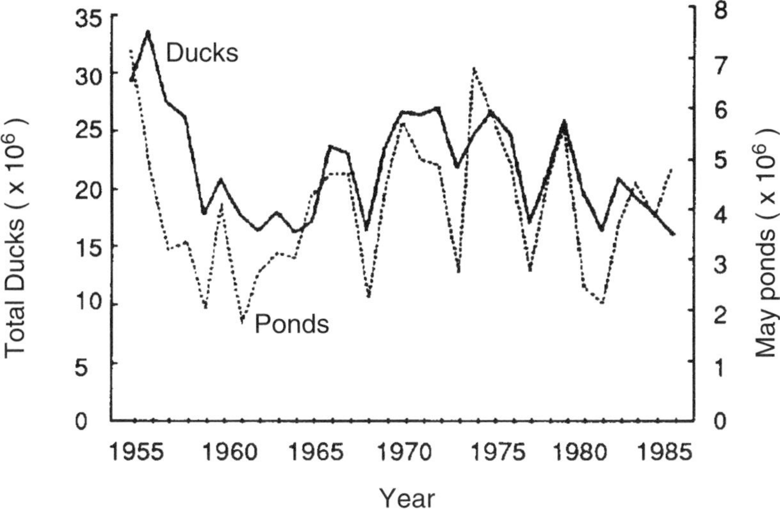

Figure 15–3. Estimated number of total ducks and ponds in May, in prairie pothole region, 1955–1985. (*From Batt et al., 1989; copyright © 1989 by Iowa State University Press, reprinted with permission*)

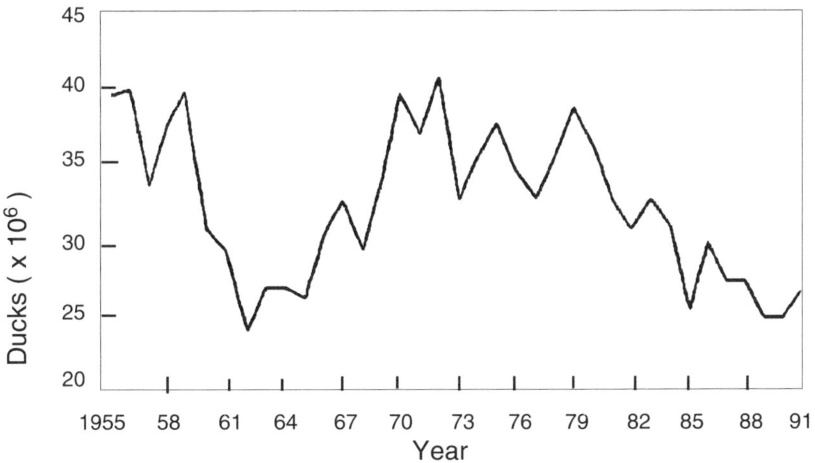

Figure 15–4. Duck breeding populations in North America, 1955–91, excluding scoters, eiders, mergansers, and oldsquaws. (*From Bortner et al., 1991*)

pintail, redhead, and canvasback (Table 15–1). Habitat degradation, both in the northern breeding grounds and in wintering areas, is considered a major cause of the decline (Harris, 1989). Gadwall populations, in contrast, have increased 22 percent over the 1955–1991 average (Bortner et al., 1991).

Hunting is closely regulated and tailored to the local region. The mallard makes up about one-third of the U.S. total of harvested ducks. About 50 percent are shot in wetlands; most of the rest are shot in agricultural fields. In Louisiana, one-third of the mallard population is killed each year. The percentage is lower for other species—about 8–13 percent. The vast flocks of geese that used to be so abundant along the Eastern Seaboard and the Gulf Coast are smaller now but are still abundant and are considered to be important as hunted species in some areas.

Fish and Shellfish

Over 95 percent of the fish and shellfish species that are harvested commercially in the United States are wetland-dependent (Feierabend and Zelanzy, 1987). The fishing industry contributed $1.7 billion to the U.S. gross national product in 1988. The degree of dependence on wetlands varies widely with species and with the type of wetland. Some important species are permanent residents; others are merely transients that feed in wetlands when the opportunity arises. Some shallow wetlands, which may exhibit several other wetland "values," may be virtually devoid of fish, whereas other types of deepwater and coastal wetlands may serve as important nursery and feeding areas. Table 15–2 lists some of the major nektonic species associated with wetlands. Table 8–6 shows dollar value of coastal fisheries. Virtually all of the freshwater species are dependent to some degree on wetlands, often spawning in marshes bordering lakes or in riparian forests during spring flooding. These species are primarily recreational, although some small local commercial fisheries exploit them. The saltwater species tend to spawn offshore, move into the coastal marsh "nursery" during their juvenile stages, and then emigrate offshore as they mature. They are often important for both commercial and recreational fisheries. The menhaden is caught only commercially, but competition between commercial and sport fishermen for shrimp, blue crab, oyster, catfish, seatrout, and striped bass can be intensive and acrimonious. Anadromous fish probably use wetlands less than the other two groups. Their most intensive use is probably by young fry that sometimes linger in estuaries and adjacent marshes on their migrations to the ocean from the freshwater streams in which they were spawned.

The relationship between the area of marsh or the marsh-water interface of coastal wetlands and the production of commercial fisheries has been illustrated for shrimp and fish harvests in Louisiana and worldwide (R. E. Turner, 1982; Gosselink, 1984; Costanza et al., 1989), blue crab production in western Florida (Lynn et al., 1981), and oyster production in Virginia (Batie and Wilson, 1978). In one example, R. E. Turner (1977, 1982) showed a direct relationship between

Table 15-2. Dominant Commercial and Recreational Wetland-Associated Fish and Shellfish

Common Name	*Scientific Name*	*Commercial Harvest, metric tons*[a]
Fresh Water		
Catfish and bullhead	*Ictalurus* spp.	16,800
Carp	*Cyprinus carpio*	11,800
Buffalo	*Istiobus* sp.	11,300
Perch	*Perca* sp., *Stizostedion* sp.	—
Pickerel	*Esox* sp.	—
Sunfish	*Lepomis* sp., *Micropterus* sp., *Pomoxis* sp.	—
Trout	*Salmo* sp., *Salvelinus* sp.	—
Anadromous		
Salmon	*Oncorhynchus* sp.	107,000
Shad and alewife	*Alosa* sp.	27,700
Striped bass	*Morone saxatilis*	5,000
Saltwater		
Menhaden	*Brevoortia* sp.	889,000
Shrimp	*Penaeus* sp.	111,000
Blue crab	*Callinectes sapidus*	63,900
Oyster	*Crassostrea* sp.	24,000
Mullet	*Mugil* sp.	15,400
Sea trout	*Cynoscion* sp.	11,300
Atlantic croaker	*Micropogonias undulatus*	9,500
Hard clam	*Mercenaria* sp.	6,800
Fluke	*Paralichthys* sp.	4,500
Soft clam	*Mya arenaria*	4,500
Bluefish	*Pomatomus saltatrix*	—
Drum	*Pogonias cromis, Sciaenops ocellata*	—
Spot	*Leiostomus xanthurus*	—

[a]Landings are 1974–75 averages
Source: After Peters et al., 1979

fish harvests and wetland area for a number of fisheries around the world, including marine, freshwater, and pond-raised (Fig. 15–5).

Analyses of fishery harvests from wetlands show the importance of the recreational catch. Although the commercial harvest is usually much better documented, several studies have shown that the recreational catch far outweighs the commercial catch for certain fisheries. Furthermore, the value to the economy of recreational fishing is usually far greater than the value of the commercial catch because sports fishermen spend more money per fish caught (they are less efficient) than their commercial counterparts. For example, DeSylva (1969) estimated that in California it costs an angler $18.11 to catch one salmon; its value from

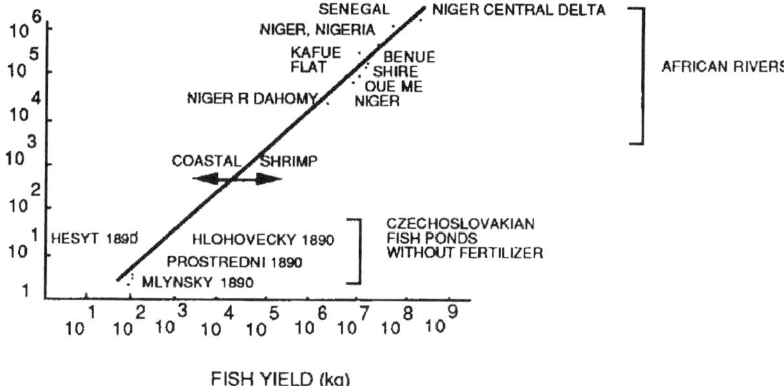

Figure 15–5. Relationships between wetland area and fish harvests. The linear slope describes the line of about 60 kg/ha yield. Pond fisheries are unfertilized, managed ponds in Czechoslovakia. African floodplain river fisheries after Welcomme (*1976*). Wetland-dependent coastal fisheries (*adapted from Turner, 1977*) are generally ten times higher than inland eccosystems. (*After Turner, 1982*)

the angler's standpoint is five times what it is to the commercial fisherman. These excess dollars feed the California economy.

Timber and Other Vegetation Harvest

The area of wetland timber in the United States is about 22 million hectares (55 million acres). Two-thirds of that area is east of the Rocky Mountains. The Mississippi River alluvial floodplain and the floodplains of rivers entering the south Atlantic are mostly deciduous wetlands, whereas the forested wetlands along the northern tier of states are primarily evergreen. The former are more extensive and potentially more valuable commercially because of the much faster growth rates in the South. R. L. Johnson (1979) estimated that the 13 million hectares of bottomland hardwood and cypress swamps in the southeastern part of the nation contained about 112 cubic meters of merchantable timber per hectare (1,600 cubic feet/acre), worth about \$620/hectare, or about \$8 billion. As timber prices climb and as the land becomes more and more valuable for agriculture, these wetlands are being clear-cut and drained (see Chap. 16), although by using sound silviculture practices, the timber industry can coexist with productive wetlands.

In addition to the timber harvest, the production of herbaceous vegetation in marshes is a potential source of energy, fiber, and other commodities. These prospects have not been explored widely in North America but are viable options elsewhere. For example, Chung (1989), Mitsch (1991 a, b), and Yan (1992) described several commercial products that are harvested from restored and natural salt marshes and freshwater marshes in China. The productivity of many wetland species (e.g., *Spartina alterniflora, Phragmites communis, Typha angustifolia, Eichhornia crassipes, Papyrus* spp.) is as great as our most vigorous agricultural crops. Ryther et al. (1979) estimated that energy put into growing wetland crops can be returned five- to ninefold in their harvest. Only about half as much energy, however, can be recovered if the crop is fermented to produce methane. The economics of commercial production has not been completely worked out, but Ryther et al. (1979) stated that a 1,000-hectare water hyacinth (*Eichhornia crassipes*) energy farm in the southern United States could produce on the order of 10^{12} BTU of methane per year and at the same time remove all the nitrogen from the wastewater from a city of about 700,000 people. It should be understood that the use of wetlands for any purpose involving the harvesting of the vegetation is bound to have a significant effect on the way the system functions. Therefore, the benefits of the harvest should be weighed against any functional values lost through the harvesting operation.

In addition to the annual production of living vegetation in wetlands, great reservoirs of buried peat exist around the world. Kivinen and Pakarinen (1981) estimated world peat resources (30 cm or more thickness) at 420 million hectares, mostly in the former Soviet Union and Canada. The United States has only about 21 million hectares, most of it in Alaska, Minnesota, and Michigan. This buried peat is, of course, a nonrenewable energy source that destroys the wetland habitat when it is mined. In the United States, peat is mined primarily for horticultural peat production, but in other parts of the world—for example, several republics of the former Soviet Union—it has been used as a fuel source for hundreds of years. It is used to generate electricity, formed into briquettes for home use, and gasified or liquefied to produce methanol and industrial fuels (see Chap. 16).

Endangered and Threatened Species

Wetland habitats are necessary for the survival of a disproportionately high percentage of endangered and threatened species. Table 15–3 summarizes the statistics but imparts no information about the particular species involved, their location, wetland habitat requirements, degree of wetland dependence, and factors contributing to their demise. Although wetlands occupy only about 3.5 percent of the land area of the United States, of the 209 animal species listed as endangered in 1986, about 50 percent depend on wetlands for survival and viability. Almost one-third of native North American freshwater fish species are endangered, threatened, or of special concern. Almost all of these were adversely affected by habitat loss (J. D. Williams and Dodd, 1979). Ernst and Brown (1989) listed 63 species of plants and 34 species of animals that are considered endangered, threatened, or candidates for listing and that occupy southern U.S. forested wetlands. Of these, amphibians and many reptiles are especially linked to wetlands. In Florida, where the number of amphibian and reptile species is

Table 15–3. Threatened and Endangered Species Associated With Wetlands

Taxon	*Number of Species Endangered*	*Number of Species Threatened*	*Percent of U.S. Total Threatened or Endangered*
Plants	17	12	28
Mammals	7	—	20
Birds	16	1	68
Reptiles	6	1	63
Amphibians	5	1	75
Mussels	20	—	66
Fish	26	6	48
Insects	1	4	38
Total	98	25	

Source: Niering 1988, from USDI, FWS, 1986

about equal to the number of mammal and breeding bird species, 18 percent of all amphibians and 35 percent of all reptiles are considered threatened or endangered or their status is unknown (Harris and Gosselink, 1990).

The fate of several endangered species is discussed below to illustrate the ecological complexity of species endangerment. Whooping cranes nest in wetlands in the Northwest Territories of Canada, in water 0.3–0.6 m deep, during the spring and summer. In the fall they migrate to the Aransas National Wildlife Refuge, Texas, stopping off in riverine marshes along the migration route. In Texas they winter in tidal marshes. All three types of wetlands are important for their survival (Williams and Dodd, 1979). The decline in the once abundant species has been attributed both to hunting and to habitat loss. The last whooping crane nest in the United States was seen in 1889 (R. S. Miller et al., 1974). In 1941 the flock consisted of 13 adults and 2 young. Since then the flock has been gradually built up to about 75 birds.

American alligator populations were reduced by hunters and poachers to such low levels that the species was declared endangered in the 1970s. Under close control, the population has subsequently been rebuilt. Alligators have a reciprocal relationship with wetlands—they depend on them, but, in return, the character of the wetland is shaped by the alligator, at least in the south Florida Everglades. As the annual dry season approaches, alligators dig "gator holes." The material thrown out around the holes forms a berm high enough to support trees and shrubs in an otherwise treeless prairie. The trees provide cover and breeding grounds for insects, birds, turtles, and snakes. The hole itself is a place where the alligator can wait out the dry period until the winter rains. It also provides a refuge for dense populations of fish and shellfish (up to 1,600/m^2). These organisms, in turn, attract top carnivores, and so the gator holes are sites of concentrated biological activity that may be important for the survival of many species.

The slackwater darter (*Etheostoma boschungi)* is an example of the specificity sometimes required by wetland species. This small fish, found in small and moderate creeks in Alabama and Tennessee, migrates 1 to 5 km upstream at certain times of the year to spawn in small marshy areas associated with water seeps. It requires shallow (2–8 cm) water and deposits its eggs on a single species of rush (*Juncus acuminatus*) in spite of the presence of other species. The larvae remain in the vicinity of these marshes for four to six weeks before returning to their home streams (Williams and Dodd, 1979).

Ecosystem Values

At the level of the whole ecosystem, wetlands have value to the public for flood mitigation, storm abatement, aquifer recharge, water-quality improvement, aesthetics, and general subsistence. Some of the ecosystem values of wetlands vary from year to year or from season to season. For example, Figure 15–6 illustrates several of the potential ecosystem values of riparian forested wetlands during flooding (spring) and dry (summer) seasons.

Flood Mitigation

Chapter 4 dealt with the importance of hydrology in determining the character of wetlands. Conversely, wetlands influence regional water-flow regimes. One way they do this is to intercept storm runoff and store storm waters, thereby changing sharp runoff peaks to slower discharges over longer periods of time (Fig. 15–7). Because it is usually the peak flows that produce flood damage, the effect of the wetland area is to reduce the danger of flooding (Novitzki, 1979; Verry and Boelter, 1979). Riverine wetlands are especially valuable in this regard. On the Charles River in Massachusetts, the floodplain wetlands were deemed so effective for flood control by the U.S. Army Corps of Engineers that they purchased them rather than build expensive flood-control structures to protect Boston (U.S. Army Corps of Engineers, 1972). The study on which the Corp's decision was made is a classic in wetland hydrologic values. It demonstrated that if the 3,400 hectares of wetlands in the Charles River Basin were drained and leveed off from the river, flood damages would increase by $17 million per year.

Bottomland hardwood forests along the Mississippi River before European settlement stored floodwater equivalent to about 60 days' river discharge. Storage capacity has now been reduced to only about 12 days (Gosselink et al.,

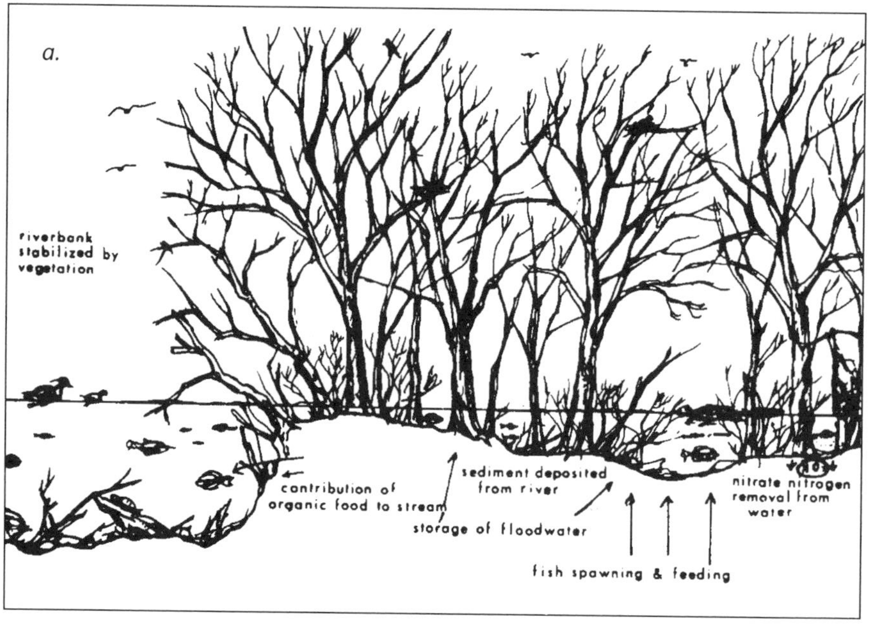

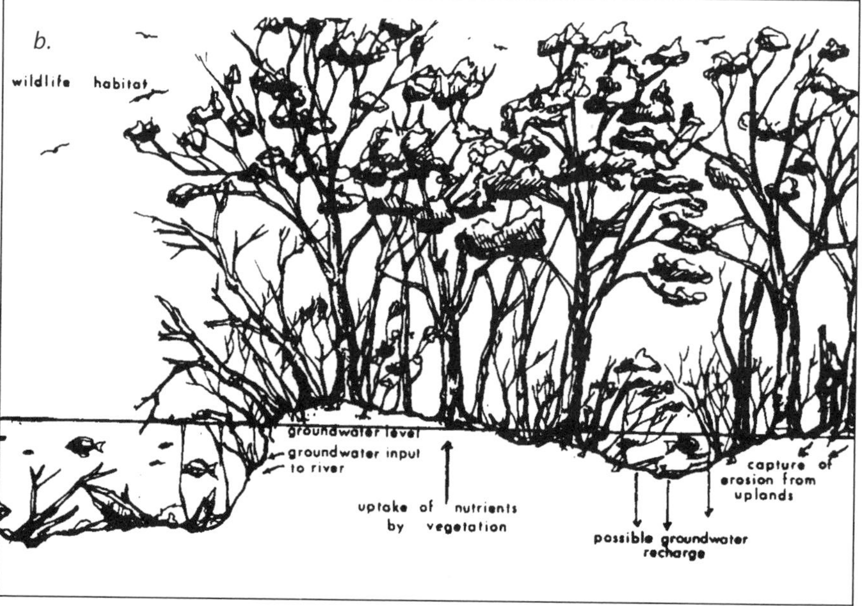

Figure 15–6. Illustration of several of the potential wetland values for riparian wetlands during *a.* flood season and *b.* dry season. (*From Mitsch et al., 1979c*)

1981) as a result of leveeing the river and draining the floodplain. The consequences—the confinement of the river to a narrow channel and the loss of storage capacity—are major reasons that flooding is increasing along the lower Mississippi River (Belt, 1975).

Novitzki (1985) analyzed the relationship between flood peaks and the percentage of basin area in lakes and wetlands. In the Chesapeake Bay drainage basin, where the wetland area was 4 percent, floodflow was only about 50 percent of that in basins containing no wetland storage (Fig. 15–8). In Wisconsin river basins that contained 40 percent lakes and wetlands, however, spring streamflow was as much as 140 percent of that in basins that do not contain storage. This apparent anomaly is probably related to a reduction in the proportion of precipitation that can infiltrate the soil and to a lack of additional storage capacity in lakes and wetlands that are at full capacity during spring floods.

The location of wetlands in the river basin can complicate the response downstream. For example, detained water in a downstream wetland of one tributary can combine with flows from another tributary to increase the flood peak rather than desynchronize flows.

A quantitative approach to the flood mitigation potential of wetlands was undertaken by Ogawa and Male (1983, 1986), who used a hydrologic simulation model to investigate the relationship between upstream wetland removal and downstream flooding. That study found that for rare floods—that is, those that experienced a 100-year recurrence interval or greater—the increase in peak streamflow was significant for all sizes of streams when wetlands were

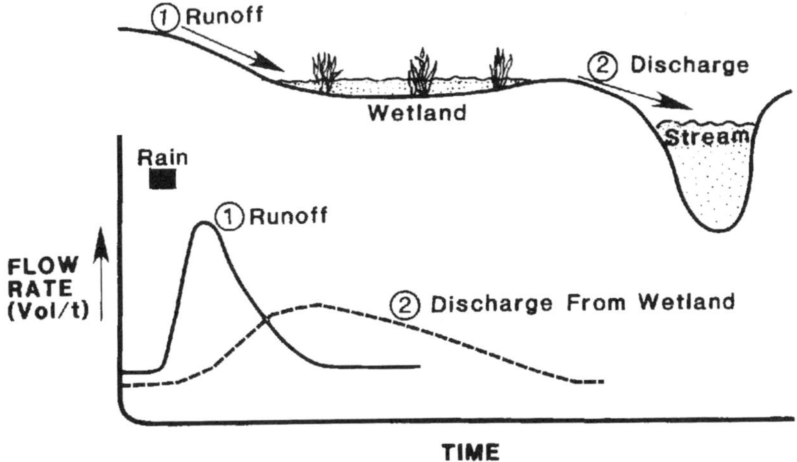

Figure 15–7 The general effect of wetlands on streamflow.

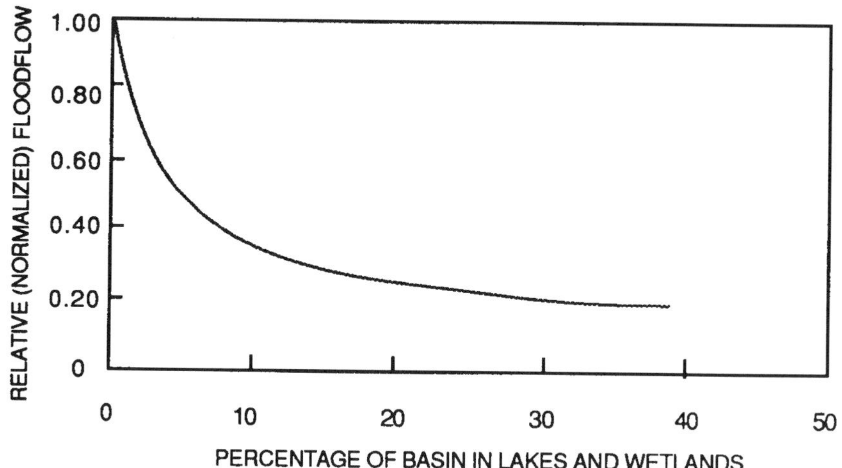

Figure 15–8. Relationship between basin storage (lake and wetland area) and floodflow in Pennsylvania streams tributary to the Chesapeake Bay. (*From Novitski, 1989, modified from Novitzki, 1985, copyright © by S. K. Majumdar*)

removed (Fig. 15–9). The authors concluded that the usefulness of wetlands in reducing downstream flooding increases with (1) an increase in wetland area, (2) the distance the wetland is downstream, (3) the size of the flood, (4) the closeness to an upstream wetland, and (5) the lack of other upstream storage areas such as reservoirs.

Storm Abatement

Coastal wetlands absorb the first fury of ocean storms as they come ashore. Salt marshes and mangrove wetlands act as giant storm buffers. This value can be seen in the context of marsh conservation versus development. Natural marshes, which sustain little permanent damage from these storms, can shelter inland developed areas. Buildings and other structures on the coast are vulnerable to storms, and damage has often been high. Inevitably, the public pays much of the cost of this damage through taxes for public assistance, rebuilding public services such as roads and utilities, and federally guaranteed insurance (Farber, 1987). Two hurricanes that struck southeastern United States in the early 1990s (Hugo in 1989; Andrew in 1992) passed through major inland wetland areas in South Carolina, Florida, and Louisiana. The damage caused to human development might have been more had remaining coastal wetlands (salt marshes and mangroves) been drained and populated.

Aquifer Recharge

Another value of wetlands related to hydrology is groundwater recharge. This function has received too little attention, and the magnitude of the phenomenon

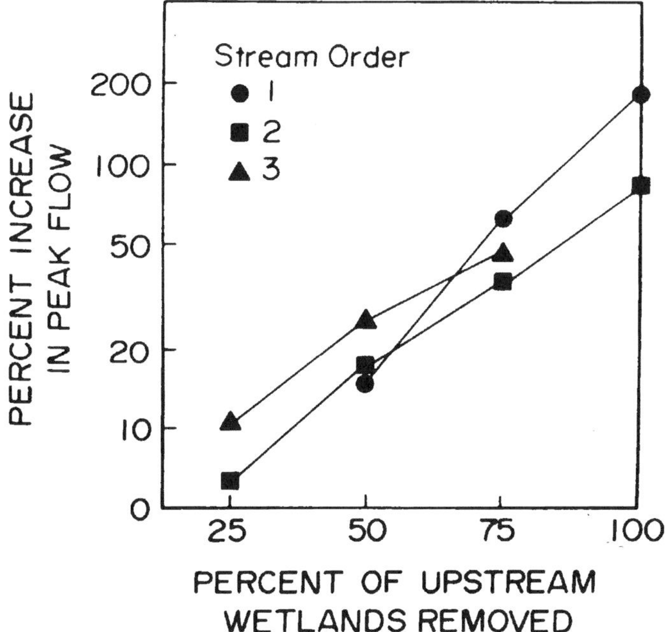

Figure 15–9. Hydrologic model simulation results showing relationships between increase in peak flood streamflow and percent wetland removal for Massachusetts watershed. Results are for various stream orders for 130-year flood. (*After Ogawa and Male, 1983*)

has not been well documented. Some hydrologists believe that although some wetlands recharge groundwater systems, most wetlands do not (Carter et al., 1979; Novitzki, 1979; Carter, 1986; Carter and Novitzki, 1988). The reason for the absence of recharge is that soils under most wetlands are impermeable (Larson, 1982). In the few studies available, recharge occurred primarily around the edges of wetlands and was related to the edge:volume ratio of the wetland. Thus recharge appears to be relatively more important in small wetlands such as prairie potholes than in large ones. These small wetlands can contribute significantly to recharge of regional groundwater (Weller, 1981). Heimburg (1984) found significant radial infiltration from cypress domes in Florida and concluded that the rate of infiltration was relative to the area of the wetland and the depth of the surficial water table. He also concluded that these wetlands represent hydrologic "highs" in the surface water table and are therefore "closely coupled to groundwater." There did not appear to be any direct percolation to deep aquifers, however.

Water Quality

Wetlands, under favorable conditions, have been shown to remove organic and inorganic nutrients and toxic materials from water that flows across them. The concept of wetlands as sinks for chemicals was discussed in detail in Chapter 5. Wetlands have several attributes that cause them to exert major influences on chemicals that flow through them, whether the chemicals are naturally added or artificially applied (Sather and Smith, 1984). These attributes include the following:

1. A reduction in water velocity as streams enter wetlands, causing sediments and chemicals sorbed to sediments to drop out of the water column;
2. A variety of anaerobic and aerobic processes in close proximity, promoting denitrification, chemical precipitation, and other chemical reactions that remove certain chemicals from the water (see Chap. 5);
3. The high rate of productivity of many wetlands that can lead to high rates of mineral uptake by vegetation and subsequent burial in sediments when the plants die;
4. A diversity of decomposers and decomposition processes in wetland sediments;
5. A high amount of contact of water with sediments because of the shallow water, leading to significant sediment-water exchange;
6. The accumulation of organic peat in many wetlands, which causes the permanent burial of chemicals.

There have also been a number of reports of efficient primary, secondary, and tertiary treatment of sewage wastewater as it flows through wetlands. More details about wetlands as wastewater and nonpoint sources of pollution control are given in Chapter 17. Where environmental circumstances are appropriate, waste organic compounds are rapidly decomposed and nitrogen is denitrified and lost to the air. Nonvolatile pollutants such as heavy metals and phosphorus accumulate under favorable conditions. When those materials saturate the ecosystem, they may begin to increase in the effluent. For these elements, permanent long-term storage depends in part on whether the wetland is accreting vertically (and thus sequestering materials in deep sediments).

Aesthetics

A real but difficult aspect of a wetland to capture is its aesthetic value, often hidden under the dry term "nonconsumptive use values," which simply means that people enjoy being out in wetlands. There are many aspects of this kind of wetland use. Wetlands are excellent "biological laboratories" where students in elementary, secondary and higher education can learn natural history first hand. Wetlands are a rich source of information about cultural heritage. The remains of prehistoric Native American villages and mounds of shells or middens have contributed to our understanding of Native American cultures and of the history of the use of our wetlands. Smardon (1979) described wetlands as visually and educationally rich environments because of their ecological interest and diversity. Their complexity makes them excellent sites for research. Many artists—the Georgia poet Sidney Lanier, the painters John Constable and John Singer

Sargent, and many others who paint and photograph wetlands—have been drawn to them. Many visitors to wetlands use hunting and fishing as excuses to experience the wildness and its solitude, expressing that frontier pioneering instinct that may lurk in all of us.

Subsistence Use

In many regions of the world, including Alaska, Canada, and several developing nations, the subsistence use of wetlands is extensive. There wetlands provide the primary resources on which village economies are based. Those societies have adapted to the local ecosystems over many generations and are integrated into them. For those cultures few alternative life-styles exist (Ellanna and Wheeler, 1986; see also Chap. 1).

Regional and Global Values

The wetlands function of maintaining water and air quality is discharged on a much broader scale than that of the wetland ecosystem itself. Wetlands may be significant factors in the global cycles of nitrogen, sulfur, methane, and carbon dioxide.

Nitrogen Cycle

The natural supply of ecologically useful nitrogen comes from the fixation of atmospheric nitrogen gas (N_2) by a small group of plants and microorganisms that can convert it into organic form. The current production of ammonia from N_2 for fertilizers is about equal to all natural fixation (Delwiche, 1970). Wetlands may be important in returning a part of this "excess" nitrogen to the atmosphere through denitrification. As discussed in Chapter 5, denitrification requires the close proximity of an aerobic and a reducing environment such as the surface of a marsh, and the denitrification rate seems to increase with the supply of nitrate. Because most temperate wetlands are the receivers of fertilizer-enriched agricultural runoff and are ideal environments for denitrification, it is likely that they are important to the world's available nitrogen balance.

Sulfur Cycle

Sulfur is another element whose cycle has been modified by humans. The atmospheric load of sulfate has been greatly increased by fossil fuel burning. It is about equally split between anthropogenic sources (104×10^{12} g/yr), chiefly caused by fossil fuel burning; and natural biogenic sources (103×10^{12} g/yr) of which salt marshes account for about 25 percent (Cullis and Hirschler, 1980; Andreae and Raembonck, 1983; Gosselink and Maltby, 1990). When sulfates are washed out of the atmosphere by rain, they acidify oligotrophic lakes and streams. When sulfates are washed into marshes, however, the intensely reducing environment of the sediment reduces them to sulfides. Some of the reduced sulfide is recycled to the atmosphere as hydrogen, methyl, and dimethyl sulfides, but most forms insoluble complexes with phosphate and metal ions (see Chap. 5). These complexes can be more or less permanently removed from circulation in the sulfur cycle.

Carbon Cycle

The carbon cycle may also be significantly affected in several ways on a global scale by wetlands. The carbon dioxide concentration in the atmosphere is steadily increasing because of the burning of fossil fuels and because the rapid clear-cutting of tropical forests results in the oxidation of organic matter in trees and soils (Woodwell et al., 1983; Detwiler and Hall, 1988; Houghton, 1990). Methane, which is released from anaerobic organic soils, may function as a sort of homeostatic regulator for the ozone layer that protects us from the deleterious effects of ultraviolet radiation (Sze, 1977), although it also contributes to the adsorption of radiant energy by the atmosphere and thus to the so-called greenhouse effect.

The huge volume of peat deposits in the world's wetlands has the potential to contribute significantly to worldwide atmospheric carbon dioxide levels, depending on the balance between draining and oxidation of the peat deposits and their formation in active wetlands. Armentano and Menges (1986) estimated that before recent human disturbances of wetlands, the net global retention of carbon in peats was 57–83 x 10^6 mt/yr, most in northern peatlands. This annual retention rate is small compared to the estimated 4,000–9,000 x 10^6 mt/yr of net primary productivity of the world's wetlands (Aselmann and Crutzen, 1989). It is also small compared to the estimated 1.4 x 10^{12} mt of organic carbon stored in the world's soils (Post et al., 1982), with about one-third of that, or 0.45 x 10^{12} mt, in boreal and subarctic peat (Gorham, 1991). There are indications that the global carbon balance of wetlands has shifted, primarily because of agricultural conversion of peatlands. By 1900 there had been no significant change in North America, but in European countries anywhere from 20 to 100 percent of the stored carbon had been lost to the atmosphere. By 1980 the total carbon shift attributed to agricultural drainage was estimated at 63–85x10^6 mt/yr, with a further 32–39x10^6 mt/yr released from peat combustion.

Gorham (1991) provided another estimate of the role of wetlands in global carbon cycling (Table 15–4). He estimated that there is a current net accumulation of carbon of 76 x 10^6 mt/yr in northern peatlands. On the other hand, oxidation attributed to drainage is 8.5–42 x 10^6 mt/yr (the low number is for long term; the high number for short term), the combustion of peat releases 26 x 10^6 mt/yr, and emissions of methane releases 46 x 10^6 mt/yr. For all wetlands in the world, Cicerone and Oremland (1988) estimated a release of 86 x 10^6 mt/yr of methane (as carbon) while Aselmann and Crutzen (1989) estimated a release of 40–160 x 10^6 mt/yr of methane from natural wetlands and another 60–140 x 10^6 mt/yr from rice paddies. The total emission of methane from all

Table 15–4. Net Carbon Fluxes to and from Boreal and Subarctic Peatlands

	Carbon flux, x 10^6 mt/yr
Undrained peatlands (3.30 x 10^6 km^2)	
Accumulated as organic carbon in peat	
Overall	96
Current	76
Released as CH_4 to atmosphere	46
Drained peatlands (0.115 x 10^6 km^2)	
Released as CO_2 by short-term drainage	42
Released as CO_2 by long-term drainage	8.5
Released as CO_2 by fuel combustion	26

Source: Gorham, 1991

sources is estimated to be 520–590 x 10^6 mt/yr. Thus wetlands, like tropical rain forests, may be shifting from being a net sink to a net source of carbon to the atmosphere. Their protection should thus be encouraged to prevent the release of yet more carbon to the atmosphere. To put these numbers in perspective, the burning of fossil fuel contributes an estimated 5,600 x 10^6 mt/yr, and the deforestation of tropical rain forests contributes an additional 400–2,800 x 10^6 mt/yr of carbon to the atmosphere (Detwiler and Hall, 1988; Houghton, 1990).

QUANTIFYING WETLAND VALUES

A number of efforts have been made to quantify the "free services" and amenities that wetlands provide for society (Wharton, 1970; Gosselink et al., 1974; Jaworski and Raphael, 1978; Mumphrey et al., 1978; Mitsch et al., 1979c; Costanza, 1984; C. W. Johnson and Linder, 1986; Leitch, 1986; Farber and Costanza, 1987; Shabman and Batie, 1988; Leitch and Ekstrom, 1989; Costanza et al., 1989; Folke, 1991). For activities that require an environmental impact statement (as required by the National Environmental Policy Act), two kinds of evaluation are involved. The first is the determination of the ecological value of the area in question—that is, the ecological quality of the site as compared with similar sites, for example, its suitability for supporting wildlife. The other component of the evaluation is a comparison of the economic value of the habitat compared to the economic value of some proposed activity that would destroy or modify it. Regardless of which kind of evaluation is required, several generic problems must be addressed.

1. *Wetlands are multiple-value systems, that is, they may be valuable for many different reasons.* Therefore, the evaluator is often faced with the problem of comparing and weighing different commodities. For example, a fresh marsh area is more valuable for waterfowl than a salt marsh area is; but the salt marsh

may be much more valuable as a fish habitat. Which is rated higher depends on the value judgment made by the evaluator, which has nothing to do with the intrinsic ecological viability of either area. This is the old "apples versus oranges" problem. The decision to some extent reflects a matter of preference. Furthermore, in most wetland evaluations, evaluators are not concerned with single commodities. Instead, they wish to apprehend the overall value of an area, that is, the value of the whole fruit basket, rather than the apples, oranges, and pears. Complexity is added when the concern is to compare the value of a natural wetland with the same piece of real estate proposed to be used for economic development—a dammed lake, a parking lot, or an oil well. In that case, the comparison is not between apples and oranges but between apples and computer chips or oranges and electrical energy. Conventional economics solves the comparison problem by reducing all commodities to a single index of value—dollars. This is difficult when some of the commodities are natural products of wetlands that do not compete in the marketplace.

2. *The most valuable products of wetlands are public amenities that have no commercial value for the private wetland owner.* The wetland owner, for example, has no control over the harvest of marsh-dependent fish that are caught in the adjacent estuary or even offshore in the ocean. The owner does not control and usually cannot capitalize on the ability of the wetland to purify waste water, and certainly cannot control its value for the global nitrogen balance. Thus there is often a strong conflict between what a private wetland owner perceives of as his or her best interests and the best interests of the public. In coastal Louisiana a marsh owner can earn revenues of perhaps $25 per acre annually from the renewable resources of his or her marsh by leasing it for hunting and fur trapping. In contrast, depending on where it is situated, a wetland may be worth hundreds of thousands of dollars per acre as a housing development or as a site for an oil well. Riparian wetlands in the Midwest and lower Mississippi River Basin yield little economic benefit to the owner for the flood mitigation and water-quality maintenance that they provide. Yet if cleared and planted in corn or soybeans, the land will provide economic benefits to the owner. Many of the current regulations that govern wetlands were initiated to protect the public's interest in privately owned wetlands (see Chap. 16).
3. *The relationship between wetland area and marginal value is complex.* Conventional economic theory states that the less there is of a commodity, the more valuable the remaining stocks. This generalization is complicated in nature because different natural processes operate on different scales. This is an important consideration in parklands and wildlife preserves, for example. Large mammals require large ranges in which to live. Small plots below a certain size will not support *any* large mammals. Thus the marginal-value generalization falls apart because the marginal value ceases to increase below a certain plot size. In fact, it becomes zero. In wetlands, the situation is even more complex because they are open ecosystems that maintain strong ties to

adjacent ecosystems. Therefore, one factor that governs the ecological value (and hence the value to society) of the wetland is its *interspersion* with other ecosystems, that is, its place in the total regional landscape. A small wetland area that supports few endemic organisms may be extremely important during critical times of the day or during certain seasons for migratory species that spend only a day or a week in the area. A narrow strip of riparian wetland along a stream that amounts to very little acreage may efficiently filter nutrients out of runoff from adjacent farmland, protecting the quality of the stream water. Its value is related to its interspersion in the landscape, not to its size. These kinds of considerations have only recently begun to be addressed in a quantitative way, and the methodology is not well developed.

4. *Commercial values are finite, whereas wetlands provide values in perpetuity. Wetland development is often irreversible.* The time frame for most human projects is from 10 to 30 years. Private entrepreneurs expect to recoup their investments and profits in projects within this time frame and seldom consider longer-term implications. Even large public-works projects such as dams for energy generation seldom are seen in terms longer than from 50 to 100 years. The destruction of natural areas, on the other hand, removes their public services forever. Often, especially for wetlands, the decision to develop is irreversible. If an upland field is abandoned, it will gradually revert to a forest; but once a wetland is drained and developed, it is usually lost forever because of associated changes in the hydrologic regime of the area. For example, in Louisiana and elsewhere, marsh areas were diked and drained for agriculture early in the twentieth century. Many of these developments have subsequently been abandoned. They did not revert to wetlands, but are now large, shallow lakes, distinguishable by their straight edges (Fig. 15–10).
5. *A comparison of economic short-term gains with wetland value in the long term is often not appropriate.* Wetland values, even when multiple functions are quantified, often cannot "compete" with short-term economic calculations of high economic-yield projects such as commercial developments or intensive agriculture. This is especially true because economic analyses typically discount the value of future amenities. The issue of wetland conservation versus development has an intergenerational component. Future generations do not compete in the marketplace, and decisions that will affect the public resources they inherit are often made without regard to their interests.
6. *Estimates of values, by their nature, are colored by the personal endowment and biases of individuals and of the society.* For example, China places a value on wetlands that is vastly different from that of the United States. There is a great deal of subjective judgment about wetlands related to different knowledge bases and different value systems. It is not surprising that wetland scientists "value" wetlands, even quantitatively, at higher levels than developers and other resource users do.

Figure 15–10. Sugar cane fields "reclaimed" from fresh marsh in coastal Louisiana. Natural marsh is in foreground. Such development projects are usually irreversible since land elevations and flooding patterns are permanently changed. (*Photograph by C. Sasser*)

A number of approaches to the valuation of wetlands have been advanced (see Lonard et al., 1981). Because of the complexities described above, there is no universal agreement about which is preferable. In part the choice depends on the circumstances. Valuations fall broadly into two classes, ecological (or functional) evaluation and economic (or monetary) evaluation. The former are generally necessary before attempting the latter, for it is the valued ecological functions that determine monetary value.

Ecological Valuation: Scaling and Weighting Approaches

The ecological-valuation approach has been widely used as a means of forming a rational basis for deciding on different management options. Probably the best developed procedures assess the relative value of wildlife habitats. E. P. Odum (1979b) described the general procedure, which is listed below.

a. Make a list of all the values that a knowledgeable person or panel can apply to the situation in question, and assign a numerical value of "1" to each.
b. Scale each factor in terms of a maximum level; for example, if 200 ducks per acre could be supported by a first-class marsh but only 100 are supported by the marsh in question, then the scaled factor is 0.5, or 50 percent of the maximum value for that item.

c. Weigh each scaled factor in proportion to its relative importance; for example, if the value 2 is considered 10 times more important to the region than the value 1, then multiply the scale value of 2 by 10.
d. Add the scaled and weighted values to obtain a value index. Because the numbers are only arbitrary and comparative, the index is most useful in comparing different wetlands or the same wetland under different management plans. It is desirable that each value judgment reflect the consensus of several "experts," for example, determined by the "Delphi method" (Dalkey, 1972).

Table 15–5 shows an example from the Habitat Evaluation Procedure (HEP) of the U. S. Fish and Wildlife Service (1980) of the application of this technique to evaluate different development plans for a cypress-gum swamp ecosystem. The present value of the swamp for a representative group of terrestrial and aquatic animals was evaluated (baseline conditions) using a habitat-suitability index (HSI) based on a range of from 0 to 1 for the optimum habitat for the species in question. The evaluation resulted in a mean terrestrial HSI of 0.8 and a mean aquatic HSI of 0.4. This baseline condition was compared with the projected habitat condition in 50 and 100 years under three projected scenarios—Plan A, Plan B, and a no-project projection. The results suggest that Plan A would be detrimental to the environment, whereas Plan B would have no effect on terrestrial habitat values and would improve aquatic ones. Whether to proceed with either of these plans is a decision that requires weighing projected environmental effects against projected economic benefits of the project.

One often neglected feature of the analysis is the effect of aggregating HSIs for different species. Although, overall, Plan B appears to be about equivalent environmentally to the no-project option, scrutiny of Table 15–5 shows that Plan B is expected to improve the habitat for swamp rabbits and large-mouthed bass but decrease its value for warblers and turtles. This kind of detailed scrutiny may be important because it indicates a change in the quality of the environment, but it is often neglected when the "apples and oranges" are combined into "fruit."

Two procedures attempt to deal with two shortcomings of the habitat evaluation discussed above by (1) evaluating all relevant goods and services (not just biotic ones) derived from the site and (2) incorporating a landscape focus. The Wetland Evaluation Technique (WET) (Adamus et al., 1987) rates a broad range of functional attributes on a scale of high, moderate, and low. The result is a list of functions, each involving a quality rating for three attributes: (1) *social significance* assesses the value of a wetland to society in terms of its economic value, strategic location (for example, upstream from an urban area that requires flood protection), or any special designations it carries (for example, habitat for an endangered species); (2) *effectiveness* is the site's capacity to carry out a function because of its physical, chemical, or biological characteristics (for example, to store flood waters); (3) *opportunity* refers to the opportunity of a wetland to perform a function to its level of capability (for example, whether the upstream

Table 15–5. Comparison of the Impact of Two Management Plans and a No-Management Control in a Cypress-Gum Swamp[a]

Species	*Baseline Condition*	*Future with Project Plan A*[b]		*Future with Project Plan B*[c]		*Future without Project*	
		50 Years	*100 Years*	*50 Years*	*100 Years*	*50 Years*	*100 Years*
			Terrestrial				
Raccoon	0.7	0.5	0.6	0.8	0.8	0.7	0.9
Beaver	0.7	0.2	0.2	0.4	0.3	0.6	0.4
Swamp Rabbit	0.7	0.2	0.2	0.8	0.8	0.7	0.4
Green Heron	0.9	0.2	0.1	0.8	0.9	0.9	1.0
Mallard	0.8	0.3	0.2	1.0	0.9	0.9	1.0
Wood Duck	0.8	0.3	0.2	0.9	1.0	1.0	1.0
Prothonotary Warbler	0.8	0.3	0.1	0.6	0.7	0.8	0.9
Snapping Turtle	0.8	0.4	0.3	0.8	0.7	0.8	0.9
Bullfrog	0.9	0.3	0.2	0.8	0.9	1.0	1.0
Total Terrestrial HSI	7.1	2.7	2.1	6.9	7.0	7.4	7.5
Mean Terrestrial HSI	0.8	0.3	0.2	0.8	0.8	0.8	0.8
			Aquatic				
Channel Catfish	0.3	0.3	0.4	0.4	0.4	0.4	0.4
Largemouth bass	0.4	0.2	0.3	0.7	0.8	0.4	0.4
Total Aquatic HSI	0.7	0.5	0.7	1.1	1.2	0.8	0.8
Mean Aquatic HSI	0.4	0.3	0.4	0.6	0.6	0.4	0.4

[a]Numbers in the tables are habitat suitability index (HSI) values,which have a maximum value of 1 for an optimal habitat.
[b]Channelization of water and clearing of swamp for agricultural development with a loss of 324 ha of wetland.
[c]Construction of levees around swamp for flood control with no loss of wetland area.
Source: Schamberger et al., 1979

watershed is capable of producing flood waters). The evaluator is charged with the task of weighting each function to get an integrated evaluation.

One of the problems with both WET and the Habitat Evaluation Procedure is that they tend to be site specific. WET deals with some contextual issues but does not reflect a landscape focus and its results are only semiquantitative. A procedure recently developed by the U.S. Fish and Wildlife Service (O'Neil et al., 1991) for bottomland hardwood forested wetland ecosystems is explicitly concerned with large-scale features and general habitat quality. Its purpose is to "numerically rate wildlife habitat quality of bottomland forests in floodplains of the southeastern United States" on a scale of 0–1.0, where "a value of 1.0 represents the types of bottomland habitats needed to support the maximum native species richness of birds, mammals, reptiles, and amphibians on a regional scale over a long time period" (O'Neil et al., 1991). The model has two components, a local one and a regional one. At the regional level, habitat fragmentation and contiguity are major considerations. The objectives focus on landscape scales, animal diversity rather than indicator species (as in the earlier Habitat Evaluation Procedure), and on long time frames rather than current status. Thus it attempts to deal with several of the problems listed in the previous section that are inherent in natural resource valuation. On the other hand, it evaluates wetlands for only one attribute—wildlife diversity—not the whole range of services provided by a wetland site.

Economic Evaluation: Common Denominator Approaches

Evaluation systems that seek to compare natural wetlands to human economic systems usually reduce all values to monetary terms (thus losing sight of the apples and oranges). Conventional economic theory assumes that in a free economy the economic benefit of a commodity is the dollar amount that the public is willing to pay for the good or service rather than be without it. This measure is formally expressed as *willingness-to-pay*, or, more accurately, *net* willingness-to-pay, "the amount society would be willing to pay to produce and/or use a good beyond that which it actually does pay" (Scodari, 1990). The principle is illustrated as follows (Scodari 1990): Suppose a fisherman were willing to pay $30 a day to use a particular fishing site but had to spend only $20 per day in travel and associated costs. The net benefit, or economic value, to the fisherman of a fishing day at the site is not the $20 expenditure but the $10 difference between what he was willing to spend and what he had to spend. If the fishing opportunity at the site were eliminated the fisherman would lose $10 worth of satisfaction fishing; the $20 cost that he would have incurred would be available to spend elsewhere. In the case of commercial goods such as harvested fish, the total value of a wetland is the sum of the net benefit to the consumer plus the net benefit to the producer (the fisherman).

Economists estimate consumer benefits through a *demand curve* and producer benefits through a *supply curve* (Fig. 15–11). When a good is sold in an open,

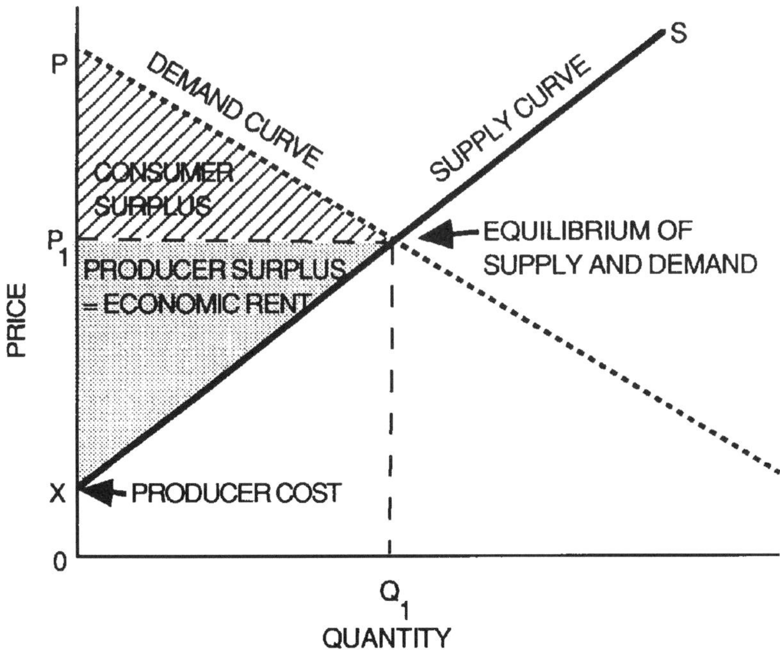

Figure 15–11. Superimposed theoretical supply and demand curves show the equilibrium price and quantity, the consumer surplus, and the economic rent.

competitive market, its market price rises as the supply volume decreases. The price of the last unit of the good purchased measures the consumer's marginal willingness to pay. For all other units of the good, however, the consumer is willing to pay more than the marginal price; that is, the scarcer an item, the more one is willing to pay for it. The marginal price also approaches the producer's marginal cost because the price must represent at a minimum the cost of production. The area under the curve bounded by the marginal price—the *consumer surplus*—represents the net benefit of the good to the consumer. Similarly, a supply curve describes the relationship of supply to price; the excess of what producers earn over production costs—the area over the good's supply curve bounded by price—is the producer surplus, or *economic rent.* Although not an exact measure of social welfare, the sum of consumer surplus and economic rent provides a useful approximation of the net benefit of a good or service (Scodari, 1990).

Although this characterization of the value of a commodity is reasonable under most conventional economic conditions, it leads to real problems in monetizing nonmarket commodities such as pure water and air and in pricing wetlands whose value in the marketplace is determined by their value as real estate, not by their "free services" to society. Consequently, attempts to monetize wetland values have generally emphasized the commercial crops from wetlands: fish, shell-

Figure 15–12. This flock of white pelicans on a coastal marsh symbolizes the difficulty of placing a dollar value on wetlands. They are not hunted so have no "sport-hunting value." Nevertheless, they are one species supported by wetlands. Have they any value? If so, how is it defined, and how quantified, in a social system in which "market values" often take precedence over any others? (*Photograph by R. Abernethy*)

fish, furs, and recreational fishing and hunting for which pricing methodologies are available (Fig. 15–12). As E. P. Odum (1979b) pointed out, this kind of pricing ignores ecosystem and global-level values related to clean air and water and other "life-support" functions. Even in the cases of market commodities from wetlands, available data are seldom adequate to develop reliable demand curves (Bardecki, 1987).

Economists recognize four more or less independent aspects of "value" that contribute to the total. These aspects are (1) use value—the most tangible portion of total value derived from identifiable direct benefits to the individual; hunting, harvesting fish, and nature study are examples; (2) social value—those amenities that accrue to a societal group rather than an individual; examples are improved water quality, flood protection, and the maintenance of the global sulfur balance; (3) option value—the value that exists for the conservation of perceived benefits for future use; (4) existence value—the benefits deriving from the simple knowledge that the valued resource exists—irrespective of whether it is ever used. For example, the capacity of an extant wetland to conserve biological diversity is an existence value.

Willingness-to-Pay Methods

In the absence of a well-developed free-market alternative, pricing methodologies have been applied. One of these, "willingness-to-pay," establishes a more or less hypothetical (contingency) market for nonmarket goods or services (Mitchell and

Carson, 1989). The evaluation of willingness-to-pay has been carried out in a number of ways (Bardecki, 1987), including:

1. *Gross expenditure.* This approach evaluates the total expenditures for a specific activity (for example, the willingness of hunters to spend dollars to travel to a site, buy equipment, and rent hunting easements) as a measure of the value of the wetland site. Aside from the fact that this method measures only one aspect of the total value of a wetland, it is beset with serious methodological difficulties (Carey, 1965; Knetsch and Davis, 1966).
2. *Travel cost.* In this method the costs of travel from different locations to a common property resource such as a wetland site are used to create a demand curve for the goods and services of that site. The total net value of the site is represented by the integral of the demand function (Hammack and Brown, 1974; Gum and Martin, 1975). This approach has been used most effectively for recreational activity. It is not effective in estimating the demand for off-site services such as downstream water-quality enhancement or flood mediation.
3. *Imputed willingness to pay.* In circumstances in which goods or services that depend on wetlands are produced and are recognized in the marketplace (for example, commercial fish harvests), the values of these goods can be interpreted as measures of society's willingness to pay for the productivity of the wetlands and hence for the wetland itself (Farber and Costanza, 1987). This method is limited to specific products or services and does not cover the entire range of wetlands values.
4. *Direct willingness to pay* (contingent value). This method uses direct surveys of consumers to generate willingness-to-pay values for the entire range of potential wetland benefits. This method can separate public from private benefits, derive marginal values rather than average ones, and deal with individual sites or the entire resource base (Bardecki, 1987). On the other hand, an individual's willingness to pay for a wetland "service" is probably strongly influenced by his understanding of the ecology and functions of wetlands. Thus the contingent-value method, like the other approaches, has both strengths and weaknesses.

Opportunity Costs

A second approach to resource evaluation in the absence of a free-market model is the *opportunity cost* approach. In general terms, the opportunity cost associated with a resource is the net worth of that resource in its best alternative use. For example, "the opportunity cost of conserving a wetland area is the net benefit which might have been derived from the best alternate use of the area which must be foregone in order to preserve it in its natural state" (Bardecki, 1987). Because determining the opportunity cost associated with wetland conservation would require the evaluation of each wetland good and service as well as the identification and valuation of the best alternative use, in practice a comprehensive evaluation of the opportunity cost of wetland conservation is far from possible. Nevertheless, it represents a useful approach to the valuation of specific wetland functions (Shabman, 1986).

Replacement Value

If one could calculate the cheapest way of replacing various services performed by a wetland and could make the case that those services would have to be replaced if the wetland were destroyed, then the figure arrived at would be the "replacement value." Some of the replacement technologies that might be necessary to replace services provided by wetland processes are listed in Table 15–6. This approach has the merit of being accepted in the world of conventional economists. For certain functions, it gives very high values compared with those of other valuation approaches discussed in this section. For example, the tertiary treatment of wastewater is extremely expensive, as is the cost of replacing the nursery function of marshes for juvenile fish and shellfish. Serious questions, however, have been raised about whether these functions would be replaced by treatment plants and fish nurseries if the wetlands were destroyed. Some ecologists and economists argue that in the long run either the services of wetlands would have to be replaced or the quality of human life would deteriorate. Other individuals argue that this assertion cannot be supported in any convincing manner.

Energy Analysis

A completely different approach uses the idea of energy flow through an ecosystem or the similar concept of "embodied energy." The concept of embodied energy (Costanza, 1980), or *emergy* (H. T. Odum, 1988, 1989), the total energy required to produce a commodity, is assumed to be a valid index of the totality of ecosystem functions and is applicable to human systems as well. Thus natural and human systems can be evaluated on the basis of one common currency. Because there is a linear relationship between embodied energy and money, the more familiar currency can also be used. Costanza (1984) and Costanza et al. (1989) showed that the economist's willingness-to-pay approach and energy analysis converge to a surprising degree, although both methods result in a great deal of uncertainty. Table 15–7 shows a comparison of some monetary values for Louisiana's coastal marshes arrived at through these two different approaches. The willingness-to-pay estimates reflect the assessment that a reasonable range of wetland value for coastal Louisiana is between \$6,000 and \$22,000 per hectare (\$2,400–\$9,000 per acre) depending on the discount rate applied to determine the present value.

The energy-analysis method is based on using the total amount of energy captured by natural ecosystems as a measure of their ability to do useful work (for nature and hence for society). Gross primary productivity (GPP) of representative

Table 15–6. Some Replacement Technologies for Societal Support Values Provided by Wetlands

Societal Support	*Replacement Technologies*
Peat Accumulation	
Accumulating and storing organic matter (peat)	Artificial fertilizers Redraining ditches
Hydrologic Function	
Maintaining drinking water quality	Water transport Pipeline to distant source
Maintaining ground water level	Well-drilling Saltwater filtering
Maintaining surface water level	Dams for irrigation Pumping water to dam Irrigation pipes and machines Water transport for domestic animals
Moderation of waterflows	Regulating gate Pumping water to stream
Biogeochemical Functions	
Processing sewage, cleansing nutrients and chemicals	Mechanical sewage treatment Sewage transport Sewage treatment plant Clear-cutting ditches and stream
Maintaining drinking water quality	Water-quality inspections Water purification plant Silos for manure from domestic animals Nitrogen filtering Water transport
Filter to coastal waters	Nitrogen reduction in sewage treatment plants
Food Chain Functions	
Providing food for humans and domestic animals	Agriculture production Import of food
Providing cover	Roofing materials
Sustaining anadromous trout populations	Releases of hatchery-raised trout Farmed salmon
Sustaining other fish species and wetland dependent flora and fauna	Work by nonprofit organizations
Species diversity; storehouse for genetic material	Replacements not possible
Bird watching, sport fishing, boating, and other recreational values	Replacement not possible
Aesthetic and spiritual values	Replacement not possible

Source: Folke, 1991

coastal marsh systems, which ranges from 48,000 to 70,000 kcal m^{-2} yr^{-1}, is converted to money units by multiplying by a conversion factor of 0.05 units fossil fuel energy/unit GPP energy and dividing by the energy/money ratio for the economy (15,000 kcal fossil fuel/1983 $). These calculations resulted in estimates of annual coastal wetland value of about $1,560 ha^{-1} yr^{-1} ($630 $acre^{-1}$ yr^{-1}), which when converted to present value for an infinite series of payments, yields the range of values of $16,000–$70,000/ha ($6,400–$28,200/acre) for the discount rates used in Table 15–7.

Costanza et al. (1989) used this range from the willingness-to-pay and energy analysis approaches to suggest that the annual loss of Louisiana coastal wetlands is costing society from $77 million to $544 million per year.

The energy-analysis method, although imprecise because of the several estimates used, is more satisfying to many wetland scientists than conventional cost-accounting methods because it is based on the inherent productivity of the ecosystem, not on perceived values that may change from generation to generation and from location to location. It is interesting to note that the energy analysis results in Table 15–7 give values higher than the willingness-to-pay approach. The sensitivity of both conventional and energy-analysis methods to the choice of a discount rate, which has been vital for decades in the outcome of cost-benefit studies, is also demonstrated in this comparison.

In analyzing other comparisons of valuation methods, Mitsch et al. (1979c) and Folke (1991) compared the use of a replacement-cost method with the use of an energy-analysis method and found them to be in a similar range. For example, the

Table 15–7. Estimates of Wetland Values in $/ha ($/acre) of Louisiana Coastal Marshes Based on Willingness-to-Pay and Energy Analysis at Two Discount Rates

	Discount Rate			
Method	*3 percent*		*8 percent*	
Willingness to Pay				
commercial fishery	$2,090	($846)	$ 783	($317)
fur trapping (muskrat and nutria)	991	(401)	373	(151)
recreation	447	(181)	114	(46)
storm protection	18,653	(7,549)	4,732	(1,915)
Total willingness-to-pay value	$22,181	($8,977)	$6,002	($2429)
Energy Analysis	$42,000–70,000 ($17,000–28,200)		$16,000–26,000 ($6,400–10,600)	
Best Estimate	$22,000–$42,000 ($9,000–$17,000)		$6,000–$16,000 ($2,400–$6,400)	

Source: Costanza et al., 1989

replacement method suggested an annual cost of from \$0.4 to \$1.1 million to replace the functions of a 2.5 km^2 peatland-lake complex on the island of Gotland, Sweden (Folke, 1991). This translates to a replacement cost of \$1,600 ha^{-1} yr^{-1} (\$650 $acre^{-1}$ yr^{-1}), with most of the cost involved in replacing the biogeochemical processes of the wetland (52–82 percent of the cost) and less involved in replacing the hydrologic processes (7–40 percent) and food chain functions (8–11 percent) (see Table 15–6) (Folke, 1991). When the energy cost of the economic replacements was compared with the energy lost when the wetland was lost, the results were remarkably similar. If the 2.5 km^2 wetland were lost, the economic-replacement cost in energy terms would range from 3.5 to 12 x 10^9 kcal/yr; the ecosystem-loss calculation ranged from 13 to 18 x 10^9 kcal/yr. Again, the energy-analysis method gave a slightly higher estimate of the energy (and hence the money) cost of wetland loss than the replacement-cost method did.

Multiple Function Approach

Values placed on wetlands using these evaluation methods have ranged from very high to low. Although few people dispute that wetlands have many and varied values, the lack of consistent, accepted methodologies for comparing them with conventional economic goods and services limits the usefulness of the estimates that have been made. Perhaps the most comprehensive attempt to evaluate wetlands for management purposes is a joint project of Wildlife Habitat Canada and Environment Canada (Manning et al., 1990). Their process of valuing a wetland development project begins with applying a multiple-function screening that incorporates a series of specific standards or benchmarks reflecting applicable societal goals associated with wetlands. Most of the standards were identified from existing legislation, stated government goals and objectives, and scientific principles. A project that fails to satisfy one or more of these goals can often be eliminated, avoiding further, detailed analysis.

Major projects passing the initial screening must subsequently be evaluated in terms of their impact on identified social values. This process involves a cost accounting of the benefits of the project compared to the cost of natural wetland goods and services lost. In this process both the willingness-to-pay and the opportunity cost methodologies are used. The results of four pilot studies showed limitations in each of the methods but also the possibility for developing a more useful process for evaluating major development projects. One key element of the evaluation process is the need to recognize and conceptually separate the relationships among the ecological functions of wetlands, the recognizable benefits to society, and the socioeconomic values that can be placed on those benefits.

Wetland Management and Protection

16

Wetland management has meant both wetland alteration and protection. In earlier times wetland drainage was considered the only policy for managing wetlands. With the recognition of wetland values, wetland protection has been emphasized by many federal and state policies. Nevertheless, significant wetland alteration continues, particularly by dredging, filling, drainage, hydrologic modification, peat mining removal for mineral extraction, and water pollution. Wetlands can also be managed in their more or less natural state for certain objectives such as fish and wildlife enhancement, agricultural and aquaculture production, water quality improvement, and flood control.

The federal government in the United States has relied on executive orders, a "no net loss" policy, and the Section 404 dredge-and-fill permit program of the Clean Water Act for wetland protection augmented by wetland protection programs in agriculture and the development of wetland delineation procedures. Some states have wetland protection laws although many more have relied on Federal regulations. Many observers believe that individual states will assume more responsibility for protecting wetlands, whereas others believe that the federal government must continue in the lead in the now highly politicized field of wetland protection. International cooperation in wetland protection, particularly through the Ramsar Convention and the North American Waterfowl Management Plan, has been emphasized in recent years as policymakers have come to realizate that wetland function knows no political boundaries.

The concept of wetland management has had different meanings at different times to different disciplines. Until the middle of the twentieth century, wetland management usually meant wetland drainage to many policymakers except for a few resource managers who maintained wetlands for hunting and fishing. Landowners were encouraged through government programs to tile and drain wetlands to make the land suitable for agriculture and other uses. Countless coastal wetlands were destroyed by dredging and filling for navigation and land development. There was little understanding of and concern for the inherent values of wetlands. The value of wetlands as wildlife habitats, particularly for waterfowl, was recognized in the first half of this century by some fish and game managers to whom wetland management often meant the maintenance of hydrologic conditions to optimize fish or waterfowl populations. Only relatively recently have other values such as those described in Chapter 15 been recognized.

Today the management of wetlands means setting several objectives, depending on the priorites of the wetland manager. In some cases, objectives such as preventing pollution from reaching wetlands and using wetlands as sites of wastewater treatment or disposal can be conflicting. Many floodplain wetlands are now managed and zoned to minimize human encroachment and maximize floodwater retention. Coastal wetlands are now included in coastal zone protection programs for storm protection and as sanctuaries and subsidies for estuarine fauna. In the meantime, wetlands continue to be altered or destroyed through drainage, filling, conversion to agriculture, water pollution, and mineral extraction.

Wetlands are now the focus of legal efforts to protect them but, as such, they are beginning to be defined by legal fiat rather than by the application of ecological principles. Protection has been implemented through a variety of policies, laws, and regulations ranging from land-use policies to zoning restrictions to enforcement of dredge-and-fill laws. In the United States, wetland protection has historically been a national initiative, but some assistance has been provided by individual states. In the international arena, agreements to protect ecologically important wetlands throughout the world have been negotiated and ratified.

AN EARLY HISTORY OF WETLAND MANAGEMENT

The early history of wetland management, a history that still influences many people today, was driven by the misconception that wetlands were wastelands that should be avoided or, if possible, drained and filled. As described by Larson and Kusler (1979), "For most of recorded history, wetlands were regarded as wastelands if not bogs of treachery, mires of despair, homes of pests, and refuges for outlaw and rebel. A good wetland was a drained wetland free of this mixture of dubious social factors." In the United States this opinion of wetlands and shallow-water environments led to the destruction of more than half of the total wetlands in the lower 48 states (Fig. 16–1; see also Chap. 3).

Some public laws actually encouraged wetland drainage. Congress passed the Swamp Land Act of 1849, which granted to Louisiana the control of all swamplands and overflow lands in the state for the general purpose of controlling floods in the Mississippi Basin. In the following year the act was extended to the states of Alabama, Arkansas, California, Florida, Illinois, Indiana, Iowa, Michigan,

Figure 16–1. This aerial photograph taken from a U-2 plane is a scene of a fresh marsh along an abandoned Mississippi River distributary, Bayou Lafourche. Residential and agricultural development has occurred on the high natural levees of this bayou. The large rectangular lake (1) was an agricultural development in the early part of the century. The levees were breached by a severe storm and it was abandoned. Below it (2) is a similar development, still in sugarcane production. The soil surface inside the levees is now about 2 m below the surrounding water level due to compaction and oxidation. Manmade canals (3) are straight and deep, natural channels (4) are tortuous and shallow. Infared imagery from 20,000 meters (65,000 feet) with 30 cm (12-inch) focal length lens. (*Photograph by NASA, Ames Research Center, Flight 78-143, October 9, 1978*)

Mississippi, Missouri, Ohio, and Wisconsin. Minnesota and Oregon were added in 1860. The act was designed to decrease federal involvement in flood control and drainage by transferring federally owned wetlands to the states, leaving to them the initiative of "reclaiming" wetlands through activities such as levee construction and drainage. By 1954, almost 100 years after the act was established, an estimated 26 million hectares (65 million acres) of land had been ceded to those 15 states for reclamation. Ironically, although the federal government passed the Swamp Land Act to get out of the flood-control business and the states sold those lands to individuals for pennies per acre, the private owners subsequently exerted great pressure on both national and state governments to protect them from floods. Further, governments are now paying enormous sums to buy those lands back for conservation purposes. Although current government policies are generally in direct opposition to the Swamp Land Act and it is now disregarded, the act cast the initial wetland policy of the United States government in the direction of wetland elimination.

Other actions led to the rapid decline of the nation's wetlands. An estimated 23 million hectares (57 million acres) of wet farmland, including some wetlands, were drained under the U.S. Department of Agriculture's Agricultural Conservation Program between 1940 and 1977 (Office of Technology Assessment, 1984). Some of the wetland-drainage activity was hastened by projects of groups such as the Depression-era WPA (Works Progress Administration) and the Soil Conservation Service (Reilly, 1979). Coastal marshes were eliminated or drained and ditched for intercoastal transportation, residential developments, mosquito control, and even for salt marsh hay production. Interior wetlands were converted primarily to provide land for urban development, road construction, and agriculture.

Typical of the prevalent attitude toward wetlands is the following quote by Norgress (1947) discussing the "value" of Louisiana cypress swamps:

> With 1,628,915 acres of cutover cypress swamp lands in Louisiana at the present time, what use to make of these lands so that the ideal cypress areas will make a return on the investment for the landowner is a serious problem of the future. . . .
>
> The lumbermen are rapidly awakening to the fact that in cutting the timber from their land they have taken the first step toward putting it in position to perform its true function—agriculture. . . .
>
> It requires only a visit into this swamp territory to overcome such prejudices that reclamation is impracticable. Millions of dollars are being put into good roads. Everywhere one sees dredge boats eating their way through the soil, making channels for drainage.
>
> After harvesting the cypress timber crop, the Louisiana lumbermen are at last realizing that in reaping the crop sown by Nature ages ago, they have left a heritage to posterity of an asset of permanent value and service—land, the true basis for wealth.

> The day of the pioneer cypress lumberman is gone, but we need today in Louisiana another type of pioneer—the pioneer who can help bring under cultivation the enormous areas of cypress cutover lands suitable for agriculture. It is important to Louisiana, to the South, and the Nation as a whole, that this be done. Would that there were some latter-day Horace Greeleys to cry, in clarion tones, to the young farmers of today, "Go South, young man; go South!"

As an example of state action leading to wetland drainage, Illinois passed the Illinois Drainage Levee Act and the Farm Drainage Act in 1879, which allowed counties to organize into drainage districts to consolidate financial resources. This action accelerated draining to the point that 27 percent of Illinois is now under some form of drainage and almost all of the original wetlands in the state (85 percent) have been destroyed. In Ohio, over 90 percent of the original wetlands were drained, partially assisted and encouraged by the then newly formed land-grant college in Columbus. That college, now The Ohio State University, still houses a "Drainage Hall of Fame."

WETLAND ALTERATION

In a sense, wetland alteration or destruction is an extreme form of wetland management. One model of wetland alteration (Fig. 16–2) assumes that three main factors influence wetland ecosystems: water level, nutrient status, and natural disturbances (Keddy, 1983). Through human activity, the modification of any one of these factors can lead to wetland alteration, either directly or indirectly. For example, a wetland can be disturbed through decreased water levels, as in draining and filling, or through increased water levels, as in downstream drainage impediments. Nutrient status can be affected through upstream flood control that decreases the frequency of nutrient inputs or through increased nutrient loading from agricultural areas.

The most common alterations of wetlands have been (1) draining, dredging, and filling of wetlands; (2) modification of the hydrologic regime; (3) highway construction; (4) mining and mineral extraction; and (5) water pollution. These wetland modifications are described in more detail below.

Wetland Conversion: Draining, Dredging, and Filling

The major cause of wetland loss in the United States continues to be conversion to agricultural use. Figure 16–3 illustrates the steady rate, interrupted by World War II, of drainage for farms since 1900 (Gosselink and Maltby, 1990). Probably about 65 percent of this land was wetland (Office of Technology Assessment, 1984). This conversion was particularly significant in the vast midwestern "breadbasket" that has provided the bulk of the grain produced on the continent

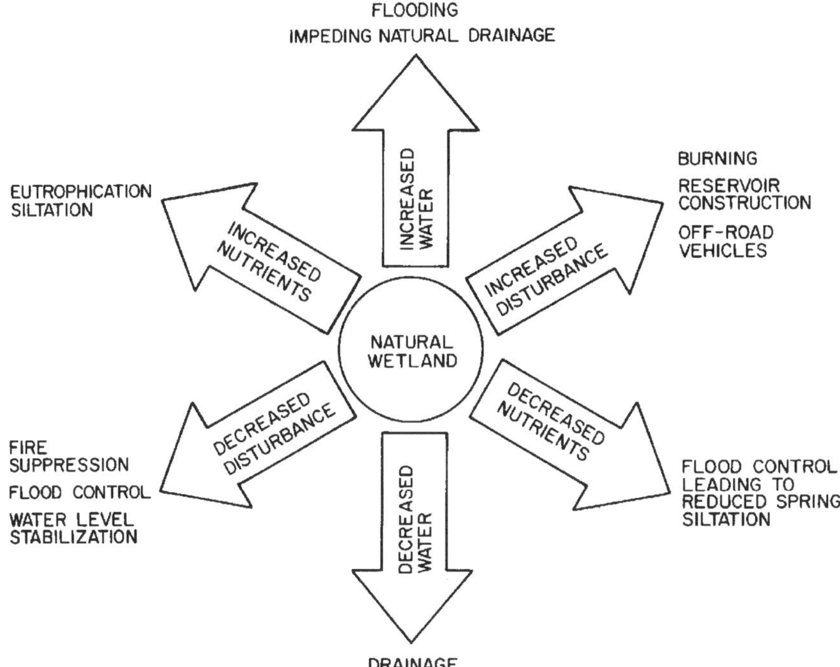

Figure 16–2. Model of human-induced impacts on wetlands, including effects on water level, nutrient status, and natural disturbance. By either increasing or decreasing any one of these factors, wetlands can be altered. (*From Keddy, 1983; copyright © 1983 by Springer-Verlag, reprinted with permission*)

(Fig. 16–4). When drained and cultivated, the fertile soils of the prairie pothole marshes and east Texas playas produce excellent crops. With ditching and modern farm equipment, it has been possible to farm these small marshes.

Since the mid-twentieth century, however, the most rapid changes have occurred in the bottomland hardwood forests of the Mississippi River alluvial floodplain. As the populations increased along the river, the floodplain was channeled and leveed so that it could be drained and inhabited. Since colonial times the floodplain provided excellent cropland, especially for cotton and sugarcane. Cultivation, however, was restricted to the relatively high elevation of the natural river levees, which flooded regularly after spring rains and upstream snowmelts but drained rapidly enough to enable farmers to plant their crops. Because the river levees were naturally fertilized by spring floods, they required no additional fertilizers to grow productive crops. One of the results of drainage and flood protection is the additional cost of fertilization. The lower parts of the floodplain, which are too wet to cultivate, were left as forests but harvested for timber. As pressure for additional cropland increased, these agriculturally marginal forests

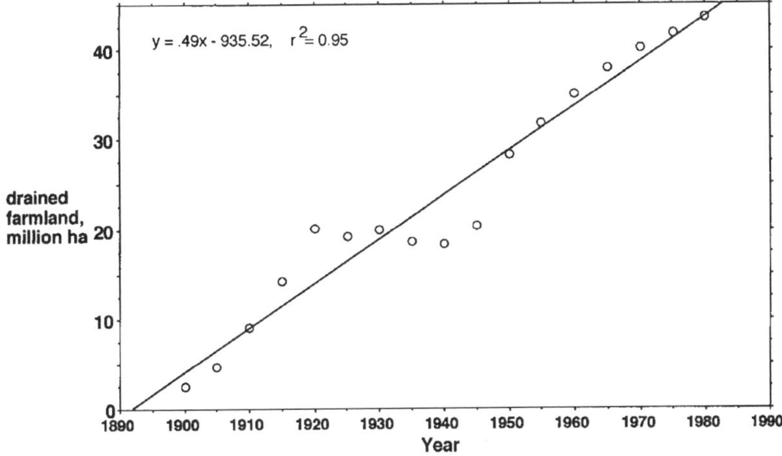

Figure 16–3. Trend of drained farmland in the United States from 1900 to 1980. (*After Gosselink and Maltby, 1990, based on data from Office of Technology Assessment, 1984*)

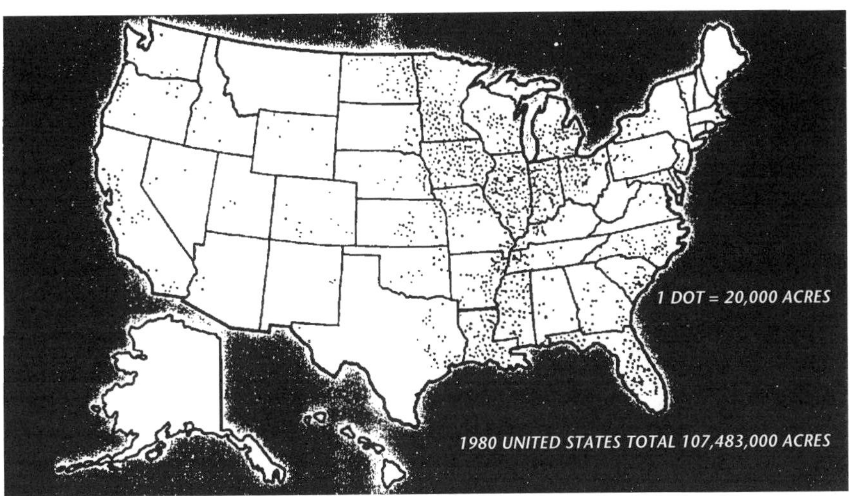

Figure 16–4. Extent and location of artificially drained agricultural land in the United States as of 1985. Each dot represents 8,000 hectares (20,000 acres). (*From Dahl, 1990*)

were clear-cut at an unprecedented rate (Fig. 16–5). This was feasible in part because of the development of soybean varieties that mature rapidly enough to be planted in June or even early July, after severe flooding has passed. Often the land thus "reclaimed" was subsequently incorporated behind flood- control levees where it was kept dry by pumps (Stavins, 1987). Clear-cutting of bottomland forests is still proceeding from north to south. Most of the available wetland has

Figure 16–5. Oblique aerial photograph of bottomland in the Tensas River Basin, Louisiana, formerly a 1-million hectare forest. In the foreground, trees have been sheared off and felled with a bulldozer blade. Above, a line of standing trees remains along a slightly wet depression. Above the standing trees, felled trees have been pushed into a line to be burned. In the distance, the cleared land has been harrowed for planting, probably to soybeans. (*Credit: Larry Harper, U.S. Army Corps of Engineers, 1981*)

been converted in Arkansas and Tennessee; Mississippi and Louisiana are experiencing large losses (Fig. 3–9).

Along the nation's coasts, especially the East and West coasts, the major cause of wetland loss is draining and filling for urban and industrial development. Compared to land converted to agricultural use, the area involved is rather small.

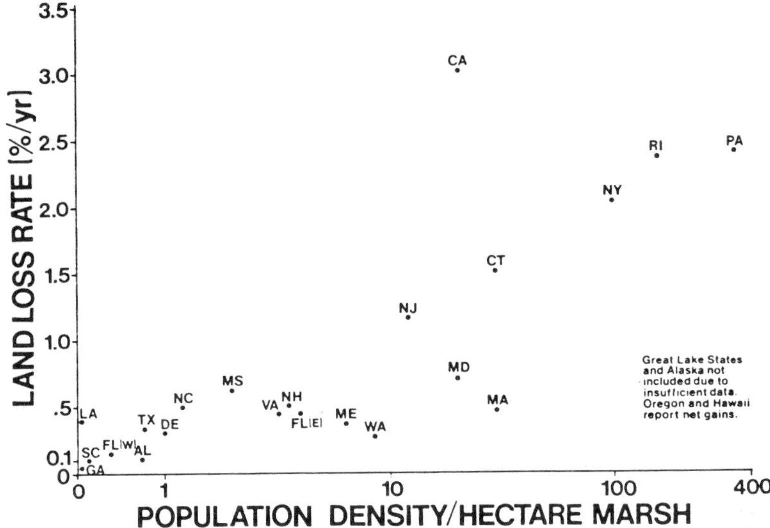

Figure 16–6. Relationship between coastal wetland loss (1954–1974) and population density for coastal counties. (*From Gosselink and Baumann, 1980; copyright © 1980 by Gebruder Borntraeger, reprinted with permission*)

Nevertheless, in some coastal states, notably California, almost all coastal wetlands have been lost. The rate of coastal wetland loss from 1954 to 1974 was closely tied to population density (Fig. 16–6). This finding underscores two facts: (1) two-thirds of the world's population lives along coasts; (2) population density puts great pressure on coastal wetlands as sites for expansion. The most rapid development of coastal wetlands occurred after World War II. In particular, several large airports were built in coastal marshes. Since the passage of federal legislation controlling wetland development, the rate of conversion has slowed.

Hydrologic Modifications

Ditching, draining, and levee building are hydrologic modifications of wetlands specifically designed to dry them out. Other hydrologic modifications destroy or change the character of thousands of hectares of wetlands annually. Usually these hydrologic changes were made for some purpose that had nothing to do with wetlands; wetland destruction is an inadvertent result. Canals, ditches, and levees are created for three primary purposes:

1. *Flood control.* Most of the canals and levees associated with wetlands are for flood control. The canals have been designed to carry floodwaters off the adjacent uplands as rapidly as possible. Normal drainage through wetlands is slow

surface sheet flow; straight, deep canals are more efficient. Ditching marshes and swamps to drain them for mosquito control or biomass harvesting is a special case designed to lower water levels in the wetlands themselves. Along most of the nation's major rivers are systems of levees constructed to prevent overbank flooding of the adjacent floodplain. Most of those levees were built by the U.S. Army Corps of Engineers after Congress passed flood-control legislation after the disastrous floods of the the 1920s and 1930s. Those levees, by separating the river from its floodplain, isolated wetlands so that they could be drained expeditiously. For example, along the lower Mississippi River the creation of levees created a demand from farmers for additional floodplain drainage. The sequence of response and demand was so predictable that farmers bought and cleared floodplain forests in anticipation of the next round of flood-control projects.

2. *Navigation and transportation.* Navigation canals tend to be larger than drainage canals. They traverse wetlands primarily to provide water-transportation access to ports and to improve transport among ports. For example, the Intracoastal Waterway was dredged through hundreds of miles of wetlands in the northern Gulf Coast. In addition, when highways were built across wetlands, fill material for the roadbed was often obtained by dredging soil from along the right-of-way, thus forming a canal parallel to the highway.
3. *Industrial activity.* Many canals are dredged to obtain access to sites within a wetland for the purpose of sinking an oil well, building a surface mine, or other kinds of development. Usually pipelines that traverse wetlands are laid in canals that are not backfilled.

The result of all of these activities can be a wetland crisscrossed with canals, especially in the immense coastal wetlands of the northern Gulf Coast (Fig. 16–7). These canals modify wetlands in a number of ecological ways by changing normal hydrologic patterns. Straight, deep canals in shallow bays, lakes, and marshes capture flow, depriving the natural channels of water. Canals are hydrologically efficient, allowing the more rapid runoff of fresh water than the normal shallow, sinuous channels do. As a result, water levels fluctuate more rapidly than they do in unmodified marshes, and minimum levels are lowered, drying the marshes. The sheet flow of water across the marsh surface is reduced by the spoil banks that almost always line a canal and by road embankments that block sheet flow. Consequently, the sediment supply to the marsh is reduced, and the water on the marsh is more likely to stagnate than when freely flooded. In addition, when deep, straight channels connect low-salinity areas to high-salinity zones, as with many large navigation channels, tidal water, with its salt, intrudes farther upstream, changing freshwater wetlands to brackish. In extreme cases, salt-intolerant vegetation is killed and is not replaced before the marsh erodes into a shallow lake. On the Louisiana coast the natural subsidence rate is high; wetlands go through a natural cycle of growth followed by decay to open bodies of water.

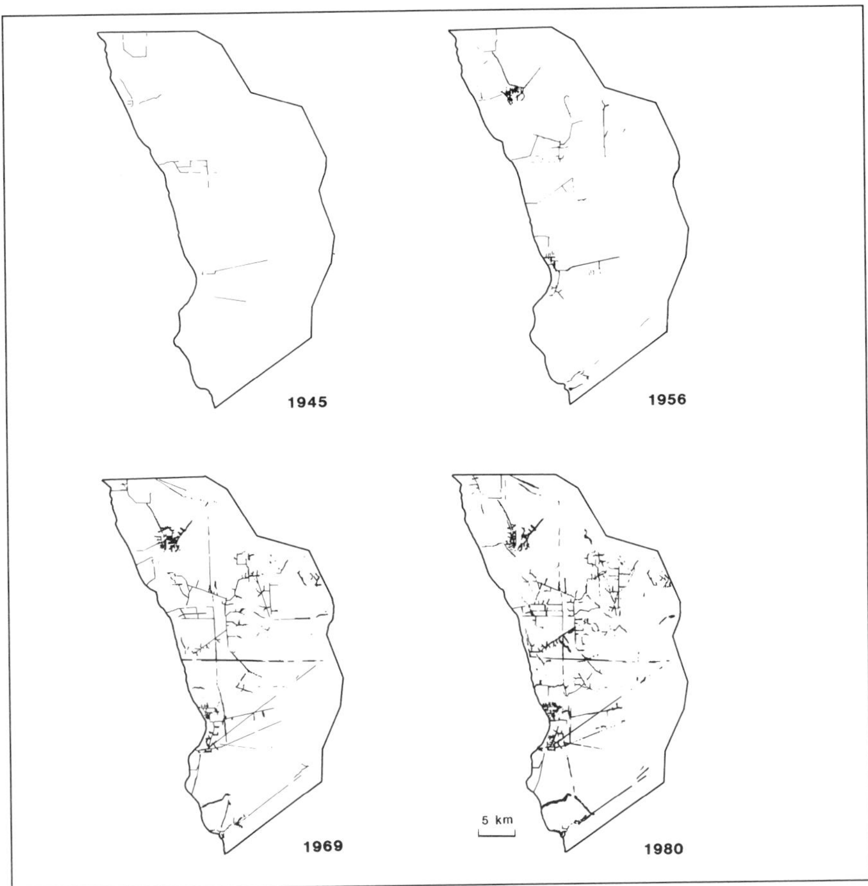

Figure 16–7. Computer images showing growth in the number and length of navigational canals constructed in the wetlands of the north central coast of the Gulf of Mexico (Barataria Bay, Louisiana) from 1945 to 1980. The concentrated nodes of canals are sites of oil fields. Each short canal segment provides access to an oil well. *(From Sasser et al., 1986; copyright © 1986 by Springer Verlag, New York, reprinted with permission)*

There canals accelerate the subsidence rate by depriving wetlands of natural sediment and nutrient subsidies.

Highway Construction

Highway construction can have a major effect on the hydrologic conditions of wetlands (see Fig. 16–8). Although few definitive studies have been able to document the extent of wetland damage caused by highways (Adamus, 1983), several studies have inferred that the major effects of highways are alteration of the

Figure 16–8. This sign is typical of the conflict between wetland protection and highway construction. Sign was near a proposed highway in central Florida. (*Photograph by W. J. Mitsch*)

hydrologic regime, sediment loading, and direct wetland removal. McLeese and Whiteside (1977) compared the effects of highways on uplands and wetlands in Michigan and found that wetlands were much more sensitive to highway construction than uplands were, particularly through the disruption of hydrologic conditions. Similarly, Clewell et al. (1976) and Evink (1980) found that highway construction in Florida led to negative effects on coastal wetlands through hydrologic isolation. The authors of the former study discovered that isolated tidal marshes became less saline and began to fill with vegetation because of the construction of a filled roadway. The authors of the latter study found that the decreased circulation that resulted from a causeway increased nutrient retention in the wetland and led to subsequent symptoms of eutrophication. Adamus (1983) concluded that the "best location for a highway that must cross a wetland is one which minimizes interference with the wetland ecosystem's most important driving forces." Other than solar energy and wind, the most important driving forces for wetlands are hydrologic, including tides, gradient currents (e.g., streamflow), runoff, and groundwater flow. The importance of protecting the hydrologic regime during highway construction is based on the contention presented in Chapter 4 that the hydrology of wetlands is the most important determinant of a wetland's structure and function.

Peat Mining

Surface peat mining has been a common activity in several European countries, particularly the lands of the former Soviet Union, since the eighteenth century. That territory accounts for almost 90 percent of peat mining in the world; most of the material derived from that process is used as a fuel for electric power production (Moore and Bellamy, 1974). In the United States, peat resources are estimated at about 63 billion tons (Table 16–1). Since its inception peat mining in the United States has been primarily undertaken for agricultural and horticultural uses and has been done on a relatively small scale. Approximately 825,000 tons of peat were mined in the United States in 1979; 77 percent came from the states of Michigan, Florida, Illinois, Indiana, and New York, in decreasing order (Carpenter and Farmer, 1981). More recent data (Table 16–1) suggest a decrease of peat mining to about 480,000 tons per year in the United States and even less (290,000 tons/yr) in Canada. Almost all of that peat was used for horticultural purposes. Mining for energy production is often proposed on a large scale for the peatlands in Minnesota and the pocosins in North Carolina. It has been estimated that Minnesota has enough peat reserves to supply its energy needs for 32 years (Williams, 1990).

Mineral and Water Extraction

Surface mining activity for materials other than peat often affects major regions of wetlands. Phosphate mining in central Florida has had a significant impact on wetlands in the region (Gilbert et al., 1981; Dames and Moore, 1983; M. T. Brown et al., 1992). Thousands of hectares of wetlands may have been lost in central Florida because of this activity alone, although the reclamation of phosphate-mined sites for wetlands is now a common practice (see Chap. 17). H. T. Odum et al. (1981) argued that "managed ecological succession" on mined sites could be an economical alternative to such expensive techniques moving earth and reclamation planting.

Surface mining of coal has also affected wetlands in some parts of the country (Brooks et al., 1985). Mitsch et al. (1983b, c) identified 46,000 hectares of wetlands, mostly bottomland hardwood forests, that could be or are being affected by surface coal mining in western Kentucky alone, whereas Cardamone et al. (1984)

Table 16–1. Estimated Reserves and Peat Production in the World

Country	Reserve, x 10^6 tons[a]	Peat Production, x 10^3 tons/yr		
		Fuel	Horticulture	Total
Former USSR	120,000	48,000	72,000	120,000
Finland	6,240	4,054	279	4,333
Ireland	2,459	3,646	196	3,842
China	27,000	480	780	1,260
Former FRG	133	170	1,077	1,247
Sweden	11,000	770	1,077	1,247
USA	62,985	—	480	480
Burundi	109	480	—	480
United Kingdom	1,500	—	370	370
Former GDR	—	—	315	315
Canada	335,000	—	294	294
Poland	1,914	—	178	178
Czechoslovakia	78	—	157	157
France	—	30	60	90
Denmark	—	—	88	88
Venezuela	—	—	60	60
Norway	2,000	1	50	51
New Zealand	—	0	6	6
Other countries	132,567	—	—	—
Total	703,021[b]	57,631	76,600	134,231

[a]ton = metric ton (mt)

[b]Approximately half of total world pool of soil carbon 1,400,000 x 10^6 ton (Post et al., 1982)

Source: Based on Immirzi et al., 1992

prescribed methods available to protect wetlands during mining or to create wetlands as part of the reclamation process (Fig. 16–9). The recognition of the potential benefits of including wetlands as part of the reclamation of coal mines has not been as widespread as one would have expected because of the strict interpretation of measures regulating the return of the land to its original contours and because of liability questions. This is in contrast to the widespread acceptance of the reclamation of wetlands on phosphorus mine sites in Florida.

In some parts of the country, the withdrawal of water from aquifers or minerals from deep mines has resulted in accelerated subsidence rates that are lowering the elevations of marshes and built-up areas alike, sometimes dramatically. For example, groundwater and mineral extraction has led to as much as 2.5 m of subsidence in northern Galveston Bay (Kreitler, 1977). Land subsidence, which can also result in the creation of lakes and wetlands, is a geologically common phenomenon in Florida. Often when excessive amounts of water are removed from the ground, underground cave-ins occur in the limestone, causing surface slumpage. Some believe that the cypress domes in north-central Florida are an indirect result of a similar natural process whereby fissures and dissolutions of underground limestone cause slight surface slumpage and subsequent wetland development.

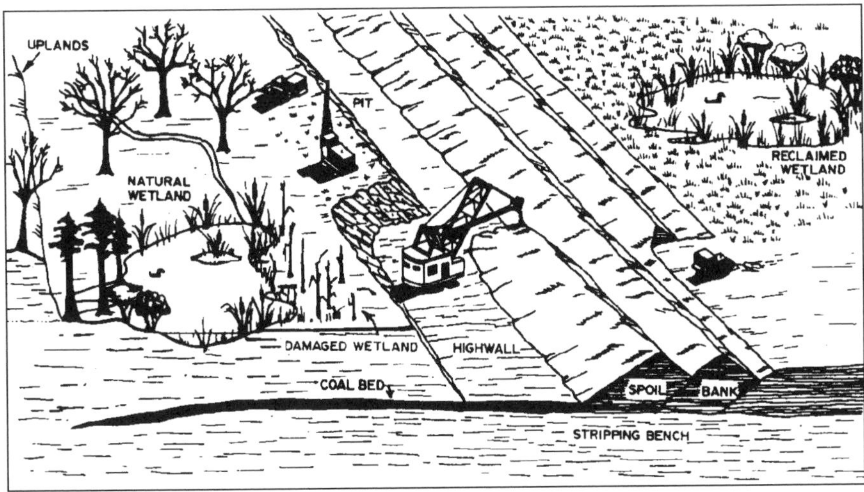

Figure 16–9. Possible use of wetlands in reclamation of coal surface mines for wildlife enhancement and for control of mine drainage. (*From Cardamone et al., 1984*)

Water Pollution

Wetlands are altered by pollutants from upstream or local runoff and, in turn, change the quality of the water flowing out of them. The ability of wetlands to cleanse water has received much attention in research and development and is discussed elsewhere (see Chapters 15 and 17 and Water Quality Management section in this chapter). The effects of polluted water on wetlands has received less attention, although there is discussion now in the United States about establishing water-quality standards for wetlands (Robb, 1992; Nichols, 1992). Many coastal wetlands are nitrogen limited; one response to nitrogen as one of the pollutants is increased productivity of the vegetation and increased standing stocks of vegetation followed by increased rates of decay of the vegetation, at least initially, and higher community respiration rates. Species composition may also change with eutrophication of wetlands. For example, increased agricultural runoff, laden with phosphorus, is believed to have caused a spread of *Typha* spp. in conservation areas that are part of the original Everglades in Florida (Koch and Reddy, 1992; Gunderson and Loftus, 1993). This, in turn, has increased fears that the phosphorus will eventually lead to invasion of *Typha* in the Everglades National Park itself replacing the natural sawgrass (*Cladium jamaicense*).

When metals or toxic organic compounds are pollutants, effects on the wetland can be dramatic. In severe cases of water pollution, wetland vegetation can be killed, as occurred when oil was spilled on a coastal marsh (J. M. Baker, 1973) or sulfates were discharged into a forested wetland (J. Richardson et al., 1983). Acid drainage from active and abandoned coal mines has been shown to affect wetlands seriously. Mitsch et al. (1983a, b, c) documented the presence of wetlands and coal surface mining adjacent to each other in western Kentucky. In many instances, waters with low pH and high iron and sulfur were discharged from these mines into or through wetlands, causing extensive ecological damage (Fig. 16–10).

In one of the most publicized and dramatic cases of water pollution of a wetland, selenium from farm runoff contaminated marshes in Kesterson National Wildlife Refuge in California's San Joaquin Valley (Presser and Ohlendorf, 1987; T. Harris, 1991). The selenium contamination led to excessive death and deformities of wildlife and to eventual "closing" of the contaminated marsh in the mid-1980s amid much controversy.

WETLAND MANAGEMENT BY OBJECTIVE

Wetlands are managed for environmental protection, for recreation and aesthetics, and for the production of renewable resources. Stearns (1978) lists 12 specific goals of wetland management: that are applicable today:

1. maintain water quality
2. reduce erosion
3. protect from floods
4. provide a natural system to process airborne pollutants
5. provide a buffer between urban residential and industrial segments to ameliorate climate and physical impact such as noise
6. maintain a gene pool of marsh plants and provide examples of complete natural communities
7. provide aesthetic and psychological support for human beings
8. produce wildlife
9. control insect populations
10. provide habitats for fish spawning and other food organisms
11. produce food, fiber, and fodder; for example, timber, cranberries, cattails for fiber
12. expedite scientific inquiry

One excellent management decision is to fence in a wetland to preserve it. Although simple, this is an act of conservation of a valuable natural ecosystem involving no substantative changes in management practices. Often, however, management has one or more specific objectives that require positive manipulation of the environment. Efforts to maximize one objective may be incompatible with the attainment of others, although in recent years most management objectives have been broadly stated to enhance a broad range of objectives. Multipurpose management generally focuses on system-level support rather than individual species. This has often been achieved indirectly through plant species manipulation because plants provide food and cover for the animals (Weller,

Figure 16–10. Wetlands, such as this riverine wetland impacted by coal mine drainage in western Kentucky, can be negatively affected by water pollution. (*Photograph by W. J. Mitsch*)

1978). When many small wetland management areas are in close proximity, different practices should be used, or the management cycle should staggered so that the different areas are not all treated the same way at the same time. Implementing a strategy of this kind would not only increase the diversity of the larger landscape but would also be attractive to wildlife.

Wildlife Enhancement

The best wetland management practices are those that enhance the natural processes of the wetland ecosystem involved. One way to accomplish this is to maintain conditions as close as possible to the natural hydrology of the wetland, including hydrologic connections with adjacent rivers, lakes, and estuaries. Unfortunately, this cannot easily be accomplished in wetlands managed for wildlife; the vagaries of nature, especially in hydrologic conditions, make planning difficult. Hence marsh management for wildlife in North America, particularly waterfowl, has often meant water level manipulation. Water level control is achieved by dikes (impoundments), weirs (solid structures in marsh outflows that maintain a minimum water level), control gates, and pumps. In general, the results of the management activity depend on how well the water level control is maintained, and control depends on the local rainfall and on the sophistication of the control structures. For example, weirs provide the poorest control; all they do is maintain a minimum water level. Pumps provide positive control of drainage or flooding depth at the desired time; and the management objectives can usually be met (Wicker et al., 1983).

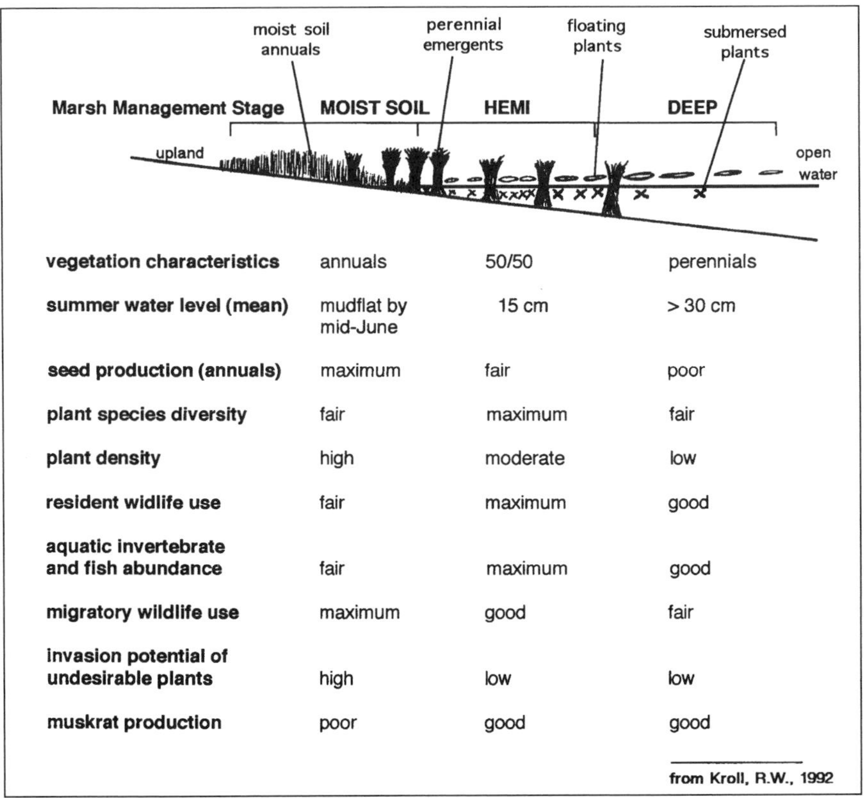

Marsh Management Stage	MOIST SOIL	HEMI	DEEP
vegetation characteristics	annuals	50/50	perennials
summer water level (mean)	mudflat by mid-June	15 cm	> 30 cm
seed production (annuals)	maximum	fair	poor
plant species diversity	fair	maximum	fair
plant density	high	moderate	low
resident widlife use	fair	maximum	good
aquatic invertebrate and fish abundance	fair	maximum	good
migratory wildlife use	maximum	good	fair
invasion potential of undesirable plants	high	low	low
muskrat production	poor	good	good

Figure 16–11. Some generalizations of water level management for vegetation, wildlife use, and other characteristics as practiced on impounded marshes near Lake Erie in northern Ohio. (*Redrawn from R. Kroll, Winous Point Shooting Club, Port Clinton, Ohio, 1992, with permission*)

To illustrate the trade-offs in wetland management for wildlife enhancement, some generalizations about water-level manipulation for the Lake Erie (Ohio) coastal marshes are shown in Figure 16–11. Maximum migratory wildlife use of the marshes happens in moist soil conditions, but these conditions are also the best for the invasion of potentially undesirable plants and are generally least favorable for the overall abundance and diversity of resident plant and animal populations. Shallow water (hemi) conditions (around 15 cm depth in summer) usually result in the highest plant species diversity and greatest fish and resident wildlife use but less migratory wildlife. Deep water conditions (>30 cm) offer the least potential for both annual emergent plants and invading, undesirable plants, but desirable migratory waterfowl use is only fair in deep water.

The set of management recommendations by Weller (1978) for prairie pothole marshes in the north-central United States and south-central Canada are other examples of multipurpose wildlife enhancement. Those practices mimic the natural cycle of the marshes (see Chap. 11). Although they may seem drastic, they are entirely natural in their consequences. In sequence, the practices are as follows:

1. When a pothole is in the open stage and there is little emergent vegetation, the cycle should be initiated by a spring drawdown. This stimulates the germination of seedlings on the exposed mud surfaces.
2. A slow increase in water level after the drawdown maintains the growth of flood-tolerant seedlings without shading them out in turbid water. Shallowly flooded areas attract dabbling ducks during the winter.
3. The drawdown cycle should be repeated for a second year to establish a good stand of emergents.
4. Low water levels should be maintained for several more seasons to encourage the growth of perennial emergents such as cattails.
5. Maintaining stable, moderate water depths for several years promotes the growth of rooted submerged aquatic plants and associated benthic fauna that make excellent food for waterfowl. During that period, the emergent vegetation will gradually die out and will be replaced by shallow ponds. When that occurs, the cycle can be initiated again, as described in (1) above.
6. Different wetland areas maintained in staggered cycles provide all stages of the marsh cycle at once, maximizing habitat diversity.

Wildlife management in coastal salt marshes such as those found in Louisiana uses a similar strategy, although the short-term cycle is not as pronounced there. Drawdowns to encourage the growth of seedlings and perennials preferred by ducks is a common practice, as is fall and winter flooding to attract dabbling ducks. As it happens, there is general agreement that stabilizing water levels is not good management, even though our society seems to feel intuitively that stability is a good thing. Wetlands thrive on cycles, especially flooding cycles, and practices that dampen these cycles also reduce wildlife productivity. Although the management practices described above enhance waterfowl production, they are generally deleterious for wetland-dependent fisheries in coastal wetlands since free access between the wetlands and the adjacent estuary is restricted; the wetlands' role in regulating water quality is also often underutilized.

Agriculture and Aquaculture

When wetlands are drained for agricultural use, they no longer function as wetlands. They are, as the local farmer says, "fast lands" removed from the effects of periodic flooding and they grow terrestrial, flood-intolerant crops. Some use is made of more or less undisturbed wetlands for agriculture, but it is minor. In New England high-salt marshes were harvested for "salt marsh hay" that was considered an excellent bedding and fodder for cattle. In fact, Russell (1976) stated that the proximity of fresh and salt hay marshes was a major factor in selecting the sites for the emergence of many towns in New England before 1650. Subsequently, marshes were ditched to allow the intrusion of tides to promote the growth of salt marsh hay (*Spartina patens*), but the extent of this practice has not been well documented. On the coast of the Gulf of Mexico where coastal marshes are firm underfoot, they are still used extensively for cattle grazing. To improve access, small embankments or raised earthen paths are constructed in those marshes.

The ancient Mexican practice of *marceno* is unique. In the freshwater wetlands of the northern coast of Mexico, small areas were cleared and planted in corn during the dry season. Those native varieties were tolerant enough to withstand considerable flooding. After harvest (or apparently sometimes before harvest), the marshes were naturally reflooded, and the native grasses were reestablished until the next dry season. This practice is no longer followed, but there has been some interest in reviving it (Orozco-Segovia, 1980).

Aquaculture usually requires more extensive manipulation of the environment than the practices mentioned above. When ponds are constructed with retaining walls or levees and pumps, little resemblance to the natural ecosystem remains. Nevertheless, attempts have been made to use estuarine-wetland areas in a more or less natural state to raise fish and shellfish. The practice with shrimp is typical. A natural marsh and pond area is enclosed by weirs, gates, or other water-control structures. Fine mesh fences allow water flux but still retain the cultured animals. Recruitment of postlarval juveniles to the aquaculture site usually occurs naturally, after which the area is sealed off and the shrimp are allowed to grow. They are harvested as they emigrate over the weirs or by seining or trawling within the enclosure. In the southern United States several commercial ventures were launched during the 1970s. None succeeded. There were too many uncertainties, including stock recruitment and predator and disease control. Historically in the United States, coastal fisheries have been considered public resources. The practice of "privatizing" coastal wetlands for shrimp culture, therefore, faces serious legal challenges.

A more successful commercial venture is crayfish farming in combination with timber production. Crayfish are an edible delicacy in the southern United States and in many foreign countries. They live in burrows in shallow flooded areas such as swamp forests and rice fields, emerging with their young early in the year to forage for food. The young grow to edible size within a few weeks and are harvested in the spring. When floodwaters retreat, the crayfish construct burrows where they remain until the next winter flood. In crayfish farms this natural cycle is enhanced by controlling water levels. An area of swamp forest is impounded; it is flooded deep during the winter and spring and drained during the summer. This cycle is ideal for crayfish, which thrive. Fish predators are controlled within the impoundments to improve the harvest. The hydrologic cycle is also favorable for forest trees. It simulates the hydrologic cycle of a bottomland hardwood forest; forest tree productivity is high, and seedling recruitment is good

Figure 16–12. Rice cultivation is an important agricultural wetland management practice throughout the world. This photo, taken in the Yangtze River Valley of China near Nanjing, shows typical rice plants being tilled early in the growing season while water is being added to the fields. (*Photo by W. J. Mitsch*)

because of the summer drawdown. Species composition tends toward species typical of bottomland hardwoods (Conner et al., 1981).

On a global scale, of course, the production of rice in managed wetlands contributes a major proportion of the world's food supply (Fig. 16–12). Aselmann and Crutzen (1989) estimated that there are approximately 1.3 million km^2 of rice paddies in the world, of which almost 90 percent are in Asia. In North America, especially in Minnesota, there are several commercial operations in the production of wild rice (*Zizania aquatica*) in wetlands.

Some rice farmers have also found that they can take advantage of the same annual flooding cycle to combine rice and crayfish production. Rice fields are drained during the summer and fall when the rice crop matures and is harvested. Then the fields are reflooded, allowing crayfish to emerge from their burrows in the rice field embankments and forage on the vegetation remaining after the rice harvest. The crayfish harvest ends when the fields are replanted with rice. When this rotation is practiced, extreme care has to be exercised in the use of pesticides.

Water Quality Enhancement

A number of studies have shown natural wetlands to be sinks for certain chemicals, particularly sediments and nutrients (see Chaps. 5 and 17). It is now common to cite the water quality role of natural wetlands in the landscape as one of the most important reasons for the protection (Fig. 16–13). The idea of applying domestic, industrial, and agricultural wastewaters, sludges, and even urban and rural runoff to wetlands to take advantage of this nutrient-sink capacity has also been explored. Many wetland treatment systems are summarized by Nichols (1983), Godfrey et al. (1985), Hammer (1989), and Cooper and Findlater (1990) and are discussed in Chapter 17. To some the idea involves wastewater treatment, to others wastewater disposal. Regardless of what it is called, wastewater recycling in wetlands is an intriguing concept involving the forging of a partnership between humanity and the ecosystem.

Much of the interest in maintaining the natural wetlands for water quality management purposes was sparked by two studies begun in the early 1970s. In one of those studies northern peatlands at Houghton Lake and other communities in Michigan were investigated by researchers from the University of Michigan for the wetlands's capacity to treat wastewater (see Richardson et al., 1978; Kadlec, 1979; Kadlec and Kadlec, 1979; Kadlec and Tilton, 1979; Tilton and

Figure 16–13. Natural wetlands, when left as major parts of the landscape, often provide water quality roles in their natural condition without much human management. In this photo, two colors of water are noted, with the dividing line approximately at the canoe. The water in the foreground is highly polluted with sediments from a watershed artificially drained by a large ditch. The clearer water in the background is from a flooding river that is passing through a natural riparian wetland. The forested wetland, with water among the trees, can be seen in the background. Picture was taken during flooding conditions on the Kankakee River near Momence, Illinois, during a typical spring flood in this river. (*Photo by W. J. Mitsch*)

Kadlec, 1979). A pilot operation for disposing of up to 380 m^3 per day (100,000 gallons per day) of secondarily treated wastewater in a rich fen at Houghton Lake led to significant reductions in ammonia nitrogen and total dissolved phosphorus as the water passed from the point of discharge. Inert materials such as chloride did not change as the wastewater passed through the wetland (Kadlec, 1979). An estimated 70 percent of ammonia nitrogen, 99 percent of nitrite and nitrate nitrogen, and 95 percent of total dissolved phosphorus were removed from the wastewater as it passed through the wetland. In 1978 the flow was increased to approximately 5,000 m^3 per day over a much larger area. Data after more than 10 years of operation at this high flow show that although the area of influence of the wastewater on the peatland has grown from 10 to 66 hectares, the effectiveness of the wetland in removing both ammonia-nitrogen and total phosphorus remained extremely high (Knight, 1990).

In a second major research effort in the 1970s to investigate water quality management in natural wetlands, wastewater was applied to several cypress domes in north-central Florida by a team of researchers from the University of Florida (Ewel, 1976; H. T. Odum et al., 1977a; Ewel and Odum, 1978, 1979, 1984). After five years of experimentation in which secondarily treated wastewater was added to the cypress domes at a rate of approximately 2.5 cm/wk (1 in/week), the results indicated that the wetland filtered nutrients, heavy metals, microbes, and viruses from the water. The productivity of the canopy pond cypress trees also increased (Nessel et al., 1982; Lemlich and Ewel, 1984). The uptake of nutrients in these systems was enhanced by a continuous cover of duckweed on the water, by the retention of nutrients in the cypress wood and litter, and by the adsorption of phosphorus onto clay and organic peat in sediments.

Wetlands that have received wastewater for a relatively long time have also been studied. Study sites have included freshwater marshes in Wisconsin (Spangler et al., 1977; Fetter et al., 1978) and forested wetlands in Florida (Boyt et al., 1977; Nessel, 1978a, b; Nessel and Bayley, 1984). All of these studies and several others have demonstrated that wetlands can serve as sinks of nutrients for several years, although their assimilation capacity can become saturated for certain chemical constituents (Kadlec and Kadlec, 1979; Richardson, 1985).

There can be other long-term benefits of using wetlands for water quality management. In the subsiding environment of Louisiana's Gulf Coast, nutrients are permanently retained in peat of wetlands receiving high nutrient wastewater as the wetland aggrades to match subsidence. In this case, wastewater discharge into a wetland can occur without saturating the system and simultaneously helps counteract the deleterious effects of subsidence (Conner and Day, 1989).

The U.S. Environmental Protection Agency (1983) summarized a number of critical technical and institutional considerations "that may act independently or jointly to influence the feasibility of using wetlands as a wastewater management alternative." Although recent policy is to discourage the use of natural wetlands as wastewater-treatment systems (Olson, 1992), these considerations are useful guides for managing wetlands for any water quality role. They include the following:

Technical Considerations

1. Other values of the wetlands such as wildlife habitat should be considered.
2. Acceptable pollutant and hydrologic loadings must be determined for the use of wetlands in wastewater management.
3. All existing wetland characteristics, including vegetation, geomorphology, hydrology, and water quality, should be well understood.
4. Site-specific analyses of wetlands, particularly as to whether they are hydrologically open or isolated, are necessary to determine their potential for wastewater management. Hydrologically isolated wetlands are likely to be altered if wastewater is applied to them, but hydrologically open wetlands are more likely to affect downstream systems.

Institutional Considerations

5. Potential conflicts over the protection and use of wetlands may arise among state agencies, federal agencies, and local groups.
6. Wastewater disposal into wetlands can often serve the dual purposes of both wetland protection and use, particularly when wetland restoration is involved.
7. State and municipal governments may be liable for damage to private wetlands from wastewater disposal. It is best for appropriate levels of government to obtain ownership or legal control of wetlands that are used for wastewater management.
8. Federal permit processes, many of which are now administered by state agencies, do not recognize wetland-disposal systems. The modification of requirements for granting permits is needed to make use of this effective method of wetland and wastewater management.

Flood Control and Groundwater Recharge

Wetlands can be managed, often passively, for their role in the hydrologic cycle. These hydrologic functions include streamflow augmentation, groundwater recharge, water supply potential, and flood protection (see Chap. 15). It is not altogether clear how well wetlands carry out these functions, nor do all wetlands perform these functions equally well. It is known, for example, that wetlands do not necessarily always contribute to low flows or recharge groundwater (Carter et al., 1979; Verry and Boelter, 1979; Carter, 1986). Some wetlands, however, should be and often are protected for their ability to hold water and slowly return it to surface and groundwater systems in periods of low water. If wetlands are impounded to retain even more water from flooding downstream areas, considerable changes in vegetation will result as the systems adapt to the new hydrologic conditions.

LEGAL PROTECTION OF WETLANDS IN THE UNITED STATES

In the early 1970s interest in wetland protection increased significantly as scientists began to identify and quantify the many values of these ecosystems. This interest in wetland protection began to be translated at the federal level in the United States into laws and public policies. Prior to this time, federal policy on wetlands was vague and often contradictory. Policies in agencies such as the U.S. Army Corps of Engineers, the Soil Conservation Service, and the Bureau of Reclamation encouraged the destruction of wetlands, whereas policies in the Department of Interior, particularly in the U.S. Fish and Wildlife Service, encouraged their protection (Kusler, 1983). Some states have also developed inland and coastal wetland laws and policies, and activity in that area appears to be increasing.

Federal Government Policies and Laws

Some of the more significant activities of the federal government that led to a more consistent wetland protection policy have included presidential orders on wetland protection and floodplain management, implementation of a dredge-and-fill permit system to protect wetlands, coastal zone management policies, and initiatives and regulations issued by various agencies. The primary wetland protection mechanisms used by the federal government are summarized in Table 16–2. Despite all of this activity related to federal wetland management, two major points are still in effect today and should be emphasized:

1. *There is no specific national wetland law.* Wetland management and protection result from the application of many laws intended for other purposes. Jurisdiction over wetlands has also been spread over several agencies, and, overall, federal policy continually changes and requires considerable interagency coordination.
2. *Wetlands have been managed under regulations related to both land use and water quality.* Neither of these approaches, taken separately, can lead to a comprehensive wetland policy. The regulatory split mirrors the scientific split noted by many wetland ecologists, a split that is personified by people who have developed expertise in either aquatic or terrestrial systems. Rarely do individuals possess expertise in both areas.

Early Presidential Orders

President Jimmy Carter issued two executive orders in May 1977 that established the protection of wetlands and riparian systems as the official policy of the federal government. Executive Order 11990, Protection of Wetlands, required all federal agencies to consider wetland protection as an important part of their policies:

Table 16–2. Major Federal Laws, Directives, and Regulations Used for the Management and Protection of Wetlands

Directive or Statute	*Date*	*Responsible Federal Agency*
Rivers and Harbors Act	1899	Army Corps of Engineers
Fish and Wildlife Coordination Act	1967	Fish and Wildlife Service
Land and Water Conservation Fund Act	1968	Fish and Wildlife Service, Bureau of Land Management, Forest Service, National Park Service
Federal Water Pollution Control Act (PL 92-500) as Amended (Clean Water Act)	1972, 1982	
Section 404—Dredge-and-Fill Permit Program		Army Corps of Engineers with assistance from Environmental Protection Agency and U.S. Fish and Wildlife Service
Section 208—Areawide Water Quality Planning		Environmental Protection Agency
Section 303—Water Quality Standards		Environmental Protection Agency
Section 401—Water Quality Certification		Environmental Protection Agency (with state agencies)
Section 402—National Pollutant Discharge Elimination System		Environmental Protection Agency (or state agencies)
Coastal Zone Management Act	1972	Office of Coastal Zone Management, Department of Commerce

continued on next page

> Each agency shall provide leadership and shall take action to minimize the destruction, loss or degradation of wetlands, and to preserve and enhance the natural and beneficial values of wetlands in carrying out the agency's responsibilities for (1) acquiring, managing, and disposing of Federal lands and facilities; and (2) providing federally undertaken, financed, or assisted construction and improvement; and (3) conducting Federal activities and programs affecting land use, including but not limited to water and related land resources planning, regulating, and licensing activities.

Executive Order 11988, Floodplain Management, established a similar federal policy for the protection of floodplains, requiring agencies to avoid activity in the floodplain wherever practicable. Furthermore, agencies were directed to revise

Table 16–2 continued

Directive or Statute	*Date*	*Responsible Federal Agency*
Flood Disaster Protection Act	1973, 1977	Federal Emergency Management Agency
Federal Aid to Wildlife Restoration Act	1974	Fish and Wildlife Service
Water Resources Development Act	1976, 1990	Army Corps of Engineers
Executive Order 11990 Protection of Wetlands	May 1977	All agencies
Executive Order 11988 Floodplain Management	May 1977	All agencies
Food Security Act, Swampbuster provisions	1985	U.S. Dept. Agriculture, Soil Conservation Service
Emergency Wetland Resources Act	1986	Fish and Wildlife Service
Wetland Delineation Manuals (various revisions)	1987, 1989, 1991	All agencies
North American Wetlands Conservation Act	1989	Fish and Wildlife Service
"No Net Loss" Policy	1988	All agencies
Wetlands Reserve Program	1991	U.S. Dept. Agriculture, Soil Conservation Service

Source: Based on data from Kusler, 1983; Environmental Defense Fund and World Wildlife Fund, 1992; Want, 1990

their procedures to consider the impact that their activities might have on flooding and to avoid direct or indirect support of floodplain development when other alternatives are available.

Both of these executive orders were significant because they set in motion a review of wetland and floodplain policies by almost every federal agency. Several agencies such as the U.S. Environmental Protection Agency and the Soil Conservation Service established policies of wetland protection prior to the issuance of these executive orders (Kusler, 1983), but many other agencies such as the Bureau of Land Management were compelled to review or establish wetland and floodplain policies (Zinn and Copeland, 1982).

No Net Loss

A more significant initiative in developing a national wetlands policy was undertaken in 1987, when a National Wetlands Policy Forum was convened by the Conservation Foundation at the request of the U.S. Environmental Protection Agency to investigate the issue of wetland management in the United States

(National Wetlands Policy Forum, 1988; Davis, 1989). The distinguished group of 20 members (which included three governors, a state legislator, state and local agency heads, the chief executive officers of environmental groups and businesses, farmers, ranchers, and academic experts) published a report that set significant goals for the nation's remaining wetlands. The forum formulated one overall objective:

> to achieve no overall net loss of the nation's remaining wetlands base and to create and restore wetlands, where feasible, to increase the quantity and quality of the nation's wetland resource base (National Wetlands Policy Forum, 1988).

The group recommended as an interim goal that the holdings of wetlands in the United States should decrease no further, and as a long-term goal that the number and quality of the wetlands should increase. In his 1988 presidential campaign and in his 1990 budget address to Congress, President George Bush echoed the "no net loss" concept as a national goal, shifting the activities of many agencies such as the Department of the Interior, the U.S. Environmental Protection Agency, the U.S. Army Corps of Engineers, and the Department of Agriculture toward achieving a unified and seemingly simple goal. Nevertheless, it was not anticipated that there would be a complete halt of wetland loss in the United States when economic or political reasons dictated otherwise. Consequently, implied in the "no net loss" concept is wetland construction and restoration to replace destroyed wetlands. The "no net loss" concept became a cornerstone of wetland conservation in the United States in the early 1990s.

The Clean Water Act Section 404 Program

The primary vehicle for wetland protection and regulation in the United States is Section 404 of the Federal Water Pollution Control Act (FWPCA amendments of 1972 (PL 92–500) and subsequent amendments (also known as the Clean Water Act). The use of Section 404 for wetland protection has been controversial and has been the subject of several court actions and revisions of regulations. The surprising point about the importance of the Clean Water Act in wetland protection is that wetlands are not directly mentioned in Section 404. This section gave authority to the Army Corps of Engineers to establish a permit system to regulate the dredging and filling of materials in "waters of the United States." At first this directive was interpreted narrowly by the Corps to apply only to navigable waters. That definition of waters of the United States was expanded to include wetlands in two 1974–1975 court decisions, *United States* v. *Holland* and *Natural Resources Defense Council* v. *Calloway*. Those decisions, along with Executive Order 11990 on Protection of Wetlands, put the Army Corps of Engineers squarely in the center of wetland protection in the United States. On

July 25, 1975, the Corps issued revised regulations for the 404 program enunciated the policy of the United States on wetlands:

> As environmentally vital areas, [wetlands] constitute a productive and valuable public resource, the unnecessary alteration or destruction of which should be discouraged as contrary to the public interest (*Federal Register*, July 25, 1975).

Wetlands were defined in those regulations to encompass coastal wetlands ("marshes and shallows and . . . those areas periodically inundated by saline or brackish waters and that are normally characterized by the prevalence of salt or brackish water vegetation capable of growth and reproduction") and freshwater wetlands ("areas that are periodically inundated and that are normally characterized by the prevalence of vegetation that requires saturated soil conditions for growth and reproduction") (*Federal Register*, July 25, 1975, as cited by Zinn and Copeland, 1982). By those actions the jurisdiction of the Corps had been extended to include 60 million hectares of wetlands, 45 percent of which are in Alaska (Zinn and Copeland, 1982). Several times since 1975, the Corps has issued revised regulations for the dredge-and-fill permit program, and in 1985 the U.S. Supreme Court, in *United States* v. *Riverside Bayview Homes,* rejected the contention that Congress did not intend to include wetland protection as part of the Clean Water Act.

The procedure for obtaining a "404 Permit" for dredge-and-fill activity in wetlands is a complex process (Fig. 16–14). The decision to issue a permit rests with the Corps' district engineer, and it must be based on a number of considerations, including conservation, economics, aesthetics, and several other factors listed in Figure 16–14. Assistance to the Corps on the dredge-and-fill permit process in wetland cases is provided by the U.S. Environmental Protection Agency, the U.S. Fish and Wildlife Service, and state agencies. The National Marine Fisheries Service also comments on coastal permit applications. The EPA has statutory authority to designate wetlands subject to permits, and also has veto power on the Corps' decisions. Some states require state permits as well as Corps permits for wetland development. The district engineer, according to Corps regulations, should not grant a permit if a wetland is identified as performing important functions for the public such as biological support, sanctuary, storm protection, flood storage, groundwater recharge, or water purification. An exception is allowed when the district engineer determines "that the benefits of the proposed alteration outweigh the damage to the wetlands resource and the proposed alteration is necessary to realize those benefits" (*Federal Register*, July 19, 1977). The effectiveness of the 404 Program has varied since the program began and has also varied from district to district.

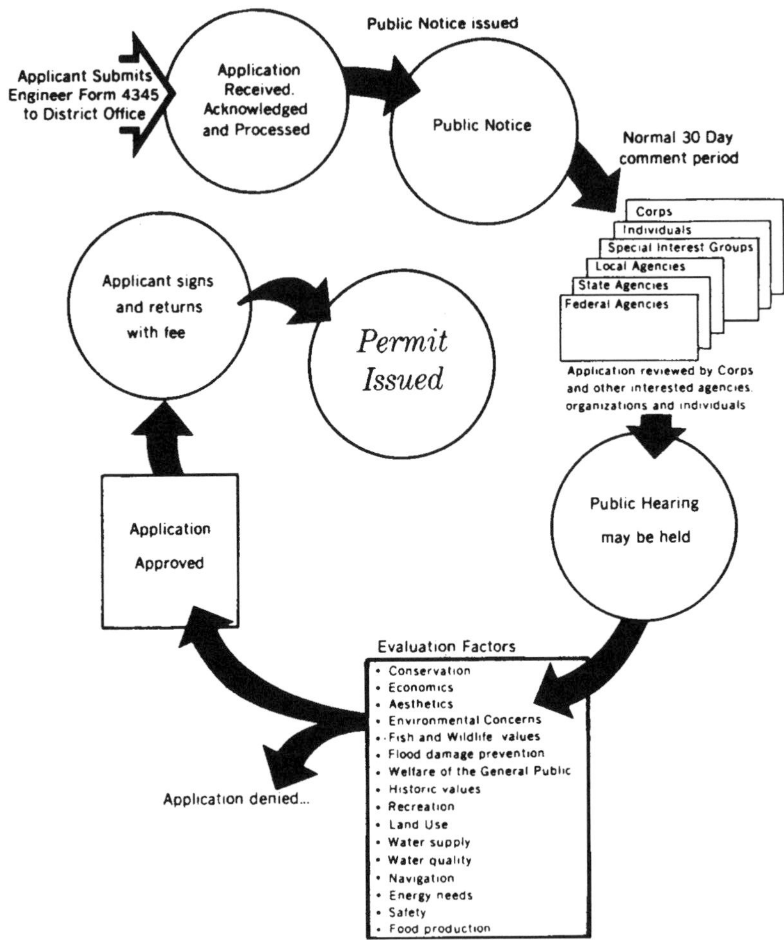

Figure 16–14. Typical U.S. Army Corps of Engineers review process for Section 404 dredge-and-fill permit request. (*From J. A. Kusler,* Our National Wetland Heritage: A Protection Guidebook; *copyright © 1983 by Environmental Law Institute, reprinted with permission*)

Swampbuster

Normal agricultural and silvicultural activities were exempted from the Section 404 permit requirements, thereby still allowing wetland drainage on farms and in commercial forests. Allowing such exemptions created conflict in the federal government: the Corps of Engineers and the Environmental Protection Agency were encouraging wetland conservation through the Clean Water Act, and the Department of Agriculture was encouraging wetland draining by providing Federal subsidies for drainage projects. The conflict ended when Congress passed, as part of the 1985 Food Security Act, the "Swampbuster" provisions that denied federal subsidies to any farm owner who knowingly converted wetlands to farmland after the act became effective. The "Swampbuster" provisions of the act drew the U.S. Soil Conservation Service into federal wetland management, primarily as an advisory agency helping farmers identify wetlands on their farms. The Soil Conservation Service also administers a Wetlands Reserve Program that was set up in 1990 to acquire easements on up to 400,000 hectares (one million acres) of agricultural land that was formerly wetland.

Wetland Delineation

To be able to determine whether a particular piece of land was a wetland and therefore if it was necessary to obtain a federal 404 permit to dredge or fill that wetland, federal agencies, beginning with the Corps of Engineers, began to develop guidelines for the demarcation of wetland boundaries in a process that came to be named *wetland delineation.* In 1987 the Army Corps of Engineers published a technical manual for wetland delineation. Subsequent to that, the Environmental Protection Agency, the Soil Conservation Service, and the U.S. Fish and Wildlife Service developed separate documents for their respective roles in wetland protection. Finally, in January 1989, after several months of negotiation, a single *Federal Manual for Identifying and Delineating Jurisdictional Wetlands* was published by the four federal agencies to unify the government's approach to wetlands. This manual specified three mandatory technical criteria, namely wetland hydrology, soils, and vegetation, for a parcel of land to be declared a wetland. The manual also provided some guidance about how to use field indicators such as water marks on trees or stains on leaves to determine recent flooding, wetland vegetation (from published lists), and hydric soils indicators such as mottling. The manual was used by developers and agencies alike to prove or disprove the presence of wetlands in the Section 404 permit process. Consulting firms specializing in wetland delineation sprung up overnight, and short courses on the methodology became very popular.

Beginning in early 1991, modifications of the manual were proposed in response to heavy lobbying by developers, agriculturalists, and industrialists for a relaxing of the wetland definitions, supposedly to lessen the regulatory burden on the private sector. A new manual was published for public comment in August 1991 but was quickly and heavily criticized for its lack of scientific credibility and unworkability (Enviromental Defense Fund and World Wildlife Fund, 1992) and it was eventually abandoned in 1992. In the meantime, the 1987 Corps technical manual is being used until some other version is adopted. Although wetland determination and delineation will remain important tools for wetland managers, the exact rules for using these tools will remain uncertain and subject to political change for many years to come.

Other Federal Activity

Several other federal laws and activities have led to wetland protection since the 1970s. The Coastal Zone Management Program, established by the Coastal Zone

Management Act of 1972, has provided up to 80 percent of matching-funds grants to states to develop plans for coastal management based on giving a high priority to protecting wetlands. The National Flood Insurance Program offers some protection to riparian and coastal wetlands by offering federally subsidized flood insurance to state and local governments that enact local regulations against development in flood-prone areas. The Clean Water Act, in addition to supporting the 404 Program, authorized six million dollars to the U.S. Fish and Wildlife Service to complete its inventory of wetlands of the United States (see Chap. 18). The Emergency Wetlands Resource Act passed by Congress in 1986 required the U.S. Fish and Wildlife Service to update its report on the status of and trends in wetlands every ten years. The first report was issued in 1982 and published one year later (Frayer et al., 1983). The first update was published in 1991 as Dahl and Johnson (1991) (see Chap. 3 for conclusions in these reports).

Table 16–3. States with Coastal Wetland Protection Programs

State	*Program*
Alabama	Permits are required for activities in coastal zone (dredging, dumping, etc.) that alter tidal movement or damage flora and fauna.
Alaska	State agencies regulate use of coastal land, waters, including offshore areas, estuaries, wetlands, tideflats, islands, sea cliffs and lagoons.
California	Permit required for development up to 1,000 yards (meters) of mean high tide; coastal zone regulated by regional regulatory boards; prohibits siting coastal-dependent developments in wetlands with some exceptions that must be permitted.
Connecticut	Permit required for all regulated activity; state inventory required.
Delaware	Permits required for all activities; has both Coastal Zone and Beach Protection Acts.
Florida	Florida Coastal Zone Management Act requires permit for erosion-control devices and excavations or erections of structures in coastal environment.
Georgia	Permits required for work in coastal salt marshes through Coastal Marshlands Protection Program.
Hawaii	County authorities issue development permits for development of coastal area with state oversight.
Louisiana	State and/or local permits required for activity in coastal wetlands. Coastal Wetland Planning, Protection and Restoration Act passed in 1990 to restore coastal wetlands.
Maine	Permits required for dredging, filling, or dumping into coastal wetlands. Comprehensive coastal/freshwater protection in Protection of Natural Resources Act.
Maryland	State permits required for activity in coastal wetlands based on Tidal Wetlands Act and Chesapeake Bay Critical Area Act.

continued on next page

The "Takings" Issue

As discussed in Chapter 15, one of the dilemmas of valuing and protecting wetlands is that the values accrue to the public at large but rarely to individual landowners who happen to have a wetland on their property. In a major ruling in June 1992 *(Lucas* v. *South Carolina),* the U.S. Supreme Court ruled that regulations denying "economically viable use of land" require compensation to the landowner, no matter how great the public interest served by the regulations (Runyon, 1993). The denial of an individual's right to use his or her property is referred to as a "taking." This case was referred back to the State of South Carolina to determine if the developer, David Lucas, was denied all economically viable use of his land (beachfront property that was rezoned by South Carolina in response to the Coastal Zone Management Act). The ultimate resolution of the takings issue in this and other pending court cases could have significant implications on Federal and state regulations of wetlands on private property.

Table 16–3 continued

State	*Program*
Massachusetts	State and local permits required for fill or alteration of coastal wetlands. Permits from local conservation commissioners.
Michigan	Permit required for development in high rule erosion areas, flood risk areas, and environmental areas of coastal Great Lakes.
Mississippi	Permits required for dredging and dumping, although there are many exemptions through Coastal Wetlands Protection Act.
New Hampshire	Permit required for dredge and fill in or adjacent to fresh and saltwater wetlands; higher priority usually given to saltwater marshes.
New Jersey	Permit required for dredging and filling; agriculture and Hackensack meadowlands exempted.
New York	Permits required for tidal wetland alteration by Tidal Wetlands Act.
North Carolina	State permit required for coastal wetland excavation or fill of estuarine waters, tidelands, or salt marshes.
Oregon	Local zoning requirements on coastal marshes and estuaries with state review.
Rhode Island	Coastal wetlands designated by order and use limited; permits required for filling, aquaculture, development activity on salt marshes.
South Carolina	Permits required for dredging, filling, and construction in coastal waters and tidelands including salt marshes.
Virginia	Wetlands Act requires permits for all activities in coastal counties with some exemptions; also 1988 Chesapeake Bay Preservation Act.
Washington	Shoreline Management Act requires local governments to adopt plans for shorelines, including wetlands; state may regulate if local government fails to do so.

Source: After Zinn and Copeland, 1982; Kusler, 1979, 1983; Want, 1990; Meeks and Runyon, 1990

State Management of Wetlands

Many individual states have issued wetland protection statutes or regulations. That activity has been described in a number of summaries, including those by Kusler (1979, 1983), Zinn and Copeland (1982), Glubiak et al. (1986), Meeks and Runyon (1990), Want (1990), and Environmental Defense Fund and World Wildlife Fund (1992). State wetland programs may become more important as the federal government attempts to delegate much of its authority to local and state governments. Kusler (1983) has suggested that although local communities may also have wetland-protection programs, states are much more probable governmental units for wetland protection for the following reasons:

1. Wetlands cross local government boundaries, making local control difficult.
2. Wetlands in one part of a watershed affect other parts that may be in different jurisdictions.
3. There is usually a lack of expertise and resources at the local level to study wetland values and hazards.
4. Many of the traditional functions of states such as fish and wildlife protection are related to wetland protection.

Many states that contain coastlines initally paid more attention to managing their coastal wetlands than to managing their inland wetlands (Kusler, 1983). This is a result of an earlier interest in coastal wetland protection at the federal level and to the development of coastal zone management programs. Table 16–3 shows some of the states that have coastal wetland protection programs. In general, those programs can be divided into those that have been based on specific coastal wetland laws and those that are designed as a part of broader regulatory programs such as coastal zone management. Several coastal states have coastal dredge-and-fill permit programs, whereas other states have specific wetland regulations administered by a state agency.

State programs for inland wetlands, although in an earlier stage of development, are more diverse, ranging from comprehensive laws to a lack of concern for inland wetlands. Comprehensive laws have been enacted in several states such as Connecticut, Rhode Island, New York, Massachusetts, Florida, New Jersey, and Minnesota (Table 16–4). Other states such as Arizona, Georgia, and Idaho have few regulations governing inland wetlands. Between these two extremes, there are many states that rely on federal-state cooperation programs or on state laws that indirectly protect wetlands. Only one state, Michigan, has assumed responsibility from the Federal government to issue Section 404 permits, although the Corps of Engineers retains the permit program in navigable waters (Meeks and Runyon, 1990) and the U.S. EPA retains Federal oversight of the program. Floodplain protection laws or scenic and wild river programs are being implemented in more than half of the states and are often effective in slow-

Table 16–4. States that Have Comprehensive Wetland Laws for Inland Waters

State	*Law*
Connecticut	Inland Wetlands and Watercourses Act
Delaware	The Wetlands Act
Florida	Henderson Wetlands Protection Act of 1984
Maine	Protection of Natural Resources Act
Maryland	Chesapeake Bay Critical Area Act
Massachusetts	Wetland Protection Act
Michigan	Goemaere–Anderson Wetland Protection Act
Minnesota	The Wetland Conservation Act of 1991
New Hampshire	Fill and Dredge in Wetlands Act
New Jersey	Freshwater Wetlands Protection Act of 1987
New York	Freshwater Wetlands Act
North Dakota	No Net Wetlands Loss Bill of 1987
Oregon	Fill and Removal Act
	Comprehensive Land Use Planning Coordination Act
Rhode Island	Freshwater Wetlands Act
Vermont	Water Resources Management Act
Wisconsin	Water Resources Development Act
	Shoreland Management Program

Source: Want, 1990; Meeks and Runyon, 1990

ing the destruction of riparian wetlands. States are also involved in wetland protection through wetland acquisition programs, conservation easement programs, preferential tax treatment for landowners who protect wetlands, and enforcement of state water quality standards as required by the Clean Water Act (Meeks and Runyon, 1990).

INTERNATIONAL WETLAND CONSERVATION

The Ramsar Convention

Intergovernmental cooperation on wetland conservation has been spearheaded by the Convention on Wetlands of International Importance, more commonly referred to as the *Ramsar Convention* because it was initially adopted at an international conference held in Ramsar, Iran, in 1971. The global treaty provides the framework for the international protection of wetlands as habitats for migratory fauna that do not observe international borders and for the benefit of human populations dependent on wetlands. The specific obligations of countries that have ratified the Ramsar Convention are the following:

1. Member countries shall formulate and implement their planning so as to promote the "wise use" of all wetlands in their territory, and develop national wetland policies.
2. Member countries shall designate at least one wetland in their territory for the "List of Wetlands of International Importance." The so-called Ramsar sites should be developed based on their "international significance in terms of ecology, botany, zoology, limnology or hydrology" (Navid, 1989). Currently there are 11 Ramsar sites comprising 1.1 million hectares in the United States, 30 sites of almost 13 million hectares in Canada, and 1 site of 47,000 hectares in Mexico. As of 1993, almost 37 million hectares at 582 wetland sites were designated as Ramsar sites in the world.
3. Member countries shall establish nature reserves at wetlands.
4. Member countries shall cooperate over shared species and development assistance affecting wetlands.

Currently 74 countries have joined the Ramsar Convention. A permanent secretariat headquartered at the International Union of Conservation of Nature and Natural Resources (IUCN) in Switzerland was established in 1987 to administer the convention, and a budget based on the United Nations scale of contributions was adopted.

North American Waterfowl Management Plan

The United States and Canada, partially as a result of collaboration begun by the Ramsar Convention, established the *North American Waterfowl Management Plan* in 1986 to conserve and restore about 2.4 million hectares of waterfowl wetland habitat in Canada and the United States. This treaty was formulated as a partial response to the steep decline in waterfowl in Canada and the United States that had become apparent in the early 1980s (see Chap. 15). This bilateral treaty is jointly administered by the U.S. Fish and Wildlife Service and the Canadian Wildlife Service but also involves public and private participation by groups such as Ducks Unlimited. The total cost of this plan was estimated to be $1.5 billion, to be paid by the two countries and private organizations. Major emphasis has been placed on sites that cross international borders, including the prairie pothole region, the lower Great Lakes-St. Lawrence River Basin, and the Middle-Upper Atlantic Coastline (Larson, 1991).

Wetland Creation and Restoration

17

Policies such as "no net loss" of wetlands and the recognition of wetland values have stimulated restoration and creation of these systems. Restored or created wetlands have specific objectives such as the mitigation of unavoidable wetland losses, wildlife enhancement, domestic wastewater treatment, coal mine drainage control, and stormwater retention and control. Although many of these constructed and restored wetlands have been successful in providing the desired results, there have been some cases of "failure" of constructed or restored wetlands generally caused by a lack of proper hydrology. Ecological engineering of wetlands is based on the concept of self-design whereby the ecosystem adapts and changes according to its physical constraints, leading to a minimum of human intervention.

Some wetland restoration involves little more than restoring the natural hydrologic conditions. Other wetland creation projects involve paying more attention to design detail. Among the hydrologic design parameters to be considered for constructing wetlands are hydroperiod, loading rates, seasonal pulses, flow patterns, and retention times. Wetland managers in the past have used mainly water depth to control the functioning of wetlands; a more comprehensive management of the flow-through characteristics of the wetland is needed. Chemical loading rates are important for wetlands designed for water pollution control and guidelines are available on nitrogen, phosphorus, and iron. Substrate characteristics such as organic content, texture, nutrients, iron, and aluminum play important roles in wetland design and construction. A wide variety of vegetation types and planting and seeding techniques are available for wetland construction. Vegetation success should be measured more by the success of the original objective of the wetland construction than by the success of individual species. Management options after wetland construction and restoration can involve plant harvesting, wildlife management, mosquito control, perturbation control, and sediment dredging. Estimates for wetland construction costs tend to be quite variable and site-specific; maintenance costs are even more difficult to determine.

Knowledge of the principles and techniques for wetland creation and restoration is one of the qualifications required of modern landscape managers and wetland scientists. The creation of wetlands in previously dry and/or nonvegetated areas or the restoration of wetlands are exciting possibilities for reversing the trend of decreasing wetland resources and for providing aesthetic and functional units to the landscape. Wetland creation and restoration can range from the relatively simple building of farmland freshwater marshes by plugging existing drainage systems to the construction of more extensive wetlands for coastal protection or wastewater treatment. Successful ecological engineering of wetlands requires that managers take advantage of increasing knowledge of wetland ecology and its principles (e.g., hydrology, biogeochemistry, adaptations, succession) to construct and restore wetlands as part of a natural landscape. In constructing and reclaiming wetlands, managers must resist the ever-present temptation to overengineer by attempting to channel natural energies that cannot be channeled and by introducing species that the design does not support. Human contribution to the design of wetlands should be kept simple, without reliance on complex technological approaches that invite failure. Boulé (1988) states that "simple systems tend to be self-regulating and self-maintaining."

There is a growing literature that reviews constructing and restoring wetlands in general (Kusler and Kentula, 1990; Marble, 1992; Kentula et al., 1992), and several reviews on construction and restoration of specific types of wetlands, including wastewater wetlands (Hammer, 1989; Cooper and Findlater, 1990; Knight, 1990), nonpoint source wetlands (Olson, 1992; Mitsch and Cronk, 1992, and Mitsch, 1992a; Kentula et al., 1992), freshwater marshes (Erwin, 1990; Hammer, 1992), tidal salt and freshwater wetlands (Broome et al., 1988; Zedler, 1988; Broome, 1990; Shisler, 1990; Thayer, 1992), mangroves (Lewis, 1990a, b), and forested wetlands (Clewell and Lea, 1990).

DEFINITIONS OF RESTORED AND CONSTRUCTED WETLANDS

Several terms are frequently used in connection with the ecological engineering of wetlands. *Wetland restoration* usually refers to the rehabilitation of wetlands that may be degraded or hydrologically altered and often involves reestablishing the vegetation. *Wetland creation* refers to the construction of wetlands where they did not exist before and can involve much more engineering of hydrology and soils. Created wetlands are also called *constructed wetlands* or *artificial wetlands*, although the last term is not preferred by many wetland scientists. Some regulators draw a distinction between constructed wetlands and created wetlands; they use the former term to designate wetlands built for wastewater or stormwater treatment and the latter term for wetlands developed on nonwetland sites to produce or replace natural habitat (Hammer, 1992). It is difficult to estimate the extent of created and restored wetlands in the United States or the world, and rates of wetland gain do not make up for the rate of wetland loss. Sometimes wetland construction or restoration can be combined with other improvements in the landscape to optimize their effectiveness. For example, reduced tillage combined with wetland restoration along streams may provide a significant benefit to the water quality of downstream systems (Loucks, 1990).

REASONS FOR CREATING AND RESTORING WETLANDS

Part of the interest in wetland creation and restoration stems from the fact that we have lost a major part of our wetland resources and their societal values (see Chaps. 3 and 15) and government policies such as "no net loss" require the replacement of wetlands for those unavoidably lost (see Chap. 16). There has also been great interest in constructing wetlands for habitat restoration or replacement, water quality enhancement, and coastal protection and restoration.

Habitat Restoration or Replacement

Mitigation Wetlands

Strict enforcement of Section 404 of the Clean Water Act by the U.S. Army Corps of Engineers (see Chap. 16) has led to the common practice of requiring that a wetland system be built to replace any wetland lost in development such as highway construction, coastal drainage and filling, or commercial development (Larson and Neill, 1987; Kusler et al., 1988). These constructed wetlands, often called *mitigation wetlands*, are built with the intent of replacing the wetland "function" lost by the development usually in the same or an adjacent watershed. They are usually designed to be at least the same size as the lost wetlands. Unfortunately, there has been little follow-up of these mitigation wetlands, and there are few satisfactory methods available to determine the "success" of a mitigation wetland in replacing the functions lost with the original wetland. Some studies suggest that there is much room for improvement in the building of mitigation wetlands. For example, Erwin (1991) found that of 40 mitigation projects in south Florida involving wetland creation and restoration, only about half of the required 430 hectares (1,058 acres) of wetlands had been constructed and that 24 of the 40 projects (60 percent) were judged to be incom-

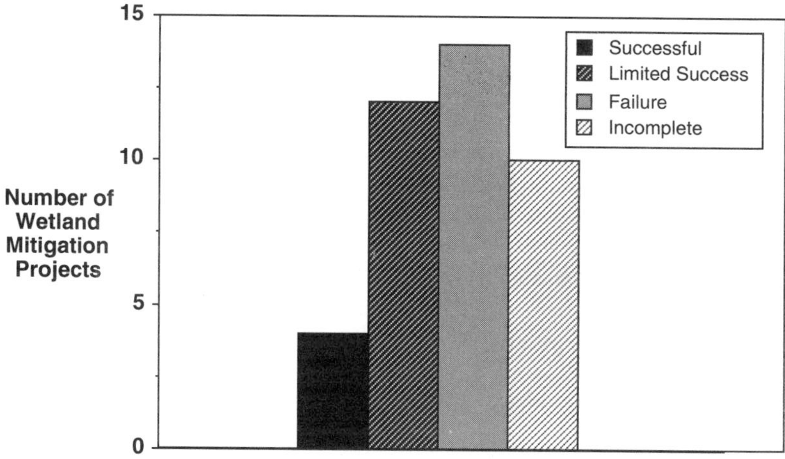

Figure 17–1. Status of 40 wetland mitigation projects in south Florida involving wetland creation, mitigation, and preservation. The average age of the projects was less than three years. Successful means that the project met all of its state goals while failure means that few goals were met or the created wetland did not have functional equivalency with a reference wetland. (*Erwin, 1991*)

plete or failures (Fig. 17–1). The most significant problems identified with the constructed wetlands were improper water levels and hydroperiod.

Farm Pond Construction and Marsh Restoration

Although the net trend in the United States is toward the destruction of inland freshwater wetlands (see Chap. 3), there has been the creation of a significant number of wetlands associated primarily with the construction and restoration of farm ponds and marshes in the Midwest. Although individually quite small (about 0.2 ha), the total number of ponds is large. The U.S. Soil Conservation Service estimated several years ago that about 50,000 ponds are constructed each year, for a 20-year total of close to 800,000 hectares of marsh created (40,000 ha/yr) from these ponds and the enlargement of existing ones (Office of Technology Assessment, 1984). Many of these ponds are built with large, shallow areas to attract waterfowl, and these shallow zones become typical pothole marshes. Dahl and Johnson (1991) determined that between the mid-1970s and the mid-1980s wetland pond areas increased by 320,000 hectares in the United States. Most of this gain (170,000 ha) resulted from building ponds on nonagricultural uplands. An additional 91,000 hectares were constructed farm ponds.

Conservation programs have encouraged individual farmers in the United States to have wetlands restored on their land. The U.S. Fish & Wildlife Service estimated that as a result of a conservation reserve program within the 1985 Food Security Act, about 36,000 hectares (90,000 acres) of wetlands were restored in the United States from 1987 to 1990 (Josephson, 1992). The north-central office of the U.S. Fish & Wildlife service estimated that they have restored more than 17,000 hectares (43,000 acres) of wetlands in voluntary cooperation with landowners in the upper midwestern United States from 1987 to 1992.

Forested Wetland Restoration

There is relatively less experience at forested wetland restoration and creation, despite that fact that these wetlands have been lost at alarming rates, particularly in the southeastern United States (see Chap. 3). Forested wetland creation and restoration are different from marsh creation and restoration because forest regeneration takes decades rather than years to complete and there is more uncertainty about the results. Forested wetland restoration is carried out on sites where hydrology and soils are largely intact. It primarily involves planting appropriate vegetation (Clewell and Lea, 1990; see Fig. 17–2). Much of this activity has centered on the Mississippi Delta region, where more than 3,700 hectares have been planted primarily with bottomland hardwood species (see Chap. 14) and to a lesser extent deepwater swamp species (see Chap. 13). Forested wetland creation that involves the engineering of an entire wetland setting has been attempted primarily in the phosphate mining region of central Florida (M. T. Brown et al., 1992). One of the largest forested wetland creation projects consisted of building 61 hectares of marsh and forested wetlands at a phosphate mine reclamation site. Fifty-five thousand trees, representing 12 wetland species, were planted (Erwin et al., 1984). The survival rate of the seedlings for the first year was 77 percent.

Water Quality Enhancement

Wastewater Wetlands

The use of wetlands for wastewater treatment (a.k.a. *constructed wetlands*) was stimulated by a number of studies in the early 1970s that demonstrated the ability of natural wetlands to remove suspended sediments and nutrients, particularly nitrogen and phosphorus, from domestic wastewater (see e.g., Nichols, 1983; Godfrey et al., 1985; Knight, 1990; see also Chaps. 5 and 16). Early ecosystem-level studies that set many of the standards for today's use of constructed wetland for domestic wastewater treatment were with forested cypress swamps in Florida (H. T. Odum et al., 1977; Boyt et al., 1977; Ewel and Odum, 1984) and peatlands in Michigan (Kadlec and Kadlec, 1979; Kadlec, 1985). Most activity involving the use of wetlands for wastewater treatment now centers on constructing new wetlands (Hammer, 1989; Knight, 1990) rather than using natural wetlands, and even sanitary engineering books have recently described wetlands as alternative natural treatment systems (Reed, 1990; Tchobanoglous and Burton, 1991). Wastewater wetland demonstrations that have been studied in

Figure 17–2. Cypress-bottomland hardwood forest restoration, including undergrowth, along a replacement headwater stream on phosphate-mined land in Florida that was reclaimed 7 years prior to this photograph. The restoration was designed and installed by A. F. Clewell, Inc. for American Cyanamid Company, Hillsborough County, Florida. (*Photo by A. F. Clewell, reprinted with permission*)

some detail include sites in Florida (Knight et al., 1987; J. Jackson, 1989), California (Gersberg et al., 1983, 1984; Gerheart et al., 1989), Arizona (Wilhelm et al., 1989), Kentucky (Steiner et al., 1987), Pennsylvania (Conway and Murtha, 1989), and North Dakota (Litchfield and Schatz, 1989). Created wetlands for processing water pollution from wastewater have been most effective for controlling organic matter, suspended sediments, and nutrients. Their effectiveness is less certain when they are used for controlling trace metals and other toxic materials not because these chemicals are not retained in the wetlands but because of concern that they might concentrate in wetland substrate and fauna (Knight, 1990).

Mine Drainage Wetlands

Acid mine drainage, with its low pH and high concentrations of iron, sulfate, and trace metals, is a major water pollution problem in many coal mining regions of the world. The use of wetlands for coal mine drainage control was probably first considered by observing volunteer *Typha* wetlands near acid seeps in a harsh environment where no other vegetation could grow. In the 1980s more than 140 wetlands were constructed in the eastern United States alone to control mine drainage (Wieder, 1989). The goal of these systems was usually the removal of iron from the water column to avoid its discharge downstream, but sulfate reduction and recovery of pH from extremely acidic conditions were also observed in many cases (e.g., Wieder and Lang, 1984; Brodie et al., 1988; Fennessy and Mitsch, 1989a, b). Design criteria for these wetlands have been developed, but they are neither consistent from site to site nor generally accepted (e.g., Brodie et al., 1988; Wieder, 1989; Fennessy and Mitsch, 1989b). Furthermore, it may not always be cost effective to construct wetlands when extremely high (>85–90 percent) iron removal efficiencies are necessary (Baker et al., 1991) and constructed wetlands do not always meet strict regulatory requirements (Wieder, 1989). Nevertheless, the use of wetlands to control this type of water pollution is viewed as a low-cost alternative to costly chemical treatment or to downstream water pollution where no other alternative is feasible.

Stormwater and Nonpoint Source Wetlands

The control of stormwater and nonpoint source pollution has been proposed as a valid application of the ecological engineering of wetlands (Livingston, 1989; Hey et al., 1989; Tourbier and Westmacott, 1989; Mitsch and Cronk, 1992; Olson, 1992). There is less experience in the design of these wetlands than with the more common wastewater wetlands, although sites receiving the equivalent of nonpoint source pollution have been constructed in northeastern Illinois (Hey et al., 1989; see Fig. 17–3), Orlando, Florida (Palmer and Hunt, 1989), and Massachusetts (Daukas et al., 1989). Experience to date has been encouraging; significant nutrients and sediments have been retained by these systems.

Coastal Protection and Restoration

A much smaller source of new wetlands has been the creation and restoration of coastal marshes and mangroves mostly in areas where considerable dredging occurs along navigable waterways. The U.S. Army Corps of Engineers annually

Figure 17–3. The Des Plaines River Wetland Demonstration Project: (*top*) experimental wetlands during construction, and *(bottom)* one year after wetland construction is completed. These wetlands, located in northeastern Illinois, have been used to investigate the role of wetlands for controlling nonpoint source pollution. (*Photos by W. J. Mitsch*)

dredges about 275 million cubic meters of material from the nation's waterways to keep them open. During the 1970s this agency initiated a large study to determine the feasibility of creating marshes out of the spoil, thus turning a liability into an asset. In addition, there has recently been much interest in creating wet-

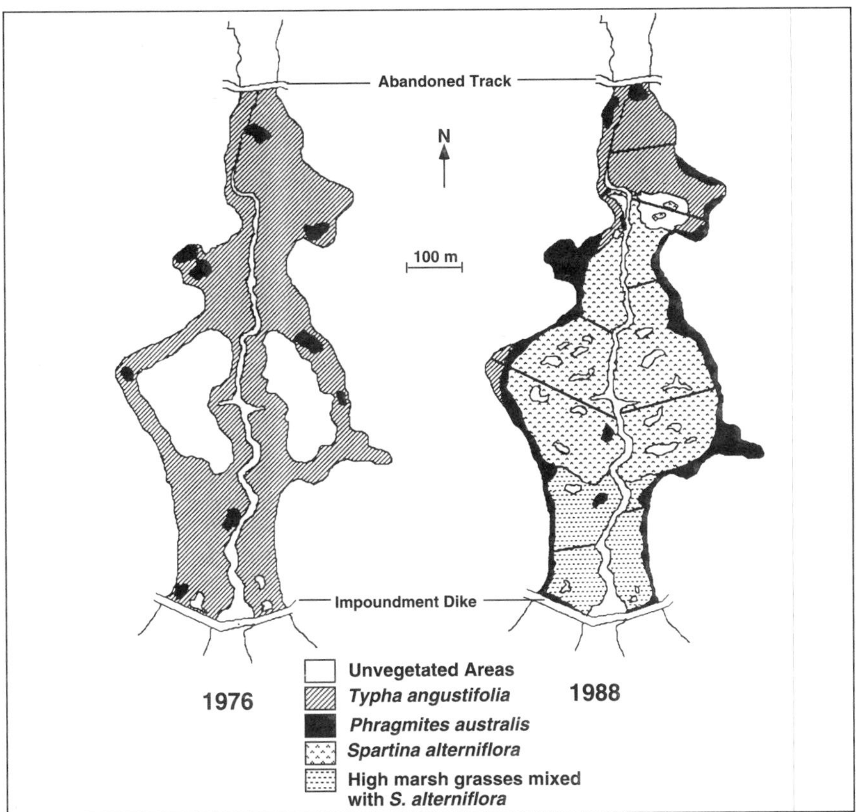

Figure 17–4. Vegetation maps of impounded salt marsh on Connecticut shoreline showing restoration to salt marsh with reintroduction of tidal flushing. Map in 1976 is prior to opening to tides, while map in 1988 indicates return of *Spartina alterniflora* after opening to tidal circulation in late 1970s and early 1980s. Lines in 1988 map indicate vegetation transects. (*From Sinicrope et al., 1990; copyright © 1990 by Estuarine Research Federation, reprinted with permission*)

lands along the Gulf Coast by diverting the sediment-rich water of the Mississippi River into shallow bays, where the sediments settle out to intertidal elevations. Both processes appear to be feasible. Some of the early pioneering work on salt marsh restoration was done in Europe (Lambert, 1964; Ranwell, 1967) and China (Chung, 1982, 1989). Much of the early work on reclaiming coastal marshes in the United States has been on the North Carolina coastline (Woodhouse, 1979; Broome et al., 1988), in the Chesapeake Bay area (Garbisch et al., 1975; Garbisch, 1977), in the salt marshes and mangroves of Florida and Puerto Rico (Lewis, 1990a, b), and along the California coastline (Zedler, 1988; Josselyn et al., 1990). In recent years much of the coastal wetland restoration has been undertaken as mitigation for coastal development projects (Zedler, 1988;

see Fig. 17–4). The passage of the Coastal Wetland Planning, Protection and Restoration Act of 1990 in Louisiana has initiated comprehensive planning aimed at the protection, restoration, and creation of millions of hectares of coastal wetlands. The plans call for building new deltas and building smaller diversions of the Mississippi River to mimic spring floods to restore subsiding marshes, restoring barrier islands, and instituting measures to protect many smaller wetlands.

WETLAND CONSTRUCTION PRINCIPLES AND GOALS

General Principles

Some general principles of ecotechnology that could be applied to the creation and restoration of wetlands (Mitsch and Cronk, 1992) are outlined below:

1. Design the system for minimum maintenance. The system of plants, animals, microbes, substrate, and water flows should be developed for self-maintenance and self-design (Mitsch and Jørgensen, 1989; H. T. Odum, 1989).
2. Design a system that utilizes natural energies such as the potential energy of streams as natural subsidies to the system. Flooding rivers and tidal circulation transport great quantities of water and nutrients in relatively short periods, subsidizing wetlands open to these flows.
3. Design the system with the hydrologic landscape and climate. Floods, droughts, and storms are to be expected, not feared.
4. Design the system to fulfill multiple goals, but identify at least one major objective and several secondary objectives.
5. Design the system as an ecotone. That may require a buffer strip around the site, but it also means that the wetland site itself will be a buffer system between upland and aquatic systems.
6. Give the system time. Wetlands do not become functional overnight. Several years may pass before plant establishment, nutrient retention, and wildlife enhancement can become optimal. Strategies that try to short-circuit ecological succession or overmanage it are doomed to failure.
7. Design the system for function, not form. If initial plantings and animal introductions fail but the overall function of the wetland, based on the fulfillment of initial objectives, is being carried out, then the wetland has not failed. The outbreak of plant diseases and the invasion of alien species are often symptomatic of other stresses and may indicate false expectations rather than ecosystem failure.
8. Do not overengineer wetland design with rectangular basins, rigid structures and channels, and regular morphology. Natural systems should be mimicked to accommodate biological systems (Brooks, 1989).

Defining Goals

The design of an appropriate wetland or series of wetlands, whether for the control of nonpoint source pollution, for a wildlife habitat, or for wastewater treatment, should start with the formulation of the overall objectives of the wetland. One view is that wetlands should be designed to maximize ecosystem longevity and efficiency and minimize cost (Girts and Knight, 1989). The most important aspect of designing a wetland is to define the goal of the wetland project (Willard et al., 1989). Among the possible goals for wetland construction are the following:

1. flood control
2. wastewater treatment (e.g., domestic wastewater or acid mine drainage)
3. stormwater or nonpoint source pollution control
4. ambient water quality improvement (e.g., instream system)
5. coastal restoration (including storm surge protection)
6. wildlife enhancement
7. fisheries enhancement
8. replacement of similar habitat (mitigation wetlands)
9. research wetland

The goal, or a series of goals, should be determined before a specific site is chosen or a wetland is designed. If several goals are identified, one must be chosen as primary.

PLACING WETLANDS IN THE LANDSCAPE

In some cases, particularly when sites are being chosen for mitigation or habitat wetlands, there is a wide array of choices of where in the landscape to locate a constructed wetland. In cases such as a coastal restoration or a wastewater wetland, there may be few choices. If a choice exists for determining a site, the following can be considered as possible locations for constructed wetlands in watersheds:

1. instream wetlands;
2. riparian wetlands;
3. multiple upstream wetlands;
4. terraced wetlands.

Each of these is described in more detail below.

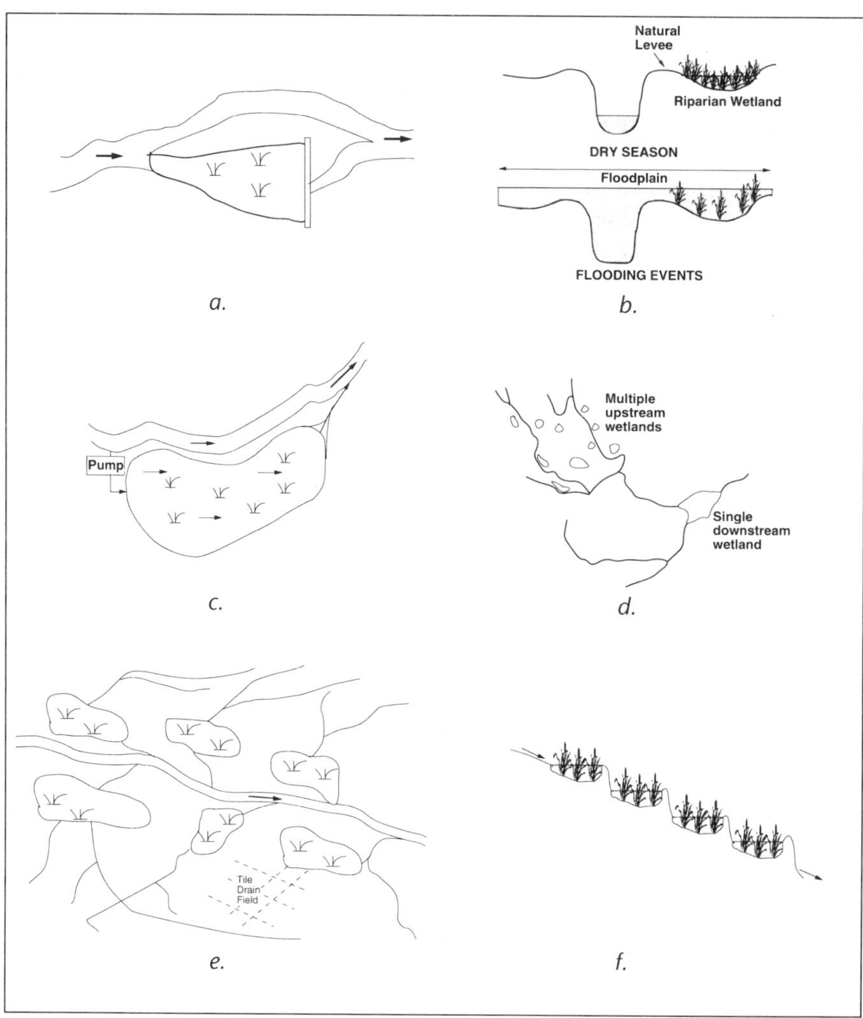

Figure 17–5. Selected locations for created or restored wetlands in a watershed including *a.* instream wetland; *b.* riparian wetland, *c.* riparian wetland with pump; *d.* multiple upstream vs. single downstream wetland, *e.* multiple upstream wetlands intercepting small streams, drainage ditches, and drainage tile, and *f.* terraced wetland for steep terrain. *(From Mitsch, 1992a; copyright © 1992 by Elsevier Science Publishers, reprinted with permission)*

Instream Wetlands

Wetlands can be designed as instream systems by adding control structures to the streams themselves or by impounding a distributary of the stream (Fig. 17–5a). Blocking an entire stream is a reasonable alternative only in low-order streams, and it is not generally cost-effective. This design is particularly vulnerable during flooding and its stability might be very unpredictable, but it has the advantage of potentially "treating" a significant portion of the water that passes that point in the stream. The maintenance of the control structure and the distributary might mean making significant management commitments to this design.

Riparian Wetlands

The restoration of entire rivers has been shown to be an elusive goal in many parts of the world because of years of drainage and channel "improvement," floodplain development, and significant loads of sediments and other nonpoint pollutants. The natural design for a riparian wetland fed primarily by a flooding stream or river (Fig. 17–5b) allows for flood events of a river to deposit sediments and chemicals on a seasonal basis in the wetland and for excess water to drain back to the stream or river. Because there are natural and also often constructed levees along major sections of streams, it is often possible to create such a wetland with minimal construction. The wetland could be designed to capture flooding water and sediments and slowly release the water back to the river after the flood passes, or to receive water from flooding and retain it through the use of flapgates.

A riparian wetland fed by a pump (Fig. 17–5c) creates the most predictable hydrologic conditions for the wetland but at an obvious extensive cost for equipment and maintenance. If it is anticipated that the primary objective of a constructed wetland is the development of a research program to determine design parameters for future wetland construction in the basin, then a series of wetlands fed by pumps is a good design. If other objectives are most important, then the use of large pumps is usually not appropriate. Small pumps may be necessary to carry a riparian wetland through drought periods. One ecological engineering project on the Des Plaines River north of Chicago in Lake County, Illinois, involves restoration of a length of a river floodplain and establishing experimental wetland basins (Hey et al., 1989; Fig. 17–3). Several experimental wetland basins (2 to 3 ha each) were constructed there using precise hydrologic controls with pumps in order to investigate the hydrologic design of wetlands subjected to high sediment loads. Results from the experimental wetlands after several years of study point to substantial productivity in all of the wetlands; most of the phosphorus and sediments delivered by the pumps were retained (Sather, 1992; Mitsch, 1992a).

Upstream Versus Downstream Wetlands

The advantages of locating several small wetlands on small streams or intercepting ditches in the upper reaches of a watershed (but not in the streams themselves) rather than creating fewer larger wetlands in the lower reaches should be considered (Fig. 17–5d). Loucks (1990) argued that a better strategy for enabling wetlands to survive extreme events is to locate a greater number of low-cost wetlands in the upper reaches of a watershed rather than to build fewer high-cost wetlands in the lower reaches. A modeling effort on flood control by Ogawa and Male (1983, 1986) suggested the opposite: The usefulness of wetlands in decreasing flooding increases with the distance the wetland is downstream.

Figure 17–5e shows a design involving the construction of multiple wetlands in the landscape to intercept small streams and drainage tiles fields. The main stream itself is not diverted, but the wetlands receive their water, sediments, and nutrients from small tributaries, swales, and overland flow. If tile drains can be located and broken upstream to prevent their discharge into tributaries, they can be made effective conduits to supply adequate water to constructed wetlands. Because tile drains are often the sources of the highest concentrations of chemicals such as nitrates from agricultural fields, the wetlands could be very effective in controlling certain types of nonpoint source pollution while creating a needed habitat in an agricultural setting.

Terraced Wetlands

Wetlands are a phenomenon of naturally flat terrain. Steeper terrain, however, is often most susceptible to high erosion and hence high contributions of suspended sediments and other chemicals. One approach is to attempt to integrate "terraced" wetlands into the landscape (Fig. 17–5f). In this case wetland basins are constructed as smaller basins that "stair-step" down steep terrain. There are a few examples of these types of wetlands, mainly coal mine drainage wetlands in the hilly Appalachian region of the eastern United States; few wetlands of this type have been constructed for wildlife habitats or other mitigation purposes.

Site Selection

Several important factors ultimately determine site selection. When the objective is defined, the appropriate site should allow for the maximum probability that the objective can be met, that construction can be done at a reasonable cost, that the system will perform in a generally predictable way, and that the long-term maintenance costs of the system are not excessive. These factors are elaborated on below (Brodie, 1989; Willard et al., 1989):

1. Find a site where wetlands previously existed or where nearby wetlands still exist. In an area such as this, the proper substrate may be present, seed sources may be on site or nearby, and the appropriate hydrologic conditions may exist. A historical meander of a stream that has been abandoned or channelized makes an excellent potential site for restoring a wetland.

Similarly, a restored tidal marsh would be difficult to establish where one had not existed before.

2. Take into account the surrounding land use and the future plans for the land. Future land-use plans such as abandoning agricultural fields to become old-field ecosystems may obviate the need for a wetland to control nonpoint runoff.
3. Undertake a detailed hydrologic study of the site including a determination of the potential interaction of groundwater with the wetland. Hydrologic conditions are paramount. Without flooding or saturated soils for at least part of the growing season, a wetland is impossible to maintain. For coastal wetlands, the tidal cycle and stages are important.
4. Find a site where natural inundation is frequent. Sites should be inspected during flood season and heavy rains, and the annual and extreme event flooding history of the site should be determined as closely as possible.
5. Inspect and characterize the soils in some detail not only to determine their permeability and depth but also to determine their chemical content. Highly permeable soils are not likely to support a wetland unless water-inflow rates are excessive.
6. Determine the quality of the groundwater, surface flows, flooding streams and rivers, and tides that may influence the site water quality. Even if the wetland is being built primarily for wildlife enhancement, chemicals in the water may be significant either to wetland productivity or to the bioaccumulation of toxic materials.
7. Evaluate on-site and nearby seed banks to ascertain their viability and response to hydrologic conditions.
8. Ascertain the availability of necessary fill material, seed, and plant stocks and access to infrastructure (e.g., roads, electricity). This is particularly important for the construction phase.
9. Determine the ownership of the land, and hence the price; these are often overriding considerations. Additional lands may need to be purchased in the future to provide a buffer zone and room for expansion.
10. For wildlife and fisheries enhancement, determine if the wetland site is along ecological corridors such as migratory flyways or spawning runs.
11. Assess site access. Public access to the site will eventually need to be controlled to avoid vandalism and personal injury. A remote site that offers possibilities of fewer mosquito complaints, lower property values, and less drastic kinds of social impact is often preferable to an urban one. Urban wetlands, however, offer intriguing possibilities for programs on wetland education for school groups and the public.
12. Ensure that an adequate amount of land is available to meet the objectives. If "aging" of a wetland, defined as an impairment of wetland function after several years of perturbation (Kadlec, 1985), is anticipated because of the inputs of sediments, nutrients, or other materials, then larger land parcels to build additional wetlands in the future should be considered.

13. Evaluate the position of the proposed wetland in the landscape. Landscapes have natural patterns that maximize the value and function of individual habitats. For example, an isolated wetland pothole functions in ways that are quite different from a wetland adjacent to a river. A forested wetland island created in an otherwise grassy or agricultural landscape will support far different species from those that inhabit a similar wetland created as part of a large forest tract.

WETLAND DESIGN

The need for rigor in designing a wetland varies widely depending on the site and application. In general, a design that can use natural processes to achieve the objectives will yield a less expensive and more satisfactory solution in the long run. On the other hand, "naturally" designed wetlands may not develop as predictably as more tightly designed systems might do. The choice of design is strongly affected by the site and the objectives. In coastal Louisiana there are now several wetland enhancement and creation projects that have simply created crevasses in the natural levee of the lower Mississippi River. These crevasses allow river water and sediment to flow into shallow estuaries and create a crevass splay or minidelta that rapidly becomes a naturally vegetated marsh. Their extent and life span can seldom be predicted, but they are inexpensive to create and they function in a natural manner because they mimic the natural geomorphic processes of the river. In the following sections we focus on much more rigidly designed wetlands in part because this kind of wetland creation requires much greater ecotechnological sophistication.

Hydrology

Hydrology is the most important variable in wetland design. If the proper hydrologic conditions are developed, the chemical and biological conditions will respond accordingly (see Chaps. 4, 5, and 6). The hydrologic conditions, in turn, depend on climate, seasonal patterns of streamflow and runoff, tides (for coastal wetlands), and possible groundwater influences. Improper hydrology leads to the failure of many created wetlands (D'Avanzo, 1989). Improper hydrologic conditions will not always correct themselves as will the more forgiving biological components of the system such as vegetation and animals. Ultimately the hydrologic conditions determine the wetland function. Several parameters used to describe the hydrologic conditions of created and restored wetlands include hydroperiod, depth, and seasonal pulses; for water quality wetlands, they include inflow rates, retention time, and basin morphology.

Hydroperiod and Depth

One of the most basic design parameters for constructed wetlands is the pattern of depth over time, a pattern called the hydroperiod (see Chap. 4). Included in this parameter is the seasonal pattern of the depth and the frequency of flooding. Coastal tidal wetlands must be designed at an elevation to allow proper tidal flushing. Bottomland forested wetlands frequently have extended periods of dry conditions during the growing season. If freshwater marsh vegetation is desired, more frequent flooding is necessary.

Wetlands that possess a variety of water depths have the most potential for developing a diversity of plants, animals, and biogeochemical processes. Deepwater areas, devoid of emergent vegetation, offer habitat for fish, can enhance nitrification as a prelude to later denitrification if nitrogen removal is desired, and can provide low velocity areas where water flow can be redistributed (Steiner and Freeman, 1989). Extremely shallow depths can provide maximum soil-water contact for certain chemical reactions such as denitrification and can accommodate a greater variety of vascular plants.

The state of Florida recommended less than 70 percent open water for created wetlands, with the rest of the wetland established with aquatic vegetation. For Maryland, specific water depth requirements for wetlands for the use of stormwater runoff treatment are (Athanas, 1987): 75 percent of the wetland should have a depth under 30 cm (50 percent less than or equal to 15 cm, and 25 percent—15 to 20 cm); 25 percent should have depths ranging from 60 cm to 100 cm.

Water levels are often controlled by outflow structures that allow the appropriate water depths for planting and controlling undesirable plants. During the start-up period of constructed wetlands, lower water levels are needed to avoid flooding new emergent plants, whereas continuous flooding is necessary for floating-leaved and submerged plants (Allen et al., 1989). Specific periods of drawdown may be necessary if wetland plants are to be started from germinating seeds. Start-up periods for the establishment of plants may take two to three years and an adequate litter-sediment compartment may take another two to three years after that (Kadlec, 1989).

Zedler (1988) cautioned that no standard hydrologic goals are warranted for salt marsh restoration. Instead, the timing and duration of tidal flushing, that is, the hydroperiod, should be specified to create the desired inundation period, circulation, and salinity that, in turn, determine the vegetation communities.

Seasonal and Year-to-Year Pulses

Seasonal patterns of river flooding are part of the river's hydroperiod and using this floading pattern offers the most natural way to replenish a riparian wetland hydrologically. For example, most of the loading of phosphorus-laden sediments to riverine wetlands occurs during the winter and spring, and a good wetland

Table 17–1. Hydrologic Loading for Wastewater Wetlands in North America and Europe

Wetland Project	*Type of Flow*	*Loading Rate, cm/day*	
		Design	*Actual*
North America			
Ann Arundel County, Md.			
freshwater wetland	surface flow	15.1	
peat wetland	percolation	10.2	
offshore wetland	surface flow	1.1	
Arcata, Calif.	surface flow	24.0	2.7
Benton, Ky.			
Cell 1	surface flow	9.5	5.3
Cell 2	surface flow	7.1	4.1
Cell 3	subsurface[a]	7.1	4.3
Brookhaven, N.Y.			
marsh	surface flow		2.3
pond	surface flow		4.5
Cannon Beach, Oreg.	surface flow		5.7
Cobalt, Ontario	surface flow		2–10
Collins, Miss.	surface	3.8	
East Lansing, Mich.	surface flow marsh/pond		1.8
Emmitsburg, Md.	subsurface		16.4
Foothills Pointe, Tenn.	subsurface	4.7	
Fort Deposit, Ala.	surface flow	1.5	
Gustine, Calif.	surface flow	3.8	
Hardin, Ky.	subsurface	5.9	
Houghton Lake, Mich.	surface flow		0.1
Incline Village, Nev.	surface flow	1.3	
Iron Bridge, Fla.	surface flow	1.6	0.6
Iselin, Pa.			
marsh	subsurface[a]	4.7	5.3
meadow	subsurface[a]	9.4	10.7
Lake Buena Vista, Fla.	surface flow		4.0
Lakeland, Fla.	surface flow	0.9	0.5

continued on next page

design takes advantage of these pulses if nutrient retention is a primary objective. These seasonal pulses are themselves not constant but vary in intensity from year to year. The rarer events, for example, the 100-year flood, could be the streamflow for which the wetland has been designed. Infrequent and nonperiodic flooding and droughts are also important for dispersing biological species to the wetland and adjusting resident species composition.

After start-up, a variable hydroperiod, exhibiting dry periods interspersed with flooding, is a natural cycle in prairie pothole freshwater marshes (Weller, 1981), and fluctuating water levels should be accepted if not encouraged (Willard et al., 1989). A fluctuating water level can often provide needed oxida-

Table 17–1 continued

Wetland Project	Type of Flow	Loading Rate, cm/day	
		Design	Actual
North America (continued)			
Listowel, Ontario	surface flow		1.3–2.0
Mountain View Sanitary, Calif.	surface flow marsh/pond	7.8	3.5
Neshaminy Falls, Pa.			
marsh	subsurface[a]	4.6	2.4
meadow	subsurface[a]	9.3	4.8
Orange County, Fla.	surface flow		1.5
Orlando, Fla.	surface flow	1.6	0.6
Paris Landing, Tenn.	subsurface	15	
Pembrook, Ky.	surface/subsurface	2.2	
Phillips High School, Bear Creek, Ala.	subsurface	3.7	
Santee, Calif.	subsurface	4.7	4.7
Silver Springs Shores, Fla.	surface flow		1.2
Vermontville, Mich.	surface flow	1.4	
Europe			
Gravesend, England	subsurface-gravel	8.2	
Marnhell, England	subsurface-soil	4.5–6.9	
Holtby, England	subsurface-soil	4.9	
Castleroe, England	subsurface-gravel and soil	4.3	
Middleton, England	subsurface-sand/gravel	8.9	
Bluther Burn, England	subsurface-flyash	6.2–10.8	
	subsurface-gravel	9.9–10.1	
Little Stretton, England	subsurface-gravel	26	
Ringsted, Denmark	subsurface-gravel	5.7	
	subsurface-clay	1.7	
Surface flow wetlands—average±standard error		5.4±1.7 (n=15)	
Subsurface flow wetlands—average±standard error		7.5±1.0 (n=23)	

[a]Plugging resulted in substantial surface flow.
Source: Watson et al., 1989

tion of organic sediments and can, in some cases, rejuvenate a system to higher levels of chemical retention (Kobriger et al., 1983; Faulkner and Richardson, 1989).

Hydrologic Inflows

Most of what is known about flow rates to wetlands (volume of water flowing into a wetland per unit area per unit time) comes from studies of designs of wetlands to treat wastewater. Table 17–1 lists the hydrologic loading rates of some of the many wastewater wetlands in North America and Europe. There are two general designs, *surface flow wetlands* as used primarily in the United States, and *subsurface flow* (or *root-zone*) *wetlands* as used primarily in Europe.

Table 17–2. Average Design Parameters for Marsh-Type Wetlands Used to Treat Coal Mine Drainage

Parameter	*Suggested Design*
Hydrology	
Loading	5 cm/day
Retention Time	> 24 hours
Iron Loading, g m^{-2} day^{-1}	2–10 (for 90 percent Fe removal)
	20–40 (for 50 percent Fe removal)
Basin Characteristics	
Depth	< 0.3 m
Cells	multiple (3 or more)
Planting Material	*Typha* sp.
Substrate Material	Organic Peat over Clay Seal

Source: Fennessy and Mitsch, 1989b

Loading rates to surface flow wetlands for wastewater treatment from small municipalities ranged from 1.4 to 22 cm/day (average = 5.4 cm/day), whereas subsurface rates varied between 1.3 and 26 cm/day (average = 7.5 cm/day). Knight (1990) reviewed several dozen wetlands constructed for wastewater treatment and found loading rates to vary between 0.7 and 50 cm/day. He recommended a rate of 2.5 to 5 cm/day for surface water systems and 6 to 8 cm/day for subsurface flow wetlands. Wile et al. (1985) recommended 2 cm/day for wastewater wetlands as optimum. M. T. Brown (1987) designed a mosaic of wetland "cells" amid forested floodplain wetlands and estimated a wastewater loading rate of 2.2 cm/day. Adamus (1990) reported an EPA guideline for loading rates of less than 5 cm/week (<0.7 cm/day) for natural wetlands receiving wastewater. This limitation was developed from the Florida state guidelines for forested wetlands which were originally developed from experiments with cypress domes in Gainesville, Florida (H. T. Odum et al., 1977a).

Hydraulic loading rates as high as 29 cm/day have been suggested for wetlands designed for acid mine drainage (Pesavento, unpublished, as cited in Watson et al., 1989), although Fennessy and Mitsch (1989b) recommended 5 cm/day as a conservative loading rate for this type of wetland (Table 17–2).

The above loading rate limitations are probably too restrictive for wetlands used for the control of nonpoint pollution and stormwater runoff, but few studies have been undertaken to determine the optimum design rates for such wetlands. The Des Plaines River Demonstration Project in northeastern Illinois (Hey et al., 1989) had initial experiments designed for loading rates of river water from 1 to 8 cm/day with pumps delivering the water from the Des Plaines River. These rates, however, were estimated from rates for comparable wastewater wetlands and may be too low for riparian wetlands receiving floodwaters.

Residence Time

As with loading rate, most of the experience with residence time of wetlands (see Chap. 4) is based on wetlands designed to treat wastewater. The optimum detention time has been suggested to be from 5 to 14 days for treatment of municipal wastewater (Wile et al., 1985; Watson and Hobson, 1989). Florida regulations on wetlands (cited in Palmer and Hunt, 1989) require that the volume in the permanent pools of the wetland must provide for a residence time of at least 14 days. M. T. Brown (1987) suggested a retention time of a riparian wetland system in Florida of 21 days in the dry season and more than 7 days in the wet season. Fennessy and Mitsch (1989b) recommended a minimum retention time of one day for acid mine drainage wetlands and much longer periods for more effective iron removal. The use of simple calculations of residence time is not always appropriate because of short-circuiting and the ineffective spreading of the waters as they pass through the wetland.

Basin Morphology

Several aspects related to the morphology of constructed wetland basins need to be considered when designing wetlands. Several states have developed guidelines for bottom profiles of constructed wetlands, particularly freshwater marshes. Florida regulations for the Orlando area (described by Palmer and Hunt, 1989) require a littoral shelf with gently sloping sides of 6:1 or flatter to a point of from 60 to 77 cm (2–2.5 feet) below the water surface. Wetland bottom slopes of less than 1 percent were recommended for wetlands built to control runoff and for wetlands in riparian settings (Bell, 1981; Kobriger et al., 1983, cited by Willard et al., 1989), whereas Steiner and Freeman (1989) suggested a substrate slope, from inlet to outlet, of 0.5 percent or less for surface flow wetlands used to control wastewater. Flow conditions should be designed so that the entire wetland is effective in nutrient and sediment retention if these are desired objectives. This may necessitate several inflow locations and a wetland configuration to avoid channelization of flows. Steiner and Freeman (1989) suggested a length-to-width ratio (L/W) of at least 10 if water is purposely introduced to the system. In riparian settings the wetland should probably be designed longitudinally to minimize flooding erosion damage to dikes built perpendicular to the flow on the floodplain.

Individual wetland cells, placed in series or parallel, often offer an effective design to create different habitats or establish different functions (Steiner and Freeman, 1989). Cells can be parallel so that alternate drawdowns can be accomplished for mosquito control or redox enhancement, or they can be in a series to enhance biological processes. Wetlands in Ohio (Fennessy and Mitsch, 1989a, b), California (Metz, 1987), and Florida (Redmond, 1981) have utilized separate cells. In other cases wetlands have been designed with cells that actively receive polluted water amid passive wetlands that receive overflow from the active cells (Best, 1987; M. T. Brown, 1987).

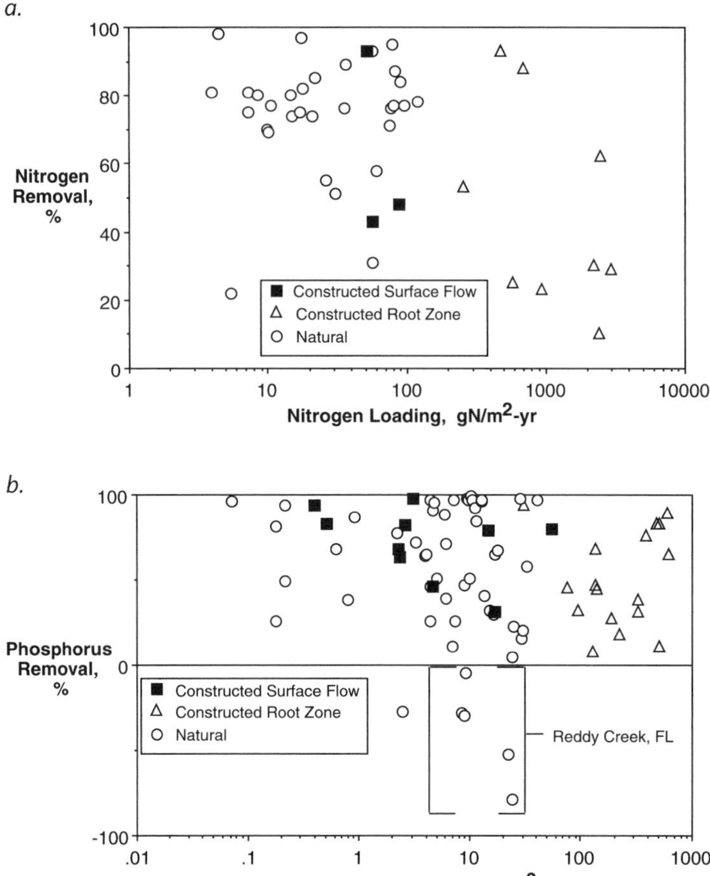

Figure 17–6. Nutrient retention in wetlands receiving wastewater or river water as a function of loading for *a.* nitrogen and *b.* phosphorus. (*Data are from Knight [1990] and unpublished data from Des Plaines River Wetland project [for P-removal].*) Reddy Creek data on phosphorus plot were for wetland in Florida that consistently exported phosphorus after wastewater was applied. (*Knight et al., 1987*) Data are shown for surface flow constructed wetlands, subsurface flow constructed wetlands (root zone), and natural wetlands. Multiple data are given for wetlands with multiple-year monitoring.

Chemical Inputs

When water flows into a wetland, it brings chemicals that may be beneficial or possibly detrimental to the functioning of that wetland. In an agricultural watershed, this inflow will include nutrients such as nitrogen and phosphorus as well as sediments and possibly trace amounts of pesticides. Wetlands in urban areas can have all of these chemicals plus other contaminants such as oils and salts. Wastewater, when added to wetlands either accidentally or on purpose, has high concentrations of nutrients and if incompletely treated, high concentrations of organic matter. At one time or another, wetlands have been subjected to all of these chemicals, and they often serve as effective sinks (see Chap. 5); yet models predicting chemical retention are still under development and estimated removals calculated by these models must be used with caution.

If the wetland is designed to retain nutrients, for example, it would be desirable to know how well that retention would occur for various nutrient inflows. Data compiled from a large number of wetland sites provide an indication of the nutrient retention of wetlands (Knight et al., 1984; Richardson and Nichols, 1985; Athanas, 1988; Faulkner and Richardson, 1989; Watson et al., 1989; Knight, 1990). Figure 17-6, compiled from some of those data sets, illustrates the percent removal of nitrogen and phosphorus for various loading rates of wastewater and nonpoint source wetlands. With much scatter, the data show generally decreasing percent retention with greater loading rates. Typically nitrogen retention in constructed wetlands is greater that 40–50 percent at inflow rates of less than 100 gN m^{-2} yr^{-1} (Fig. 17–6a). Much of that nitrogen retention can be assumed to be caused by denitrification. Phosphorus retention is more erratic and site specific (Fig. 17–6b) and depends on the chemical characteristics of the wetland substrate (see below). One set of data on Figure 17–6b showed a wetland in Florida with consistent phosphorus export (negative removal percent); wastewater was applied to phosphorus-saturated peat soils with insufficient retention capacity to accomplish additional phosphorus removal (R. Knight, personal communication).

Subsurface flow wastewater wetlands subjected to higher loading rates retain more mass (Table 17–3). Surface flow wastewater wetlands, in turn, have generally much higher sediment and phosphorus retention than natural and restored wetlands receiving nonpoint source waters do because of their much higher loading rates. The limited data for natural and nonpoint source wetlands in Table 17–3 suggest that mass retention in natural riparian wetlands is about the same as in restored nonpoint source wetlands but that the efficiency of retention is lower in natural systems.

Although most evaluations of the efficiency of wetlands have been concerned with this capacity to remove nutrients, sediments, and organic carbon, there is some literature on other chemicals such as iron (e.g., Wieder, 1989; Fennessy and Mitsch, 1989a, b) and selected metals such as cadmium, chromium, copper, lead, mercury, nickel, and zinc (reviewed by Richardson and Nichols, 1985). Metals are often easily sequestered by wetland soils or biota or both. McArthur (1989) reported on a case study in Florida where lead was reduced by 83 percent, zinc by 70 percent, and total solids by 55 percent as stormwater passed through a 0.4 hectare (0.9 acre) pond-wetland complex. The accumulation of selenium became an issue in marshes in the Kesterson National Wildlife Refuge

Table 17–3. Comparison of Natural, Nonpoint Source with Wastewater Wetland Sediment and Phosphorus Retention

	Sediments			*Phosphorus*		
Wetland Site	*Loading* $g\ m^{-2}\ yr^{-1}$	*Retention* $g\ m^{-2}\ yr^{-1}$	*Percent retention*	*Loading* $gP\ m^{-2}\ yr^{-1}$	*Retention* $gP\ m^{-2}\ yr^{-1}$	*Percent retention*
Natural Wetlands						
Heron Pond, southern Ill. forested wetland	15,000	447	3	80.2	3.6	4.5
Old Woman Creek, coastal Lake Erie marsh	—	—	—	8	0.8	10
Restored and Created Wetlands— Nonpoint Source Control						
Des Plaines River Wetlands—Oct. 1989–Sept 1990						
Wetland 3—high flow	926	816	88	2.3	1.5	63
Wetland 4—low flow	218	203	93	0.5	0.4	83
Wetland 5—high flow	889	800	90	2.3	1.6	68
Des Plaines River Wetlands–Apr.–Sept. 1991						
Wetland 3—high flow	1955	1840	94	2.7	2.2	82
Wetland 4—low flow	328	317	97	0.44	0.41	94
Wetland 5—high flow	2127	2096	98	3.6	3.55	98
RANGE—Nonpoint Source Wetlands	**218–2127**	**203–2096**	**88–98**	**0.4–3.6**	**0.4–3.5**	**63–98**
Created Wetlands - Wastewater Treatment–Surface Flow						
Arcata, Calif.—high flow	6520	5570	85	—	—	—
Arcata, Calif.—medium flow	3260	2760	85	—	—	—
Arcata, Calif.—low flow	1630	1360	84	—	—	—
Emittsburg, Md.	1760	1270	73	—	—	—
Iselin, Pa.	1640	1560	95	56	45	80
Listowel 4, Ontario—high concentration	520	480	93	15	12	79
Listowel 3, Ontario—low concentration	107	65	61	4.7	2.1	46
Santee, Calif.—bulrush	980	960	98	—	—	—
Santee, Calif.—cattail	980	880	90	—	—	—
Santee, Calif.—reed	980	840	86	—	—	—
RANGE—Surface Wastewater Wetlands	**107–6520**	**65–5570**	**61–98**	**4.7–56**	**2.1–45**	**46–80**

Table 17–3 continued

Wetland Site	*Sediments*			*Phosphorus*		
	Loading $g\ m^{-2} \cdot yr^{-1}$	*Retention* $g\ m^{-2} \cdot yr^{1}$	*Percent retention*	*Loading* $gP\ m^{-2}\ yr^{-1}$	*Retention* $gP\ m^{-2}\ yr^{-1}$	*Percent retention*
Created Wetlands—Wastewater Treatment–Subsurface Flow						
Bluther Burn, England (4 different substrates)	3610–5880	3030–4930	84–87	390–631	296–540	65–89
Castleroe, England	1680	1100	65	190	51	27
Gravesend, England	1500	730–1020	49–68	139	62–95	45–68
Holtby, England	3650	3250	89	131	11	8
RANGE—Subsurface Wastewater Wetlands	**1500–5880**	**1100–4930**	**49–89**	**131–631**	**11–540**	**8–89**

Source: Knight, 1990; Mitsch, 1992a

in California when it began to accumulate in biota there as a result of the inflow of irrigation waters for several years. As a result, the Kesterson marshes were found to be a threat to fish and wildlife in the region and were dewatered (Hammer, 1992).

Substrate/Soils

The substrate is important to the overall function of a constructed wetland and is the primary medium supporting rooted vegetation. The soil or subsoil must also have permeability low enough to retain standing water or saturated soils. If a wetland is designed to improve water quality, the substrate will play a significant role in its ability to retain certain chemicals. The sediments retain certain chemicals and provide the habitat for micro-and macro-flora and fauna that are involved in several chemical transformations. The substrate is not as significant for the retention of suspended organic matter and solids (along with the chemicals adsorbed to sediment particles), because their retention is based primarily on net deposition (sedimentation minus resuspension) that results from the slow velocities characteristic of wetlands.

Some of the characteristics of substrates that do play a role in the design of wetlands are reviewed here.

Organic Content

The organic content of soils has some significance for the retention of chemicals in a wetland. Mineral soils generally have lower cation exchange capacity than organic soils do; the former is dominated by various metal cations, and the latter is dominated by the hydrogen ion (see Chap. 5). Organic soils can therefore remove some contaminants (e.g., certain metals) through ion exchange and can enhance nitrogen removal by providing an energy source and anaerobic conditions appropriate for denitrification. Organic matter in wetland soils generally varies between 15 and 75 percent (Faulkner and Richardson, 1989), with higher concentrations in peat-building systems such as bogs and fens and lower concentrations in open wetlands such as riparian bottomland wetlands subject to mineral sedimentation or erosion. When wetlands are constructed, especially subsurface flow wetlands, organic matter such as composted mushrooms, peat, or detritus is often added in one of the layers. For construction of many wetlands, organic soils are avoided because they are low in nutrients, can cause low pH, and often provide inadequate support for rooted aquatic plants (Allen et al., 1989).

Soil Texture

Clay material, although often favored for surface flow wetlands to prevent percolation of water to groundwater, may also limit root and rhizome penetration and may be impermeable to water for plant roots. In that case loam soils are preferable. Sandy soils, although generally low in nutrients, do anchor plants and allow water to reach the plant roots readily (Allen et al., 1989). The use of local soils, underlain with impermeable clay to prevent downward percolation, is often the best design. If on-site top soils are to be returned to the wetland, adequate temporary storage should be provided (Willard et al., 1989). If clay is not available on site, it may be advisable to add a layer to slow percolation.

Subsurface Flow and Gravel

Wetlands for wastewater treatment are designed for either surface flow over the substrate or subsurface flow through the substrate (Fig. 17–7). Surface flow systems are generally less effective in removing pollutants but are closer in design to natural wetlands. Subsurface flow wetlands have frequently been used, particularly in Europe, for treating wastewater but their long-term effectiveness has been questioned. Wieder et al. (1989) suggested that subsurface flow through artificial wetlands can be through soil media (*root-zone method*) or through rocks or sand (*rock-reed filters*), with the flow in both cases 15 to 30 cm below the surface. Gravel is added to the substrate of artificial wetlands to provide a relatively high permeability that allows water to percolate into the root zone of the plants where microbial activity is high (Gersberg et al., 1983, 1984, 1986, 1989). In a survey of several hundreds of wetlands build in Europe for sewage treatment in rural settings, Cooper and Hobson (1989) reported that gravel has been used in combination with soil but that the substrate remains the greatest uncertainty in artificial reed (*Phragmites*) wetlands in Europe. Gravel can be silica based or limestone based; the former has much less capacity for phosphorus retention (Cooper and Hobson, 1989). A more recent evaluation of the European-design subsurface wetlands indicated that they decreased in hydrologic conductivity after a few years and became essentially overloaded surface flow wetlands (Steiner and Freeman, 1989). One study, which investigated six substrates for their effectiveness in controlling acid mine drainage, concluded that all performed in a similar fashion (Brodie et al., 1988).

Depth and Layering of Substrate

The depth of substrate is an important design consideration for wastewater and mine drainage wetlands, particularly those that use subsurface flow. The depth of substrate is of less concern for surface water wetlands than for subsurface flow wetlands. All wetlands should have an adequate layer of clay materials at an appropriate depth if downward percolation is not desired. Meyer (1985) described a layered substrate in wetlands to control stormwater runoff as having the following materials from the base upward: 60 cm of 1.9 cm limestone; 30 cm of 2 mm crushed limestone to raise pH and aid in the precipitation of dissolved heavy metals and phosphate; 60 cm of coarse to medium sand as filter, and 50 cm organic soil. A common substrate depth for subsurface flow wetlands

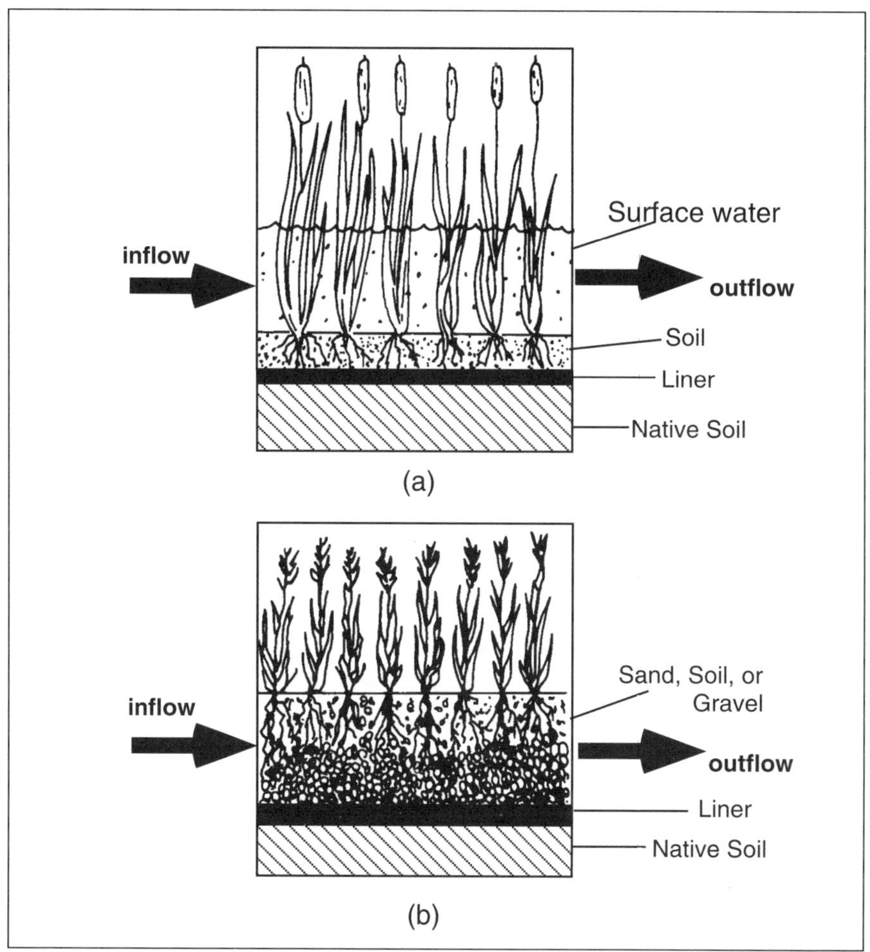

Figure 17–7. Cross-sections of types of wetlands used to treat wastewater including *a.* surface flow wetlands (also known as free water surface [FWS] wetlands) and *b.* subsurface flow wetlands (also know as vegetated submerged bed [VSB] wetlands, root zone wetlands, or rock-reed filters). *(From Knight, 1990; copyright © 1990 by Water Environment Federation, Alexandria, Virginia, reprinted with permission)*

is 60 cm (Steiner and Freeman, 1989). The depth of suitable substrate should be adequate to support and hold vegetation roots (see Vegetation below).

Nutrients

Although exact specifications of nutrient conditions in substrate necessary to support aquatic plants are not well known (Allen et al., 1989), low nutrient levels characteristic of organic, clay, or sandy soils can cause problems in initial plant growth. Although fertilization may be necessary in some cases to establish plants and enhance growth, it should be avoided if possible in wetlands that eventually will be used as sinks for macronutrients. When fertilization is required, slow-release granular and tablet fertilizers are often useful (Allen et al., 1989).

Iron and Aluminum

When soils are submerged and anoxic conditions result, iron is reduced from the ferric (Fe^{+++}) to the ferrous (Fe^{++}) ions, releasing phosphorus that was previously held as ferric phosphate compounds. The Fe-P compound can be a significant source of phosphorus to overlying and interstitial waters (Faulkner and Richardson, 1989) after flooding and anaerobic conditions occur. Phosphorus can also be retained by wetlands as oxides and hydroxides of iron and aluminum (Richardson, 1985). There is apparent disagreement in the literature about whether the phosphorus is retained by ligand exchange or by precipitation (Faulkner and Richardson, 1989). Nevertheless, the iron and aluminum content of a wetland soil exerts significant influences on the ability of that wetland, whether constructed or natural, to retain nutrients such as phosphorus.

Vegetation

Types

The species of vegetation types to be introduced to created and restored wetlands depend on the type of wetland desired, the region, and the climate as well as the design characteristics described above. Table 17–4 summarizes some of the plant species used for wetland creation and restoration projects. Common plants used for freshwater marshes include soft-stem bulrush (*Scirpus validus*), cattails (*Typha latifolia* and *Typha angustifolia*), sedges (*Carex* spp.), and floating-leaved aquatic plants such as white water lily (*Nymphaea odorata*) and spatterdock (*Nuphar luteum*). Submersed plants are not as common in wetland design, and their propagation is often hampered by turbidity.

For coastal salt marshes, *Spartina alterniflora* is the primary choice for coastal marsh restoration (Broome et al., 1988) although both *Spartina townsendii* and *S. anglica* have been used to restore salt marshes in Europe (Ranwell, 1987) and in China (Chung, 1982; 1989). The details of successful coastal wetland creation are, of course, site specific, but several generalizations seem to be valid in most situations (H. K. Smith, 1980; Seneca, 1980):

1. Sediment elevation seems to be the most critical factor determining the successful establishment of vegetation and the plant species that will survive. The site must be intertidal and, in general, the upper half of the intertidal zone is more rapidly vegetated than lower elevations. Sediment composition does not seem to be a critical factor in colonization by plants unless the

Table 17–4. Selected Plant Species Planted in Created and Restored Wetlands

Scientific Name	*Common Name*
Freshwater Marsh—Emergent	
Acorus calamus	sweet flag
Cladium jamaicense	sawgrass
Carex spp.	sedges
Eleocharis spp.	spike rush
Glyceria spp.	manna grass
Hibiscus spp.	rose mallow
Iris pseudacorus	yellow iris
Iris versicolor	blue iris
Juncus effusus	soft rush
Leersia oryzoides	rice cutgrass
Panicum virgatum	switchgrass
Peltandra virginica	arrow arum
Phalaris arundinacea	reed canary grass
Phragmites australis[a]	giant reed
Polygonum spp.	smartweed
Pontederia cordata	pickerelweed
Sagittaria rigida	duck potato
Sagittaria latifolia	duck potato; arrowhead
Saururus cernuus	lizard's tail
Scirpus acutus	hard-stem bulrush
Scirpus americanus	three-square bulrush
Scirpus cyperinus[a]	woolgrass
Scirpus fluviatilis	river bulrush
Scirpus validus[a]	soft-stem bulrush
Sparganium eurycarpum	giant burreed
Spartina pectinata	prairie cordgrass
Typha angustifolia[a]	narrow-leaved cattail
Typha latifolia[a]	wide-leaved cattail
Zizania aquatica	wild rice
Freshwater Marsh—Submersed	
Ceratophyllum demersum	coontail
Elodea nuttallii	waterweed
Myriophyllum aquaticum	milfoil
Potamogeton pectinatus	Sago pondweed
Vallisneria spp.	wild celery; tape grass
Freshwater Marsh—Floating	
Azolla caroliniana	water fern
Eichhornia crassipes	water hyacinth
Hydrocotyle umbellata	water pennywort
Lemna spp.	duckweed
Nymphaea odorata	fragrant white water lily
Nuphar luteum	spatterdock
Pistia stratiotes	water lettuce
Salvinia rotundifolia	floating moss
Wolffia sp.	water meal

continued

Table 17–4 continued

Scientific Name	*Common Name*
Bottomlands/Forested Wetland	
Acer rubrum	red maple
Acer floridanum	Florida maple
Acer saccharinum	silver maple
Alnus spp.	alder
Carya illinoensis	pecan
Celtis occidentalis	hackberry
Cephalanthus occidentalis	buttonbush
Cornus stolonifera	red-osier dogwood
Fraxinus caroliniana	water ash
Fraxinus pennsylvanica	green ash
Gordonia lasianthus	loblolly bay
Liquidambar styracifula	sweet gum
Platanus occidentalis	sycamore
Populus deltoides	cottonwood
Quercus falcata var. *pagodaefolia*	cherrybark oak
Quercus nigra	water oak
Quercus nuttallii	Nuttall oak
Quercus phellos	willow oak
Salix spp.	willow
Ulmus americana	American elm
Deepwater Swamp	
Nyssa aquatica	swamp tupelo
Nyssa sylvatica var. *biflora*	black gum
Taxodium distichum	bald cypress
Taxodium distichum var. *nutans*	pond cypress
Salt Marsh	
Distichlis spicata	spike grass
Salicornia sp.	saltwort
Spartina alterniflora	cordgrass (eastern United States)
Spartina anglica	cordgrass (Europe; China)
Spartina foliosa	cordgrass (western United States)
Spartina patens	salt meadow grass
Spartina townsendii	cordgrass (Europe)
Mangrove	
Rhizophora mangle	red mangrove
Avicennia germinans	black mangrove
Laguncularia racemosa	white mangrove

[a]Commonly planted in wastewater wetlands

deposits are almost pure sand that is subject to rapid dessication at the upper elevations. Another important requirement is protection of the site from high wave energy. It is difficult or impossible to establish vegetation at high-energy sites.
2. Most deposits seem to revegetate naturally if the elevation is appropriate and the wave energy is moderate. Sprigging live plants has been accomplished successfully in a number of cases, and seeding also has been successful in the upper half of the intertidal zone.
3. Good stands can be established during the first season of growth, although sediment stabilization does not occur until after two seasons. Within four years, successfully planted sites are indistinguishable from natural marshes.

Forested wetland restoration and creation usually involve the establishment of seedlings. In the Southeast, deciduous hardwood species typical of bottomland forests are planted. They include Nuttall oak (*Quercus nuttallii*), cherrybark oak (*Q. falcata* var. *pagodaefolia*), willow oak (*Q. phellos*), water oak (*Q. nigra*), cottonwood (*Populus deltoides*), sycamore (*Platanus occidentalis*), green ash (*Fraxinus pennsylvanica*), sweetgum (*Liquidambar styracifula*), and pecan (*Carya illinoensis*). There is less use of deepwater plants such as bald cypress (*Taxodium distichum*) and water tupelo (*Nyssa aquatica*) (Clewell and Lea, 1990). In Florida, a diverse variety of wetland oaks, bays, gums, ashes, and pines are also used in forested wetland restoration (Erwin et al., 1984; M. T. Brown et al., 1992).

Exotic or Undesirable Plant Species

In some cases, certain plants are viewed as desirable or undesirable for reasons such as their value to wildlife or their aesthetics. Reed grass (*Phragmites australis*) is often favored in constructed wetlands in Europe (Cooper and Hobson, 1989) but is not generally used in the United States. Some plants are considered undesirable in wetlands because they are aggressive competitors. In the southern United States, the floating aquatic plant water hyacinth (*Eichhornia crassipes*) and alligator weed (*Alternanthera philoxeroides*) and, in the northern United States, the emergent purple loostrife (*Lythrum Salicaria*) are considered undesirable alien plants in wetlands, although water hyacinth has frequently been evaluated for its role in nutrient retention (e.g., Mitsch, 1977; Ma and Yan, 1989). Throughout the United States, cattail (*Typha* spp.) is championed by some and disdained by others, for it is a rapid colonizer but of limited wildlife value (W. E. Odum, 1987).

Natural Succession versus Horticulture

An important general consideration of wetland design is whether plant material is going to be allowed to develop naturally from some initial seeding and planting or whether continuous horticultural selection for desired plants will be imposed. To develop a wetland that will ultimately be a low-maintenance one, natural successional processes need to be allowed to proceed. Often this means some initial period of invasion by undesirable species, but if proper hydrologic conditions are imposed, those invasions may be temporary. The best strategy is to introduce, by seeding and planting, as many choices as possible to allow natural processes to sort out the species and communities in a timely fashion (H. T. Odum, 1989; Dunn, 1989). Providing some help to this selection process, for example, selective weeding, may be necessary in the beginning, but ultimately the system needs to survive with its own successional patterns unless significant labor-intensive management can be used. W. E. Odum (1987) distinguished freshwater wetland succession from coastal saltwater wetland succession by stating that "in many freshwater wetland sites it may be an expensive waste of time to plant species which are of high value to wildlife. . . . It may be wiser to simply accept the establishment of disturbance species as a cheaper although somewhat less attractive solution."

Sinicrope et al. (1990) described the natural restoration of a 20-hectare impounded salt marsh in Connecticut over ten years (see Fig. 17–4). In that case, a wetland had been isolated from tidal flushing for many years and had become dominated by *Typha angustifolia*. In the late 1970s and early 1980s, several major culverts were installed to reintroduce tidal flushing, leading to a significant decline in *T. angustifolia* from 74 percent to 16 percent cover and a recovery of *Spartina alterniflora* from <1 percent cover to 45 percent cover. *Phragmites australis*, which tolerates brackish conditions, did not decrease as expected but increased from 6 percent to 17 percent cover and was generally found along the edges of the marsh in a stunted (0.3 to 1.5 m tall) condition. A relatively simple alternation of the hydrologic conditions reestablished a salt marsh.

Planting Techniques

Plants can be introduced to a wetland by transplanting roots, rhizomes, tubers, seedling, or mature plants; by broadcasting seeds obtained commercially or from other sites; by importing substrate and its seed bank from nearby wetlands; or by relying completely on the seed bank of the original site. If planting stocks rather than site seed banks are used, it is most desirable to choose plants from wild stock rather than nurseries because the former are generally better adapted to the environmental conditions that they will face in constructed wetlands. The plants should come from nearby if possible and should be planted within 36 hours of collection. If nursery plants are used, they should be from the same general climatic conditions and should be shipped by express service to minimize losses (Allen et. al., 1989).

Esry and Cairns (1989) described three cells of a marsh designed for stormwater runoff. They were planted differently to determine the effectiveness and hardiness of different plant covers. The first cell was planted with locally obtained sawgrass (*Cladium*), and the second cell was planted with bulrush (*Scirpus*) obtained in nurseries. Both of these species were hand planted in rows. The third cell was planted with pickerelweed (*Pontederia*) that had been gathered in a nearby farm pond. The plants of the first and the third cells have taken hold and thrived. The second cell still contains some bulrushes, but their overall survival was limited. In this cell there was rapid volunteer growth of duckweed (*Lemna*) and emergent plants.

For emergent plants, the use of planting materials with at least 20 to 30 cm stems was recommended (Tomljanovich and Perez, 1989), and whole plants, rhizomes, or tubers rather than seeds have been most successful. In temperate climates, both fall and spring planting times are possible for certain species, but spring plantings are generally more successful. Garbisch (1989) suggested that spring planting is desirable to minimize the destructive grazing of plants in the winter by migratory animals and to avoid the uprooting of the new plants by ice.

Transplanting plugs or cores (8–10 cm in diameter) from existing wetlands is another technique that has been used with success, for it brings seeds, shoots, and roots of a variety of wetland plants to the newly constructed wetland (Kobriger et al., 1983; Allen et al., 1989). M. T. Brown (1987) gave planting recommendations for a Florida site used for the renovation of treated effluent. He suggested that marshes be planted at densities to ensure rapid colonization, adequate seed source, and effective competition with *Typha* spp. Specifically this could mean from 2,000 to 5,000 plants/hectare (800 to 2000 plants/acre). M. T. Brown (1987) and Willard et al. (1989) found that a varied bed form adds to the diversity of the vegetation.

Another example of specific vegetation guidelines was provided by Athanas (1987) and Livingston (1989) for wetland construction in the state of Maryland for stormwater management. Specific recommendations were given for plant species: Planting was to include at least two aggressive wetland species (primary species) and three secondary species planted in smaller numbers than the primary species (Table 17–5). *Typha* spp. and *Phragmites australis* (= *communis*) are considered aggressive plants but without much value to wildlife. For this reason, they were not included on the list of desirable species. *Peltandra virginica* is an example of a secondary species that is less aggressive (probably because it depends more on seed germination than on vegetative propagation) but seems to have good wildlife value.

The planting guidelines given are listed below.

1. Primary species should cover 30 percent of the shallow zone and be spaced at 1 meter (3-feet) intervals.
2. These species should be in four monospecific areas.
3. One hundred clumps per hectare (40 clumps/acre) should be distributed over the rest of the wetland.
4. For secondary species, plant 125 individuals per hectare (50 per acre), and for

Table 17-5. Recommended Wetland Plant Species for Stormwater Management in Maryland

Primary Species
Sagittaria latifolia
Scirpus americanus
Scirpus validus

Secondary Species
Acorus Calamus
Cephalanthus occidentalis
Hibiscus moscheutos
Hibiscus laevis
Leersia oryzoides
Nuphar luteum
Peltandra virginica
Pontederia cordata
Saururus cernuus

Undesirable Species
Typha latifolia
Typha angustifolia
Phragmites australis

Source: Athanas, 1987; and Livingston, 1989

each species, plant 25 clumps of 5 individuals per hectare (10 per acre) close to edge of the wetland, but space clumps as far apart as possible to segregate species.

Seeding Techniques

If seeds and seed banks are used for wetland vegetation, several precautions must be taken. The seed bank should be evaluated for seed viability and species present (van der Valk, 1981). The use of seed banks from other nearby sites can be an effective way to develop wetland plants in a constructed wetland if the hydrologic conditions in the new wetland are similar. Weller (1981) stated that seed bank transplants were successful for many different species, including sedges (*Carex* spp.), *Sagittaria* sp., *Scirpus acutus*, *S. validus*, and *Typha* spp. The disruption of the wetland site where the seed bank is obtained must also be considered.

When seeds are used directly to vegetate a wetland, they must be collected when they are ripe and stratified if necessary. If commercial stocks are used, the purity of the seed stock must be determined (Garbisch, 1989; Willard et al., 1989). The seeds can be added with commercial drills or by broadcasting from the ground, watercraft, or aircraft. Seed broadcasting is most effective when there is little to no standing water in the wetland.

Management After Construction

Plant Harvesting

The harvesting of plants generally does not result in a great quantity of chemicals, for example nutrients, being removed from the system unless the material is harvested several times in a growing season. In some cases plant harvesting may be advisable as part of the routine maintenance of the constructed wetland. Wieder et al. (1989) suggested plant harvesting as a mechanism to alter the effect that plants have on the ecosystem, usually by putting it back into an earlier stage of succession when net growth may be greater. Plant harvesting may also be necessary to control mosquitoes, reduce congestion in the water, change the residence time of the basin, and allow for greater plant diversity. Yan Jingsong (personal communication, Nanjing, China, 1989) stated that the practice in China is to harvest the shoots of emergent plants regularly, thereby increasing the size, strength, and number of remaining shoots and encouraging vegetative reproduction. Suzuki et al. (1989) stated that harvesting *Phragmites* twice during the growing season—once at peak nutrient content and the second at the end of plant growth—leads to a maximum removal of nitrogen and phosphorus.

Sometimes other plant management techniques such as drawdowns followed by burning are used as means of controlling the invasion of woody vegetation if that invasion is considered undesirable (Warners, 1987). When the controlled burning of wetlands is used, the wetland manager needs to consider the impact on wildlife (Willard et al., 1989) as well as the potential reintroduction of inorganic nutrients to the water column from the sediments when water is again added.

Wildlife Management

Although the development of wildlife is a welcomed and often desired aspect of created wetlands, beaver and muskrat can burrow into dikes, or create obstructions to flow and remove vegetation (Tomljanovich and Perez, 1989); sand, gravel, or wire screening can be used to discourage burrowing. In other cases, animals grazing on newly planted perennial herbs and seedlings are particularly destructive. Garbisch (1989) suggested that the timing of planting is important, especially when migratory animals are involved in destructive grazing in the winter. Similarly, deeper wetlands often become havens for undesirable fish such as carp that can cause excessive turbidity and uproot vegetation.

Weller (1981) recommended 50 percent open water in Midwest United States basin wetlands to encourage wildlife, and Brinson et al. (1981b) suggested creating diverse habitats with live and dead vegetation, islands, and floating structures. In many cases of wetland construction, wildlife enhancement begins within a few years after construction. At a constructed wetland at Pintail Lake in Arizona, the area's waterfowl population increased dramatically. By the second year of use, duck nest density had increased 97 percent over the first year (Wilhelm et al., 1989). Considerable increase in avian activity was also noted at the Des Plaines River Wetlands Demonstration Project in northeastern Illinois (Fig. 17–3). Migrating waterfowl increased from 3 to 15 species and from 13 to 617 individuals between 1985 (preconstruction) and 1990 (one year after water was introduced to the wetlands). The number of wetland-dependent breeding birds increased from 8 to 17 species and two state-designated endangered birds, the least bittern and the yellow-headed blackbird, nested at the site after wetland construction (Hickman and Mosca, 1991).

Mosquitoes

Mosquitoes are controlled in constructed wetlands in California by changing the conditions of the wetland (e.g., hydrology) to inhibit mosquito larvae development or by using chemical or biological control (Martin and Eldridge, 1989). Wieder et al. (1989) and others have propossed mosquito control by fish, especially the air-gulping mosquito fish (*Gambusia affinis*). Apparently little is known about the role that water quality may perform in controlling mosquitoes. Bacterial insecticides (e.g., *Bacillus sphaericus*) and the fungus *Lagenidium giganteum* are known pathogens of mosquito larvae, but they have not been extensively tested (Martin and Eldridge, 1989). Boxes for swallows and bats have also been used to control adult mosquito populations at constructed wetlands.

Perturbation Control

The average conditions used in the design of a wetland do not reflect the actual conditions, where seasonal and less frequent perturbations are uncertain in frequency and magnitude but may require some response (Willard and Hiller, 1989; Brooks, 1989; Girts and Knight, 1989). Willard and Hiller (1989) recommended that wetlands be designed and planned for the worst-case conditions of perturbations but that a balance be maintained among form, function, and persistence. It is useful to remember that wetland design is an inexact science and that perturbations may change the original design (e.g., selected plant species) to something else. If wetland functions, particularly those that are related to the objectives of the wetland, remain intact, changes in species and forms will not be so significant.

Sediment Dredging

This is an optional management technique in constructed wetlands. Its use depends on whether the wetlands are filling in with sediments, thus shortening their effective lives, and whether sediment accumulation is viewed as an undesirable feature relative to the objectives of the wetland. Dredging is generally an expensive operation and one that should not be attempted frequently in the life of a constructed wetland. Dredging not only removes sediments but also removes the seed bank and rooted plants. The process of dredging sediments from a constructed wetlands may require a regulatory permit even if it is done to "improve" wetland function. The best approach, unless dredging is unavoidable,

Table 17–6. Construction Costs of Wetlands for Various Uses in the United States

Wetland	*State*	*Use*	*Area, ha*	*Cost,* *$/hectare*	*$/acre*	*Source*
Ballona Wetland	California	habitat, recreation	87.4	$70,100	$28,400	Metz, 1987
Greenwood Urban Wetland	Florida	stormwater runoff	11.0	$51,500	$20,800	Palmer and Hunt, 1989
Lake Jackson Restoration	Florida	urban runoff	4.0	$199,500	$80,700[a]	Esry and Cairns, 1989
Santee Marsh	California	wastewater treatment	0.1	$1,820,000	$737,000[b]	Gersberg et al., 1989
Iselin Marsh/Pond/Meadow	Pennsylvania	wastewater treatment	0.2	$2,080,000	$842,000[b]	Conway and Murtha, 1989
Pintail Lake	Arizona	wastewater treatment	20.2	$73,800	$30,000	Wilhelm et al., 1989
Jacques Marsh	Arizona	wastewater treatment	18.0	$75,300	$30,500	Wilhelm et al., 1989
Kash Creek (Impoundment 3)	Alabama	acid mine drainage	0.4	$84,200	$34,000	Brodie et al., 1989a
SIMCO Mine	Ohio	acid mine drainage	0.2	$480,000	$194,000	Kolbash and Murphy, unpub. data
Widows Creek Steam Plant	Alabama	ash pond seepage	0.5	$69,800	$28,200	Brodie et al., 1989b
Kingston	Alabama	ash pond seepage	0.9	$142,100	$57,500	Brodie et al., 1989b
Bolivar Peninsula	Texas	disposal site for dredge	8.0	$34,100	$13,800	Newling, 1982
Windmill Point	Virginia	disposal site for dredge	8.0	$25,300	$10,300	Newling, 1982
Blue River Reclamation Project	Colorado	riparian restoration	12.0	$41,300	$16,700	Roesser, 1988
		Average		**$374,800**	**$152,000**	
		Median		**$74,500**	**$30,200**	

[a]includes area on which an impoundment and filter were built
[b]cost reflects entirely artificial wetland and includes a good deal of plumbing
Source: Mitsch and Cronk, 1992

is to "accept the [sediment] accumulation as a natural part of wetland dynamics" (Willard et al., 1989).

Economics

Construction Costs

The construction of a new wetland involves careful consideration of a number of criteria, including a realistic look at cost. The amount of funding available, the period of time for which it is available, and the limits and rules concerning its expenditure are questions to be dealt with early in the planning stages of a constructed wetland (Newling, 1982). Tomljanovich and Perez (1989) recommended that an estimate of the cost of a new wetland's development should include the following items:

1. engineering plan;
2. preconstruction site preparation;
3. construction (labor, equipment, materials, supervision, indirect and overhead charges).

To this list should be added the cost of adequate monitoring to determine whether construction objective are being met.

The cost of wetland construction varies widely and depends on location, type, and objectives of the wetland as well as the maintenance required (Table 17–6). Factors that add to the cost variation include access to the site, substrate characteristics, cost of protective structures, local labor rates, and the availability of equipment (Newling, 1982). Compared to many other systems of water-quality improvement, wetlands are relatively inexpensive to build. As the data in Table 17–6 indicate, however, some wetlands that require human and technological intervention such as the Santee Marsh in California and the Iselin wastewater wetland plant in Pennsylvania are extremely costly to construct. The Pintail Lake and Jacques Marsh sites in Arizona were fairly inexpensive to build because they were constructed in preexisting but dry lake basins. Digging and basin formation were not necessary at those sites, and the natural formations helped to bring down the construction costs (Wilhelm et al., 1989).

Operational Costs

Operating and maintenance costs vary according to the wetland's use and to the amount and complexity of mechanical parts and plumbing that the wetland contains. Fewer data on these operational costs are available. A pump, filter, impoundment tank, and piping add considerably to both the construction and the maintenance costs of a wetland. Wetlands fed by natural runoff or by water that enters the site from adjacent waterways using only the force of gravity are far less expensive to maintain than highly mechanized wetlands.

Classification and Inventory of Wetlands

18

Wetlands have been classified since the early 1900s, beginning with the peatland classifications of Europe and North America. Regional wetland classifications and inventories have been developed for several states. Some of these have classified wetlands according to their vegetative life forms, whereas others also use the hydrologic regime. Classifications based on environmental forcing functions, particularly hydrologic flow, offer a promising approach. More recently, classifications based on wetland function and value have been explored.

The U.S. Fish and Wildlife Service has been involved in two major wetland classifications and inventories, one completed in 1954, and one begun in the mid-1970s but not yet completed for the entire United States. The early classification described 20 wetland types based on flooding depth, dominant forms of vegetation, and salinity regimes. "Classification of Wetlands and Deepwater Habitats of the United States" uses a hierarchical approach based on systems, subsystems, classes, subclasses, dominance types, and special modifiers to define wetlands and deepwater habitats precisely. Wetland inventories are carried out at many different scales with several different imageries and with both aircraft and satellite platforms. Eventually wetlands in the entire United States will be mapped at a scale of 1:24,000 for most areas and at 1:100,000 or smaller for a few areas.

In order to deal realistically with wetlands on a regional scale, wetland managers have found it necessary both to define the different types of wetlands that exist and to determine their extent and distribution. Those activities, the first called *wetland classification*, and the second called a *wetland inventory,* are valuable undertakings for both the wetland scientist and the wetland manager. Classifications and inventories of wetlands have been developed for the entire United States, for Canada, and, in individual states, provinces, and regions for many purposes over the past century. Some of the earliest efforts were undertaken for the purpose of finding wetlands that could be drained for human use; later classifications and inventories centered on the desire to compare different types of wetlands in a given region, often for their value to waterfowl. The protection of multiple ecological values of wetlands is the most recent purpose and now the most common reason for wetland classification and inventory. Recognition of wetland "value" has led some to now seek wetland classifications based on priorities for protection, with highest protection afforded to those wetlands with the greatest value. Like other techniques, classifications and inventories are valuable only when the user is familiar with their scope and limitations.

GOALS OF WETLAND CLASSIFICATION

Several attempts have been made to classify wetlands into categories that follow their structural and functional characteristics. These classifications depend on a well-understood general definition of wetlands (see Chap. 2), although a classification contains definitions of individual wetland types. A primary goal of wetland classifications, according to Cowardin et al. (1979), "is to impose boundaries on natural ecosystems for the purposes of inventory, evaluation, and management." These authors identified four major objectives of a classification system:

1. to describe ecological units that have certain homogeneous natural attributes;
2. to arrange these units in a system that will aid decisions about resource management;
3. to identify classification units for inventory and mapping;
4. to provide uniformity in concepts and terminology.

The first objective deals with the important task of grouping ecosystems that have similar characteristics in much the same way that taxonomists categorize species in taxonomic groupings. The wetland attributes that are frequently used to group and compare wetlands include vegetation type, plant or animal species, or hydrologic conditions.

The second objective, to aid wetland managers, can be met in several ways when wetlands are classified. Classifications (which are definitions of different types of wetlands) enable the wetland manager to deal with wetland regulation and protection in a consistent manner from region to region and from one time to the next. Classifications also enable the wetland manager to pay selectively more attention to those types of wetlands that are most threatened or functionally the most valuable to a given region.

The third and fourth objectives, to provide consistency in the formulation and use of inventories, mapping, concepts, and terminology, are also important in wetland management. The use of consistent terms to define particular types of wetlands is needed in the field of wetland science (see Chap. 3). These terms should then be applied uniformly to wetland inventories and mapping so that different regions can be compared and so that there will be a common understanding of wetland types among wetland scientists, wetland managers, and wetland owners.

HISTORY OF WETLAND CLASSIFICATION

Peatland Classifications

Many of the earliest wetland classifications were undertaken for the northern peatlands of Europe and North America. An early peatland classification in the United States, developed by Davis (1907), described Michigan bogs according to three criteria: (1) the landform on which the bog was established such as shallow lake basins or deltas of streams; (2) the method by which the bog was developed such as from the bottom up or from the shores inward; and (3) the surface vegetation such as tamarack or mosses. Based on the work of Weber (1908), Potonie (1908), Kulczynski (1949), and others in Europe, Moore and Bellamy (1974)

Table 18–1. Hydrologic Classification of European Peatlands

A. *Rheophilous Mire*—Peatland influenced by groundwater derived from outside the immediate watershed

- Type 1—Continuously flowing water that inundates the peatland surface
- Type 2—Continuously flowing water beneath a floating mat of vegetation
- Type 3—Intermittent flow inundating the mire surface
- Type 4—Intermittent flow of water beneath a floating mat of vegetation

B. *Transition Mire*—Peatland influenced by groundwater derived solely from the immediate watershed

- Type 5—Continuous flow of water
- Type 6—Intermittent flow of water

C. *Ombrophilous Mire*

- Type 7—Peatland never subject for flowing groundwater

Source: After Moore and Bellamy, 1974, based on Bellamy, 1968

described seven types of peatlands based on flow-through conditions (Table 18–1). Three general categories, called rheophilous, transition, and ombrophilous, describe the degree to which peatlands are influenced by outside drainage. These categories are discussed in more detail in Chapter 12 in the more modern terminology of minerotrophic, transition, and ombotrophic peatlands.

Most peatlands, of course, are limited to northern temperate climes and do not include all or even most types of wetlands in North America. These classifications, however, served as models for more inclusive classifications. They are significant because they combined the chemical and physical conditions of the wetland with the vegetation description to present a balanced approach to wetland classification.

Circular 39 Classification

In the early 1950s, the U.S. Fish and Wildlife Service recognized the need for a national wetlands inventory to determine "the distribution, extent, and quality of the remaining wetlands in relation to their value as wildlife habitat" (Shaw and Fredine, 1956). A classification was developed for that inventory (Martin et al., 1953), and the results of both the inventory and the classification scheme were published in U.S. Fish and Wildlife Circular No. 39 (Shaw and Fredine, 1956). Twenty types of wetlands were described under four major categories:

1. Inland fresh areas
2. Inland saline areas
3. Coastal freshwater areas
4. Coastal saline areas

In each of the four categories, the wetlands were arranged in order of increasing water depth or frequency of inundation. A brief description of the site characteristics of the 20 wetland types is given in Table 18–2.

Types 1 through 8 are freshwater wetlands that include bottomland hardwood forests (Type 1), infrequently flooded meadows (Type 2), freshwater nontidal marshes (Types 3 and 4), open water less than 2 m deep (Type 5), shrub-scrub swamps (Type 6), forested swamps (Type 7), and bogs (Type 8). Types 9 through 11 are inland wetlands that have saline soils. They are defined according to the degree of flooding. Types 12 through 14 are wetlands that, although freshwater, are close enough to the coast to be influenced by tides. Types 15 through 20 are coastal saline wetlands that are influenced by both salt water and tidal action. These include salt flats and meadows (Types 15 and 16), true salt marshes (Types 17 and 18), open bays (Type 19), and mangrove swamps (Type 20).

This wetland classification was the most widely used in the United States until 1979, when the present National Wetlands Inventory classification was

Table 18–2. Circular 39 Wetland Classification by U.S. Fish and Wildlife Service

Type Number	*Wetland Type*	*Site Characteristics*
	Inland Fresh Areas	
1.	Seasonally flooded basins or flats	Soil covered with water or waterlogged during variable periods, but well drained during much of the growing season; in upland depressions and bottomlands
2.	Fresh meadows	Without standing water during growing season; waterlogged to within a few inches of surface
3.	Shallow fresh marshes	Soil waterlogged during growing season; often covered with 15 cm or more of water
4.	Deep fresh marshes	Soil covered with 15 cm to 1 m of water
5.	Open fresh water	Water less than 2 m deep
6.	Shrub swamps	Soil waterlogged; often covered with 15 cm or more of water
7.	Wooded swamps	Soil waterlogged; often covered with 30 cm of water; along sluggish streams, flat uplands, shallow lake basins
8.	Bogs	Soil waterlogged; spongy covering of mosses
	Inland Saline Areas	
9.	Saline flats	Flooded after periods of heavy precipitation; waterlogged within few cm of surface during the growing season
10.	Saline marshes	Soil waterlogged during growing season; often covered with 0.7 to 1 m of water; shallow lake basins
11.	Open saline water	Permanent areas of shallow saline water; depth variable
	Coastal Fresh Areas	
12.	Shallow fresh marshes	Soil waterlogged during growing season; at high tide as much as 15 cm of water; on landward side, deep marshes along tidal rivers, sounds, deltas
13.	Deep fresh marshes	At high tide covered with 15 cm to 1 m water; along tidal rivers and bays
14.	Open fresh water	Shallow portions of open water along fresh tidal rivers and sounds
	Coastal Saline Areas	
15.	Salt flats	Soil waterlogged during growing season; sites occasionally to fairly regularly covered by high tide; landward sides or islands within salt meadows and marshes
16.	Salt meadows	Soil waterlogged during growing season; rarely covered with tide water; landward side of salt marshes
17.	Irregularly flooded salt marshes	Covered by wind tides at irregular intervals during the growing season; along shores of nearly enclosed bays, sounds, etc.

continued

Table 18–2 continued

Type Number	Wetland Type	Site Characteristics
18.	Regularly flooded salt marshes	Covered at average high tide with 15 cm or more of water; along open ocean and along sounds
19.	Sounds and bays	Portions of saltwater sounds and bays shallow enough to be diked and filled; all water landward from average low-tide line
20.	Mangrove swamps	Soil covered at average high tide with 15 cm to 1 m of water; along coast of southern Florida

Source: After R. L. Smith, 1980 after Shaw and Fredine, 1956

Table 18–3. Classification of Natural Ponds and Lakes in the Glaciated Prairie Region

Class	Circular 39 Type
I—Ephemeral Ponds	1
II—Temporary Ponds	1, 2
III—Seasonal Ponds and Lakes	3, 4
IV—Semipermanent Ponds and Lakes	3, 4, 5, 10, 11
Subclasses A and B—cover type 1	3
Subclasses A and B—cover type 2	4
Subclasses A and B—cover types 3 and 4	4, 5
Subclasses C, D, and E—cover types 1 and 2	10
Subclasses C, D, and E—cover types 3 and 4	11
V—Permanent Ponds and Lakes	5, 11
Subclass B—cover types 3 and 4	5
Subclasses C, D, and E—cover types 3 and 4	11
VI—Alkali Ponds and Lakes	9
VII—Fen (alkaline bog) ponds	8

[a]Compared to Circular 39 wetland classification, Shaw and Fredine, 1956 (see Table 18–2)
Source: After Stewart and Kantrud, 1971

adopted. It is still referred to today by some wetland managers and is regarded by many as elegantly simple compared with its successor. It primarily used the life forms of vegetation and the depth of flooding to identify of wetland type. Salinity was the only chemical parameter used, and although wetland soils were addressed in the Circular 39 publication, they were not used to define wetland types.

Adaptations to Circular 39

Two additional attempts to classify wetlands using categories similar to those in Circular 39 are noteworthy. The prairie potholes of the upper Midwest were included in a classification by Stewart and Kantrud (1971, 1972). A list of the classes used and a comparison with Circular 39 types are shown in Table 18–3.

Table 18–4. Classes and Subclasses of Freshwater Wetlands in the Glaciated Northeastern United States

Wetland Class	*Wetland Subclass*	
Open Water	(OW-1)	Vegetated
	(OW-2)	Nonvegetated
Deep Marsh	(DM-1)	Dead woody
	(DM-2)	Shrub
	(DM-3)	Sub-shrub
	(DM-4)	Robust
	(DM-5)	Narrow-leaved
	(DM-6)	Broad-leaved
Shallow Marsh	(SM-1)	Robust
	(SM-2)	Narrow-leaved
	(SM-3)	Broad-leaved
	(SM-4)	Floating-leaved
Seasonally Flooded Flats	(SF-1)	Emergent
	(SF-2)	Shrub
Meadow	(M-1)	Ungrazed
	(M-2)	Grazed
Shrub Swamp	(SS-1)	Sapling
	(SS-2)	Bushy
	(SS-3)	Compact
	(SS-4)	Aquatic
Wooded Swamp	(WS-1)	Deciduous
	(WS-2)	Evergreen
Bog	(BG-1)	Shrub
	(BG-2)	Wooded

Source: From Golet and Larson, 1974

Seven classes were used, all based on vegetation zones that occupied the deepest portion of each pothole basin. Each zone, in turn, was defined according to its hydrologic characteristics (e.g., temporary pond). Six subclasses denoted variations in salinity, and four cover types referred to the spatial patterns of the emergent vegetation.

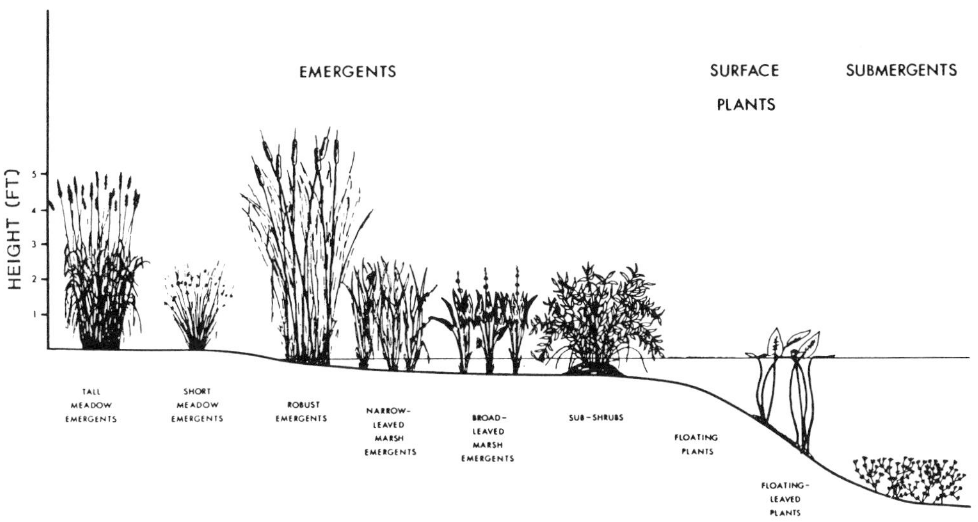

Figure 18–1. Subforms of vegetation used to classify wetlands of glaciated northeastern United States as described in Table 18–4. (*From Golet and Larson, 1974*)

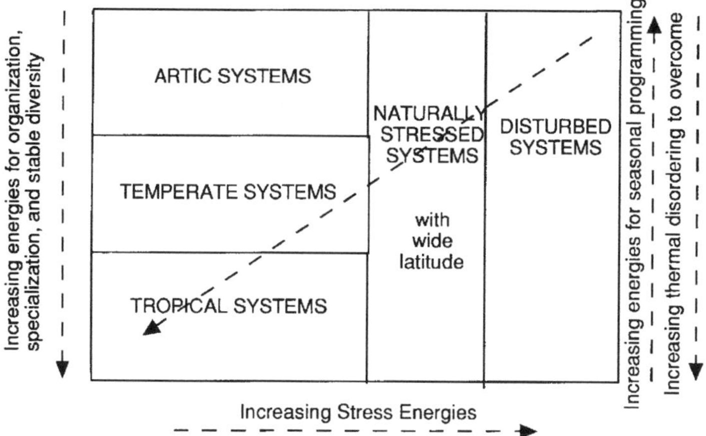

Figure 18–2. Coastal ecosystem classification system based on latitude (and hence solar energy) and major stresses. (*After H. T. Odum et al., 1974*)

Golet and Larson (1974) detailed a freshwater wetland classification for the glaciated northeastern United States based on categories similar to those in Circular 39, Types 1 through 8 (Table 18–4). This classification introduced 24 subclasses based on 18 possible life forms, as shown in Figure 18–1. In addition, five wetland size categories, seven site types, eight cover types, three vegetative interspersion types, and six surrounding habitat types were used to refine the definition of wetlands.

Coastal Wetland Classification

A classification and functional description of coastal ecosystems, was developed by H. T. Odum et al. (1974) in "Coastal Ecological Systems of the United States." The significance of this classification was its approach to categorizing ecosystems according to their major forcing functions (e.g., seasonal programming of sunlight and temperature) and stresses (e.g., ice) (Figure 18–2). Coastal wetland types in this classification include salt marshes and mangrove swamps. Salt marshes, found in the Type C category of natural temperate ecosystems with seasonal programming, have "light tidal regimes" and "winter cold" as forcing function and stress, respectively. Mangrove swamps are classified as Type B (natural tropical ecosystems) because they have abundant light, show little stress, and reflect little seasonal programming. Three additional classes, Type A (naturally stressed systems of wide latitudinal range), Type D (natural arctic ecosystems with ice stress), and Type E (emerging new systems associated with man), were included in this classification. The last class, which includes new systems formed by pollution such as pesticides and oil spills, is still an interesting concept that could be applied to other wetland classifications.

Table 18–5. Classification of Forested Wetlands in Florida

1. *Cypress Ponds (Domes)—Still-Water*
   - Acid water ponds
   - Hard water ponds
   - Pasture ponds
   - Enriched ponds
2. *Other Nonstream Swamps*
   - Gum pond (swamp)
   - Lake border swamp
   - Dwarf cypress
   - Bog swamp (Okefenokee Swamp)
   - Bay swamp
   - Shrub bog
   - Herb bog
   - Seepage swamp
   - Hydric hammock (North Florida type)
   - South Florida hammock
   - Melaleuca swamp
3. *Cypress Strand—Slowly Flowing Water*
4. *River Swamps and Floodplains*
   - Alluvial river swamps
   - Blackwater river and creek swamps
   - Backswamp
   - Spring run swamp
   - Tidewater swamp
5. *Saltwater Swamps—Mangroves*
   - Riverine black mangroves
   - Fringe red mangroves
   - Overwash red mangroves
   - Scrub mangroves

Source: After Wharton et al., 1976

Florida Wetland Classification

Other classifications of wetlands have further developed the approach of using the forcing functions to define wetlands. Wharton et al. (1976) described the forested swamps of Florida according their hydrologic inputs. The major types in this classification are given in Table 18–5. The wetlands are arranged according to water flow. The first two classes, "cypress ponds" and "other nonstream swamps," involve wetlands that experience little water inflow except precipitation and groundwater in some cases. The third class, "cypress strands," includes slowly flowing water typical of the southern Florida river basins, whereas "river swamps and floodplains" are more typical of continuously flooded alluvial swamps and periodically flooded riparian forests. The fifth class, "saltwater

Table 18–6. Freshwater Classification According to Hydrologic Regime

Wetland Type	*Hydropulse*	*Example*
1. Raised—Convex	Seasonal precipitation and capillarity	Ombrotrophic bog
2. Meadow	Seasonal precipitation, capillarity; little upstream inflow	Blanket bog
3. Sunken—Concave	Seasonal precipitation and upstream inflow	Fen
4. Lotic	Seasonal precipitation, runoff, groundwater, and flowthrough	Fen; reed marsh
5. Tidal	Tides	Salt marsh
6. Lentic	Variable or seasonal Overbank flooding	Riparian wetland

Source: After Gosselink and Turner, 1978

swamps—mangroves," are those forested wetlands that are affected by tides and salt water.

Hydrodynamic Energy Gradient Classification

A similar approach to classifying wetlands according to their sources and velocity of water flow was developed as a "Classification of Wetland Systems on a Hydrodynamic Energy Gradient" by Gosselink and Turner (1978). The classification is summarized in Table 18–6. As described by the authors, "In general flow rate, or other indications of hydrologic energy such as renewal time or frequency of flooding, increases from raised convex wetlands to lentic and tidal wetlands." Raised-convex systems are primarily ombrotrophic bogs that are fed only by precipitation and capillary action, whereas lentic wetlands include riparian bottomlands that receive seasonal flooding pulses. This classification, although applied only to nonforested freshwater wetlands, has applicability to wetlands in general. It offers a useful approach toward incorporating wetland functions into a wetland classification. Classification schemes based on functional processes, however, are difficult to delimit because they are not necessarily identified by different vegetation associations or other easily mapped features. Hence these classification schemes are seldom used for inventory purposes.

WETLAND CLASSIFICATION FOR VALUE

More recently there has been considerable interest in classifying wetlands for their value to society in order to simplify the issuance of permits for wetlands activities; that is, high-value wetlands would presumably receive more protection than low-value wetlands. Figure 18–3 illustrates some of the difficulties of using this approach to classify a bottomland forest riparian system. We pose the

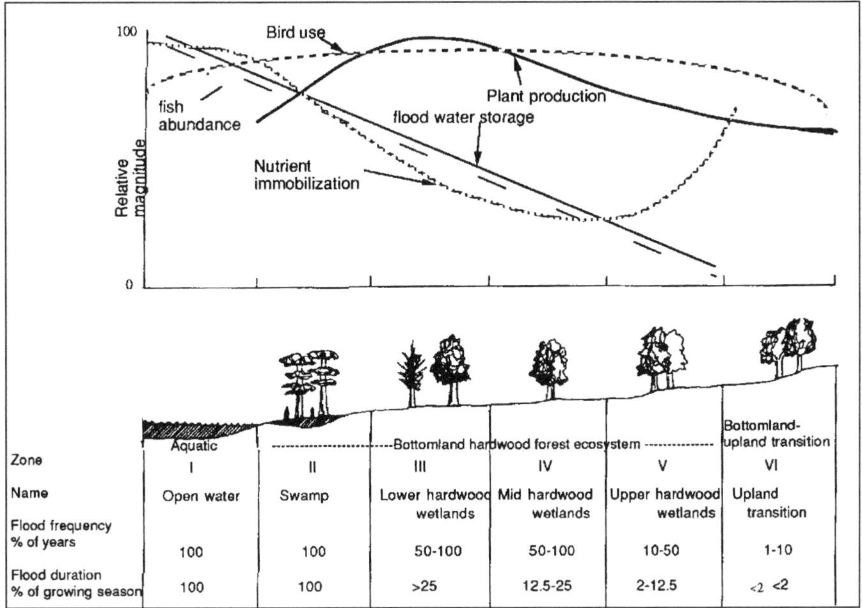

Figure 18–3. A simplified representation of a section across a bottomland hardwood forest from stream to upland, showing how various functions of interest to humans change across the transect. In actuality, the area is a complex spatial pattern of intermixed zones, not a linear gradient. This is a multiple value resource; different ecological processes peak in different zones, and these peaks are not directly related to their value to humans.

question of whether one can classify this floodplain into high-value wetlands and wetlands of lesser value. This is not a trivial question because the wetlands status and hence the protection of the upper end of the bottomland zone is an issue of serious concern to developers, farmers, and agency regulators. Floodplains are multiple-value resources; as Figure 18–3 shows, different ecological processes peak in different zones, and those peaks are not directly related to their value to humans. For example, although most flood water is stored at low elevations in the floodplain, the highest zone is important because it helps moderate large infrequent floods that do the most damage. Because tree growth rates are highest in the seasonally flooded zone (III), that zone might be considered most valuable for timber harvest. In fact, however, cypress, which is found primarily in zone II, is a more valuable timber crop because of its superior rot and insect resistance. The example illustrates some of the difficulties of classifying wetlands according to value:

1. A wetland is a multiple-value system (see Chap. 15). Any categorization by relative value implies that priorities have been set that trade off one value for

another. Who makes these decisions? They are not technical decisions; rather, they involve human and societal judgments.
2. Value is not linearly related to function. It is influenced by human perceptions that may change as development pressure, population density, political interests, and other human factors change.
3. Wetland value classification involves, either implicitly or explicitly, some sort of risk assessment, for example, the weight to give to a wetland's value to moderate rare but severe floods.
4. Categorization by value also involves decisions about scale. Conventional economics places a higher value on 1 hectare in a 10-hectare forest than on 1 hectare in a 100,000-hectare forest. But for a bear, a 10-hectare forest patch is useless: It cannot survive in such a small area. If a bear population is a desired component of a wetland system, it is necessary to elevate the priority assigned to large, unbroken forest tracts.

Because of the complexity of ecological patterns and functions, the classification of wetlands involves tradeoffs among competing social values that imply an earlier commitment to certain goals (for example, to farming over conservation, or to hunting over fishing).

PRESENT UNITED STATES WETLAND CLASSIFICATION

The U.S. Fish and Wildlife Service began a rigorous wetland inventory of the nation's wetlands in 1974. Because this inventory was designed to fulfill several scientific and management objectives, a new classification scheme, broader than Circular 39, was developed and finally published in 1979 as a "Classification of Wetlands and Deepwater Habitats of the United States" (Cowardin et al., 1979). Because wetlands were found to be continuous with deepwater ecosystems, both categories were addressed in this classification. It is thus a comprehensive classification of all continental aquatic and semi-aquatic ecosystems. As described in that publication,

> This classification, to be used in a new inventory of wetlands and deepwater habitats of the United States, is intended to describe ecological taxa, arrange them in a system useful to resource managers, furnish units for mapping, and provide uniformity of concepts and terms. Wetlands are defined by plants (hydrophytes), soils (hydric soils), and frequency of flooding. Ecologically related areas of deep water, traditionally not considered wetlands, are included in the classification as deepwater habitats.

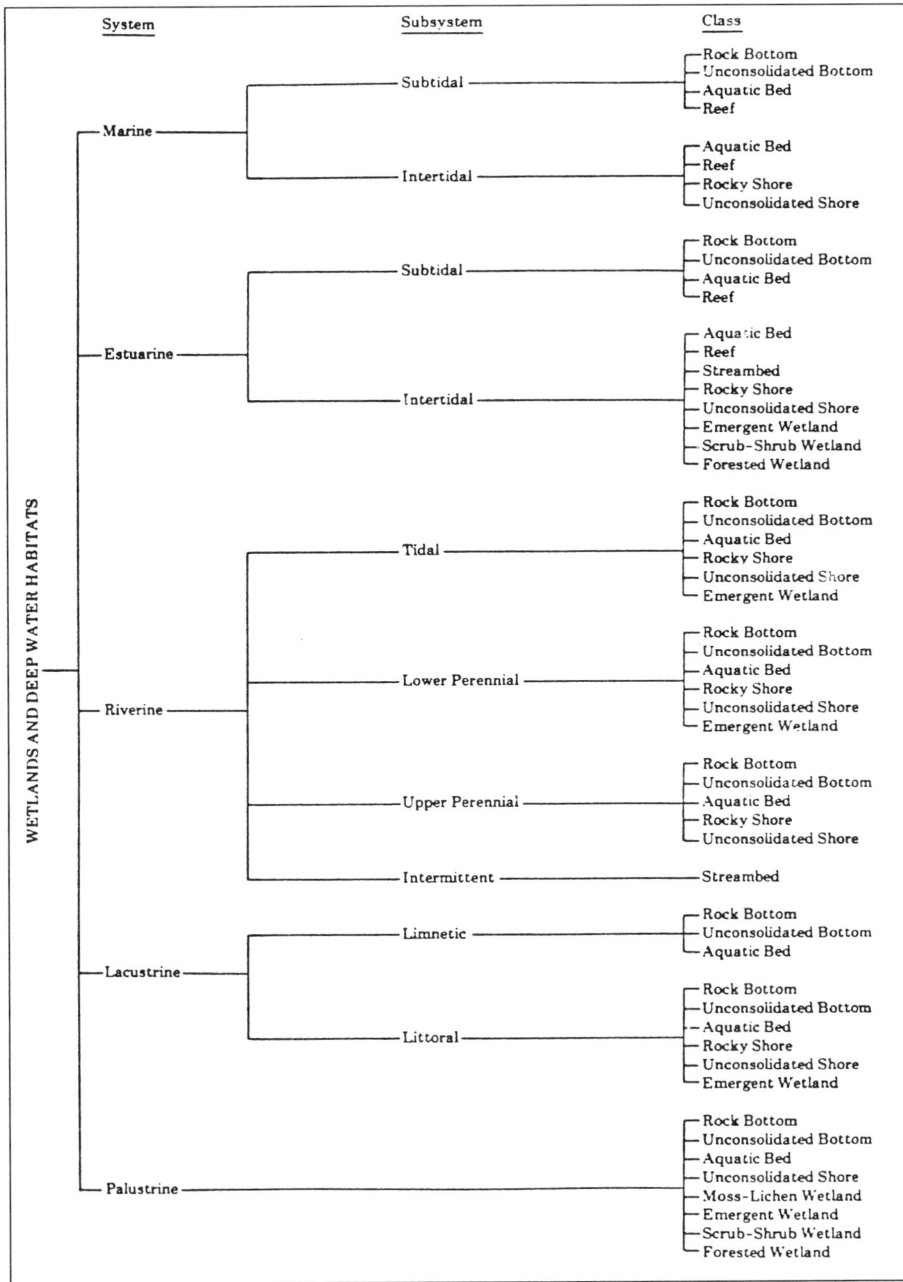

Figure 18–4. Wetland and deepwater habitat classification heirarchy showing system, subsystems, and classes. (*From Cowardin et al., 1979*)

This classification is based on a hierarchical approach analogous to taxonomic classifications used to identify plant and animal species. The first three levels of the classification hierarchy are given in Figure 18–4. The broadest level is *systems*: "a complex of wetlands and deepwater habitats that share the influence of similar hydrologic, geomorphologic, chemical, or biological factors." These broad categories include the following:

1. *Marine*—open ocean overlying the continental shelf and its associated high-energy coastal line.
2. *Estuarine*—deepwater tidal habitats and adjacent tidal wetlands that are usually semi-enclosed by land but have open, partially obstructed, or sporadic access to the ocean and in which ocean water is at least occasionally diluted by freshwater runoff from the land.
3. *Riverine*—wetland and deepwater habitats contained within a channel with two exceptions: (1) wetlands dominated by trees, shrubs, persistent emergents, emergent mosses, or lichens, and (2) habitats with water containing ocean-derived salts in excess of 0.5 parts per thousand.
4. *Lacustrine*—wetlands and deepwater habitats with all of the following characteristics: (1) situated in a topographic depression or a dammed river channel; (2) lacking trees, shrubs, persistent emergents, emergent mosses, or lichens with greater than 30 percent areal coverage; and (3) total area exceeds 8 hectare (20 acres). Similar wetland and deepwater habitats totaling less than 8 hectare are also included in the Lacustrine system when an active wave-formed or bedrock shoreline feature makes up all or part of the boundary or when the depth in the deepest part of the basin exceeds 2 m (6.6 feet) at low water.
5. *Palustrine*—All nontidal wetlands dominated by trees, shrubs, persistent emergents, emergent mosses or lichens, and all such wetlands that occur in tidal areas where salinity stemming from ocean-derived salts is below 0.5 parts per thousand. It also includes wetlands lacking such vegetation but with all of the following characteristics: (1) area less than 8 hectares; (2) lack of active wave-formed or bedrock shoreline features; (3) water depth in the deepest part of the basin of less than 2 m at low water; (4) salinity stemming from ocean-derived salts of less than 0.5 parts per thousand.

Several *subsystems*, as shown in Figure 18–4, give further definition to the systems. These include the following:

1. *subtidal*—the substrate is continuously submerged;
2. *intertidal*—the substrate is exposed and flooded by tides; this includes the splash zone;
3. *tidal*—for riverine systems, the gradient is low and the water velocity fluctuates under tidal influence.
4. *lower perennial*—riverine systems with continuous flow, low gradient, and no tidal influence;

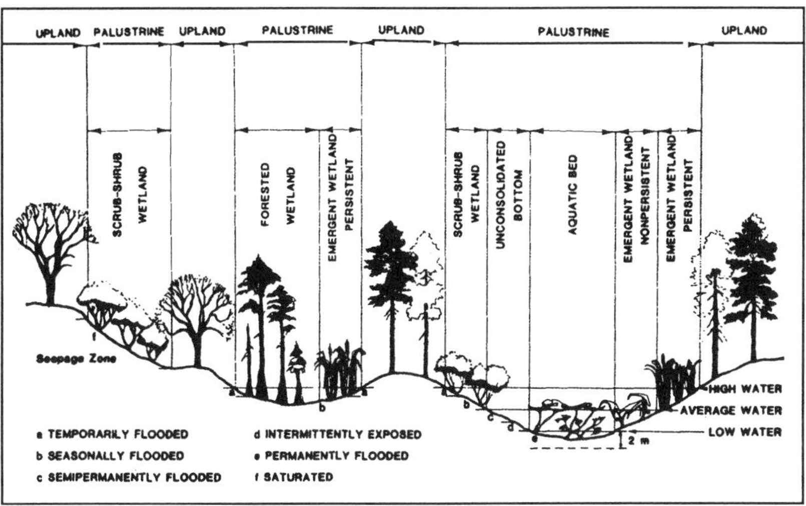

Figure 18–5. Features and examples of wetland classes and hydrologic modifiers in the Palustrine System. (*From Cowardin et al., 1979*)

5. *upper perennial*—riverine systems with continuous flow, high gradient, and no tidal influence;
6. *intermittent*—riverine systems in which water does not flow for part of the year;
7. *limnetic*—all deepwater habitats in lakes;
8. *littoral*—wetland habitats of a lacustrine system that extends from shore to a depth of 2 m below low water or to the maximum extent of nonpersistent emergent plants.

The *class* of a particular wetland or deepwater habitat describes the general appearance of the ecosystem in terms of either the dominant vegetation or the substrate type. When more than 30 percent cover by vegetation is present, a vegetation class is used (e.g., shrub-scrub wetland). When less than 30 percent of the substrate is covered by vegetation, then a substrate class is used (e.g., unconsolidated bottom). Definitions and examples of most of the classes in this classification system are given in Table 18–7. The typical demarcation of many of the classes of palustrine systems is shown in Figure 18–5.

Most inland wetlands fall into the Palustrine system, in the classes moss-lichen, emergent, scrub-shrub, or forested wetland. Coastal wetlands are classified in the same classes within the estuarine system, and intertidal subsystem. Only nonpersistent emergent wetlands are classified into other systems.

Further descriptions of the wetlands and deepwater habitats are possible through the use of *subclasses, dominance types*, and *modifiers*. Subclasses such as "persistent" and "nonpersistent" give further definition to a class such

Table 18–7. Classes, Subclasses, and Examples of Dominance Types for Wetland and Deepwater Habitat Classification by U.S. Fish and Wildlife Service

Class		*Examples of Dominance Type*		
Subclass	*Definition*	*Marine/Estuarine*	*Lacustrine/Riverine*	*Palustrine*
Rock Bottom				
Bedrock	Bedrock covers 75% or more of surface	Lobster *(Homarus)*	Brook leech *(Helobdella)*	—
Rubble	Stones and boulders cover more than 75% of surface	Sponge *(Hippospongia)*	Chironomids	Water penny *(Psephenus)*
Unconsolidated Bottom				
Cobble-gravel	At least 25% of particles smaller than stones and less than 30% vegetation cover	Clam *(Mya)*	Mayfly *(Baetis)*	Oligochaete worms
Sand	At least 25% sand cover and less than 30% vegetation cover	Wedge shell *(Donax)*	Mayfly *(Ephemerella)*	Sponge *(Eunapius)*
Mud	At least 25% silt and clay, although coarser sediments can be intermixed; less than 25% vegetation	Scallop *(Placopecten)*	Freshwater mollusk *(Anodonta)*	Fingernail clam *(Pisidium)*
Organic	Unconsolidated material predominantly organic matter and less than 25% vegetation cover	Clam *(Mya)*	Sewage worm *(Tubifex)*	Oligochaete worms
Aquatic Bed				
Algal	Algae growing on or below surface of water	Kelp *(Macrocystis)*	Stonewort *(Chara)*	Stonewort *(Chara)*
Aquatic moss	Aquatic moss growing at or below the surface	—	Moss *(Fissidens)*	—
Rooted vascular	Rooted vascular plants growing submerged or as floating-leaved	Turtlegrass *(Thalassia)*	Water lily *(Nymphaea)*	Ditch grasses *(Ruppia)*
Floating vascular	Floating vascular plants growing on water surface	—	Water hyacinth *(Eichhornia crassipes)*	Duckweed *(Lemna)*

Table 18–7 continued

Class / *Subclass*	*Definition*	*Examples of Dominance Type*		
		Marine/Estuarine	*Lacustrine/Riverine*	*Palustrine*
Reef	Ridgelike or moundlike structures formed by sedentary invertebrates			
Coral		Coral *(Porites)*	—	—
Mollusk		Eastern oyster *(Crassostrea virginica)*	—	—
Worm		Reefworm *(Sabellaria)*	—	—
Streambed	Intermittent streams (riverine system) or systems completely dewatered at low tide			
Bedrock	Bedrock covers 75% or more of surface	—	Mayfly *(Ephemerella)*	—
Rubble	Stones, boulders, and bedrock cover more than 75% of channel	—	Fingernail clam *(Pisidium)*	—
Cobble-gravel	At least 25% of substrate smaller than stones	Blue mussel *(Mytilus)*	Snail *(Physa)*	—
Sand	Sand particles predominate	Ghost shrimp *(Callianassa)*	Snail *(Lymnea)*	—
Mud	Silt and clay predominate	Mud snail *(Nassarius)*	Crayfish *(Procambarus)*	—
Organic	Peat or muck predominates	Mussel *(Modiolus)*	Oligochaete worms	—

continued

Table 18–7 continued

Class / *Subclass*	*Definition*	*Examples of Dominance Type*		
		Marine/Estuarine	*Lacustrine/Riverine*	*Palustrine*
Rocky Shore	High-energy habitats that lie exposed to wind-driven waves or strong currents			
Bedrock	Bedrock covers 75% or more of surface	Acorn barnacle *(Chthamalus)*	Liverwort *(Marsupella)*	—
Rubble	Stones, boulders, and bedrock cover more than 75% of surface	Mussel *(Mytilus)*	Lichens	—
Unconsolidated Shore	Landforms such as beaches, bars, and flats that have less than 30% vegetation and are found adjacent to unconsolidated bottoms			
Cobble-gravel	At least 25% of particles smaller than stones	Periwinkle *(Littorina)*	Mollusk *(Elliptio)*	—
Sand	At least 25% sand	Wedge shell *(Donax)*	Fingernail clam *(Pisidium)*	—
Mud	At least 25% silt and clay	Fiddler crab *(Uca)*	Fingernail clam *(Pisidium)*	—
Organic	Unconsolidated material, predominantly organic matter	Fiddler crab *(Uca)*	Chironomids	—
Vegetated	Nontidal shores exposed for sufficient time to colonize annuals or perennials	—	Cocklebur *(Xanthium)*	Summer cypress *(Kochia)*
Moss-Lichen Wetland				
Moss	Mosses cover substrate other than rock; emergents, shrubs and trees cover less than 30% of area	—	—	Peatmoss *(Sphagnum)*
Lichen	Lichens cover substrate other than rock; emergents, shrubs and trees cover less than 30% of area	—	—	Reindeer moss *(Cladonia)*

continued

Table 18–7 continued

Class / *Subclass*	*Definition*	*Examples of Dominance Type*		
		Marine/Estuarine	*Lacustrine/Riverine*	*Palustrine*
Emergent Wetland	Erect, rooted, herbaceous aquatic plants			
Persistent	Species that normally remain standing until the beginning of the next growing season	Cordgrass *(Spartina)*	—	Cattail *(Typha)*
Nonpersistent	No obvious sign of emergent vegetation at certain seasons	Samphire *(Salicornia)*	Wild rice *(Zizania)*	Pickerelweed *(Pontederia)*
Scrub-Shrub Wetland	Dominated by wood vegetation less than 6 m tall			
Broad-leaved deciduous		Marsh elder *(Iva)*	—	Buttonbush *(Cephalanthus)*
Needle-leaved deciduous		—	—	Dwarf cypress *(Taxodium)*
Broad-leaved evergreen		Mangrove *(Rhizophora)*	—	Fetterbush *(Lyonia)*
Needle-leaved evergreen		—	—	Stunted pond pine *(Pinus serotina)*
Dead		—	—	—
Forested Wetland	Woody vegetation 6 m or taller			
Broad-leaved deciduous		—	—	Red maple *(Acer rubrum)*
Needle-leaved deciduous		—	—	Bald cypress *(Taxodium distichum)*
Broad-leaved evergreen		Mangrove *(Rhizophora)*	—	Red bay *(Persea)*
Needle-leaved evergreen		—	—	Northern white cedar *(Thuja occidentalis)*
Dead		—	—	—

Source: Cowardin et al., 1979

Table 18–8. Modifiers Used in Wetland and Deepwater Habitat Classification by U.S. Fish and Wildlife Service

Water Regime Modifiers (Tidal)

Subtidal—substrate permanently flooded with tidal water
Irregularly exposed—land surface exposed by tides less often than daily
Regularly flooded—alternately floods and exposes land surfaces at least daily
Irregularly flooded—land surface flooded less often than daily

Water Regime Modifiers (Nontidal)

Permanently flooded—water covers land surface throughout year in all years
Intermittently exposed—surface water present throughout year except in years of extreme drought
Semipermanently flooded—surface water persists throughout growing season in most years. When surface water is absent, water table is at or near surface
Seasonally flooded—surface water is present for extended periods, especially in early growing season but is absent by the end of the season
Saturated—substrate is saturated for extended periods during growing season but surface water is seldom present
Temporarily flooded—surface water is present for brief periods during growing season but water table is otherwise well below the soil surface
Intermittently flooded—substrate is usually exposed but surface water is present for variable periods with no seasonal periodicity

Salinity Modifiers

Marine and Estaurine	Riverine, Lacustrine, and Palustrine	Salinity, (parts per thousand)
Hyperhaline	Hypersaline	>40
Euhaline	Eusaline	30-40
Mixohaline (brackish)	Mixosaline	0.5-30
Polyhaline	Polysaline	18.0-30
Mesohaline	Mesosaline	5.0-18
Oligohaline	Oligosaline	0.5-5
Fresh	Fresh	<0.5

pH Modifiers

Acid	pH less than 5.5
Circumneutral	pH 5.5-7.4
Alkaline	pH greater than 7.4

Soil Material Modifiers[a]

Mineral
(1) Less than 20% organic carbon and never saturated with water for more than a few days, or
(2) Saturated or artificially drained and has
(a) less than 18% organic carbon if 60% or more is clay
(b) less than 12% organic carbon if no clay
(c) a proportional content of organic carbon between 12 and 18% if clay content is between 0 and 60%.

Organic
Other than mineral as described above.

[a]U.S. Soil Conservation Service, 1975
Source: Cowardin et al., 1979

as emergent vegetation (Table 18–7). Type refers to a particular dominant plant species (e.g., bald cypress, *Taxodium distichum*, for a needle-leaved deciduous forested wetland) or a dominant sedentary or sessile animal species (e.g., eastern oyster, *Crassostrea virginica*, for a mollusk reef). Modifiers (Table 18-8) are used after classes and subclasses to describe more precisely the water regime, the salinity, the pH, and the soil. For many wetlands, the description of the environmental modifiers adds a great deal of information about their physical and chemical characteristics. Unfortunately, those parameters are difficult to measure consistently in large-scale surveys such as inventories.

WETLAND INVENTORIES

Once wetlands have been defined and classified, an important question for management becomes "for any given region, how many and what types of wetlands are there?" Determining the extent of various types of wetlands in a region is accomplished by an inventory. An inventory can be made of a small watershed, a county or a parish, an entire state or a province, or an entire nation. Whatever the size of the area to be surveyed, the inventory must be based on some previously defined classification and should be constructed to meet the needs of specific users of information on wetlands.

Remote Sensing of Wetlands

Platform

The use of remote sensing platforms, particularly aircraft at various altitudes and satellites, has proved to be an effective way of gathering data for large-scale wetland surveys. The choice of which platform to use depends on the resolution required, the area to be covered, and the cost of the data collection. Low-altitude aircraft surveys offer a relatively inexpensive and fairly effective way to survey small areas. High-altitude aircraft offer much greater coverage in each scene (photograph) and may be less expensive per unit area than low-altitude aircraft when costs of photo interpretation are included.

A third alternative, monitoring wetlands from satellites in orbit, has potential and has been an alternative since the launching of the first of the LANDSAT satellites in 1972. Imagery from satellites covering most of the world is available at a reasonable cost. Wetland scientists working on the National Wetlands Inventory (see below) found, however, even for this large-scale inventory, that "LANDSAT could not provide the desired level of detail without what appeared to be an excessive amount of collateral data such as aerial photographs and field work" (Nyc and Brooks, 1979). A comparison between LANDSAT and high-

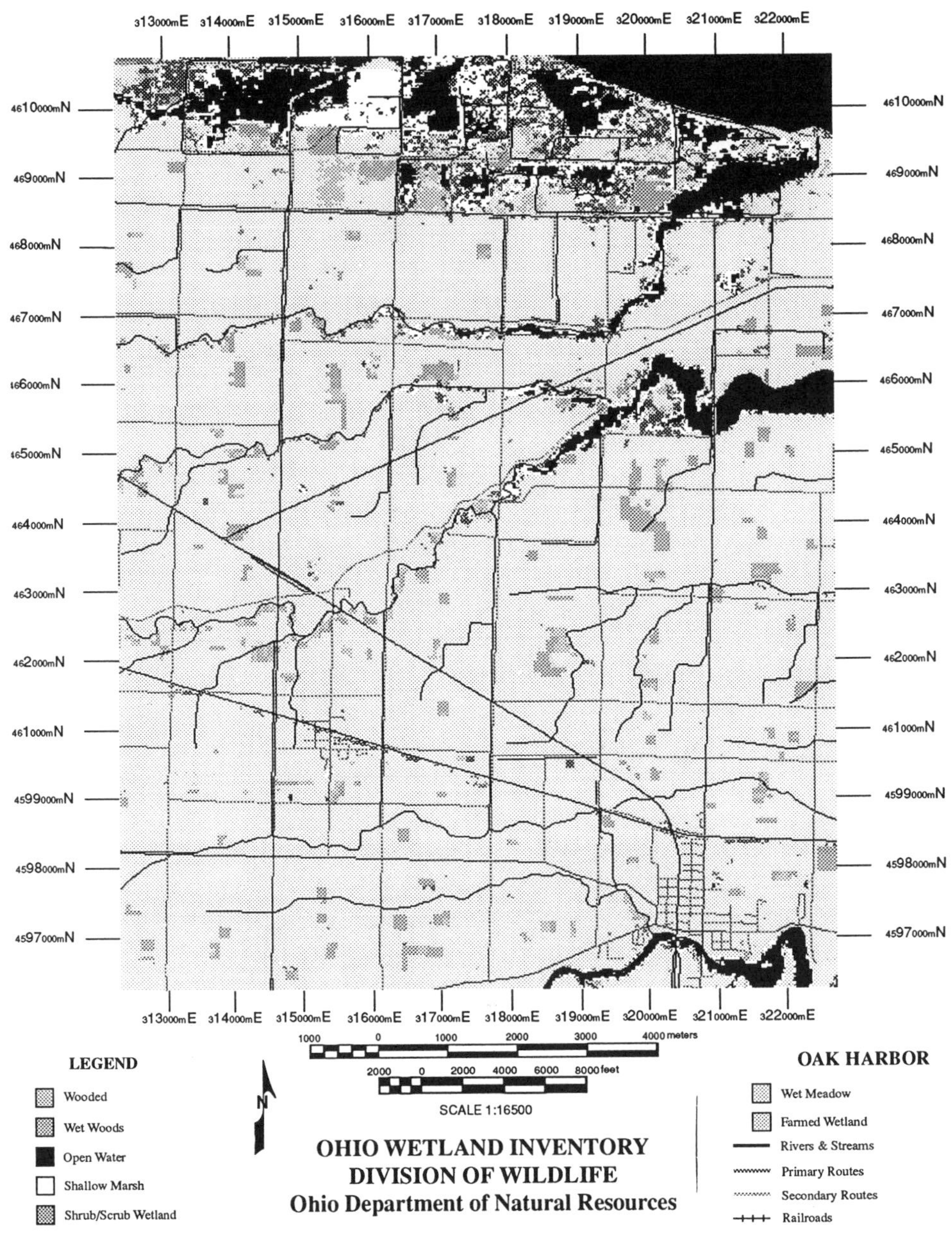
313000mE
314000mE
315000mE
316000mE
317000mE
318000mE
319000mE
320000mE
321000mE
322000mE
4610000mN
469000mN
468000mN
467000mN
466000mN
465000mN
464000mN
463000mN
462000mN
461000mN
4599000mN
4598000mN
4597000mN
1000 0 1000 2000 3000 4000 meters
2000 0 2000 4000 6000 8000 feet
SCALE 1:16500
LEGEND
Wooded
Wet Woods
Open Water
Shallow Marsh
Shrub/Scrub Wetland
N
OHIO WETLAND INVENTORY
DIVISION OF WILDLIFE
Ohio Department of Natural Resources
OAK HARBOR
Wet Meadow
Farmed Wetland
Rivers & Streams
Primary Routes
Secondary Routes
Railroads

altitude photography in the prairie pothole region revealed that 61 of the small pothole wetlands were not identified by LANDSAT and that only 3 wetland types could be identified, compared with 15 types by high-altitude photography. In other cases, satellite data, when used in conjunction with other data bases such as hydric soil maps, have provided a useful approach for inventorying wetlands in large areas. The state of Ohio, for example, has adopted that approach to inventory the 107,000 km^2 state (Fig. 18–6).

Imagery

In addition to choosing the remote sensing platform, the wetland scientist or manager has the choice of several types of imagery from different types of sensors. Color photography and color-infrared photography have been popular for many years for wetland inventories from aircraft, although black-and-white photography has been used with some success. R. R. Anderson (1969, 1971) experimented with a mix of photography and scanners to determine the best remote sensing imagery for the delineation of coastal salt marsh communities along the Patuxent River and the Chesapeake Bay. He found that color-infrared film provided good definition of plant communities and that infrared scanning was best for observing temperature differentials in the marsh. The Earth Satellite Corporation (1972) used both black–and–white–infrared and color-infrared to distinguish different *Spartina* communities in New Jersey's coastal wetlands.

High-resolution multispectral scanner imagery from low-altitude plane flights has also been used with some success to map wetlands. For example, Jensen et al. (1983, 1984) were able to map a nontidal wetland in South Carolina by using a classification scheme that had been modified slightly from the class and subclass components of the National Wetlands Inventory. They achieved a classification accuracy of 83 percent over a diverse 4,000 hectare area. Because of the potential for significant savings compared with manual mapping from aerial photographs, this could be a useful approach.

Second generation satellite imagery from the "Thematic Mapper" provides multispectral data at 10–30 m resolution (see Fig. 18–6). The choice of imagery, however, is governed in part by a balance between required resolution and the amount of data that can be processed. Improving resolution from 100 m to 30 m, for example, would increase data-processing requirements by 10 times for each channel processed. This requirement imposes both a spatial and a temporal limit on the use of satellite imagery. If the final maps could be used by public agencies and local landowners to delineate wetlands for regulatory purposes, the additional cost of increased resolution might be supportable.

Figure 18–6 (*opposite page*). Ohio wetland inventory map prepared by the Ohio Department of Natural Resources, Division of Wildlife, from supervised classification of LANDSAT 5 Thematic Mapper (TM) satellite imagery (30 m resolution), hydric soil data, and limited on-site evaluation. (*Courtesy of Ohio Department of Natural Resources, Division of Wildlife*)

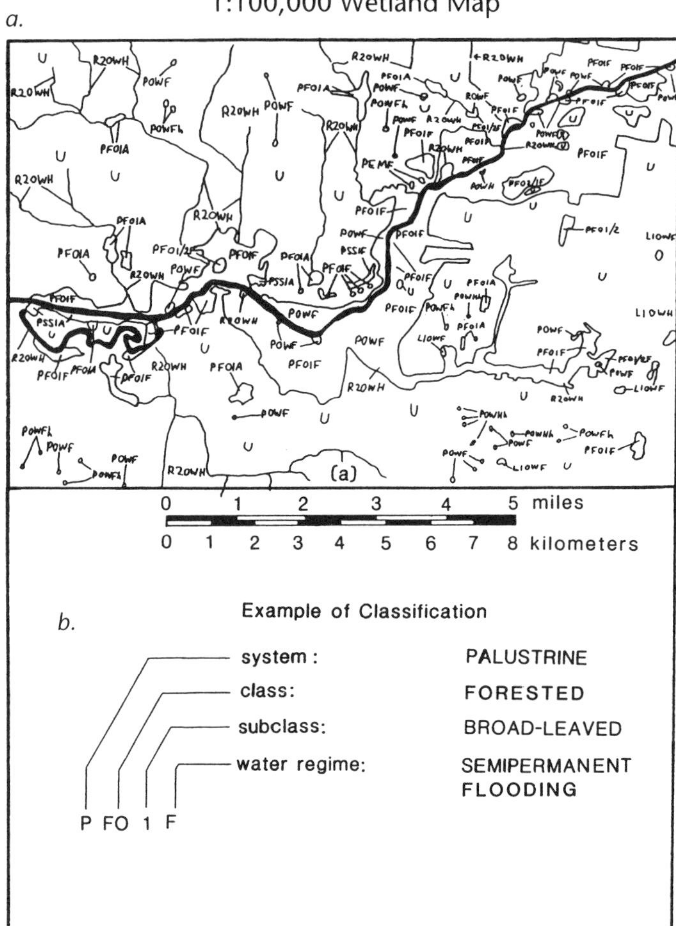

Figure 18–7. Sample of mapping technique used at 1:100,000 scale by National Wetlands Inventory showing *a.* portion of map (redrawn), and *b.* example of classification notation. (*After Dyersburg, Tennessee, 1:100,000 wetland map provided courtesy of National Wetlands Inventory, U.S. Fish and Wildlife Service*)

The National Wetlands Inventory

The Cowardin et al. (1979) classification scheme has provided the basic mapping units for the United States National Wetlands Inventory (NWI) being carried out by the U.S. Fish and Wildlife Service. For the National Wetlands Inventory, aerial photography at scales ranging from 1:60,000 to 1:130,000 is the primary source of data, with color-infrared photography providing the best delineation of

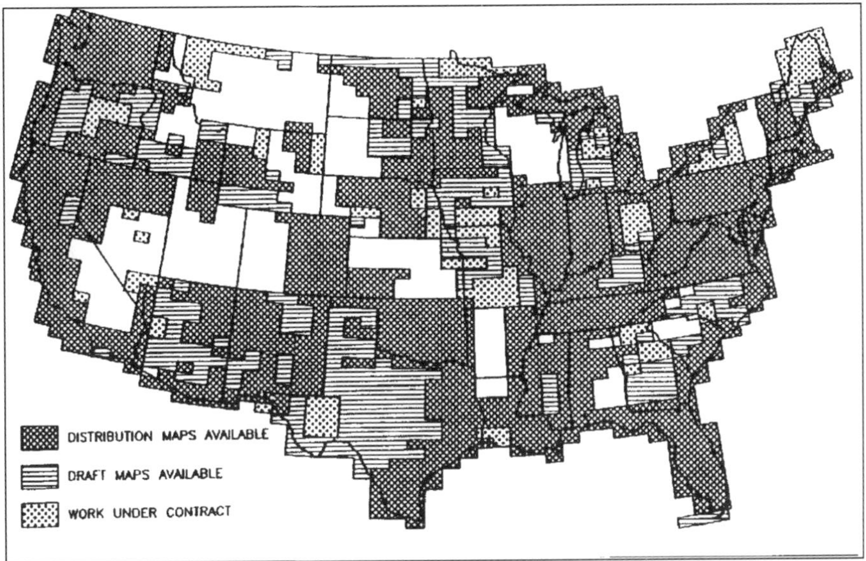

Figure 18–8. Status of National Wetlands Inventory, April 1991. (*From Wilen, 1991*)

wetlands (Wilen and Pywell, 1981; Tiner and Wilen, 1983). The National Wetlands Inventory is using 1:60,000 color infrared photography for most of its mapping. Photointerpretation and field reconnaissance are then used to define wetland boundaries according to the wetland classification system. The information is summarized on 1:24,000 and 1:100,000 maps, using an alphanumeric system as illustrated in Figure 18–7.

By late 1991, 70 percent of the lower 48 states, 22 percent of Alaska, and all of Hawaii had been mapped (Wilen, 1991). It is anticipated that the entire conterminous United States will be mapped by 1998 (Fig. 18–8). Initial areas for mapping were selected based on agency needs and requests from states and other government units. Fourteen percent of U.S. wetlands maps have been computerized into digital geographic information systems (GIS), including New Jersey, Delaware, Illinois, Maryland, Washington, and Indiana and parts of Virginia, Minnesota, and South Carolina.

The NWI program is now also producing "Status and Trends" reports on U.S. wetlands, the most recent documenting wetlands losses in the past 200 years (Dahl, 1990) and wetlands status and trends from the mid-1970s to the mid-1980s (Dahl and Johnson, 1991; see Chap. 3). Data for these reports were derived from a sample of U.S. wetlands and will be revised every ten years. Other products from NWI include "map reports" for each 1:100,000 scale wetland map, a wetland plant database, and state wetland reports. The entire wetland inventory, at the level of detail envisioned for NWI, will take many years to complete. Once it is completed and the database is computerized, a continuous and more vigilant monitoring of our nation's wetland resources will be possible.

Although National Wetlands Inventory wetland maps can provide much important information about the general location of wetlands, the determination of wetland boundaries to within a tolerance of less than a meter for regulatory purposes will probably always require on-site evaluation. In part this simply reflects the limit of resolution of the 1:24,000 scale maps currently used by NWI. The width of a fine line on these maps represents about 5 m on the ground. Improving the scale to 1:12,000 would theoretically double the resolution but would require four times as many maps to cover the same area.

References

Abell, D. L., ed., 1984, *California Riparian Systems–Protection, Management and Restoration for the 1990s.* General Technical Report PSW–110, Pacific Southwest Forest and Range Experiment Station, U.S. Department of Agriculture Forest Service, Berkeley, Ca.

Abernethy, R. K., 1986, *Environmental Conditions and Waterfowl Use of a Backfilled Pipeline Canal,* M.S. Thesis, Louisiana State Univ., Baton Rouge. 125p.

Abernethy, V., and R. E. Turner, 1987, U.S. forested wetlands: 1940–1980, *Bioscience* **37:**721–727.

Adams, D. A. 1963, Factors influencing vascular plant zonation in North Carolina salt marshes, *Ecology* **44:**445-456.

Adams. D. D., 1978, *Habitat Development Field Investigations: Windmill Point Marsh Development Site, James River, Virginia; App. F: Environmental Impacts of Marsh Development with Dredged Material: Sediment and Water Quality, vol. II: Substrate and Chemical Flux Characteristics of a Dredged Material Marsh,* U.S. Army Waterways Experiment Station Tech. Rep. D-77-23, Vicksburg, Miss., 72p.

Adams, D. F., S. O. Farwell, E. Robinson, M. R. Pack, and W. L. Bamesberger, 1981, Biogenic sulfur source strengths, *Environmental Science and Technology* **15:**1493–1498.

Adams, D. O., and S. F. Yang, 1977, Methionine metabolism in apple tissue. Implication of 5-adenosylmethionine as an intermediate in the conversion of methionine to ethylene, *Plant Physiol.* **60:**892–896.

Adams, D. O., and S. F. Yang, 1979, Ethylene biosynthesis: identification of 1-amonocyclopropane-1-carbvoxylic acid as an intermediate in the conversion of methionine to ethylene, *Proc. National Acad. Sci. USA* **76:**170–174.

Adams, G. D., 1988, Wetlands of the prairies of Canada, in *Wetlands of Canada,* National Wetlands Working Group, ed., Environment Canada and Polyscience Publications, Inc., Ottawa, Ontario, and Montreal, Quebec, pp. 155–198.

Adamus, P. R., 1983, *A Method for Wetland Functional Assessment,* vol. I *Critical Review and Evaluation Concepts,* and vol. Il *FHWA Assessment Method,* U.S. Department of Transportation, Federal Highway Administration Reports FHWA-IP-82-23 and FHWA-IP-82-24, Washington, D.C., 176p. and 134p.

Adamus, P. R., 1990, *Wetlands and Water Quality: EPA's Research and Monitoring Implementation Plan for the Years 1989–1994,* Environmental Research Laboratory, Office of Research and Development, USEPA, Corvallis, Oregon. USEPA Office of Wetlands Protection, USEPA Environmental Research Laboratory, and Roy F. Weston, Inc., 44p.

Adamus, P. R., E. J. Clairain, R. D. Smith, and R. E. Young, 1987, *Wetland Evaluation Technique* (WET), vol. II: *Methodology*, Department of the Army, Waterways Experiment Station, Vicksburg, Miss., Operational Draft.

Adamus, P., and Brandt., 1990, *Impacts on Quality of Inland Wetlands of the United States: A Survey of Indicators, Techniques, and Applications of Community Level Biomonitoring Data,* U.S. Environmental Protection Agency, EPA/600/3–90/073, Washington, D.C., 406p.

Allam, A. I., and J. P. Hollis, 1972, Sulfide inhibition of oxidases in rice roots, *Phytopathology* **62:**634–639.

Allen, H. H., G. J. Pierce, and R. Van Wormer, 1989, Considerations and techniques for vegetation establishment in constructed wetlands, in *Constructed Wetlands for Wastewater Treatment,* D. A. Hammer, ed., Lewis Publishers, Inc., Chelsea, Mich., pp. 405–416.

Allen, J. R. L., 1992, Tidally influenced marshes in the Severn Estuary, southwest Britain, in *Saltmarshes: Morphodynamics, Conservation and Engineering Signficance*, J. R. L. Allen and K. Pye, eds. Cambridge Univ. Press, England pp. 123–147.

Allen, J. R. L., and K. Pye, eds., 1992, *Saltmarshes: Morphodynamics, Conservation and Engineering Signficance,* Cambridge Univ. Press, Cambridge, England, 184 pp.

Anderson, C. M., and M. Treshow, 1980, A review of environmental and genetic factors that affect height in *Spartina alterniflora*, Loisel (Salt marsh and grass), *Estuaries* **3:**168–176.

Anderson, F. O., 1976, Primary production in a shallow water lake with special reference to a reed swamp, *Oikos* **27:**243–250.

Anderson, R. C., and J. White, 1970, A cypress swamp outlier in southern Illinois, *Illinois, State Acad. Sci. Trans.* **63:**6–13.

Anderson, R. R.,1969, *The Use of Color lnfrared Photography and Thermal lmagery in Marshland and Estuarine Studies,* NASA Earth Resources Aircraft Program .Status Review III:40-1-40-29.

Anderson, R. R., 1971, *Multispectral Analysis of Aquatic Ecosystems in Chesapeake Bay,* 7th International Symposium on Remote Sensing of Environment Proc., Univ. of Michigan, Ann Arbor, **3:**2217–2227.

Andreae, M. O., and H. Raembonck, 1983, Di-methyl sulfide in the surface ocean and the marine atmosphere: a global view, *Science* **221:**744–747.

Aneja, V. P., J. H. Overton, Jr., and A. P. Aneja, 1981, Emission survey of biogenic sulfur flux from terrestrial surfaces, *J. Air Pollution Control Association* **31:**256–258.

Antlfinger, A. E., and E. L. Dunn, 1979, Seasonal patterns of CO_2 and water vapor exchange of three salt-marsh succulents, *Oecologia* (Berl.) **43:**249–260.

Armentano, T. V., 1990, Soils and ecology: tropical wetlands, in *Wetlands: A Threatened Landscape*, M. Williams, ed., Basil Blackwell, Oxford, pp. 115–144.

Armentano, T. V., and E. S. Menges, 1986, Patterns of change in the carbon balance of organic soil wetlands of the termperate zone, *J. of Ecology* **74:**755–774.

Armstrong, W., 1964, Oxygen diffusion from the roots of some British bog plants, *Nature* **204:**801–802.

Armstrong, W., 1975, Waterlogged soils, in *Environment and Plant Ecology,* J. R. Etherington, ed., Wiley, London, pp. 181–218.

Armstrong, W., 1978, Root aeration in the wetland condition, in *Plant Life in Anaerobic Environments,* D. D. Hook and R. M. M. Crawford, eds., Ann Arbor Science Publishers, Inc., Ann Arbor, Mich. pp. 269–297.

Aselmann, I., and P. J. Crutzen, 1989, Global distribution of natural freshwater wetlands and rice paddies, their net primary productivity, seasonality and possible methane emissions, *J. Atmospheric Chemistry* **8:**307–358.

Atchue, J. A., III, H. G. Marshall, and F. P. Day, Jr. ,1982, Observations of phytoplankton composition from standing water in the Great Dismal Swamp, *Castanea* **47:**308–312.

Atchue, J. A., III, F. P. Day, Jr., and H. G. Marshall, 1983, Algae dynamics and nitrogen and phosphorus cycling in a cypress stand in the seasonally flooded Great Dismal Swamp, *Hydrobiologia* **106:**115–122.

Athanas, C., 1987, *Guidelines for Constructing Wetland Stormwater Basins,* Maryland Department of Natural Resources. 23p.

Athanas, C., 1988, Wetlands creation for stormwater treatment, in *Wetlands: Increasing our Wetland Resources,* J. Zelazny and J. S. Feierabend, eds., *Proceedings of the Conference Wetlands: Increasing our Wetland Resources,* Washington, D.C., October 4–7, 1987, Corporate Conservation Council, National Wildlife Federation, Washington, D.C., pp 61–66.

Atlas, R. M., and R. Bartha, 1981, *Microbial Ecology: Fundamentals and Applications,* Addison-Wesley, Reading, Mass., 560p.

Auble, G. T., 1991, Modeling wetland and riparian vegetation change, in *Wetlands and River Corridor Management*, J. A. Kusler and S. Daly, eds., Association of Wetland Managers, Berne, N.Y., pp. 399–403.

Auble, G. T., B. C. Patten, R. W. Bosserman, and D. B. Hamilton, 1982, A hierarchical model to organize integrated research on the Okefenokee Swamp, in *Ecosystem Dynamics in Freshwater Wetlands and Shallow Water Bodies,* vol. II, SCOPE and UNEP, Centre of International Projects, Moscow, USSR, pp. 203–217.

Auble, G. T., J. Friedman, and M. L. Scott, in press, Relating riparian vegatation to existing and future streamflows, *Ecol. Applications*.

Aurand, D., and F. C. Daiber, 1973, Nitrate and nitrite in the surface waters of two Delaware salt marshes, *Chesapeake Sci.* **14:**105–111.

Aust, W. M., R. Lea, and J. D. Gregory, 1991, Removal of floodwater sediments by a clearcut tupelo-cypress wetland, *Water Resources Bulletin* **27:**111–116.

Babcock, H. M., 1968, The phreatophyte problem in Arizona, *Ariz. Watershed Symp. Proc.* **12:**34–36.

Baker, J. M., 1973, Recovery of salt marsh vegetation from successive oil spillages, *Environ. Pollut.* **4:**223–230.

Baker, K., M. S. Fennessy, and W. J. Mitsch, 1991, Designing wetlands for controlling coal mine drainage: an ecologic-economic modelling approach, *Ecological Economics* **3:**1–24.

Baker, W. L., 1989, Macro- and Micro-scale influences on riparian vegetation in Western Colorado, *Annals of the Asoc. Amer. Geographers* **79:**65–78.

Baker-Blocker, A., T. M. Donahue, and K. H. Mancy, 1977, Methane flux from wetlands, *Tellus* **29:**245–250.

Ball, M. C., 1980, Patterns of secondary succession in a mangrove forest in south Florida, *Oecologia* **44:**226–235.

Balling, S. S., and V. H. Resh, 1983, The influence of mosquito control recirculation ditches on plant biomass, production, and composition in two San Francisco Bay salt marshes, *Estuarine Coastal Shelf Sci.* **16:**151–161.

Balogh, G. R., and T. A. Bookhout, 1989, Purple loosestrife (*Lytrum salicaria*) in Ohio's Lake Erie marshes, *Ohio J. Sci.* **89**:62–64.

Bardecki, M. J., 1987, *Wetland Evaluation: Methodology Development and Pilot Area Selection*, Canadian Wildlife Service and Wildlife Habitat Canada, Toronto, Canada, Report 1.

Bares, R. H., and M. K. Wali, 1979, Chemical relations and litter production of *Picia mariana* and *Larix laricina* stands on an alkaline peatland in northern Minnesota, *Vegetatio* **40:**79–94.

Barsdate, R. J., and V. Alexander, 1975, The nitrogen balance of Arctic tundra: pathways, rates, and environmental implications, *J. Environ. Qual.* **4:**111–117.

Bartlett, C. H., 1904, *Tales of Kankakee Land,* Charles Scribner's Sons, New York, 232p.

Bartsch, I., and T. R. Moore, 1985, A preliminary investigation of primary production and decomposition in four peatlands near Schefferville, Quebec, *Can. J. Bot.* **63:**1241–1248.

Batie, S. S., and J. R. Wilson, 1978, Economic values attributable to Virginia's coastal wetlands as inputs in oyster production, *S. J. Agric. Economics* **10:**111–117.

Batt, B. D. J., M. G. Anderson, C. D. Anderson, and F. D. Caswell, 1989, The use of prairie potholes by North American ducks, in *Northern Prairie Wetlands*, A. G. van der Valk, ed., Iowa State Univ. Press, Ames, Iowa, pp. 204–227.

Baumann, R. H., J. W. Day, Jr., and C. A. Miller, 1984, Mississippi deltaic wetland survival: sedimentation versus coastal submergence, *Science* **224:**1093–1095.

Bay, R. R., 1967, Groundwater and vegetation in two peat bogs in northern Minnesota, *Ecology* **48:**308–310.

Bay, R. R., 1969, Runoff from small peatland watersheds, *J. Hydrology* **9:**90–102.

Bazilevich, N. I., L. Ye. Rodin, and N. N. Rozov, 1971, Geophysical aspects of biological productivity, *Soviet Geog.* **12:**293–317.

Bazilevich, N. I., and A. A. Tishkov, 1982, Conceptual balance model of chemical element cycles in a mesotrophic bog ecosystem, in *Ecosystem Dynamics in Freshwater Wetlands and Shallow Water Bodies,* vol. II. SCOPE and UNEP, Centre of International Projects, Moscow, USSR, pp. 236–272.

Beadle, L. C., 1974, *The Inland Waters of Tropical Africa*, Longman, London,

Beaumont, P., 1975, Hydrology, in *River Ecology,* B. Whitton, ed., Blackwell, Oxford, pp. 1–38.

Beck, L. T., 1977, *Dlstribution and Relative Abundance of Freshwater Macroinverebrates of the Lower Atchafalaya River Basin,* Louisiana, Master's thesis, Louisiana State University, Baton Rouge.

Bedinger, M. S., 1979, Relation between forest species and flooding, in *Wetland Functions and Values: The State of Our Understanding,* P. E. Greeson, J. R. Clark, and J. E. Clark, eds., American Water Resources Assoc., Minneapolis, Minn., pp. 427–435.

Bedinger, M. S., 1981, Hydrology of bottomland hardwood forests of the Mississippi Embayment, in *Wetlands of Bottomland Hardwood Forests,* J. R. Clark and J. Benforado, eds., Elsevier, Amsterdam, pp. 161–176.

Beeftink, W. G., 1977a, Salt marshes, in *The Coastline,* R. S. K. Barnes, ed., Wiley, New York, pp. 93–121.

Beeftink, W. G., 1977b, The coastal salt marshes of Western and Northern Europe: An ecological and phytosociological approach, in *Ecosystems of the World 1: Wet Coastal Ecosystems.*, V. J. Chapman, ed., Elsevier Science Publishing Co., New York, pp. 109–149.

Beeftink. W. G., and J. M. Gehu, 1973, Spartinetea maritimae, in *Prodrome des Groupements Vegetaux d'Europe,* lieferung 1, R. Tuxen, ed., J. Cramer Verlag, Lehre, pp. 1–43.

Bell, H. E., 1981, *Illinois Wetlands: Their Value and Management,* State of Illinois Institute of Natural Resources, Document Number 81/33, Springfield, Illinois. 133p.

Bellamy, D. J. and J. Rieley, 1967, Some ecological statistics of a "miniature bog," *Oikos* **18:**33–40.

Bellamy, D. J., 1968, *An Ecological Approach to the Classification of the Lowland Mires in Europe,* in 3rd Internat. Peat Congress Proc. Quebec, Canada, pp. 74–79.

Belt, C. B., Jr., 1975, The 1973 flood and man's constriction of the Mississippi River, *Science* **189:**681–684.

Berggren, T. J., and J. T. Lieberman, 1977, Relative contributions of Hudson, Chesapeake, and Roanoke striped bass, *Morone saxatilis,* to the Atlantic coast fishery, *Fish. Bull.* **76:**335–345.

Berkeley, E., and D. Berkeley, 1976, Man and the Great Dismal, *Virginia J. Science* **27:**141–171.

Bernard, J. M., and B. A. Solsky, 1977, Nutrient cycling in a *Carex lacustris* wetland, *Can. J. Bot.* **55:**630–638.

Bernatowicz, S., S. Leszczynski, and S. Tyczynska, 1976, The influence of transpiration by emergent plants on the water balance of lakes, *Aquatic Botany* **2:**275–288.

Bertani, A., I. Bramblila, and F. Menegus, 1980, Effect of anaerobiosis on rice seedlings: Growth, metabolic rate, and rate of fermentation products, *J. Exp. Bot.* 3:325–331.

Bertness, M. D., 1992, The ecology of a New England Salt marsh, *American Scientist* **80:**260–268.

Beschel, R. E., and P. J. Webber, 1962, Gradient analysis in swamp forests, *Nature* **194:**207–209.

Best, R. G., 1987, Natural wetlands—southern environment: wastewater to wetlands, where do we go from here? in *Aquatic Plants for Water Treatment and Resource Recovery,* K. R. Reddy and W. H. Smith, eds., *Proceedings of the Conference Research and Application of Aquatic Plants for Water Treatment and Resource Recovery,* Orlando, Florida. Magnolia Publishing, Inc., Orlando, Fla., pp. 99–119.

Bhowmik, N. G., A. P. Bonini, W. C. Bogner, and R. P. Byrne, 1980, Hydraulics of flow and sediment transport to the Kankakee River in Illinois, *Illinois State Water Survey Rep. of Investigation 98,* Champaign, Ill.,170p.

Biggs, R. B., and D. A. Flemer, 1972, The flux of particulate carbon in an estuary, *Mar. Biol.* **12:**11–17.

Bishop, J. M., J. G. Gosselink, and J. M. Stone, 1980, Oxygen consumption and hemolymph osmolality of brown shrimp, *Penaeus aztecus, Fish. Bull.* **78:**741–757.

Black, C. C., Jr., 1973, Photosynthetic carbon fixation in relation to net CO_2 uptake, *Ann. Rev. Plant Physiol.* **24:**253–286.

Blem, C. R., and L. B. Blem, 1975, Density, biomass, and energetics of the bird and mammal populations of an Illinois deciduous forest, *Illinois Acad. Sci. Trans.* **68:**156–184.

Blom, C. W., G. M. Bogemann, P. Laan, A. J. M. van der Sman, H. M. van der Steeg, and L. A. Voesenek. et al., 1990, Adaptations to flooding in plants from river areas, *Aquatic Botany* **38:**29–47.

Boelter, D. H., and E. S. Verry, 1977, *Peatland and Water in the Northern Lake States,* General Technical Report NC-31, U.S. Dept. of Agriculture, Forestry Service, North Central Experiment Station, St. Paul, Minn.

Bolin, E. G., and F. S. Guthery, 1982, Playa, irrigation and wildlife in West Texas, *47th N. Amer. Wildl. Nat. Resour. Conf. Trans.* **47:**528–541.

Bolin, E. G., L. H. Smith, and H. L. Scramm, Jr., 1989, Playa lakes: prairie wetlands of the Southern High Plains, *Bioscience* **39:**615–623.

Bonasera, J., J. Lynch, and M. A. Leck, 1979, Comparison of the allelopathic potential of four marsh species, *Torrey Bot. Club Bull.* **106:**217–222.

Bormann, F. H., G. E. Likens, and J. S. Eaton, 1969, Biotic regulation of particulate and solution losses from a forested ecosytem, *BioScience* **19:**600–610.

Bortner, J. B., F. A. Johnson, G. W. Smith, and R. E. Trost, 1991, *1991 Status of Waterfowl and Fall Flight Forecast*, Canadian Wildlife Service and U.S. Fish and Wildlife Service, Office of Migratory Bird Management, Laurel, Md.

Bosserman, R. W., 1983a, Dynamics of physical and chemical parameters in Okefenokee Swamp, *J. Freshwater Ecology* **2:**129–140.

Bosserman, R. W., 1983b, Elemental composition of *Utricularia*—periphyton ecosystems from Okefenokee Swamp, *Ecology* **64:**1637–1645.

Boto, K. G., and J. S. Bunt, 1981, Tidal export of particulate organic matter from a northern Australian mangrove system, *Estuarine, Coastal and Shelf Science* **13:**247–255.

Boto, K. G., and J. T. Wellington, 1984, Soil characteristics and nutrient status in a northern Australian mangrove forest, *Estuaries* **7:**61–69.

Boulé, M.E. 1988. Wetland Creation and Enhancement in the Pacific Northwest, in *Proceedings of the Conference Wetlands: Wetlands: Increasing our Wetland Resources,* J. Zelazny and J. S. Feierabend, ed., Washington D.C. Corporate Conservation Council, National Wildlife Federation, Washington, D.C., pp. 130–136.

Boyd, C. E., 1970, Losses of mineral nutrients during decomposition of *Typha latifolia*, *Arch. Hydrobiol.* **66:**511–517.

Boyd, C. W., 1971, The dynamics of dry matter and chemical substances in a *Juncus effusus* population, *Amer. Midland Nat.* **86:**28–45.

Boyt, F. L., S. E. Bayley, and J. Zoltek, 1977, Removal of nutrients from treated wastewater by wetland vegetation, *Journal of the Water Pollution Control Federation.* **49:**789–799.

Brackke, F. H., ed., 1976, *Impact of Acid Precipitation on Forests and Freshwater Ecosystems in Norway*, SNSF Report 6, Oslo-As, 111p.

Bradbury, I. K., and J. Grace, 1983, Primary production in wetlands, in *Ecosystems of the World 4A. Mires: Swamp. Bog, Fen and Moor,* A. J. P. Gore, ed., Elesvier, Amsterdam, Netherlands, pp. 285–310.

Bradford, K. J., and S. F. Yang, 1980, Xylem transport of 1-amino-cyclopropane-1-carboxylic acid, an ethylene precursor in waterlogged tomato plants, *Plant Physiol.* **65:**322–326.

Bradley, C. E., and D. G. Smith, 1986, Plains cottonwood recruitment and survival on a prairie meandering river floodplain, Milk River, southern Alberta and northern Montana, *Can. J. Bot.* **64:**1433–1442.

Brady, N. C., 1974, *The Nature and Properties of Soils,* 8th ed., Macmillan, New York, 639p.

Brakenridge, G. R., 1988, River flood regime and floodplain stratigraphy, in *Flood Geomorphology*, V. R. Baker, R. C. Kochel, and P. C. Patton, eds., John Wiley and Sons, New York, pp. 139–156.

Brinson, M. M., 1977, Decomposition and nutrient exchange of litter in an alluvial swamp forest, *Ecology* **58:**601–609.

Brinson, M. M., H. D. Bradshaw, R. N. Holmes, and J. B. Elkins, Jr., 1980, Litterfall, stemflow, and throughfall nutrient fluxes in an alluvial swamp forest, *Ecology* **61:**827–835.

Brinson, M. M., A. E. Lugo, and S. Brown, 1981a, Primary productivity, decomposition and consumer activity in freshwater wetlands, *Annu. Rev. Ecol. Systematics* **12:**123–161.

Brinson, M. M., B. L. Swift, R. C. Plantico, and J. S. Barclay, 1981b, *Riparian Ecosystems: Their Ecology and Status,* U.S. Fish and Wildlife Service, Biol. Serv. Prog., FWS/OBS-81/17, Washington, D.C., 151p.

Brinson, M. M., M. D. Bradshaw, and E. S. Kane, 1984, Nutrient assimilative capacity of an alluvial floodplain swamp, *J. Applied Ecology* **21:**1041–1057.

Britsch, L. D., and E. B. Kemp, 1990, *Land Loss Rates, Report 1, Mississippi River Deltaic Plain,* U.S. Army Corps of Engineers, Waterways Expt. Sta., Vicksburg, Miss., COE Tech Report GL–90–2.

Brock, J. H., 1985, Physical characteristics and pedogenesis of soils in riparian habitats along the upper Gila River basin, in *Riparian Ecoystems and Their Management: Reconciling Conflicting Uses*, R. R. Johnson et al., eds., Forest Service, U.S. Department of Agriculture, Rocky Mountain Forest and Range Experiment Station, Fort Collins, Colo., General Technical Report RM–120, pp. 49–53.

Brodie, G. A., 1989, Selection and evaluation of sites for constructed wastewater treatment wetlands, in *Constructed Wetlands for Wastewater Treatment,* D. A. Hammer, ed., Lewis Publishers, Inc., Chelsea, Mich., pp. 307–318.

Brodie, G. A., D. A. Hammer, and D. A. Tomljanovich, 1988, An evaluation of substrate types in constructed wetland acid drainage treatment systems, in *Mine Drainage and Surface Mine Reclamation,* vol. I: *Mine Water and Mine Waste,* United States Department of the Interior, Pittsburgh, Pa., pp. 389–398.

Brodie, G. A., D. A. Hammer, and D. A. Tomljanovich, 1989a, Treatment of acid drainage with a constructed wetland at the Tennessee Valley Authority 950 coal mine, in *Constructed Wetlands for Wastewater Treatment*, D. A. Hammer, ed., Lewis Publishers, Inc., Chelsea, Mich., pp. 201–210.

Brodie, G. A., D. A. Hammer, and D. A. Tomljanovich, 1989b, Constructed wetlands for treatment of ash pond see page, in *Constructed Wetlands for Wastewater Treatment,* D. A. Hammer, ed., Lewis Publishers, Inc., Chelsea, Mich., pp. 211–220.

Brody, M., and E. Pendleton, 1987, *FORFLO: A Model to Predict Changes in Bottomland Hardwood Forests*, Office of Information Transfer, U.S. Fish Wildl. Service, Fort Collins, Colo.

Brooks, R. P., 1989, Wetland and waterbody restoration and creation associated with mining, in *Wetland Creation and Restoration,* J. A. Kusler and M. E. Kentula, eds., Island Press, Washington, D.C., pp. 529–548.

Brooks, R. P., D. E. Samuel, and J. B. Hill, eds., 1985, *Wetlands and Water Management on Mined Lands,* Proceedings of a Conference Oct. 23–24,1985, The Pennsylvania State University, University Park, Pa., 393p.

Broome, S. W., 1990, Creation and restoration of tidal wetlands of the southeastern United States, in *Wetland Creation and Restoration,* J. A. Kusler and M. E. Kentula, eds., Island Press, Washington, D.C., pp. 37–72.

Broome, S. W., W. W. Woodhouse, Jr., and E. D. Seneca, 1975a, The relationship of mineral nutrients to growth of *Spartina alterniflora* in North Carolina: I. Nutrient status of plants and soils in natural stands, *Soil Sci. Soc. Amer. Proc.* **39**(2):295–301.

Broome, S. W., W. W. Woodhouse, Jr., and E. D. Seneca, 1975b, The relationship of mineral nutrients to growth of *Spartina alterniflora* in North Carolina: II. The effects of N, P and Fe fertilizers, *Soil Sci. Soc. Amer. Proc.* **39**:301–307.

Broome, S. W., E. D. Seneca, and W. W. Woodhouse, Jr., 1988, Tidal salt marsh restoration, *Aquatic Botany* **32:**1–22.

Brown, C. A., 1984, Morphology and biology of cypress trees, in *Cypress Swamps,* K. C. Ewel and H. T. Odum, eds., Univ. Presses of Florida, Gainesville, pp. 16–24.

Brown, M. T., 1987, *Conceptual Design for a Constructed Wetlands System for the Renovation of Treated Effluent,* Report from the Center for Wetlands, University of Florida, 18p.

Brown, M. T., 1988, A simulation model of hydrology and nutrient dynamics in wetlands, *Computers, Environment and Urban Systems* **12**(4):221–237.

Brown, M. T., R. E. Tighe, T. R. McClanahan, and R. W. Wolfe, 1992, Landscape reclamation at a Central Florida phosphate mine, *Ecological Engineering* **1:**323–354.

Brown, S. L., 1978, *A Comparison of Cypress Ecosystems in the Landscape of Florida,* Ph.D. dissertation, University of Florida, 569p.

Brown, S. L., 1981, A comparison of the structure, primary productivity, and transpiration of cypress ecosystems in Florida, *Ecol. Monogr.* **51:**403–427.

Brown, S. L., 1984, The role of wetlands in the Green Swamp, in *Cypress Swamps,* K. C. Ewel and H. T. Odum, eds., Univ. Presses of Florida, Gainesville, pp. 405–415.

Brown, S. L., 1990, Structure and dynamics of basin forested wetlands in North America, in *Forested Wetlands, Ecosystems of the World,* A. E. Lugo, M. M. Brinson, and S. L. Brown, eds., Elsevier, Amsterdam, pp.171–199.

Brown, S. L., M. M. Brinson, and A. E. Lugo, 1979, Structure and function of riparian wetlands, in *Strategies for Protection and Management of Floodplain Wetlands and Other Riparian Ecosystems,* R. R. Johnson and J. F. McCormick, eds., U.S. Forest Service Gen. Tech. Report WO-12, Washington, D.C., pp. 17–31.

Brown, S. L., and A. E. Lugo, 1982, A comparison of structural and functional characteristics of saltwater and freshwater forested wetlands, in *Wetlands—Ecology and Management,* B. Gopal, R. E. Turner, R. G. Wetzel, and D. F. Whigham, eds., National Institute of Ecology and International Scientific Publications, Jaipur, India, pp. 109–130.

Brown, S. L., and D. L. Peterson, 1983, Structural characteristics and biomass production of two Illinois bottomland forests, *Amer. Midl. Nat.* **110:**107–117.

Brown, S. L., E. W. Flohrschutz, and H. T. Odum, 1984, Structure, productivity, and phosphorus cycling of the scrub cypress ecosystem, in *Cypress Swamps*, K. C. Ewel and H. T. Odum, eds., University Presses of Florida, Gainesville, pp. 304–317.

Brundage, H. M., III, and R. E. Meadows, 1982, Occurrence of the endangered shortnose sturgeon, *Acipenser brevirostrum,* in the Delaware River estuary, *Estuaries* **5:**203–208.

Bryan, C. F., D. J. DeMost, D. S. Sabins, and J. P. Newman, Jr., 1976, *A Limnological Survey of the Atchafalaya Basin,* Annual report, Louisiana Cooperative Fishery Research Unit, School of Forestry and Wildlife Manage., Louisiana State University, Baton Rouge, 285p.

Bunnell, F. L., and P. Dowding, 1974, ABISKO-A generalized decomposition model for comparisons between tundra sites, in *Soil Organisms and Decomposition in Tundra,* A. J. Holding, O. W. Heal, S. F. McLean, Jr. and P. U. Flanagan, eds., Tundra Biome Steering Committee, Stockholm, Sweden, pp. 227–247.

Bunnell, F. L., and K. A. Scoullar, 1975, ABISKO II: A computer simulation model of carbon flux in tundra ecosystems, in *Structure and Function of Tundra Ecosystems,* T. Rosswall and O. W. Heal, eds., Ecol. Bull. (Stockholm) **20:**425–448.

Burdick, D. M., and I. A. Mendelssohn, 1990, Relationship between anatomical and metabolic responses to soil waterlogging in the coastal grass *Spartina patens*. *J. Exptl. Bot.* **41**:223–228.

Buresh, R. J., M. E. Casselman, and W. H. Patrick, Jr., 1980a, Nitrogen fixation in flooded soil systems: a review, *Adv. Agron.* **33:**149–192.

Buresh, R. J., R. D. Delaune, and W. H. Patrick, Jr., 1980b, Nitrogen and phosphorus distribution and utilization by *Spartina alterniflora* in a Louisiana Gulf Coast Marsh, *Estuaries* **3:**111–121.

Burkham, D. E., 1972, Channel changes of the Gila River, Safford Valley, Arizona, 1846–1970, *Geol. Surv. Prof. Pap. (U.S.)* **665–G:**1–24.

Burkholder, P. R., 1956, Studies on the nutritive value of *Spartina* grass growing in the marsh areas of coastal Georgia, *Bull. Torrey Bot. Club* **83:**327–334.

Burkholder, P. R., and G. H. Bornside, 1957, Decomposition of marsh grass by anaerobic marine bacteria, *Torrey Bot. Club Bull.* **84:**366–383.

Burns, L. A., 1978, *Productivity, Biomass, and Water Relations in a Florida Cypress Forest,* Ph.D. dissertation, University of North Carolina.

Burton, J. D., and P. S. Liss, 1976, *Estuarine Chemistry,* Academic Press, London, 229p.

Cairns, J., Jr., M. M. Brinson, R. L. Johnson, W. B. Parker, R. E. Turner, and P. V. Winger, 1981, Impacts associated with southeastern bottomland hardwood forest ecosystems, in *Wetlands of Bottomland Hardwood Forests,* J. R. Clark and J. Benforado, eds., Elsevier, Amsterdam, pp. 303–332.

Cameron, C. C., 1970, Peat deposits of northeastern Pennsylvania, *U.S. Geological Survey Bull. 1317-A,* 90p.

Cameron, C. C., 1973, Peat, in United States Mineral Resources, *U.S. Geol. Survey Prof. Paper No. 820,* pp. 505–513.

Capone, D. G., and R. P. Kiene, 1988, Comparison of microbial dynamics in marine and freshwater sediments: contrasts in anaerobic carbon metabolism, *Limnology and Oceanography* **33:**725–749.

Cardamone, M. A., J. R. Taylor, and W. J. Mitsch, 1984, *Wetlands and Coal Surface Mining: A Management Handbook—with Particular Reference to the Illinois Basin of the Eastern Interior Coal Region,* Center for Environmental Science and Management, University of Louisville, Kentucky, 99p.

Carey, O. L., 1965, The economics of recreation: progress and problems, *West. Econ. J.* **8:**169–173.

Carpenter, E. J., C. D. Van Raalte, and I. Valiela, 1978, Nitrogen fixation by algae in a Massachusetts salt marsh, *Limnol. Oceanogr.* **23:**318–327.

Carpenter, J. M., and G. T. Farmer, 1981, *Peat Mining: An Initial Assessment of Wetland Impacts and Measures to Mitigate Adverse Impacts,* U.S. EPA PB 82-130766, Washington, D.C., 61p.

Carter, M. R., L. A. Burns, T. R. Cavinder, K. R. Dugger, P. L. Fore, D. B. Hicks, H. L. Revells, T. W. Schmidt, 1973, *Ecosystem Analysis of the Big Cypress Swamp and Estuaries,* U.S. EPA 904/9-74-002, Region IV, Atlanta.

Carter, V., 1986, An overview of the hydrologic concerns related to wetlands in the United States, *Can. J. Bot.* **64:**364–374.

Carter, V., M. S. Bedinger, R. P. Novitzki, and W. O. Wilen, 1979, Water resources and wetlands, in *Wetland Functions and Values: The State of Our Understanding,* P. E. Greeson, J. R. Clark, and J. E. Clark, eds., American Water Resources Assoc., Minneapolis, Minn., pp. 344–376.

Carter, V., and R. P. Novitzki, 1988, Some comments on the relation between ground water and wetlands, in *The Ecology and Management of Wetlands,* D. D. Hook et al., eds., vol. 1: *Ecology of Wetlands,* Timber Press, Portland, Oregon, pp. 68–86.

Casparie, W. A., and J. G. Streefkerk, 1992, Climatological, stratigraphic and palaeoecological aspects of mire development, in *Fens and Bogs in the Netherlands: Vegetation, History, Nutrient Dynamics and Conservation,* J. T. A. Verhoeven, ed., Kluwer Academic Publishers, Dordrecht, Netherlands, pp. 81–129.

Castro, M. S., and F. E. Dierberg, 1987, Biogenic hydrogen sulfide emissions from selected Florida wetlands, *Water, Air, and Soil Pollution* **33:**1–13.

Cely, J., 1974, Is the Beidler Tract in Congaree Swamp Virgin?, in *Congaree Swamp: Greatest Unprotected Forest on the Continent*, Sierra Club Publication, Columbia, S.C.

Chabreck, R. H., 1972, Vegetation, water and soil characteristics of the Louisiana coastal region, *Louisiana State Univ. Agr. Exp. Stn. Bull.* 664, Baton Rouge, 72p.

Chabreck, R. H., 1979, Wildlife harvest in wetlands of the United States, in *Wetland Function and Values: The State of Our Understanding,* P. E. Greeson, J. R. Clark, and J. E. Clark, eds., American Water Resources Assoc., Minneapolis, Minn., pp. 618–631.

Chabreck, R. H., 1988, *Coastal Marshes: Ecology and Wildlife Management*, Univ. of Minnesota Press, Minneapolis, Minn., 138p.

Chalmers, A. G., R. G. Wiegert, and P. L. Wolf, 1985, Carbon balance in a salt marsh: interactions of diffusive export, tidal deposition and rainfall-caused erosion., *Estuarine Coastal Shelf Sci.* **21:**757–771.

Chamie, J. P., and C. J. Richardson, 1978, Decomposition in northern wetlands, in *Freshwater Wetlands-Ecological Processes and Management Potential,* R. E. Good, D. F. Whigham, and R. L. Simpson, eds., Academic Press, New York, pp. 115–130.

Chapman, S. B., 1965, The ecology of Coom Rigg Moss, Northumberland. III. Some water relations of the bog system, *J. Ecology* **53:**371–384.

Chapman, V. J., 1938, Studies in salt marsh ecology I-III, *J. Ecology* **26:**144–221.

Chapman, V. J., 1940, Studies in salt marsh ecology VI-VII, *J. Ecology* **28:**118–179.

Chapman, V. J., 1960, *Salt Marshes and Salt Deserts of the World,* Interscience, New York, 392p.

Chapman, V. J., 1974, Salt marshes and salt deserts of the world, in *Ecology of Halophytes,* R. J. Reimold and W. H. Queen, eds., Academic Press, New York, pp. 3–19.

Chapman, V. J., 1975, The salinity problem in general; its importance and distribution with special reference to natural halophytes, in *Plants in Saline Environments*, A. Poljakoff-Mayber and J. Gale, eds., Ecological Studies No. 15, Springer-Verlag, New York, pp. 7–24.

Chapman, V. J., 1976a, *Coastal Vegetation,* 2nd ed., Pergamon Press, Oxford, 292p.

Chapman, V. J., 1976b, *Mangrove Vegetation,* J. Cramer, Vaduz, Germany, 447p.

Chapman, V. J., 1977, *Wet Coastal Ecosytems,* Elsevier, Amsterdam, 428p.

Chow, V. T., ed., 1964, *Handbook of Applied Hydrology,* McGraw-Hill, New York, 1453p.

Chung, C. H., 1982, Low marshes, China, in *Creation and Restoration of Coastal Plant Communities,* R. R. Lewis III, ed., CRC Press, Inc., Boca Raton, Florida., pp. 131–145.

Chung, C. H., 1983, Geographical distribution of *Spartina anglica* Hubbard in China, *Bull. Mar. Sci.* **33:**753–758.

Chung, C. H., 1985, The effects of introduced *Spartina* grass on coastal morphology in China, *Z. Geomorph. N.F. Suppl. Bd.* **57:**169–174.

Chung, C. H., 1989, Ecological engineering of coastlines with salt marsh plantations, in *Ecological Engineering: An Introduction to Ecotechnology*, W. J. Mitsch and S. E. Jørgensen, eds., Wiley, New York, pp. 255–289.

Cicerone, R. J., and R. J. Oremland, 1988, Biogeochemical aspects of atmospheric methane, *Global Biogeochem. Cycles* **2:**299–327.

Cintrón, G., A. E. Lugo, and R. Martinez, 1985, Structural and functional properties of mangrove forests, in *The Botany and Natural History of Panama,* IV *Series: Monographs in Systematic Botany,* vol. 10, W. G. D'Arcy and M. D. Correa, eds., Missouri Botanical Garden, St. Louis, pp. 53–66.

Clark, J. E., 1979, Fresh water wetlands: habitats for aquatic invertebrates, amphibians, reptiles, and fish, in *Wetland Functions and Values: The State of Our Understanding.* P. E. Greeson, J. R. Clark, and J. E. Clark, eds., American Water Resources Association, Minneapolis, Minn., pp. 330–343.

Clark, J. R., and J. Benforado, eds., 1981, *Wetlands of Bottomland Hardwood Forests,* Elsevier, Amsterdam, 401p.

Clements, F. E., 1916. Plant Succession, *Carnegie Institution of Washington,* Pub. 242, 512p.

Clements, F. E., 1924, *Methods and Principles of Paleo-ecology,* Carnegie Institution of Washington, Washington, D.C., Yearbook 32.

Clewell, A. F., L. F. Ganey, Jr., D. P Harlos, and E. R. Tobi, 1976, *Biological Effects of Fill Roads across Salt Marshes,* Florida Dept. of Transportation Report FL–E.R-1-76, Tallahassee, Fl.

Clewell, A. F. and R. Lea, 1990, Creation and restoration of forested wetland vegetation in the southeastern United States, in *Wetland Creation and Restoration,* J. A. Kusler and M. E. Kentula, eds., Island Press, Washington, D.C., pp. 195–230.

Clymo, R. S., 1963, Ion exchange in *Sphagnum* and its relation to bog ecology, *Ann. Bot. (Lond.) N.S.* **27:**309–324.

Clymo, R. S., 1965, Experiments on breakdown of *Sphagnum* in two bogs, *J. Ecology* **53:**747–758.

Clymo, R. S., 1970, The growth of *Sphagnum:* methods of measurement, *J. Ecology* **58:**13–49.

Clymo, R. S., 1983, Peat, in *Mires: Swamp, Bog, Fen, and Moor,* Ecosystems of the World 4A, A. J. P. Gore, ed., Elsevier, Amsterdam, pp. 159–224.

Clymo, R. S., and E. J. F. Reddaway, 1971, Productivity of *Sphagnum* (bog-moss) and peat accumulation, *Hydrobiologia (Bucharest)* **12:**181–192.

Clymo, R. S., and P. M. Hayward, 1982, The ecology of *Sphagnum,* in *Bryophyte Ecology,* A. J. E. Smith, ed., Chapman and Hall, London, pp. 229–289.

Cohen, A. D., D. J. Casagrande, M. J. Andrejko, and G. R. Best, eds., 1984, *The Okefenokee Swamp: Its Natural History, Geology, and Geochemistry.* Wetland Surveys, Los Alamos, New Mexico, 709p.

Coles, B., and J. Coles, 1989, *People of the Wetlands, Bogs, Bodies and Lake-dwellers,* Thames and Hudson, New York, N.Y., 215p.

Conner, W. H., and J. W. Day, Jr., 1976, Productivity and composition of a bald cypress-water tupelo site and a bottomland hardwood site in a Louisiana swamp, *Amer. J. Bot.* **63:**1354–1364.

Conner, W. H., J. G. Gosselink, and R. T. Parrondo, 1981, Comparison of the vegetation of three Louisiana swamp sites with different flooding regimes, *Amer. J. Bot.* **68:**320–331.

Conner, W. H., and J. W. Day, Jr., 1982, The ecology of forested wetlands in the southeastern United States, in *Wetlands: Ecology and Management,* B. Gopal, R. E. Turner, R. G. Wetzel, and D. F. Whigham, eds., National Institute of Ecology and International Scientific Publications, Jaipur, India, pp. 69–87.

Conner, W. H., and J. W. Day, eds., 1987, *The Ecology of Barataria Basin, Louisiana: An Estuarine Profile,* U.S. Fish Wildl. Serv., Washington, D.C., vol. *Biol. Rep.* **85**(7.13), 165p.

Conner, W. H., and J. W. Day, Jr., 1989, *A Use Attainability Analysis of Wetlands for Receiving Treated Municipal and Small Industry Wastewater: A Feasibility Study Using Baseline Data from Thibodaux, La.,* prepared for La. Dept. of Environmental Quality, Center for Wetland Resources, LSU, Baton Rouge, La.

Conner, W. H., and K. Flynn, 1989, Growth and survival of baldcypress (*Taxodium distichum* [L.] Rich.) planted across a flooding gradient in a Louisiana bottomland forest, *Wetlands* **9:**207–217.

Conway, T. E., and J. M. Murtha, 1989, The Iselin Marsh Pond Meadow, in *Constructed Wetlands for Wastewater Treatment,* D. A. Hammer, ed., Lewis Publishers, Inc., Chelsea, Mich., pp. 139–144.

Cooper, A. W., 1974, Salt marshes, in *Coastal Ecological Systems of the United States,* vol. II, H. T. Odum, B. J. Copeland, and E. A. McMahan, eds., The Conservation Foundation, Washington, D.C., pp. 55–96.

Cooper, J. R., J. W. Gilliam, and T. C. Jacobs, 1986, Riparian areas as a control of nonpoint pollutants, in *Watershed Research Perspectives,* D. L. Correll, ed., Smithsonian Institution Press, Washington, D.C., pp. 166–192.

Cooper, J. R., and J. W. Gilliam, 1987, Phosphorous redistibution from cultivated fields into riparian areas, *Soil Science Society of America Journal* **51:**1600–1604.

Cooper, J. R., J. W. Gilliam, R. B. Daniels, and W. P. Robarge, 1987, Riparian areas as filters for agricultural sediment, *Soil Science of America Journal* **51:**416–420.

Cooper, P. F., and J. A. Hobson, 1989, Sewage treatment by reed bed systems: the present situation in the United Kingdom, in *Constructed Wetlands for Wastewater Treatment,* D. A. Hammer, ed., Lewis Publishers, Inc. Chelsea, Mich., pp. 153–172.

Cooper, P. F., and B. C. Findlater, eds., 1990, *Constructed Wetlands in Water Pollution Control,* Pergamon Press, Oxford, England, 605p.

Cooper, W. S.,1913, The climax forest of Isle Royale, Lake Superior, and its development, *Bot. Gaz.* **55:**1–44, 115–140, 189–235.

Corlett, R. T., 1986, The mangrove understory: some additional observations, *Journal of Tropical Ecology* **2:**93–94.

Correll, D. L., ed., 1986, *Watershed Research Perspectives*, Smithsonian Institution Press, Washington, D.C., Smithsonian Environmental Research Center, Edgewater, Md., 421p.

Costanza, R., 1980, Embodied energy and economic evaluation, *Science* **210:**1219–1224.

Costanza, R., 1984, Natural resource valuation and management: toward ecological economics, in *Integration of Economy and Ecology—An Outlook for the Eighties,* A. M. Jansson, ed., Univ. of Stockholm Press, Stockholm, Sweden, pp. 7–18.

Costanza, R., F. H. Sklar, M. L. White, and J. W. Day, Jr., 1988, A dynamic spatial simulation model of land loss and marsh succession in coastal Louisiana, in *Wetland Modelling,* W. J. Mitsch, S. E. Jørgensen, and Milan Straskraba, eds., Elsevier, Amsterdam, Netherlands, pp. 99–114.

Costanza, R., S. C. Farber, and J. Maxwell, 1989, Valuation and management of wetland ecosystems, *Ecological Economics* **1:**335–361.

Costanza, R., F. H. Sklar, and M. L. White, 1990, Modeling coastal landscape dynamics. *Bioscience* **40**(2):91–107.

Costlow, J. D., C. G. Boakout, and R. Monroe, 1960, The effect of salinity and temperature on larval development of *Sesarma cinereum* (Bosc.) reared in the laboratory, *Biol. Bull.* **118:**183–202.

Cowardin, L. M., V. Carter, F. C. Golet, and E. T. LaRoe, 1979, *Classificaion of Wetlands and Deepwater Habitats of the United States,* U.S. Fish & Wildlife Service Pub. FWS/OBS-79/31, Washington, D.C., 103p.

Cowardin, L. M., D. S. Gilmer, and C. W. Shaiffer, 1985, Mallard recruitment in the agricultural environment of North Dakota, *Wildlife Monogr.* **92:**1–37.

Cowles, H. C., 1899, The ecological relations of the vegetation on the sand dunes of Lake Michigan, *Bot. Gaz.* **27:**95–117,167–202, 281–308, 361–369.

Cowles, S., 1975, *Metabolism Measurements in a Cypress Dome,* Master's thesis, University of Florida, Gainesville, 275p.

Cragg, J. B., 1961, Some aspects of the ecology of moorland animals, *J. Anim. Ecol.* **30:**205–234.

Craighead, F. C., 1971, *The Trees of South Florida,* University of Miami Press, Coral Gables, Florida, 212p.

Crill, P. M., K. B. Bartlett, R. C. Harriss, E. Gorham, E. S. Verry, D. I. Sebacher, L. Madzar, and W. Sanner, 1988, Methane flux from Minnesota peatlands, *Global Biogeochemical Cycles* **2:**371–384.

Crisp, D. T., 1966, Input and output of minerals for an area of Pennine moorland: the importance of precipitation, drainage, peat erosion, and animals, *J. Appl. Ecol.* **3:**327–348.

Cronk, J. K., and W. J. Mitsch, The effects of hydrology on the water column primary productivity of four constructed freshwater wetlands, unpublished manuscript.

Crum, H., 1988, *A Focus on Peatlands and Peat Mosses*, The University of Michigan Press, Ann Arbor, Mich., 306p.

Cullis, C. F., and M. M. Hirschler, 1980, Atmospheric sulfur: natural and man-made sources, *Atmospheric Environment* **14:**1263–1278.

Cummins, K. W., 1974, Stream ecosystem structure and function, *Bioscience* **24:**631–641.

Curtis, J. T., 1959, *The Vegetation of Wisconsin,* University of Wisconsin Press, Madison, 657p.

Cushing, E. J., 1963, *Late-Wisconsin Pollen Stratigraphy in East-central Minnesota,* Ph.D. dissertation, Univ. of Minnesota, Minneapolis.

Cypert, E., 1961, The effect of fires in the Okefenokee Swamp in 1954 and 1955, *Am. Midl. Nat.* **66:**485–503.

Cypert, E., 1972, The origin of houses in the Okefenokee prairies, *Am. Midl. Nat.* **87:**448–458.

D'Avanzo, C., 1989, Long-term evaluation of wetland creation projects, in *Wetland Creation and Restoration: The Status of the Science,* J. A. Kusler and M. E. Kentula, eds., USEPA. Corvallis, Oreg., pp. 75–84.

Dabel, C. V., and F. P. Day, Jr., 1977, Structural comparison of four plant communities in the Great Dismal Swamp, Virginia, *Torrey Bot. Club Bull.* **104:**352–360.

Dacey, J. W. H., 1980, Internal winds in water lilies: an adaption for life in anaerobic sediments, *Science* **210:**1017–1019.

Dachnowski-Stokes, A. P., 1935, Peat land as a conserver of rainfall and water supplies, *Ecology* **16:**173–177.

Dahl, T. E., 1990, *Wetlands Losses in the United States, 1780s to 1980s*, U.S. Department of the Interior, Fish and Wildlife Service, Washington, D.C., 21p.

Dahl, T. E., and C. E. Johnson, 1991, *Wetlands Status and Trends in the Conterminous United States Mid-1970s to mid-1980s,* U.S. Department of Interior, Fish and Wildlife Service, Washington, D.C., 28p.

Daiber, F. C., 1977, Salt marsh animals: distributions related to tidal flooding, salinity and vegetation, in *Wet Coastal Ecosystems,* V. J. Chapman, ed., Elsevier, Amsterdam, pp. 79–108.

Daiber, F. C., 1982, *Animals of the Tidal Marsh,* Van Nostrand Reinhold, New York, 442p.

Dalkey, N. C., 1972, *Studies in the Quality of Life: Delphi and Decision Making,* Lexington Books, Lexington, Mass., 161p.

Dames and Moore, 1983, *A Survey of Wetland Reclamation Projects in the Florida Phosphate Industry,* final report prepared for Florida Institute of Phosphate Research, Bartow, Fla., 59p. plus appen.

Damman, A. W. H., 1978, Distribution and movement of elements in ombrotrophic peat bogs, *Oikos* **30:**480–495.

Damman, A. W. H., 1979, Geographic Patterns in Peatland Development in Eastern North America, in *Proceedings of the International Symposium of Classification of Peat and Peatlands,* Hyytiala, Finland, International Peat Society, pp. 42–57.

Damman, A. W. H., 1990, Nutrient status of ombrotrophic peat bogs, *Aquilo, Ser. Bot.* **28:**5–14.

Damman, A. W. H., and T. W. French, 1987, The ecology of peat bogs of the glaciated Northeastern United States: a community profile, *U.S. Fish Wildl. Serv. Biol. Rep.* **85**(7.16):100.

Darnell, R., 1976, Impacts of construction activities in wetlands of the United States, *U. S. Environmental Protection Agency* EPA-600/3-76-045, Corvallis, Or., 393p.

Dauffenbach, L., R. M. Atlas, and W. J. Mitsch, 1981, A computer simulation model of the fate of crude petroleum spills in Arctic tundra ecosystems, in *Energy and Ecological Modelling,* W. J. Mitsch, R. W. Bosserman and J. M. Klopstek, eds., Elsevier, Amsterdam, pp. 145–155.

Daukas, P., D. Lowry, and W. W. Walker, Jr., 1989, Design of wet detention basins and constructed wetlands for treatment of stormwater runoff from a regional shopping mall in Massachusetts, in *Constructed Wetlands for Wastewater Treatment,* D. A. Hammer, ed., Lewis Publishers, Inc., Chelsea, Mich., pp. 686–694.

Davis, A. M., 1984, Ombrotrophic peatlands in Newfoundland, Canada: Their origins, development and trans-Atlantic affinities, *Chem. Geol.* **44:**287–309.

Davis, C. A., 1907, Peat, essays on its origin, uses, and distribution in Michigan, in *Report State Board Geological Survey Michigan for 1906,* pp. 95–395.

Davis, C. B., and A. G. van der Valk, 1978a, The decomposition of standing and fallen litter of *Typha glauca* and *Scirpus fluviatilis, Can. J. Bot.* **56:**662–675.

Davis, C. B., and A. G. van der Valk, 1978b, Litter decomposition in prairie glacial marshes, in *Freshwater Wetlands: Ecological Processes and Management Potential,* R. E. Good, D. F. Whighan, and R. L. Simpson, ed., Academic Press, New York, pp. 99–113.

Davis, D. G., 1989, No net loss of the nation's wetlands: a goal and a challenge, *Water Environment and Technology* **4:5**13–514.

Davis, J. H., 1940, The ecology and geologic role of mangroves in Florida, *Carnegie Institution, Washington Publ. No. 517,* pp. 303–412.

Davis, J. H., 1943, The natural features of southern Florida, especially the vegetation and the Everglades, *Florida Geol. Survey Bull. 25,* 311p.

Davis, L. V., and I. E. Gray, 1966, Zonal and seasonal distribution of insects in North Carolina salt marshes, *Ecol. Monogr.* **36:**275–295.

Davis, S. N., and R. J. M. DeWiest, 1966, *Hydrogeology,* Wiley, New York, 463p.

Dawes , C. J., 1981, *Marine Botany,* Wiley, New York, 628p.

Day, F. P., Jr., 1982, Litter decomposition rates in the seasonally flooded Great Dismal Swamp, *Ecology* **63:**670–678.

Day, F. P., 1987, Production and decay in a *Chamaecyparis thyoides* swamp in southeastern Virginia, in *Atlantic White Cedar Wetlands,* A. Laderman, ed., Westview Press, Boulder, Col., pp. 123–132.

Day, F. P., Jr., and C. V. Dabel, 1978, Phytomass budgets for the Dismal Swamp ecosystem, *Virginia J. Sci.* **29:**220–224.

Day, F. P., S. K. West, and E. G. Tupacz, 1988, The influence of ground-water dynamics in a periodically flooded ecosystem, the Great Dismal Swamp, *Wetlands* **8:**1–13.

Day, F. P., Jr., J. P. Megonigal, and L. C. Lee, 1989, Cypress root decomposition in experimental wetland mesocosms, *Wetlands* **9:**263–282.

Day, J. H., 1981, *Estuarine Ecology: With Particular Reference to Southern Africa,* A. A. Balkema, Rotterdam, 411p.

Day, J. W., Jr., W. G. Smith, P. Wagner, and W. Stowe, 1973, *Community Structure and Carbon Budget in a Salt Marsh and Shallow Bay Estuarine System in Louisiana,* Center for Wetland Resources, Louisiana State University, Baton Rouge, Sea Grant Publ. LSU-SG-72-04, 30p.

Day, J. W., Jr., T. J. Butler, and W. G. Conner, 1977, Productivity and nutrient export studies in a cypress swamp and lake system in Louisiana, in *Estuarine Processes,* vol. II, M. Wiley, ed., Academic Press, New York, pp. 255–269.

Day, J. W., Jr., W. H. Conner, F. Ley-Lou, R. H. Day, and A. M. Navarro, 1987, The productivity and composition of mangrove forests, Laguna de Términos, Mexico, *Aquatic Botany* **27:**267–284.

Day, J. W., Jr., C. A. S. Hall, W. M. Kemp, and A. Yánez-Arancibia, 1989, *Estuarine Ecology*, Wiley, New York, 558p.

Day, R. T., P. A. Keddy, J. M. NcNeil, and T. J. Carleton, 1988, Fertility and disturbance gradients: a summary model for riverine marsh vegetation, *Ecol.* **69:**1044–1054.

de la Cruz, A. A., 1974, Primary productivity of coastal marshes in Mississippi, *Gulf Res. Rep.* **4:**351–356.

de la Cruz, A. A., 1978, Primary production processes: summary and recommendations, in *Freshwater Wetlands: Ecological Processes and Management Potential,* R. E. Good, D. F. Whigham, and R. L. Simpson, eds., Academic Press, New York, pp. 79–86.

DeBusk, W. F., and K. R. Reddy, 1987, Removal of floodwater nitrogen in a cypress swamp receiving primary wastewater effluent, *Hydrobiologia* **153:**79–86.

Deghi, G. S., K. C. Ewel, and W. J. Mitsch, 1980, Effects of sewage effluent application on litterfall and litter decomposition in cypress swamps, *J. Appl. Ecol.* **17:**397–408.

Delaune, R. D., R. J. Buresh, and W. H. Patrick, Jr., 1979, Relationship of soil properties to standing crop biomass of *Spartina alterniflora* in a Louisiana marsh, *Estuarine Coast. Mar. Sci.* **8:**477–487.

Delaune, R. D., and W. H. Patrick, Jr., 1979, Nitrogen and phosphorus cycling in a Gulf Coast salt marsh, in *Estuarine Perspectives,* V. S. Kennedy, ed., Academic Press, New York, pp. 143–151.

Delaune, R. D., C. M. Reddy, and W. H. Patrick, Jr., 1981, Accumulation of plant nutrients and heavy metals through sedimentation processes and accretion in a Louisiana salt marsh, *Estuaries* **4:**328–334.

Delaune, R. D., C. J. Smith, and W. H. Patrick, 1983a, Methane release from Gulf coast wetlands, *Tellus* **35B:**8–15.

Delaune, R. D., C. J. Smith, and W. H. Patrick, Jr., 1983b, Nitrogen losses from a Louisiana Gulf Coast salt marsh, *Est. Coast. Shelf Sci.* **17:**133–142.

Delaune, R. D., R. H. Baumann, and J. G. Gosselink, 1983c, Relationships among vertical accretion, coastal submergence, and erosion in a Louisiana Gulf Coast marsh, *J. Sed. Petrol.* **53:**147–157.

Delaune, R. D., and C. W. Lindau, 1990, Fate of added 15N labelled nitrogen in a *Sagittaria lancifolia* L. Gulf Coast marsh, *J. Freshwater Ecol.* **5**(3):429–431.

Delmas, R., J. Baudey, J. Servant, and Y. Baziard, 1980, Emissions and concentrations of hydrogen sulfide in the air of the tropical forest of the Ivory Coast and of temperate regions in France, *J. Geophysical Research* **85:**4468–4474.

Delwiche, C. C., 1970, The nitrogen cycle, *Sci. Am.* **223:**137–146.

Demaree, D., 1932, Submerging experiments with *Taxodium, Ecology* **13:**258–262.

Dennis, J. G., and P. L. Johnson, 1970, Shoot and rhizome-root standing crops of tundra vegetation at Barrow, Alaska, *Arct. Alp. Res.* **2:**253–266.

Dennis, J. V., 1988, *The Great Cypress Swamps*, Louisiana State Univ. Press, Baton Rouge, LA, 142p.

Dennis, W., and W. Batson, 1974, The floating log and stump communities in the Santee Swamp of South Carolina, *Castanea* **39:**166–170.

Dent, D. L., 1986, *Acid Sulphate Soils: a Baseline for Research and Development,* ILRI Publication 39, Wageningen, Netherlands.

Dent, D. L., 1992, Reclamation of acid sulphate soils, in *Soil Restoration, Advances in Soil Science,* Vol. 17, R. Lal and B. A. Stewart, eds., Springer-Verlag, New York, pp. 79–122.

Derksen, A. J., 1989, Autumn movements of underyearling northern pike, *Esox lucius*, from a large Manitoba marsh, *Can. Field Nat.* **103**(3):429–431.

DeRoia, D. M., and T. A. Bookhout, 1989, Spring feeding ecology of teal on the Lake Erie marshes (abstract), *Ohio J. Sci.* **89**(2):3.

DeSylva, D. P., 1969, Trends in marine sport fisheries research, *Am. Fish. Soc. Trans.* **98:**151–169.

Detwiler, R. P., and C. A. S. Hall, 1988, Tropical forests and the global carbon cycle, *Science* **239:**42–47.

Diaz, R. J.,1977, *The Effects of Pollution on Benthic Communities of the Tidal James River, Virginia,* Ph.D. dissertation, University of Virginia, 149p.

Dierberg, F. E., and P. L. Brezonik, 1983a, Tertiary treatment of municipal wastewater by cypress domes, *Water Research* **17:**1027–1040.

Dierberg, F. E., and P. L. Brezonik, 1983b, Nitrogen and phosphorus mass balances in natural and sewage-enriched cypress domes, *Journal of Applied Ecology* **20:**323–337.

Dierberg, F. E., and P. L. Brezonik, 1984, Nitrogen and phosphorus mass balances in a cypress dome receiving wastewater, in *Cypress Swamps,* K. C. Ewel and H. T. Odum, eds., University Presses of Florida, Gainesville, pp. 112–118.

Dierberg, F. E., and P. L. Brezonik, 1985, Nitrogen and phosphorus removal by cypress swamp sediments, *Water, Air and Soil Pollution* **24:**207–213.

Diers, R., and J. L. Anderson, 1984, Part I, Development of soil mottling, in *Soil Survey Horizons*, winter 1984, pp. 9–12.

Dolan, T. J., S. E. Bayley, J. Zoltek, Jr., and A. Hermann, 1981, Phosphorus dynamics of a Florida freshwater marsh receiving treated wastewater, *J. Appl. Ecol.* **18:**205–219.

Dopson, J. R., P. W. Greenwood, and R. L. Jones, 1986, Holocene forest and wetland vegetation dynamics at Barrington Tops, New South Wales, *J. Biogeography* **13:**561–585.

Dorffling, K., D. Tietz, J. Streich, and M. Ludewig, 1980, Studies on the role of abscisic acid in stomatal movements, in *Plant Growth Substances 1979*, F. Skoog, ed., Springer-Verlag, Berlin, pp. 274–285.

Dorge, C. L., W. J. Mitsch, and J. R. Wiemhoff, 1984, Cypress wetlands in southern Illinois, in *Cypress Swamps*, K. C. Ewel and H. T. Odum, eds., University Presses of Florida, Gainesville, pp. 393–404.

Douglas, M. S., 1947, *The Everglades: River of Grass,* Ballantine, New York, 308p.

Doyle, G. J., 1973, Primary production estimates of native blanket bog and meadow vegetation growing on reclaimed peat at Glenamoy, Ireland, in *Primary Production and Production Processes, Tundra Biome,* L. C. Bliss and F. E. Wielgolaski, eds., Tundra Biome Steering Committee, Edmonton and Stockholm, pp. 141–151.

Driscoll, C. T., G. E. Likens, L. O. Hedin, J. S. Eaton, and F. H. Bormann, 1989, Changes in the chemistry of surface waters, *Environmental Science and Technology* **23:**137–143.

DuBarry, A. P., Jr., 1963, Germination of bottomland tree seed while immersed in water, *Journal of Forestry* **61:**225–226.

Duever, M. J., 1984, Environmental factors controlling plant communities of the Big Cypress Swamp, in *Environments of South Florida: Present and Past II,* P. J. Gleason, ed., Miami Geological Society, Coral Gables, Fla., pp. 127–137.

Duever, M. J., 1988, Hydrologic processes for models of freshwater wetlands, in *Wetland Modelling,* W. J. Mitsch, M. Straskraba, and S. E. Jørgensen, eds., Elsevier Science Publishers, Amsterdam pp. 9–39.

Duever, M. J., 1990, Hydrology, in *Wetlands and Shallow Continental Water Bodies,* B. C. Patten, ed., vol. 1, SPB Academic Publishing, The Hague, Netherlands, pp. 61–89.

Duever, M. J., J. E. Carlson, and L. A. Riopelle, 1984, Corkscrew Swamp: a virgin cypress strand, in *Cypress Swamps*, K. C. Ewel and H. T. Odum, eds., University Presses of Florida, Gainesville, pp. 334–348.

Duever, M. J., J. E. Carlson, J. F. Meeder, L. C. Duever, L. H. Gunderson, L. A. Riopelle, T. R. Alexander, R. L Myers, and D. P. Spangler, 1986, *The Big Cypress National Preserve*, Research Report no. 8, National Audubon Society, New York.

Dunbar, J. B., 1990, *Land Loss Rates, Report 2, Louisiana Chenier Plain*, U.S. Army Corps of Engineers, Waterways Experiment Sta., Vicksburg, Miss., COE Tech Report GL–90–2.

Dunbar, J. B., L. D. Britsch, and E. B. Kemp, 1992, *Land Loss Rates, Report 3, Louisiana Coastal Plain.*, U.S. Army Corps of Engineers, Waterways Experiment Station, Vicksburg, Miss., Technical Report GL–90–2.

Dunn, M. L., 1978, *Breakdown of Freshwater Tidal Marsh Plants,* M.S. Thesis, Univ. Virginia, Charlottesville.

Dunn, W. J., 1989, Wetland succession—What is the appropriate paradigm?, in *Wetlands Concerns and Successes,* D. W. Fisk, ed., *Proceedings of the Conference American Water Resources Association: Wetlands Concerns and Successes*, Tampa, Florida. American Water Resources Association. Tampa, Fla., pp. 473–488.

Dunne, T., and L. B. Leopold, 1978, *Water in Environmental Planning,* W. H. Freeman and Company, New York, 818p.

Durno, S. E., 1961, Evidence regarding the rate of peat growth, *J. Ecol.* **49:**347–351.

Dvorak, J., 1978, Macrofauna of invertebrates in helohyte communities, in *Pond Littoral Ecosystems,* D. Dykyjova and J. Kvet, eds., Springer-Verlag, Berlin, pp. 389–392.

Dykyjova, D., and J. Kvet, eds., 1978, *Pond Littoral Ecosystems,* Springer-Verlag, Berlin, 464p.

Dykyjova, D., and J. Kvet, 1982, Mineral nutrient economy in wetlands of the Trebon Basin Biosphere Reserve, Czechoslovakia, in *Wetlands—Ecology and Management,* B. Gopal, R. E. Turner, R. G. Wetzel and D. F. Whigham, eds., National Institute of Ecology and International Scientific Publications, Jaipur, India, pp. 335–355.

Earth Satellite Corporation, 1972, Aerial multiband wetlands mapping, *Photogramm. Eng.* **38:**1188–1189.

Eggelsmann, R., 1963, *Die Potentielle und Aktuelle Evaporation eines Seeklimath-ochmoores,* Internat. Assoc. Sci. Hydrol. Publication No. 62 pp. 88–97.

Egler, F. E., 1952, Southeast saline Everglades vegetation, Florida, and its management, *Veg, Acta, Geobot.* **3:**213–265.

Eisenlohr, W. S., 1966, Water loss from a natural pond through transpiration by hydrophytes, *Water Resour. Res.* **2:**443–453.

Eisenlohr, W. S., 1972, *Hydrologic Investigations of Prairie Potholes in North Dakota, 1958–1969,* U.S. Geol. Survey, Washington, D.C., Paper 585.

Elder, J. F., 1985, Nitrogen and phosphorus speciation and flux in a large Florida river-wetland system, *Water Resour. Res.* **21:**724–732.

Elder, J. F., and H. C. Mattraw, Jr., 1982, Riverine transport of nutrient and detritus to the Apalachicola Bay estuary, Florida, *Water Resour. Bull.* **18:**849–856.

Ellanna, L. J., and P. C. Wheeler, 1986, Subsistence use of wetlands in Alaska, in *Alaska: Regional Wetland Functions, Proceedings of a Workshop Held at Anchorage, Alaska, May 28–29, 1986,* Public. no. 90–1, A. G. van der Valk and J. Hall, eds., The Environmental Institute, Univ. Massachusetts, Amherst, Mass., pp. 85–103.

Ellison, R. L., and M. M. Nichols, 1976, Modern and holocene foraminifera in the Chesapeake Bay region, *Mar. Sed. Spec. Publ.* **1:**131–151.

Emery, K. O., and E. Uchupi, 1972, Western North Atlantic Ocean: topography, rocks, structure, water, life, and sediments, *Am. Assoc. Petroleum Geologists, Mem. 17,* 532p.

Environmental Defense Fund and World Wildlife Fund, 1992, *How Wet Is a Wetland? The Impact of the Proposed Revisions to the Federal Wetlands Delineation Manual,* Environmental Defense Fund and World Wildlife Fund, Washington, D.C., 175p.

Ernst, J. P., and V. Brown, 1989, Conserving endangered species on southern forested wetlands, in *Proceedings of the Symposium: The Forested Wetlands of the Southern United States,* D. D. Hook and R. Lea, eds., U.S. Department of Agriculture, Forest Service, Southeastern Forest Experiment Station, Asheville, N.C., pp. 135–145.

Ernst, W. H. O., 1990, Ecophysiology of plants in waterlogged and flooded environments, *Aquatic Bot.* **38:**73–90.

Erwin, K. L., 1990, Freshwater marsh creation and restoration in the Southeast, in *Wetland Creation and Restoration,* J. A. Kusler and M. E. Kentula, eds., Island Press, Washington, D.C., pp. 233–265.

Erwin, K. L., 1991, *An Evaluation of Wetland Mitigation in the South Florida Water Management District,* Volume I, Final Report to South Florida Water Management District, West Palm Beach, Florida, 124p.

Erwin, K. L., G. R. Best, W. J. Dunn, and P. M. Wallace, 1984, Marsh and forested wetland reclamation of a central Florida phosphate mine, *Wetlands* **4:**87–104.

Esry, D. H., and D. J. Cairns, 1989, Overview of the Lake Jackson Restoration Project with artificially created wetlands for treatment of urban runoff, in *Wetlands Concerns and Successes,* D. W. Fisk, ed., *Proceedings of the Conference American Water Resources Association: Wetlands Concerns and Successes,* Tampa, Florida. American Water Resources Association. Tampa, Fla., pp. 247–257.

Etherington, J. R., 1983, *Wetland Ecology,* Edward Arnold, London, 67p.

Evink, G. L., 1980, *Studies of Causeways in the Indian River, Florida,* Florida Dept. of Transportation Rept. FL-ER-7-80, Tallahassee, Fla., 140p.

Ewel, K. C., 1976, Effects of sewage effluent on ecosystem dynamics in cypress domes. in *Freshwater Wetlands and Sewage Effluent Disposal,* D. L. Tilton, R. H. Kadlec, and C. J. Richardson, eds., University of Michigan, Ann Arbor, pp. 169-195.

Ewel, K. C., 1990, Swamps, in *Ecosystems of Florida,* R. L. Myers and J. J. Ewel, eds., University of Central Florida Press, Orlando, pp. 281–323.

Ewel, K. C., and W. J. Mitsch, 1978, The effects of fire on species composition in cypress dome ecosystems, *Florida Sci.* **41:**25-31.

Ewel, K. C., and H. T. Odum, 1978, Cypress swamps for nutrient removal and wastewater recycling, in *Advances in Water and Wastewater Treatment Biological Nutrient Removal,* M. B Wanielista and W. W. Eckenfelder, Jr., eds., Ann Arbor Sci. Pub., Inc., Ann Arbor, Mich. pp. 181–198.

Ewel, K. C., and H. T. Odum, 1979, Cypress domes: nature's tertiary treatment filter, in *Utilization of Municipal Sewage Effluent and Sludge on Forest and Disturbed Land,* W. E. Sopper and S. N. Kerr, eds., The Pennsylvania State University Press, University Park, pp. 103–114.

Ewel, K. C., and H. T. Odum, eds., 1984, *Cypress Swamps,* University Presses of Florida, Gainesville, Fla., 472p.

Ewel, K. C., and L. P. Wickenheiser, 1988, Effects of swamp size on growth rates of cypress (*Taxodium distichum*) trees, *American Midland Naturalist* **120:**362–370.

Ewel, K. C., and J. E. Smith, 1992, Evapotranspiration from Florida pondcypress swamps, *Water Resources Bull.* **28:**299–304.

Ewing, K., and K. A. Kershaw, 1986, Vegetation patterns in James Bay coastal marshes, I. Environmental factors on the south coast, *Can. J. Bot.* **64:**217–226.

Faber, P. A., E. Keller, A. Sands, and B. M. Masser, 1989, *The Ecology of Riparian Habitats of the Southern California Coastal Region: a Community Profile*, U.S. Fish Wildl. Serv., Washington, D.C., Biol. Rep. **85**(7.27), 152p.

Fail, J. L., M. N. Hamzah, B. L. Haines, and R. L. Todd, 1989, Above and belowground biomass, production, and element accumulation in riparian forests of an agricultural watershed, in *Proceedings of the Symposium: The Forested Wetlands of the Southern United States,* Asheville, N.C., D. D. Hook, and R. Lea, ed. U.S. Department of Agriculture, Forest Service, Southeastern Forest Experiment Station, pp.193–223.

Farber, S., 1987, The value of coastal wetlands for protection of property against hurricane wind damage, *J. Environ. Econ. and Manage.* **14:**143–151.

Farber, S., and R. Costanza, 1987, The economic value of wetlands systems, *J. Environmental Manage.* **24:**41–51.

Faulkner, S. P., and C. J. Richardson, 1989, Physical and chemical characteristics of freshwater wetland soils, in *Constructed Wetlands for Wastewater Treatment,* D. A. Hammer, ed., Lewis Publishers, Chelsea, Michigan, pp. 41–72.

Faulkner, S. P., W. H. Patrick, Jr., and R. P. Gambrell, 1989, Field techniques for measuring wetland soil parameters, *Soil Science Society of America Journal* **53:**883–890.

Feierabend, S. J., and J. M. Zelazny, 1987, *Status Report on our Nations's Wetlands*, National Wildlife Federation, Washington, D.C., 50p.

Feminella, J. W., and V. H. Resh, 1989, Submersed macrophytes and grazing crayfish: an experimental study of herbivory in a California freshwater marsh, *Holarctic Ecol.* **12:**1–8.

Fenchel, T., 1969, The ecology of marine microbenthos. IV. Structure and function of the benthic ecosystem, its chemical and physical factors and the microfauna communities with special reference to the ciliated protozoa., *Ophelia* **6:**1–182.

Fennessy, M. S., 1991, *Ecosystem Development in Restored Riparian Wetlands,* Ph.D. Dissertation, The Ohio State University, Columbus, 201p.

Fennessy, M. S., and W. J. Mitsch, 1989a, Design and use of wetlands for renovation of drainage from coal mines, in *Ecological Engineering: An Introduction to Ecotechnology,* W. J. Mitsch and S. E. Jørgensen, eds., Wiley, New York, pp. 231–253.

Fennessy, M. S., and W. J. Mitsch, 1989b, Treating coal mine drainage with an artificial wetland. *Research Journal Water Pollution Control Federation,* **61**:1691–1701.

Fetter, C. W., Jr., W. E. Sloey, and F. L. Spangler, 1978, Use of a natural marsh for wastewater polishing, *J. Water Pollution Control Fed.* **50:**290–307.

Field, D. W., A. J. Reyer, P. V. Genovese, and B. D. Shearer, 1991, *Coastal Wetlands of the United States,* National Oceanic and Atmospheric Administration and U.S. Fish Wildl. Serv., Washington, D.C.,

Finlayson, M., and M. Moser, eds., 1991, *Wetlands,* Facts On File, Oxford, England, 224p.

Fitter, A. H., and R. K. M. Hay, 1987, *Environmental Physiology of Plants*, Academic Press, New York, 423p.

Fleming, M., G. Lin, and L. da S. L. Sternberg, 1990, Influence of mangrove detritus in an estuarine ecosystem, *Bulletin of Marine Science* **47:**663–669.

Flores-Verdugo, F., J. W. Day, Jr., and R. Briseno-Duenas, 1987, Structure, litter fall, decomposition, and detritus dynamics of mangroves in a Mexican coastal lagoon with an ephemeral inlet, *Marine Ecology—Progress Series* **35:**83–90.

Flores-Verdugo, F., F. Gonzáles-Farías, O. Ramírez-Flores, F. Amezcua-Linares, A. Yánez-Arancibia, M. Alvarez-Rubio, and J. W. Day, Jr. 1990, Mangrove ecology, aquatic primary productivity, and fish community dynamics in the Teacapán-Agua Brava lagoon-estuarine system (Mexican Pacific), *Estuaries* **13:**219–230.

Folke, C., 1991, The societal value of wetland life-support, in *Linking the Natural Environment and the Economy*, C. Folke and T. Kaberger, eds., Kluwar Academic Publ., Dordrecht, the Netherlands, pp. 141–171.

Forman, R. T. T., and M. Godron, 1986, *Landscape Ecology*, Wiley, New York.

Forrest, G. I., and R. A. H. Smith, 1975, The productivity of a range of blanket bog types in the Northern Pennines, *J. Ecol.* **63:**173–202.

Foster, D. R., G. A. King, P. H. Glaser, and H. E. Wright, 1983, Origin of string patterns in boreal peatlands, *Nature* **306:**256–258.

Fox, T. C., R. A. Kennedy, and A. A. Alani, 1988, Biochemical adaptations to anoxia in barnyard grass, in *The Ecology and Management of Wetlands, vol. 1: Ecology of Wetlands.*, D. D. Hook et al., ed., Timber Press, Portland, Oregon, 2 vol., pp. 359–372.

Frayer, W. E., T. J. Monahan, D. C. Bowden, and F. A. Graybill, 1983, *Status and Trends of Wetlands and Deepwater Habitat in the Conterminous United States, 195Os to 1970s,* Dept. of Forest and Wood Sciences, Colorado State University, Fort Collins, 32p.

Frazer, C., J. R. Longcore, and D. G. McAuley, 1990a, Habitat use by postfledging American black ducks in Maine and New Brunswick, *J. Wildl. Manage.* **54**(3):451–459.

Frazer, C., J. R. Longcore, and D. G. McAuley, 1990b, Home range and movements of postfledging American ducks in eastern Maine, *Can. J. Zool.* **68:**1228–1291.

Fredrickson, L. H., 1979, Lowland hardwood wetlands: current status and habitat values for wildlife, in *Wetland Functions and Values: The State of Our Understanding,* P. E. Greeson, J. R. Clark, and J. E. Clark, eds., American Water Resources Assoc., Minneapolis, Minn., pp. 296–306.

Fredrickson, L. H., and F. A. Reid, 1990, Impacts of hydrologic alteration on management of freshwater wetlands, in *Management of Dynamic Ecosystems, North Central Section,* J. M. Sweeney, ed., The Wildlife Society, West Lafayette, Indiana, pp. 71–90.

Freiberger, H. J., 1972, *Streamflow Variation and Distribution in the Big Cypress Watershed During Wet and Dry Periods,* Map Series 45, Bureau of Geology, Florida Department of Natural Resources, Tallahassee, Florida.

Frey, R. W., and P. B. Basan, 1985, Coastal salt marshes, in *Coastal Sedimentary Environments.*, R. A. Davis, Jr., ed., Springer-Verlag, Inc., New York, pp. 225–301.

Frissell, C. A., 1986, *A Hierarchical Stream Habitat Classification System: Development and Demonstration,* M.S. Thesis, Oregon State Univ., Corvallis.

Frissell, C. A., W. J. Liss, C. E. Warren, and M. D. Hurley, 1986, A hierarchical framework for stream habitat classification: viewing streams in a watershed context, *Environmental Management* **10:**199–214.

Futyma, R. P., 1988, Fossil pollen and charcoal analysis in wetland development studies at Indiana Dunes National Lakeshore, in *Interdisciplinary Approaches to Freshwater Wetlands Research,* D. A. Wilcox, ed., Michigan State University Press, East Lansing, Mich., pp. 11–23.

Gaddy, L. L., 1978, Congaree: forest of giants, *American Forests* **84:**51–53.

Gallagher, J. L., 1978, Decomposition processes: summary and recommendations, in *Freshwater Wetlands: Ecological Processes and Management Potential.* R. E. Good, D. F. Whigham, and R. L. Simpson, eds., Academic Press, New York, pp. 145–151.

Gallagher, J. L., and F. C. Daiber, 1974, Primary production of edaphic algae communities in a Delaware salt marsh, *Limnol. Oceanogr.* **19:**390–395.

Gallagher, J. L., R. J. Reimold, R. A. Linthurst, and W. J. Pfeiffer, 1980, Aerial production, mortality, and mineral accumulation-export dynamics in *Spartina alterniflora* and *Juncus roemerianus* plant stands in a Georgia salt marsh, *Ecology* **61:**303–312.

Gambrell, R. P., and W. H. Patrick, Jr., 1978, Chemical and microbiological properties of anaerobic soils and sediments, in *Plant Life in Anaerobic Environments,* D. D. Hook and R. M. M. Crawford, eds., Ann Arbor Sci. Pub. Inc., Ann Arbor, Mich., pp. 375–423.

Garbisch, E. W., 1977, *Recent and Planned Marsh Establishment Work Throughout the Contiguous United States: A Survey and Basic Guidelines.* CR D–77–3, U.S. Army Corps of Engineers Waterways Experiment Station, Vicksburg, Mississippi.

Garbisch, E. W., 1989, Wetland enhancement, restoration, and construction, in *Wetlands Ecology and Conservation: Emphasis in Pennsylvania,* S. K. Majumdar, ed., The Pennsylvania Academy of Science. Easton, Pa., pp. 261–275.

Garbisch, E. W., P. B. Woller, and R. J. McCallum, 1975, *Salt Marsh Establishment and Development,* Tech. Memo. 52, U.S. Army Coastal Engineering Research Center, Fort Belvoir, Virginia.

Gardner, L. R., 1975, Runoff from an intertidal marsh during tidal exposure: regression curves and chemical characteristics, *Limnol. Oceanogr.* **20:**81–89.

Gemborys, S. R., and E. J. Hodgkins, 1971, Forests of small stream bottoms in the coastal plain of southwestern Alabama, *Ecology* **52:**70–84.

Gerheart, R. A., F. Klopp, and G. Allen, 1989, Constructed free surface wetlands to treat and receive wastewater: pilot project to full scale, in *Constructed Wetlands for Wastewater Treatment,* D. A. Hammer, ed., Lewis Publishers, Inc., Chelsea, Mich., pp. 121–137.

Gerritsen, J., and H. S. Greening, 1989, Marsh seed banks of the Okefenokee Swamp: effects of hydrologic regime and nutrients, *Ecology* **70:** 750–763.

Gersberg, R. M., B. V. Elkins, and C. R. Goldman, 1983, Nitrogen removal in artifical wetlands, *Water Resources,* **17**(9):1009–1014.

Gersberg, R. M., B. V. Elkins, and C. R. Goldman, 1984, Use of artificial wetlands to remove nitrogen from wastewater, *Journal of the Water Pollution Control Federation,* **56**(2):152–156.

Gersberg, R. M., B. V. Elkins, S. R. Lyon, and C. R. Goldman, 1986, Role of aquatic plants in wastewater treatment by artificial wetlands, *Water Resources,* **20**(3):363–368.

Gersberg, R. M., S. R. Lyon, R. Brenner, and B. V. Elkins, 1989, Integrated wastewater treatment using artificial wetlands: a gravel marsh case study, in *Constructed Wetlands for Wastewater Treatment,* D. A. Hammer, ed., Lewis Publishers, Inc. Chelsea, Mich., pp. 145–152.

Gilbert, T., T. King, and B. Barnett, 1981, *An Assessment of Wetland Habitat Establishment at a Central Florida Phosphate Mine Site,* U.S. Fish and Wildlife Service Pub. FWS/OBS-81/45, Atlanta, Ga., 96p.

Gilman, K., 1982, Nature conservation in wetlands: two small fen basins in western Britain, in *Ecosystem Dynamics in Freshwater Wetlands and Shallow Water Bodies,* vol. I, D. O. Logofet and N. K. Luckyanov, eds., SCOPE and UNEP Workshop, Center of International Projects, Moscow, USSR, pp. 290–310.

Ginsburg, M., L. Sachs, and B. Z. Ginsburg, 1971, Ion metabolism in a halobacterium II. Ion concentration in cells at different levels of metabolism, *J. Membr. Biol.* **5:**78–101.

Girts, M. A., and R. M. Knight, 1989, *Operations Optimization,* in *Constructed Wetlands for Wastewater Treatment,* D. A. Hammer, ed., Lewis Publishers, Inc., Chelsea, Mich., pp. 417–430.

Giurgevich, J. R., and E. L. Dunn, 1978, Seasonal patterns of CO_2 and water vapor exchange of *Juncus roemerianus* Scheele in a Georgia salt marsh, in *Am. J. Bot.* **65:**502–510.

Giurgevich, J. R., and E. L. Dunn, 1979, Seasonal patterns of CO_2 and water vapor exchange of the tall and short forms of *Spartina alterniflora* Loisel in a Georgia salt marsh, *Oecologia* **43:**139–156.

Glaser, P. H., 1983a, *Carex exilis* and *Scirpus cespitosus* var. callosus in patterned fens in northern Minnesota, *Mich. Bot.* **22:**22–26.

Glaser, P. H., 1983b, *Eleocharis rostellata* and its relation to spring fens in Minnesota, *Mich Bot.* **22:**19–21.

Glaser, P. H., 1983c, Vegetation patterns in the North Black River peatland, northern Minnesota, *Can. J. Bot.* **61:**2085–2104.

Glaser, P. H., 1987, *The Ecology of Patterned Boreal Peatlands of Northern Minnesota: A Community Profile.,* U.S. Fish Wildl. Serv. Rep., Report 85 (7.14), Washington, D.C., 98p.

Glaser, P. H., and G. A. Wheeler, 1980, The development of surface patterns in the Red Lake peatland, northern Minnosota, in *Proc. Sixth Int. Peat Congress,* Duluth, Minn., 31–35.

Glaser, P. H., G. A. Wheeler, E. Gorham, and H. E. Wright Jr., 1981, The patterned mires of the Red Lake peatland, northern Minnosota: vegetation, water chemistry, and landforms, in *J. Ecology* **69:**575–599.

Glaser, P. H., J. A. Janssens, and D. I. Siegel, 1990, The response of vegetation to chemical and hydrological gradients in the Lost River peatland, northern Minnesota, in *J. Ecology* **78:**1021–1048.

Gleason, H. A., 1917, The structure and development of the plant association, *Torrey Bot. Club Bull.* **44:**463–481.

Glob, P. V., 1969, *The Bog People: Iron Age Man Preserved,* trans. by R. Bruce-Mitford, Cornell University Press, Ithaca, New York, 200p.

Glooschenko, V., and P. Grondin, 1988, Wetlands of eastern temperate Canada, in *Wetlands of Canada,* National Wetlands Working Group, ed., Environment Canada and Polyscience Publications, Inc., Ottawa, Ontario and Montreal, Quebec, pp. 199–248.

Glubiak, P. G., R. H. Nowka, and W. J. Mitsch, 1986, Federal and state management of inland wetlands: are states ready to assume control? *Environ. Manage.* **10:**145–156.

Godfrey, P. J., E. R. Kaynor, S. Pelczarski, and J. Benforado, eds., 1985, *Ecological Considerations in Wetlands Treatment of Municipal Wastewaters,* Van Nostrand Reinhold Company, New York, 474p.

Godwin, H., 1981, *The Archives of the Peat Bogs,* Cambridge University Press, Cambridge, England, 229p.

Goldman, C. R., 1961, The contribution of alder trees (*Alnus tenuifolia*) to the primary production of Castle Lake, California, *Ecology* **42:**282–288.

Golet, F. C., and J. S. Larson, 1974, *Classification of Freshwater Wetlands in the Glaciated Northeast,* U.S. Fish and Wildlife Service Resources Publ. 116, Washington, D.C., 56p.

Golley, F. B., H. T. Odum, and R. F. Wilson, 1962, The structure and metabolism of a Puerto Rican red mangrove forest in May, *Ecology* **43:**9–19.

Gomez, M. M., and F. P Day, Jr., 1982, Litter nutrient content and production in the Great Dismal Swamp, *Am. J. Bot.* **69:**1314–1321.

Good, R. E., D. F. Whigham, and R. L. Simpson, eds., 1978, *Freshwater Wetlands: Ecological Processes and Management Potential,* Academic Press, New York, 378p.

Good, R. E., N. F. Good, and B. R. Frasco, 1982, A review of primary production and decomposition dynamics of the belowground marsh component, in *Estuarine Comparisons,* V. S. Kennedy, ed., Academic Press, New York, pp. 139–157.

Goodman, G. T., and D. F. Perkins, 1968, The role of mineral nutrients in *Eriophorum* communities IV. Potassium supply as a limiting factor in an *E. vaginatum* community, *J. Ecology* **56**:685-696.

Goodman, P. J., and W. T. Williams, 1961, Investigations into "die-back," *J. Ecology* ***49****:391-398.*

Gopal, B., R. E. Turner, R. G. Wetzel, and D. F. Whigham, eds., 1982, *Wetlands—Ecology and Management,* National Institute of Ecology and International Scientific Publications, Jaipur, India, 514p.

Gopal, B., and V. Masing, 1990, Biology and ecology, in *Wetlands and Shallow Continental Water Bodies,* B. C. Patten, ed., SPB Academic Publishing, the Hague, Netherlands, pp. 91– 239.

Gore, A. J. P.. ed., 1983a. *Ecosystems of the world, vol.4A, Mires: Swamp, Bog, Fen, and Moor,* Elsevier, Amsterdam, 440p.

Gore, A. J. P., 1983b. Introduction, in *Ecosystems of the World,* vol. 4B, Mires: Swamp, Bog, Fen, and Moor, Elsevier, Amsterdam, pp. 1–34.

Gore, A. J. P., and S. E. Allen, 1956, Measurement of exchangeability and total cation content for H^+, Na^+, K^+, Mg^+, Ca^+, and iron in high level blanket peat, *Oikos* **7:**48–55.

Gorham, E., 1956, The ionic composition of some bogs and fen waters in the English lake district, *J. Ecol.* **44:**142–152.

Gorham, E., 1961, Factors influencing supply of major ions to inland waters, with special references to the atmosphere, *Geol. Soc. Amer. Bull.* **72:**795–840.

Gorham, E., 1967, *Some Chemical Aspects of Wetland Ecology,* Technical Mem. Committee on Geotechnical Research, National Research Council of Canada, No. 90, pp. 20–38.

Gorham, E., 1974, The relationship between standing crop in sedge meadows and summer temperature, *J. Ecol.* **62:**487–491.

Gorham, E., 1987, The Natural and Anthropogenic Acidification of Peatlands, in *Effects of Atmospheric Pollutants on Forests, Wetlands, and Agricultural Ecosystems,* T. C. Hutchinson and K. M. Meema, eds., Springer-Verlag, Berlin, pp. 493–512.

Gorham, E.,1991, Northern peatlands: role in the carbon cycle and probable responses to climatic Warming, *Ecol. Applications* **1:**182–195.

Gorham, E., S. E. Bayley, and D. W. Schindler, 1984a, Ecological effects of acid deposition upon peatlands: a neglected field in "acid-rain" research, *Can. J. Fish Aquatic Sci.* **41:**1256–1268.

Gorham, E., S. J. Eisenreich, J. Ford, and M. V. Santelmann, 1984b, The chemistry of bog waters, in *Chemical Processes in Lakes*, W. Strum, ed., Wiley, New York, pp. 339–363.

Gosselink, J. G., 1984, *The Ecology of Delta Marshes of Coastal Louisiana: A Community Profile,* U.S. Fish and Wildlife Service, Biological Services FWS/OBS-84/09, Washington, D.C., 134p.

Gosselink, J. G., and C. J. Kirby, 1974, Decomposition of salt marsh grass, *Spartina alterniflora* Loisel, *Limnol. Oceanogr.* **19:**825–832.

Gosselink, J. G., E. P. Odum, and R. M. Pope, 1974, *The Value of the Tidal Marsh,* Center for Wetland Resources Publ. LSU-SG-74-03, Louisiana State University, Baton Rouge, 30p.

Gosselink, J. G., and R. E. Turner, 1978, The role of hydrology in freshwater wetland ecosystems, in *Freshwater Wetlands: Ecological Processes and Management Potential,* R. E. Good, D. F. Whigham, and R. L. Simpson, eds., Academic Press, New York, pp. 63–78.

Gosselink, J. G., and R. H. Baumann, 1980, Wetland inventories: wetland loss along the United States coast, *Z. Geomorphol.,* N.F. Suppl.-Bd., **34:**173–187.

Gosselink, J. G., W. H. Conner, J. W. Day, Jr., and R. E. Turner, 1981, Classification of wetland resources: land, timber, and ecology, in *Timber Harvesting in Wetlands,* B. D. Jackson and J. L. Chambers, eds., Div. of Cont. Ed., Louisiana State Univ., Baton Rouge, pp. 28–48.

Gosselink, J. G., R. Hatton, and C. S. Hopkinson, 1984, Relationship of organic carbon and mineral content to bulk density in Louisiana marsh soils, *Soil Sci.* **137:**177–180.

Gosselink, J. G., et al., 1990a, Landscape conservation in a forested wetland watershed, *Bioscience* **40:**588–600.

Gosselink, J. G., L. C. Lee, and T. A. Muir, eds., 1990b, *Ecological Processes and Cumulative Impacts: Illustrated by Bottomland Hardwood Wetland Ecosystems*, Lewis Publishers, Chelsea, Mich., 708p.

Gosselink, J. G., M. M. Brinson, L. C. Lee, and G. T. Auble, 1990c, Human activites and ecological processes in bottomland hardwood ecosystems: the report of the ecosystem workgroup in *Ecological Processes and Cumulative Impacts: Illustrated by Bottomland Hardwood Wetland Ecosystems*, J. G. Gosselink, L. C. Lee, and T. A. Muir, eds., Lewis Publishers, Chelsea, Mich., pp. 549–598.

Gosselink, J. G., and E. Maltby, 1990, Wetland losses and gains, in *Wetlands: A Threatened Landscape*, M. Williams, ed., Basil Blackwell Ltd., Oxford, England, 296–322 pp.

Graf, W. L., 1988, Definition of flood plains along arid-region rivers, in *Flood Geomorphology*, V. R. Baker, R. C. Kochel, and P. C. Patton, eds., Wiley, New York, pp. 231–259.

Grant, J., and U. V. Bathmann, 1987, Swept away: resuspension of bacterial mats regulates benthic-pelagic exchange of sulfur, *Science* **236:**1472–1474.

Grant, R. R., and R. Patrick, 1970, Tinicum Marsh as a water purifier, in *Two Studies of Tinicum Marsh, Delaware and Philadelphia Counties, Pa.,* J. McCormick, R. R. Grant, Jr., and R. Patrick, eds., The Conservation Foundation, Washington, D.C., pp. 105–131.

Gray, A. J.,1992, Saltmarshes plant ecology: zonation and succession revisited, in *Saltmarshes: Morphodynamics, Conservation and Engineering Significance,* Cambridge Univ. Press, Cambridge, England, pp. 63–79.

Gray, L. C., O. E. Baker, F. J. Marschner, B. O. Weitz, W. R. Chapline, W. Shepard, and R. Zon, 1924, *The Utilization of our Lands for Crops, Pasture, and Forests,* in U.S. Dept. of Agriculture Yearbook, 1923, Government Printing Office, Washington. D.C., p. 415–506.

Greenwood, D. J., 1961, The effect of oxygen concentration on the decomposition of organic materials in soil, *Plant and Soil* **14:**360–376.

Greeson, P. E., J. R. Clark, and J. E. Clark, eds., 1979, *Wetland Functions and Values: The State of Our Understanding,* Proceedings of National Symposium on Wetlands, Lake Buena Vista, Florida, American Water Resources Assoc. Tech. Publ. TPS 79-2, Minneapolis, Minn., 674p.

Gregory, S. V., F. J. Swanson, W. A. McKee, and K. W. Cummins, 1991, An ecosystem perspective of riparian zones, *Bioscience* **41:**540–551.

Grigal, D. F., 1985, Sphagnum production in forested bogs of northern Minnesota, *Can. J. Bot.* **63:**1204–1207.

Grigal, D. F., C. C. Buttlema, and L. K. Kernik, 1985, Productivity of the woody strata of forested bogs in northern Minnesota, *Can. J. Bot.* **63:**2416–2424.

Gross, W. J., 1964, Trends in water and salt regulation among aquatic and amphibious crabs, *Biol. Bull.* **127:**447-466.

Gum, R. L., and W. E. Martin, 1975, Problems and solutions in estimating the demand for and value of rural outdoor recreation, *Am. J. Agric. Econ.* **57:**558–566.

Gunderson, L. H., and W. T. Loftis, 1993, The Everglades, in *Biodiversity of the Southeastern United States: Lowland Terrestrial Communities*, W. H. Martin, S. G. Boyce, and A. C. E. Echternacht, eds., Wiley, New York, pp. 199–255.

Guthery, F. S., J. M. Pates, and F. A. Stormer, 1982, Characterization of playas of the north-central Llano Estacado in Texas, *47th Am. Wildl. Nat. Resour. Conf. Trans.* **47:**516–527.

Hackney, C. T., and A. A. de la Cruz, 1980, Insitu decomposition of roots and rhizomes of two tidal marsh plants, *Ecology* **61:**226–231.

Haines, B. L., and E. L. Dunn, 1976, Growth and resource allocation responses of *Spartina alterniflocra* Loisel to three levels of NH_4-N, Fe, and NaCI in solution culture, *Bot. Gaz.* **137:**224–230.

Haines, E. B., 1979, Interactions between Georgia salt marshes and coastal waters: a changing paradigm, in *Ecological Processes in Coastal and Marine Systems,* R. J. Livingston, ed., Plenum Press, New York, pp. 35–46.

Haines, E. B., A. G. Chalmers, R. B. Hanson, and B. Sherr, 1977, Nitrogen pools and fluxes in a Georgia salt marsh, in *Estuarine Processes,* vol. 2, M. Wiley, ed., Academic Press, New York, pp. 241–254.

Hall, F R., R. J. Rutherford, and G. L. Byers, 1972, *The Influence of a New England Wetland on Water Quantity and Quality,* New Hampshire Water Resource Center Research Report 4, Univ. of New Hampshire, Durham, 51p.

Hall, H. D., and V. W. Lambou, 1990, The ecological significance to fisheries of bottomland hardwood ecosystems: values, detrimental impacts, and assessment: the report of the fisheries workgroup, in *Ecological Processes and Cumulative Impacts: Illustrated by Bottomland Hardwood Wetland Ecosystems,* J. G. Gosselink, L. C. Lee, and T. A. Muir, eds., Lewis Publishers, Chelsea, Mich., pp. 481–531.

Hall, T. F., and W. T. Penfound, 1939, A phytosociological study of a cypress-gum swamp in southern Louisiana, *Am. Midl. Nat.* **21:**378–395.

Hammack, J., and G. M. Brown, 1974, *Waterfowl and Wetlands: Toward Bioeconomic Analysis*, Johns Hopkins Univ. Press, Baltimore, Md.

Hammer D. A., ed., 1989, *Constructed Wetlands for Wastewater Treatment,* Lewis Publishers, Inc., Chelsea, Mich., 831p.

Hammer D. A., 1992, *Creating Freshwater Wetlands,* Lewis Publishers, Inc., Chelsea, Mich., 298p.

Hammer, D. E., and R. H. Kadlec, 1983, *Design Principles for Wetland Treatment Systems,* U.S. Environmental Protection Agency, Ada, Oklahoma, EPA–600/2–83–26, 244p.

Harris, L. D., 1988, The nature of cumulative impacts on biotic diversity of wetland vertebrates, *Environmental Management* **12:**675–693.

Harris, L. D., 1989, The faunal significance of fragmentation of southeastern bottomland forests, in *Proceedings of the Symposium: the Forested Wetlands of the Southern United States,* Asheville, N.C., D. D. Hook and R. Lea, eds., U.S. Department of Agriculture, Forest Service, Southeastern Forest Experiment Station, 126–134.

Harris, L. D., and J. G. Gosselink, 1990, Cumulative impacts of bottomland hardwood forest conversion on hydrology, water quality, and terrestrial wildlife, in *Ecological Processes and Cumulative Impacts Illustrated by Bottomland Hardwood Wetland Ecosystems*, J. G. Gosselink, L. C. Lee, and T. A. Muir, eds., Lewis Publishers, Chelsea, Mich., pp. 259–322.

Harris, R. R., 1988, Associations between stream valley geomorphology and riparian vegetation as a basis for landscape analysis in the Eastern Sierra Nevada, California, USA, *Environ. Manage.* **12:**219–228.

Harris, T., 1991, *Death in the Marsh*, Island Press, Washington, D.C., 245p.

Harriss, R. C., D. I. Sebacher, and F. P Day, Jr., 1982, Methane flux in the Great Dismal Swamp, *Nature* **297:**673–674.

Hatton, R. S., 1981, *Aspects of Marsh Accretion and Geochemistry: Barataria Basin, La.,* Master's thesis, Louisiana State University, Baton Rouge.

Havill, D. C., A. Ingold, and J. Pearson, 1985, Sulfide tolerance in coastal halophytes, *Vegetatio* **62:**279–285.

Heal, O. W., H. E. Jones, and J. B. Whittaker, 1975, Moore House, U.K., in *Structure and Function of Tundra Ecosystems,* T. Rosswall and O. W. Heal, eds., Ecol. Bull. 20, Swedish Natural Science Research Council, Stockholm, pp. 295–320.

Heald, E. J., 1969. *The Production of Organic Detritus in a South Florida Estuary,* Ph.D. dissertation, University of Miami, 110p.

Heald, E. J., 1971, *The Production of Organic Detritus in a South Florida Estuary,* University of Miami Sea Grant Technical Bulletin No. 6, Coral Gables, Fla., 110p.

Heard, R. W., 1982, *Guide to Common Tidal Marsh Invertebrates of the Northeastern Gulf of Mexico*, Mississippi–Alabama Sea Grant Consortium, Univ. of South Alabama, Mobile, Ala., MASGP-79–004.

Heilman, P E., 1968, Relationship of availability of phosphorus and cations to forest succession and bog formation in interior Alaska, *Ecology* **49:**331–336.

Heimburg, K., 1984, Hydrology of north-central Florida cypress domes, in *Cypress Swamps,* K. C. Ewel and H. T. Odum, eds., University Presses of Florida, Gainesville, pp. 72–82.

Heinle, D. R., and D. A. Flemer, 1976, Flows of materials between poorly flooded tidal marshes and an estuary, *Mar. Biol.* **35:**359–373.

Heinselman, M. L., 1963, Forest sites, bog processes, and peatland types in the glacial Lake Agassiz Region, Minnesota, *Ecol. Monogr.* **33:**327–374.

Heinselman, M. L., 1970, Landscape evolution and peatland types, and the Lake Agassiz Peatlands Natural Area, Minnesota, *Ecol. Monogr.* **40:**235–261.

Heinselman, M. L., 1975, Boreal peat lands in relation to environment, in *Coupling of Land and Water Systems,* A. D. Hasler, ed., Ecological Studies No. 10, Springer-Verlag, New York, pp. 93–103.

Hem, J. D., 1970, *Study and Interpretation of the Chemical Characteristics of Natural Water,* U.S.G.S. Water Supply Paper 1473, Washington, D.C.

Hemond, H. F., 1980, Biogeochemistry of Thoreau's Bog, Concord, Mass., *Ecol. Monogr.* **50:**507–526.

Hemond, H. F., 1983, The nitrogen budget of Thoreau's bog, *Ecology* **64:**99–109.

Hemond, H. F., and R. Burke, 1981, A device for measurement of infiltration in intermittently flooded wetlands, *Limnol. Oceanogr.* **26:**795-800.

Hemond, H. F., and J. L. Fifield, 1982, Subsurface flow in salt marsh peat: a model and field study, *Limnol. Oceanogr.* **27:**126-136.

Hemond, H. F., and J. C. Goldman, 1985, On non-Darcian water flow in peat, in *Journal of Ecology* **73:**579–584.

Hemond, H. F., T. P. Army, W. K. Nuttle, and D. G. Chen, 1987, Element cycling in wetlands: interactions with physical mass transport, in *Sources and Fates of Aquatic Pollutants*, R. A. Hites and S. J. Eisenreich, eds., American Chemical Society, Chicago, Illinois, pp. 519–540.

Herdendorf, C. E., 1987, *The Ecology of the Coastal Marshes of Western Lake Erie: A Community Profile*, U.S. Fish and Wildlife Service, Biological Report 85(7.9), Washinghton, D.C., 171p. plus microfiche appendices.

Hern, S. C., V. W. Lambou, and J. R. Butch, 1980, *Descriptive Water Quality for the Atchafalaya Basin,* Louisiana, EPA-600/4-80-614, U.S. Environmental Protection Agency, Las Vegas, Nev., 168p.

Hewlett, J. D., and A. R. Hibbert, 1967, Factors affecting the response of small watersheds to precipitation in humid areas, in *International Symposium on Forest Hydrology.* W. E. Sopper and H. W. Lull, eds., Pergamon Press, New York, pp. 275–290.

Hey, D. L., M. A. Cardamone, J. H. Sather, and W. J. Mitsch, 1989, Restoration of riverine wetlands: the Des Plaines river wetlands demonstration project, in *Ecological Engineering: An Introduction to Ecotechnology,* W. J. Mitsch and S. E. Jørgensen, eds., Wiley, New York, pp. 159–183.

Hickman, S. C., and V. J. Mosca, 1991, *Improving Habitat Quality for Migratory Waterfowl and Nesting Birds: Assessing the Effectiveness of the Des Plaines River Wetlands Demonstration Project,* Technical Paper no. 1, Wetlands Research, Inc., Chicago, Ill., 13p.

Higgins, K. F., 1977, Duck nesting in intensively farmed areas of North Dakota, *J. Wildl. Manage.* **41:**232–242.

Hill, B. H., 1985, The breakdown of macrophytes in a reservoir wetland, *Aquatic Bot.* **21:**23–31.

Hill, F. B., V. P. Aneja, and R. M. Felder, 1978, A technique for measurement of biogenic sulfur emission fluxes, *Environ. Sci. Health* **13:**199–225.

Hill, P. L., 1983, *Wetland-stream Ecosystems of the Western Kentucky Coalfield: Environmental Dlsturbance and the Shaping of Aquatic Community Structure,* Ph.D. dissertation, University of Louisville, Kentucky, 290p.

Ho, C. L., and S. Schneider, 1976, Water and sediment chemistry, Chapter VI.l, in *Louisiana Offshore Oil Port: Environmental Baseline Study,* vol. IV, Gosselink, J. G., R. Miller, M. Hood, and L. M. Bahr, Jr., eds., LOOP, Inc., New Orleans, La.

Hofstetter, R. H. ,1983, Wetlands in the United States, in *Ecosystems of the World,* vol. 4B, *Mires: Swamp, Bog, Fen and Moor,* A. J. P. Gore, ed., Elsevier, Amsterdam, pp. 201–244.

Hogg, E. H., and R. W. Wein, 1987, Growth dynamics of floating *Typha* mats: seasonal transformation and internal deposition of organic material, *Oikos* **50:**197–205.

Hogg, E. H., and R. W. Wein, 1988a, The contribution of *Typha* components to floating mat buoyancy, *Ecol.* **69**:1035–1031.

Hogg, E. H., and R. W. Wein, 1988b, Seasonal change in gas content and buoyancy of floating *Typha* mats, *J. Ecol.* **76:**1055–1068.

Hollis, J.,1967, *Toxicant Diseases of Rice,* Louisiana Agr. Exp. Station, Bull. 614, Louisiana State University, Baton Rouge.

Hook, D. D., and J. R. Scholtens, 1978, Adaptations and flood tolerance of tree species, in *Plant Life in Anaerobic Environments,* D. D. Hook and R. M. M. Crawford, eds. Ann Arbor Science, Ann Arbor, Mich., pp. 299–331.

Hook, D. D., et al., eds., 1988, *The Ecology and Management of Wetlands,* vols. 1 and 2, Croom Helm, London, and Timber Press, Portland, Ore., 592p. and 394p.

Hook, D. D., and R. Lea, eds., 1989, *Proceedings of the Symposium: The Forested Wetlands of the Southern United States*, Southeastern Forest Experiment Station, U.S. Department of Agriculture, Forest Service., Asheville, N.C., vol. Gen. Tech. Rep. SE–50, 168p.

Hopkinson, C. S., Jr., and J. W. Day, Jr., 1977, A model of the Barataria Bay salt marsh ecosystem, in *Ecosystem Modeling in Theory and Practice,* C. A. J. Hall and J. W. Day, Jr., eds.,Wiley, New York, pp. 235–265.

Hopkinson, C. S., Jr., and J. W. Day, Jr., 1980*a*, Modeling the relationship between development and storm water and nutrient runoff, *Environ. Manage.* **4:**315–324.

Hopkinson, C. S., Jr., and J. W. Day, Jr., 1980*b*, Modeling hydrology and eutrophication in a Louisiana swamp forest ecosystem, *Environ. Manage.* **4:**325–335.

Hopkinson, C. S., Jr., J. G. Gosselink, and F. T. Parrondo, 1980, Production of coastal Louisiana marsh plants calculated from phenometric techniques, *Ecology* **61:**1091–1098.

Houghton, R. A., 1990, The global effects of tropical deforestation, *Environmental Science and Technology* **24:**414–424.

Howarth, R. W., and J. M. Teal, 1979, Sulfate reduction in a New England salt marsh, *Limnol. Oceanogr.* **24:**999–1013.

Howarth, R. W., and J. M. Teal, 1980, Energy flow in a salt marsh ecosystem: the role of reduced inorganic sulfur compounds, *Am. Nat.* **116:**862–872.

Howarth, R. W., and A. Giblin, 1983, Sulfate reduction in the saltmarshes at Sapelo Island, Georgia, *Limnol. and Oceanogr.* **28:**70–82.

Howarth, R. W., A. Giblin, J. Gale, B. J. Peterson, and G. W. Luther, III, 1983, Reduced sulfur compounds in the pore waters of a New England salt marsh, *Environ. Biogeochem. Ecology Bull.* (Stockholm) **35:**135–152.

Howes, B. L., R. W. Howarth, J. M. Teal, and I. Valiela, 1981, Oxidation-reduction potentials in a salt marsh: spatial patterns and interactions with primary production, *Limnol. Oceanogr.* **26:**350–360.

Howes, B. L., J. W. H. Dacey, and G. M. King, 1984, Carbon flow through oxygen and sulfate reduction pathways in salt marsh sediments, *Limnol. Oceanogr.* **29:**1037–1051.

Howes, B. L., J. W. Dacey, and J. M. Teal, 1985, Annual carbon mineralization and below-ground production of *Spartina alterniflora* in a New England salt marsh, *Ecology* **66:**595–605.

Hsu, S-A., M. E. C. Giglioli, P. Reiter, and J. Davies, 1972, Heat and water balance studies on Grand Cayman, *Carib. J. Sci.* **12**(1–2):9–22.

Hudec, K., and K. Stastny, 1978, Birds in the reedswamp ecosystem, in *Pond Littoral Ecosystems,* D. Dykyjova and J. Kvet, eds., Springer-Verlag, Berlin, pp. 366–375.

Huenneke, L. F., and R. R. Sharitz, 1986, Microsite abundance and distribution of woody seedlings in a South Carolina cypress-tupelo swamp, *Am. Mid. Nat.* **115:**328–335.

Huffman, R. T., and S. W. Forsythe, 1981, Bottomland hardwood forest communities and their relation to anaerobic soil conditions, in *Wetlands of Bottomland Hardwood Forests,* J. R. Clark and J. Benforado, eds., Elsevier, Amsterdam, pp. 187–196.

Huffman, R. T., and R. E. Lonard, 1983, Successional patterns on floating vegetation mats in a southwestern Arkansas bald cypress swamp, *Castanea* **48**:73–78.

Hupp, C. R., and E. E. Morris, 1990, A dendrogeomorphic approach to measurement of sedimentation in a forested wetland, Black Swamp, Arkansas, *Wetlands* **10:**107–124.

Hupp, C. R., and D. E. Bazemore, 1993, Temporal and spatial patterns of wetland sedimentation, West Tennessee, *J. Hydrology* **141**:179–196.

Hustedt, F., 1955, Marine littoral diatoms of Beaufort, North Carolina, *Duke Univ. Mar. Stn. Bull.* **6**:1–67.

Hutchinson, G. E., 1973, Eutrophication: the scientific background of a contemporary practical problem, *Am. Sci.* **61**:269–279.

Hyatt, R. A., and G. A. Brook, 1984, Groundwater flow in the Okefenokee Swamp and hydrologic and nutrient budgets for the period August, 1981 through July, 1982, in *The Okefenokee Swamp: Its Natural History, Geology, and Geochemistry,* A. D. Cohen, D. J. Casagrande, M. J. Andrejko, and G. R. Best, eds., Wetland Surveys, Los Alamos, N.M., pp. 229–245.

Immirzi, C. P., E. Maltby, and R. S. Clymo, 1992, *The Global Status of Peatlands and Their Role in Carbon Cycling*, Wetland Ecosystems Research Group, Department of Geography, Univ. Exeter, prepared for Friends of the Earth, London, 145 pp.

Ingram, H., 1957, Microorganisms resisting high concentrations of sugar or salt, *Soc. Cen. Microbiolog. Symp.* **7**:90–135.

Ingram, H. A. P.,1967, Problems of hydrology and plant distribution in mires, *J. Ecol.* **55**:711–724.

Ingram, H. A. P., 1982, Size and shape in raised mire ecosystems: a geophysical model, *Nature* **297**:300–303.

Ingram, H. A. P., 1983, Hydrology, in *Ecosystems of the World, vol.4A, Mires: Swamp, Bog, Fen, and Moor,* A. J. P. Gore, ed., Elsevier, Amsterdam, pp. 67–158.

Ingram, H. A. P., D. W. Rycroft, and D. J. A. Williams, 1974, Anomalous transmission of water through certain peats, *J. Hydrology* **22**:213–218.

Ingvorsen, K., and T. D. Brock, 1982. Electron flow via sulfate reduction and methanogenesis in the anaerobic hypolimnion of Lake Mendota, *Limnol. and Oceanogr.* **27**:559–564.

Ivanov, K. E., 1981, *Water Movement in Mirelands,* translated from Russian by A. Thomson and H. A. P. Ingram, Academic Press, London, 276p.

Jackson, J., 1989, Man-made wetlands for wastewater treatment: two case studies, in *Constructed Wetlands for Wastewater Treatment,* D. A. Hammer, ed., Lewis Publishers, Inc., Chelsea, Mich., pp. 574–580.

Jackson, M. B., 1982, Ethylene as a growth promoting hormone under flooded conditions, in *Plant Growth Substances*, P. F. Waring, ed., Academic Press, London, pp. 291–301.

Jackson, M. B., 1985, Ethylene and responses of plants to soil waterloggong and submergence, in *Annual Rev. Plant Physiol.* **36**:145–174.

Jackson, M. B., 1988, Involvement of the hormones ethylene and abscisic acid in some adaptive responses of plants to submergence, soil waterlogging and oxygen shortage, in *The Ecology and Management of Wetlands. Volume 1: Ecology of Wetlands*, D. D. Hook et al., ed., Timber Press, Portland, Oregon, 2 vol., pp. 373–382.

Jackson, M. B., 1990, Hormones and developmental change in plants subjected to submergence or soil waterlogging, *Aquatic Bot.* **38**:49–71.

Jackson, M. B., C. M. Dobson, B. Herman, and A. Merryweather, 1984, Modification of 3,5-diiodo-4-hydroxybenzoic acid activity and stimulation of ethylene production by small concentrations of oxygen in the root environment, *Plant Growth Regul.* **2**:251–62.

Jacobs, T. C., and J. W. Gilliam, 1985, Riparian losses of nitrate from agricultural drainage waters, *J. Environmental Quality* **14**:472–478.

Janzen, D. H., 1985, Mangroves: where's the understory? *Journal of Tropical Ecology* **1**:89–92.

Jaworski, E., and C. N. Raphael, 1978, *Fish, Wildlife and Recreational Values of Michigan's Coastal Wetlands,* prepared by Department of Geography and Geology, Eastern Michigan University, Ypsilanti, Michigan, for Great Lakes Shorelines Sect., Div. of Land Resources Programs, Mich. Dept. of Nat. Resources, Lansing, 209p.

Jensen, J. R., E. J. Christensen, and R. Sharitz, 1983, in *Renewable Resources Management—Application of Remote Sensing* (Proceedings) American Society Photogrammetry, Falls Church, Va., pp. 318–336.

Jensen, J. R., E. J. Christensen, and R. Sharitz, 1984, Nontidal wetland mapping in South Carolina using airborne multi-spectral scanner data, *Remote Sensing Environ.* **16**:1–12.

Jiménez, J. A., A. E. Lugo, and G. Cintrón, 1985, Tree mortality in mangrove forests, *Biotropica* **17**:177–185.

Johnson, C. W., 1985, *Bogs of the Northeast*, University Press of New England, Hanover, N.H., 269p.

Johnson, C. W., in press, Dams and riparian forests: Case study from the Upper Missouri River. *Rivers.*

Johnson, C. W., and R. L. Linder, 1986, An economic valuation of South Dakota wetlands as a recreation resource for resident hunters, *Landscape Journal* **5:**33–38.

Johnson, D. C., 1942, *The Origin of the Carolina Bays,* Columbia University Press, New York, 341p.

Johnson, F. L., and D. T. Bell, 1976, Plant biomass and net primary production along a flood-frequency gradient in a streamside forest, *Castanea* **41:**156–165.

Johnson, R. L., 1979, Timber harvests from wetlands, in *Wetland Functions and Values: The State of Our Understanding,* P. E. Greeson, J. R. Clark, and J. E. Clark, eds., American Water Resource Assoc., Minneapolis, Minn., pp. 598–605.

Johnson, R. R., 1979, The lower Colorado River; a western system, in *Strategies for the Protection and Management of Floodplain Wetlands and Other Riparian Ecosystems,* R. R. Johnson and J. F. McCormick, tech. coord., Proceedings of the Symposium, Callaway Gardens, Georgia, December 11-13, 1978, U.S. Forest Service General Technical Report WO-12, Washington, D.C., pp. 41–55.

Johnson, R. R., and J. F. McCormick, (tech. coord.) 1979, *Strategies tor the Protection and Management of Floodplain Wetlands and Other Riparian Ecosystems,* Proceedings of the Symposium, Callaway Gardens, Georgia. December 11-13, 197/8, U.S. Forest Service General Technical Report WO-12, Washington, D.C., 410p.

Johnson, R. R., C. D. Ziebell, D. R. Patton, P. F. Ffolliott, and R. H. Hamre, eds., 1985, *Riparian Ecosystems and Their Management: Reconciling Conflicting Uses,* U.S. Department of Agriculture, Forest Service, Washington, D.C., General Technical Report RM–120.

Johnson, R. R., and C. H. Lowe, 1985, On the development of riparian ecology, in *Riparian Ecosystems and Their Management,* R. R. Johnson et al., eds., U.S. Department of Agriculture, Forest Service, General Tech. Report RM–120, Washington, D.C., pp. 112–116.

Johnson, W. B., C. E. Sasser, and J. G. Gosselink, 1985, Succession of vegetation in an evolving river delta, Atchafalaya Bay, Louisiana, *J. Ecology* **73:**973–986.

Johnson, W. C., R. L. Burgess, and W. R. Keammerer, 1976, Forest overstory vegetation and environment on the Missouri River floodplain in North Dakota, *Ecol. Monogr.* **46:**59–84.

Johnson, W. C., T. L. Sharik, R. A. Mayes, and E. P. Smith, 1987, Nature and cause of zonation discreteness around glacial prairie marshes, *Can. J. Bot.* **65**(8):1622–1632.

Johnston, C. A., 1991, Sediment and nutrient retention by freshwater wetlands: effects on surface water quality, *Critical Reviews in Environmental Control* **21:**491–565.

Johnston, C. A., G. D. Bubenzer, G. B. Lee, F. W. Madison, and J. R. McHenry, 1984, Nutrient trapping by sediment deposition in a seasonally flooded lakeside wetland, *J. Environmental Quality* **13:** 283–290.

Johnston, C. A., N. E. Detenbeck, and G. J. Niemi, 1990, The cumulative effect of wetlands on stream water quality and quantity: a landscape approach, *Biogeochemistry* **10:**105–141.

Jordan, T. E., D. L. Correll, W. T. Peterjohn, and D. E. Weller, 1986, Nutrient flux in a landscape: the rhode river watershed and receiving waters, in *Watershed Research Perspectives,* D. L. Correll, ed., Smithsonian Institution Press, Washington, D.C., pp. 57–76.

Josephson, J., 1992, Status of wetlands, *Environmental Science and Technology* **26:**422.

Josselyn, M., 1983, *The Ecology of San Francisco Bay Tidal Marshes: A Community Profile,* U.S. Fish and Wildlife Service FWS/OBS-83/23, Slidell, La., 102p.

Josselyn, M., J. Zedler, and T. Griswold, 1990, Wetland mitigation along the Pacific coast of the United States, in *Wetland Creation and Restoration,* J. A. Kusler and M. E. Kentula, eds., Island Press, Washington, D.C., pp. 3–36.

Junk, W. J., 1970, Investigations on the ecology and production biology of the "floating meadows" (Paspalo-Echinochloetum) on the middle Amazon: I. The floating vegetation and its ecology, *Amazonia* **2:**449–495.

Junk, W. J., 1982, Amazonian floodplains: their ecology, present and potential use, in *Ecosystem Dynamics in Freshwater Wetlands and Shallow Water Bodies,* vol. 1., D. O. Logofet and N. N. Luckanov, eds., SCOPE and UNEP Workshop, Center of International Projects, Moscow, USSR, pp. 98–126.

Junk, W. J., P. B. Bayley, and R. E. Sparks, 1989, The flood pulse concept in river-floodplain systems, in *Proceedings of the International Large River Symposium, Can. Spec. Publ. Fish Aquat. Sci. 106,* D. P. Dodge, ed. [Dept. of Fisheries and Oceans, Ottawa, Ontario, Canada] pp. 110–127.

Kaatz, M. R., 1955, The Black Swamp: a study in historical geography, *Ann. Assoc. Amer. Geogr.* **35:**1–35.

Kadlec, R. H., 1979, Wetlands for tertiary treatment, in *Wetland Functions and Values: The State of Our Understanding,* P. E. Greeson, J. R. Clark, and J. E. Clark, eds., American Water Resources Assoc., Minneapolis, Minn., pp. 490–504.

Kadlec, R. H., 1983, The Bellaire wetlands: wastewater alteration and recovery, *Wetlands* **3:**44–63.

Kadlec, R. H., 1985, Aging phenomenon in wastewater wetlands, in *Ecological Considerations in Wetland Treatment of Municipal Wastewaters,* P. J. Godfrey et al., eds., Van Nostrand Reinhold, New York, pp. 338–350.

Kadlec, R. H., 1989, Hydrologic factors in wetland water treatment, in *Constructed Wetlands for Wastewater Treatment,* D.A. Hammer, ed., Lewis Publishers, Chelsea, Mich., pp. 21–40.

Kadlec, R. H., and D. L. Tilton, 1979, The use of freshwater wetlands as a tertiary wastewater treatment alternative, *CRC Crit. Rev. Environ. Control* **9:**185–212.

Kadlec, R. H., and J. A. Kadlec, 1979, Wetlands and water quality, in *Wetland Functions and Values: The State of Our Understanding,* P. E. Greeson, J. R. Clark, and J. E. Clark, eds., American Water Resources Association, Minneapolis, Minn., pp. 436–456.

Kadlec, R. H., R. B. Williams, and R. D. Scheffe, 1988, Wetland evapotranspiration in temperate and arid climates, in *The Ecology and Management of Wetlands,* Timber Press, Portland, Ore., D. D. Hook et al., eds., pp. 146–160.

Kadlec, R. H., and X.-M. Li, 1990, Peatland ice/water quality, *Wetlands* **10:**93–106.

Kale, H. W., 1965, *Ecology and Bioenergetics of the Long-billed Marsh Wren,* Telmatodytes palustris griseus *(Brewster) in Georgia Salt Marshes,* Nuttall Ornithol. Club. Publ. 5, Cambridge.

Kangas, P. C., and A. E. Lugo, 1990, The distribution of mangroves and saltmarsh in Florida, *Tropical Ecology* **31:**32–39.

Kantrud, H. A., and R. E. Stewart, 1977, Use of natural basin wetlands by breeding waterfowl in North Dakota, *J. Wildl. Manage.* **41:**243–253.

Kantrud, H. A., J. B. Millar, and A. G. van der Valk, 1989, Vegetation of wetlands of the prairie pothole region, in *Northern Prairie Wetlands,* A. G. van der Valk, ed., Iowa State University Press, Ames, Iowa, pp. 132–187.

Kaplan, W., I. Valiela, and J. M. Teal, 1979, Denitrification in a salt marsh ecosystem, *Limnol. Oceanogr.* **24:**726–734.

Kaswadji, R. F., J. G. Gosselink, and R. E. Turner, 1990, Estimation of primary production using five different methods in a *Spartina alterniflora* salt marsh, *Wetlands Ecology and Management* **1:**57–64.

Kauffman, J. B., and W. C. Kreuger, 1984, Livestock impacts on riparian ecosystems and streamside management implications—a review, *J. Range Management* **37:**685–691.

Kawase, M., 1981, Effects of ethylene on aerenchyma development, *Am. J. Bot.* **68:**61–65.

Keammerer, W. R., W. C. Johnson, and R. L. Burgess, 1975, Floristic analysis of the Missouri River bottomland forests in North Dakota, *Can. Field Nat.* **89:**5–19.

Keddy, P A., 1983, Freshwater wetland human-induced changes: indirect effects must also be considered, *Environ. Manage.* **7:**299–302.

Keddy, P. A., 1992, Water level fluctuations and wetland conservation, in *Wetlands of the Great Lakes,* Proceedings of an International Symposium, J. Kusler and R. Smandon, eds., Association of State Wetland Managers, Berne, New York, pp. 79–91.

Keddy, P. A., and A. A. Reznicek, 1986, Great Lakes vegetation dynamics: the role of fluctuating water levels and buried seeds, *J. Great Lakes Res.* **12:**25–36.

Kemp, G. P., and J. W. Day, Jr., 1984, Nutrient dynamics in a Louisiana swamp receiving agricultural runoff, in *Cypress Swamps,* K. C. Ewel and H. T. Odum, eds., University Presses of Florida, Gainesville, pp. 286–293.

Kemp, G. P., W. H. Conner, and J. W. Day, Jr., 1985, Effects of flooding on decomposition and nutrient cycling in a Louisiana swamp forest, *Wetlands* **5:**35–51.

Kentula, M. E., R. P. Brooks, S. E. Gwin, C. C. Holland, A. D. Sherman, and J. C. Sifness, 1992, *Wetlands: An Approach to Improving Decision Making in Wetland Restoration and Creation,* Island Press, Washington, D.C., 151p.

Kilham, P., 1982, The biogeochemistry of bog ecsosystems and the chemical ecology of *Sphagnum, Mich. Bot.* **21:**159–168.

King, D. R., and G. S. Hunt, 1967, Effect of carp on vegetation in a Lake Erie marsh, *J. Wildl. Manage.* **31**(1):181–188.

Kinler, N. W., G. Linscombe, and P. R. Ramsey, 1987, Nutria, in *Wild Furbearer Management and Conservation in North America,* M. Novak, J. A. Balen, M. E. Obbard, and B. Mallocheds, eds., Ministry of Natural Resources, Ontario, Canada, pp. 326–343.

Kirby. C. J., and J. G. Gosselink, 1976, Primary production in a Louisiana Gulf Coast *Spartina alterniflora* marsh, *Ecology* **57:**1052–1059.

Kirk, P. W., Jr., 1979, *The Great Dismal Swamp,* University Press of Virginia, Charlottesville, Va., 427p.

Kitchens, W. M ., Jr., J. M. Dean, L. H. Stevenson, and J. M. Cooper, 1975, The Santee Swamp as a nutrient sink, in *Mineral Cycling in Southeastern Ecosystems,* F G. Howell, J. B. Gentry, and M. H. Smith, eds., ERDA Symposium Series 740513, USGPO. Washington, D.C., pp. 349–366.

Kivinen, E., and P. Pakarinen, 1981, Geographical distribution of peat resources and major peatland complex types in the world, *Annals Acad. Sciencia Fennicae, Series A, Geology–Geography* **132:**1–28.

Klarer, D. M., and D. F. Millie, 1989, Amelioration of storm-water quality by a freshwater estuary, *Archiv für Hydrobiologie* **116:**375–389.

Klopatek, J. M., 1974, *Production of Emergent Macrophytes and Their Role in Mineral Cycling Within a Freshwater Marsh,* Master's thesis, University of Wisconsin—Milwaukee.

Klopatek, J. M., 1978, Nutrient dynamics of freshwater riverine marshes and the role of emergent macrophytes, in *Freshwater Wetlands: Ecological Processes and Management Potential,* R. E. Good, D. F. Whigham, and R. L. Simpson, eds., Academic Press, New York, pp. 195–216.

Knetsch, J. L., and R. K. Davis, 1966, Comparisons of methods for recreation evaluation, in *Water Research,* A. Kneese and S. Smith, eds., Johns Hopkins Univ. Press, Baltimore, Md., pp. 125–142.

Knight, A. W., and R. L. Bottorff, 1984, The importance of riparian vegetation to stream ecosystems, in *California Riparian Systems: Ecology, Conservation, and Productive Management*, R. F. Warner and K. M. Hendrix, eds., Univ. California Press, Berkeley, pp. 160–167.

Knight, R. L., 1990, Wetland Systems, in *Natural Systems for Wastewater Treatment, Manual of Practice FD-16,* Water Pollution Control Federation, Alexandria, Va., pp. 211–260.

Knight, R. L., B. H. Winchester, and J. C. Higman, 1984, Carolina bays—feasibility for effluent advanced treatment and disposal, *Wetlands* **4:**177–204.

Knight, R. L., T. W. McKim, and H. R. Kohl, 1987, Performance of a natural wetland treatment system for wastewater management, *Journal Water Pollution Control Federation* **59:**746–754.

Knopf, F. L., and M. L. Scott, 1990, Altered flows and created landscapes, in *Management of Dynamic Ecosystems*, J. M. Sweeney, ed., North Cent. Section, The Wildl. Soc., West Lafayette, Indiana, pp. 47–70.

Kobriger, N. P., T. V. Dupuis, W. A. Kreutzberger, F. Stearns, G. Guntenspergen, and J. R. Keough, 1983, *Guidelines for the Management of Highway Runoff on Wetlands,* National Research Council Transportation Research Board, Washington D.C. National Cooperative Highway Research Program Report 264.

Koch, M. S., and I. A. Mendelssohn, 1989, Sulphide as a soil phytotoxin: differential responses in two marsh species, *J. Ecology* **77:**565–578.

Koch, M. S., I. A. Mendelssohn, and K. L. McKee, 1990, Mechanism for the hydrogen sulfide–induced growth limitation in wetland macrophytes, *Limnol. Oceanogr.* **35:**399–408.

Koch, M. S., and K. R. Reddy, 1992, Distribution of soil and plant nutrients along a trophic gradient in the Florida Everglades, *Soil Science Soc. America J.* **56:**1492–1499.

Kologiski, R. L., 1977, *The phytosociology of the Green Swamp, North Carolina,* North Carolina Agri. Exp. Sta. Tech. Bull. 20, Raleigh, N.C., 101p.

Kortekaas, W. M., E. Van der Maarel, and W. G. Beeftink, 1976, A numerical classification of European *Spartina* communities, *Vegetatio* **33:**51–60.

Kovalchik, B. L., 1987, *Riparian Zone Associations of the Deschutes, Ochoco. Fremont, and Winema National Forests*, U.S. Department of Agriculture, Forest Service, Pacific Northwest Region, Region 6, Ecology Technical Paper 279–287.

Kramer, P J., W. S. Riley, and T. T. Bannister, 1952, Gas exchange of cypress knees, *Ecology* **33:**117–121.

Kratz, T. K., and C. B. DeWitt, 1986, Internal factors controlling peatland-lake ecosystem development, *Ecology* **67:**100–107.

Kreitler, C. W., 1977, Faulting and land subsidence from groundwater and hydrocarbon production, Houston-Galveston, Texas, in *Second International Symposium on Land Subsidence Proceedings,* International Association of Hydrological Sciences Publ. No. 121, pp. 435–446.

Krieger, K., D. M. Klarer, and R. Heath, eds., 1992, Coastal wetlands of the Laurentian Great Lakes: current knowledge and research needs. *J. Great Lakes Research* **18**(4)525–699.

Kroll, R., 1992, *Marsh Management Generalizations* (unpubl. sketch), Winous Point Shooting Club, Port Clinton, Ohio.

Kruczynski, W. L., C. B. Subrahmanyam, and S. H. Drake, 1978, Studies on the plant community of a North Florida salt marsh, Part I, Primary production, *Bull. Mar. Sci.* **28:**316–334.

Kuenzler, E. J., 1974, Mangrove swamp systems, in *Coastal Ecological Systems of the United States,* vol. 1, H. T. Odum, B. J. Copeland, and E. A. McMahan, eds., The Conservation Foundation, Washington, D.C., pp. 346–371.

Kuenzler, E. J., and N. J. Craig, 1986, Land use and nutrient yields of the Chowan River watershed, in *Watershed Research Perspectives*, D. L. Correll, ed., Smithsonian Institution Press, Washington, D.C., pp. 57–76.

Kuenzler, E. J., P J. Mulholland, L. A. Yarbro, and L. A. Smock, 1980, *Distributions and Budgets of Carbon, Phosphorus, Iron and Manganese in a Floodplain Swamp Ecosystem,* Water Resources Research Institute of North Carolina Report No. 17. Raleigh, N.C., 234p.

Kulzynski S., 1949, *Peat Bogs of Polesie,* Acad. Pol. Sci. Mem., Series B, No. 15, 356p.

Kurz, H., 1928, Influence of *Sphagnum* and other mosses on bog reactions, *Ecology* **9:**56–69.

Kurz, H., and D. Demaree, 1934, Cypress buttresses in relation to water and air, *Ecology* **15:**36–41.

Kurz. H., and K. A. Wagner, 1953, Factors in cypress dome development, *Ecology* **34:**17–164.

Kushlan, J. A., 1989, Avian use of fluctuating wetlands, in *Freshwater Wetlands and Wildlife,* DOE Symposium Series no. 61, R. R. Sharitz and J. W. Gibbons, eds., USDOE Office of Scientific and Technical Information, Oak Ridge, Tennessee, pp. 593–604.

Kushlan, J. A., 1990, Freshwater marshes, in *Ecosystems of Florida,* R. L. Myers and J. J. Ewel, eds., University of Central Florida Press, Orlando, pp. 324–363.

Kushlan, J. A., 1991, The Everglades, in *The Rivers of Florida*, R. J. Livingston, ed., Springer-Verlag, New York, pp. 121–142.

Kushner, D. J., 1978, Life in high salt and solute concentrations: halophilic bacteria, in *Microbial Life in Extreme Environments,* Kushner, D. J., ed., Academic Press, New York, 465p.

Kusler, J. A., 1979, *Strengthening State Wetland Regulations,* U.S. Fish and Wildlife Service FWS/OBS-79/98, Washington, D.C.

Kusler, J. A., 1983, *Our National Wetland Heritage: A Protection Guidebook,* Environmental Law Institute, Washington, D.C., 167p.

Kusler, J. A., and J. Montanari, eds., 1978, *Proceedings of the National Wetland Protection Symposium, Reston, Virginia, June 6-8,1977,* U.S. Fish and Wildlife Service FWS/OBS-78/97, Washington, D.C., 255p.

Kusler, J. A., M. L. Quammen, and G. Brooks, eds., 1988, *Mitigation of Impacts and Losses,* Proceedings of The National Wetland Symposium, New Orleans, Louisiana, Omnipress, Madison, WI. 460p.

Kusler, J. A., and M. E. Kentula, eds., 1990, *Wetland Creation and Restoration: The Status of the Science,* Island Press, Washington, D.C., 594p.

Kvet, J., and S. Husak, 1978, Primary data on biomass and production estimates in typical stands of fishpond littoral plant communities, in *Pond Littoral Ecosystems,* D. Dykyjova and J. Kvet, eds., Springer-Verlag, Berlin, pp. 211–216.

Laanbroek, H. J., 1990, Bacterial cycling of minerals that affect plant growth in waterlogged soils: a review, *Aquatic Botany* **38:**109–125.

Laderman, A. D., 1989, *The Ecology of the Atlantic White Cedar Wetlands: A Community Profile*, U.S. Fish Wildl. Serv., Biol. Report **85**(7.21), 114p.

Lambers, H., E. Steingrover, and G. Smakman, 1978, The significance of oxygen transport and of metabolic adaptation in flood tolerance of *Senecio* species, *Physiol. Plant* **43:**277–281.

Lambert, J. M., 1964. The *Spartina* story, *Nature* **204:**1136–1138.

Lambou, V. W., 1965, The commercial and sport fisheries of the Atchafalaya Basin floodway, 17th Annual Conference of S.E. Associated Game and Fish Commissioners Proc., pp. 256–281.

Lambou, V. W., 1990, Importance of bottomland forest zones to fishes and fisheries: a case history, in *Ecological Processes and Cumulative Impacts: Illustrated by Bottomland Hardwood Wetland Ecosystems*, J. G. Gosselink, L. C. Lee, and T. A. Muir, eds., Lewis Publishers, Chelsea, Mich., pp. 125–193.

Larcher, W., 1991, *Physiological Plant Ecology,* Reprinted 2nd ed., Springer-Verlag, New York, 303p.

Larsen, J. A., 1982, *Ecology of the Northern Lowland Bogs and Conifer Forests,* Academic Press, New York, 307p.

Larson, J. S., 1982, Wetland value assessment—state of the art, in *Wetlands: Ecology and Management,* B. Gopal, R. E. Turner, R. G. Wetzel, and D. F. Whigham, eds., Natural Institute of Ecology and International Scientific Publications, Jaipur, India, pp. 417–424.

Larson, J. S., 1991, North America, in *Wetlands,* M. Finlayson, and M. Moser, ed., Facts on File, Oxford, England, pp. 57–84.

Larson, J. S., and J. A. Kusler, 1979, Preface, in *Wetland Functions and Values: The State of Our Understanding,* P. E. Greeson, J. R. Clark, and J. E. Clark, eds., American Water Resources Association, Minneapolis, Minn.

Larson, J. S., M. S. Bedinger, C. F. Bryan, S. Brown, R. T. Huffman, E. L. Miller, D. G. Rhodes, and B. A. Touchet, 1981, Transition from wetlands to uplands in southeastern bottomland hardwood forests, in *Wetlands of Bottomland Hardwood Forests,* J. R. Clark and J. Benforado, eds., Elsevier, Amsterdam, pp. 225–273.

Larson, J. S., and C. Neill, ed. 1987, *Mitigating Freshwater Wetland Alterations in the Glaciated Northeastern United States: An Assessment of the Science Base,* Proceedings of the Conference, the University of Massachusetts, Amherst, 143p.

Leck, M. A., 1979, Germination behavior of *Impatiens capensis* Meerb. (Balsaminaceae), *Bartonia* **46:**1–11.

Leck, M. A., and K. J. Graveline, 1979, The seed bank of a freshwater tidal marsh, in *Am. J. Bot.* **66:**1006–1015.

Leck, M. A., and R. L. Simpson, 1987, Seed bank of a freshwater tidal wetland: turnover and relationship to vegetation change, *Amer. J. Bot.* **74:**360–370.

Lee, G. F., E. Bentley, and R. Amundson, 1975, Effect of marshes on water quality, in *Coupling of Land and Water Systems,* A. D. Hasler, ed., Springer-Verlag, New York, pp. 105–127.

Lee, J. K., R. A. Park, and P. W. Mausel, 1991, Application of geoprocessing and simulation modeling to estimate impacts of sea level rise on northeastern coast of Florida, *Photogrammetric Engineering and Remote Sensing* **58:**1579–1586.

Lee, L. C., 1983, *The Floodplain and Wetland Vegetation of Two Pacific Northwest River Ecosystems,* Ph.D. Thesis, Univ. Washington, Seattle.

Lee, L. C., T. M. Hinckley, and M. L. Scott, 1985, Plant water status relationships among major floodplain sites of the Flathead River, Montana, *Wetlands* **5:**15–34.

Lee, R., 1980, *Forest Hydrology,* Columbia University Press, New York, 349p.

Lee, S. Y., 1989, Litter production and turnover of the mangrove *Kandelia candel* (L.) Druce in a Hong Kong tidal shrimp pond, *Estuarine, Coastal and Shelf Science* **29:**75–87.

Lee, S. Y., 1990, Primary productivity and particulate organic matter flow in an estuarine mangrove-wetland in Hong Kong, *Marine Biology* **106:**453–463.

Lee, S. Y., 1991, Herbivory as an ecological process in a *Kandelia candel* (Rhizophoraceae) mangal in Hong Kong, *J. Tropical Ecology* **7:**337–348.

Lefeuvre, J. C., ed., 1989, *Conservation and Development: The Sustainable Use of Wetland Resources,* published abstracts of the Third INTECOL Wetlands Conference, Rennes, France, September 1988, Laboratoire d'Evolution des Systems Naturels et Modifiés, Paris.

Lefor, M. W., and W. C. Kennard, 1977, *Inland Wetland Definitions,* University of Connecticut Institute of Water Resources Report No. 28, 63p.

Leitch, J. A., 1986, Economics of prairie wetland values, *North Dakota Acad. Science* **40:**44.

Leitch, J. A., 1989, Politicoeconomic overview of prairie potholes, in *Northern Prairie Wetlands,* A. G. van der Valk, ed., Iowa State University Press, Ames, Iowa, pp. 3–14.

Leitch, J. A., and L. E. Danielson, 1979, *Social, Economic, and Institutional Incentives to Drain or Preserve Prairie Wetlands*, Department of Agricultural and Applied Economics, University of Minnesota, St. Paul, 78p.

Leitch, J. A., and B. L. Ekstrom, 1989, *Wetland Economics and Assessment: An Annotated Bibliography*, Garland Publishing, New York.

Lemlich, S. K., and K. C. Ewel, 1984, Effects of wastewater disposal on growth rates of cypress trees, *Journal of Environmental Quality* **13:**602–604.

Leopold, L. B., M. G. Wolman, and J. E Miller, 1964, *Fluvial Processes in Geomorphology,* W. H. Freeman, San Francisco, 522p.

Levitt, J., 1980, *Responses of Plants to Environmental Stresses, vol. II, Water, Radiation, Salt, and Other Stresses,* Academic Press, New York, 607p.

Lewis, R. R., 1990a, Creation and restoration of coastal plain wetlands in Florida, in *Wetland Creation and Restoration,* J. A. Kusler and M. E. Kentula, eds., Island Press, Washington, D.C., pp. 73–101.

Lewis, R. R., 1990b, Creation and restoration of coastal plain wetlands in Puerto Rico and the U.S. Virgin Islands, in *Wetland Creation and Restoration,* J. A. Kusler and M. E. Kentula, ed., Island Press, Washington, D.C., pp. 103–123.

Lieth, H., 1975, Primary production of the major units of the world, in *Primary Productivity of the Biosphere,* H. Lieth and R. H. Whitaker, eds., Springer-Verlag, New York, pp. 203–215.

Likens, G. E., F. H. Bormann, R. S. Pierce, J. S. Eaton, and N. M. Johnson, 1977, *Biogeochemistry of a Forested Ecosystem,* Springer-Verlag, New York, 146p.

Likens, G. E., F. H. Bormann, R. S. Pierce, and J. S. Eaton, 1985, The Hubbard Brook Valley, in: *An Ecosystem Approach to Aquatic Ecology: Mirror Lake and Its Environment,* G. E. Likens, ed., Springer-Verlag, New York, pp. 9–39.

Linacre, E., 1976, Swamps, in *Vegetation and the Atmosphere,* vol. 2, *Case Studies,* J. L. Monteith, ed., Academic Press, London, pp. 329–347.

Lindeman, R. L., 1941, The developmental history of Cedar Creek Lake, Minnesota, *Am. Midl. Nat.* **25:**101–112.

Lindeman, R. L., 1942, The trophic-dynamic aspect of ecology, *Ecology* **23:**399–418.

Lindsay, W. L., 1979, *Chemical Equilibria in Soils.* Wiley, New York, 449p.

Linsley, R. K., and J. B. Francini, 1979, *Water Resources Engineering,* 3rd ed., McGraw–Hill, New York, 716p.

Linthurst, R. A., and R. J. Reimold, 1978, An evaluation of methods for estimating the net aerial primary productivity of estuarine angiosperms, *J. Appl. Ecol.* **15:**919–931.

Lippson, M. A. J., M. S. Haire, A. F. Holland, F. Jacobs, J. Jensen, R. L. Moran-Johnson, T. T. Polgar, and W. A. Richkus, 1979, *Environmental Atlas of the Potomac Estuary,* Williams and Heintz Map Corp., Washington, D.C.

Lipschultz, F., 1981, Methane release from a brackish intertidal salt-marsh embayment of Chesapeake Bay, Maryland, *Estuaries* **4:**143–145.

Litchfield, D. K. and D. D. Schatz, 1989, Constructed Wetlands for Wastewater Treatment at Amoco Oil Company's Mandan, North Dakota, Refinery, in *Constructed Wetlands for Wastewater Treatment,* D. A. Hammer, ed., Lewis Publishers, Inc. Chelsea, Mich., pp. 101–119.

Little, E. L., Jr., 1971, *Atlas of United States Trees,* vol. I, Conifers and important hardwoods, Misc. Pub. No. 1146, U.S. Department of Agriculture—Forest Service, USGPO, Washington, D.C.

Littlehales, B., and W. A. Niering, 1991, *Wetlands of North America*, Thomasson-Grant, Charlottesville, Virginia, 160p.

Littlejohn C. B., 1977., An analysis of the role of natural wetlands in regional water management, in *Ecosystem Modeling in Theory and Practice,* C. A. S. Hall and J. W. Day, Jr., eds., Wiley, New York, pp. 451–476.

Livingston, D. A., 1963, Chemical composition of rivers and lakes, *U. S. Geological Survey Prof Paper 440G,* 64p.

Livingston, E. H., 1989, Use of wetlands for urban stormwater management, in *Constructed Wetlands for Wastewater Treatment,* D.A. Hammer, ed., Lewis Publishers, Inc., Chelsea, Mich., pp. 253–264.

Livingstone, D. C., and D. G. Patriquin, 1981, Belowground growth of *Spartina alterniflora* Loisel: habit, functional biomass and non-structural carbohydrates, *Estuarine Coastal Shelf Sci.* **12:**579–588.

Loftin, M. K., L. A. Toth, and J. T. B. Obeysekera, 1990, *Kissimmee River Restoration: Alternative Plan Evaluation and Preliminary Design Report*, South Florida Water Management District, West Palm Beach, Florida, 148p. plus appen.

Logofet, D. O. and N. K. Luckyanov, eds., 1982, *Ecosystem Dynamics in Freshwater Wetlands and Shallow Water Bodies,* Proceeedings Workshop Misk, Pinsk, and Tskhaltoubo, USSR, July 12–26, 1981, Scientific Committee on Problems of the Environment and United Nations Environment Programme, Center of International Projects, Moscow, USSR, Vol. I and II. 312p. and 424p.

Logofet, D. O., and G. A. Alexandrov, 1984, Modelling of matter cycle in a mesotrophic bog ecosystem. 1. Linear analysis of carbon environs, *Ecol. Modelling* **21:**247–258.

Logofet, D. O., and G. A. Alexandrov, 1988, Interference between mosses and trees in the framework of a dynamic model of carbon and nitrogen cycling in a mesotrophic bog ecosystem, in *Wetland Modelling*, W. J. Mitsch, M. Straskraba, and S. E. Jørgensen, eds., Elsevier, Amsterdam, pp. 55–66.

Lonard, R. T., E. J. Clairain, R. T. Huffman, J. W. Hardy, L. D. Brown, P E. Ballard, and J. W. Watts, 1981, *Analysis of Methodologies Used for the Assessment of Wetland Values,* Environmental Laboratory, U.S. Army Corps of Engineers, Waterways Expt. Station, Vicksburg, Miss., 68p. plus appendices.

Loucks, O. L., 1990, Restoration of the pulse control function of wetlands and its relationship to water quality objectives, in *Wetland Creation and Restoration,* J. A. Kusler, and M. E. Kentula, eds., Island Press, Washington, D.C., pp. 467–477.

Lowrance, R., R. Todd, J. Fail, Jr., O. Hendrickson, Jr., R. Leonard, and L. Asmussen, 1984, Riparian forests as nutrient filters in agricultural watersheds, *BioScience* **34:**374–377.

Lugo, A. E., 1980, Mangrove ecosystems: successional or steady state? *Biotropica* (supplement) **12:**65–72.

Lugo, A. E., 1981, The island mangroves of Inagua, *J. Nat. History* **15:**845–852.

Lugo, A. E., 1986, Mangrove understory: an expensive luxury? *Journal of Tropical Ecology* **2:**287–288.

Lugo, A. E., 1988, The mangroves of Puerto Rico are in trouble, *Acta Cientifica* **2:**124.

Lugo, A. E., 1990a, *Mangroves of the Pacific Islands: Research Opportunities*, U.S. Department of Agriculture Forest Service, Pacific Southwest Research Station, General Technical Report PSW–118, Berkeley, California, 13p.

Lugo, A. E., 1990b, Fringe wetlands, in *Forested Wetlands, Ecosystems of the World 15*, A. E. Lugo, M. Brinson, and S. Brown, eds. Elsevier, Amsterdam, pp. 143–169.

Lugo, A. E., and S. C. Snedaker, 1974, The ecology of mangroves, *Ann. Rev. Ecol. Systematics* **5:**39–64.

Lugo, A. E., G. Evink, M. M. Brinson, A. Broce, and J. C. Snedaker, 1975, Diurnal rates of photosynthesis, respiration, and transpiration in mangrove forests in South Florida, in *Tropical Ecological Systems—Trends in Terrestrial and Aquatic Research,* F. B. Golley and E. Medina, eds., Springer-Verlag, New York, pp. 335–350.

Lugo, A. E., M. Sell, and S. C. Snedaker, 1976, Mangrove ecosystem analysis, in *Systems Analysis and Simulation in Ecology,* vol. IV, B. C. Patten, ed., Academic Press, New York, pp. 113–145.

Lugo, A. E., and C. Patterson–Zucca, 1977, The impact of low temperature stress on mangrove structure and growth, *Trop. Ecol.* **18:**149–161.

Lugo, A. E., G. Cintrón, and C. Goenaga, 1981, Mangrove ecosystems under stress, in *Stress Effects on Natural Ecosystems*, G. W. Barrett and R. Rosenberg, eds., Wiley, New York, pp. 129–153.

Lugo, A. E., S. L. Brown, and M. M, Brinson, 1988, Forested wetlands in freshwater and salt-water environments, *Limnology and Oceanography* **33:**894–909.

Lugo, A. E., M. M. Brinson, and S. L. Brown, eds., 1990, *Forested Wetlands, Ecosystems of the World 15*, Elsevier, Amsterdam, 527p.

Lunz, J. D., T. W. Zweigler, R. T. Huffman, R. J. Diaz, E. J. Clairain, and L. J. Hunt, 1978, *Habitat Development Field Investigations Windmill Point Marsh Development Site, James River Virginia; summary report,* U.S. Army Waterways Exp. Station Tech. Rep. D-79-23, Vicksburg, Miss., 116p.

Lurssen, K., K. Naumann, and R. Schroder, 1979, 1-Aminocylopropane-1-carboxylic acid—an intermediate of the ethylene biosynthesis in higher plants, *Z. Pflanzenphysiol.* **92:**285–294.

Lynn, G. D., P. D. Conroy, and F. J. Prochaska, 1981, Economic valuation of marsh areas for marine production processes (Florida), *J. Environ. Econ. Manage.* **8:**175–186.

Lynn, L. M., and E. F. Karlin, 1985, The vegetation of the low-shrub bogs of northern New Jersey and adjacent New York: ecosystems at their southern limit, *Bull. Torrey Bot. Club* **112:**436–444.

Ma, S., and J. Yan, 1989, Ecological engineering for treatment and utilization of wastewater, in *Ecological Engineering: An Introduction to Ecotechnology,* W. J. Mitsch and S. E. Jørgensen, eds., John Wiley and Sons, Inc., New York, N.Y., pp. 185–218.

Macbeth Division of Kollmoggen Instruments Corporation, 1992, *Munsell Soil Color Charts,* Macbeth Division of Kollmoggen Instruments Corporation, Newburgh, New York, 10p. and charts.

MacDonald, K. B., 1977, Plant and animal communities of Pacific North American salt marshes, in *Ecosystems of the World,* vol. 1, Wet Coastal Ecosystems, B. J. Chapman, ed., Elsevier, Amsterdam, pp. 167–191.

MacMannon, M., and R. M. M. Crawford, 1971, A metabolic theory of flooding tolerance; the significance of enzyme distribution and behavior, *New Phytol.* **10:**299–306.

Macnae, W., 1963, Mangrove swamps in South Africa, *J. Ecol.* **51:**1–25.

Madden, C. J., J. W. Day, Jr., and J. M. Randall, 1988. Freshwater and marine coupling in estuaries of the Mississippi River deltaic plain, *Limnol. and Oceanogr.* **33:**982–1004.

Maguire, C. and P. O. S. Boaden, 1975, Energy and evolution in the thiobios: an extrapolation from the marine gastrotrich *Thiodasys sterreri, Cahiers de Biol. Mar.* **16:**635–646.

Mahall, B. E., and R. B. Park, 1976a, The ecotone between *Spartina foliosa* Trin. and *Salicornia virginica* L. in salt marshes of northern San Francisco Bay. 1. Biomass and production, *J. Ecol.* **64:**421–433.

Mahall, B. E., and R. B. Park, 1976b, The ecotone between *Spartina foliosa* Trin. and *Salicornia virginica* L. in salt marshes of northern San Francisco Bay. II. Soil water and salinity. *J. Ecol.* **64:**793–809.

Mahall, B. E., and R. B. Park, 1976c, The ecotone between *Spartina foliosa* Trin. and *Salicornia virginica* L. in salt marshes of northern San Francisco Bay, III. Soil aeration and tidal immersion, *J. Ecol.* **64:**811–819.

Maki, T. E., A. J. Weber, D. W. Hazel, S. C. Hunter, B. T. Hyberg, D. M. Flinchum, J. P. Lollis, J. B. Rognstad, and J. D. Gregory, 1980, *Effects of Stream Channelization on Bottomland and Swamp Forest Ecosystems,* North Carolina Water Resources Research Institute Report 80-147, Raleigh, N.C., 135p.

Malmer, N., 1962, Studies on mire vegetation in the Archaean area of southwestern Gotland, (South Sweden). 1. vegetation and habitat conditions on the Akahuit mire, *Opera Botanica* **7:**1–322.

Malmer, N., 1975, Development of bog mires, in *Coupling of Land and Water Systems,* Ecology Studies No. 10, A. D. Hasler, ed., Springer-Verlag, New York, pp 85–92.

Malmer, N., and H. Sjors, 1955, Some determinations of elementary constituents in mire plants and peat, *Bot. Not.* **108:**46–80.

Malone, M., and I. Ridge, 1983, Ethylene-induced growth and proton excretions in the aquatic plant *Nymphoides peltata., Planta* **157:**71–73.

Maltby, E., 1986, *Waterlogged Wealth: Why Waste the World's Best Wet Places?* Earthscan, Washington, D.C., 200p.

Maltby, E., and R. E. Turner, 1983, Wetlands of the world, *Geog. Mag.* **55:**12–17.

Maltby, E., P. J. Dugan, and J. C. Lefeuvre, eds., 1992, *Conservation and Development: The Sustainable Use of Wetland Resources,* IUCN, Gland, Switzerland, 219p.

Manning, E., M. Bardecki, and W. Bond, 1990, Measuring the value of renewable resources: The case of wetlands, in *The Common Property Conference, Duke University, Durham, N.C.,* Sept. 26–28, 1990.

Mantel, L. H., 1968, The foregut of *Gecarcinus lateralis* as an organ of salt and water balance, *Am. Zool.* **8:**433–442.

Marble, A. D., 1992, *A Guide to Wetland Functional Design,* Lewis Publishers, Chelsea, Mich., 222p.

Marois, K. C., and K. C. Ewel, 1983, Natural and management-related variation in cypress domes, *Forest Sci.* **29:**627–640.

Martin, A. C., N. Hutchkiss, F. M. Uhler, and W. S. Bourn, 1953, *Classification of Wetlands of the United States,* U.S. Fish and Wildlife Service Special Science Report—Wildlife 20, Washington, D.C., 14p.

Martin, C. V., and B. F. Eldridge, 1989, California's experience with mosquitoes in aquatic wastewater treatment systems, in *Constructed Wetlands for Wastewater Treatment,* D. A. Hammer, ed., Lewis Publishers, Inc. Chelsea, Mich., pp. 393–398.

Martini, I. P., D. W. Cowell, and G. M. Wickware, 1980, Geomorphology of Southwestern James Bay: A Low Energy, Emergent Coast in *The Coastline of Canada,* S. B. McCann, ed., Pap. Geol. Surv. Can., pp. 293–301.

Mason, C. F., and R. J. Bryant, 1975, Production, nutrient content and decomposition of *Phragmites communis* Trin. and *Typha angustifolia* L., *Ecol.* **63:**71–95.

Matthews, E. 1990, *Global distribution of forested wetlands.* Addendum to Forested Wetlands, A. E. Lugo, M. Brinson, and S. Brown, eds., Elsevier, Amsterdam.

Matthews, E. and I. Fung, 1987, Methane emissions from natural wetlands: global distribution, area, and environmental characteristics of sources, *Global Biogeochemical Cycles,* **1:**61–86.

Mattoon, W. R., 1915, *The Southern Cypress,* U.S. Department of Agriculture Bulletin 272.

Mattoon, W. R., 1916, Water requirements and growth of young cypress, *Soc. Am. Foresters Proc.* **11:**192–197.

Mazda, Y., Y. Sato, S. Sawamoto, H. Yokochi, and E. Wolanski, 1990, Links between physical, chemical and biological processes in Bashita-minato, a mangrove swamp in Japan, *Estuarine, Coastal and Shelf Science* **31:**817–833.

McArthur, B.H., 1989, The use of isolated wetlands in Florida for stormwater treatment, in *Wetlands Concerns and Successes,* D. W. Fisk, ed., Proceedings of the Conference, Tampa, Florida, American Water Resources Association, Bethesda, Md., pp. 185–194.

McArthur, J. V., 1989, Aquatic and terrestrial linkages: floodplain functions, in *Proceedings of the Symposium: The Forested Wetlands of the Southern United States, Asheville, N.C.,* D. D. Hook and R. Lea, eds., U.S. Department of Agriculture, Forest Service, Southeastern Forest Experiment Station, pp. 107–116.

McArthur, J. V., and G. R. Marzolf, 1987, Changes in soluble nutrients of prairie riparian vegetation during decomposition on a floodplain, *Amer. Mid. Nat.* **117:**26–34.

McCaffrey, R. J., 1977, *A Record of the Accumulation of Sediment and Trace Metals in a Connecticut, U.S.A. Salt Marsh,* Ph.D. dissertation, Yale University, 156p.

McComb, A. J., and P. S. Lake, 1990, *Australian Wetlands*, Angus and Robertson, London, 258p.

Mclntosh, R. P., 1980, The background and some current problems of theoretical ecology *Synthese* **43:**195–255.

Mclntosh, R. P., 1985, *The Background of Ecology, Concept and Theory,* Cambridge University Press, Cambridge, 383p.

McKay, D., P. J. Leinonen, J. C. K. Overall, and B. R. Wood, 1975, The behavior of crude oil spilled on snow, *Arctic* **28:**9–20.

McKee, K. L., and I. A. Mendelssohn, 1989, Response of a freshwater marsh plant community to increased salinity and increased water level, *Aquat. Bot.* **34**(4):301–316.

McKee, K. L., I. A. Mendelssohn, and M. W. Hester, 1988, Reexamination of pore water sulfide concentrations and redox potentials near the aerial roots of *Rhizophora mangle* and *Avicennia Germinans*, *Amer. J. Bot.* **75:**1352–1359.

McKinley, C. E., and F. E. Day, Jr., 1979, Herbaceous production in cut-burned, uncut- burned, and control areas of a *Chamaecyparls thyoides* (L.) BSP (Cupressaceae) stand in the Great Dismal Swamp, *Torrey Bot. Club Bull.* **106:**20–28.

McKnight, D., E. Thurman, R. Wershaw, and H. Hemond, 1985, Biogeochemistry of aquatic humic substances in Thoreau's Bog, Concord, Mass., *Ecology* **66**:1339–1352.

McKnight, J. S., D. D. Hook, O. G. Langdon, and R. L. Johnson, 1981, Flood tolerance and related characteristics of trees of the bottomland forests of the southern United States, in *Wetlands of Bottomland Hardwood Forests,* J. R. Clark and J. Benforado, eds., Elsevier, Amsterdam, pp. 29–69.

McLaughlin, D. B., and H. J. Harris, 1990, Aquatic insect emergence in two Great Lakes marshes, *Wetlands Ecol. Manage.* **1:**111–121.

McLeese, R. L., and E. P. Whiteside, 1977, Ecological effects of highway construction upon Michigan woodlots and wetlands: soil relationships, *J. Environ. Qual.* **6:**467–471.

McLeod, K. W., L. S. Donovan, and N. J. Stumpff, 1988, Responses of woody aeedlings to elevated flood water temperatures, in *The Ecology and Management of Wetlands,* vol 1: *Ecology of Wetlands*, D. D. Hook et al., ed., Timber Press, Portland, Ore., 2 vol., pp. 441–451.

McNaughton, S. J., 1966, Ecotype function in the *Typha* community-type, *Ecol. Monogr.* **36:**297–325.

McNaughton, S. J., 1968, Autotoxic feedback in relation to germination and seedling growth in *Typha latifolia, Ecology* **49:**367–369.

Meade, R., and R. Parker, 1985, Sediment in Rivers of the United States in *National Water Summary*, U.S.G.S. Water Supply Paper 2275, Washington, D.C., pp. 49–60.

Medina, E., E. Cuevas, M. Popp, and A. E. Lugo, 1990, Soil salinity, sun exposure, and growth of *Acrostichum aureum*, the mangrove fern, *Botanical Gazette* **151:**41–49.

Meeks, G., and L. C. Runyon, 1990, *Wetlands Protection and the States*, National Conference of State Legislatures, Denver, Colo., 26p.

Megonigal, J. P., and F. P. Day, Jr., 1988, Organic matter dynamics in four seasonally flooded forest communities of the Dismal Swamp, *Amer. J. Bot.* **75:**1334–1343.

Mehlich, A., 1972, Uniformity of soil test results as influenced by volume weight, *Comm. on Soil Sci. and Plant Analysis* **4:**475–486.

Mendelssohn, I. A., 1979, Nitrogen metabolism in the height forms of *Spartina alterniflora* in North Carolina, *Ecology* **60:**574–584.

Mendelssohn, I. A., K. L. McKee, and W. H. Patrick, Jr., 1981, Oxygen deficiency in *Spartina alterniflora* roots: metabolic adaptation to anoxia, *Science* **214:**439-441.

Mendelssohn, I. A., and M. L. Postek, 1982, Elemental analysis of deposits on the roots of *Spartina alterniflora* Loisel, *Am. J. Bot.* **69:**904–912.

Mendelssohn, I. A., K. L. McKee, and M. L. Postek, 1982, Sublethal stresses controlling *Spartina alterniflora* productivity, in *Wetlands Ecology and Management,* B. Gopal, R. E. Turner, R. G. Wetzel, and D. F Whigham, eds., National Institute of Ecology and International Science Publications, Jaipur, India, pp. 223–242.

Mendelssohn, I. A., and D. M. Burdick, 1988, The relationship of soil parameters and root metabolism to primary production in periodically inundated soils, in *The Ecology and Management of Wetlands.* vol 1: *Ecology of Wetlands*, D. D. Hook et al., eds., Timber Press, Portland, Ore., 2 vol., pp. 398–428.

Metz, E. D., 1987, Guidelines for planning and designing a major wetlands restoration project: Ballona Wetland case study, in *Proceedings of the Conference Wetlands: Increasing our Wetland Resources,* J. Zelazny and J. S. Feierabend, eds., Washington, D.C., Corporate Conservation Council and National Wildlife Federation, Washington, D.C., pp. 80–87.

Metzler, K., and R. Rosza, 1982, Vegetation of fresh and brackish tidal marshes in Connecticut, *Connecticut Bot. Soc. Newsletter* **10:**2–4.

Meyer, A. H., 1935, The Kankakee "Marsh" of northern Indiana and Illinois, *Michigan Acad. Sci. Arts & Letters Papers* **21:**359–396.

Meyer, J. L., 1985, A detention basin/artificial wetland treatment system to renovate stormwater runoff from urban, highway, and industrial areas, *Wetlands.* **5:**135–145.

Middleton, B. A., 1990, Effect of water depth and clipping frequency on the growth and survival of four wetland plant species, *Aquat. Bot.* **37:**189–196.

Millar, J. B., 1971, Shoreline-area as a factor in rate of water loss from small sloughs, *J. Hydrology* **14:**259–284.

Miller, P. C., 1972, Bioclimate, leaf temperature, and primary production in red mangrove canopies in south Florida, *Ecology* **53:**22–45.

Miller, P. C., and L. L. Tieszen, 1972, A preliminary model of processes affecting primary productivity in the Arctic tundra, *Arctic Alp. Res.* **4:**1–18.

Miller, P. C., B. D. Collier, and F. L. Bunnell, 1975, Development of ecosystem modelling in the Tundra Biome, in *Systems Analysis and Simulation in Ecology,* Vol. III, B. C. Patten, ed., Academic Press, New York, pp. 95–115.

Miller, P C., W. A. Stoner, and L. L. Tieszen, 1976, A model of stand photosynthesis for the wet meadow tundra at Barrow, Alaska, *Ecology* **57:**411–430.

Miller, R. S., D. B. Botkin, and R. Mendelssohn, 1974, The whooping crane (*Grus americana*) population of North America, *Biol. Conserv.* **6:**106–111.

Minnesota Department of Natural Resources, 1984, *Recommendations for the Protection of Ecologically Significant Peatlands in Minnesota*, Minnesota Department of Natural Resources, Minneapolis, 55p.

Minshall, G. W., R. C. Peterson, K. W. Cummins, T. L. Bott, J. R. Sedall, C. E. Cushing, and R. L. Vannote, 1983, Interbiome comparison of stream ecosystem dynamics, *Ecol. Monogr.* **53:**1–25.

Minshall, G. W., et al., 1985, Development in stream ecosystem theory, *Can. J. Fish. Aquat. Sci.* **37:**130–137.

Minshall, G. W., W. S. E. Jensen, and W. S. Platts, 1989, *The Ecology of Stream and Riparian Habitats of the Great Basin Region: A Community Profile,* U.S. Fish Wildl. Serv. Biol. Rep. **85(7.24):**142.

Mitchell, D. S., and B. Gopal, 1991, Invasion of Tropical Freshwaters by Alien Aquatic Plants, in *Ecology of Biological Invasion in the Tropics,* P. S. Ramakrishnan, ed., pp. 139–154.

Mitchell, J. G., R. Gehman, and J. Richardson, 1992, Our disappearing wetlands. *National Geographic* **182**(4):3–45.

Mitchell, R. C., and R. T. Carson, 1989, *Using Surveys to Value Public Goods: The Contingent Valuation Method*, Resources for the Future, Washington, D.C.

Mitsch, W. J., 1977, Water hyacinth (*Eichhornia crassipes*) nutrient uptake and metabolism in a north-central Florida marsh, *Archiv. fur Hydrobiologia,* **81:**188–210.

Mitsch, W. J., 1979, Interactions between a riparian swamp and a river in southern Illinois, in *Strategies for the Protection and Management of Floodplain Wetlands and Other Riparian Ecosystems,* R. R. Johnson and J. F. McCormick, tech. coords., Proceedings Symposium, Callaway Gardens, Dec. 1978, U.S. Forest Service General Technical Report WO-12, Washington, D.C., pp. 63–72.

Mitsch, W. J., 1983, Ecological models for management of freshwater wetlands, in *Application of Ecological Modeling in Environmental Management,* Part B, S. E. Jørgensen and W. J. Mitsch, eds., Elsevier, Amsterdam, pp. 283–310.

Mitsch, W. J., 1984, Seasonal patterns of a cypress dome pond in Florida, in *Cypress Swamps,* K. C. Ewel and H. T. Odum, eds., University Presses of Florida, Gainesville, pp. 25–33.

Mitsch, W. J., 1988, Productivity-hydrology-nutrient models of forested wetlands, in *Wetland Modelling,* W. J. Mitsch, M. Straskraba and S. E. Jørgensen, eds., Elsevier, Amsterdam, pp. 115–132.

Mitsch, W. J., ed., 1989, *Wetlands of Ohio's Coastal Lake Erie: A Hierarchy of Systems,* Ohio Sea Grant Program, Columbus, Ohio. NTIS, OHSU–BS–007, 186p.

Mitsch, W. J., 1991a, Ecological engineering: approaches to sustainability and biodiversity in the U.S. and China, in *Ecological Economics: The Science and Management of Sustainability*, R. Costanza, ed., Columbia University Press, New York, pp. 428–448.

Mitsch, W. J., 1991b, Ecological engineering: the roots and rationale of a new ecological paradigm, in *Ecological Engineering for Wastewater Treatment,* C. Etnier and B. Guterstam, eds., Bokskogen, Gothenburg, Sweden, pp.19–37.

Mitsch, W. J., 1992a, Landscape design and the role of created, restored, and natural riparian wetlands in controlling nonpoint source pollution, *Ecological Engineering* **1:**27–47.

Mitsch, W. J., 1992b, Combining ecosystem and landscape approaches to Great Lakes wetlands, *Journal of Great Lakes Research* **18:**552–570.

Mitsch, W. J., C. L. Dorge, and J. R. Wiemhoff, 1977, *Forested Wetlands for Water Resource Management in Southern Illinois.* Research Report No. 132, Illinois Univ. Water Resources Center, Urbana, Ill., 225p.

Mitsch, W. J., and K. C. Ewel, 1979, Comparative biomass and growth of cypress in Florida wetlands, *Am. Midl. Nat.* **1O1:**417–426.

Mitsch, W. J., C. L. Dorge, and J. R. Wiemhoff, 1979a, Ecosystem dynamics and a phosphorus budget of an alluvial cypress swamp in southern Illinois, *Ecology* **60:**1116–1124.

Mitsch, W. J., W. Rust, A. Behnke, and L. Lai, 1979b, *Environmental Observations of a Riparian Ecosystem during Flood Season,* Research Report 142, Illinois Univ. Water Resources Center, Urbana, Ill., 64p.

Mitsch, W. J., M. D. Hutchison, and G. A. Paulson, 1979c, *The Momence Wetlands of the Kankakee River in Illinois—An Assessment of Their Value,* Illinois Institute of Natural Resources, Doc 79/17, Chicago, 55p.

Mitsch, W. J., J. W. Day, Jr., J. R. Taylor, and C. Madden, 1982, Models of North American freshwater wetlands, *Int. J. Ecol. Environ. Sci.* **8:**109–140.

Mitsch, W. J., J. R. Taylor, and K. B. Benson, 1983a, Classification, modelling and management of wetlands—a case study in western Kentucky, in *Analysis of Ecological Systems: State-of-the-art in Ecological Modelling,* W. K. Lauenroth, G. V. Skogerboe, and M. Flug, eds., Elsevier, Amsterdam. pp. 761–769.

Mitsch, W. J., J. R. Taylor, K. B. Benson, and P. L. Hill, Jr., 1983b, *Atlas of Wetlands in the Principal Coal Surface Mine Region of Western Kentucky,* U.S. Fish and Wildlife Service Report FWS/OBS 82/72, Washington, D.C., 135p.

Mitsch, W. J., J. R. Taylor, K. B. Benson, and P. L. Hill, Jr., 1983c, Wetlands and coal surface mining in western Kentucky—a regional impact assessment, *Wetlands* **3:**161–179.

Mitsch, W. J., and W. G. Rust, 1984, Tree growth responses to flooding in a bottomland forest in northeastern Illinois, *Forest Sci.* **30:**499–510.

Mitsch, W. J., M. Straskraba, and S. E. Jørgensen, eds., 1988, *Wetland Modelling*, Elsevier, Amsterdam, 227p.

Mitsch, W. J., and S. E. Jørgensen, 1989, Introduction to ecological engineering, in *Ecological Engineering: An Introduction to Ecotechnology,* W. J. Mitsch and S. E. Jørgensen, eds., Wiley, New York, pp. 3–12.

Mitsch, W. J., and B. C. Reeder, 1991, Modelling nutrient retention of a freshwater coastal wetland: estimating the roles of primary productivity, sedimentation, resuspension and hydrology, *Ecological Modelling* **54:**151–187.

Mitsch, W. J., J. R. Taylor, and K. B. Benson, 1991, Estimating primary productivity of forested wetland communities in different hydrologic landscapes, *Landscape Ecology* **5:**75–92.

Mitsch, W. J., and J. K. Cronk, 1992, Creation and restoration of wetlands: some design consideration for ecological engineering, in *Advances in Soil Science,* R. Lal and B. A. Stewart, eds., vol. 17 — *Soil Restoration,* Springer-Verlag, New York, pp. 217–259.

Mitsch, W. J., and B. C. Reeder, 1992, Nutrient and hydrologic budgets of a Great Lakes coastal freshwater wetland during a drought year, *Wetlands Ecology and Management* **1(4):**211–223.

Mitsch, W. J., ed., 1994, *Global Wetlands: Old World and New,* Elsevier, Amsterdam, 967 pp.

Mohanty, S. K., and R. N. Dash, 1982, The chemistry of waterlogged soils, in *Wetlands—Ecology and Management,* B. Gopal, R. E. Turner, R. G. Wetzel, and D. F. Whigham, eds., Natural Institute of Ecology and International Scientific Publications, Jaipur, India, pp. 389–396.

Monk, C. D., and T. W. Brown, 1965, Ecological consideration of cypress heads in northcentral Florida, *Am. Midl. Nat.* **74:**126–140.

Montague, C. L., S. M. Bunker, E. B. Haines, M. L. Pace, and R. L. Wetzel, 1981, Aquatic macroconsumers, in *The Ecology of a Salt Marsh,* L. R. Pomeroy and R. G. Wiegert, eds., Springer-Verlag, New York, pp. 69–85.

Montague, C. L., and R. G. Wiegert, 1990, Salt Marshes, in *Ecosystems of Florida,* R. L. Myers and J. J. Ewel, eds., University of Central Florida Press, Orlando, Fla., pp. 481–516.

Montz, G. N., and A. Cherubini, 1973, An ecological study of a bald cypress swamp in St. Charles Parish, Louisiana, *Castanea* **38:**378–386.

Moon, G. J., B. F. Clough, C. A. Peterson, and W. G. Allaway, 1986, Apoplastic and symplastic pathways in *Avicennia marina* (Forsk.) Vierh. roots revealed by fluorescent tracer dyes, *Aust J. Plant Physiol.* **13:**637–48.

Moore, D. R. J., and P. A. Keddy, 1989, The relationship between species richness and standing crop in wetlands: the importance of scale, *Vegetatio* **79:**99–106.

Moore, D. R. J., P. A. Keddy, C. L. Gaudet, and I. C. Wisheu, 1989, Conservation of wetlands: do infertile wetlands deserve a higher priority? *Biol. Conserv.* **47:**203–217.

Moore, P. D., ed., 1984, *European Mires,* Academic Press, London, 367p.

Moore, P. D., and D. J. Bellamy, 1974, *Peatlands,* Springer-Verlag, New York, 221p.

Moore, T. R., 1989, Growth and net production of Sphagnum at five fen sites, subarctic eastern Canada, *Can. J. Bot.* **67:**1203–1207.

Moorhead, K. K., and K. R. Reddy, 1988, Oxygen transport through selected aquatic macrophytes, *J. Environ. Qual.* **17:**138–142.

Morris, J. T., 1982, A model of growth responses by *Spartina alterniflora* to nitrogen limitation, *Journal of Ecology* **70:**25–42.

Morris, J. T., 1984, Effects of oxygen and salinity on ammonium uptake by *Spartina alterniflora* Loisel and *Spartina patens* (Aiton), *J. Exp. Mar. Biol. Ecol.* **78:**87–98.

Morris, J. T., and J. W. H. Dacey, 1984, Effects of O_2 on ammonium uptake and root respiration by *Spartina alterniflora., Am. J. Bot.* **71:**979–985.

Morris, J. T., and W. B. Bowden, 1986, A mechanistic, numerical model of sedimentation, mineralization, and decomposition for marsh sediments, *Soil Sci Soc. America J.* **50:**96–105.

Morris, J. T., B. Kjerfve, and J. M., Dean, 1990, Dependence of estuarine productivity on anomalies in mean sea level, *Limnol. Oceanogr.* **35:**926–930.

Morss, W. L., 1927, The plant colonization of marshlands in the estuary of the River Nith, *J. Ecol.* **15:**310–343.

Mortimer, C. H., 1941-1942, The exchange of dissolved substances between mud and water in lakes, *J. Ecol.* **29:**280–329; **30:**147–201.

Mulholland, P. J., 1979, *Organic Carbon in a Swamp-Stream Ecosystem and Export by Streams in Eastern North Carolina,* Ph.D. Dissertation, University of North Carolina.

Mulholland, P. J., and E. J. Kuenzler, 1979, Organic carbon export from upland and forested wetland watersheds, *Limnol. Oceanogr.* **24:**960–966.

Mumphrey, A. J., J. S. Brooks, T. D. Fox, L. B. Fromberg, R. J. Marak, and J. D. Wilkinson, 1978, *The Valuation of Wetlands in the Barataria Basin,* Urban Studies Institute, University of New Orleans, New Orleans.

Naiman, R. J., 1988, Animal influences on ecosystem dynamics, *BioScience* **38:**750–752.

Naiman, R. J., T. Manning, and C. A. Johnston, 1991, Beaver population fluctuations and tropospheric methane emissions in boreal wetlands, *Biogeochemistry* **12:**1–15.

National Research Council, 1992, *Restoration of Aquatic Ecosystems,* National Academy Press, Washington, D.C., 552p.

National Wetlands Policy Forum, 1988, *Protecting America's Wetlands: An Action Agenda,* Conservation Foundation, Washington, D.C., 69p.

National Wetlands Working Group, 1988, *Wetlands of Canada,* Ecological Land Classification Series, no. 24, Environment Canada, Ottawa, Ontario, and Polyscience Publications, Inc., Montreal, Quebec, 452p.

Nature Conservancy, 1992, *The Forested Wetlands of the Mississippi River: An Ecosystem in Crisis*, The Nature Conservancy, Baton Rouge, LA, 25 pp.

Navid, D., 1989, The international law of migratory species: the Ramsar Convention, *Natural Resources Journal* **29:**1001–1016.

Neill, C., 1990a, Effects of nutrients and water levels on emergent macrophytes biomass in a prairie marsh, *Can. J. Bot.* **68:**1007–1014.

Neill, C., 1990b, Effects of nutrients and water levels on species composition in prairie whitetop (*Scolochloa festucacea*) marshes, *Can. J. Bot.* **68:**1015–1020.

Neill, C., 1990c, Nutrient limitation of hardstem bulrush (*Scirpus acutus* Muhl.) in a Manitoba interlake region marsh, *Wetlands* **10**(1):69–75.

Neill, C., and L. A. Deegan, 1986, The effect of Mississippi River delta lobe development on the habitat composition and diversity of Louisiana coastal wetlands, *Amer. Midl. Naturalist* **116:**296–303.

Nelson, J. W., J. A. Kadlec, and H. R. Murkin, 1990a, Seasonal comparisons of weight loss of two types of *Typha glauca* Godr. leaf litter, *Aquatic Bot.* **37:**299–314.

Nelson, J. W., J. A. Kadlec, and H. R. Murkin, 1990b, Response by macroinvertebrates to cattail litter quality and timing of litter submergence in a northern prairie marsh, *Wetlands* **10:**47–60.

Nelson, U. C., 1947, Woodland caribou in Minnesota, *J. Wildl. Manage,* **11:**283–284.

Nessel, J. K., 1978a, Phosphorus cycling, productivity and community structure in the Waldo cypress strand, in *Cypress Wetlands For Water Management, Recycling, and Conservation,* 4th annual report, H. T. Odum and K. C. Ewel, eds., Center for Wetlands, University of Florida, Gainesville, pp. 750–801.

Nessel, J. K., 1978b, *Distribution and Dynamics of Organic Matter and Phosphorus in a Sewage Enriched Cypress Strand*, M.S. Thesis, University of Florida, Gainesville, 159p.

Nessel, J. K., K. C. Ewel, and M. S. Burnett, 1982, Wastewater enrichment increases mature pond cypress growth rates, *Forest Sci.* **28:**400–403.

Nessel, J. K., and S. E. Bayley, 1984, Distribution and dynamics of organic matter and phosphorus in a sewage-enriched cypress swamp, in *Cypress Swamps,* K. C. Ewel and H. T. Odum, eds., University Presses of Florida, Gainesville, pp. 262–278.

Nester, E. W., C. E. Roberts, B. J. McCarthy, N. N. Pearsall, 1973, *Microbiology— Molecules, Microbes, and Man,* Holt, Rinehart, and Winston, Inc., New York, 719p.

Newling, C. J., 1982, Feasibility report on a Santa Ana River marsh restoration and habitat development project, in *Habitat Development at Eight Corps of Engineers Sites; Feasibility and Assessment,* M. C. Landin, ed., U.S. Army Engineer Waterways Experiment Station Environmental Laboratory, Vicksburg, Miss., pp. 45–84.

Nichols, A. B., 1992, Wetlands are getting more respect, *Water Environment and Technology* (Nov. 1992), pp. 46–51.

Nichols, D.S., 1983, Capacity of natural wetlands to remove nutrients from wastewater, *J. Water Pollution Control Federation* **55:**495–505.

Niering, W. A., 1985, *Wetlands,* Alfred A. Knopf, New York, 638p.

Niering, W. A., 1988, Endangered, threatened and rare wetland plants and animals of the continental United States, in *The Ecology and Management of Wetlands, Vol. 1: Ecology of Wetlands*, D. D. Hook, and et al., ed., pp. 227–238.

Niering, W. A., 1989, Wetland vegetation development, in *Wetlands Ecology and Conservation: Emphasis in Pennsylvania,* S. K. Majumdar, R. P. Brooks, F. J. Brenner and J. R.W. Tiner, eds., The Pennsylvania Academy of Science, Easton, pp. 103–113.

Niering, W. A., and R. S. Warren, 1977, Salt marshes, in *Coastal Ecosystem Management,* J. R. Clark, ed., Wiley, New York, pp. 697–702.

Niering, W. A., and R. S. Warren, 1980, Vegetation patterns and processes in New England salt marshes, *Bioscience* **30:**301–307.

Nixon, S. W., 1980, Between coastal marshes and coastal waters—a review of twenty years of speculation and research on the role of salt marshes in estuarine productivity and water chemistry, in *Estuarine and Wetland Processes,* P. Hamilton and K. B. MacDonald, eds., Plenum, New York, pp. 437–525.

Nixon, S. W., 1982, *The Ecology of New England High Salt Marshes: A Community Profile*, U.S. Fish and Wildlife Service, Office of Biological Services, Washington, D.C., FWS/OBS-81/55.

Nixon, S. W., ed., 1988, *Comparative Ecology of Freshwater and Marine Ecosystems, Limnol. and Oceanogr.* **33**(4)Part 2:649–1025.

Nixon, S. W., and C. A. Oviatt, 1973, Ecology of a New England salt marsh, *Ecol. Monogr.* **43:**463–498.

Nixon, S. W., and V. Lee, 1986, *Wetlands and Water Quality*, Wetlands Research Program, Tech. Rept. y-86-2, U.S. Army Engineers Waterway Experiment Station, Vicksburg, MS.

Nordquist, G. E., and E. C. Birney, 1980, *The Importance of Peatland Habitats to Small Mammals in Minnesota*, Peat Program, Minn. Dept. Nat. Resources, Minneapolis.

Norgress, R. E., 1947, The history of the cypress lumber industry in Louisiana, *Louisiana Hist. Q.* **30:**979–1059.

Novacek, J. M., 1989, The water and wetland resources of the Nebraska Sandhills, in *Northern Prairie Wetlands*, A. G. van der Valk, ed., Iowa State University Press, Ames, Iowa, pp. 340–384.

Novak, M., J. A., Balen, M. E. Obbard, and B. Mallocheds, eds., 1987, *Wildlife Furbearer Management and Conservation in North America,* Ontario Trappers Assn., Ontario, Canada, 1150 p.

Novitzki, R. P., 1979, Hydrologic characteristics of Wisconsin's wetlands and their influence on floods, stream flow, and sediment, in *Wetland Functions and Values: The State of Our Understanding,* P. E. Greeson, J. R. Clark, J. E. Clark, eds., American Water Resource Association, Minneapolis, Minn., pp. 377–388.

Novitzki, R. P., 1985, The effects of lakes and wetlands on flood flows and base flows in selected northern and eastern states, in *Proceedings of a Conference—Wetlands of the Chesapeake*, H. A. Groman, et al., eds., Environmental Law Institute, Washington, D.C., pp. 143–154.

Novitzki, R. P., 1989, Wetland hydrology, in *Wetlands Ecology and Conservation: Emphasis in Pennsylvania*, S. K. Majumdar, R. P. Brooks, F. J. Brenner, and R. W. Tiner, eds., Pennsylvania Academy of Science, Easton, Pa., pp. 47–64.

Nyc, R., and P. Brooks, 1979, *National Wetlands Inventory Project: Inventorying the Nation's Wetlands with Remote Sensing,* paper presented at Corp of Engineers Remote Sensing Symposium, Reston, Virginia, 29–31 October, 1979, 11p.

O'Brien, A. L., 1988, Evaluating the cumulative effects of alteration on New England wetlands, *Environmental Management* **12:** 627–636.

O'Neil, L. J., T. M. Pullen, and R. L. Schroeder, 1991, *A Wildlife Community Habitat Evaluation Model for Bottomland Hardwood Forests in the Southeastern United States*, U.S. Department of Interior, Fish and Wildl. Serv., Washington, D.C., Biological Report 91(x).

O'Neil, T., 1949, *The Muskrat in the Louisiana Coastal Marsh*, Louisiana Department of Wildlife and Fisheries, New Orleans, LA, 152p.

Odum, E. P., 1961, The role of tidal marshes in estuarine production, *N.Y. State Cons.* **15**(6):12–15.

Odum, E. P., 1969, The strategy of ecosystem development, *Science* **164:**262–270.

Odum, E. P., 1971, *Fundamentals of Ecology,* 3rd ed., W. B. Saunders Co., Philadelphia. 544p.

Odum, E. P., 1979a, Ecological importance of the riparian zone, in *Strategies For Protection and Management of Floodplain Wetlands and Other Riparian Ecosystems,* R. R. Johnson and J. F. McCormick, tech. coords., Proceedings of the Symposium, Callaway Gardens, Georgia, Dec. 11–13,1978, U.S. Forest Service General Technical Report WO-12, Washington, D.C., pp. 2–4.

Odum, E. P., 1979b, The value of wetlands; a hierarchical approach, in *Wetland Functions and Values: The State of Our Understanding,* P. E. Greeson, J. R. Clark, and J. E. Clark, eds., American Water Resources Assoc., Bethesda, Md., pp. 1–25.

Odum, E. P., 1980, The status of three ecosystem-level hypotheses regarding salt marsh estuaries: tidal subsidy, outwelling, and detritus-based food chains, in *Estuarine Perspectives*, V. S. Kennedy, ed., Academic Press, New York, pp. 485–495.

Odum, E. P., 1981, Foreword, in *Wetlands of Bottomland Hardwood Forests,* J. R. Clark and J. Benforado, eds., Elsevier, Amsterdam, pp. xi–xiii.

Odum. E. P., and A. A. de la Cruz, 1967, Particulate organic detritus in a Georgia salt marsh-estuarine ecosystem, in *Estuaries,* G. H. Lauff, ed., American Association for the Advancement of Science, Washington, D.C., pp. 383–388.

Odum, H. T., 1951, The Carolina Bays and a Pleistocene weather map, *Amer. J. Sci.* **250**:262–270.

Odum, H. T., 1982, Role of wetland ecosystems in the landscape of Florida, in *Ecosystem Dynamics in Freshwater Wetlands and Shallow Water Bodies,* Vol. II. D. O. Logofet and N. K. Luckyanov, eds., SCOPE and UNEP Workshop, Center of International Projects, Moscow, USSR, pp. 33–72.

Odum, H. T., 1983, *Systems Ecology—An Introduction,* Wiley, New York, 644p.

Odum, H. T., 1984, Summary: cypress swamps and their regional role, in *Cypress Swamps,* K. C. Ewel and H. T. Odum, eds., University Presses of Florida, Gainesville. pp. 416–443.

Odum, H. T., 1988, Self-organization, transformity, and information, *Science* **242:**1132–1139.

Odum, H. T., 1989, Ecological engineering and self-organization, in *Ecological Engineering*, W. J. Mitsch and S. E. Jørgensen, eds., Wiley, New York, pp. 79–101.

Odum, H. T., B. J. Copeland, and E. A. McMahan, eds., 1974, *Coastal Ecological Systems of the United States,* The Conservation Foundation, Washington, D.C., 4 vols.

Odum, H. T., K. C. Ewel, W. J. Mitsch, and J. W. Ordway, 1977a, Recycling Treated Sewage Through Cypress Wetlands in Florida, in *Wastewater Renovation and Reuse,* F. M. D'Itri, ed., Marcel Dekker, Inc., New York, N.Y., pp. 35–67.

Odum, H. T., W. M. Kemp, M. Sell, W. Boynton, and M. Lehman, 1977b, Energy analysis and coupling of man and estuaries, *Environ. Manage.* **1:**297–315.

Odum, H. T., P. Kangas, G. R. Best, B. T. Rushton, S. Leibowitz, J. R. Butner, and T. Oxford, 1981, *Studies on Phosphate Mining, Reclamation, and Energy,* Center for Wetlands, University of Florida, Gainesville, 142p.

Odum, W. E.. 1970, *Pathways of Energy Flow in a South Florida Estuary,* Ph.D. Dissertation, University of Miami, Coral Gables, Florida, 162p.

Odum, W. E., 1987, Predicting ecosystem development following creation and restoration of wetlands, in *Wetlands: Increasing our Wetland Resources,* J. Zelazny and J. S. Feierabend, eds., Proceedings of the Conference Wetlands: Increasing our Wetland Resources, Washington D.C., Corporate Conservation Council, National Wildlife Federation, Washington, D.C., pp. 67–70.

Odum, W. E., 1988, Comparative ecology of tidal freshwater and salt marshes, *Ann. Rev. Ecol. Syst.* **19:**147–176.

Odum, W. E., and E. J. Heald, 1972, Trophic analyses of an estuarine mangrove community, *Bulletin of Marine Science* **22:**671–738.

Odum, W. E., and M. A. Heywood, 1978, Decomposition of intertidal freshwater marsh plants, in *Freshwater Wetlands; Ecological Processes and Management Potential,* R. E. Good, D. G. Whigham, and R. L. Simpson. eds., Academic Press, New York, pp. 89–97.

Odum, W. E., J. S. Fisher, and J. C. Pickral, 1979, Factors controlling the flux of particulate organic carbon from estuarine wetlands, in *Ecological Processes in Coastal and Marine Systems,* R. J. Livingston, ed., Plenum, New York, pp. 69–80.

Odum, W. E., C. C. McIvor, and T. J. Smith, III, 1982, *The Ecology of the Mangroves of South Florida: a Community Profile*, U.S. Fish & Wildlife Service, Office of Biological Services, Technical Report FWS/OBS 81–24, Washington, D.C.

Odum, W. E., T. J. Smith III, J. K. Hoover, and C. C. McIvor, 1984, *The Ecology of Tidal Freshwater Marshes of the United States East Coast: A Community Profile,* U.S. Fish and Wildlife Service, FWS/OBS-87/17, Washington, D.C., 177p.

Odum, W. E., and C. C. McIvor, 1990, Mangroves, in *Ecosystems of Florida*, R. L. Myers and J. J. Ewel, eds., University of Central Florida Press, Orlando, pp. 517–548.

Office of Technology Assessment. 1984. *Wetlands: Their Use and Regulation,* OTA, U.S. Congress, OTA–O–206,Washington, D.C., 208p.

Ogaard, L. A., J. A. Leitch, D. F. Scott, and W. C. Nelson, 1981, *The Fauna of the Prairie Wetlands: Research Methods and Annotated Bibliography,* Research Report No. 86, North Dakota State University, Agric. Exp. Station, Fargo, North Dakota, 2p.

Ogawa, H., and J. W. Male, 1983, *The Flood Mitigation Potential of Inland Wetlands,* University of Massachusetts, Amherst, Water Resources Research Center Publication no. 138, 164p.

Ogawa, H., and J. W. Male, 1986, Simulating the flood mitigation role of wetlands, *J. Water Resource Planning and Management* **112:**114–128.

Oliver, J. D., and T. Legovic, 1988, Okefenokee marshland before, during, and after nutrient enrichment by a bird rookery, *Ecol. Mod.* **43**(3-4):195–223.

Olson, R. K., ed., 1992, The role of created and natural wetlands in controlling nonpoint source pollution, *Ecological Engineering* **1**:1–170.

Omernik, J. M., 1977, *Nonpoint Source-Stream Nutrient Level Relationships: A Nationwide Study*, EPA–600/3–79–105, Corvallis Environmental Research Laboratory, U.S. EPA, Corvallis, Oregon.

Orozco-Segovia, A. D. L., 1980, One option for the use of marshes of Tabasco, Mexico, in *Wetlands Restoration and Creation: Proceedings of the 7th Annual Conference, 1979 May 16-17, Tampa, Florida,* D. P. Cole, ed., pages 209–218. Available from Hillsborough Community College, Tampa, Florida.

Otte, L. J., 1981, *Origin, Development and Maintenance of Pocosin Wetlands of North Carolina*, North Carolina Department of Natural Resources and Community Development, Raleigh, Unpublished Report to the North Carolina Natural Heritage Program.

Overbeck, F., and H. Happach, 1957, Uber das Wachstum und den Wasserhaushalt einiger Hochmoor-Sphagnum, *Flora,* Jena **144:**335–402.

Padgett, W. G., A. P. Youngblood, and A. H. Winward, 1989, *Riparian Community Type Classification of Utah and Southeastern Idaho*, Intermountain Region, U.S. Department Agriculture Forest Service, R4-Ecol-89–01.

Pakarinen, P., 1978, Production and nutrient ecology of three *Sphagnum* species in southern Finnish raised bogs, *Annales Botanici Fennici* **15:**15–26.

Pakarinen, P., and E. Gorham, 1983, Mineral Element Composition of *Sphagnum fuscum* peats collected from Minnesota, Manitoba, and Ontario, in *Proceedings of the International Symposium on Peat Utilization, Bemidji, Minnesota*, Bemidji State Univ. Center for Environmental Studies, Bemidji, Minnesota, 417–429.

Pallis, M., 1915, The structural history of Plav: the floating fen of the delta of the Danube, *J. Linn. Soc. Bot.* **43:**233–290.

Palmer, C. N., and J. D. Hunt, 1989, Greenwood urban wetland a manmade strormwater treatment facility, in *Wetlands Concerns and Successes,* D. W. Fisk, ed., Proceedings of the Conference American Water Resources Association, Tampa, Florida, pp. 1–10.

Park, R. A., J. K. Lee, P. W. Mausel, and R. C. Howe, 1991, Using remote sensing for modeling the impacts of sea level rise, *World Resource Review* **3**:184–205.

Parsons, K. E., and A. A. de la Cruz, 1980, Energy flow and grazing behavior of conocephaline grasshoppers in a *Juncus roemerianus* marsh, *Ecology* **61:**1045–1050.

Patrick, R., J. Cairns, Jr., and S. S. Roback, 1967, An ecosystematic study of the fauna and flora of the Savannah River, *Acad. Nat. Sci. Philadelphia Proc.* **118:**109–407.

Patrick, W. H., Jr., 1981, Bottomland soils, in *Wetlands of Bottomland Hardwood Forests,* J. R. Clark and J. Benforado, eds., Elsevier, Amsterdam, pp. 177–185.

Patrick, W. H., Jr., and R. D. Delaune, 1972, Characterization of the oxidized and reduced zones in flooded soil, *Soil Sci. Soc. Am. Proc.* **36:**573–576.

Patrick, W. H., Jr., and R. A. Khalid, 1974, Phosphate release and sorption by soils and sediments: effect of aerobic and anaerobic conditions, *Science* **186:**53–55.

Patrick, W. H., Jr., and K. R. Reddy, 1976, Nitrification-denitrification reactions in flooded soils and water bottoms: dependence on oxygen supply and ammonium diffusion, *J. Environ. Quality* **5:**469–472.

Patten, B. C., ed., 1990, *Wetlands and Shallow Continental Water Bodies,* vol. 1. *Natural and Human Relationships,* SPB Academic Publishing, the Hague, Netherlands, 759p.

Patterson, S. G., 1986, *Mangrove Community Boundary Interpretation and Detection of Areal Changes in Marco Island, Florida: Application of Digital Image Processing and Remote Sensing Techniques*, U.S. Fish & Wildlife Service, Office of Biological Services, Report 86(10).

Pearlstein, L., H. McKellar, and W. Kitchens, 1985, Modelling the impacts of a river diversion on bottomland forest communities in the Santee River floodplain, South Carolina, *Ecol. Mod.* **29:**283–302.

Pearsall, W. H., 1920, The aquatic vegetation of the English lakes, *J. Ecology* **8:**163–201.

Pearson, J., and D. C. Havill, 1988, The effect of hypoxia and sulfide on culture grown wetland and non-wetland plants. 2. Metabolic and physiological changes, *J. Experimental Botany* **39:**431–439.

Pedersen, A., 1975, Growth measurements of five *Sphagnum* species in South Norway, *Norwegian J. of Botany* **22:**277–284.

Pederson, R. L., and L. M. Smith, 1988, Implications of wetland seed bank research: a review of Great Britain and prairie marsh studies, in *Interdisciplinary Approaches to Freshwater Wetlands Research*, D. A. Wilcox, ed., Michigan State University Press, East Lansing, Mich., pp. 81–95.

Pelikan, J., 1978, Mammals in the reedswamp ecosystem, in *Pond Littoral Ecosystems,* D. Dykyjova and J. Kvet, eds., Springer-Verlag, Berlin, pp. 357–365.

Penfound, W. T., 1952, Southern swamps and marshes, *Bot. Rev.* **18:**413–446.

Penfound, W. T., and E. S. Hathaway, 1938, Plant communities in the marshlands of southeastern Louisiana, *Ecol. Monogr.* **8:**1–56.

Penfound, W. T., and T. T. Earle, 1948, The biology of the water hyacinth, *Ecol. Monogr.* **18:**447–472.

Penman, H. L., 1948, Natural evaporation from open water, bare soil, and grass, *Proc. Royal Society of London* **93:**120–145.

Peterjohn, W. T., and D. L. Correll, 1984, Nutrient dynamics in an agricultural watershed: observations on the role of a riparian forest, *Ecology* **65:**1466–1475.

Peters, D. S., D. W. Ahrenholz, and T. R. Rice, 1979, Harvest and value of wetland associated fish and shellfish, in *Wetland Functions and Values: The State of Our Understanding,* Greeson, P. E., J. R. Clark, and J. E. Clark, eds., American Water Resources Assoc., Minneapolis, Minn., pp. 606–617.

Peverly, J. H., 1982, Stream transport of nutrients through a wetland, *J. Environ. Quality* **11:**38–43.

Pezeshki, S. R., S. W. Matthews, and R. D. Delaune, 1991, Root cortex structure and metabolic responses of *Spartina patens* to soil redox conditions, *Environmental and Experimental Bot.* **31:**91–97.

Pfeiffer, W. J., and R. G. Wiegert, 1981, Grazers on *Spartina* and their predators, in *The Ecology of a Salt Marsh,* L. R. Pomeroy and R. G. Wiegert, eds., Springer-Verlag, New York, pp. 87–112.

Phillips, J. D., 1989, Fluvial sediment storage in wetlands, *Water Resources Bulletin* **25:**867–873.

Phipps, R. L., 1979, Simulation of wetlands forest vegetation dynamics, *Ecol. Mod.* **7:**257–288.

Phipps, R. L., and L. H. Applegate, 1983, Simulation of management alternatives in wetland forests, in *Application of Ecological Modelling in Environmental Management,* Part B, S. E. Jørgensen and W. J. Mitsch, eds., Elsevier, Amsterdam, pp. 311–339.

Pjavchenko, N. J., 1982, Bog ecosystems and their importance in nature, in *Proceedings of International Workshop on Ecosystems Dynamics in Wetlands and Shallow Water Bodies,* Vol. 1, D. O. Logofet and N. K. Luckyanov, eds., SCOPE and UNEP Workshop, Center for International Projects, Moscow, USSR, pp. 7–21.

Pokorny, I., O. Lhotsky, P. Denny, and E. G. Turner, eds., 1987, Waterplants and wetland processes, *Archiv für Hydrobiologie,* Heft 27, E. Schweizerbart'sche Verlagsbuchhandlung, Stuttgart, Germany, 265p.

Pomeroy, L. R., 1959, Algae productivity in salt marshes of Georgia, *Limnol. Oceanogr.* **4:**386–397.

Pomeroy, L. R., L. R. Shenton, R. D. Jones, and R. J. Reimold, 1972, Nutrient flux in estuaries, in *Nutrients and Eutrophication,* G. E. Likens, ed., Am. Soc. Limnol. Oceanogr. Special Symposium, Allen Press, Lawrence, Kans., pp. 274–291.

Pomeroy, L. R., K. Bancroft, J. Breed, R. R. Christian, D. Frankenberg, J. R. Hall, L. G. Maurer, W. J. Wiebe, R. G. Wiegert, and R. L. Wetzel, 1977, Flux of organic matter through a salt marsh, in *Estuarine Processes,* vol. II, M. Wiley, ed., Academic Press, New York, pp. 270–279.

Pomeroy, L. R., and R. G. Wiegert, ed., 1981, *The Ecology of a Salt Marsh,* Springer-Verlag, New York, 271p.

Pomeroy, L. R., W. M. Darley, E. L. Dunn, J. L Gallagher, E. B. Haines, and D. M. Witney, 1981, Primary production, in *Ecology of a Salt Marsh,* L. R. Pomeroy and R. G. Wiegert, eds., Springer-Verlag, New York pp. 39–67.

Ponnamperuma, F. N., 1972, The Chemistry of Submerged Soils, *Adv. Agron.* **24:**29–96.

Pool, D. J., A. E. Lugo, and S. C. Snedaker, 1975, Litter production in mangrove forests of southern Florida and Puerto Rico, in *Proceedings of the International Symposium on Biology and Management of Mangroves,* G. Walsh, S. Snedaker, and H. Teas, eds., Institute of Food and Agriculture Science, University of Florida, Gainesville, pp. 213–237.

Porcher, R., 1981, The vascular flora of the Francis Beidler Forest in Four Holes Swamp, Berkeley and Dorchester Counties, South Carolina, *Castanea* **46:**248–80.

Post, W. M., W. R. Emmanuel, P. J. Zinke, and A. G. Stangenberger, 1982, Soil carbon pools and world life zones, *Nature* **298:**156–159.

Potonie, R., 1908, *Aufbau und Vegetation der Moore Norddeutschlands,* Englers. Bot. Jahrb. 90, Leipzig.

Powell, S. W., and F. P. Day, 1991, Root production in four communities in the Great Dismal Swamp, *Amer. J. Bot.* **78:**288–297.

Prentki, R. T., T. D. Gustafson, and M. S. Adams, 1978, Nutrient movements in lakeshore marshes, in *Freshwater Wetlands: Ecological Processes and Management Potential,* R. E. Good, D. F. Whigham, and R. L. Simpson, eds., Academic Press, New York, pp. 169–194.

Presser, T. S., and H. M. Ohlendorf, 1987, Biogeochemical cycling of selenium in the San Joaquin Valley, *Environmental Management* **11:**805–821.

Price, J. S., and M. K. Woo, 1988, Origin of salt in coastal marshes of Hudson and James bays., *Can. J. Earth Sci.* **25:**145–147.

Pride, R. W., F. W. Meyer, and R. N. Cherry, 1966, Hydrology of Green Swamp area in central Florida, *Florida Div. Geology Rept. Inv.* 42. Talahassee, Fla., 137p.

Prince, H. H., and F. M. D'Itri, eds., 1985, *Coastal Wetlands,* Lewis Publishers, Inc., Chelsea, Mich., 286p.

Prouty, W. F., 1952, Carolina Bays and their origin, *Geol. Soc. Amer. Bull.* **63:**167–224.

Puriveth, P., 1980, Decomposition of emergent macrophytes in a Wisconsin marsh, *Hydrobiol.* **72:**231–242.

Putnam, J. A., G. M. Furnival, and J. S. McKnight, 1960, *Management and Inventory of Southern Hardwoods,* USDA Agricultural Handbook 181, Washington, D.C., 1O2p.

Puustjarvi, V., 1957, On the base status of peat soils, *Acta Agric. Scand.* **7:**190–223.

Rabinowitz, D., 1978, Dispersal properties of mangrove propagules, *Biotropica* **10:**47–57.

Radforth, N. W., 1962, Organic terrain and geomorphology, *Can. Geographer* **71:**8–11.

Radforth, N. W., and C. O. Brawner, eds., 1977, *Muskeg and the Northern Environment in Canada,* University of Toronto Press, Toronto, Canada, 399p.

Ragotzkie, R. A., L. R. Pomeroy, J. M. Teal, and D. C. Scott, ed., 1959, *Proceedings Salt Marsh Conference, Marine Institute, Univ. Georgia, Sapelo Island, Georgia,* Marine Institute, Univ. Georgia, Athens, Ga., 133p.

Rainey, G. B., 1979, Factors affecting nutrient chemistry distribution in Louisiana coastal marshes, Master's Thesis, Louisiana State University, Baton Rouge.

Ranwell, D. S., 1967, World resources of *Spartina townsendii* and economic use of *Spartina* marshland, *Coastal Zone ManagementJournal* **1:**65–74.

Ranwell, D. S., 1972, *Ecology of Salt Marshes and Sand Dunes,* Chapman and Hall, London, 258p.

Ranwell, D. S., E. C. F. Bird, J. C. E. Hubbard, and R. E. Stebbings, 1964, *Spartina* salt marshes in Southern England. V. Tidal submergence and chlorinity in Poole Harbour, *J. Ecology* **52:**627–641.

Raskin, I., and H. Kende, 1983, Regulation of growth in rice seedlings, *J. Plant Growth Regulation* **2:**193–203.

Raskin, I., and H. Kende, 1984, Regulation of growth in stem sections of deep-water rice, *Planta* **160:**66–72.

Reader, R. J., 1978, Primary production in Northern bog marshes, in *Freshwater Wetlands: Ecological Processes and Management Potential,* R. E. Good, D. F. Whigham, and R. L. Simpson, eds., Academic Press, New York, pp. 53–62.

Reader, R. J., and J. M. Stewart, 1971, Net primary productivity of the bog vegetation in southeastern Manitoba, *Can. J. Bot.* **49:**1471–1477.

Reader, R. J., and J. M. Stewart, 1972, The relationship between net primary production and accumulation for a peatland in southeastern Manitoba, *Ecology* **53:**1024–1037.

Reddy, K. R., and W. H. Patrick, Jr., 1984, Nitrogen transformations and loss in flooded soils and sediments, *CRC Crit. Rev. Environ. Control* **13:**273–309.

Reddy, K. R., T. C. Feijtel, and W. H. Patrick, Jr., 1986, Effect of Soil Redox Conditions on Microbial Oxidation of Organic Matter, in: *The Role of Organic Matter in Modern Agriculture,* Y. Chen and Y. Avnimelech, eds., Martinus Nijhoff Publishers, Dordrecht, pp. 117–156.

Reddy, K. R., and D. A. Graetz, 1988, Carbon and nitrogen dynamics in wetland soils, in: *The Ecology and Management of Wetlands,* vol. I, D. D. Hook et al., eds., Timber Press, Portland, Ore., pp. 307–318.

Redfield, A. C., 1965, Ontogeny of a salt marsh estuary, *Science* **147:**50–55.

Redfield, A. C., 1972, Development of a New England salt marsh, *Ecol. Monogr.* **42:**201–237.

Redman, F. H., and W. H. Patrick, Jr., 1965, *Effect of Submergence on Several Biological and Chemical Soil Properties,* Louisiana Agricultural Exp. Station, Bull. No. 592, Louisiana State University, Baton Rouge.

Redmond, A.M., 1981, Considerations for design of an artificial marsh for use in stormwater renovation, in *Wetlands Restoration and Creation,* R. H. Stovall, ed., Proceedings of the Eighth Annual Conference on Wetlands Restoration and Creation, Hillsborough Community College, Tampa, Fla., pp. 189–199.

Reed, S. C., ed., 1990, *Natural Systems for Wastewater Treatment,* Manual of Practice FD–16, Water Pollution Control Federation, Washington, D.C., 270p.

Reeder, B. C., and W. J. Mitsch, 1990, *What Is a Great Lakes Coastal Wetland Worth? A Bibliography,* Ohio Sea Grant College Program, Columbus, Ohio, Technical Series Report OHSU-TS-007, 18p.

Reilly, W. K., 1979, Can science help save interior wetlands? in *Wetland Functions and Values: The State of our Understanding,* P. E. Greeson, J. R. Clark, and J. E. Clark, eds., American Water Resources Assoc., Minneapolis, Minn., pp. 26–30.

Reimold, R. J., 1972, The movement of phosphorus through the marsh cord grass, *Spartina alterniflora* Loisel, *Limnol. Oceanogr.* **17:**606–611.

Reimold, R. J., 1974, Mathematical modeling—*Spartina*, in *Ecology of Halophytes,* R. J. Reimold and W. M. Queen, eds., Academic Press, New York, pp. 393–406.

Reimold, R. J., and F. C. Daiber, 1970, Dissolved phosphorus concentrations in a natural salt marsh of Delaware, *Hydrobiologia* **36:**361–371.

Reiners, W. A., 1972, Structure and energetics of three Minnesota forests, *Ecol. Monogr.* **42:**71–94.

Reppert, R. T., W. Sigleo, E. Stakhiv, L. Messman, and C. Meyers, 1979, *Wetland Values: Concepts and Methods for Wetlands Evaluation,* U.S. Army Corps of Engineers, Institute for Water Resources, Fort Belvoir, Va., IWR Res. Rep. 79-R-1, 109p.

Richardson, C. J., 1979, Primary productivity values in freshwater wetlands, in *Wetland Functions and Values: The State of Our Understanding,* P. E. Greeson, J. R. Clark, and J. E. Clark, eds., American Water Resources Assoc., Minneapolis, Minn., pp. 131–145.

Richardson, C. J., ed., 1981, *Pocosin Wetlands,* Hutchinson Ross Publishing Co., Stroudsburg, Pa., 364p.

Richardson, C. J.,1983, Pocosins: vanishing wastelands or valuable wetlands? *BioScience* **33:**626–633.

Richardson, C. J., 1985, Mechanisms controlling phosphorus retention capacity in freshwater wetlands, *Science* **228:**1424–1427.

Richardson, C. J., W. A. Wentz, J. P. M. Chamie, J. A. Kadlec, and D. L. Tilton, 1976, Plant growth, nutrient accumulation and decomposition in a central Michigan peatland used for effluent treatment, in *Freshwater Wetlands and Sewage Effluent Disposal,* D. L. Tilton, R. H. Kadlec, and C. J. Richardson, eds., University of Michigan, Ann Arbor, Mich., pp. 77–117.

Richardson, C. J., D. L. Tilton, J. A. Kadlec, J. P. M. Chamie, and W. A. Wentz, 1978, Nutrient dynamics of northern wetland ecosystems, in *Freshwater Wetlands— Ecological Processes and Management Potential,* R. E. Good, D. F. Whigham, and R. L. Simpson, eds., Academic Press, New York, pp. 217–241.

Richardson, C. J., R. Evans, and D. Carr, 1981, Pocosins: an ecosystem in transition, in *Pocosin Wetlands,* C. J. Richardson, ed., Hutchinson Ross Publishing Co., Stroudsburg, Pa., pp. 3–19.

Richardson, C. J., and D. S. Nichols, 1985, Ecological analysis of wastewater management criteria in wetland ecosystems, in *Ecological Considerations in Wetlands Treatment of Municipal Wastewaters,* P. J. Godfrey, E. R. Kaynor, S. Pelczarski and J. Benforado, eds., Van Nostrand Reinhold, New York, pp. 351–391.

Richardson, C. J., and P. E. Marshall, 1986, Processes controlling movement, storage, and export of phosphorus in a fen peatland, *Ecol. Monog.* **56:**279–302.

Richardson, J., P. A. Straub, K. C. Ewel, and H. T. Odum, 1983, Sulfate-enriched water effects on a floodplain forest in Florida, *Environ. Manage.* **7:**321–326.

Rigg, G. B., 1925, Some sphagnum bogs of the North Pacific coast of America, *Ecology* **6:**260–278.

Riley, J. P., and G. Skirrow, 1975, *Chemical Oceanography,* 2nd ed., vol. 2, Academic Press, New York, 647p.

Risotto, S., and R. E. Turner, 1985, Annual fluctuations in the abundance of the comercial fisheries of the Mississippi River and tributaries, *North Amer. J. Fish. Management* **4:**557–574.

Ritchie, S. A., 1990, A simulation model of water depth in mangrove basin forests, *J. American Mosquito Control Association* **6:**213–222.

Robb, D. M., 1989, *Diked and Undiked Freshwater Coastal Marshes of Western Lake Erie.* Master's thesis, The Ohio State University, Columbus, Ohio, 145p.

Robb, D. M., 1992, The role of wetland water quality standards in nonpoint source pollution control strategies, *Ecological Engineering* **1:**143–148.

Roberts, J. K. M., 1988, Cytoplasmic acidosis and flooding in crop plants, in *The Ecology and Management of Wetlands.* vol 1, *Ecology of Wetlands*, D. D. Hook, ed., Timber Press, Portland, Ore., pp. 392–397.

Robertson, P. A., 1987, The woody vegetation of Little Black Slough: an undisturbed upland-swamp forest in southern Illinois, in *Proc. of the Central Hardwood Forest Conference VI, Knoxville,* Tenn., R. L. Hays, F. W. Woods, and H. DeSelm, eds., Univ. Tennessee, Department of Forestry, Wildlife and Fisheries, Dept of Botany; USDA Forest Service, Southern Forest Experiment Station., pp. 353–367.

Rochefort, L., D. H. Vitt, and S. E. Bayley, 1990, Growth, production, and decomposition dynamics of *Sphagnum* under natural and experimentally acidified conditions, *Ecology* **71:**1986–2000.

Roe, H. B., and Q. C. Ayres, 1954, *Engineering for Agricultural Drainage,* McGraw-Hill, New York, 501p.

Roesser, J. C., 1988, The Blue River Reclamation Project, in *Proceedings of the Conference Restoration, Creation and Management of Wetland and Riparian Ecosystems in the American West,* K. M. Mutz, D. J. Cooper, M. L. Scott and L. K. Miller, eds., PIC Technologies, Denver, Col., pp. 94–101.

Romanov, V. V., 1968, *Hydrophysics of Bogs,* trans. from Russian by N. Kaner, ed. by Prof. Heimann, Israel Program for Scientific Translation, Jerusalem. Available from Clearinghouse for Federal Scientific and Technical Information, Springfield, Virginia, 299p.

Rood, S. B., and J. M. Mahoney, 1990, Collapse of riparian poplar forests downstream from dams in western prairies: probable causes and prospects for mitigation, *Environ. Manage.* **14:**451–464.

Rosswall, T., et al., 1975, Structure and function of tundra ecosystems, *Ecological Bulletin (Sweden)* **20:**265–294.

Roulet, N. T., R. Ash, and T. R. Moore, 1992, Low boreal wetlands as a source of atmospheric methane, *J. Geophysical Research* **97:**3739–3749.

Rowe, J. S., 1972, *Forest Regions of Canada*, Canadian Forest Service, Publ. No. 1300, 172p.

Ruebsamen, R. N., 1972, *Some Ecological Aspects of the Fish Fauna of a Louisiana Intertidal Pond System,* Master's thesis, Louisiana State University, Baton Rouge, 80p.

Runyon, L. C., 1993, *The Lucas Court Case and Land-Use Planning*, National Conference of State Legislators, Denver, Colorado, Supplement to State Legislatures vol. 1, No. 10 (March).

Russell, H. S., 1976, *A Long, Deep Furrow. Three Centuries of Farming in New England.* University Press of New England, Hanover, N.H., 671p.

Ruttner, F., 1963, *Fundamentals of Limnology,* 3rd ed., University of Toronto Press, Canada, 295p.

Rycroft, D. W., D. J. A. Williams, and H. A. E Ingram, 1975, The transmission of water through peat. I. Review, *J. Ecol.* **63:**535–556.

Rykiel, E. J., Jr., 1977, *Toward Simulation and Systems Analysis of Nutrient Cycling in the Okefenokee Swamp, Georgia,* Technical Completion Report, USDI/OWRT Project No. A-06GA, University of Georgia, Athens, Ga., 139p.

Rykiel, E. J., Jr., 1984, General hydrology and mineral budgets for Okefenokee Swamp: ecological significance, in *The Okefenokee Swamp: Its Natural HIstory, Geology, and Geochemistry,* A. D. Cohen, D. J. Casagrande, M. J. Andrejko, and G. R. Best, eds., Wetland Surveys, Los Alamos, N. Mex., pp. 212–228.

Rymal, D. E., and G. W. Folkerts, 1982, Insects associated with pitcher plants (*Sarracenia*, Sarraceniaceae), and their relationship to pitcher plant conservation: a review, *J. Ala. Acad. Sci.* **53:**131–151.

Ryther, J. H., R. A. Debusk, M. D. Hanisak, and L. D. Williams, 1979, Freshwater macrophytes for energy and waste water treatment, in *Wetland Functions and Values: The State of Our Understanding,* P. E. Greeson, J. R. Clark, and J. E. Clark, eds., American Water Resources Association, Minneapolis, Minn., pp. 652–660.

Sass, R. L., F. M. Fisher, and P. A. Harcombe, 1990, Methane production and emission in a Texas rice field, *Global Biogeochemical Cycles* **4:**47–68.

Sasser, C. E., 1977, *Distribution of Vegetation in Louisiana Coastal Marshes as Response to Tidal Flooding,* Master's thesis, Louisiana State University, Baton Rouge, 40p.

Sasser, C. E., G. W. Peterson, D. A. Fuller, R. K. Abernathy, and J. G. Gosselink, 1982, *Environmental Monitoring Program, Louisiana Offshore Oil Port Pipeline,* 1981 Annual Report, Coastal Ecology Laboratory, Center for Wetland Resources, Louisiana State University, Baton Rouge, 299p.

Sasser, C. E., and J. G. Gosselink, 1984, Vegetation and primary production in a floating freshwater marsh in Louisiana, *Aquatic Bot.* **20:**245–255.

Sasser, C. E., M. D. Dozler, J. G. Gosselink, and J. M. Hill, 1986, Spatial and Temporal Changes in Louisiana's Baratarin Basin Marshes, 1945-1980, *Environmental Management* **10:**671–680.

Sasser, C. E., J. G. Gosselink, and G. P. Shaffer, 1991, Distribution of nitrogen and phosphorus in a Louisiana freshwater floating marsh, *Aquatic Bot.* **41:**317–331.

Sather, J. H., ed., 1992, *Intensive Studies of Wetland Functions: 1990–91 Research Summary of the Des Plaines River Wetland Demonstration Project*, Project Technical Paper No. 2, Wetlands Research Inc., Chicago, Il., 11p.

Sather, J. H., and R. D. Smith, 1984, *An Overview of Major Wetland Functions and Values,* Western Energy and Land Use Team, U.S. Fish and Wildlife Service, FWS/ 0BS-84/18, Washington, D.C., 68p.

Savage, H., 1956, *River of the Carolinas: The Santee*, Rinehart, New York, N.Y., 435p.

Savage, H., 1983, *The Mysterious Carolina Bays,* Univ. South Carolina Press, Columbia, S.C., 121p.

Schaeffer-Novelli, Y., G. Cintrón-Molero, R. R. Adaime, and T. M. de Camargo, 1990, Variability of mangrove ecosystems along the Brazilian coast, *Estuaries* **13:**204–218.

Schamberger, M. L., C. Short, and A. Farmer, 1979, Evaluation wetlands as a wildlife habitat, in *Wetland Functions and Values: The State of Our Understanding,* P. E. Greeson, J. R. Clark, and J. E. Clark, eds., American Water Resources Assoc., Minneapolis, Minn., pp. 74–83.

Schat, H., 1984, A comparative ecophysiological study on the effects of waterlogging and submergence on dune slack plants: growth, survival and mineral nutrition in sand culture experiments, *Oecologia (Berl.)* **62:**279–86.

Scheffe, R. D., 1978, *Estimation and Prediction of Summer Evapotranspiration from a Northern Wetland,* Master's thesis, University of Michigan, Ann Arbor, 69p.

Schlesinger, W. H., 1978, Community structure, dynamics, and nutrient ecology in the Okefenokee Cypress Swamp-Forest, *Ecol. Monogr.* **48:**43–65.

Schlesinger, W. H., and B. F. Chabot, 1977, The use of water and minerals by evergreen and deciduous shrubs in Okefenokee Swamp, *Bot. Gazette* **138:**490–497.

Schlosser, I. J., and J. R. Karr, 1981a, Water quality in agricultural watersheds: impact of riparian vegetation during base flow, in *Water Resour. Bull.* **17:**233–240.

Schlosser, I. J., and J. R. Karr, 1981b, Riparian vegetation and channel morphology impact on spatial patterns of water quality in agricultural watersheds, *Environ. Manage.* 5:233–243.

Schneider, R. L., and R. R. Sharitz, 1986, Seed bank dynamics in a southeastern riverine swamp, *Am. J. Bot.* **73:**1022–1030.

Schneider, R. L. and R. R. Sharitz, 1988, Hydrochory and regeneration in a bald cypress-water tupelo swamp forest, *Ecology* **69:**1055–1063.

Scholander, P. F. 1968, How mangroves desalinate seawater, *Physiologica Plantarum* **21:**251–261.

Scholander, P. F., L. van Dam, and S. I. Scholander, 1955, Gas exchange in the roots of mangroves, in *Am. J. Bot.* **42:**92–98.

Scholander, P. F., H. T. Hammel, E. D. Bradstreet, and E. A. Hemmingsen, 1965, Sap pressure in vascular plants, in *Science* **148:**339–346.

Scholander, P. F., E. D. Bradstreet, H. T. Hammel, and E. A. Hemmingsen, 1966, Sap concentrations in naiophytes and some other plants, in *Plant Physiol.* **41:**529–532.

Schubel, J. R., and D. W. Pritchard, 1990, Great Lakes estuaries—phooey, *Estuaries* **13:**508–509.

Schumacher, T. E., and A. J. M. Smucker, 1985, Carbon transport and root respiration of split root systems of *Phaseolus vulgaris* subjected to short term localized anoxia, in *Pl. Physiol.* **78:**359–364.

Schumm, S. A., and R. W. Lichty, 1963, *Channel Widening and Floodplain Construction Along Cimarron River in Southwestern Kansas,* U.S. Geological Survey, Washington, D.C., Professional Paper 352-D.

Schutz, H., A. Holzapfel–Pschorn, R. Conrad, H. Rennenberg, and W. Seiler, 1989, A three year continuous record on the influence of daytime, season and fertilizer treatment on methane emission rates from an Italian rice paddy field, *J. Geophysical Res.* **94:**16405–16416.

Schwintzer, C. R., 1983, Nonsymbiotic and symbiotic nitrogen fixation in a weakly minerotrophic peatland, *Am. J. Bot.* **70:**1071–1078.

Scodari, P. F., 1990, *Wetlands Protection: The Role of Economics,* Environmental Law Institute, Washington, D.C., 89p.

Sedell, J. R., P. A. Bisson, and J. A. June, 1980, Ecology and habitat requirements of fish populations in South Fork Hoh River, Olympic National Park, in *Proc. 2d Conf. Scientific Res. in National Parks,* NPS/ST-80/02/7 **7:**47–63.

Sedell, J. R., and K. J. Luchessa, 1982, Using the historical record as an aid to salmonid habitat enhancement, in *Proceedings of a Symposium on Acquisition and Utilization of Aquatic Habitat Inventory Information,* N. B. Armantrout, ed., Western Divsion, American Fisheries Soc., Portland, Oreg., pp. 210–223.

Sedell, J. R., F. H. Everest, and F. J. Swanson, 1982, Fish habitat and streamside management: past and present, in *Proceedings of the Society of American Foresters Annual Meeting,* 1981, Bethesda, Md., *Soc. Amer. Foresters* 244–255.

Sedell, J. R., and J. L. Froggatt, 1984, Importance of streamside forests to large rivers: the isolation of the Willamette River, Oregon, USA, from its floodplain by snagging and streamside forest removal, in *Verhandlungen, Internationale Vereinigung fur Theoretische und Augewandte Limnologie* **22:**1828–1834.

Sedell, J. R., and W. S. Duval, 1985, *Influence of Forest and Rangeland Managment on Anadromous Fish Habitat in Western North America: 5. Water Transportation and Storage of Logs*, Pacific Northwest Forest and Range Experiment Station, U.S. Dept. Agriculture, Portland, Oreg., General Technical Report PNW-186.

Seiler, W. A., A. Holzapfel–Pschorn, R. Conrad, and D. Scharffe, 1984, Methane emission from rice paddies, *J. Atmospheric Chemistry* **1**:214–268.

Sell, M. G., 1977, *Modeling the Response of Mangrove Ecosystems to Herbicide Spraying, Hurricanes, Nutrient Enrichment and Economic Development,* Ph.D. dissertation, University of Florida, 389p.

Sellers, R. A., 1973, Mallard releases in understocked prairie pothole habitat, *J. Wildl. Manage.* **37:**10–32.

Seneca, E. D., 1980, Techniques for creating salt marshes along the east coast, in *Rehabilitation and Creation of Selected Coastal Habitats,* J. C. Lewis and E. W. Bunce, eds., U.S. Fish and Wildlife Service, Biol. Service Program, FWS/OBS-80/ 27, Washington, D.C., pp. 1–5.

Shabman, L. A., and S. S. Batie, 1988, *Socioeconomic Values of Wetlands: Literature Review,* 1970–1985, Army Engineer Waterways Experiment Station, Vicksburg, Miss., Technical Report Y-88.

Shabman, L., 1986, The contribution of economics to wetlands valuation and management, in *Proceedings of the National Wetlands Assessment Symposium.*, J. A. Kusler and P. Reixinger, eds., Association of State Wetland Managers Tech. Report, pp. 9–13.

Shaffer, G. W., C. E. Sasser, J. G. Gosselink, and M. Rejmanek, 1992, Vegetation dynamics in the emergent Atchafalaya delta, Louisiana, USA, *J. Ecology* **80:**in press.

Sharitz, R. R., and J. W. Gibbons, eds., 1982, *The Ecology of Southeastern Shrub Bogs (Pocosins) and Carolina Bays: A Community Profile*, U.S. Fish Wildl. Serv., Div. of Biol. Services, Washington, D.C., FWS/OBS–82/04.

Sharitz, R. R., and J. W. Gibbons, 1989, *Freshwater Wetlands and Wildlife*, U.S. Dept. of Energy, NTIS, Springfield, Virginia, 1265p.

Sharitz, R. R., and W. J. Mitsch, 1993, Southern floodplain forests, in *Biodiversity of the Southeastern United States: Lowland Terrestrial Communities*, W. H. Martin, S. G. Boyce, and A. C. E. Echternacht, eds., Wiley, New York, pp. 311–372.

Shaver, G. R., and J. M. Melillo, 1984, Nutrient budgets of marsh plants: efficiency concepts and relation to availability, *Ecology* **65:**1491–1510.

Shaw, S. P., and C. G. Fredine, 1956, *Wetlands of the United States, Their Extent, and Their Value for Waterfowl and Other Wildlife,* U.S. Department of Interior, Fish and Wildlife Service, Circular 39, Washington, D.C., 67p.

Shew, D. M., R. A. Linthurst, and E. D. Seneca, 1981, Comparison of production computation methods in a southeastern North Carolina *Spartina alterniflora* salt marsh. *Estuaries* **4**:97–109.

Shisler, J. K., 1990, Creation and restoration of coastal wetlands of the northeastern United States, in *Wetland Creation and Restoration,* Kusler, J. A., and M. E. Kentula, eds., Island Press, Washington, D.C., pp. 143–170.

Shjeflo, J. B., 1968, Evapotranspiration and the water budget of prairie potholes in North Dakota, *U.S. Geological Survey Prof. Paper. 585-B,* 49p.

Siegel, D. I., 1983, Ground water and evolution of patterned mires, glacial Lake Agissiz peatlands, northern Minnesota, *J. Ecol.* **71:**913–921.

Siegel, D. I., 1988a, A review of the recharge-discharge function of wetlands, in *The Ecology and Management of Wetlands,* D.D. Hook et al., eds., vol. 1: *Ecology of Wetlands,* Timber Press, Portland, Oregon, pp. 59–67.

Siegel, D. I., 1988b, Evaluating cumulative effects of disturbance on the hydrologic function of bogs, fens, and mires, *Environmental Management* **12:**621–626.

Siegley, C. E., R. E. J. Boerner, and J. M. Reutter, 1988, Role of the seed bank in the development of vegetation on a freshwater marsh created from dredge spoil, *J. Great Lakes Res.* **14:**267–276.

Sikora, J. P., W. B. Sikora, C. W. Erkenbrecher, and B. C. Coull, 1977, Significance of ATP, carbon and caloric content of meiobenthic nematodes in partitioning benthic biomass, *Mar. Biol.* **44:**7–14.

Sikora, L. J., and D. R. Keeney, 1983, Further aspects of soil chemistry under anaerobic conditions, in *Ecosystems of the World: Mires, Swamp, Bog, Fen and Moor*, A. J. P. Gore, ed., Elsevier Scientific Publ. Co., Amsterdam, pp. 247–256.

Sikora, W. B., 1977, *The Ecology of* Palaemonetes pugio *in a Southeastern Salt Marsh Ecosystem with Particular Emphasis on Production and Trophic Relationship,* Ph.D. dissertation, University of South Carolina, 122p.

Silvola, J., and I. Hanski, 1979, Carbon accumulation in a raised bog, *Oecologia (Berlin)* **37:**285–295.

Simpson, R. L., D. F. Whigham, and R. Walker, 1978, Seasonal patterns of nutrient movement in a freshwater tidal marsh, in *Freshwater Wetlands: Ecological Processes and Management Potential,* R. E. Good, D. F. Whigham, and R. L. Simpson, eds., Academic Press, New York, pp. 243–257.

Simpson, R. L., R. E. Good, R. Walker, and B. R. Frasco, 1981, *Dynamics of Nitrogen, Phosphorus and Heavy Metals in Delaware River Freshwater Tidal Wetland,* Final Tech. Completion Report, Corvallis Environmental Research Laboratory, Environmental Protection Agency, Corvallis, Oregon, 192p.

Simpson, R. L., R. E. Good, M. A. Leck, and D. F. Whigham, 1983, The ecology of freshwater tidal wetlands, *BioScience* **33:**255–259.

Singer, P. C., and W. Stumm, 1970, Acidic mine drainage: the rate-determining step, *Science* **167:**1121–1123.

Sinicrope, T. L., P. G. Hine, R. S. Warren, and W. A. Niering, 1990, Restoring of an impounded salt marsh in New England, *Estuaries* **13:**25–30.

Sjors, H., 1961, Bogs and fens on Attawapiskat River, Northern Ontario, *Nat. Mus. Can. Bull.* 186, 133p.

Sklar, F. H., and W. H. Conner, 1979, Effects of altered hydrology on primary production and aquatic animal populations in a Louisiana swamp forest, in *Proceedings of the 3rd Coastal Marsh and Estuary Management Symposium,* J. W. Day, Jr., D. D. Culley, Jr., R. E. Turner, and A. T. Humphrey, Jr., eds., Louisiana State University, Division of Continuing Education, Baton Rouge, pp. 101–208.

Sklar, F. H., R. Costanza, and J. W. Day, Jr., 1985, Dynamic spatial simulation modelling of coastal wetland habitat succession, *Ecol. Mod.* **29:**261–281.

Sloey, W. E., F. L. Spangler, and C. W. Fetter, Jr., 1978, Management of freshwater wetlands for nutrient assimilation, in *Freshwater Wetlands: Ecological Processes and Management Potential,* R. E. Good, D. F. Whigham, and R. L. Simpson, eds., Academic Press, New York, pp. 321–340.

Small, E., 1972, Ecological significance of four critical elements in plants of raised *Sphagnum* peat bogs, *Ecology* **53:**498–503.

Smalley, A. E., 1960, Energy flow of a salt marsh grasshopper population, *Ecology* **41:**672–677.

Smardon, R. C., 1979, Visual-cultural values of wetlands, in *Wetland Functions and Values: The State of Our Understanding,* P. E. Greeson, J. R. Clark, and J. E. Clark, eds., American Water Resources Assoc., Minneapolis, Minn., pp. 535–544.

Smart, R. M., and J. W. Barko, 1980, Nitrogen nutrition and salinity tolerance of *Distichlis spicata* and *Spartina alterniflora, Ecology* **61:**630–638.

Smith, A. M., and T. ap Rees, 1979, Pathways of carbohydrate fermentation in the roots of marsh plants, *Planta* **146:**327–334.

Smith, C. J., R. D. Delaune, and W. H. Patrick, Jr., 1982, Carbon and nitrogen cycling in a *Spartina alterniflora* salt marsh, in *The Cycling of Carbon, Nitrogen, Sulfur and Phosphorus in Terrestrial and Aquatic Ecosystems,* J. R. Freney and I. E. Galvally, eds., Springer-Verlag, New York, pp. 97–104.

Smith, C. S., M. S. Adams, and T. D. Gustafson, 1988, The importance of belowground mineral element stores in cattails (*Typha latifolia* L.), *Aquat. Bot.* **30:**343–352.

Smith, H. K., 1980, Coastal habitat development in the dredged material research program, in *Rehabilitation and Creation of Selected Coastal Habitats,* J. C. Lewis and E. W. Bunce, eds., U.S. Fish and Wildlife Service, Biol. Service Program, FWS/OBS-80/27, pp. 117–125.

Smith, L. M., and J. A. Kadlec, 1985, Predictions of vegetation change following fire in a Great Salt Lake marsh, *Aquatic Botany* **21:**43–51.

Smith, R. A., R. B. Alexander, and M. G. Wolman, 1987, Water-quality trends in the nation's rivers, *Science* **235:**1607–1615.

Smith, R. C., 1975, Hydrogeology of the experimental cypress swamps, in *Cypress Wetlands for Water Management, Recycling and Conservation,* H. T. Odum and K. C. Ewel, eds., Second Annual Report to NSF and Rockefeller Foundation, Center for Wetlands, University of Florida, Gainesville, pp. 114–138.

Smith, R. L., 1980, *Ecology and Field Biology,* 3rd ed., Harper and Row, New York, 835p.

Smith, R. L., and M. J. Klug, 1981, Electron donors utilized by sulfate-reducing bacteria in eutrophic lake sediments, *Appl. Environ. Microbiol.* **42:**116–121.

Smith, T. J. I., and W. E. Odum, 1981, The effects of grazing by snow geese on coastal salt marshes, *Ecology* **62:**98–106.

Smits, A. J. M., R. M. J. C. Kleukers, C. J. Kok, and A. G. van der Velde, 1990a, Alcohol dehydrogenase isozymes in the roots of some nymphaeid and isoetid macrophytes. Adaptations to hypoxic sediment conditions? *Aquatic Botany* **38:**19–27.

Smits, A. J. M., P. Laan, R. H. Thier, and A. G. van der Velde, 1990b, Root aerenchyma, oxygen leakage patterns and alcoholic fermentation ability of the roots of some nymphaeid and isoetid macrophytes in relation to the sediment type of their habitat, *Aquatic Bot.* **38:**3–17.

Smock, L. A., and K. L. Harlowe, 1983, Utilization and processing of freshwater wetland macrophytes by the detritivore *Asellus forbesi*, *Ecology* **64:**1556–1565.

Snedaker, S. C., 1989, Overview of ecology of mangroves and information needs for Florida Bay, *Bulletin of Marine Science* **44:**341–347.

Solem, T., 1986, Age, origin and development of blanket mires in soer-Troendelag, central Norway, *Boreas* **15:**101–115.

Soper, E. K., 1919, The peat deposits of Minnesota, *Minn. Geol. Surv. Bull.* **16:**1–261.

Spangler, F. L., C. W. Fetter, Jr., and W. E. Sloey, 1977, Phosphorus accumulation- discharge cycles in marshes, *Water Resour. Bull.* **13:**1191–1201.

Sprunt, A., Jr., 1961, Emerald kingdom, *Audubon Mag.,* January–February 1961, pp. 25–40.

Stanek, W., and I. A. Worley, 1983, A terminology of virgin peat and peatlands, in *Proc. of the International Symposium on Peat Utilization,* Bemidji, Minn., C. H. Fuchsman and S. A. Spigarelli, eds., pp. 75–102.

Stavins, R., 1987, *Conversion of Forested Wetlands to Agricultural Uses: Executive Summary*, Environmental Defense Fund, New York, 72p.

Stearns, F., 1978, Management potential: summary and recommendations, in *Freshwater Wetlands: Ecological Processes and Management Potential,* R. E. Good, D. F. Whigham, and R. L. Simposon, eds., Academic Press, New York, pp. 357–363.

Steever, E. Z., R. S. Warren, and W. A. Niering, 1976, Tidal energy subsidy and standing crop production of *Spartina alterniflora, Estuarine Coast. Mar Sci.* **4:**473–478.

Steiner, G. R., J. T. Watson, D. Hammer, and D. F, Harker, Jr., 1987, Municipal wastewater treatment with artificial wetlands—a TVA/Kentucky demonstration, in *Aquatic Plants for Wastewater Treatment and Resource Recovery,* K. R. Reddy and W. H. Smith, eds., Magnolia Pub., Inc., Orlando, Florida, p. 923.

Steiner, G. R., and R. J. Freeman, Jr., 1989, Configuration and substrate design considerations for constructed wetlands for wastewater treatment, in *Constructed Wetlands for Wastewater Treatment,* D. A. Hammer, ed., Lewis Publishers, Inc., Chelsea, Mich., pp. 363–378.

Stelzer, R., and A. Lauchli, 1978, Salt and flooding tolerance of *Puccinellia peisonis,* III. Distribution and localization of ions in the plant, *Z. Pflanzenphysiol* **88:**437–448.

Stephenson, T. D., 1990, Fish reproductive utilization of coastal marshes of Lake Ontario near Toronto, ***J. Great Lakes Res.*** **16:**71–81.

Steudler, P. A., and B. J. Peterson, 1984, Contribution of gaseous sulfur from salt marshes to the global sulphur cycle, *Nature* **311:**455–457.

Stevenson, F. W., 1986, *Cycles of Soil*, Wiley, New York.

Stevenson, J. S., D. R. Heinle, D. A. Flemer, R. J. Small, R. A. Rowland, and J. F. Ustach, 1977, Nutrient exchanges between brackish water marshes and the estuary, in *Estuarine Processes.* vol. II, M. Wiley, ed., Academic Press, New York, pp. 219–240.

Steward, K. K., 1990, Aquatic weed problems and management in the eastern United States, in *Aquatic Weeds: The Ecology and Management of Nuisance Aquatic Vegetation,* A. H. Pieterse, and K. J. Murphy, eds., Oxford University Press, New York, pp. 391–405.

Stewart, G. R., and M. Popp, 1987, The ecophysiology of mangroves, in *Plant Life in Aquatic and Amphibious Habitats,* R. M. M. Crawford, ed., Spec Publ. British Ecological Society, vol. 5, pp. 333–345.

Stewart, R. E., 1962, *Waterfowl Populations in the Upper Chesapeake Region,* U.S. Fish and Wildlife Service Spec. Sci. Rep. Wildl. Research Pub. 65, 208p.

Stewart, R. E., and C. S. Robbins, 1958, Birds of Maryland and the District of Columbia, *U.S. Fish and Wildlife Service North America Fauna Service Research Pub.* 62, 401p.

Stewart, R. E., and H. A. Kantrud, 1971, Classification of natural ponds and lakes in the glaciated prairie region, *U.S. Fish and Wildlife Service Research Pub.* 92, 57p.

Stewart, R. E., and H. A. Kantrud, 1972, Vegetation of the prairie potholes, North Dakota, in relation to quality of water and other environmental factors, *U.S. Geological Survey Prof.* Paper 85-D, 36p.

Stewart, R. E., and H. A. Kantrud, 1973, Ecological distribution of breeding water-fowl populations in North Dakota, *J. Wildl. Manage.* **37:**39–50.

Stockwell, S. S., and M. L. Hunter, 1985, *Distribution and abundance of birds, amphibians and reptiles, and small mammals in peatlands of central Maine.*, Maine Department of Inland Fisheries and Wildlife, Augusta, 89p.

Stout, J. P., 1978, *An Analysis of Annual Growth and Productivity of* Juncus roemerianus *Scheele and* Spartina alterniflora *Loisel in Coastal Alabama,* Ph.D. thesis, Univ. of Alabama, Tuscaloosa.

Stout, J. P., 1984, The Ecology of Irregularly Flooded Salt Marshes of the Northeastern Gulf of Mexico: A Community Profile, *U.S. Fish Wildl. Serv. Biol. Rep.,* 85(7.1), Washington, D.C., 98p.

Stromberg, J. S., and D. C. Patten, 1990, Riparian vegetaton instream flow requirements: A case study from a diverted stream in the eastern Sierra Nevada, California, USA, *Environ. Manage.* **14:**185–194.

Stroud, L. M., 1976, *Net Primary Production of Belowground Material and Carbohydrate Patterns in Two Height Forms of* Spartina alterniflora *in Two North Carolina Marshes,* Ph.D. dissertation, North Carolina State University, Raleigh.

Stuckey, R. L., 1980, Distributional history of *Lythrum Salicaria* (purple loosestrife) in North America, *Bartonia* **47:**3–20.

Stumm. W., and J. J. Morgan, 1970, *Aquatic Chemistry: An Introduction Emphasizing Chemical Equilibria in Natural Waters,* Wiley, New York, 583p.

Sullivan, M. J., 1978, Diatom community structure taxonomic and statistical analyses of a Mississippi salt marsh, *J. Phycol.* **14:**468–475.

Sullivan, M. J., and F. C. Daiber, 1974, Response in production of cordgrass *Spartina alterniflora* to inorganic nitrogen and phosphorus fertilizer, *Chesapeake Sci.* **15:**121–123.

Suzuki, T., W. G. A. Nissanka, and Y. Kurihara, 1989, Amplification of total dry matter, nitrogen and phosphorus removal from stands of *Phragmites australis* by harvesting and reharvesting regenerated shoots, in *Constructed Wetlands for Wastewater Treatment,* D. A. Hammer, ed., Lewis Publishers, Inc., Chelsea, Mich., pp. 530–535.

Svensson, B. H., and T. Rosswall, 1980, Energy flow through the subarctic mire at Stordalen, *Ecol. Bull. (Stockholm)* **30:**283–301.

Swanson, G. A., and H. F. Duebbert, 1989, Wetland habitats of waterfowl in the prairie pothole region, in *Northern Prairie Wetlands,* A. G. van der Valk, ed., Iowa State Univ. Press, Ames, Iowa, pp. 228–267.

Swarzenski, C., E. M. Swenson, C. E. Sasser, and J. G. Gosselink, 1991, Marsh mat flotation in the Louisiana Delta Plain, *J. Ecology* **79:**999–1011.

Szaro, R. C., 1989, Riparian forest and scrubland community types of Arizona and New Mexico, *Desert Plants* **9:**69–139.

Sze, N. D., 1977, Anthropogenic CO_2 emissions: implications for the atmospheric CO_2-OH-CH_4 cycle, *Science* **195:**673–675.

Tarnocai, C., 1979, Canadian wetland registry, in *Proceedings of a Workshop on Canadian Wetlands Environment,* C. D. A. Rubec and F. C. Pollett, eds., Canada Land Directorate, Ecological Land Classification Series, No. 12., pp. 9–38.

Taylor, J. R., 1985, *Community Structure and Primary Productivity of Forested Wetlands in Western Kentucky*, Ph.D. Dissertation, University of Louisville, Louisville, Kentucky, 139p.

Taylor, J. R., M. A. Cardamone, and W. J. Mitsch, 1990, Bottomland hardwood forests: their functions and values, in *Ecological Processes and Cumulative Impacts: Illustrated by Bottomland Hardwood Wetland Ecosystems*, J. G. Gosselink, L. C. Lee, and T. A. Muir, eds., Lewis Publishers, Chelsea, Mich., pp. 13–86.

Tchobanoglous, G., and F. L. Burton, 1991, *Wastewater Engineering: Treatment, Disposal, and Reuse,* 3d ed., McGraw-Hill, New York, 1334p.

Teal, J. M., 1958, Distribution of fiddler crabs in Georgia salt marshes, *Ecology* **39:**18–19.

Teal, J. M., 1962, Energy flow in the salt marsh ecosystem of Georgia, *Ecology* **43:**614–624.

Teal, J. M., 1986, *The Ecology of Regularly Flooded Salt Marshes of New England: A Community Profile*, U.S. Fish Wildl. Serv., Washington, D.C., Biol. Rep. 85(7.4), 61p.

Teal, J. M., and J. W. Kanwisher, 1966, Gas transport in the marsh grass *Spartina alterniflora, J. Exp. Botany* **17:**355–361.

Teal, J. M., and M. Teal, 1969, *Life and Death of the Salt Marsh*, Little, Brown, Boston, 278p.

Teal, J. M., I. Valiela, and D. Berla, 1979, Nitrogen fixation by rhizosphere and free-living bacteria in salt marsh sediments, *Limnol. Oceanogr.* **24:**126–132.

Teas, H. J., and R. J. McEwan, 1982, An epidemic dieback gall disease of *Rhizophora* mangroves in Gambia, West Africa, *Plant Disease* **66:**522–523.

Terasmae, J., 1977, Postglacial history of Canadian muskeg, in *Muskeg and the Northern Environment in Canada*, N. W. Radforth and C. O. Brawner, eds., University of Toronto Press, Toronto, Canada, pp. 9–30.

Tharp, M. L., 1978, *Modeling Major Perturbations on a Forest Ecosystem,* M.S. thesis, U. Tennessee, Knoxville.

Thayer, G. W., ed., 1992, *Restoring the Nation's Marine Environment*, Maryland Sea Grant College, College Park, Maryland, 716p.

Thibodeau, F. R., and N. H. Nickerson, 1986, Differential oxidation of mangrove substrate by *Avicennia germinas* and *Rhizophora mangle*, *American Journal of Botany* **73:**512–516.

Thomas, B., 1976, *The Swamp,* Norton, New York, 223p.

Thompson, E., 1983, Origin of Surface Patterns in a Subarctic Peatland in *Proc. International Symposium on Peat Utilization,* Bemidji, Minn., C. H. Fuchsman and S. A. Spigarelli, ed.

Thorp, J. H., E. M. McEwan, M. F. Flynn, and F. R. Hauer, 1985, Invertebrate colonization of submerged wood in a cypress-tupelo swamp and blackwater stream, *American Midland Naturalist* **113:**56–68.

Tilton, D. L., 1977, Seasonal growth and foliar nutrients of *Larix laricina* in three wetland ecosystems, *Can. J. Bot.* **55:**1291–1298.

Tilton, D. L., and R. H. Kadlec, 1979, The utilization of a freshwater wetland for nutrient removal from secondarily treated wastewater effluent, *J. Environ. Qual.* **8:**328–334.

Tiner, R. W., 1984, *Wetlands of the United States: Current Status and Recent Trends,* National Wetlands Inventory, Fish and Wildlife Service, U.S. Department of Interior, Washington, D.C., 58p.

Tiner, R. W., and B. O. Wilen, 1983, *The U.S. Fish and Wildlife Services National Wetlands Inventory Project,* unpublished report, U.S. Fish and Wildlife Service, Washington, D.C., 19p.

Titus, J. G., et al. 1991, Greenhouse effect and sea level rise: the cost of holding back the sea, *Coastal Management* **19:**171–204.

Todd. D. K.,1964, Groundwater, in *Handbook of Applied Hydrology,* V. T. Chow, ed., McGraw-Hill, New York, pp. 13-1–13-55.

Tomlinson, P. B., 1986, *The Botany of Mangroves*, Cambridge University Press, London, 413p.

Tomljanovich, D. A., and O. Perez, 1989, Constructing the wastewater treatment wetland—some factors to consider, in *Constructed Wetlands for Wastewater Treatment,* D. A. Hammer, ed., Lewis Publishers, Inc., Chelsea, Mich., pp. 399–404.

Tourbier, J. T. and R. Westmacott, 1989, Looking good: the use of natural methods to control urban runoff, *Urban Land* (April 1989):32–35.

Train, E., and F. P Day. Jr., 1982, Population age structure of tree species in four communities in the Great Dismal Swamp, *Castanea* **47:**1–16.

Transeau, E. N., 1903, On the geographic distribution and ecological relations of the bog plant societies of northern North America, *Bot. Gaz.* **36:**401-420.

Tupacz, E. G., and F. P. Day, 1990, Decomposition of roots in a seasonally flooded swamp ecosystem, *Aquatic Botany* **37:**199–214.

Turner, M. G., 1989, Landscape ecology: the effect of pattern of process, *Ann. Rev. Ecol. Syst.* **20:**171–197.

Turner, R. E., 1976, Geographic variations in salt marsh macrophyte production: a review, *Contr. Mar. Sci.* **20:**47–68.

Turner, R. E., 1977, Intertidal vegetatian and commercial yields of penaeid shrimp, *Amer. Fish. Soc. Trans.* **106:**411–416.

Turner, R. E., 1978, Community plankton respiration in a salt marsh estuary and the importance of macrophytic leachates, *Limnol. Oceanogr.* **23:**442–451.

Turner, R. E., 1982, Protein yields from wetlands, in *Wetlands: Ecology and Management*, B. Gopal, ed., National Institute of Ecology and International Scientific Publications, Jaipur, India, pp. 405–415.

Turner, R. E., 1988a, Secondary production in riparian wetlands, *Trans. 53rd North American Wildl. and Nat. Res. Conf.*, pp. 491–501.

Turner, R. E., 1988b, Fish and fisheries of inland wetlands, *Water Qual. Bull.* **13:**7–9, 13.

Turner, R. E., S. W. Forsythe, and N. J. Craig, 1981, Bottomland hardwood forest land resources of the southeastern United States, in *Wetlands of Bottomland Hardwood forests,* J. R. Clark and J. Benforado, eds., Elsevier, Amsterdam, pp. 13–28.

Twilley, R. R., 1982, *Litter Dynamics and Organic Carbon Exchange in Black Mangrove* (Avicennia germinans) *Basin Forests in a Southwest Florida Estuary*, Ph.D. dissertation, University of Florida, Gainesville.

Twilley, R. R., 1985, The exchange of organic carbon in basin mangrove forests in a southwest Florida estuary, *Estuarine, Coastal and Shelf Science* **20:**543–557.

Twilley, R. R., 1988, Coupling of mangroves to the productivity of estuarine and coastal waters, in *Coastal-Offshore Ecosystem: Interactions, Lecture Notes on Coastal and Estuarine Studies,* vol. 22, B.O. Jansson, ed., Springer-Verlag, Berlin, pp. 155–180.

Twilley, R. R., A. E. Lugo, and C. Patterson-Zucca, 1986, Litter production and turnover in basin mangrove forests in southwest Florida, *Ecology* **67:**670–683.

Tyler, P. A., 1976, Lagoon of Islands, Tasmania—Deathknell for a unique ecosystem? *Biol. Conservation* **9:**1–11.

U.S. Army Corps of Engineers, 1972, *Charles River Watershed, Massachusetts,* New England Division, Waltham, Mass., 65p.

U.S. Department of Interior, Bureau of Reclamation, 1984, *Water Measurement Manual,* 2d ed., revised reprint, U.S. Government Printing Office, Washington, D.C., 327p.

U.S. Environmental Protection Agency, 1983, *Freshwater Wetlands for Wastewater Management: Environmental Impact Statement—Phase I Report,* EPA 904/9-83- 107, U.S. EPA Region IV, Atlanta, Ga., 380p.

U.S. Fish and Wildlife Service, 1980, *Habitat Evaluation Procedures (HEP)*, Division of Ecological Services, Fish and Wildl. Serv., Washington, D.C., ESM 102.

U.S. Soil Conservation Service, 1975, *Soil Taxonomy: A Basic System of Soil Classification For Making and Interpreting Soil Surveys,* U.S. Soil Conservation Service Agric. Handbook 436, Washington, D.C., 754p.

U.S. Soil Conservation Service, 1987, *Hydric Soils of the United States.* in cooperation with the National Technical Committee for Hydric Soils, Washington, D.C.

Urban, N. R., 1983, *The Nitrogen Cycle in a Forested Bog Watershed in Northern Minnesota,* M.S. Thesis, Univ. of Minnesota, Minneapolis.

Urban, N. R., and S. J. Eisenreich, 1988, Nitrogen cycling in a forested Minnesota bog, *Can J. Bot.* **66:**435–449.

Urban, N. R., S. J. Eisenreich, and E. Gorham, 1985, Proton cycling in bogs: geographic variation in northeastern North America, in *Proceedings NATO Advanced Research Workshop on the Effects of Acid Deposition on Forest, Wetland, and Agricultural Ecosystems,* Toronto, May 13–17, 1985, T. C. Hutchinson and K. Meema, eds., Springer-Verlag, New York, pp. 577–598.

Ursin, M. J., 1972, *Life in and around the Salt Marsh,* T. Y. Crowell Co., New York, 110p.

Valiela, I., 1984, *Marine Ecological Processes,* Springer-Verlag, New York, 546p.

Valiela, I., J. M. Teal, and W. Sass, 1973, Nutrient retention in salt marsh plots experimentally fertilized with sewage sludge, *Estuarine Coast. Mar. Sci.* **1:**261–269.

Valiela, I., and J. M. Teal, 1974, Nutrient limitation in salt marsh vegetation, in *Ecology of Halophytes,* R. J. Reimold and W. H. Queen, eds., Academic Press, New York, pp. 547–563.

Valiela, I., J. M. Teal, and N. Y. Persson, 1976, Production and dynamics of experimentally enriched salt marsh vegetation: belowground biomass, *Limnol. Oceanogr.* **21:**245–252.

Valiela, I., J. M. Teal, S. Volkmann, D. Shafer, and E. J. Carpenter, 1978, Nutrient and particulate fluxes in a salt marsh ecosystem: tidal exchanges and inputs by precipitation and groundwater, *Limnol. Oceanogr.* **23:**798–812.

Valiela, I., and J. M. Teal, 1979, The nitrogen budget of a salt marsh ecosystem, *Nature* **280:**652–656.

Van Engel, W. A., and E. B. Joseph, 1968, *Characterization of Coastal and Estuarine Fish Nursery Grounds as Natural Communities,* Final Rep. Bur. Commercial Fisheries, Va. Inst. Marine Science, Glocester Point, Va., 43p.

Van Raalte, C. D., and I. Valiela, 1976, Production of epibenthic salt marsh algae: light and nutrient limitation. *Limnol. Oceanogr.* **21**:862–872.

van der Valk, A. G., 1981, Succession in wetlands: a Gleasonian approach, *Ecology* **62**:688–696.

van der Valk, A. G., 1982, Succession in temperate North American wetlands, in *Wetlands: Ecology and Management,* B. Gopal, R. E. Turner, R. G. Wetzel, and D. F. Whigham, eds., National Institute for Ecology and International Science Publications, Jaipur, India, pp. 169–179.

van der Valk, A. G., ed., 1989, *Northern Prarie Wetlands*, Iowa State University Press, Ames, 400p.

van der Valk, A. G., and C. B. Davis, 1978a, Primary production of prairie glacial marshes, in *Freshwater Wetlands: Ecological Processes and Management Potential,* R. E. Good, D. F. Whigham, and R. L. Simpson, eds., Academic Press, New York, pp. 21–37.

van der Valk, A. G., and C. B. Davis, 1978b, The role of seed banks in the vegetation dynamics of prairie glacial marshes, *Ecology* **59**:322–335.

van der Valk, A. G., C. B. Davis, J. L. Baker, and C. E. Beer, 1979, Natural freshwater wetlands as nitrogen and phosphorus traps for land runoff, in *Wetland Functions and Values: The State of Our Understanding,* P. E. Greeson, J. R. Clark, and J. E. Clark, eds., American Water Resources Assoc., Minneapolis, Minn., pp. 457–467.

van der Valk, A. G., J. M. Rhymer, and H. R. Murkin, 1991, Flooding and the decomposition of litter of four emergent plant species in a prairie wetland, *Wetlands* **11**:1–16.

Vannote, R. L., G. W. Minshall, K. W. Cummins, J. R. Sedell, and C. E. Cushing, 1980, The river continuum concept, *Can. J. Fish. Aquat. Sci.* **37**:130–137.

Vartapetian, B. B., 1988, Ultrastructure studies as a means of evaluating plant tolerance to flooding, in *The Ecology and Management of Wetlands.* vol 1, *Ecology of Wetlands*, D. D. Hook, ed., Timber Press, Portland, Ore., 2 vol., pp. 452–466.

Vega, A., and K. C. Ewel, 1981, Wastewater effects on a waterhyacinth marsh and adjacent impoundment, in *Environmental Management* **5**:537–541.

Verhoeven, J. T. A., 1986, Nutrient dynamics in minerotrophic peat mires, *Aquatic Botany* **25**:117–137.

Verhoeven, J. T. A., ed., 1992, *Fens and Bogs in the Netherlands: Vegetation, History, Nutrient Dynamics and Conservation,* Kluwer Academic Publishers, Dordrecht, The Netherlands, 490p.

Vermeer, J. G., and F. Berendse, 1983, The relationship between nutrient availability, shoot biomass and species richness in grassland and wetland communities, *Vegetatio* **53**:121–126.

Vernberg, F. J., 1981. Benthic macrofauna, in *Funcional Adaptations of Marine Organisms,* F. J. Vernberg and W. B. Vernberg, eds., Academic Press, New York, pp. 179–230.

Vernberg, W. B., and F. J. Vernberg, 1972, *Environmental Physiology of Marine Animals,* Springer-Verlag, New York, 346p.

Vernberg, W. B., and B. C. Coull, 1981, Meiofauna, in *Functional Adaptations of Marine Organisms,* F. J. Vernberg and W. B. Vernberg, eds., Academic Press, New York, pp. 147–177.

Vernon, R. 0.,1947, Cypress domes, *Science* **105**:97–99.

Verry, E. S., and D. H. Boelter, 1979, Peatland hydrology, in *Wetland Functions and Values: The State of Our Understanding,* P. E. Greeson, J. R. Clark, and J. E. Clark, eds., American Water Resources Assoc., Minneapolis, Minn., pp. 389–402.

Verry, E. S., and D. R. Timmons, 1982, Waterborne nutrient flow through an upland-peatland watershed in Minnesota, *Ecology* **63**:1456–1467.

Viereck, L. A., and E. L. Little Jr., 1972, *Alaska trees and shrubs*, U.S. Department of Agriculture, Washington, D.C., Handbook 410, 265p.

Visser, J. M., 1989, *The Impact of Vertebrate Herbivores on the Primary Production of Sagittaria Marshes in the Wax Lake Delta, Atchafalaya Bay, Louisiana,* Ph.D. thesis, Louisiana State University, Baton Rouge, 88p.

Vitt, D. H., P. Achuff, and R. E. Andrus, 1975a, The vegetation and chemical properties of patterned fens in the Swan Hills, north central Alberta., *Can. J. Bot.* **53**:2776–2795.

Vitt, D. H., H. Crum, and J. A. Snider, 1975b, The vertical zonation of *Sphagnum* species in hummock-hollow complexes in northern Michigan, *Mich. Bot.* **14**:190–200.

Voigts, D. K., 1976, Aquatic invertebrate abundance in relation to changing marsh vegetation, *Am. Midl. Nat.* **95**:312–322.

Wadsworth, J. R., Jr., 1979, *Duplin River Tidal System: Sapelo Island, Georgia.* Map reprinted in Jan. 1982 by the University of Georgia Marine Institute, Sapelo Island, Georgia.

Wainscott, V. J., C. Bartley, and P. Kangas, 1990, Effect of muskrat mounds on microbial density on plant litter, *Am. Midl. Nat.* **123:**399–401.

Walker, D., 1970, Direction and rate in some British post-glacial hydroseres, in *Studies in the Vegetational HIstory of the British Isles,* D. Walker and R. G. West, eds.. Cambridge University Press, Cambridge, pp. 117–139.

Walmsley, M. E., 1977, Physical and Chemical Properties of Peat in *Muskeg and the Northern Environment of Canada*, N. W. Radforth and C. O. Brawner, ed., Univ. of Toronto Press, Toronto, pp. 82–129.

Walsh, T., and T. A. Barry, 1958, The chemical composition of some Irish peats, *Royal Irish Acad. Proc.* **59:**305–328.

Walter, H., 1973, *Vegetation of the Earth,* Springer-Verlag, New York, 237p.

Walters, C., L. Gunderson, and C. S. Holling, 1992, Experimental policies for water management in the Everglades, *Ecological Applications* **2:**189–202.

Wample, R. L., and D. M. Reid, 1979, The role of endogenous auxins and ethylene in the formation of adventitious roots and hypocotyl hypertrophy in flooded sunflower plants (*Helianthus annuus* L.), *Physiol. Plant.* **45:**219–226.

Want, W. L., 1990, *Law of Wetlands Regulation* (Release #1/7/90), Clark-Boardman Company, Ltd., New York.

Warner, D., and D. Wells, 1980, *Bird Population Structure and Seasonal Habitat Use as Indicators of Environment Quality of Peatlands*, Minn. Dep. Nat. Resour., Minn. Peat Program, 84p.

Warners, D. P., 1987, Effects of burning on sedge meadow studied, *Restoration and Management Notes* **5**(2):90–91.

Watson, J. T., and J. A. Hobson, 1989, Hydraulic design considerations and control structures for constructed wetlands for wastewater treatment, in *Constructed Wetlands for Wastewater Treatment,* D. A. Hammer, ed., Lewis Publishers, Inc., Chelsea, Mich., pp. 379–392.

Watson, J. T., S. C. Reed, R. H. Kadlec, R. L. Knight, and A. E. Whitehouse, 1989, Performance expectations and loading rates for constructed wetlands, in *Constructed Wetlands for Wastewater Treatment,* D. A. Hammer, ed., Lewis Publishers, Inc., Chelsea, Mich., pp. 319–361.

Webb, J., and M. B. Jackson, 1986, A transmission and cryo-scanning electron microscopy study of the formation of aerenchyma (cortical gas-filled space) in adventitious roots of rice (*Oryza sativa*)., *J. Exp. Bot.* **37:**832–41.

Weber, C. A., 1908, Aufbau und Vegetation der Moore Norddeutschlands, *Engler's Bot. Jahrb.* **40**(Suppl):19–34.

Webster, J. R., and G. M. Simmons, 1978, Leaf breakdown and invertebrate colonization on a reservoir bottom, *Verin. Limnol.* **20:**1587–1596.

Welcomme, R. L., 1976, The Role of African Flood Plains in Fisheries in *Proc. Internat. Conf. Conservation of Wetlands and Waterfowl*, M. Smart, ed., Intern. Waterfowl Res. Bureau, Slimbridge, pp. 332–335.

Welcomme, R. L., 1979, *Fisheries Ecology of Floodplain Rivers*, Longman, New York,

Weller, M. W., 1978, Management of freshwater marshes for wildlife, in *Freshwater Wetlands: Ecological Processes and Management Potential,* R. E. Good, D. F. Whigham, and R. L. Simpson, eds., Academic Press, New York, pp. 267–284.

Weller, M. W., 1981, *Freshwater Marshes,* University of Minnesota Press, Minneapolis, Minn., 146p.

Weller, M. W., 1987, *Freshwater Marshes, Ecology and Wildlife Management*, 2d ed., Univ. of Minnesota Press, Minneapolis, Minn., 165p.

Weller, M. W., and C. S. Spatcher, 1965, Role of habitat in the distribution and abundance of marsh birds, *Iowa State University Agric. and Home Economics Exp. Station Spec. Rep. 43,* Ames, Iowa, 31p.

Welling, C. H., R. L. Pederson, and A. G. van der Valk, 1988a, Recruitment from the seed bank and the development of zonation of emergent vegetation during a drawdown in a prairie wetland, *J. Ecology* **76:**483–496.

Welling, C. H., R. L. Pederson, and A. G. van der Valk, 1988b, Temporal patterns in recruitment from the seed bank during drawdowns in a prairie wetland, *J. Appl. Ecol.* **25:**99–1007.

Wells, B. W., 1928, Plant communities of the coastal plain of North Carolina and their successional relations, *Ecology* **9:**230–242.

Wells, E. D., and S. Zoltai, 1985, The Canadian system of wetland classification and its application of circumboreal wetlands, Proc. Field Symposium on Classification of Mire Vegetation, *Aquilo (Bot.)* **21:**45–52.

Wells, J. T., S. J. Chinburg, and J. M. Coleman, 1982, *Development of the Atchafalaya River Deltas: Generic Analysis*, Coastal Studies Institute, Louisiana State University, Baton Rouge, prepared for U.S. Army Engineers, Waterways Experiment Station, Vicksburg, Mississippi,

Welsh, S. L., N. D. Atwood, L. C. Higgins, and S. Goodrich, 1987, *A Utah Flora.* Great Basin Naturalist Memoir no. 9, Brigham Young Univ., Provo, Utah.

Werme, C. E., 1981, *Resource Partitioning in the Salt Marsh Fish Community,* Ph.D. dissertation, Boston University, Boston, 126p.

West, R. G., 1964, Inter-relations of ecology and quaternary paleobotany, *J. Ecology* (suppl.) **52:**47–57.

Wetzel, R. G., and D. F. Westlake, 1969, Periphyton, in *A Manual on Methods for Measuring Primary Productivity in Aquatic Environments,* R. A. Vollenweider, ed., IBP Handbook 12, Blackwell, Oxford, England, pp. 33–40.

Wetzel, R. L., and S. Powers, 1978, Habitat development field investigations, Windmill Point marsh development site, James River, Virginia App. D: environmental impacts of marsh development with dredged material: botany, soil, aquatic biology, and wildlife, *U.S. Army Waterways Experiment Sta. Tech. Rep.* D-77-2., Vicksburg, Miss., 292p.

Wharton, C. H., 1970, *The Southern River Swamp—A Multlple-Use Environment,* Bureau of Business and Economic Research, Georgia State University, Atlanta, 48p.

Wharton, C. H., 1978, *The Natural Environments of Georgia*, Georgia Dept. Natural Resources, Atlanta, Ga., 227p.

Wharton, C. H., H. T. Odum, K. Ewel, M. Duever, A. Lugo, R. Boyt, J. Bartholomew, E. DeBellevue, S. Brown, M. Brown, and L. Duever, 1976, *Forested Wetlands of Florida—Their Management and Use,* Center for Wetlands, University of Florida, Gainesville, 421p.

Wharton, C. H., V. W. Lambou, J. Newsom, P. V. Winger, L. L. Gaddy, and R. Mancke, 1981, The fauna of bottomland hardwoods in southeastern United States, in *Wetlands of Bottomland Hardwood Forests,* J. R. Clark and J. Benforado, eds., Elsevier, Amsterdam, pp. 87–100.

Wharton, C. H., W. M. Kitchens, E. C. Pendleton, and T. W. Sipe, 1982, The ecology of bottomland hardwood swamps of the southeast: a community profile, *U.S. Fish and Wildlife Service, Biological Services Program FWS/OBS-81/37,* 133p.

Wheeler, B. D., 1980, Plant communities of rich-fen systems in England and Wales, *J. Ecology* **68:**365–395.

Wheeler, B. D., and K. E. Giller, 1982, Species richness of herbaceous fen vegetation in Broadland, Norfolk, in relation to the quantity of above-ground plant material, *J. Ecol.* **70:**179–200.

Whigham, D. F., and R. L. Simpson, 1975, *Ecological Studies of the Hamilton Marshes,* Progress report for the period June 1974-January 1975, Rider College, Biology Department, Lawrenceville, N.J.

Whigham, D. F., and R. L. Simpson, 1977, Growth, mortality, and biomass partitioning in freshwater tidal wetland populations of wild rice *(Zizania aquatica var. aquatica), Torrey Bot. Club Bull.* **104:**347–351.

Whigham, D. F., J. McCormick, R. E. Good, and R. L. Simpson, 1978, Biomass and primary production in freshwater tidal wetlands of the Middle Atlantic Coast, in *Freshwater Wetlands: Ecological Processes and Management Potential,* R. E. Good. D. F. Whigham, and R. L. Simpson, eds., Academic Press, New York, pp. 3–20.

Whigham, D. F., and S. E. Bayley, 1979, Nutrient dynamics in freshwater wetlands, in *Wetland Functions and Values: The State of Our Understanding,* P. E. Greeson, J. R. Clark, and J. E. Clark, eds., American Water Resources Assoc., Minneapolis, Minn.. pp. 468–478.

Whigham, D. F., R. L. Simpson, and M. A. Leck, 1979, The distribution of seeds, seedlings, and established plants of arrow arum (*Peltandra virginica* (L.) Kunth) in a freshwater tidal wetland, *Torrey Bot. Club Bull.* **106:**193–199.

Whigham, D. F., R. L. Simpson, and K. Lee, 1980, *The Effect of Sewage Effluent on the Structure and Function of a Freshwater Tidal Wetland,* Water Resources Research Institute Report, Rutgers University, New Brunswick, N.J., 160p.

Whigham, D. F., C. Chitterling, B. Palmer, and J. O'Neill, 1986, Modification of Runoff from Upland Watersheds—the Influence of a Diverse Riparian Ecosystem in *Watershed Research Perspectives*, D. L. Correll, ed., Smithsonian Institution Press, Washington, D.C., pp. 305–332.

White, D. A., T. E. Weiss, J. M. Trapani, and L. B. Thien, 1978, Productivity and decomposition of the dominant salt marsh plants in Louisiana, *Ecology* **59:**751–759.

Whitehead, D. R., 1972, Developmental and environmental history of the Dismal Swamp, *Ecol. Monogr.* **42:**301–315.

Whiting, G. J., H. N. McKellar, and T. G. Wolaver, 1989, Nutrient exchange between a portion of vegetated saltmarsh and the adjoining creek, *Limnol. Oceanogr.* **34**:463–473.

Whitney, D. E., G. M. Woodwell, and R. W. Howarth, 1975, Nitrogen fixation in Flax Pond: a Long Island salt marsh, *Limnol. Oceanogr.* **20:**640–643.

Whitney, D. M., A. G. Chalmers, E. B. Haines, R. B. Hanson, L. R. Pomeroy, and B. Sherr, 1981, The cycles of nitrogen and phosphorus, in *The Ecology of a Salt Marsh,* L. R. Pomeroy and R. G. Weigert, eds., Springer-Verlag, New York, pp. 163–18I .

Whittaker, R. H., 1967, Gradient analysis of vegetation, *Biol. Rev.* **42:**207–264.

Whooten, H. H., and M. R. Purcell, 1949, *Farm Land Development: Present and Future by Clearing, Drainage, and Irrigation,* U.S. Department of Agriculture, Circular 825, Washington, D.C.

Wicker, K. M., D. Davis, and D. Roberts, 1983, *Rockefeller State Wildlife Refuge and Game Preserve: Evaluation of Wetland Management Techniques,* Coastal Environments, Inc., Baton Rouge, La.

Wicker, K. M., G. C. Castille, D. J. Davis, S. M. Gagliano, D. W. Roberts, D. S. Sabins, and R. A. Weinstein, 1982, *St. Bernard Parish: A Study in Wetland Management,* Coastal Environments, Inc., Baton Rouge, La., 132p.

Widdows, J., B. L. Bayne, D. R. Livingstone, R. I. E. Newell, and E. Donkin, 1979, Physiological and biochemical responses of bivalve mollusks to exposure to air, *Comparative Biochemistry and Physiology* **62** Part A(2):301–308.

Wiebe, W. J., and L. R. Pomeroy, 1972, Micro-organisms and their association with aggregates and detritus in the sea: a microscopic study, *1st Ital. Iydrobiol. Mem.* **29**(Suppl.):325–352.

Wiebe, W. J., R. R. Christian, J. A. Hansen, G. King, B. Sherr, and G. Skyring, 1981, Anaerobic respiration and fermentation, in *The Ecology of a Salt Marsh,* L. R. Pomeroy and R. G. Wiegert, eds., Springer-Verlag, New York, pp. 137–159.

Wieder, R. K., 1989, A survey of constructed wetlands for acid coal mine drainage treatment in the Eastern United States, *Wetlands* **9:**299–315.

Wieder, R. K., and G. E. Lang, 1982, Modifications of acid mine drainage in a freshwater wetland, in *Symposium on wetlands of the unglaciated Appalachian region,* B. R. McDonald, ed., West Virginia University, Morgantown, pp. 43–53.

Wieder, R. K., and G. E. Lang, 1983, Net primary production of the dominant bryophytes in a *Sphagnum*-dominated wetland in West Virginia, *Bryologist* **86:**280–286.

Wieder, R. K., and G. E. Lang, 1984, Influence of wetlands and coal mining on stream water chemistry, *Water, Air, and Soil Pollution* **23:**381–396.

Wieder, R. K., G. Tchobanoglous, and R. W. Tuttle, 1989, Preliminary considerations regarding constructed wetlands for wastewater treatment, in *Constructed Wetlands for Wastewater Treatment,* D. A. Hammer, ed., Lewis Publishers, Chelsea, Mich., pp. 297–306.

Wiegert, R. G., 1986, Modeling spatial and temporal variability in a salt marsh: sensitivity to rates of primary production, tidal migration and microbial degradation, in *Estuarine Variability*, D. A. Wolfe, ed., Academic Press, New York, pp. 405–426.

Wiegert, R. G., R. R. Christian, J. L. Gallagher, J. R. Hall, R. D. H. Jones, and R. L. Wetzel, 1975, A preliminary ecosystem model of a Georgia salt marsh, in *Estuarine Research,* Vol. 1, L. E. Cronin, ed., Academic Press, New York, pp. 583–601.

Wiegert, R. G., and R. L. Wetzel, 1979, Simulation experiments with a fourteen compartment model of a *Spartina* salt marsh, in *Marsh-Estuarine Systems Simulations,* R. F. Dame, ed., University of South Carolina Press, Columbia, pp. 7–39.

Wiegert, R. G., R. R. Christian, and R. L. Wetzel, 1981, A model view of the marsh, in *The Ecology of a Salt Marsh,* L. R. Pomeroy and R. G. Wiegert, eds., Springer-Verlag, New York, pp. 183–218.

Wiegert, R. G., and B. J. Freeman, 1990, *Tidal Salt Marshes of the Southeast Atlantic Coast: A Community Profile*, U. S. Department of Interior, Fish and Wildl. Service, Washington, D.C., Biological Report 85(7.29).

Wiemhoff, J. R., 1977, *Hydrology of a Southern Illinois Cypress Swamp,* Master's thesis, Illinois Institute of Technology, Chicago. Il. 98p.

Wilcox, D. A., R. A. Shedlock, and W. H. Hendrickson, 1986, Hydrology, water chemistry, and ecological relations in the raised mound of Cowles Bog, *J. Ecology* **74:**1103–1117.

Wilde, S. A., and G. W. Randall, 1951, Chemical characteristics of groundwater in forest and marsh soils of Wisconsin, *Wisc. Acad. Sci. Arts Lett. Trans.* **40:**251–259.

Wile, I., G. Miller, and S. Black, 1985, Design and Use of Artificial Wetlands, in *Ecological Considerations in Wetland Treatment of Municipal Wastewaters,* P. J. Godfrey, E. R. Kaynor, S. Pelczarski and J. Benforado, eds., Van Nostrand Reinhold, New York, pp. 26–37.

Wilen, B. O., and H. R. Pywell, 1981, *The National Wetlands Inventory,* paper presented at In-Place Resource Inventories: Principles and Practices—a national workshop, Orono, Maine, Aug. 9–14,10p.

Wilen, B., 1991, Fact sheets and information, National Wetlands Inventory, St. Petersburg, Fla.

Wilhelm, M., S. R. Lawry, and D. D. Hardy, 1989, Creation and management of wetlands using municipal wastewater in northern Arizona: a status report, in *Constructed Wetlands for Wastewater Treatment,* D. A. Hammer, ed., Lewis Publishers, Inc., Chelsea, Mich., pp. 179–185.

Willard, D. E., and A. K. Hiller, 1989, Wetland dynamics: considerations for restored and created wetlands, in *Wetland Creation and Restoration,* J. A. Kusler and M. E. Kentula, eds., Island Press, Washington, D.C., pp. 459–466.

Willard, D. E., V. M. Finn, D. A. Levine, and J. E. Klarquist, 1989, Creation and restoration of riparian wetlands in the agricultural midwest, in *Wetland Creation and Restoration,* J. A. Kusler and M. E. Kentula, eds., Island Press, Washington, D.C., pp. 327–337.

Williams, G. P., and M. G. Wolman, 1984, *Downstream Effects of Dams on Alluvial Rivers,* U.S. Geological Survey, Washington, D.C., Professional Paper 1286.

Williams, J. D., and C. K. Dodd, Jr., 1979, Importance of wetlands to endangered and threatened species, in *Wetland Functions and Values: The State of Our Understanding,* P E. Greeson, J. R. Clark, and J. E. Clark, eds., American Water Resources Assoc., Minneapolis, Minn., pp. 565–575.

Williams, M., ed., 1990, *Wetlands: A Threatened Landscape,* Basil Blackwell, Inc., Oxford, 419p.

Williams, R. B., 1962, *The Ecology of Diatom Populations in a Georgia Salt Marsh,* Ph.D. thesis, Harvard University, Cambridge, Mass.

Williams, R. B., and M. B. Murdock, 1972, Compartmental analysis of the production of *Juncus roemerianus* in a North Carolina salt marsh, *Chesapeake Sci.* **13:**69–79.

Willis, C. N., and W. J. Mitsch, Effects of hydrologic and nutrient variability on emergence and growth of aquatic macrophytes, unpub. manuscript.

Wilson, C. L., and W. E. Loomis, 1967, *Botany,* 4th ed., Holt, Rinehart and Winston, New York, 626p.

Wilson, K. A., 1962, *North Carolina Wetlands: Their Distribution and Management,* North Carolina Wildlife Resources Commission, Raleigh, N.C., 169p.

Wilson, L. R., 1935, Lake development and plant succession in Vilas County, Wisconsin, Part 1, The medium hard water lakes, *Ecol. Monogr.* **5:**207–247.

Winter, T. C., 1988, A conceptual framework for assessing cumulative impacts on the hydrology of nontidal wetlands, *Environmental Management* 12:605–620.

Winter, T. C., 1989, *Hydrologic Studies of Wetlands in the Northern Prairie,* in *Northern Prairie Wetlands,* A.G. van der Valk, ed., Iowa State University Press, Ames, Iowa, pp. 16–54.

Winter, T. C., and M. R. Llamas, eds., 1993, Hydrogeology of wetlands, Special Issue, *J. Hydrology* 141: 1-269.

Wolaver, T. G., and J. D. Spurrier, 1988, The exchange of phosphorus between a euhaline vegetated marsh and the adjacent tidal creek, *Est. Coast. Shelf Shic.* **26:**203–214.

Woodhouse, W. W., Jr., 1979, *Building Salt Marshes Along the Coasts of the Continental United States,* U.S. Army, Coastal Engineering Research Center, Fort Belvoir, Virginia, Special Report 4.

Woodwell, G. M., 1956, *Phytosociology of Coastal Plain Wetlands of the Carolinas,* Master's thesis, Duke University, Durham, N.C., 52p.

Woodwell, G. M., and D. E. Whitney, 1977, Flax Pond ecosystem study: exchanges of phosphorus between a salt marsh and the coastal waters of Long Island Sound, *Mar. Biol.* **41:**1–6.

Woodwell, G. M., D. E. Whitney, C. A. S. Hall, and R. A. Houghton, 1977, The Flax Pond ecosystem study: exchanges of carbon in water between a salt marsh and Long Island Sound, *Limnol. Oceanogr.* **22:**833–838.

Woodwell, G. M., C. A. S. Hall, D. E. Whitney, and R. A. Houghton, 1979, The Flax Pond ecosystem study: exchanges of inorganic nitrogen between an estuarine marsh and Long Island Sound, *Ecology* **60:**695–702.

Woodwell, G. M., J. E. Hobbie, R. A. Houghton, J. M. Melillo, B. Moore, B. J. Peterson, and G. R. Shaver, 1983, Global deforestation: contribution to atmospheric carbon dioxide, *Science* **222:**1081–1086.

Wright, A. H., and A. A. Wright, 1932, The habits and composition of vegetation of Okefenokee Swamp, Georgia, *Ecol. Monogr.* **2:**109–232.

Wright, J. 0., 1907, *Swamp and Overflow Lands in the United States,* U.S. Department of Agriculture Circular 76, Washington, D.C.

Yan, J., 1992, Ecological techniques and their applications with some case studies in China, *Ecological Engineering* **1:**261–285.

Yarbro, L. A., 1979, *Phosphorus Cycling in the Creeping Swamp Floodplain Ecosystem and Exports from the Creeping Swamp Watershed,* Ph.D. dissertation, University of North Carolina, Chapel Hill.

Yarbro, L. A., 1983, The influence of hydrologic variations on phosphorus cycling and retention in a swamp stream ecosystem, in *Dynamics of Lotic Ecosystems,* T. D. Fontaine and S. M. Bartell, eds., Ann Arbor Science, Ann Arbor, Mich., pp. 223–245.

Yates, R. F. K., and F. R. Day, Jr., 1983, Decay rates and nutrient dynamics in confined and unconfined leaf litter in the Great Dismal Swamp, *Am. Midl. Nat.* **110:**37–45.

Zedler, J. B., 1980, Algae mat productivity: comparisons in a salt marsh, *Estuaries* 3:122–131.

Zedler, J. B., 1982, *The Ecology of Southern California Coastal Salt Marshes: A Community Profile*, U.S. Fish Wildl. Serv., Biol. Services Program, Washington, D.C., FWS/OBS-81/54.

Zedler, J. B., 1988, Salt marsh restoration: lessons from California, in *Rehabilitating Damaged Ecosystems,* J. Cairns, ed., vol. I, CRC Press, Boca Raton, Florida, pp. 123–138.

Zedler, P. H., 1987, *The Ecology of Southern California Vernal Pools: A Community Profile,* Biological Report 85(7.11), U.S. Fish and Wildlife Service, Washington, D.C., NTIS.

Zieman, J. C., and W. E. Odum, 1977, *Modeling of Ecological Succession and Production in Estuarine Marshes,* U.S. Army Engineers Waterways Exp. Station Tech. Rep. D-77-35, Vicksburg, Miss.

Zimmerman, R. J., T. J. Minellos, D. L. Smith, and J. Kostera, 1990, *The Use of Juncus and Spartina Marshes by Fisheries Species in Lavaca Bay, Texas, with Reference to Effects of Floods*, National Marine Fisheries Service, National Oceanographic and Atmospheric Agency, Washington, D.C., NOAA Tech. Memo. NMFS-SEFC–251.

Zinn, J. A., and C. Copeland, 1982, *Wetland Management,* Congressional Research Service, The Library of Congress, Washington, D.C., 149p.

Ziser, S. W., 1978, Seasonal variations in water chemistry and diversity of the phytophilia macroinvertebrates of three swamp communities in southeastern Louisiana, *Southwest Nat.* **23:**545–562.

Zoltai, S. C., 1979, An outline of the wetland regions of Canada, in *Proceedings of a Workshop on Canadian Wetlands,* C. D. A. Rubec and F. C. Pollett, eds., Environment Canada, Lands Directorate, Ecological Land Classifications Series, No. 12, Saskatoon, Saskatchewan, pp. 1–8.

Zoltai, S. C., 1988, Wetland environments and classification, in *Wetlands of Canada, Ecological Land Classification Series, no. 24.,* National Wetlands Working Group, ed., Environment Canada, Ottawa, Ontario, and Polyscience Publications, Inc., Montreal, Quebec, pp. 1–26.

Zoltai, S. C., and F. C. Pollet, 1983, Wetlands in Canada: their classification, distribution, and use, in *Mires: Swamp, Bog, Fen and Moor,* A. J. P. Gore, ed., Elsevier, Amsterdam, pp. 245–268.

Zoltai, S. C., S. Taylor, J. K. Jeglum, G. F. Mills, and J. D. Johnson, 1988, Wetlands of boreal Canada, in *Wetlands of Canada, Ecological Land Classification Series, no. 24.*, National Wetlands Working Group, ed., Environment Canada, Ottawa, Ontario, and Polyscience Publications, Inc., Montreal, Quebec, pp. 97–154.

Index

Numbers printed in **bold type** refer to pages where the terms are defined.